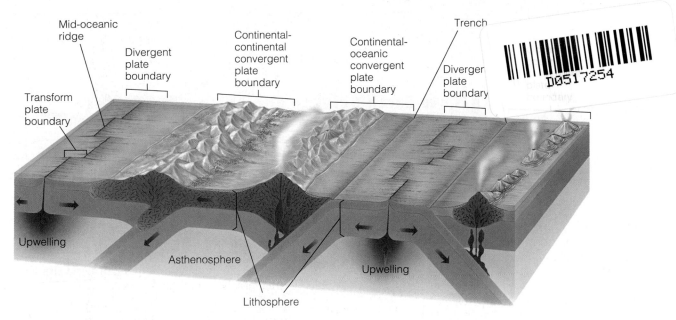

Three Principal Types of Plate Boundaries (Figure 1.16)

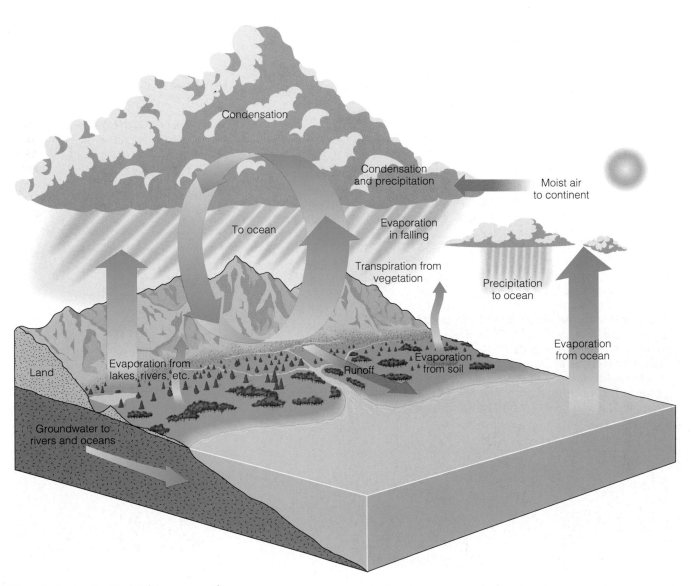

The Hydrologic Cycle (Figure 15.6)

Books in the Brooks/Cole Earth Science and Astronomy Series:

Internet and Media Products in the Brooks/Cole Earth Science Series:

Blue Skies™, College Edition, CD-ROM, SAMSON
Brooks/Cole's Earth Science Resource Center:
    http://www.brookscole.com/geo
Earth Online: An Internet Guide for Earth Science, RITTER
In-Terra-Active™ 2.0: Physical Geology Interactive
    Student CD-ROM, BROWN AND DUNNING
The Brooks/Cole Geolink 3.0: A Geology and
    Oceanography Presentation Tool
The Brooks/Cole Astrolink 2.0: An Astronomy
    Presentation Tool
The Geology Workbook for the Web, BLACKERBY
Earth Systems Today CD-ROM, SAMSON

Earth Science

Earth Science Today, MURPHY, NANCE
Earth Science, DUTCH, MONROE, MORAN
Earth Online: An Internet Guide for Earth Science, RITTER

Meteorology

Meteorology Today, 6th, AHRENS
Essentials of Meteorology, 2nd, AHRENS
Meteorology Today for Scientists and Engineers,
    2nd, STULL

Astronomy

Horizons: Exploring the Universe, 6th, SEEDS
Foundations of Astronomy, 1999 edition, SEEDS
Universe: Origins and Evolution, SNOW AND BROWNSBERGER
Astronomy: The Cosmic Journey, 5th, HARTMANN AND IMPEY
Moons and Planets, an Introduction to Planetary
    Science, 4th, HARTMANN
Astronomy on the Web, KURLAND
Current Perspectives in Astronomy and Physics, POMPEA
The Universe Revealed, IMPEY, HARTMANN
Great Ideas for Teaching Astronomy, POMPEA
Astronomy: The Solar System and Beyond, SEEDS

Physical Geology

Physical Geology: Exploring the Earth, 4th, MONROE AND WICANDER
Essentials of Geology, 2nd, WICANDER AND MONROE

Historical Geology

Historical Geology: Evolution of Earth and Life Through
    Time, 3rd, WICANDER AND MONROE

Physical and Historical Geology

The Changing Earth, 2nd, MONROE AND WICANDER

Environmental Geology

Geology and the Environment, 2nd, PIPKIN AND TRENT

Geology Supplements

Current Perspectives in Geology, 2000 edition,
    MCKINNEY/MCHUGH/MEADOWS

Oceanography

Essentials of Oceanography, 2nd, GARRISON
Oceanography: An Invitation to Marine Science, 3rd,
    GARRISON
Oceanography: An Introduction, 5th, INGMANSON AND
    WALLACE
Introduction to Ocean Science, SEGAR
Marine Life and the Sea, MILNE

Physical Geology

Fourth Edition

Exploring the Earth

James S. Monroe

Reed Wicander

Central Michigan University

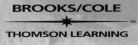

BROOKS/COLE

THOMSON LEARNING

Australia • Canada • Mexico • Singapore • Spain • United Kingdom • United States

BROOKS/COLE

THOMSON LEARNING

Sponsoring Editor: *Nina Horne*
Editorial Assistant: *Rebecca Eisenman*
Marketing Team: *Rachel Alvelais, Carla Martin-Falcone*
Marketing Assistant: *Mandie Houghan*
Assistant Editor: *Sam Subity*
Production Editor: *Keith Faivre*
Production Service: *Nancy Shammas, New Leaf Publishing Services*
Manuscript Editor: *Linda Purrington*
Developmental Editor: *Marie Carigma-Sambilay*

Cover Design: *Vernon T. Boes*
Cover Illustration: *William Scott Jennings*
Interior Design: *Delgado Design*
Permissions Editor: *Lillian Campobasso*
Photo Researcher: *Christine Johnson*
Print Buyer: *Vena Dyer*
Typesetting: *The Beacon Group, Inc.*
Printing and Binding: *Transcontinental Printing, Inc.*

For more information about this or any other Brooks/Cole product, contact:
BROOKS/COLE
511 Forest Lodge Road
Pacific Grove, CA 93950 USA
www.brookscole.com
1-800-423-0563 (Thomson Learning Academic Resource Center)

Printed in Canada

10 9 8 7 6 5 4 3 2 1

Library of Congress Cataloging-in-Publication Data

Monroe, James S. (James Stewart), [date].
    Physical geology : exploring the earth / James S. Monroe, Reed Wicander.—4th ed.
        p. cm.
    Includes bibliographical references and index.
    ISBN 0-534-57222-7
    1. Physical geology. I. Wicander, Reed, [date]. II. Title.

QE28.2.M655 2000
550—dc21

00-063037

About the Authors

James S. Monroe

James S. Monroe is professor emeritus of geology at Central Michigan University where he taught physical geology, historical geology, prehistoric life, and stratigraphy and sedimentology since 1975. He has co-authored several textbooks with Reed Wicander and has interests in Cenozoic geology and geologic education.

Photo courtesy of Melanie Wicander

Reed Wicander

Reed Wicander is a geology professor at Central Michigan University where he teaches physical geology, historical geology, prehistoric life, and invertebrate paleontology. He has co-authored several geology textbooks with James S. Monroe. His main research interests involve various aspects of Paleozoic palynology, specifically the study of acritarchs, on which he has published many papers. He is a past president of the American Association of Stratigraphic Palynologists and currently a councillor of the International Federation of Palynological Societies.

About the Cover

Your textbook cover features a painting of the Grand Canyon by William Scott Jennings called *The Awakening*. We chose this painting for the cover because the Grand Canyon is not only symbolic of geology, but many of the basic principles of geology can be illustrated within its walls. The Grand Canyon is also linked to human experience in many different ways, which is an important theme of this book. (The Grand Canyon also serves as a background photo for all the chapter openers.)

Major John Wesley Powell, a Civil War veteran who lost his right arm in the battle of Shiloh, was the first geologist to explore the Grand Canyon region, leading a group of hardy explorers in fragile wooden boats down the uncharted Colorado River through the Grand Canyon in 1869. President Theodore Roosevelt named the Grand Canyon a national monument in 1908, and Congress upgraded it to a national park in 1919. The Grand Canyon has become such a popular tourist site that cars are now banned from many of the roads in an attempt to reduce pollutants.

When we stand on the rim and look down into the Grand Canyon, we are really looking far back in time, all the way back to the early history of our planet. More than one billion years of history are preserved in the rocks of the Grand Canyon, indicating episodes of mountain building as well as periods of transgressions and regressions of shallow seas. Three of the six fundamental principles of relative dating are also illustrated by the sequence of rocks exposed in the Grand Canyon. The Grand Canyon did not come into being by some catastrophic event, but rather by the simple action of the Colorado River eroding its channel over millions of years.

Considering the importance of the Grand Canyon to geology and its role in human history, it thus seems most appropriate that the cover of this book should be a painting of the Grand Canyon in all its grandeur.

Contents

Chapter 3

Igneous Rocks and Intrusive Igneous Activity 60

Chapter 4

Volcanism 90

Chapter 5

Weathering, Erosion, and Soil   126

Chapter 6

Sediment and Sedimentary Rocks   156

Chapter 7

Metamorphism and Metamorphic Rocks   190

Chapter 8

Geologic Time   214

Chapter 9

Earthquakes 248

Chapter 10

Earth's Interior 288

Chapter 11

The Seafloor   314

Chapter 12

Plate Tectonics: A Unifying Theory   346

Chapter 13

Deformation, Mountain Building, and the Evolution of Continents 380

Chapter 14

Mass Wasting 422

Chapter 15

Running Water 454

Chapter 16

Groundwater 494

Chapter 17

Glaciers and Glaciation 532

Chapter 18

The Work of Wind and Deserts 572

Chapter 19

Shorelines and Shoreline Processes 602

Chapter 20

A History of the Universe, Solar System, and Planets 640

Preface

Earth is a dynamic planet that has changed continuously during its 4.6 billion years of existence. The size, shape, and geographic distribution of the continents and ocean basins have changed through time, as have the atmosphere and biota. We have become increasingly aware of how fragile our planet is and, more importantly, how interdependent all of its various systems are. We have learned that we cannot continually pollute our environment and that our natural resources are limited and, in most cases, nonrenewable. Furthermore, we are coming to realize how central geology is to our everyday lives. For these and other reasons, geology is one of the most important college or university courses a student can take.

Physical Geology: Exploring the Earth is designed for a one-semester introductory course in geology that serves both majors and nonmajors in geology and the Earth sciences. One of the problems with any introductory science course is that students are overwhelmed by the amount of material that must be learned. Furthermore, most of the material does not seem to be linked by any unifying theme and does not always appear to be relevant to their lives.

The goals of this book are to provide students with a basic understanding of geology and its processes and, more importantly, with an understanding of how geology relates to the human experience: that is, how geology affects not only individuals, but society in general. With these goals in mind, we introduce the major themes of the book in the first chapter to provide students with an overview of the subject and to enable them to see how the various systems of Earth are interrelated. We also discuss the economic and environmental aspects of geology throughout the book rather than treating these topics in separate chapters. In this way students can see, through relevant and interesting examples, how geology impacts our lives.

NEW FEATURES IN THE FOURTH EDITION

The fourth edition has undergone considerable rewriting and updating to produce a book that is easier to read with a high level of current information and many new photographs, figures, prologues, and perspectives. Drawing on the comments and suggestions of reviewers, we have incorporated many new features into this edition.

New material in this edition of *Physical Geology: Exploring the Earth* includes an expanded section on Earth Systems and an added emphasis on the systems approach throughout the book. There is updated information in every chapter, particularly such recent events as the volcanic eruptions on Montserrat Island (Chapter 1), the earthquakes in Turkey (Chapter 9), and the flooding and landslides in Venezuela (Chapter 14). The chapters on surface processes (Chapters 14–19) are still largely descriptive, but many of the sections of these chapters have been rewritten to emphasize the systems approach in discussing the dynamic nature of these processes.

Updated information on mineral and energy resources and environmental issues has been added to many chapters, such as a perspective on the precious metals (Chapter 2), chemical reactions in the atmosphere that yield sulfuric acid (Chapter 5), additional information on oil shales and tar sands (Chapter 6), and dam failures and their resulting floods (Chapter 15).

Other important changes include a number of new Prologues, such as the volcanic eruptions on Montserrat Island (Chapter 1), granitic rocks and their occurrence at various national parks (Chapter 3), the rocks and fossils in John Day Fossil Beds, Oregon (Chapter 6), the earthquakes in Turkey (Chapter 9), and flooding and landslides in Venezuela (Chapter 14). New Perspectives also appear, such as columnar jointing and unusual volcanoes (Chapter 4), geologic time and climate change (Chapter 8), the opening and closing of oceans (Chapter 11), paleogeographic reconstrucitons and maps (Chapter 12), the geologic evolution of San Francisco (Chapter 13), Waterton Lake National Park, Alberta, and Glacier National Park, Montana (Chapter 17), the geologic history of Uluru and Kata Tijuta, Australia (Chapter 18), and the Outer Banks of North Carolina (Chapter 19).

Many photographs in the third edition have been replaced, including many of the chapter opening photographs. In addition, a number of photographs within the chapters have been enlarged to enhance their visual impact.

We feel the rewriting and updating done in the text as well as the addition of new photographs greatly improves the fourth edition by making it easier to read and comprehend, as well as a more effective teaching tool. Additionally, improvements have been made in the ancillary package that accompanies the book.

TEXT ORGANIZATION

Plate tectonic theory is the unifying theme of geology and this book. This theory has revolutionized geology because it provides a global perspective of Earth and allows geologists to treat many seemingly unrelated geologic phenomena as part of a total planetary system. Because plate tectonic theory is so important, it is introduced in Chapter 1 and is discussed in most subsequent chapters in terms of the subject matter of that chapter.

Another theme of this book is that Earth is a complex, dynamic planet that has changed continually since its origin some 4.6 billion years ago. We can better understand this complexity by using a systems approach in the study of Earth. As mentioned earlier, we have expanded the section on Earth Systems in Chapter 1 and emphasized this approach throughout the book.

We have organized *Physical Geology: Exploring the Earth* into several informal categories. Chapter 1 is an introduction to geology and Earth Systems, its relevance to the human experience, plate tectonic theory, the rock cycle, geologic time and uniformitarianism, and the origin of the solar system and Earth. Chapters 2–7 examine Earth's materials (minerals and igneous, sedimentary, and metamorphic rocks) and the geologic processes associated with them, including the role of plate tectonics in their origin and distribution. Chapter 8 discusses geologic time, introduces several dating methods, and explains how geologists correlate rocks. Chapters 9–13 deal with the related topics of Earth's interior, the seafloor, earthquakes, deformation and mountain building, and plate tectonics. Chapters 14–19 cover Earth's surface processes. Chapter 20 discusses the origin of the universe and solar system and Earth's place in the evolution of these larger systems.

We have found that presenting the material in this order works well for most students. We know, however, that many instructors prefer an entirely different order of topics, depending on the emphasis in their course. We have therefore written this book so that instructors can present the chapters in any order that suits the needs of their course.

CHAPTER ORGANIZATION

All chapters have the same organizational format. Each chapter opens with a photograph that relates to the chapter material, a detailed outline that engages the students by having many of the headings as questions, a Chapter Objectives outline, followed by a Prologue, which is intended to stimulate interest in the chapter by discussing some aspect of the material.

The text is written in a clear informal style, making it easy for students to comprehend. Numerous color diagrams and photographs complement the text, providing a visual representation of the concepts and information presented.

Each chapter contains at least one Perspective that presents a brief discussion of an interesting aspect of geology or geological research. What Would You Do? boxes are a new feature in each chapter. These boxes are designed to encourage critical thinking by the students as they attempt to solve a hypothetical problem or issue on the local, national, and global level. Each chapter has at least two What Would You Do? boxes. Mineral and energy resources are discussed in the final sections of a number of chapters to provide interesting, relevant information in the context of the chapter topics.

The end-of-chapter materials begin with a concise review of important concepts and ideas in the Chapter Summary. The Important Terms, which are printed in boldface type in the chapter text, are listed at the end of each chapter for easy review, and a full glossary of important terms appears at the end of the text. The Review Questions are another important feature of this book; they include multiple-choice questions with answers as well as short answer and essay questions, and thought-provoking and quantitative questions under the Points to Ponder heading. Many new multiple-choice, short answer, essay, and Points to Ponder questions have been added in each chapter for this edition. Each chapter concludes with World Wide Web Activities that list World Wide Web sites students can visit to get additional information about the topics covered in that chapter, and a CD-ROM Exploration. In the previous three editions, a list of Additional Readings appeared at the end of each chapter. In the fourth edition, however, this list has been deleted from the text and moved to the World Wide Web. For students interested in pursuing a particular topic, an up-to-date list of Additional Readings for each chapter is available at **http://www.brookscole.com/geo**

FOR STUDENTS

CURRENT PERSPECTIVES IN GEOLOGY, 2000 EDITION
Michael McKinney, Kathleen McHugh, and Susan Meadows (University of Tennessee, Knoxville)
This book of 42 current readings is designed to supplement any geology textbook and is ideal for instructors who include a writing component in their course. The

articles are culled from a number of popular science magazines (such as *American Scientist, National Wildlife, Discover, Science, New Scientist,* and *Nature*). Available for sale to students or bundled at a discount with any Brooks/Cole geology text.
0-534-37213-9

ESSENTIAL STUDY SKILLS FOR SCIENCE STUDENTS
Daniel Chiras (University of Colorado–Denver)
Designed to accompany any introductory science text. It offers tips on improving your memory, learning more quickly, getting the most out of lectures, preparing for tests, producing first-rate term papers, and improving critical thinking skills. For just one dollar extra, it can be bundled with every student copy of the text.
0-534-37595-2

EARTH SYSTEMS TODAY™: THE BROOKS/COLE EARTH SCIENCES CD-ROM FOR STUDENTS
Create a landslide. Predict a volcanic eruption. Locate the epicenter of an earthquake. Or explore other key topics in geology through some 50 interactive exercises on CD-ROM. In each case, students can manipulate variables and data and view the results of their selections. Over 35 minutes of full motion video clips, plus animations, help illustrate difficult concepts. This CD-ROM offers an easy-to-use, graphically oriented interface, a searchable glossary, and user-tracking for professors to monitor student progress.
0-534-37733-5

GEOLOGY WORKBOOK FOR THE WEB
Bruce Blackerby (California State University–Fresno)
Over twenty chapters designed to encourage students to explore geology-related sites on the Internet. Chapter exercises, on topics from Minerals to Cenozoic Earth, focus most heavily on physical geology. The workbook comes three-hole punched with perforated pages for easy tearing, so worksheets can be handed in as assignments.
0-314-21072-5

EARTH ONLINE
Michael Ritter (University of Wisconsin–Stevens Point)
An inexpensive, hands-on Internet guide written for the novice. It provides a tool for students to get "up and running" on the Internet with homework exercises, lab exercises, Web searches, and more. To keep the book as useful as possible, the author maintains an Earth Online home page with exercises, tips, new links, and constant updates of the exercises and reference sites. Access it through the Brooks/Cole Earth Science Resource Center at http://www.brookscole.com/geo/.
0-534-51707-2

STUDY GUIDE
Christopher Haley (Virginia Wesleyan College)
A valuable tool to help you excel in this course! Filled with sample test questions, learning objectives, useful analogies, key terms, drawings and figures, and vocabulary reviews to guide your study.
0-534-37520-0

ACKNOWLEDGMENTS

As the authors, we are, of course, responsible for the organization, style, and accuracy of the text, and any mistakes, omissions, or errors are our responsibility. The finished product is the culmination of many years of work during which we received numerous comments and advice from many geologists who reviewed parts of the text. We wish to express our sincere appreciation to the reviewers who reviewed the third edition and made many helpful and useful comments that led to the improvements seen in this fourth edition.

Steven R. Dent, *Northern Kentucky University*
Michael R. Forrest, *Rio Hondo Community College*
René De Hon, *Northeast Louisiana University*
Robert B. Jorstad, *Eastern Illinois University*
David T. King, *Auburn University*
Peter L. Kresan, *University of Arizona*
Gary D. Rosenberg, *Indiana University–Purdue University at Indianapolis*
Darrel W. Schmitz, *Mississippi State University*
David G. Towell, *Indiana University*

We would also like to thank the reviewers of the second edition whose comments and suggestions greatly improved the third edition.

Gary Allen, *University of New Orleans*
Richard Beck, *Miami University*
Roger Bilham, *University of Colorado, Boulder*
John P. Buchanan, *Eastern Washington University*
Jeffrey B. Connelly, *University of Arkansas at Little Rock*
Peter Copeland, *University of Houston*
Rachael Craig, *Kent State University*
P. Johnathan Patchett, *University of Arizonia*
Gary Rosenberg, *Indiana University–Purdue University at Indianapolis*
Chris Sanders, *Southeast Missouri State University*
Paul B. Tomascak, *University of Maryland at College Park*
Harve S. Waff, *University of Oregon*

In addition, we thank those individuals who reviewed the first edition and made many useful and insightful comments that were incorporated into the second edition of this book.

Theodore G. Benitt, *Nassau Community College*
Bruce A. Blackerby, *California State University–Fresno*
Gerald F. Brem, *California State University–Fullerton*
Ronald E. Davenport, *Louisiana Tech University*
Isabella M. Drew, *Ramapo College of New Jersey*
Jeremy Dunning, *Indiana University*
Norman K. Grant, *Miami University*
Norris W. Jones, *University of Wisconsin–Oshkosh*
Patricia M. Kenyon, *University of Alabama*
Peter L. Kresan, *University of Arizona*
Dave B. Loope, *University of Nebraska–Lincoln*
David N. Lumsden, *Memphis State University*
Steven K. Reid, *Morehead State University*
M.J. Richardson, *Texas A&M University*
Charles J. Ritter, *University of Dayton*
Gary D. Rosenberg, *Indiana University–Purdue University at Indianapolis*
J. Alexander Speer, *North Carolina State University*
James C. Walters, *University of Northern Iowa*

Lastly, we would like to provide thanks to the individuals who reviewed the first edition of this book in manuscript form. Their comments helped make the book a success for students and instructors alike.

Gary C. Allen, *University of New Orleans*
R. Scott Babcock, *Western Washington University*
Kennard Bork, *Denison University*
Thomas W. Broadhead, *University of Tennessee at Knoxville*
Anna Buising, *California State University at Hayward*
F. Howard Campbell III, *James Madison University*
Larry E. Davis, *Washington State University*

Noel Eberz, *California State University at San Jose*
Allan A. Ekdale, *University of Utah*
Stewart S. Farrar, *Eastern Kentucky University*
Richard H. Fluegeman, Jr., *Ball State University*
William J. Fritz, *Georgia State University*
Kazuya Fujita, *Michigan State University*
Norman Gray, *University of Connecticut*
Jack Green, *California State University at Long Beach*
David R. Hickey, *Lansing Community College*
R. W. Hodder, *University of Western Ontario*
Cornelis Klein, *University of New Mexico*
Lawrence W. Knight, *William Rainey Harper College*
Martin B. Lagoe, *University of Texas at Austin*
Richard H. Lefevre, *Grand Valley State University*
I. P. Martini, *University of Guelph, Ontario*
Michael McKinney, *University of Tennessee at Knoxville*
Robert Merrill, *California State University at Fresno*
Carleton Moore, *Arizona State University*
Alan P. Morris, *University of Texas at San Antonio*
Harold Pelton, *Seattle Central Community College*
James F. Petersen, *Southwest Texas State University*
Katherine H. Price, *DePauw University*
William D. Romey, *St. Lawrence University*
Gary Rosenberg, *Indiana University–Purdue University at Indianapolis*
David B. Slavsky, *Loyola University of Chicago*
Edward F. Stoddard, *North Carolina State University*
Charles P. Thornton, *Pennsylvania State University*
Samuel B. Upchurch, *University of South Florida*
John R. Wagner, *Clemson University*

We also wish to thank Kathy Benison, Richard V. Dietrich (Professor Emeritus), David J. Matty, Jane M. Matty, Wayne E. Moore (Professor Emeritus), and Sven Morgan of the Geology Department, and Bruce M. C. Pape of the Geography Department of Central Michigan University, as well as Eric Johnson (Hartwick College, New York) and Stephen D. Stahl (SUNY, College at Oneonta, New York) for providing us with photographs and answering our questions concerning various topics. We also thank Pam Iacco of the Geology Department, whose general efficiency was invaluable during the preparation of this book. Theresa Barber, Justin Reuter, and Adrienne Sims, also of the Geology Department, helped with some of the clerical tasks associated with this book. We are also grateful for the generosity of the various agencies and individuals from many countries who provided photographs.

Special thanks must go to Nina Horne, acquisitions editor at Brooks/Cole Thomson Learning, who saw this edition through to completion. She was assisted by her editorial associate, John-Paul Ramin. We are equally indebted to our production editor, Keith Faivre, whose attention to detail and consistency is greatly appreciated. We thank Nancy Shammas, production manager for all her help and Linda Purrington for her copyediting skills. We also thank Lillian Campobasso, who coordinated the permissions portion of the text, and Astor Indexers who did the index. Because geology is largely a visual science, we extend special thanks to Carlyn Iverson, who rendered the reflective art, and to the artists at Precision Graphics, who were responsible for much of the rest of the art program. We also thank the artists at Magellan Geographix, who rendered many of the maps. They did an excellent job, and we enjoyed working with them.

As always, our families were very patient and encouraging when much of our spare time and energy were devoted to this book. We again thank them for their continued support and understanding.

Developing Critical Thinking and Study Skills

INTRODUCTION

College is a demanding and important time, a time when your values will be challenged, and you will try out new ideas and philosophies. You will make personal and career decisions that will affect your entire life. One of the most important lessons you can learn in college is how to balance your time among work, study, and recreation. If you develop good time management and study skills early in your college career, you will find that your college years will be successful and rewarding.

This section offers some suggestions to help you maximize your study time and develop critical thinking and study skills that will benefit you, not only in college, but throughout your life. While mastering the content of a course is obviously important, learning how to study and to think critically is, in many ways, far more important. Like most things in life, learning to think critically and study efficiently will initially require additional time and effort, but once mastered, these skills will save you time in the long run.

You may already be familiar with many of the suggestions and may find that others do not directly apply to you. Nevertheless, if you take the time to read this section and apply the appropriate suggestions to your own situation, we are confident that you will become a better and more efficient student, find your classes more rewarding, have more time for yourself, and get better grades. We have found that the better students are usually also the busiest. Because these students are busy with work or extracurricular activities, they have had to learn to study efficiently and manage their time effectively.

One of the keys to success in college is avoiding procrastination. While procrastination provides temporary satisfaction because you have avoided doing something you did not want to do, in the long run it leads to stress. While a small amount of stress can be beneficial, waiting until the last minute usually leads to mistakes and a subpar performance. By setting clear, specific goals and working toward them on a regular basis, you can greatly reduce the temptation to procrastinate. It is better to work efficiently for short periods of time than to put in long, unproductive hours on a task, which is usually what happens when you procrastinate.

Another key to success in college is staying physically fit. It is easy to fall into the habit of eating junk food and never exercising. To be mentally alert, you must be physically fit. Try to develop a program of regular exercise. You will find that you have more energy, feel better, and study more efficiently.

GENERAL STUDY SKILLS

Most courses, and geology in particular, build upon previous material, so it is extremely important to keep up with the coursework and set aside regular time for study in each of your courses. Try to follow these hints, and you will find you do better in school and have more time for yourself:

- Develop the habit of studying on a daily basis.

- Set aside a specific time each day to study. Some people are day people, and others are night people. Determine when you are most alert and use that time for study.

- Have an area dedicated for study. It should include a well-lighted space with a desk and the study materials you need, such as a dictionary, thesaurus, paper, pens, and pencils, and a computer if you have one.

- Study for short periods and take frequent breaks, usually after an hour of study. Get up and move around and do something completely different. This will help you stay alert, and you'll return to your studies with renewed vigor.

- Try to review each subject every day or at least the day of the class. Develop the habit of reviewing lecture material from a class the same day.

- Become familiar with the vocabulary of the course. Look up any unfamiliar words in the glossary of your textbook or in a dictionary. Learning the language of the discipline will help you learn the material.

GETTING THE MOST FROM YOUR NOTES

If you are to get the most out of a course and do well on exams, you must learn to take good notes. Taking good notes does not mean you should try to write down every word your professor says. Part of being a good note taker is knowing what is important and what you can safely leave out.

Early in the semester, try to determine whether the lecture will follow the textbook or be predominantly new material. If much of the material is covered in the textbook, your notes do not have to be as extensive or detailed as when the material is new. In any case, the following suggestions should make you a better note taker and enable you to derive the maximum amount of information from a lecture:

- Regardless of whether the lecture discusses the same material as the textbook or supplements the reading assignment, read or scan the chapter the lecture will cover *before* class. This way you will be somewhat familiar with the concepts and can listen critically to what is being said rather than trying to write down everything. Later a few key words or phrases will jog your memory about what was said.

- Before each lecture, briefly review your notes from the previous lecture. Doing this will refresh your memory and provide a context for the new material.

- Develop your own style of note taking. Do not try to write down every word. These are notes you're taking, not a transcript. Learn to abbreviate and develop your own set of abbreviations and symbols for common words and phrases: for example, w/o (without), w (with), = (equals), ∧ (above or increases), ∨ (below or decreases), < (less than), > (greater than), & (and), u (you).

- Geology lends itself to many abbreviations that can increase your note-taking capability: for example, pt (plate tectonics), ig (igneous), meta (metamorphic), sed (sedimentary), rx (rock or rocks), ss (sandstone), my (million years), and gts (geologic time scale).

- Rewrite your notes soon after the lecture. Rewriting your notes helps reinforce what you heard and gives you an opportunity to determine whether you understand the material.

- By learning the vocabulary of the discipline before the lecture, you can cut down on the amount you have to write—you won't have to write down a definition if you already know the word.

- Learn the mannerisms of the professor. If he or she says something is important or repeats a point, be sure to write it down and highlight it in some way. Students have told me (RW) that when I stated something twice during a lecture, they knew it was important and probably would appear on a test. (They were usually right!)

- Check any unclear points in your notes with a classmate or look them up in your textbook. Pay particular attention to the professor's examples, which usually elucidate and clarify an important point and are easier to remember than an abstract concept.

- Go to class regularly and sit near the front of the class if possible. It is easier to hear and see what is written on the board or projected onto the screen, and there are fewer distractions.

- If the professor allows it, tape record the lecture, but don't use the recording as a substitute for notes. Listen carefully to the lecture and write down the important points; then fill in any gaps when you replay the tape.

- If your school allows it, and if they are available, buy class lecture notes. These are usually taken by a graduate student who is familiar with the material; typically they are quite comprehensive. Again use these notes to supplement your own.

- Ask questions. If you don't understand something, ask the professor. Many students are reluctant to do this, especially in a large lecture hall, but if you don't understand a point, other people are probably confused as well. If you can't ask questions during a lecture, talk to the professor after the lecture or during office hours.

GETTING THE MOST OUT OF WHAT YOU READ

The old adage that "you get out of something what you put into it" is true when it comes to reading textbooks. By carefully reading your text and following these suggestions, you can greatly increase your understanding of the subject:

- Look over the chapter outline to see what the material is about and how it flows from topic to topic. If you have time, skim through the chapter before you start to read in depth.

- Pay particular attention to the tables, charts, and figures. They contain a wealth of information in abbreviated form and illustrate important concepts and ideas. Geology, in particular, is a visual science, and the figures and photographs will help you visualize what is

being discussed in the text and provide actual examples of features such as faults or unconformities.

- As you read your textbook, highlight or underline key concepts or sentences, but make sure you don't highlight everything. Make notes in the margins. If you don't understand a term or concept, look it up in the glossary.

- Read the chapter summary carefully. Be sure you understand all the key terms, especially those in boldface or italic type. Because geology builds on previous material, it is imperative that you understand the terminology.

- Go over the end-of-chapter questions. Write your answers as if you were taking a test. Only when you see your answer in writing will you know if you really understood the material.

- Access the latest geologic information on the Internet. The end-of-chapter World Wide Web Activities will enhance your understanding of the chapter concepts and the way geologic information is disseminated today. Knowing how to search the Internet is an essential skill.

DEVELOPING CRITICAL THINKING SKILLS

Few things in life are black and white, and it is important to be able to examine an issue from all sides and come to a logical conclusion. One of the most important things you will learn in college is to think critically and not accept everything you read and hear at face value. Thinking critically is particularly important in learning new material and relating it to what you already know. Although you can't know everything, you can learn to question effectively and arrive at conclusions consistent with the facts. Thus, these suggestions for critical thinking can help you in all your courses:

- Whenever you encounter new facts, ideas, or concepts, be sure you understand and can define all of the terms used in the discussion.

- Determine how the facts or information was derived. If the facts were derived from experiments, were the experiments well executed and free of bias? Can they be repeated? The controversy over cold fusion is an excellent example. Two scientists claimed to have produced cold fusion reactions using simple experimental laboratory apparatus, yet other scientists have never been able to achieve the same reaction by repeating the experiments.

- Do not accept any statement at face value. What is the source of the information? How reliable is the source?

- Consider whether the conclusions follow from the facts. If the facts do not appear to support the conclusions, ask questions and try to determine why they don't. Is the argument logical or is it somehow flawed?

- Be open to new ideas. After all, the underlying principles of plate tectonic theory were known early in this century yet were not accepted until the 1970s despite overwhelming evidence.

- Look at the big picture to determine how various elements are related. For example, how will constructing a dam across a river that flows to the sea affect the stream's profile? What will be the consequences to the beaches that will be deprived of sediment from the river? One of the most important lessons you can learn from your geology course is how interrelated the various systems of Earth are. When you alter one feature, you affect numerous other features as well.

IMPROVING YOUR MEMORY

Why do you remember some things and not others? The reason is that the brain stores information in different ways and forms, making it easy to remember some things and difficult to remember others. Because college requires that you learn a vast amount of information, any suggestions that can help you retain more material will help you in your studies:

- Pay attention to what you read or hear. Focus on the task at hand and avoid daydreaming. Repetition of any sort will help you remember material. Review the previous lecture before going to class, or look over the last chapter before beginning the next. Ask yourself questions as you read.

- Use mnemonic devices to help you learn unfamiliar material. For example, the order of the Paleozoic periods (Cambrian, Ordovician, Silurian, Devonian, Mississippian, Pennsylvanian, and Permian) of the geologic time scale can be remembered by the phrase, **Campbell's Onion Soup Does Make Peter Pale**, or the order of the Cenozoic Epochs (Paleocene, Eocene, Oligocene, Miocene, Pliocene, and Pleistocene) can be remembered by the phrase, **Put Eggs On My Plate Please**. Using rhymes can also be helpful.

- Look up the roots of important terms. If you understand where a word comes from, its meaning will be easier to remember. For example, *pyroclastic* comes from *pyro,* meaning "fire," and *clastic,* meaning "broken pieces." Hence a pyroclastic rock is one formed by volcanism and composed of pieces of other rocks. We have provided the roots of many important terms throughout this text to help you remember their definitions.

- Outline the material you are studying. This practice will help you see how the various components are interrelated. Learning a body of related material is much easier than learning unconnected and discrete facts. Looking for relationships is particularly helpful in geology because so many things are interrelated. For example, plate tectonics explains how mountain building, volcanism, and earthquakes are all related. The rock cycle relates the three major groups of rocks to each other and to subsurface and surface processes (Chapter 1).

- Use deductive reasoning to tie concepts together. Remember that geology builds on what you learned previously. Use that material as your foundation and see how the new material relates to it.

- Draw a picture. If you can draw a picture and label its parts, you probably understand the material. Geology lends itself very well to this type of memory device because so much is visual. For example, instead of memorizing a long list of glacial terms, draw a picture of a glacier and label its parts and the type of topography it forms.

- Focus on what is important. You can't remember everything, so focus on the important points of the lecture or the chapter. Try to visualize the big picture and use the facts to fill in the details.

PREPARING FOR EXAMS

For most students, tests are the critical part of a course. To do well on an exam, you must be prepared. These suggestions will help you focus on preparing for examinations:

- The most important advice is to study regularly rather than try to cram everything into one massive study session. Get plenty of rest the night before an exam, and stay physically fit to avoid becoming susceptible to minor illnesses that sap your strength and lessen your ability to concentrate on the subject at hand.

- Set up a schedule so that you cover small parts of the material on a regular basis. Learning some concrete examples will help you understand and remember the material.

- Review the chapter summaries. Construct an outline to make sure you understand how everything fits together. Drawing diagrams will help you remember key points. Make flash cards to help you remember terms and concepts.

- Form a study group, but make sure your group focuses on the task at hand, not on socializing. Quiz each other and compare notes to be sure you have covered all the material. We have found that students dramatically improved their grades after forming or joining a study group.

- Write the answers to all the Review Questions. Before doing so, however, become thoroughly familiar with the subject matter by reviewing your lecture notes and reading the chapter. Otherwise, you will spend an inordinate amount of time looking up answers.

- If you have any questions, visit the professor or teaching assistant. If review sessions are offered, be sure to attend. If you are having problems with the material, ask for help as soon as you have difficulty. Don't wait until the end of the semester.

- If old exams are available, look at them to see what is emphasized and what types of questions are asked. Find out whether the exam will be all objective or all essay or a combination. If you have trouble with a particular type of question (such as multiple choice or essay), practice answering questions of that type—your study group or a classmate may be able to help.

TAKING EXAMS

The most important thing to remember when taking an exam is not to panic. This, of course, is easier said than done. Almost everyone suffers from test anxiety to some degree. Usually, it passes as soon as the exam begins, but in some cases, it is so debilitating that an individual does not perform as well as he or she could. If you are one of those people, get help as soon as possible. Most colleges and universities have a program to help students overcome test anxiety or at least keep it in check. Don't be afraid to seek help if you suffer test anxiety. Your success in college depends to a large extent on how well you perform on exams, so by not seeking help, you are only hurting yourself. In addition, the following suggestions may be helpful:

- First of all, relax. Then look over the exam briefly to see its format and determine which questions are worth the most points. If it helps, quickly jot down any information you are afraid you might forget or particularly want to remember for a question.

- Answer the questions that you know the best first. Make sure, however, that you don't spend too much time on any one question or on one that is worth only a few points.

- If the exam is a combination of multiple choice and essay, answer the multiple-choice questions first. If you are not sure of an answer, go on to the next one. Sometimes the answer to one question can be found in another question. Furthermore, the multiple-choice questions may contain many of the facts needed to answer some of the essay questions.

- Read the question carefully and answer only what it asks. Save time by not repeating the question as your opening sentence to the answer. Get right to the point. Jot down a quick outline for longer essay questions to make sure you cover everything.

- If you don't understand a question, ask the examiner. Don't assume anything. After all, it is your grade that will suffer if you misinterpret the question.

- If you have time, review your exam to make sure you covered all the important points and answered all the questions.

- If you have followed our suggestions, by the time you finish the exam, you should feel confident that you did well and will have cause for celebration.

CONCLUDING COMMENTS

We hope that the suggestions we have offered will be of benefit to you, not only in this course but throughout your college career. Though it is difficult to break old habits and change a familiar routine, we are confident that following these suggestions will make you a better student. Furthermore, many of the suggestions will help you work more efficiently, not only in college, but also throughout your career. Learning is a lifelong process that does not end when you graduate. The critical thinking skills that you learn now will be invaluable throughout your life, both in your career and as an informed citizen.

Chapter 1

The Soufriere Hills volcano is Montserrat Island's dominant landmark.

OBJECTIVES

At the end of this chapter, you will have learned that

- Geology is the study of Earth.
- Earth is a complex, integrated system.
- Geology plays an important role in the human experience as well as affecting us as individuals and as members of society and nation-states.

- The solar system and planets evolved from a turbulent, rotating cloud of material surrounding the embryonic Sun.
- Theories are based on the scientific method.
- Plate tectonic theory revolutionized geology.

- The rock cycle illustrates the interrelationships between Earth's internal and external processes and shows how and why the three major rock groups are related to each other.
- An appreciation of geologic time and of the principle of uniformitarianism is central to understanding the evolution of Earth and its biota.

Understanding Earth: An Introduction to Physical Geology

Prologue

Montserrat Island's tranquility was shattered on July 18, 1995, when the Soufriere Hills volcano (from the French word *soufre*, meaning sulphur or brimstone), located at the south end of the island, began erupting, spewing hot ash, steam, gases, and rocks skyward. Although this volcano had been dormant during historic times, its recent outburst is merely the latest in a long history of volcanic eruptions on this Caribbean island. By the end of August 1997, Plymouth—the capital city of Montserrat—was in virtual ruins from the devastating effects of massive ash flows and airborne ash. Its 4,000 residents evacuated; Plymouth became a ghost town.

During August 1997, government officials ordered the evacuation of not only the southern part of Montserrat, where the Soufriere Hills volcano is located, but also the central portion of the island, because of the continued threats of a possible cataclysmic eruption. By September 9, 1997, the southern two-thirds of the island were completely off limits to everyone, and those still remaining on the island had been urged to move as far north as possible. Continued eruptions during September destroyed Montserrat's airport, as well as many villages. The ash flow that buried part of the airport's runway continued into the sea, heating the water to temperatures high enough to kill many of the fish and lobsters in the bay.

By the end of 1997, eruptions had buried many towns, various lava domes had grown on the Soufriere Hills volcano and collapsed, sending volcanic debris flowing down the hillsides. Ash continued to spew forth from these eruptions, sometimes forming ash plumes over 14,000 meters high, not only covering the immediate area surrounding the volcano, but also falling on neighboring islands. Eruptions and ash flows have continued into 2000 with no evidence that the volcano will become dormant any time in the near future.

Why have we chosen the eruption of the Soufriere Hills volcano on Montserrat as an introduction to geology? The reason is that it illustrates many of the themes addressed in this book, such as how geology relates to the human experience, how Earth's various subsystems and cycles are interrelated, and why Earth is considered a dynamic planet.

Montserrat and other islands of the Lesser Antilles in the Caribbean Sea are part of a chain of volcanic islands resulting from the collision between two plates, the South American plate and the Caribbean plate. As one plate descends into Earth's interior and begins melting, the molten rock produced begins rising to the surface, where it may erupt on the surface as a volcano, or in this case, the island of Montserrat. In addition, the collision between plates produces earthquakes, which can also have devastating effects.

As you read this book, keep in mind that the different topics you are studying concern parts of dynamic interrelated systems, and are not isolated pieces of information. Volcanic eruptions such as occurred on Montserrat are the result of complex interactions involving Earth's interior and surface. These eruptions not only have an immediate effect on the surrounding area, but also contribute to climatic changes that affect the entire planet.

INTRODUCTION TO EARTH SYSTEMS

A major benefit of the space age has been the ability to look back from space and view our planet in its entirety. Every astronaut has remarked in one way or another on how Earth stands out as an inviting oasis in the otherwise black void of space (Figure 1.1). We are able to see not only the beauty of our planet, but also its fragility and the role humans play in shaping and modifying its environment. The pollution of the atmosphere, oceans, and many of our lakes and streams, as well as the denudation of huge areas of tropical forests, plus scars from strip mining and the depletion of the ozone layer are all visible in the satellite images beamed back from space and attest to the impact humans have had on Earth (Figure 1.2).

A theme of this book is that Earth is a complex, dynamic planet that has changed continually since its origin some 4.6 billion years ago. These changes and the present-day features we observe are the result of interactions between the various interrelated internal and external Earth subsystems and cycles. In fact, Earth is unique among the planets of our solar system in that it supports life and has oceans of water, a hospitable atmosphere, and a variety of climates. It is ideally suited for life as we know it because of a combination of factors, including its distance from the Sun and the evolution of its interior, crust, oceans, and atmosphere. Over time, changes in Earth's atmosphere, oceans, and to some extent, its crust have been influenced by life processes. In turn, these physical changes have affected the evolution of life.

Figure 1.1
Apollo 17 view of Earth. Almost the entire coastline of Africa is clearly shown in this view, with Madagascar visible off its eastern coast. The Arabian Peninsula can be seen at the northeastern edge of Africa, and the Asian mainland is on the horizon toward the northeast. The present location of continents and ocean basins is the result of plate movement. The interaction of plates through time has affected the physical and biological history of Earth.

Figure 1.2
Denudation of a large area of tropical forest in Costa Rica.

If you view Earth as a whole, you can see innumerable interactions occurring between its various components. Furthermore, these components do not act in isolation but are interconnected—as one part of the system changes, it affects the other parts of the system. The system concept is a useful way to study a complex subject, such as Earth, because you can divide the whole into smaller components that you can easily study without losing sight of how the components all fit together.

You can better understand the complexity of Earth by thinking of it as a system. A **system** is a combination of related parts that interact in an organized fashion (Figure 1.3). Information, materials, and energy entering the system from the outside are *inputs,* whereas information, materials, and energy that leave the system are *outputs.* An automobile is a good example of a system. Its various subsystems include engine, transmis-

sion, steering, and brakes. These subsystems are interconnected in such a way that a change in any one of them affects the others. The main input into the automobile system is gasoline, and its outputs are movement, heat, and pollutants.

You can examine Earth in much the same way you look at an automobile—that is, as a system of interconnected components that interact and affect each other in many ways. The principal subsystems of Earth are the *atmosphere, hydrosphere, biosphere, lithosphere, mantle,* and *core* (Figure 1.4). The complex interactions among these subsystems result in a dynamically changing body that exchanges matter and energy and recycles them into different forms (Table 1.1). The rock cycle is an excellent example of how the interaction between Earth's internal and external processes recycles Earth materials to form the three major rock groups (see Figure 1.17, page 20). Likewise, the movement of plates has profoundly affected the formation of landscapes, the distribution of mineral resources, and atmospheric and oceanic circulation patterns, which in turn have affected global climate changes.

Also, don't forget that humans are part of the Earth system, and our presence alone affects this system to some extent. Actions we take can produce wide-ranging effects of which we might not be aware (see Perspective 1.1, page 8). Thus an understanding of geology, and science in general, is of great importance. If the human species is to survive, we must understand how the various Earth systems work and interact with each other, and how our actions affect the delicate balance between these systems.

When people discuss and debate such environmental issues as acid rain, the greenhouse effect and global

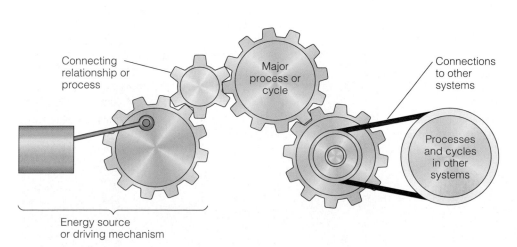

Figure 1.3
A series of gears can be used to illustrate how some of Earth's systems and processes interact. Pistons and driving rods represent energy sources or driving mechanisms, large gears represent important processes or cycles, and small gears represent connecting processes or relationships. Pulleys are used to show relationships to other systems. Although gears are a useful way to represent systems diagrammatically, remember that real Earth systems are far more complex.

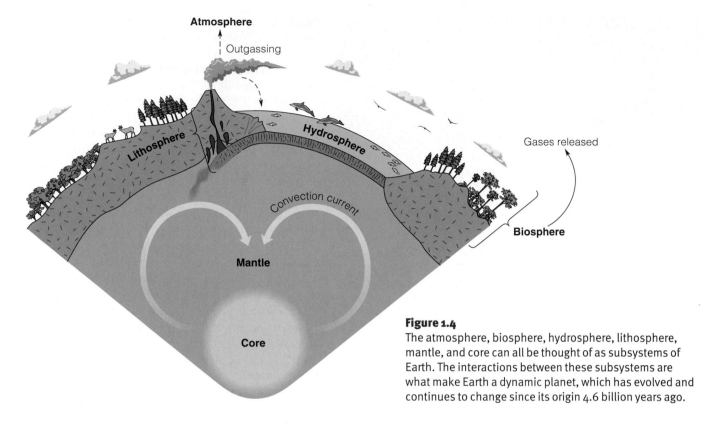

Figure 1.4
The atmosphere, biosphere, hydrosphere, lithosphere, mantle, and core can all be thought of as subsystems of Earth. The interactions between these subsystems are what make Earth a dynamic planet, which has evolved and continues to change since its origin 4.6 billion years ago.

warming, and the depleted ozone layer, it is important to remember that these effects are not isolated, but part of the larger Earth system. Furthermore, remember that Earth goes through time cycles of much longer duration than humans are used to. Although they may have disastrous short-term effects on the human species, global warming and cooling are also part of a longer-term cycle that has resulted in many glacial advances and retreats during the past 1.6 million years. Because of their geologic perspective, geologists can make vital contributions to the debate on global warming. They can study long-term trends by analyzing deep-sea sediments, ice cores, changes in sea level during the geologic past, and the distribution of plants and animals through time.

Table 1.1

Interactions Between Earth's Principal Subsystems

	ATMOSPHERE	HYDROSPHERE	BIOSPHERE	LITHOSPHERE
ATMOSPHERE	Interaction among various air masses	Surface currents driven by wind Evaporation	Gases for respiration Dispersal of spores, pollen, and seeds by wind	Weathering by wind erosion Transport of water vapor for precipitation of rain and snow
HYDROSPHERE	Input of water vapor and stored solar heat	Hydrologic cycle	Water for life	Precipitation Weathering and erosion
BIOSPHERE	Gases from respiration	Removal of disolved materials by organisms	Global ecosystems Food cycles	Modification of weathering and erosion processes Formation of soil
LITHOSPHERE	Input of stored solar heat Landscapes affect air movements	Source of solid and dissolved materials	Source of mineral nutrients Modification of ecosystems by plate movements	Plate tectonics

WHAT IS GEOLOGY?

What is geology and what do geologists do? **Geology,** from the Greek *geo* and *logos,* is defined as the study of Earth. It is generally divided into two broad areas—physical geology and historical geology. **Physical geology** is the study of Earth materials, such as minerals and rocks, as well as the processes operating within Earth and on its surface. **Historical geology** examines the origin and evolution of Earth, its continents, oceans, atmosphere, and life.

The discipline of geology is so broad that it is subdivided into many different fields or specialties. Table 1.2 shows many of the diverse fields of geology and their relationship to the sciences of astronomy, physics, chemistry, and biology.

Nearly every aspect of geology has some economic or environmental relevance. Many geologists are involved in exploration for mineral and energy resources, using their specialized knowledge to locate the natural resources on which our industrialized society is based. As the demand for these nonrenewable resources increases, geologists are applying the basic principles of geology in increasingly sophisticated ways to help focus their attention on areas with a high potential for economic success (Figure 1.5).

Although locating mineral and energy resources is extremely important, geologists are also being asked to use their expertise to help solve many environmental problems. Some geologists are involved in finding groundwater for the ever-burgeoning needs of communities and industries or in monitoring surface and underground water pollution and suggesting ways to clean it up. Geologic engineers help find safe locations for dams, waste-disposal sites, and power plants and design earthquake-resistant buildings.

Geologists are also involved in making short- and long-range predictions about earthquakes and volcanic eruptions and the potential destruction that may result.

Figure 1.5
Geologists use computers and other sophisticated technology in their search for petroleum and other natural resources.
Source: © Jean Miele/The Stock Market/TSM.

Table 1.2

Specialties of Geology and Their Broad Relationship to the Other Sciences

SPECIALTY	AREA OF STUDY	RELATED SCIENCE
Geochronology	Time and history of Earth	Astronomy
Planetary geology	Geology of the planets	
Paleontology	Fossils	Biology
Economic geology	Mineral and energy resources	
Environmental geology	Environment	
Geochemistry	Chemistry of Earth	Chemistry
Hydrogeology	Water resources	
Mineralogy	Minerals	
Petrology	Rocks	
Geophysics	Earth's interior	Physics
Structural geology	Rock deformation	
Seismology	Earthquakes	
Geomorphology	Landforms	
Oceanography	Oceans	
Paleogeography	Ancient geographic features and locations	
Stratigraphy/sedimentology	Layered rocks and sediments	

Perspective 1.1

The Aral Sea

The Aral Sea in the Central Asian desert region of the former Soviet Union (Figure 1) is a continuing environmental and human disaster. In 1960 it was the world's fourth largest lake. Since then its size has steadily decreased from 67,000 km² to about 34,000 km², and its volume has fallen from 1090 km³ to 300 km³, so that today it is only the sixth largest lake. Presently it is divided into two separate basins, a small northern basin and a larger southern basin. At its present rate of reduction, the Aral Sea could disappear within the next 30 years.

What could have caused such a disaster? For thousands of years, the Aral Sea was fed by two large rivers, the Amu Dar'ya and Syr Dar'ya. The headwaters of both rivers begin in the mountains more than 2000 km to the southeast and flow northward through the Kyzyl Kum and Kara Kum deserts into the Aral Sea. Until the early part of this century, a balance existed between the water supplied by the Amu Dar'ya and the Syr Dar'ya rivers and the rate of evaporation of the Aral Sea, which has no outlet. In 1918, however, it was decreed that waters from the two rivers supplying the Aral Sea would be diverted to irrigate millions of acres of cotton so that the Soviet Union could become self-sufficient in cotton production. As a result of this decision, the Aral Sea has become an ecological and environmental nightmare affecting some 35 million people.

To gain an appreciation of the magnitude of this disaster, consider that as late as the 1960s the Aral Sea was home to 24 native species of fish and supported a major fishing industry (Figure 2). The town of Muynak was the major fishing port and produced 3% of the Soviet Union's annual catch. The fishing industry provided about 60,000 jobs, with approximately 10,000 fishermen working out of Muynak. Today, Muynak is more than 20 km from the shoreline of the Aral Sea, and there are no native fish species left in the sea.

As the Aral Sea shrinks, vast areas of the former sea bottom are exposed. This new sediment, far from being rich and productive, contains large amounts of sodium chloride and sodium sulfate. The concentration of

Figure 1
Location (below) and 1985 space shuttle view of the Aral Sea (right).

salts is so high that only one plant species grows in it. It is too salty for anything else.

As the winds blow across the near-barren land, salt and dust are picked up and carried throughout the Aral region. These salts cause great damage to the cotton crops and other vegetation of the Aral basin and also exact a heavy toll on humans. The dry, salty dust has caused respiratory and eye diseases to increase dramatically during the past 30 years, as well as the reported number of cases of throat cancer. In addition, the drinking water supply has become so polluted that many people suffer from intestinal disorders.

As a result of the diversion of water from the Amu Dar'ya and the Syr Dar'ya rivers and the resulting reduction in the Aral Sea, desertification has become a major problem in the Aral basin. This has caused a change in the weather patterns of the region, so it is now colder in the winter and hotter and drier in the summer.

Despite the damage done to the region by the dying of the Aral Sea, irrigation for cotton production continues. However, the soil in many of the fields has become increasingly salty due to lower stream and groundwater levels, requiring more irrigation to maintain the same production levels. This means that even more water is diverted from the Aral Sea, causing it to shrink still further.

Can the Aral Sea be saved? Only by a concerted and cooperative effort by the countries that comprise the Aral basin. Realizing this, the now independent states of Kazakhstan, Uzbekistan, Kyrgyzstan, Tadzhikistan, and Turkmenistan took the first steps in 1992 by signing a formal agreement on sharing the waters of the Amu Dar'ya and Syr Dar'ya. In addition, these states entered into discussions to create a council to oversee and coordinate the management of the basin's resources.

Although full restoration of the Aral Sea is probably not possible, maintaining its current surface level and even raising the surface level of the small northern Aral Sea are feasible. If this could be accomplished, salinity levels would be reduced, making it possible to reintroduce native fish species and revive commercial fishing.

Figure 2
Fishing boats lie abandoned in the dry seabed that was once part of the Aral Sea. As recently as the 1960s, the Aral Sea supported a large fishing industry, but today the water is too saline for any fish to survive. Consequently, frozen fish are shipped in from the Pacific for processing in plants that are now more than 20 km from the Aral Sea.

In addition, they are working with civil defense planners to help draw up contingency plans should such natural disasters occur.

As this brief survey illustrates, geologists follow a wide variety of pursuits. As the world's population increases and greater demands are made on Earth's limited resources, the need for geologists and their expertise will become even greater.

HOW IS GEOLOGY RELATED TO THE HUMAN EXPERIENCE?

Many people are surprised at the extent to which we depend on geology in our everyday lives and also at the numerous references to geology in the arts, music, and literature. Rocks and landscapes are realistically represented in many sketches and paintings. Examples by famous artists include Leonardo da Vinci's *Virgin of the Rocks* and *Virgin and Child with Saint Anne,* Giovanni Bellini's *Saint Francis in Ecstasy* and *Saint Jerome,* and Asher Brown Durand's *Kindred Spirits* (Figure 1.6).

In the field of music, Ferde Grofé's *Grand Canyon Suite* was, no doubt, inspired by the grandeur and timelessness of Arizona's Grand Canyon and its vast rock exposures (Figure 1.7). The rocks on the island of Staffa in the Inner Hebrides provided the inspiration for Felix Mendelssohn's famous *Hebrides* Overture.

References to geology abound in *The German Legends of the Brothers Grimm.* Jules Verne's *Journey to the Center of the Earth* describes an expedition into Earth's interior (see Chapter 10 Prologue). On one level, the poem "Ozymandias" by Percy B. Shelley deals with the fact that nothing lasts forever and even solid rock eventually disintegrates under the ravages of time and weathering. References to geology can even be found in comics, two of the best known being *B.C.* by Johnny Hart and *The Far Side* by Gary Larson (Figure 1.8).

Geology has also played an important role in history. Wars have been fought for the control of such natural resources as oil, gas, gold, silver, diamonds, and other valuable minerals. Empires throughout history have risen and

Figure 1.6
Kindred Spirits by Asher Brown Durand (1849) realistically depicts the layered rocks occurring along gorges in the Catskill Mountains of New York State. Durand was one of numerous artists of the 19th-century Hudson River School, who were known for their realistic landscapes. This painting was done to show Durand conversing with the recently deceased Thomas Cole, the original founding force of the Hudson River School.

Figure 1.7
Ferde Grofé's *Grand Canyon Suite* pays homage to the grandeur of Arizona's Grand Canyon.

"You know, I used to like this hobby ... But shoot! Seems like *everybody's* got a rock collection."

Figure 1.8
References to geology are frequently found in comics as this *Far Side* cartoon by Gary Larson illustrates.

Figure 1.9
As these headlines from various newspapers indicate, geologic events affect our everyday lives.

fallen on the distribution and exploitation of natural resources. The configuration of Earth's surface, or its topography, which is shaped by geologic agents, plays a critical role in military tactics. Natural barriers such as mountain ranges and rivers have frequently served as political boundaries.

HOW DOES GEOLOGY AFFECT OUR EVERYDAY LIVES?

Destructive volcanic eruptions, devastating earthquakes, disastrous landslides, large sea waves, floods, and droughts make headlines because they wreck many people's lives (Figure 1.9). Although we cannot prevent most of these natural disasters, the more we know about them the better we can predict—and perhaps control—the severity of their impact.

Equally important, but not always as well understood or appreciated, is the connection between geology and economic and political power. The distribution of mineral and energy resources is unequal, and no country is self-sufficient for all its mineral and energy needs. Throughout history, people have fought wars to secure needed mineral and energy resources. We need look no further than 1990–1991 to see that the United States became involved in the Gulf War largely because of the need to protect U.S. oil interests in that region. Mineral and energy availability and needs often shape foreign policy. The sanctions that the United States imposed on South Africa in 1986, for example, did not include most of the important minerals we had been importing and needed for our industrialized society, such as platinum-group minerals. Many foreign policies and treaties develop from the need to acquire and maintain adequate mineral and energy resources.

Most readers of this book will not become professional geologists. Everyone, however, should have a basic understanding of the geologic processes that ultimately affect us all. Such an understanding is important so that you can avoid, for example, building in an area prone to landslides or flooding. Who wants to watch their dream house slide down a hillside or clean up after a flood?

You may also become involved in geologic decisions in other ways—for example, as a member of a planning board or as a property owner with mineral rights. In such cases, you must have a knowledge of geology to make informed decisions. Furthermore, many professionals must deal with geologic issues as part of their jobs. For example, lawyers are becoming more involved in issues ranging from ownership of natural resources to environmental impact reports. As government takes a greater role in environmental issues and regulations, members of Congress have increased the number of staff people devoted to studying environment and geology-related issues.

Most people are unaware of the extent to which geology affects their lives. For many people, the connection between geology and such well-publicized problems as nonrenewable energy and mineral resources, let alone waste disposal and pollution, is simply too far removed or too complex to fully appreciate. But consider for a moment just how dependent we are on geology in our daily routines.

Much electricity for appliances comes from burning coal, oil, or natural gas, or from uranium consumed in nuclear-generating plants. Geologists locate the coal, petroleum, and uranium. The copper or other metal wires through which electricity travels are manufactured from materials found as the result of mineral exploration. The buildings we live and work in owe their very existence to geologic resources. A few examples are the concrete foundation (concrete is a mixture of clay, sand, or gravel, and limestone), the drywall (made largely from the mineral gypsum), the windows (the mineral quartz is the principal ingredient in the manufacture of glass), and the metal or plastic plumbing fixtures inside the building (the metals are from ore deposits, and the plastics are most likely manufactured from petroleum distillates of crude oil).

Furthermore, when we go to work, our cars and public transport are powered and lubricated by some type of petroleum by-product and are constructed of metal alloys and plastics. And the roads or rails we ride over are made of geologic materials, such as gravel, asphalt, concrete, or steel. All these items are the result of processing geologic resources.

As individuals and societies, the standard of living we enjoy directly depends on the consumption of geologic materials. Therefore, we need to be aware of geology and of how our use and misuse of geologic resources may affect the delicate balance of nature and irrevocably alter our culture as well as our environment.

The concept of **sustainable development** has received increasing attention, particularly since the United Nations Conference on Environment and Development met in Rio de Janeiro, Brazil, during the summer of 1992. This important concept links satisfying basic human needs with safeguarding our environment to ensure continued economic development. By redefining "wealth" to include such natural resources as clean air and water, as well as productive land, appropriate measures can be taken to ensure future generations have suf-

ficient natural resources to maintain and improve their standard of living.

If we are to have a world in which poverty is not widespread, then we must develop policies that encourage management of our natural resources along with continuing economic development. A growing global population will mean increased demand for food, water, and natural resources, particularly nonrenewable mineral and energy resources. Geologists will play an important role in locating the needed resources, as well as in protecting the environment for future generations.

WHAT WOULD YOU DO?

The concept of sustainable development links satisfying basic human needs with safeguarding our environment to ensure continued economic development. The standard of living we enjoy directly depends on the consumption of geologic materials. As the president of a large multinational mining company, discuss how you might balance the need to extract valuable ore deposits and turn a profit for your company versus the need to safeguard the environment, particularly if these deposits are located in an underdeveloped country with no environmental protection laws.

GLOBAL GEOLOGIC AND ENVIRONMENTAL ISSUES FACING HUMANKIND

Most scientists would argue that the greatest environmental problem facing the world today is overpopulation (Figure 1.10). With the world's population passing six billion people in 1999, it is projected that the world's population will grow by at least another billion people during the next two decades, bringing Earth's human population to more than seven billion. Although this may not seem a *geologic* problem per se, remember that these people must be fed, housed, and clothed—and all with minimal impact on the environment. Some of this population growth will take place in areas that are already at risk from such geologic hazards as earthquakes, volcanic eruptions, and mass wasting. Safe and adequate water supplies must be found and kept from being polluted. More oil, gas, coal, and alternative energy resources must be discovered and used to provide the energy to fuel the economies of nations with ever-increasing populations. New mineral resources must be found. In addition, to decrease dependency on

Figure 1.10
Overpopulation is the greatest environmental problem facing the world today. Until the world's increasing population is brought under control, people will continue to strain Earth's limited resources. A bathing fair at Pushkar, Rajasthan, India.

new sources of these materials, ways to reduce usage and to reuse materials must also be found.

The effects of overpopulation on the global ecosystem are varied. For many of the poor and nonindustrialized countries, the problem is too many people and not enough food. For the more developed and industrialized countries, too many people are rapidly depleting both the nonrenewable and renewable natural resource base. And in the most industrialized countries, people are producing more pollutants than the environment can safely recycle on a human time scale. In all cases, environmental imbalance created by a human population is exceeding Earth's carrying capacity.

One result of environmental imbalance is global warming caused by the greenhouse effect. Carbon dioxide is a by-product of respiration and the burning of organic material. As such, it is a component of the global ecosystem and is constantly recycled as part of the carbon cycle. The concern in recent years over the increase in atmospheric carbon dioxide has to do with its role in the greenhouse effect. The recycling of carbon dioxide between the crust and atmosphere is an important climatic regulator because carbon dioxide—as well as other gases such as methane, nitrous oxide, chlorofluorocarbons, and water vapor—allow sunlight to pass through them but trap the heat reflected back from Earth's surface. Heat is thus retained, causing the temperature of Earth's surface and, more importantly, the atmosphere to increase, producing the greenhouse effect.

Until the Industrial Revolution began during the mid-18th century, human contribution to the global temperature pattern was negligible. With industrialization and its accompanying burning of tremendous amounts of fossil fuels, carbon dioxide levels in the atmosphere have been steadily increasing since about 1880. In fact, atmospheric levels of carbon dioxide are currently almost 30% higher than a century ago, and nitrous oxide levels have climbed 15% and methane has increased 100% since the Industrial Revolution. At their current yearly rate of increase, greenhouse gas concentrations will probably triple from their present concentrations within the next 100 years, unless something is done to reduce their production.

The first steps in reducing greenhouse-causing emissions were taken in 1997 in Kyoto, Japan, where a treaty was negotiated calling for the United States and 37 other industrialized nations to reduce their greenhouse emissions by 2008. According to the Kyoto Protocol, the United States is required to cut carbon dioxide emissions by 7% below 1990 levels, and the European Union and Japan must reduce their carbon dioxide emissions to 8% and 6%, respectively, below previous levels. However, only 13 of the 84 nations signing the Kyoto Protocol have ratified it, and most major industrialized nations are unlikely to do so until at least 2001.

In addition to industrial carbon dioxide emissions, research also indicates that deforestation of large areas, particularly in the tropics, is another cause of increased levels of carbon dioxide (Figure 1.2). This is because plants use carbon dioxide in photosynthesis and thus remove it from the atmosphere. With a decrease in the global vegetation cover, less carbon dioxide is removed from the atmosphere.

Because of the increase in human-produced greenhouse gases during the last 200 years, many scientists are concerned that a global warming trend has already begun and will result in severe climatic shifts. Most computer models based on the current rate of increase in greenhouse gases show Earth warming as a whole by about 3°C during the next 100 years. Such a temperature change will be uneven, however, with the greatest warming occurring in the higher latitudes.

As a consequence of this warming, rainfall patterns and growing conditions will shift dramatically. Under such a climate change, countries in high and middle latitudes such as Canada and Europe will see an increase in crop yields, whereas crop yields in lower latitudes will generally decrease. With continued global warming, mean sea level will also rise as icecaps and glaciers melt and contribute their water to the world's oceans. Scientists predict that by the 2050s, sea level will rise 21 cm, increasing the number of people at risk from flooding in coastal areas by approximately 20 million.

We cannot leave the subject of global warming without pointing out that many scientists are not convinced that the global warming trend is the direct result of increased human activity related to industrialization.

They point out that although there has been an increase in greenhouse gases, there is still uncertainty about their rate of generation and rate of removal and about whether the rise in global temperature during the past century is the result of normal climatic variations through time or the result of human activity. Furthermore, they note that even if there is a general global warming during the next century, it is not certain that the dire predictions made by proponents of the global warming theory will come true. Earth, as we know, is a remarkably complex system, with many feedback mechanisms and interconnections throughout its various subsystems and cycles. It is very difficult to predict all the consequences that global warming would have for atmospheric and oceanic circulation patterns.

THE ORIGIN OF THE SOLAR SYSTEM AND THE DIFFERENTIATION OF EARLY EARTH

According to the currently accepted theory for the origin of the solar system (Figure 1.11), interstellar material in a spiral arm of the Milky Way Galaxy condensed and began collapsing. As this cloud gradually collapsed under the influence of gravity, it flattened and

Figure 1.11
The currently accepted theory for the origin of our solar system involves (a) a huge nebula condensing under its own gravitational attraction, then (b) contracting, rotating, and (c) flattening into a disk, with the Sun forming in the center and eddies gathering up material to form planets. As the Sun contracted and began to visibly shine, (d) intense solar radiation blew away unaccreted gas and dust until finally, (e) the Sun began burning hydrogen and the planets completed their formation.

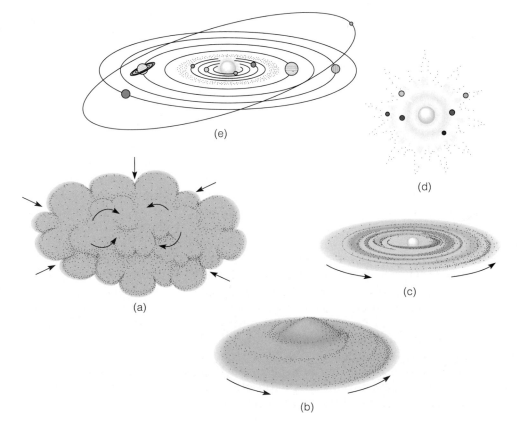

began rotating counterclockwise, with about 90% of its mass concentrated in the central part of the cloud. The rotation and concentration of material continued, and an embryonic Sun formed, surrounded by a turbulent, rotating cloud of material called a *solar nebula.*

The turbulence in this solar nebula formed localized eddies where gas and solid particles condensed. During the condensation process, gaseous, liquid, and solid particles began accreting into ever-larger masses called *planetesimals* that eventually became true planetary bodies. While the planets were accreting, material that had been pulled into the center of the nebula also condensed, collapsed, and was heated to several million degrees by gravitational compression. The result was the birth of a star, our Sun.

Some 4.6 billion years ago, enough material eventually gathered together in one of the turbulent eddies that swirled around the early Sun to form the planet Earth. Scientists think that this early Earth was rather cool, so the accreting elements and nebular rock fragments were solids rather than gases or liquids. This early Earth is also thought to have been of generally uniform composition and density throughout (Figure 1.12a). It was composed mostly of compounds of silicon, iron, magnesium, oxygen, aluminum, and smaller amounts of all other chemical elements. Subsequently, when Earth underwent heating, this homogeneous composition disappeared (Figure 1.12b), and the result was a differentiated planet consisting of a series of concentric layers of differing composition and density (Figure 1.12c). This differentiation into a layered planet is probably the most significant event in Earth history. Not only did it lead to the formation of a crust and eventually to continents, but it was also probably responsible for the emission of gases from the interior, which eventually led to the formation of the oceans and the atmosphere.

WHY IS EARTH A DYNAMIC PLANET?

Earth is a dynamic planet that has continuously changed during its 4.6-billion-year existence. The size, shape, and geographic distribution of continents and ocean basins have changed through time; the composition of the atmosphere has evolved; and life-forms existing today differ from those that lived during the past. We can easily visualize how mountains and hills are worn down by erosion and how landscapes are changed by the forces of wind, water, and ice. Volcanic eruptions and earthquakes reveal an active interior, and folded and fractured rocks indicate the tremendous power of Earth's internal forces.

Earth consists of three concentric layers: the core, the mantle, and the crust (Figure 1.13). This orderly division results from density differences between the layers as a function of variations in composition, temperature, and pressure.

The **core** has a calculated density of 10 to 13 grams per cubic centimeter (g/cm^3) and occupies about 16% of Earth's total volume. Seismic (earthquake) data indicate that the core consists of a small, solid inner region and a larger, apparently liquid, outer portion. Both are thought to consist largely of iron and a small amount of nickel.

The **mantle** surrounds the core and comprises about 83% of Earth's volume. It is less dense than the core (3.3–5.7 g/cm^3) and is thought to be composed largely of *peridotite,* a dark, dense igneous rock containing abundant iron and magnesium. The mantle can be divided into three distinct zones based on physical characteristics. The lower mantle is solid and forms most of the volume of Earth's interior. The **asthenosphere** surrounds the lower mantle. It has the same

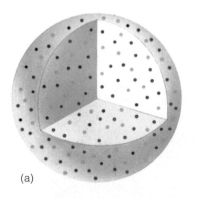
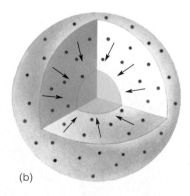
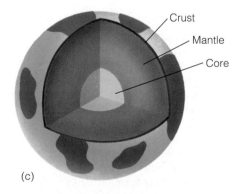

Crust
Mantle
Core

(a)   (b)   (c)

Figure 1.12
(a) Early Earth was probably of uniform composition and density throughout. (b) Heating of the early Earth reached the melting point of iron and nickel, which, being denser than silicate minerals, settled to Earth's center. At the same time, the lighter silicates flowed upward to form the mantle and the crust. (c) In this way, a differentiated Earth formed, consisting of a dense iron–nickel core, an iron-rich silicate mantle, and a silicate crust with continents and ocean basins.

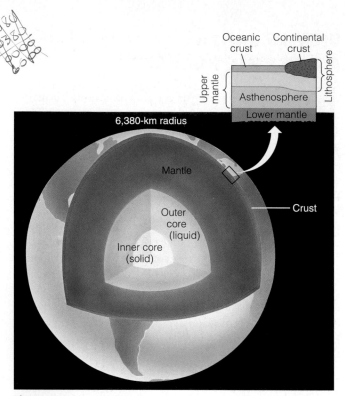

Figure 1.13

A cross section of Earth, illustrating the core, mantle, and crust. The enlarged portion shows the relationship between the lithosphere (composed of the continental crust, oceanic crust, and solid upper mantle) and the underlying asthenosphere and lower mantle.

composition as the lower mantle but behaves plastically and slowly flows. Partial melting within the asthenosphere generates *magma* (molten material), some of which rises to the surface because it it less dense than the rock from which it was derived. The upper mantle consists of the asthenosphere and the overlying solid mantle rocks up to the base of the crust. The solid portion of the upper mantle and the over-

lying crust constitute the **lithosphere,** which is broken into numerous individual pieces called **plates** that move over the asthenosphere as a result of underlying *convection cells* (Figure 1.14). Interactions of these plates are responsible for such phenomena as earthquakes, volcanic eruptions, and the formation of mountain ranges and ocean basins.

The **crust,** the outermost layer of Earth, consists of two types. *Continental crust* is thick (20–90 km), has an average density of 2.7 g/cm³, and contains considerable silicon and aluminum. *Oceanic crust* is thin (5–10 km), denser than continental crust (3.0 g/cm³), and is composed of the igneous rock *basalt*.

Since the widespread acceptance of plate tectonic theory more than 25 years ago, geologists have viewed Earth from a global perspective in which all its subsystems and cycles are interconnected. Thus, the distribution of mountain chains, major fault systems, volcanoes, and earthquakes, the origin of new ocean basins, the movement of continents, and several other geologic processes and features are all interrelated.

GEOLOGY AND THE FORMULATION OF THEORIES

The term **theory** has various meanings. In colloquial usage, it means a speculative or conjectural view of something—hence the widespread belief that scientific theories are little more than unsubstantiated wild guesses. In scientific usage, however, a theory is a coherent explanation for one or several related natural phenomena supported by a large body of objective evidence. From a theory are derived predictive statements that can be tested by observations and/or experiments so that their validity can be assessed. The law of universal

Figure 1.14

Earth's plates are thought to move as a result of underlying mantle convection cells in which warm material from deep within Earth rises toward the surface, cools, and then, upon losing heat, descends back into the interior. The movement of these convection cells is thought to be the mechanism responsible for the movement of Earth's plates, as shown in this diagrammatic cross section.

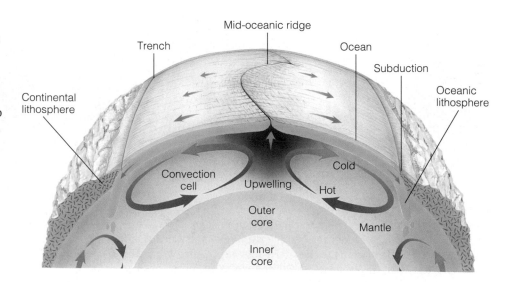

gravitation is an example of a theory describing the attraction between masses (an apple and Earth and in the popularized account of Newton and his discovery).

Theories are formulated through the process known as the **scientific method.** This method is an orderly, logical approach that involves gathering and analyzing the facts or data about the problem under consideration. Tentative explanations, or **hypotheses,** are then formulated to explain the observed phenomena. Next the hypotheses are tested to see if what they predict actually occurs in a given situation. Finally, if one of the hypotheses is found, after repeated tests, to explain the phenomena, then the hypothesis is proposed as a theory. Remember, however, that in science even a theory is still subject to further testing and refinement as new data become available.

The fact that a scientific theory can be tested and is subject to such testing separates science from other forms of human inquiry. Because scientific theories can be tested, they have the potential of being supported or even proved wrong. Accordingly, science must proceed without any appeal to beliefs or supernatural explanations, not because such beliefs or explanations are necessarily untrue but because we have no way to investigate them. For this reason, science makes no claim about the existence or nonexistence of a supernatural or spiritual realm.

Each scientific discipline has certain theories that are of particular importance for that discipline. In geology, the formulation of plate tectonic theory has changed the way geologists view Earth. Geologists now view Earth history in terms of interrelated events that are part of a global pattern of change. Before plate tectonic theory was generally accepted, numerous hypotheses were proposed and tested. Thus, the evolution of this theory illustrates the scientific method at work (see Perspective 1.2).

PLATE TECTONIC THEORY

The acceptance of **plate tectonic theory** is recognized as a major milestone in the geologic sciences. It is comparable to the revolution caused by Darwin's theory of evolution in biology. Plate tectonics has provided a framework for interpreting the composition, structure, and internal processes of Earth on a global scale. It has led to the realization that the continents and ocean basins are part of a lithosphere–atmosphere–hydrosphere (water portion of the planet) system that evolved together with Earth's interior (Table 1.3).

According to plate tectonic theory, the lithosphere is divided into plates that move over the asthenosphere (Figure 1.15). Zones of earthquake activity, volcanic

activity, or both mark most plate boundaries. Along these boundaries, plates diverge, converge, or slide sideways past each other.

At **divergent plate boundaries,** plates move apart as magma rises to the surface from the asthenosphere (Figure 1.16). The magma solidifies to form rock, which attaches to the moving plates. The margins of divergent plate boundaries are marked by mid-oceanic ridges in oceanic crust, such as the Mid-Atlantic Ridge, and are recognized by linear rift valleys, where newly forming divergent boundaries occur beneath continental crust (Figure 1.15).

Plates move toward one another along **convergent plate boundaries** where one plate sinks beneath another plate along what is known as a **subduction zone** (Figure 1.16). As the plate descends into Earth, it becomes hotter until it melts, or partially melts, thus generating a magma. As this magma rises, it may erupt at Earth's surface, forming a chain of volcanoes. The Andes Mountains on the west coast of South America are a good example of a volcanic mountain range formed as a result of subduction of the Nazca plate beneath the South American plate along a convergent plate boundary (Figure 1.15).

Transform plate boundaries are sites where plates slide sideways past each other (Figure 1.16). The San Andreas fault in California is a transform plate boundary separating the Pacific plate from the North American

Table 1.3

Plate Tectonics and Earth Systems	
SOLID EARTH	Plate tectonics is driven by convection in the mantle and in turn drives mountain-building and associated igneous and metamorphic activity.
ATMOSPHERE	Arrangement of continents affects solar heating and cooling, and thus winds and weather systems. Rapid plate spreading and hot-spot activity may release volcanic carbon dioxide and affect global climate.
HYDROSPHERE	Continental arrangement affects ocean currents. Rate of spreading affects volume of mid-oceanic ridges and hence sea level. Placement of continents may contribute to onset of ice ages.
BIOSPHERE	Movement of continents creates corridors or barriers to migration, the creation of ecological niches, and transport of habitats into more-or-less favorable climates.
EXTRATERRESTRIAL	Arrangement of continents affects free circulation of ocean tides and influences tidal slowing of Earth's rotation.

Source: Adapted by permission from Stephen Dutch, James S. Monroe, and Joseph Moran, Earth Science (Minneapolis/St. Paul: West Publishing Co.).

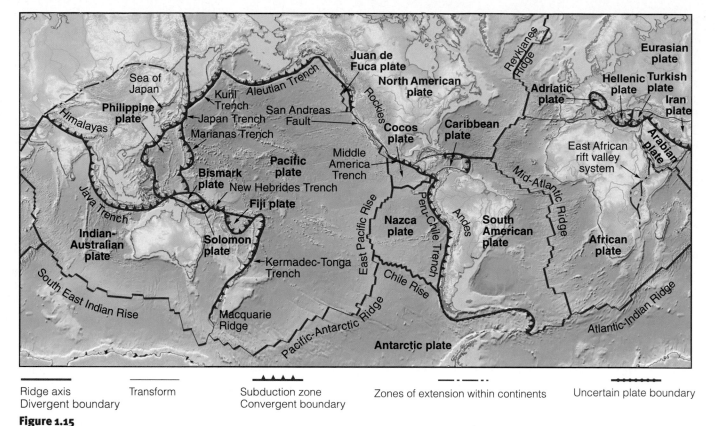

Ridge axis
Divergent boundary

Transform

Subduction zone
Convergent boundary

Zones of extension within continents

Uncertain plate boundary

Figure 1.15
Earth's lithosphere is divided into rigid plates of various sizes that move over the asthenosphere.

plate (Figure 1.15). The earthquake activity along the San Andreas fault results from the Pacific plate moving northward relative to the North American plate.

A revolutionary concept when it was proposed in the 1960s, plate tectonic theory has had significant and far-reaching consequences in all fields of geology because it provides the basis for relating many seemingly unrelated geologic phenomena. Besides being responsible for the major features of Earth's crust, plate movements also affect the formation and distribution of

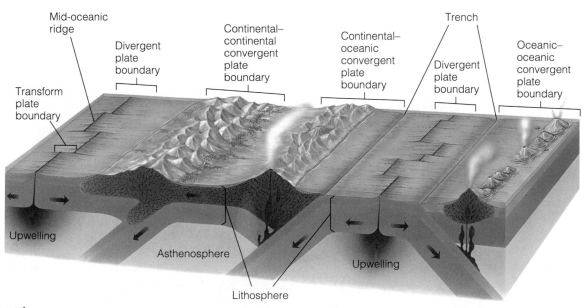

Figure 1.16
An idealized cross section illustrating the relationship between the lithosphere and the underlying asthenosphere and the three principal types of plate boundaries: divergent, convergent, and transform.

Perspective 1.2

The Formulation of Plate Tectonic Theory

The idea that continents moved during the past goes back to the time when people first noticed that the margins of eastern South America and western Africa looked as if they fit together. Geologists also noticed that similar or identical fossils occur on widely separated continents, that the same types of rocks from the same time period are found on different continents, and that ancient rocks and features indicating former glacial conditions occur in today's tropical areas. In 1912 Alfred Wegener, a German meteorologist, proposed the hypothesis of *continental drift* to explain and synthesize these myriad facts.

Wegener stated that at one time all continents were united into a single supercontinent which later broke apart, with individual continents drifting to their current locations. The continental drift hypothesis explained why the shorelines of different continents fit together, how different mountain ranges were once part of a larger continuous mountain range, why the same fossil animals and plants are found on different continents, and why rocks indicating glacial conditions are now found on continents located in the tropics.

Wegener's hypothesis and its predictions could be tested by asking what type of rocks or fossils one would expect to find at a given location on a continent if that continent was in the tropics 180 million years ago. To test the continental drift hypothesis, researchers only had to go into the field and examine the rocks and fossils for a particular time period on any continent to see if they indicated what the hypothesis predicted for the proposed location of that continent. In almost all cases, the data fit the hypothesis. However, Wegener's hypothesis did not explain how continents moved over oceanic crust and what the driving mechanism of continental movement was.

During the late 1950s and early 1960s, new data about the seafloor emerged that enabled geologists to propose the hypothesis of *seafloor spreading*. This hypothesis suggested that continents and segments of oceanic crust move together as single units, and that some type of thermal convection cell system operating within Earth was the mechanism responsible for plate movements.

Seafloor spreading and continental drift were then combined into a single hypothesis in which moving rigid plates are composed of continental and/or oceanic crust and the underlying upper mantle. In this hypothesis, plates move away from mid-oceanic ridges where new crust is formed, and toward oceanic trenches where crust is consumed or destroyed, and mountain chains are formed adjacent to the oceanic trenches.

According to this combined hypothesis, Europe and North America should be steadily moving away from each other at a rate of up to several centimeters per year. Precise measurements of continental positions by satellites have verified this, thus confirming the validity of the plate movement hypothesis.

Furthermore, if plates are moving away from mid-oceanic ridges as predicted by the combined hypothesis, then rocks of the oceanic crust should be progressively older with increasing distance from mid-oceanic ridges. To test this prediction, deep-sea sediment and oceanic crust were drilled as part of a massive scientific study of ocean basins. Analysis of the oceanic crust and the layer of sediment immediately above it showed that the age of the oceanic crust does indeed increase with distance from the mid-oceanic ridges, and that the oldest oceanic crust is adjacent to the continental margins.

With the confirmation of these and other predictions of the combined hypothesis, most geologists accept that the hypothesis is correct and therefore call it the *plate tectonic theory*. Its acceptance has been widespread because of the overwhelming evidence supporting it as well as explaining the relationship among many seemingly unrelated geologic features and events.

Earth's natural resources, as well as influence the distribution and evolution of the world's biota.

Plate tectonic theory has been particularly crucial in interpreting Earth history. For example, the Appalachian Mountains in eastern North America and the mountain ranges of Greenland, Scotland, Norway, and Sweden are not the result of unrelated mountain-building episodes but rather are part of a larger mountain-building event that involved the closing of an ancient "Atlantic Ocean" and the formation of the supercontinent Pangaea about 245 million years ago.

THE ROCK CYCLE

Arock is an aggregate of **minerals,** which are naturally occurring, inorganic, crystalline solids that have definite physical and chemical properties. Minerals are composed of elements such as oxygen, sili-

con, and aluminum, and elements are made up of atoms, the smallest particles of matter that still retain the characteristics of an element. More than 3500 minerals have been identified and described, but only about a dozen make up the bulk of the rocks in the crust (see Table 2.7).

Geologists recognize three major groups of rocks—*igneous, sedimentary,* and *metamorphic*—each of which is characterized by its mode of formation. Each group contains a variety of individual rock types that differ from one another on the basis of composition or texture (the size, shape, and arrangement of mineral grains).

The **rock cycle** is a way of viewing the interrelationships between Earth's internal and external processes (Figure 1.17). It relates the three rock groups to each other; to surficial processes such as weathering, transportation, and deposition; and to internal processes such as magma generation and metamorphism. Plate movement is the mechanism responsible for recycling rock materials and therefore drives the rock cycle.

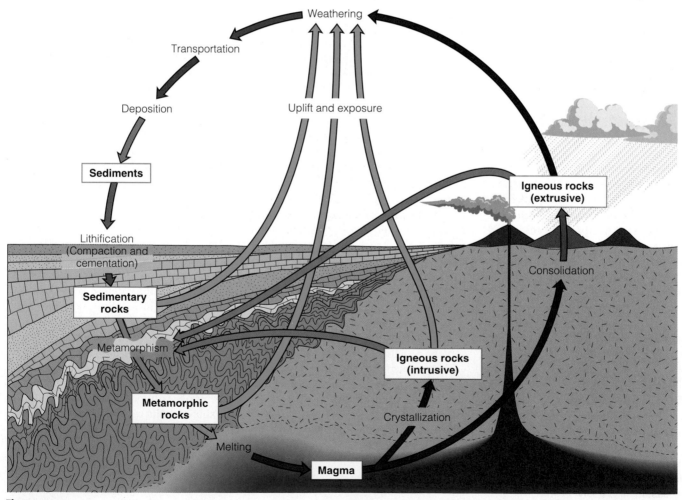

Figure 1.17
The rock cycle showing the interrelationships between Earth's internal and external processes and how each of the three major rock groups is related to the others.

Igneous rocks result from the crystallization of magma, or the accumulation and consolidation of volcanic ejecta such as ash. As a magma cools, minerals crystallize, and the resulting rock is characterized by interlocking mineral grains. Magma that cools slowly beneath the surface produces *intrusive igneous rocks* (Figure 1.18a); magma that cools at the surface produces *extrusive igneous rocks* (Figure 1.18b).

Rocks exposed at Earth's surface are broken into particles and dissolved by various weathering processes. The particles and dissolved material may be transported by wind, water, or ice and eventually deposited as *sediment*. This sediment may then be compacted or cemented into sedimentary rock.

Sedimentary rocks originate by consolidation of rock fragments, precipitation of mineral matter from solution, or compaction of plant or animal remains (Figure 1.18c and d). Because sedimentary rocks form at or near Earth's surface, geologists can make inferences about the environment in which they were deposited, the type of transporting agent, and perhaps even something about the source from which the sediments were derived (see Chapter 6). Accordingly, sedimentary rocks are especially useful for interpreting Earth history.

Metamorphic rocks result from the alteration of other rocks, usually beneath the surface, by heat, pressure, and the chemical activity of fluids. For example, marble, a rock preferred by many sculptors and builders, is a metamorphic rock produced when the agents of metamorphism are applied to the sedimentary rock limestone or dolostone. Metamorphic rocks are either *foliated* (Figure 1.18e) or *nonfoliated* (Figure 1.18f). Foliation, the parallel alignment of minerals due to pressure, gives the rock a layered or banded appearance.

(a) Granite

(b) Basalt

(c) Conglomerate

(d) Limestone

(e) Gneiss

(f) Quartzite

Figure 1.18
Hand specimens of common igneous (a, b), sedimentary (c, d), and metamorphic (e, f) rocks. (a) Granite, an intrusive igneous rock. (b) Basalt, an extrusive igneous rock. (c) Conglomerate, a sedimentary rock formed by the consolidation of rock fragments. (d) Limestone, a sedimentary rock formed by the extraction of mineral matter from seawater by organisms or by the inorganic precipitation of the mineral calcite from seawater. (e) Gneiss, a foliated metamorphic rock. (f) Quartzite, a nonfoliated metamorphic rock.

The Rock Cycle and Plate Tectonics

Interactions among plates determine, to a certain extent, which of the three rock groups will form (Figure 1.19). For example, weathering and erosion produce sediments that are transported by agents such as running water from the continents to the oceans, where they are deposited and accumulate. These sediments, some of which may be lithified and become sedimentary rock, become part of a moving plate along with the underlying oceanic crust. When plates converge, heat and pressure generated along the plate boundary may lead to igneous activity and metamorphism within the descending oceanic plate, thus producing various igneous and metamorphic rocks.

Some of the sediment and sedimentary rock is subducted and melts, whereas other sediments and sedimentary rocks along the boundary of the nonsubducted plate are metamorphosed by the heat and pressure generated along the converging plate boundary. Later, the mountain range or chain of volcanic islands formed along the convergent plate boundary will once again be weathered and eroded, and the new sediments will be transported to the ocean to begin yet another cycle.

The interrelationship between the rock cycle and plate tectonics is a good example of how Earth's various subsystems and cycles are all interrelated. Heating within Earth's interior produces convection cells that power the movement of plates and of magma, which forms intrusive and extrusive igneous rocks. Movement along plate boundaries may result in volcanic activity, earthquakes, and in some cases, mountain building. The interactions among atmosphere, hydrosphere, and biosphere contribute to the weathering of rocks exposed on Earth's surface. Plates descending back into Earth's interior generate heat and pressure, which may lead to metamorphism and the generation of magma, and to yet another recycling of materials.

WHAT IS GEOLOGIC TIME AND UNIFORMITARIANISM?

An appreciation of the immensity of geologic time is central to understanding the evolution of Earth and its biota. Indeed, time is one of the main aspects that sets geology apart from the other sciences, except astronomy. Most people have difficulty comprehending geologic time because they tend to think in terms of the human perspective—seconds, hours, days, and years. Ancient history is what occurred hundreds or even thousands of years ago. When geologists talk of ancient geologic history, however, they are referring to events that happened hundreds of millions or even billions of years ago. To a geologist, recent geologic events are those that occurred within the last million years or so.

As we discussed earlier in this chapter, it is important to remember that Earth goes through cycles of much longer duration than the human perspective of time. Because of their geologic perspective of time and how the various Earth subsystems and cycles are interrelated, geologists can make valuable contributions to many of the current environmental debates, such as those involving global warming and sea level changes.

The **geologic time scale** resulted from the work of many 19th-century geologists who pieced together information from numerous rock exposures and constructed a sequential chronology based on changes in

Figure 1.19
Plate tectonics and the rock cycle. The cross section shows how the three major rock groups—igneous, metamorphic, and sedimentary—are recycled through both the continental and oceanic regions.

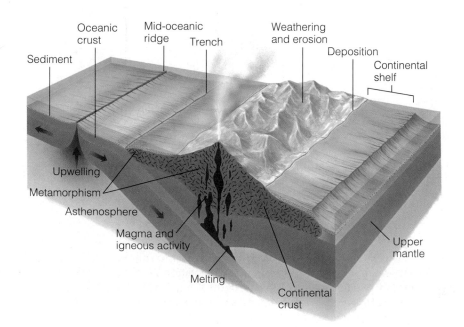

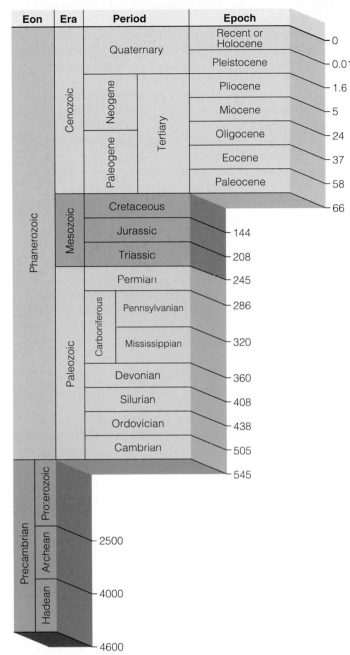

Figure 1.20
The geologic time scale. Numbers to the right of the columns are ages in millions of years before the present.

Uniformitarianism is a powerful principle that allows us to use present-day processes as the basis for interpreting the past and for predicting potential future events. We should keep in mind that uniformitarianism does not exclude such sudden or catastrophic events as volcanic eruptions, earthquakes, landslides, or flooding. These are processes that shape our modern world, and, in fact, some geologists view the history of Earth as a series of such short-term or punctuated events. Such a view is certainly in keeping with the modern principle of uniformitarianism.

Furthermore, uniformitarianism does not require that the rates and intensities of geologic processes be constant through time. We know that volcanic activity was more intense in North America 5 to 10 million years ago than it is today, and that glaciation has been more prevalent during the last several million years than in the previous 300 million years.

What uniformitarianism means is that even though the rates and intensities of geologic processes have varied during the past, the physical and chemical laws of nature have remained the same. Although Earth is in a dynamic state of change and has been ever since it formed, the processes that have shaped it during the past are the same ones operating today.

WHY IS THE STUDY OF GEOLOGY IMPORTANT?

The most important lesson to be learned from the study of geology is that Earth is an extremely complex planet in which interactions are taking place between its various subsystems and have been for the past 4.6 billion years. If the human species is to survive, we must understand how the various subsystems work and interact with each other, and more importantly, how our actions affect the delicate balance among these systems.

The study of geology is more than learning lots of facts about Earth. Geology is an integral part of our lives. The standard of living that we as individuals and societies enjoy and hope to pass on to the next generation directly depends on the consumption of natural resources and interaction with the environment. An appreciation of geology and its relationship to the environment is critical if we, as a species, are to continue to exist on this planet.

As you study the various topics discussed in this book, keep in mind the themes discussed in this chapter and how, like the parts of a system, they are interrelated. By relating each chapter's topic to its place in the Earth system, you will gain a greater appreciation of why geology is so integral to our lives.

Earth's biota through time. Subsequently, with the discovery of radioactivity in 1895 and the development of various radiometric dating techniques, geologists have since been able to assign absolute age dates in years to the subdivisions of the geologic time scale (Figure 1.20).

One of the cornerstones of geology is the **principle of uniformitarianism.** It is based on the premise that present-day processes have operated throughout geologic time. Therefore, to understand and interpret geologic events from evidence preserved in rocks, we must first understand present-day processes and their results.

Bruce Babbitt

Science: Preserving Our Heritage

Bruce Babbitt graduated from the University of Notre Dame where he majored in geology and later received a master's degree in geophysics from the University of Newcastle in England. He had planned on a career in mining until a trip to Bolivia in 1962. As he says, the poverty he saw there "crystallized my social awareness." Instead of mining, he became a lawyer, graduating from Harvard Law School. He returned to his home state of Arizona where he became the state's attorney general in 1974. He subsequently served two terms as governor. In 1993 President Bill Clinton named Babbitt to be Secretary of the Department of the Interior.

The following essay is based on Secretary Babbitt's statement before the Committee on Natural Resources of the U.S. House of Representatives in February 1993 and his speech to the National Press Club in April 1993.

I grew up in Flagstaff, Arizona, a small town in the West. In those days Flagstaff was the center of a small regional economy dependent on the management of natural resources—mining, forestry, reclamation, and grazing—and on decisions made about those resources at the Department of the Interior. Our horizons, then as now, were dominated by the Grand Canyon National Park and the Navajo Indian Reservation. Our opportunities and our problems, then as now, were inextricably intertwined with the management of the federal

and Indian lands of the West.

I learned early on that the development of the West was guided, for good or ill, by the policies set by Congress and the federal government and administered by the Department of the Interior. The role of the department as land manager, natural resources steward, wildlife conservator, parks curator, and trustee were not abstract notions. Each decision made by the department echoes through the economy, politics, and quality of life in the West.

But the Interior Department is not just the "Department of the West." The department's minerals management responsibilities extend from the outer continental shelf of Alaska to George's Bank; its Office of Surface Mining has nationwide regulatory responsibilities; and the National Park Service plays an increasingly important role in offering recreational opportunities to city dwellers in the East as well as to citizens across the rest of America.

I believe we have a particular responsibility toward the great parks of the nation. And when I say "great parks," I mean all parks because each in its own right inspires its visitors. We must care for the crown jewels of the park system, such as Yellowstone, Grand Canyon, Yosemite, Acadia, and the Great Smokies. At the same time, we must care for our urban parks, such as Gateway and Golden Gate.

To this end, we successfully fought to protect the extraordinary wildlife and ecological resources of the California desert. This area is truly one of our national treasures. The abundance of wildlife and the amazing ecological diversity of the desert deserve to be protected for future generations of Americans. At the same time, these areas need to remain accessible and available as a place where citizens today—particularly those whose lives are mainly spent in the metropolitan regions of our country—can achieve the encounter with their natural heritage that is the fundamental purpose of our national parks and wilderness areas.

My goal is to improve the management of the nation's natural resources

and to balance needed development with a renewed emphasis on stewardship and conservation so that the United States can meet its needs in the 21st century. One way we will accomplish this is through the use of market principles in resource allocation. For one thing, this is an issue of fairness; it is simply unreasonable to say to the American people: Everyone should pay their fair share—except miners, timber companies, ranchers, and reclamation water users. Second, the move to market has important environmental benefits—market pricing of resources encourages conservation and the efficient allocation of limited resources.

One priority will be reform of the Mining Law of 1872. The law has been tinkered with before, but many problems remain—such as disposal of valuable public resources for nominal fees, inadequate environmental regulation, and lack of secure tenure for mineral exploration. The mining industry, other users of the public lands, and above all the American people will benefit if we have a modern mining law—one that takes full account of the public interest in the lands and minerals owned by the American people.

The transition to market principles for resource allocation also has major implications for the development and use of water in the West. The search for more water from ever-more distant sources with greater environmental destruction is a time-honored western tradition that must now give way to a simple reality: There is plenty of water developed and available in the West if only we will allow market principles to replace heavy-handed bureaucratic allocations. By pricing water at its true cost and thereby encouraging its conservation and wise use, there will be plenty of water for everyone.

Above all, then, we are set on creating a new American land ethic. We will bring old and true economic values to a new and urgent cause: the imperative to live more lightly and productively on the land—our land. ∎

Chapter Summary

1. Geology, the study of Earth, is divided into two broad areas: Physical geology is the study of Earth materials as well as the processes that operate within and on Earth's surface; historical geology examines the origin and evolution of Earth, its continents, oceans, atmosphere, and life.

2. Earth can be viewed as a system of interconnected components that interact and affect each other. The principal subsystems of Earth are the atmosphere, hydrosphere, biosphere, lithosphere, mantle, and core. Earth is a dynamic planet that continually changes because of interactions among its various subsystems and cycles.

3. Geology is part of the human experience. We can find examples of it in the arts, music, and literature. A basic understanding of geology is also important for dealing with the many environmental problems and issues facing society.

4. Geologists engage in a variety of occupations, the main one being exploration for mineral and energy resources. They are also becoming increasingly involved in environmental issues and making short- and long-range predictions of the potential dangers from such natural disasters as volcanic eruptions and earthquakes.

5. About 4.6 billion years ago, the solar system formed from a rotating cloud of interstellar matter.

Eventually, as this cloud condensed, it collapsed under the influence of gravity and flattened into a rotating disk. Within this rotating disk, the Sun, planets, and moons formed from the turbulent eddies of nebular gases and solids.

6. Earth is differentiated into layers. The outermost layer is the crust, which is divided into continental and oceanic portions. Below the crust is the solid portion of the upper mantle. The crust and the solid portion of the upper mantle, or lithosphere, overlie the asthenosphere, a zone that slowly flows. The asthenosphere is underlain by the solid lower mantle. Earth's core consists of an outer liquid portion and an inner solid portion.

7. The lithosphere is broken into a series of plates that diverge, converge, and slide sideways past one another.

8. The scientific method is an orderly, logical approach that involves gathering and analyzing facts about a particular phenomenon, formulating hypotheses to explain the phenomenon, testing the hypotheses, and finally proposing a theory. A theory is a testable explanation for some natural phenomenon that has a large body of supporting evidence.

9. Plate tectonic theory provides a unifying explanation for many geologic features and events. The interaction between plates is responsible for volcanic eruptions,

earthquakes, the formation of mountain ranges and ocean basins, and the recycling of rock material.

10. Igneous, sedimentary, and metamorphic rocks comprise the three major groups of rocks. Igneous rocks result from the crystallization of magma or the consolidation of volcanic ejecta. Sedimentary rocks are formed by the consolidation of rock fragments, precipitation of mineral matter from solution, or compaction of plant or animal remains. Metamorphic rocks are produced from other rocks, generally beneath Earth's surface, by heat, pressure, and chemically active fluids.

11. The rock cycle illustrates the interactions between internal and external Earth processes and shows how the three rock groups are interrelated.

12. Time sets geology apart from the other sciences, except astronomy, and an appreciation of the immensity of geologic time is central to understanding Earth's evolution. The geologic time scale is the calendar geologists use to date past events.

13. The principle of uniformitarianism is basic to the interpretation of Earth history. This principle holds that the laws of nature have been constant through time and that the same processes operating today have operated throughout the past, although at different rates.

Important Terms

asthenosphere
convergent plate boundary
core
crust
divergent plate boundary
geologic time scale
geology
historical geology
hypothesis

igneous rock
lithosphere
mantle
metamorphic rock
mineral
physical geology
plate
plate tectonic theory
principle of uniformitarianism

rock
rock cycle
scientific method
sedimentary rock
subduction zone
sustainable development
system
theory
transform plate boundary

Review Questions

1. A combination of related parts interacting in an organized fashion is:

   a. _____ a theory;
   b. _____ a cycle;
   c. _____ uniformitarianism;
   d. _____ a system;
   e. _____ a hypothesis.

2. The study of Earth materials is:

   a. _____ environmental geology;
   b. _____ physical geology;
   c. _____ historical geology;
   d. _____ structural geology;
   e. _____ economic geology.

3. The interaction between the atmosphere, hydrosphere, and biosphere is a major contributor to:

   a. _____ the generation of magma;
   b. _____ mountain building;
   c. _____ weathering of Earth materials;
   d. _____ metamorphism;
   e. _____ plate movement.

4. Plates are composed of:

   a. _____ the crust and upper mantle;
   b. _____ the asthenosphere and the solid portion of the upper mantle;
   c. _____ the crust and asthenosphere;
   d. _____ continental and oceanic crust only;
   e. _____ the core and mantle.

5. Plate movement is thought to be the result of:

   a. _____ density differences between the mantle and core;
   b. _____ rotation of the mantle around the core;
   c. _____ gravitational forces;
   d. _____ the Coriolis effect;
   e. _____ convection cells.

6. Which of the following statements about a scientific theory is *not* true?

   a. _____ it is an explanation for some natural phenomenon;
   b. _____ it is a conjecture or guess;
   c. _____ it has a large body of supporting evidence;
   d. _____ it is testable;
   e. _____ predictive statements can be derived from it.

7. Rocks resulting from the crystallization of magma are:

   a. _____ igneous;
   b. _____ metamorphic;
   c. _____ sedimentary;
   d. _____ migmatites;
   e. _____ answers (a) and (d).

8. Which of the following is an example of a divergent plate boundary?

   a. _____ the Andes Mountains of South America;
   b. _____ the San Andreas fault;
   c. _____ the Mid-Atlantic Ridge;
   d. _____ the Himalayas;
   e. _____ the Hawaiian Islands.

9. The premise that present-day processes have operated throughout geologic time is the principle of:

   a. _____ continental drift;
   b. _____ uniformitarianism;
   c. _____ Earth systems;
   d. _____ plate tectonics;
   e. _____ scientific deduction.

10. According to the currently accepted theory for the origin of the solar system:

    a. _____ a huge nebula collapsed under its own gravitational attraction;
    b. _____ the nebula formed a disk with the Sun in the center;
    c. _____ planetesmials accreted from gaseous, liquid, and solid particles;
    d. _____ all of the previous;
    e. _____ none of the previous.

11. The driving force of the rock cycle is:

    a. _____ uniformitarianism;
    b. _____ plate tectonics;
    c. _____ gravitational forces;
    d. _____ density differences between Earth's interior and surface;
    e. _____ volcanism.

12. Why is it important that everyone have a basic understanding of geology, even if they aren't going to become geologists?

13. Describe how you would use the scientific method to formulate a hypothesis explaining the similarity of mountain ranges on the East Coast of North America and those in England, Scotland, and the Scandinavian countries. How would you test your hypothesis?

14. Discuss what is meant by the statement "The health and well-being of the world's economy are completely dependent on geologic resources."

15. Discuss how the three major layers of Earth differ from each other and why the differentiation into a layered planet is probably the most significant event in Earth history.

16. Describe the various ways in which geology affects our everyday lives.

17. Explain how the principle of uniformitarianism allows for catastrophic events.

18. Discuss why plate tectonic theory is a unifying theory of geology.

19. Discuss the relationship between overpopulation and geology and how they are related.

20. Explain the advantage of using a system's approach to the study of Earth.

Points to Ponder

1. Propose a hypothesis that might have predated plate tectonic theory, explaining the similar distribution of volcanoes and earthquakes.

2. Why is an accurate geologic time scale particularly important for geologists in examining changes in global temperatures during the past?

3. Explain why a knowledge of geology would be useful in planning a military campaign against another country.

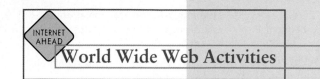

For these website addresses, along with current updates and exercises, log on to
http://www.brookscole.com/geo/

➤ EARTH SCIENCE RESOURCES ON THE INTERNET

This site, maintained by the Geology Department at the University of North Carolina at Chapel Hill, is an excellent starting place to learn how to find geoscience items on the World Wide Web. As you will see, a tremendous amount of information, graphics, and maps is available, and this site will get you started finding many of them and practicing your Net skills.

1. Read the *Starting up* and *How to find things* sections.
2. Visit several of the geology sites mentioned to see the variety of information, videos, and images available.
3. Visit the home page of several of the different geologic societies and organizations to see what the societies do and what information is available from them.
4. Check the home page of several of the different schools and universities to find information on course offerings, faculty, and requirements to earn a degree. Compare the home pages and information available from different universities around the world.

➤ ONLINE RESOURCES FOR EARTH SCIENTISTS (ORES)

This large and comprehensive Earth science resource list is operated by Bill Thoen and Ted Smith and provides links to many other resource lists. It is continually updated and is certainly a site worth visiting.

1. Check out the various headings under *Contents* to see what is available in the different geoscience and related areas, and explore the various links to other resource lists.
2. When you find a particular link interesting and would like to return to it in the future, make a "Bookmark" of the page (that is, save its URL). This way, when you want to go back to the site, all you have to do is find its name in the "Bookmark Menu" and click on it, and your browser automatically connects you to that site.

➤ WEST'S GEOLOGY DIRECTORY

This comprehensive directory of geologic Internet links is an excellent place to find links on virtually any geologic topic. The Contents-Index lists numerous topics. Just click on a topic, and a list of sites, which you can then click on, will appear. The links are regularly updated, and sites are checked, reviewed, and rated where possible. In addition, there is also a link to field trip guides and bibliographies.

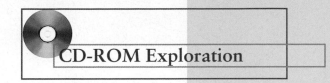

CD-ROM Exploration

➤ Exploring your *Earth Systems Today* CD-ROM will add to your understanding of the material in this chapter.

TOPIC: CURRENT NEWS

MODULE: EARTH TODAY

Explore activities in this module to see if you can discover the following for yourself:

Go to the News Panel and select "Earth Alert" or one of the other options and examine at least two of the world's "current hot spots" of geological activity. Comment on the geological aspects of these current news stories.

To what extent was each disaster you investigated predictable by geologists, and to what extent was each unpredictable?

If you were a planner or engineer with some geology background, what would you suggest for future mitigation of damage and possible loss of life in such disasters?

What other places on Earth are susceptible to disasters like the ones you just studied? What is it about the geology of those places that is related to their common risk of disaster?

Chapter 2

A spectacular example of the mineral tourmaline from the Himalaya Mine, San Diego County, California. Many museums in the United States and several other countries display mineral specimens from this mine.

OBJECTIVES

At the end of this chapter, you will have learned that:

- Atoms bond to form elements and compounds.
- Geologists have a specific definition for the term *mineral*.
- You can distinguish minerals from other naturally occurring and manufactured substances.

- Minerals are incredibly varied.
- Silicate and carbonate minerals, among others, constitute important mineral groups.
- Physical properties of minerals, related to their composition and structure, are used to identify minerals.
- Minerals form in various ways and under varied conditions.

- Certain minerals are important as constituents of rocks, and minerals differ qualitatively from rocks.
- Various minerals and rocks are important natural resources that are essential to industrialized societies.

Minerals

Prologue

T
he color, brilliance, and other properties of some minerals, rocks, and fossils have fascinated people for thousands of years. Several of these objects have served as religious symbols, talismans, or have been worn, carried, applied externally, or ingested for their presumed mystical or curative powers. Archaeological evidence indicates that people in Spain and France were carving objects from bone, ivory, horn, and various stones at least 75,000 years ago. By 3400 B.C., the ancient Egyptians were making ornaments from rock crystal (colorless quartz), amethyst (purple quartz), lapis lazuli (a rock composed of a variety of minerals), and several other stones. They mined turquoise more than 5000 years ago, and by 1650 B.C. they were mining emeralds from near the Red Sea.

Most gemstones are minerals—more rarely, rocks—that are cut and/or polished for jewelry. To qualify as a gemstone, a mineral or rock must be appealing for some reason. The most desirable quality is beauty, but brilliance, durability, and scarcity are important as well. Intense color is a preferred quality in many gemstones, but most gem-quality diamonds are colorless and valued in part because of their brilliance. Gemstones are even more desirable when some kind of lore is associated with them. According to legend, diamond wards off evil spirits, sickness, and floods. Topaz was once thought to avert mental disorders; ruby was believed to preserve its owner's health and warn of imminent bad luck. A symbol of immortality, a preserver of chastity, and a cure for dysentery are qualities attributed to emerald. Relating gemstones to birth month gives them even more appeal to many people (Table 2.1).

Gemstones are actually the raw materials for gems. That is, gems have been fashioned in some way from gemstones by shaping, polishing, or cutting. Many minerals and rocks are attractive but fail to qualify as gemstones. Cut and polished fluorite is beautiful, but is fairly common and too soft to be durable. Diamond, in contrast, meets most of the criteria for a gemstone, although not all diamonds are of gem quality; most, in fact, are used in industrial applications. Only about two dozen minerals and rocks are used as gemstones, some of which are considered *precious,* and others are referred to as *semiprecious.* Diamond (Figure 2.1), ruby, and sapphire (red and blue varieties of the mineral corundum), and emerald (a bright green variety of the mineral beryl) are the precious gemstones; precious opal is also sometimes included in this group. Several of the semiprecious gemstones are garnet, jade, tourmaline, topaz, peridot (olivine), aquamarine (light bluish green beryl), turquoise (Figure 2.2), and several varieties of quartz, such as amethyst, rose quartz, agate, and tiger's eye.

Also included among the semiprecious gemstones is an organic substance known as *amber,* and *pearl,* which is produced by organisms. Amber is a fossil resin from coniferous trees that commonly contains insects (Figure 2.3). Even though amber is neither a mineral nor a rock, people nevertheless consider it a semiprecious gemstone. Best known from the Baltic Sea region of Europe, sun-worshiping cultures believed that amber, with its golden translucence resembling the sun's brilliance, had mystical powers.

Table 2.1

Birthstones

Month	Birthstone	Mineral	Symbolizes
January	Garnet	Garnet	Constancy
February	Amethyst	Purple quartz	Sincerity
March	Aquamarine	Light green, blue-green beryl	Courage
	Bloodstone	Greenish chalcedony	
April	Diamond	Diamond	Innocence
May	Emerald	Beryl and several other green minerals	Love, success
June	Pearl	Dense, spherical white or light-colored calcareous concretion produced by organisms, especially mollusks	Health, longevity
	Alexandrite	Greenish chrysoberyl	
	Moonstone	Type of potassium feldspar	
July	Ruby	Red corundum	Contentment
August	Peridot	Pale, clear yellowish-green olivine	Married happiness
	Sardonyx	Gem variety of chalcedony	
September	Sapphire	Blue transparent corundum	Clear thinking
October	Opal	Opal	Hope
	Tourmaline	Tourmaline	
November	Topaz	Topaz	Fidelity
December	Turquoise	Turquoise	Prosperity
	Zircon	Zircon	

Mollusks produce pearl, which is essentially ready to use when found; it needs no shaping or polishing. A mollusk, such as a clam or oyster, forms pearl by depositing successive layers of tiny mineral crystals around some irritant. Most pearls are lustrous white, but some are silver gray, green, or black.

Transparent gemstones are most often cut to yield small, polished plane surfaces known as *facets,* which enhance the quality of reflected light. Gem cutters maximize the brilliance of diamond by faceting the gem so that it reflects as much light as possible (Figure 2.1). Opaque and translucent gemstones are rarely faceted. Rather, they are cut and polished into dome-shaped stones known as cabochons to emphasize their most interesting features, or they are simply polished by tumbling.

Most people are surprised to learn that diamond, the most sought-after gemstone, and the mineral graphite consist of the same chemical element, carbon. Other than composition, though, they share little in common; gem-quality diamond is colorless and the hardest mineral, whereas graphite is gray and so soft

that it can be scratched by a fingernail. The way the carbon atoms are arranged in these two minerals accounts for these differences. In diamond, the atoms bond tightly to one another in a three-dimensional framework, but in graphite the atoms form sheets that are held together by weak bonds. Incidentally, many people think the true test of a diamond is whether or not it will scratch glass. Diamond will indeed scratch glass, but so will several other minerals, including common quartz.

"Cutting" a diamond, the hardest substance known, is actually done by several processes. Diamond possesses four internal planes of weakness, or what are known as *cleavage planes,* so that if cleaved perfectly along these planes a diamond yields a "stone" shaped like two pyramids placed base to base. Large diamonds are commonly preshaped by cleaving them into smaller pieces that are then further shaped by sawing and grinding with diamond dust. A diamond's value is determined by evaluating its color (colorless specimens are most desirable), clarity (lack of internal flaws), cut, and carat, which is the measure

Figure 2.1
Their hardness, brilliance, beauty, durability, and scarcity make diamonds the most sought-after gemstones. The pendant in this necklace, housed in the Smithsonian Institution, is the Victoria Transvaal diamond from South Africa.

Figure 2.2
Turquoise—a sky blue, blue-green, or light green hydrated copper aluminum phosphate—is a semiprecious gemstone used for jewelry and as a decorative stone.

of a precious stone's weight (1 carat = 200 milligrams). The Victoria Transvaal diamond in Figure 2.1 weighs 68 carats, and the world's most famous blue diamond, the Hope diamond, is 45.5 carats.

Gemstones and the gems derived from them have been used for many purposes, including as personal symbols of wealth and power as well as for their presumed mystical powers as good luck charms and cures for various ailments. Many people own small gems, but most of the truly magnificent ones are in museums, collections of crown jewels, or religious objects.

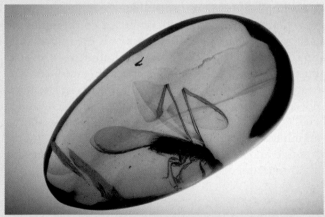

Figure 2.3
Insect preserved in amber. Even though amber is an organic substance, it is nevertheless valued as a semiprecious gemstone.

INTRODUCTION

In the Prologue we used the term *mineral* without really defining it. The term is used in a variety of ways, as, for example, a designation for dietary substances essential for good nutrition, such as calcium, iron, potassium, and magnesium. These are actually chemical elements and not minerals, at least in the geologic sense. Substances that are neither animal nor vegetable are also commonly called *minerals*. This usage implies that minerals are inorganic, which is correct, but not all inorganic substances are minerals. For instance, water and water vapor are not minerals even though they are composed of the same chemical elements as ice, which is a mineral, because minerals must be solids rather than liquids or gases. Geologists define a **mineral** as a naturally occurring, inorganic, crystalline solid. *Crystalline* means that minerals have an ordered internal arrangement of atoms. Minerals also have a narrowly defined chemical composition and characteristic physical properties such as color, hardness, and density. We will examine each part of this definition later in this chapter.

Most rocks are made up of one or more minerals, so minerals are the building blocks of rocks. Minerals as rock constituents are important not only to geologists, but also for other reasons. Ore deposits, for example, are natural concentrations of minerals or rocks that can be extracted and used for some purpose such as the manufacture of steel and aluminum. Gemstones (including diamonds, topaz, and emeralds) are minerals, glass is manufactured from sand composed of the mineral quartz, and phosphate-bearing minerals are mined and used for ceramics, fertilizers, and metallurgy. In short, industrialized societies depend on finding and using a variety of mineral resources including iron, copper, gold, and many others. One reason for the economic success of the United States and Canada is their abundant mineral resources. Furthermore, many minerals are attractive and eagerly sought out by private collectors and for museum displays (Figure 2.4).

WHAT DOES MATTER CONSIST OF?

Matter is defined as anything that has mass and occupies space. Accordingly, water, plants, animals, the atmosphere, and minerals and rocks are composed of matter. Three phases or states of matter are recognized*—*solids, liquids,* and *gases,* all of which are

---

*Actually, scientists recognize a fourth state of matter known as *plasma,* an ionized gas as in fluorescent and neon lights, and matter in the sun and stars.

Figure 2.4
Mineral display at the California Academy of Sciences in San Francisco.

important in geology (Figure 2.5, Table 2.2). Liquids such as surface water and groundwater and atmospheric gases are important geologic agents that will be discussed in later chapters. Here, though, we are interested mostly in solids, because by definition minerals are solids.

Elements and Atoms

Our discussion of minerals begins with a consideration of elements and atoms because all matter is made up of chemical **elements,** each of which is composed of tiny particles known as **atoms.** Atoms are the smallest units of matter that retain the characteristics of a particular element. That is, they cannot be split into substances of different composition. Ninety-two naturally occurring elements have been discovered, some of which are listed

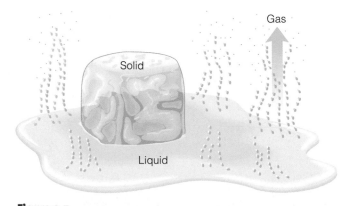

Figure 2.5
The three states of matter are shown by water as a solid (ice), a liquid, and a gas (water vapor). Only ice is a mineral, because all minerals are solids and ice has a specific internal arrangement of atoms.

Table 2.2

Phases or States of Matter

PHASE	CHARACTERISTICS	EXAMPLES
Solid	Rigid substance that retains its shape unless distorted by a force	Minerals, rocks, iron, wood, ice
Liquid	Flows easily and conforms to the shape of the containing vessel; has a well-defined upper surface and greater density than a gas	Water, lava, wine, blood, gasoline
Gas	Flows easily and expands to fill all parts of a containing vessel; lacks a well-defined upper surface; is compressible	Helium, nitrogen, air, water vapor

Table 2.3

Symbols, Atomic Numbers, and Electron Configurations for Some of the Naturally Occurring Elements

ELEMENT	SYMBOL	ATOMIC NUMBER	NUMBER OF ELECTRONS IN EACH SHELL			
			1	2	3	4
Hydrogen	H	1	1			
Helium	He	2	2			
Lithium	Li	3	2	1		
Beryllium	Be	4	2	2		
Boron	B	5	2	3		
Carbon	C	6	2	4		
Nitrogen	N	7	2	5		
Oxygen	O	8	2	6		
Fluorine	F	9	2	7		
Neon	Ne	10	2	8		
Sodium	Na	11	2	8	1	
Magnesium	Mg	12	2	8	2	
Aluminum	Al	13	2	8	3	
Silicon	Si	14	2	8	4	
Phosphorus	P	15	2	8	5	
Sulfur	S	16	2	8	6	
Chlorine	Cl	17	2	8	7	
Argon	Ar	18	2	8	8	
Potassium	K	19	2	8	8	1
Calcium	Ca	20	2	8	8	2

in Table 2.3, and several more have been made in laboratories (see the Periodic Table of Elements in Appendix B). All naturally occurring elements and most artificially produced ones have a name and a symbol, such as oxygen (O), aluminum (Al), and potassium (K).

At the center of an atom is a tiny **nucleus** made up of one or more particles known as **protons** with a positive electrical charge, and **neutrons,** which are electrically neutral (Figure 2.6). The nucleus is only about 1/100,000 of the diameter of an atom, yet it contains virtually all of the atom's mass. **Electrons,** particles with a negative electrical charge, orbit rapidly around the nucleus at specific distances in one or more **electron shells.** The electrons determine how an atom interacts with other atoms, but the nucleus determines how many electrons an atom has, because the positively charged protons attract and hold negatively charged electrons in their orbits.

The number of protons in its nucleus determines an atom's identity and its **atomic number.** Hydrogen (H), for instance, has 1 proton in its nucleus and thus has an atomic number of 1. The nuclei of helium (He) atoms possess 2 protons, whereas those of carbon (C) have 6, and uranium (U) have 92, so their atomic numbers are 2, 6, and 92, respectively. Atoms are also characterized by their **atomic mass number,** which is the sum of protons and neutrons in the nucleus (electrons contribute negligible mass to atoms). However, atoms of the same chemical element might have different atomic mass numbers because the number of neutrons can vary. All carbon (C) atoms have 6 protons—otherwise they would not be carbon—but the number of neutrons can be 12, 13, or 14. Thus, we can recognize three types of carbon, each with a different atomic mass number, or what are known as *isotopes* (Figure 2.7).

These isotopes of carbon, or those of any other element, behave the same chemically—carbon 12 and

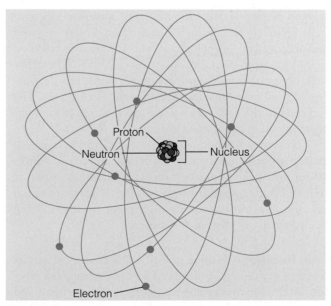

Figure 2.6
The structure of an atom. The dense nucleus consisting of protons and neutrons is surrounded by a cloud of orbiting electrons.

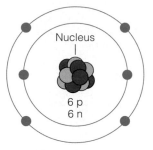
^{12}C (Carbon 12)

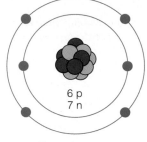
^{13}C (Carbon 13)

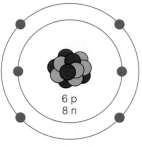
^{14}C (Carbon 14)

Figure 2.7
Schematic representation of the isotopes of carbon. Carbon has an atomic number of 6 and an atomic mass number of 12, 13, or 14, depending on the number of neutrons (n) in its nucleus.

carbon 14 are both present in carbon dioxide (CO_2), for example. However, some isotopes are radioactive, meaning that they spontaneously decay or change to other stable elements. Carbon 14 is radioactive, whereas both carbon 12 and carbon 13 are stable. Radioactive isotopes are important in determining the ages of rocks (see Chapter 8).

Bonding and Compounds

The interaction of electrons around atoms can result in two or more atoms joining together, a process known as **bonding.** If atoms of two or more different elements bond, the resulting substance is a **compound.** Gaseous oxygen consists only of oxygen atoms and is thus an element, whereas the mineral quartz, consisting of silicon and oxygen atoms, is a compound. Most minerals are compounds, although gold, silver, and several others are important exceptions.

To understand bonding, it is necessary to delve deeper into the structure of atoms. Recall that negatively charged electrons orbit the nuclei of atoms in electron shells. With the exception of hydrogen, which has only one proton and one electron, the innermost electron shell of an atom contains only two electrons. The other shells contain various numbers of electrons, but the outermost shell never has more than eight (Table 2.3). The electrons in the outermost shell are those that are usually involved in chemical bonding.

Two types of chemical bonds, *ionic* and *covalent,* are particularly important in minerals, and many minerals contain both types of bonds. Two other types of chemical bonds, *metallic* and *van der Waals,* are much less common but extremely important in determining the properties of some useful minerals.

Ionic Bonding   Notice in Table 2.3 that most atoms have fewer than eight electrons in their outermost electron shell. However, some elements, including neon and argon, have complete outer shells containing eight electrons; because of this electron configuration, these elements, known as the *noble gases,* do not react readily with other elements to form compounds. Interactions

among atoms tend to produce electron configurations similar to those of the noble gases. That is, atoms interact so that their outermost electron shell is filled with eight electrons, unless the first shell (with two electrons) is also the outermost electron shell, as in helium.

One way that the noble gas configuration can be attained is by the transfer of one or more electrons from one atom to another. Common salt is composed of the elements sodium (Na) and chlorine (Cl), each of which is poisonous, but when combined chemically they form the compound sodium chloride (NaCl), better known as the mineral halite. Notice in Figure 2.8a that sodium has 11 protons and 11 electrons; thus the positive electrical charges of the protons are exactly balanced by the negative charges of the electrons, and the atom is electrically neutral. Likewise, chlorine with 17 protons and 17 electrons is electrically neutral (Figure 2.8a). But neither sodium nor chlorine has 8 electrons in its outermost electron shell; sodium has only 1, whereas chlorine has 7. To attain a stable configuration, sodium loses the electron in its outermost electron shell, leaving its next shell with 8 electrons as the outermost one (Figure 2.8a). Sodium now has one fewer electrons (negative charge) than it has protons (positive charge), so it is an electrically charged **ion** and is symbolized Na^{+1}.

The electron lost by sodium is transferred to the outermost electron shell of chlorine, which had 7 electrons to begin with. The addition of one more electron gives chlorine an outermost electron shell of 8 electrons, the configuration of a noble gas. But its total number of electrons is now 18, which exceeds by 1 the number of protons. Accordingly, chlorine also becomes an ion, but it is negatively charged (Cl^{-1}). An **ionic bond** forms between sodium and chlorine because of the attractive force between the positively charged sodium ion and the negatively charged chlorine ion (Figure 2.8a).

In ionic compounds, such as sodium chloride (the mineral halite), the ions are arranged in a three-dimensional framework that results in overall electrical neutrality. In halite, sodium ions are bonded to chlorine ions on all sides, and chlorine ions are surrounded by sodium ions (Figure 2.8b).

In addition to containing ions of a single element, minerals commonly contain tightly bonded, charged

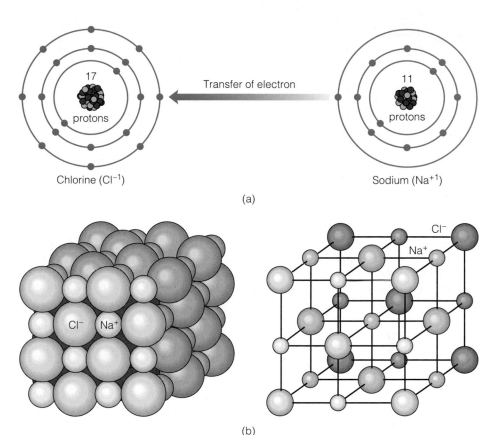

Chlorine (Cl^{-1})

Transfer of electron

Sodium (Na^{+1})

(a)

Figure 2.8
(a) Ionic bonding. The electron in the outermost shell of sodium is transferred to the outermost electron shell of chlorine. Once the transfer has occurred, sodium and chlorine are positively and negatively charged ions, respectively. (b) The crystal structure of sodium chloride, the mineral halite. The diagram on the left shows the relative sizes of sodium and chlorine ions, and the diagram on the right shows the locations of the ions in the crystal structure.

(b)

groups of different atoms known as *radicals*. Even though composed of different elements, these radicals behave as single units within minerals. A good example is the *carbonate radical*, in which three oxygen atoms and one carbon atom bond to form a unit designated CO_3, which acts as an ion with a -2 charge (Figure 2.9). Other common radicals in minerals with their formulas and charges, are *sulfate* (SO_4, -2), *hydroxyl* (OH, -1), and *silicate* (SiO_4, -4).

Covalent Bonding
Covalent bonds form between atoms when their electron shells overlap and electrons are shared. For example, atoms of the same element, such as oxygen in oxygen gas, cannot bond by transferring electrons from one atom to another. And carbon (C), which forms the minerals graphite and diamond, has four electrons in its outermost electron shell (Figure 2.10a). If these four electrons were transferred

to another carbon atom, the atom receiving the electrons would have the noble gas configuration of eight electrons in its outermost electron shell, but the atom contributing the electrons would not.

In such situations, adjacent atoms share electrons by overlapping their electron shells. A carbon atom in diamond, for instance, shares all four of its outermost electrons with a neighbor to produce a stable noble gas configuration (Figure 2.10a).

Covalent bonds are not restricted to substances composed of atoms of a single kind. Among the most common minerals, the silicates (discussed later in this chapter), the element silicon forms partly covalent and partly ionic bonds with oxygen.

Metallic and van der Waals Bonds
Metallic bonding results from an extreme type of electron sharing. The electrons of the outermost electron shell of metals such

Figure 2.9
Many minerals contain radicals, which are complex groups of atoms tightly bonded together. The silicate and carbonate radicals are particularly common in many minerals, such as quartz (SiO_2) and calcite ($CaCO_3$).

Carbonate
CO_3 (-2)

Hydroxyl
OH (-1)

Sulfate
SO_4 (-2)

Silicate
SiO_4 (-4)

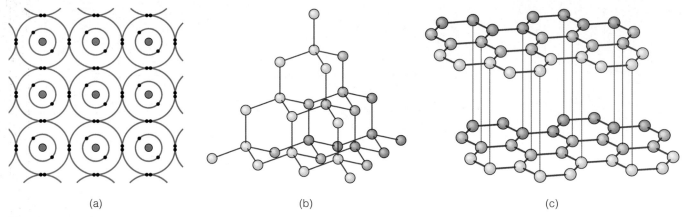

Figure 2.10
(a) Covalent bonds formed by adjacent atoms sharing electrons in diamond. (b) Carbon atoms in diamond are covalently bonded to form a three-dimensional framework. (c) Covalent bonding is also found in graphite, but here the carbon atoms are bonded together to form sheets that are held to one another by van der Waals bonds. The sheets themselves are strong, but the bonds between sheets are weak.

as gold, silver, and copper readily move about from one atom to another. This electron mobility accounts for the fact that metals have a metallic luster (their appearance in reflected light), provide good electrical and thermal conductivity, and can be easily reshaped. Only a few minerals possess metallic bonds, but those that do are very useful; copper, for example, is used for electrical wiring because of its high electrical conductivity.

Some electrically neutral atoms and molecules* have no electrons available for ionic, covalent, or metallic bonding. They nevertheless have a weak attractive force between them when in proximity. This weak attractive force is a *van der Waals* or *residual bond.* The carbon atoms in the mineral graphite are covalently bonded to form sheets, but the sheets are weakly held together by van der Waals bonds (Figure 2.10c). This type of bonding makes graphite useful for pencil leads; when a pencil is moved across a piece of paper, small pieces of graphite flake off along the planes held together by van der Waals bonds and adhere to the paper.

WHAT ARE MINERALS?

Before we discuss minerals in more detail, let us recall our formal definition: A mineral is a naturally occurring, inorganic, crystalline solid, with a narrowly defined chemical composition and characteristic physical properties. The next sections examine each part of this definition.

---

*A molecule is the smallest unit of a substance having the properties of that substance. A water molecule (H_2O), for example, possesses two hydrogen atoms and one oxygen atom.

Naturally Occurring Inorganic Substances

The criterion "naturally occurring" excludes from minerals all substances manufactured by humans. Accordingly, most geologists do not regard synthetic diamonds and rubies and a number of other artificially synthesized substances as minerals. This criterion is particularly important to those who buy and sell gemstones, most of which are minerals, because some human-made substances are very difficult to distinguish from natural gem minerals.

Some geologists think the term *inorganic* in the mineral definition is superfluous. It does remind us that animal matter and vegetable matter are not minerals. Nevertheless, some organisms, including corals, clams, and a number of other animals, construct their shells of the compound calcium carbonate ($CaCO_3$), which is either the mineral aragonite or calcite.

Mineral Crystals

By definition minerals are **crystalline solids,** in which the constituent atoms are arranged in a regular, three-dimensional framework, as in the mineral halite (Figure 2.8b). Under ideal conditions, such as in a cavity, mineral crystals can grow and form perfect crystals that possess planar surfaces (crystal faces), sharp corners, and straight edges (Figure 2.11). In other words, the regular geometric shape of a well-formed mineral crystal is the exterior manifestation of an ordered internal atomic arrangement. Not all rigid substances are crystalline solids; natural and manufactured glass lack the ordered arrangement of atoms and are said to be *amorphous,* meaning "without form."

As early as 1669, the well-known Danish scientist Nicholas Steno determined that the angles of intersection

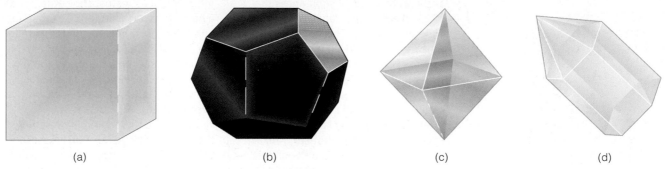

(a)   (b)   (c)   (d)

Figure 2.11
Mineral crystals develop in a variety of shapes, several of which are shown here. (a) Cubic crystals are typical of the minerals halite, galena, and pyrite. (b) Dodecahedron crystals such as those of garnet have 12 sides. (c) Diamond has octahedral, or 8-sided, crystals. (d) A prism terminated by a pyramid is found in quartz.

of equivalent crystal faces on different specimens of quartz are identical. Since then, the *constancy of interfacial angles* has been demonstrated for many other minerals, regardless of their size, shape, or geographic occurrence (Figure 2.12). Steno postulated that mineral crystals are composed of very small, identical building blocks and that the arrangement of these blocks determines the external form of the crystals. Such regularity of the external form of minerals must surely mean that external crystal form is controlled by internal structure.

In many cases, numerous minerals grow in proximity, as in a cooling lava flow, and do not have an opportunity to develop well-formed crystals. Even though well-formed mineral crystals are not common, all minerals of a given species have the same internal atomic structure.

Crystalline structure can be demonstrated even in minerals lacking obvious crystals. For example, many minerals possess a property known as *cleavage,* meaning that they break or split along closely spaced, smooth planes. The fact that these minerals can be split along such smooth planar surfaces indicates that the mineral's internal structure controls such breakage. The behavior of light and X-ray beams transmitted through minerals also provides compelling evidence for an orderly arrangement of atoms within minerals.

Chemical Composition of Minerals

Mineral composition is generally shown by a chemical formula, which is a shorthand way of indicating the numbers of atoms of different elements composing a mineral. The mineral quartz consists of one silicon (Si) atom for every two oxygen (O) atoms, and thus has the formula SiO_2; the subscript number indicates the number of atoms. Orthoclase is composed of one potassium, one aluminum, three silicon, and eight oxygen atoms, so its formula is $KAlSi_3O_8$. A few minerals are composed of

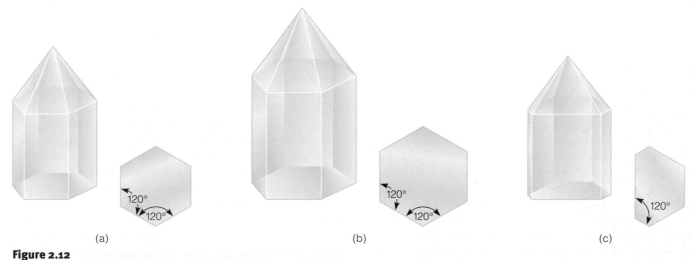

(a)   (b)   (c)

Figure 2.12
Side views and cross sections of three quartz crystals showing the constancy of interfacial angles. (a) A well-shaped crystal. (b) A larger crystal. (c) A poorly shaped crystal. The angles formed between equivalent crystal faces on different specimens of the same mineral are the same regardless of the size or shape of the specimens.

a single element. Known as **native elements,** they include such minerals as graphite and diamond, both of which are composed of carbon (C), silver (Ag), platinum (Pt), and gold (Au) (see Perspective 2.1).

The definition of a mineral contains the phrase *a narrowly defined chemical composition* because some minerals actually have a range of compositions. For many minerals, the chemical composition is constant: Quartz is always composed of silicon and oxygen (SiO_2), and halite contains only sodium and chlorine (NaCl). Other minerals have a range of compositions because one element can substitute for another if the atoms of two or more elements are nearly the same size and the same charge. Notice in Figure 2.13 that iron and magnesium atoms are about the same size; therefore, they can substitute for one another. The chemical formula for the mineral olivine is $(Mg,Fe)_2SiO_4$, meaning that, in addition to silicon and oxygen, it may contain only magnesium, only iron, or a combination of both. As a matter of fact, the term *olivine* is usually applied to minerals containing both iron and magnesium, whereas forsterite is olivine with only magnesium (Mg_2SiO_4) and olivine with only iron is fayalite (Fe_2SiO_4). A number of other minerals also have ranges of compositions, so these are actually mineral groups with several members.

Physical Properties of Minerals

The last criterion in our definition of a mineral, *characteristic physical properties,* refers to such properties as

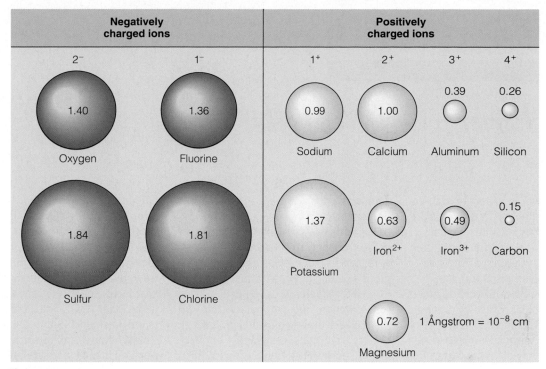

Figure 2.13
Electrical charges and relative sizes of ions common in minerals. The numbers within the ions are the radii shown in Ångstrom units.

hardness, color, and crystal form. These properties are controlled by composition and structure. We will have more to say about physical properties of minerals later in this chapter.

HOW MANY MINERALS ARE KNOWN?

More than 3500 minerals have been identified and described, but only a few—perhaps two dozen—are particularly common. Considering that 92 naturally occurring elements have been discovered, one might think that an extremely large number of minerals could be formed, but several factors limit the number of possible minerals. For one thing, many combinations of elements are chemically impossible; no compounds are composed of only potassium and sodium or of silicon and iron, for example. Another important factor restricting the number of common minerals is that only eight chemical elements make up the bulk of Earth's crust (Table 2.4). Oxygen and silicon constitute more than 74% (by weight) of the crust and nearly 84% of the atoms available to form compounds. By far the most common minerals in the crust consist of silicon and oxygen, combined with one or more of the other elements listed in Table 2.4.

MINERAL GROUPS RECOGNIZED BY GEOLOGISTS

Objects such as rocks and minerals can be classified in many ways, such as by color, size, shape, density, composition, structure, or a combination of such features. In any case, all classification schemes share two characteristics: (1) They systematically categorize similar objects and (2) their purpose is to convey information.

Table 2.4

Common Elements in Earth's Crust

ELEMENT	SYMBOL	PERCENTAGE OF CRUST (BY WEIGHT)	PERCENTAGE OF CRUST (BY ATOMS)
Oxygen	O	46.6%	62.6%
Silicon	Si	27.7	21.2
Aluminum	Al	8.1	6.5
Iron	Fe	5.0	1.9
Calcium	Ca	3.6	1.9
Sodium	Na	2.8	2.6
Potassium	K	2.6	1.4
Magnesium	Mg	2.1	1.8
All others		1.5	0.1

Perspective 2.1

The Precious Metals

As one would expect from their name, the precious metals are desirable for some reason. The best known is gold, which is valued for its pleasing appearance, the ease with which it can be reshaped, and its durability and scarcity. Silver and platinum-group metals are also included among the precious metals, and these too are sought out for several reasons, including a number of industrial applications. Gold and platinum rarely combine with other elements, so they are usually found as native elements. Silver, however, can be found as either a native element or a compound such as the mineral argentite (Ag_2S).

Among the hundreds of minerals used by humans, gold is certainly among the most highly prized and sought after. This deep yellow mineral (Figure 1a) has been mined for at least 6000 years, and archaeological evidence indicates that people 40,000 years ago in Spain possessed small quantities of gold. It has been the

cause of feuds and wars and one incentive for European exploration of the Americas, especially in Central and South America. Gold is too heavy for use in tools or weapons and too soft to hold a cutting edge, so it has been prized for jewelry, ritual objects, a symbol of wealth, and a monetary standard. Much of the gold used in North America still goes into jewelry, but it is also used in the chemical industry, gold plating, electrical circuitry, dentistry, and glass making. South Africa was the 1998 world leader in gold production, with the United States a distant second, followed by Australia, China, and Canada. Most U.S. production comes from mines in Nevada, California, South Dakota, and Alaska; mines in Ontario, followed by Quebec and Alberta, account for most Canadian production.

In the United States, gold was first profitably mined in North Carolina in 1801 and in Georgia in 1829, but the truly spectacular discovery was made in California in 1848, which culminated in the gold rush of 1849. Thousands of people flocked to California to find riches, but only a few found what they

sought. Nevertheless, more than $200 million in gold was mined during the five years of the gold rush proper (1848–1853). In 1876, another gold rush took place following a report by Lt. Col. George Armstrong Custer that "gold in satisfactory quantities can be obtained in the Black Hills." The flood of miners in the Black Hills (South Dakota), the Holy Wilderness of the Sioux Indians, resulted in the Indian War, during which Custer and some 260 of his men were annihilated at the Battle of the Little Bighorn in Montana in June 1876.

Canada, too, has had its gold rushes. The first discovery came in 1850 in the Queen Charlotte Islands along the Pacific Coast, and by 1858 about 10,000 people were panning for gold there. However, the greatest Canadian gold rush was between 1897 and 1899 when about 35,000 people traveled to the remote, hostile Klondike region of the Yukon Territory. Dawson City grew so rapidly that during the winter of 1897 hundreds of people had to be evacuated because of food shortages. As in other gold rushes, merchants made out better than miners, most of whom barely eked out a living.

(a)

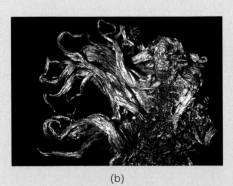
(b)

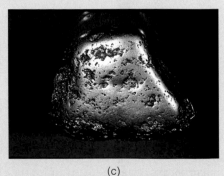
(c)

Figure 1
(a) Specimen of gold from Grass Valley, California. (b) Silver. (c) Platinum.

Certainly the lure of gold is not new, even in the Americas. Spanish explorers during the 1500s and 1600s believed stories of a fabled city of gold in South America known as El Dorado ("The Gilded One"), and the explorer Coronado went as far north as Kansas in quest of Gran Quivira, where people reportedly ate from plates made of gold. In the southwestern United States one still hears rumors of lost gold mines in the Superstition Mountains. Probably the most famous is the Lost Dutchman's Mine, which was actually "discovered" by a German immigrant, Jacob Waltz, between 1863 and 1886. Waltz did find gold, but he died in 1891 without revealing where it came from. Whether the legend of the Lost Dutchman's Mine is based on fact or fantasy remains unresolved, but rumors of its existence have enticed many people to search for it.

The photography industry uses most of the silver in North America for silver halide film, a market that will surely diminish as digital imaging becomes more common. Silver also finds uses in jewelry, flatware, surgical instruments, and backing for mirrors (Figure 1b). The United States, second only to Mexico in silver production, must import more than half the silver it needs, most from Canada, where British Columbia is the largest producer. Most silver in the United States is mined in Nevada followed by Alaska, Arizona, and Idaho.

Probably the most famous area of silver mining in North America is the now defunct Comstock Lode at Virginia City, Gold Hill, and Silver City, Nevada, which was active from 1859 until 1898 (Figure 2). Virginia City, with a present economy based entirely on tourism, had a population of more than 30,000 at its peak. One of its better-known residents, at least for a time, was the author Samuel Clemens, better known as Mark Twain.

The Comstock Lode was discovered in 1857 when miners Pat McLaughlin and Peter O'Reilly found gold. However, the mine was named for Henry Comstock, who convinced McLaughlin and O'Reilly that the gold was on his property. In any case, the deposit yielded more than $300 million during the first 20 years of mining. It was the richest silver deposit ever discovered on this continent. In fact, Nevada became a state in 1864 even though it had too few people to qualify for statehood. This fact was overlooked, however, because gold and silver from its Comstock Lode was helpful in maintaining the U.S. economy during the Civil War.

Even though silver continues to be an important mineral commodity, its value is considerably less than that of gold; in May 2000 gold sold for $275.15 per ounce, whereas silver commanded a price of only $4.99 per ounce. Silver production from U.S. mines in 1998 is valued at $338 million as opposed to the $4 billion in gold mined in the same year.

The platinum-group metals include *platinum, palladium, rhodium, ruthenium, iridium,* and *osmium* (Figure 1c). However, only palladium and platinum are used in large quantities, although the others have various uses as well. In the United States, only one mine in Montana produced platinum-group metals during 1998, but some metals were derived as by-products of copper refining in Texas and Utah. The largest producer by far is South Africa, with Russia a distant second, followed by the United States and Canada.

The automotive industry is the largest user of platinum-group metals, as oxidation catalysts in catalytic converters. These metals are also used in the chemical industry, for cancer and chemotherapy, and some platinum alloys are used for jewelry. It is interesting to note that a thin layer of iridium in sedimentary rocks at the boundary between the Cretaceous and Tertiary Periods is cited as partial evidence for the theory that dinosaur extinctions were caused by a meteorite impact.

(a)

(b)

Figure 2
(a) The Yellowjacket Mine in the Comstock Lode at Gold Hill, Nevada. The tall, wooden structure is the mine's headframe, where equipment and miners were lowered into a mine shaft. (b) Hydrothermal (hot water) solutions invaded fractured igneous rocks (gray) where silver- and gold-bearing quartz (milky white) formed. Even though the Comstock Lode was a remarkably rich deposit, ore was found at fewer than 100 of about 16,000 mining claims.

Geologists recognize mineral classes or groups, each of which contains members sharing the same negatively charged ion or radical; Table 2.5 lists several of these mineral groups.

The Silicate Minerals

Because silicon and oxygen are the two most abundant elements in the crust, it is not surprising that many minerals contain these elements. A combination of silicon and oxygen is known as **silica,** and the minerals containing silica are **silicates.** Quartz (SiO_2) is pure silica because it is composed entirely of silicon and oxygen. But most silicates have one or more additional elements, as in orthoclase ($KAlSi_3O_8$) and olivine [$(Mg,Fe)_2SiO_4$]. Silicate minerals include about one-third of all known minerals, but their abundance is even more impressive when one considers that they make up perhaps 95% of Earth's crust.

The basic building block of all silicate minerals is the **silica tetrahedron,** which consists of one silicon atom and four oxygen atoms (Figure 2.14); the silica radical in Figure 2.9. These atoms are arranged so that the four oxygen atoms surround a silicon atom, which occupies the space between the oxygen atoms; thus a four-faced pyramidal structure is formed. The silicon atom has a positive charge of 4, and each of the four oxygen atoms has a negative charge of 2, resulting in a radical with a total negative charge of 4 $(SiO_4)^{-4}$.

Because the silica tetrahedron has a negative charge, it does not exist in nature as an isolated ion group; rather, it combines with positively charged ions or shares it oxygen atoms with other silica tetrahedra. In the simplest silicate minerals, the silica tetrahedra exist as single units bonded to positively charged ions. In minerals containing isolated tetrahedra, the silicon to oxygen ratio is 1:4, and the negative charge of the silica ion is balanced by positive ions (Figure 2.15a). Olivine [$(Mg,Fe)_2SiO_4$], for example, has either two magnesium (Mg^{+2}) ions, two iron (Fe^{+2}) ions, or one of each to offset the -4 charge of the silica ion.

Table 2.5

Some of the Mineral Groups Recognized by Geologists

Mineral Group	Negatively Charged Ion or Radical	Examples	Composition
Carbonate	$(CO_3)^{-2}$	Calcite	$CaCO_3$
		Dolomite	$CaMg(CO_3)_2$
Halide	Cl^{-1}, F^{-1}	Halite	$NaCl$
		Fluorite	CaF_2
Native element	—	Gold	Au
		Silver	Ag
		Diamond	C
		Graphite	C
Oxide	O^{-2}	Hematite	Fe_2O_3
		Magnetite	Fe_3O_4
Silicate	$(SiO_4)^{-4}$	Quartz	SiO_2
		Potassium feldspar	$KAlSi_3O_8$
		Olivine	$(Mg,Fe)_2SiO_4$
Sulfate	$(SO_4)^{-2}$	Anhydrite	$CaSO_4$
		Gypsum	$CaSO_4\cdot2H_2O$
Sulfide	S^{-2}	Galena	PbS
		Pyrite	FeS_2

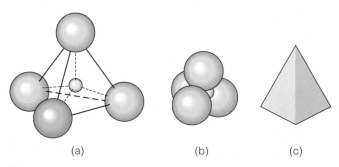

(a) (b) (c)

Figure 2.14

The silica tetrahedron. (a) Expanded view showing oxygen atoms at the corners of a tetrahedron and a small silicon atom at the center. (b) View of the silica tetrahedron as it really exists with the oxygen atoms touching. (c) The silica tetrahedron represented diagramatically; the oxygen atoms are at the four points of the tetrahedron.

Silica tetrahedra may also be arranged so that they join together to form chains of indefinite length (Figure 2.15b). Single chains, as in the pyroxene minerals, form when each tetrahedron shares two of its oxygens with an adjacent tetrahedron; the result is a silicon to oxygen ratio of 1:3. Enstatite, a pyroxene-group mineral, reflects this ratio in its chemical formula $MgSiO_3$. Individual chains, however, possess a net -2 electrical charge, so they are balanced by positive ions, such as Mg^{+2}, that link parallel chains together (Figure 2.15b).

			Formula of negatively charged ion group	Silicon to oxygen ratio	Example
(a)	Isolated tetrahedra	▲	$(SiO_4)^{-4}$	1:4	Olivine
(b)	Continuous chains of tetrahedra	Single chain	$(SiO_3)^{-2}$	1:3	Pyroxene group
		Double chain	$(Si_4O_{11})^{-6}$	4:11	Amphibole group
(c)	Continuous sheets		$(Si_4O_{10})^{-4}$	2:5	Micas
(d)	Three-dimensional networks	Too complex to be shown by a simple two-dimensional drawing	$(SiO_2)^0$	1:2	Quartz

Figure 2.15

Structures of some of the common silicate minerals shown by various arrangements of silica tetrahedra. (a) Isolated tetrahedra. (b) Continuous chains. (c) Continuous sheets. (d) Networks. The arrows adjacent to single-chain, double-chain, and sheet silicates indicate that these structures continue indefinitely in the directions shown.

The amphibole group of minerals is characterized by a double-chain structure in which alternate tetrahedra in two parallel rows are cross-linked (Figure 2.15b). The formation of double chains results in a silicon to oxygen ratio of 4:11, so each double chain possesses a -6 electrical charge. Mg^{+2}, Fe^{+2}, and Al^{+2} are usually involved in linking the double chains together.

In sheet structure silicates, three oxygens of each tetrahedron are shared by adjacent tetrahedra (Figure 2.15c). Such structures result in continuous sheets of silica tetrahedra with silicon to oxygen ratios of 2:5. Continuous sheets also possess a negative electrical charge satisfied by positive ions located between the sheets. This particular structure accounts for the characteristic sheet structure of the *micas*, such as biotite and muscovite, and the *clay minerals*.

Three-dimensional networks of silica tetrahedra form when all four oxygens of the silica tetrahedra are shared by adjacent tetrahedra (Figure 2.15d). Such sharing of oxygen atoms results in a silicon to oxygen ratio of 1:2, which is electrically neutral. Quartz is a common framework silicate.

Ferromagnesian Silicates Among the silicate minerals are those containing iron (Fe), magnesium (Mg), or both, known as **ferromagnesian silicates.** These minerals are commonly dark colored and more dense than nonferromagnesian silicates. Some of the common ferromagnesian silicate minerals are olivine, the pyroxenes, the amphiboles, and biotite (Figure 2.16). Olivine, an olive green mineral, is common in some igneous rocks but uncommon in most other

(a) Olivine

(b) Augite

(c) Hornblende

(d) Biotite mica

Figure 2.16

Common ferromagnesian silicates. (a) Olivine. (b) Augite, a pyroxene-group mineral. (c) Hornblende, an amphibole-group mineral. (d) Biotite mica.

rock types. The pyroxenes and amphiboles are actually mineral groups, but the varieties augite and hornblende are the most common. Biotite mica is a common, dark-colored ferromagnesian silicate with a distinctive sheet structure.

Nonferromagnesian Silicates   The **nonferromagnesian silicates,** as their name implies, lack iron and magnesium, are generally light colored, and are less dense than ferromagnesian silicates (Figure 2.17). The most common minerals in the crust are nonferromagnesian silicates known as *feldspars*. Feldspar is a general name, however, and two distinct groups are recognized, each of which includes several species. The *potassium feldspars,* represented by microcline and orthoclase ($KAlSi_3O_8$), are common in igneous, metamorphic, and some sedimentary rocks. Like all feldspars, microcline and orthoclase have two internal planes of weakness along which they break or cleave.

The second group of feldspars, the *plagioclase feldspars,* range from calcium-rich ($CaAl_2Si_2O_8$) to sodium-rich ($NaAlSi_3O_8$) varieties. They possess the characteristic feldspar cleavage and typically are white or cream to medium gray. Plagioclase cleavage surfaces commonly show numerous distinctive, closely spaced, parallel lines called *striations*.

Quartz (SiO_2), a very abundant nonferromagnesian silicate (see Perspective 2.2), is common in the three major rock groups, especially in such rocks as granite, gneiss, and sandstone. A framework silicate, it can usually be recognized by its glassy appearance and hardness (Figure 2.17a).

Another fairly common nonferromagnesian silicate is muscovite, which is a mica. Like biotite it is a sheet silicate, but muscovite is typically nearly colorless (Figure 2.17d), whereas biotite is dark colored. Various clay minerals also possess the sheet structure typical of the micas, but their crystals are so small that they can be seen only with extremely high magnification. These clay minerals are important constituents of several types of rocks and are essential components of soils.

(a) Quartz

(b) Orthoclase

(c) Plagioclase

(d) Muscovite

Figure 2.17
Common nonferromagnesian silicates. (a) Quartz. (b) The potassium feldspar orthoclase. (c) Plagioclase feldspar. (d) Muscovite.

Carbonate Minerals

Carbonate minerals are those containing the negatively charged carbonate radical $(CO_3)^{-2}$. Examples include calcium carbonate ($CaCO_3$) as the minerals *aragonite* or *calcite* (Figure 2.18a). Aragonite is unstable and commonly changes to calcite, the main constituent of the sedimentary rock *limestone*. A number of other car-

Figure 2.18
(a) Calcite ($CaCO_3$) is the most common carbonate mineral. (b) The sulfide mineral galena (PbS) is the ore of lead. (c) Gypsum ($CaSO_4 \cdot 2H_2O$) is a common sulfate mineral. (d) Halite (NaCl) is a good example of a halide mineral.

(a) Calcite

(b) Galena

(c) Gypsum

(d) Halite

bonate minerals are known, but only one of these need concern us: *Dolomite* [$CaMg(CO_3)_2$] is formed by the chemical alteration of calcite by the addition of magnesium. Sedimentary rock composed of the mineral dolomite is *dolostone* (see Chapter 6).

Other Mineral Groups

Besides silicates and carbonates, several other mineral groups are recognized (see Table 2.5). In the oxides, an element is combined with oxygen as in *hematite* (Fe_2O_3). Hematite and another iron oxide known as *magnetite* (Fe_3O_4) are both commonly present in small quantities in a variety of rocks. Rocks containing high concentrations of hematite and magnetite, such as those in the Lake Superior region of Canada and the United States, are sources of iron ores for the manufacture of steel.

The sulfides have a positively charged ion combined with sulfur (S^{-2}), such as in the mineral *galena* (PbS), which contains lead (Pb) and sulfur (Figure 2.18b). Sulfates contain an element combined with the complex sulfate radical (SO_4)$^{-2}$; *gypsum* ($CaSO_4 \cdot 2H_2O$) is a good example (Figure 2.18c). The halides contain halogen elements such as chlorine (Cl^{-1}) and fluorine (F^{-1}); examples include the minerals *halite* (NaCl) (Figure 2.18d) and fluorite (CaF_2).

HOW ARE MINERALS IDENTIFIED?

All minerals possess characteristic physical properties that are determined by their internal structure and chemical composition. Many physical properties are remarkably constant for a given mineral species, but some, especially color, may vary. Although a professional geologist may use sophisticated techniques in studying and identifying minerals, most common minerals can be identified by using the following physical properties (see Appendix C).

Color and Luster

For many minerals, color varies because of minute amounts of impurities (Figure 1, Perspective 2.2). The fact that some minerals are found in a variety of colors is distressing to beginning students because the most obvious mineral property is not particularly useful for identification. However, some generalizations about color can be made. Ferromagnesian silicates are typically black, brown, or dark green, although olivine is olive green (Figure 2.16a). Nonferromagnesian silicates, on the other hand, can vary considerably in color, but are

only rarely dark (Figure 2.17). Furthermore, for those minerals that have the appearance of metals, color is rather consistent.

Luster (not to be confused with color) is the appearance of a mineral in reflected light. Two basic types of luster are recognized: *metallic* and *nonmetallic* (Figure 2.19). They are distinguished by observing the quality of light reflected from a mineral and determining if it has the appearance of a metal or a nonmetal. Several types of nonmetallic luster are recognized, including glassy or vitreous, greasy, waxy, brilliant (as in diamond), and dull or earthy.

Crystal Form

As previously noted, mineral crystals are not common, so many mineral specimens you encounter will not show the perfect crystal form typical of that mineral species (Figures 2.11 and 2.20). Keep in mind, however, that even though crystals may not be apparent, minerals nevertheless possess the atomic structure

(a)   (b)

Figure 2.19
Hematite (a) has the appearance of a metal and is said to have a metallic luster, whereas orthoclase (b) has a nonmetallic luster.

(a)

(b)

(c)

Figure 2.20
Mineral crystals. (a) Cubic crystals of fluorite. (b) Calcite crystal. (c) Blade-shaped crystals of barite.

Perspective 2.2

Quartz—Common, Interesting, Attractive, and Useful

According to one estimate, Earth's crust consists of 51% feldspars, 24% ferromagnesian silicates (pyroxenes, amphiboles, and biotite), 12% quartz, and 13% other minerals. So quartz is common to begin with, but it is far more common than this list would indicate. Indeed, quartz is by far the most common mineral in the sand on beaches, in stream channels, in desert sand dunes, and on barrier islands adjacent to seashores. Three factors account for this abundance: (1) It is abundant in common rocks such as granite and gneiss, (2) quartz is resistant to mechanical breakdown, and (3) it is chemically stable, meaning that quartz does not decompose easily. In any event, in most peoples' experience it will be the most commonly encountered mineral.

Given that it is so common, and anyone who has looked closely at sand knows that individual sand grains are not particularly impressive, what is so interesting about quartz? It has several practical uses, but its interest lies in the fact that it varies from colorless, well-formed crystals to numerous color varieties and textural types. Several varieties of quartz are used as semi-precious stones in jewelry and as decorative stones. Sand deposits composed mostly of quartz are called *silica sands* and are used in the manufacture of glass, optical equipment, abrasives such as sandpaper, the manufacture of steel alloys, and for a variety of other industrial applications.

During the Middle Ages, transparent quartz crystals were believed to be ice, frozen so solidly that they would not melt (Figure 1). In fact, the term *crystal* is derived from a Greek word meaning "ice." Even today, *crystal* refers not only to mineral crystals but also to clear, colorless glass of high quality, such as crystal ware, crystal chandeliers, or the transparent glass or plastic cover of a watch or clock dial. Colorless quartz in particular has been used as a semi-precious stone for jewelry. The term *rhinestone* originally referred to transparent quartz crystals used for jewelry made in Germany. *Herkimer "diamonds"* are simply colorless quartz crystals from Herkimer County, New York. In the past, large, transparent quartz crystals were shaped into spheres for use as fortune-tellers' crystal balls.

Color varieties of quartz include milky quartz, which is commonly found as well-formed crystals. A milky quartz crystal 3.5 m long, 1.7 m in diameter, and weighing 11.8 metric tons was discovered in Siberia. Amethyst (purple), citrine (yellow to orange), rose quartz (pale pink to deep rose), and smoky quartz (smoky brown to black) are color varieties of quartz (Figure 1). Several other varieties of quartz referred to as *cryptocrystalline* have such tiny crystals that they can be detected only when highly magnified. These varieties include agate, which shows color banding (Figure 1); flint (usually brownish, gray, or black); jasper (red, yellow, or brown); and several others.

---

that would have yielded well-formed crystals if they had developed under ideal conditions.

Some minerals do typically occur as crystals. For example, 12-sided crystals of garnet are common, as are 6- and 12-sided crystals of pyrite. Minerals that grow in cavities or are precipitated from circulating hot water (hydrothermal solutions) in cracks and crevices in rocks also commonly occur as crystals.

Crystal form can be a useful characteristic for mineral identification, but a number of minerals have the same crystal form. Pyrite (FeS_2), galena (PbS), and halite (NaCl) all occur as cubic crystals, but they can usually be easily identified by other properties such as color, luster, hardness, and density.

Cleavage and Fracture

Cleavage is a property of individual mineral crystals. Not all minerals possess cleavage, but those that do tend to break, or split, along a smooth plane or planes of weakness determined by the strength of the bonds within the mineral crystal. Cleavage can be characterized in terms of quality (perfect, good, poor), direction, and angles of intersection of cleavage planes. Biotite, a

In addition to the pleasing appearance of some quartz, the mineral has the property of *piezoelectricity* ("pressure" electricity). When pressure is applied to a quartz crystal, an electric current is generated. If an electric current is applied to a quartz crystal, as by a watch's battery, the crystal expands and contracts extremely rapidly and regularly (about 100,000 times per second). Quartz crystal clocks were first developed in 1928, and now quartz watches and clocks are commonplace. Even inexpensive quartz timepieces are very accurate: Precision-manufactured quartz clocks used in astronomical observations do not gain or lose more than 1 second in 10 years.

An interesting historical note is that during World War II (1939–1945) the United States had difficulty obtaining Brazilian quartz crystals needed for making radios. This shortage prompted the development of artificially synthesized quartz, and now most of the quartz used in watches and clocks is synthetic.

(a) Colorless quartz

Figure 1
Varieties of quartz.

(b) Smoky quartz

(c) Amethyst

(d) Citrine

(e) Milky quartz

(f) Rose quartz

(g) Agate

common ferromagnesian silicate, has perfect cleavage in one direction (Figure 2.21a). The fact that biotite preferentially cleaves along a number of closely spaced, parallel planes is related to its structure; it is a sheet silicate with the sheets of silica tetrahedra weakly bonded to one another by iron and magnesium ions.

Feldspars possess two directions of cleavage that intersect at right angles (Figure 2.21b), and the mineral halite has three directions of cleavage, all of which intersect at right angles (Figure 2.21c). Calcite also possesses three directions of cleavage, but none of the intersection angles is a right angle, so cleavage fragments of calcite are rhombohedrons (Figure 2.21d). Minerals with four directions of cleavage include fluorite and diamond (Figure 2.21e). Ironically, diamond, the hardest mineral, can be easily cleaved. A few minerals such as sphalerite, an ore of zinc, have six directions of cleavage (Figure 2.21f).

Cleavage is an important diagnostic property of minerals, and its recognition is essential in distinguishing between some minerals. The pyroxene mineral augite and the amphibole mineral hornblende, for example, look much alike: both are generally dark green to black, have the same hardness, and possess two directions

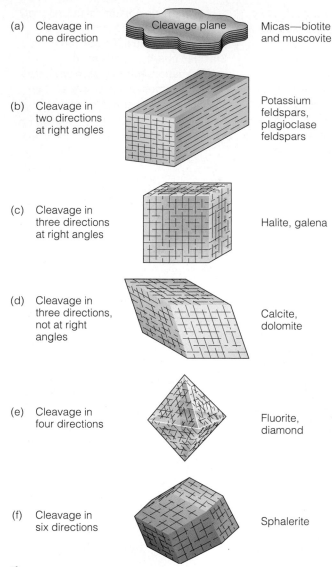

(a) Cleavage in one direction — Cleavage plane — Micas—biotite and muscovite

(b) Cleavage in two directions at right angles — Potassium feldspars, plagioclase feldspars

(c) Cleavage in three directions at right angles — Halite, galena

(d) Cleavage in three directions, not at right angles — Calcite, dolomite

(e) Cleavage in four directions — Fluorite, diamond

(f) Cleavage in six directions — Sphalerite

Figure 2.21
Several types of mineral cleavage. (a) One direction. (b) Two directions at right angles. (c) Three directions at right angles. (d) Three directions, not at right angles. (e) Four directions. (f) Six directions.

of cleavage. But the cleavage planes of augite intersect at about 90 degrees, whereas the cleavage planes of hornblende intersect at angles of 56 degrees and 124 degrees (Figure 2.22).

In contrast to cleavage, *fracture* is mineral breakage along irregular surfaces, indicating that planes of weakness are absent. Any mineral can be fractured if enough force is applied, but the fracture surfaces will not be smooth. Fracture surfaces are commonly uneven or conchoidal (smoothly curved).

Hardness

Hardness is the resistance of a mineral to abrasion. The Austrian geologist Friedrich Mohs devised a relative hardness scale for 10 minerals. He arbitrarily assigned a hardness value of 10 to diamond, the hardest mineral known, and lesser values to the other minerals. Relative hardness can be determined easily by using the Mohs hardness scale (Table 2.6). Quartz will scratch fluorite but cannot be scratched by fluorite, gypsum can be scratched by a fingernail, and so on. Hardness is controlled mostly by internal structure. Both graphite and diamond are composed of carbon, but the former has a hardness of 1 to 2, whereas the latter has a hardness of 10.

Specific Gravity

The *specific gravity* of a mineral is the ratio of its weight to the weight of an equal volume of water. A mineral with a specific gravity of 3.0 is three times as heavy as water. Like all ratios, specific gravity is not expressed in units such as grams per cubic centimeter—it is a dimensionless number.

Specific gravity varies in minerals, depending on their composition and structure. Among the common silicates, the ferromagnesian silicates have specific gravities ranging from 2.7 to 4.3, whereas the nonferromagnesian silicates vary from 2.6 to 2.9. Obviously, the ranges of values overlap somewhat, but for the most

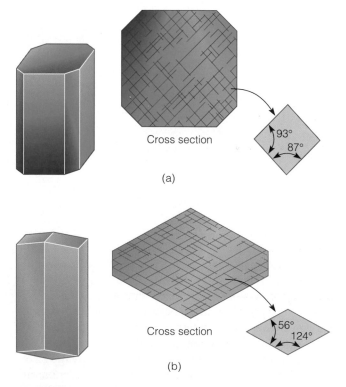

Cross section — 93° 87° — (a)

Cross section — 56° 124° — (b)

Figure 2.22
Cleavage in augite and hornblende. (a) Augite crystal and cross section of crystal showing cleavage. (b) Hornblende crystal and cross section of crystal showing cleavage.

Table 2.6

HARDNESS	MINERAL	HARDNESS OF SOME COMMON OBJECTS
10	Diamond	
9	Corundum	
8	Topaz	
7	Quartz	
		Steel file (6 1/2)
6	Orthoclase	
		Glass (5 1/2–6)
5	Apatite	
4	Fluorite	
3	Calcite	Copper penny (3)
		Fingernail (2 1/2)
2	Gypsum	
1	Talc	

Mohs Hardness Scale

part ferromagnesian silicates have greater specific gravities than nonferromagnesian silicates. In general, the metallic minerals, such as galena (7.58) and hematite (5.26), are heavier than nonmetals. Pure gold has a specific gravity of 19.3, making it about two and one-half times more dense than lead. Structure as a control of specific gravity is illustrated by the native element carbon (C). The specific gravity of graphite varies from 2.09 to 2.33; that of diamond is 3.5.

Other Properties Used in Identifying Minerals

A number of other physical properties characterize some minerals. Talc has a distinctive soapy feel, graphite writes on paper, halite tastes salty, and magnetite is magnetic (Figure 2.23). Calcite possesses the property of *double refraction,* meaning that an object when viewed through a transparent piece of calcite will have a double image (Figure 2.23c). Some minerals are plastic and, when bent into a new shape, will retain that shape; others are flexible and, if bent, will return to their original position when the forces that bent them are removed.

A simple chemical test to identify the minerals calcite and dolomite involves applying a drop of dilute hydrochloric acid to the mineral specimen. If the mineral is calcite, it will react vigorously with the acid and release carbon dioxide, which causes the acid to bubble or effervesce. Dolomite, in contrast, will not react with hydrochloric acid unless it is powdered.

WHERE AND HOW DO MINERALS ORIGINATE?

Thus far we have discussed the composition, structure, and physical properties of minerals but have not fully addressed how they originate. A common phenomenon accounting for the origin of some minerals is the cooling of molten rock material known as *magma* (magma that reaches the surfaces is called *lava*). As magma or lava cools, minerals crystallize and grow, thereby determining the mineral composition of various igneous rocks such as basalt (dominated by ferromagnesian silicates) and granite (dominated by nonferromagnesian silicates) (see Chapter 3). Hot water solutions derived from magma commonly invade cracks and crevasses in adjacent rocks, and from these solutions a variety of minerals might crystallize, some of economic importance. Many well-formed quartz crystals are found in these so-called hydrothermal veins. Minerals also originate when water in hot springs cools (see Chapter 16), and when hot, mineral-rich water discharges onto the seafloor at hot springs known as "black smokers" (see Chapter 11).

Dissolved materials in seawater, more rarely lake water, commonly combine to form minerals such as halite (NaCl), gypsum ($CaSO_4 \cdot 2H_2O$), and several others when the water evaporates. Aragonite and/or calcite,

(a) Graphite

(b) Magnetite

(c) Calcite

Figure 2.23
Various mineral properties. (a) Graphite, the mineral from which pencil leads are made, writes on paper. (b) Magnetite, an important iron ore, is magnetic. (c) Transparent pieces of calcite show double refraction.

both varieties of calcium carbonate ($CaCO_3$), might also form from evaporating water, but most originates when organisms such as clams, oysters, corals, and floating microorganisms use this compound to construct their shells. And a few plants and animals use silicon dioxide (SiO_2) for their skeletons, which accumulate as mineral matter on the seafloor when the organisms die (see Chapter 6).

Some clay minerals form when chemical processes compositionally and structurally alter other minerals, such as feldspars (see Chapter 5), and others originate when rocks are changed during metamorphism (see Chapter 7). In fact, the agents that cause metamorphism—heat, pressure, and chemically active fluids—are responsible for the origin of many minerals. A few minerals even originated when gasses such as hydrogen sulfide (H_2S) and sulfur dioxide (SO_2) react at volcanic vents to produce sulfur (Figure 2.24).

Figure 2.24
Native sulphur crystals forming around a gas vent on Kilauea Volcano, Hawaii.

ROCK-FORMING MINERALS

Rocks are generally defined as aggregates of one or more minerals. Two important exceptions to this definition are natural glass such as obsidian (see Chapters 3 and 4) and the sedimentary rock coal (see Chapter 6). Although many minerals are present in various kinds of rocks, only a few varieties are common enough to be designated as **rock-forming minerals.** Most of the others constitute such a small proportion of rocks that they can be disregarded in identification and classification; these are generally called *accessory minerals.*

We have already emphasized that Earth's crust is composed largely of silicate minerals. This being the case, one would suspect that most rocks are also composed of silicate minerals, and this is correct. Only a few of the hundreds of known silicates are common in rocks, however, although many are accessories. The common rock-forming silicates are summarized in Table 2.7. The common igneous rock basalt, for instance, is made up mostly of ferromagnesian silicates such as pyroxene and olivine, and plagioclase feldspar, a nonferromagnesian silicate. Granite, another igneous rock, is composed predominantly of potassium feldspar and quartz, both of which are nonferromagnesian silicates (Figure 2.25). Both of these rocks contain a variety of accessory minerals, most of which are silicates as well. The minerals just mentioned are also common in metamorphic rocks, and many sedimentary rocks are composed of quartz, feldspars, and various clay minerals.

(a) Granite

Figure 2.25
The igneous rock granite is composed largely of potassium feldspar and quartz, lesser amounts of plagioclase feldspar, and accessory minerals such as biotite mica. (a) Hand specimen of granite. (b) Photomicrograph showing the various minerals.

(b) Photomicrograph of granite

Table 2.7

Rock-Forming Minerals

MINERAL	COMPOSITION	PRIMARY OCCURRENCE
FERROMAGNESIAN SILICATES		
Olivine	$(Mg,Fe)_2SiO_4$	Igneous, metamorphic rocks
Pyroxene group		
Augite most common	Ca, Mg, Fe, Al silicate	Igneous, metamorphic rocks
Amphibole group		
Hornblende most common	Hydrous* Na, Ca, Mg, Fe, Al silicate	Igneous, metamorphic rocks
Biotite	Hydrous K, Mg, Fe silicate	All rock types
NONFERROMAGNESIAN SILICATES		
Quartz	SiO_2	All rock types
Potassium feldspar group		
Orthoclase, microcline	$KAlSi_3O_8$	All rock types
Plagioclase feldspar group	Varies from $CaAl_2S_2O_8$ to $NaAlSi_3O_3$	All rock types
Muscovite	Hydrous K, Al silicate	All rock types
Clay mineral group	Varies	Soils and sedimentary rocks
CARBONATES		
Calcite	$CaCO_3$	Sedimentary rocks
Dolomite	$CaMg(CO_3)_2$	Sedimentary rocks
SULFATES		
Anhydrite	$CaSO_4$	Sedimentary rocks
Gypsum	$CaSO_4 \cdot 2H_2O$	Sedimentary rocks
HALIDES		
Halite	NaCl	Sedimentary rocks

Contains elements of water in some kind of union.

The most common nonsilicate rock-forming minerals are the two carbonates, calcite ($CaCO_3$) and dolomite [$CaMg(CO_3)_2$], the primary constituents of the sedimentary rocks limestone and dolostone, respectively. Among the sulfates and halides, gypsum ($CaSO_4 \cdot 2H_2O$) and halite (NaCl), respectively, are the only common rock-forming minerals.

MINERAL RESOURCES AND RESERVES

As noted in the Introduction, the United States and Canada, both highly industrialized nations, enjoy considerable economic success because they have abundant natural resources. But what is a resource? Geologists at the U.S. Geological Survey define a **resource** as "A concentration of naturally occurring solid, liquid, or gaseous material in or on Earth's crust in such form and amount that economic extraction of a commodity from the concentration is currently or potentially feasible."

Resources consist of metals or *metallic resources,* sand, gravel, crushed stone, sulfur, salt, and a variety of other *nonmetallic resources,* and *energy resources,* including oil, natural gas, coal, and uranium. An important distinction must be made between a resource, which is the total amount of a commodity whether discovered or undiscovered, and a **reserve,** which is that part of the resource base that can be economically extracted. Liquid oil can be extracted from rock known as oil shale, for instance, so it is part of the resource base, but at present it cannot be recovered economically.

What constitutes a resource as opposed to a reserve depends on several factors. Iron-bearing minerals are present in many rocks, but in quantities or ways that make their recovery uneconomical. As a matter of fact, most economic concentrations of minerals are mined in only a few areas; 75% of all metals mined in the world come from about 150 locations. Geographic location also might determine what is considered a reserve. A mineral concentration in a remote region may not be mined because transportation costs are too high, and what may be considered a resource in the United States and Canada may be mined in a developing country where labor costs are low.

Figure 2.26

(a) Shovels such as this one at the Tilden Mine in northern Michigan load trucks with as much as 75 tons of iron ore. The ore is crushed and processed to produce iron pellets (b) measuring about 1 cm for shipment to steel mills.

(a)

(b)

The market price of a commodity is, of course, important in evaluating a potential resource. From 1935 to 1968, the U.S. government maintained the price of gold at $35 per troy ounce (3.1 g). When this restriction was removed and the price of gold became subject to supply and demand, the price rose (reaching an all-time high of $843 per troy ounce during January 1980). As a result, many marginal deposits became reserves and many abandoned mines were reopened.

Technological innovations can also change the status of a resource. By the time of World War II (1939–1945), the rich iron ores of the Great Lakes region of the United States and Canada had been depleted. However, the development of a method for separating the iron from unusable rocks and shaping it into pellets ideal for use in blast furnaces made it feasible to mine rocks containing less iron (Figure 2.26). Most of the iron ore mined in North America comes from mines in Newfoundland and Quebec, Canada, and from Minnesota and Michigan. Nearly all of it is shaped into pellets before it is shipped to steel mills.

Most people are aware that numerous commodities are extracted or mined and refined for various uses, but have little knowledge about their occurrence, methods of recovery, and economics. Geologists are instrumental in finding various recoverable commodities, but engineers are involved in the extraction processes, and ultimately people with training in business and economics make the decision of whether a commodity is worth extracting or not. In short, extraction must yield a profit. More than $40 billion in mineral commodities, other than oil, natural gas, and coal, were mined in the United States in 1997, and Canadian mineral production for 1998 totaled about $17 billion (Canadian dollars) (Tables 2.8 and 2.9).

Everyone knows that various metals such as gold and iron ore are resources, as are oil and natural gas. However, some quite common minerals are also important to the economies of industrialized nations. Quartz, gypsum, clay minerals, and feldspars are good examples. Pure quartz sand is used in the manufacture of glass and optical instruments as well as sandpaper and steel alloys. Wallboard or drywall is made of gypsum, clay minerals are needed to make ceramics and paper, and feldspars such as orthoclase are used for porcelain, ceramics, enamel, and glass.

Direct access to mineral resources is essential for industrialization and the high standard of living enjoyed by many nations. The United States and Canada are fortunate to have abundant resources, but the amount of resources used in North America since Europeans settled this continent has steadily increased. Each resident of North America now uses about 14 metric tons of resources per year, a large part of which is bulk items such as sand and gravel, cement, and crushed stone. It is no exaggeration to say that industrialized societies are totally dependent on mineral resources, but, unfortunately, many such resources are being used much faster than they form. Mineral resources are *nonrenewable,* meaning that once a reserve has been depleted, new deposits or suitable substitutes, if available, must be found.

Adequate supplies of some mineral resources are available for the indefinite future (sand and gravel and crushed stone, for example), but many others are either limited or must be imported. For some essential metals, the United States depends totally on imports. No cobalt or nickel was mined in this country during 1998, and all manganese, an essential element in manufacturing steel, is imported from Gabon, Australia, Mexico, and Brazil. More than half the crude oil used in the United States is imported, much from the Middle East where more than 50% of the proven reserves exist. A poignant reminder of our dependence on the availability of resources was the U.S. response to the takeover of Kuwait by Iraq during August 1990.

Even though the United States is resource rich, its large population and high standard of living make it dependent on imports of a number of mineral commodities for part or all of its needs. Canada, in contrast, is more self-reliant, meeting most of its own domestic mineral and energy needs. Nevertheless, it must import

Table 2.8

Nonfuel Mineral Production in the Ten U.S. States with the Highest Value for 1997 (Oil, natural gas, and coal are not included, otherwise the ranking would be different.)

RANK	STATE	VALUE (THOUSANDS)	PERCENTAGE OF U.S. TOTAL	TOP THREE MINERALS IN ORDER OF VALUE
1	Arizona	3,540,000	8.74	Copper, sand and gravel, cement*
2	Nevada	3,270,000	8.08	Gold, copper, silver
3	California	3,040,000	7.51	Cement, sand and gravel, boron
4	Florida	1,830,000	4.53	Phosphate rock, crushed stone, cement
5	Texas	1,790,000	4.43	Cement, crushed stone, sand and gravel
6	Georgia	1,680,000	4.15	Clays, crushed stone, cement
7	Minnesota	1,680,000	4.14	Iron ore, sand and gravel, crushed stone
8	Utah	1,680,000	4.14	Copper, gold, magnesium metal
9	Michigan	1,660,000	4.10	Iron ore, cement, sand and gravel
10	Missouri	1,310,000	3.23	Crushed stone, lead, cement

*Cement is manufactured from mineral and rock products including limestone, clay, and silica. It is usually combined with an aggregate such as sand or gravel to form concrete.

Source: U.S. Geological Survey Minerals Yearbook.

Table 2.9

Nonfuel Mineral Production by the Five Canadian Provinces with the Highest Value for 1998

RANK	PROVINCE	VALUE (THOUSANDS—CANADIAN DOLLARS)	PERCENTAGE OF CANADIAN TOTAL	TOP MINERALS IN ORDER OF VALUE
1	Ontario	4,994,000	29	Copper, gold, nickel
2	Quebec	3,443,000	20	Copper, gold, zinc
3	Saskatchewan	2,176,000	12.8	Potash, uranium
4	British Columbia	1,918,000	11.3	Copper, gold, zinc
5	Manitoba	909,000	5.3	Nickel, copper, zinc

Source: Canadian Minerals Yearbook.

phosphate, chromium, manganese, and bauxite, the ore of aluminum. Canada also produces more crude oil and natural gas than it uses, and is among world leaders in producing and exporting uranium.

Most of the largest and richest mineral deposits have probably already been discovered, and in some cases depleted. To ensure continued supplies of essential minerals and energy resources, geologists and other scientists, government agencies, and leaders of business and industry continually assess the status of resources in view of changing economic and political conditions and developments in science and technology. In the following chapters, we will discuss the origin and distribution of various mineral and energy resources and reserves.

WHAT WOULD YOU DO?

Your country is running short of a natural resource essential to its industrial productivity. Your choices to remedy this situation are (1) find more deposits of the resource in your country; (2) develop a technology that can use a substitute; or (3) import the commodity from a politically unstable nation. What factors would you evaluate to determine which of these three alternatives is most cost effective? Might factors other than cost be a consideration?

Chapter Summary

1. All matter is composed of chemical elements, each of which consists of atoms. Individual atoms have a nucleus, containing protons and neutrons, and electrons that circle the nucleus in electron shells.

2. Atoms are characterized by their atomic number (the number of protons in the nucleus) and their atomic mass number (the number of protons plus the number of neutrons in the nucleus).

3. Bonding is the process whereby atoms are joined to other atoms. If atoms of different elements are bonded, they form a compound. Ionic and covalent bonds are most common in minerals, but metallic and van der Waals bonds also occur in a few.

4. Most minerals are compounds, but a few, including gold and silver, are composed of a single element and are called *native elements*.

5. All minerals are crystalline solids, meaning that they possess an orderly internal arrangement of atoms.

6. Some minerals vary in chemical composition because atoms of different elements can substitute for one another provided that the electrical charge is balanced and the atoms are of about the same size.

7. Of the more than 3500 known minerals, most are silicates. Ferromagnesian silicates contain iron (Fe) and magnesium (Mg), and nonferromagnesian silicates lack these elements.

8. In addition to silicates, several other mineral groups are recognized, including carbonates, oxides, sulfides, sulfates, and halides.

9. The physical properties of minerals such as color, hardness, cleavage, and crystal form are controlled by composition and structure.

10. A few minerals are common enough constituents of rocks to be designated rock-forming minerals. Most rock-forming minerals are silicates, but the carbonates calcite and dolomite are also common.

11. Many resources are concentrations of minerals of economic importance.

12. Any commodity that can be or has the potential to be extracted for some use is a resource. Metals constitute metallic resources, whereas commodities such as sand, gravel, and salt are nonmetallic resources, and oil, natural gas, coal, and uranium are energy resources.

13. Reserves are that part of the resource base that can be extracted economically. The status of a resource versus a reserve depends on several factors, including market price, geographic location, and developments in science and technology.

14. The United States, although rich in resources, must import part or all of some essential mineral and energy commodities. Canada is more self-reliant, but it too must import some commodities such as aluminum ore.

Important Terms

atom
atomic mass number
atomic number
bonding
carbonate mineral
cleavage
compound
covalent bond
crystalline solid
electron

electron shell
element
ferromagnesian silicate
ion
ionic bond
mineral
native element
neutron
nonferromagnesian silicate
nucleus

proton
reserve
resource
rock
rock-forming mineral
silica
silica tetrahedron
silicate

1. The term *crystalline* in the definition of a mineral means that a mineral:

   a. _____ has an atomic number of at least 92;

   b. _____ possesses more protons than neutrons;

   c. _____ is characterized by physical properties such as hardness and cleavage;

   d. _____ has an orderly internal arrangement of atoms;

   e. _____ is an essential constituent of obsidian.

2. Minerals that break along smooth internal planes of weakness have the property known as:

   a. _____ cleavage;

   b. _____ double refraction;

   c. _____ specific gravity;

   d. _____ ionic bonding;

   e. _____ streak.

3. The two common rock-forming minerals from the carbonate group are:

   a. _____ hematite and magnetite;

   b. _____ calcite and dolomite;

   c. _____ clay and feldspar;

   d. _____ halite and gypsum;

   e. _____ coal and obsidian.

4. The atoms of the noble gases do not react to form compounds because they have:

   a. _____ a positive electrical charge;

   b. _____ too many neutrons;

   c. _____ eight electrons in their outermost electron shell;

   d. _____ two directions of cleavage that intersect at right angles;

   e. _____ a deficiency of silica and an excess of iron and magnesium.

5. The number of protons plus neutrons in an atom's nucleus determines its:

   a. _____ resistance to abrasion;

   b. _____ value as a resource;

   c. _____ frequency of bonding with other atoms;

   d. _____ chemical formula;

   e. _____ atomic mass number.

6. The two most abundant elements in Earth's crust are:

   a. _____ iron and calcium;

   b. _____ iridium and platinum;

   c. _____ oxygen and silicon;

   d. _____ potassium and magnesium;

   e. _____ hydrogen and nitrogen.

7. The weight of a mineral compared to that of an equal volume of water is known as its:

   a. _____ specific gravity;

   b. _____ hardness;

   c. _____ atomic number;

   d. _____ electrical charge;

   e. _____ cleavage.

8. Which group of minerals in the following list contains only ferromagnesian silicates?

   a. _____ quartz-calcite-iron;

   b. _____ potassium feldspar-mica-biotite;

   c. _____ olivine-pyroxene-amphibole;

   d. _____ halite-gypsum-coal;

   e. _____ granite-hematite-galena.

9. The ferromagnesian silicate olivine has the chemical formula $(Mg,Fe)_2SiO_4$, which means that:

   a. _____ magnesium is heavier than iron;

   b. _____ magnesium and iron must be present in equal amounts;

   c. _____ it is a sheet silicate with weak bonds between the sheets;

   d. _____ magnesium and iron can substitute for one another;

   e. _____ magnesium is more abundant than iron in Earth's crust.

10. In what type of bonding are electrons transferred from one atom to another?

    a. _____ ionic;

    b. _____ tetrahedral;

    c. _____ van der Waals;

    d. _____ silicate;

    e. _____ carbonate.

11. A pyramidal structure consisting of four oxygen atoms and one silicon atom is known as the:

    a. _____ oxygen-silicon ion;

    b. _____ oxide building block;

    c. _____ silica tetrahedron;

    d. _____ basic amphibole;

    e. _____ carbonate pyramid.

12. The silicon ion has a positive charge of 4, and oxygen has a negative charge of 2. Thus, the ion group (SiO_4) has a:

    a. _____ positive charge of 2;

    b. _____ negative charge of 2;

    c. _____ negative charge of 1;

    d. _____ positive charge of 4;

    e. _____ negative charge of 4.

13. Why is a distinction made between rock-forming minerals and accessory minerals? Also, list three silicate and two carbonate minerals and one sulfate rock-forming mineral.

14. How does the fact that all specimens of a given mineral species have the same angles between the same crystal faces indicate that these minerals are crystalline?

15. What accounts for the fact that some minerals, such as plagioclase feldspars, have a range of chemical compositions? Give an example from the ferromagnesian silicates.

16. A hypothetical atom has 19 protons, 14 neutrons, and 12 electrons. What is this atom's atomic number, atomic mass number, and electrical charge?

17. Some minerals possess the physical property known as *cleavage*. Describe cleavage and explain what controls it. Also, explain why cleavage fragments of calcite and halite, both with three directions of cleavage, are differently shaped.

18. How do compounds and native elements differ? Give two examples along with chemical formulas of minerals that are compounds and two minerals that are native elements.

19. How does a crystalline solid differ from a liquid and a gas?

20. Compare ionic and covalent bonding.

21. Considering the possible combinations of naturally occurring elements, why are so few minerals actually common components of rocks?

22. Diamond and graphite are both composed of carbon (C) but differ considerably in all physical properties. Explain how they differ and why.

23. In addition to silicates, name three other major mineral groups and explain how minerals are assigned to these groups.

24. How do silicate minerals differ from minerals in other major mineral groups? What are the two subgroups of silicates, and what are the criteria for assigning minerals to either subgroup?

Points to Ponder

1. How would the color and density of a rock composed mostly of nonferromagnesian silicates differ from one made up primarily of ferromagnesian silicates?

2. Why must the United States, a natural resource–rich nation, import most or all of some of the resources it needs? What are some resources that are in especially short supply in this country, and what problems does a dependence on imports create?

3. Where would be some good places to look for mineral crystals?

4. Explain how the composition and structure of minerals control such mineral properties as hardness, cleavage, color, and specific gravity.

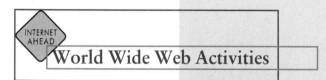

World Wide Web Activities

For these Web site addresses, along with current updates and exercises, log on to
http://www.brookscole.com/geo/

➤ SMITHSONIAN GEM & MINERAL COLLECTION

This site contains a variety of photos and information about various minerals in the Smithsonian National Museum of Natural History.

➤ MINERALOGY AND PETROLOGY RESEARCH ON THE WEB

This site contains links to mineralogy and petrology journals, professional societies, databases, research groups, and mineral collecting and commercial sites.

1. Go to the *Mineralogical Databases* section, and click on the *Alphabetical Mineral Reference* heading. It will take you to a listing of minerals and information about them. Look up information about the common rock-forming mineral feldspar. How many different types of feldspar are there? What are the different uses of feldspar? What are the major differences between orthoclase and plagioclase?

2. Click on the *Gem Database* link found at the beginning of *Alphabetical Mineral Reference*. Look up your birthstone or any gem that interests you. What did you learn about it that you did not already know?

➤ AMETHYST GALLERIES

Amethyst Galleries maintains this site, which has a wealth of information on and images of hundreds of minerals. It lists minerals alphabetically by name and class and has minerals in interesting groups such as gemstones and birthstones. A full text search for mineral identification by keyword is also available.

WEST'S GEOLOGY DIRECTORY

This site was introduced in Chapter 1 as a comprehensive directory of geologic Internet links and an excellent place to find links on virtually any geologic topic. Scroll down and click *Minerals,* and then click on *Minerals and Mineralogy.* Here you can find information on all common minerals and many others, including gemstones and jewelry. Click on any of the mineral group names listed. For example, go to the *mica group* of minerals:

1. What are the uses of biotite, muscovite, and lepidolite? Which one of these minerals is least common?
2. What types of rocks are these minerals found in?

Also go to *quartz.* What types of quartz are listed, and what are their properties?

THE MINERAL GALLERY

This site has images of dozens of minerals, along with information on their chemistry, physical properties, uses, associations with other minerals, and locales where they are found. Click on *sphalerite.* What is its chemical formula, its hardness, and uses? Click on *malachite* and see where it is found, the type of rock it is found in, and its uses. Click on *topaz* and give its chemical formula, specific gravity, and uses? Also, what is the origin of the name *topaz?*

U.S. GEOLOGICAL SURVEY MINERAL RESOURCE SURVEYS PROGRAM

Maintained by the U.S. Geological Survey (USGS), this site contains information on the USGS Mineral Resource Survey Program, Fact Sheets, Contacts, and numerous links to a variety of sources on mineral resources.

1. Go to the *Contacts* section. How would you use this section if you were interested in learning more about asbestos?
2. Go to the *Publications and Information Products* section of the *Mineral Information* link under the *Links* section. How would you use this section if you were interested in finding out about the supply and demand of gold in the world?

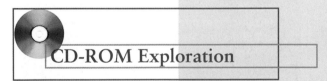

CD-ROM Exploration

Exploring your *Earth Systems Today* CD-ROM will add to your understanding of the material in this chapter.

TOPIC: EARTH'S MATERIALS

MODULE: ATOMS AND CRYSTAL STRUCTURE

Explore activities in this module to see if you can discover the following for yourself:

Use the animation in this module to gain an understanding of how atoms in magma behave and ultimately bond to make minerals. What bonds are the first to form and why? What mineral forms first? Why?

Use the graphics on mineral structures to observe how silicate minerals are assembled out of separate atoms. How are silica tetrahedra linked to each structure?

Chapter 3

A lava flow in Hawaii cooling to form the igneous rock basalt. This particular kind of lava flow with a smooth, ropy surface is called pahoehoe. It advances as a series of small lobes.

OBJECTIVES

At the end of this chapter, you will have learned that:

■ Most magma is composed of silicon and oxygen with smaller amounts of many other chemical elements.

■ Temperature and composition are the most important controls on the mobility of magma.

■ Magma originates below the surface, mostly at divergent and convergent plate boundaries.

■ A variety of processes can account for changes in the composition of a magma.

■ Composition and texture are properties of igneous rocks used to identify them.

■ Igneous textures indicate whether igneous rocks are extrusive or intrusive.

■ Intrusive igneous bodies known as *plutons* form when magma cools below the surface.

Igneous Rocks and Intrusive Igneous Activity

Prologue

Nearly everyone is familiar with the term *granite,* although few know that geologists define granite as an igneous rock with specific amounts of quartz, potassium feldspar, and plagioclase feldspar. Granite and several similar-appearing rocks, collectively referred to as *granitic,* are common, especially in the inner parts of mountain ranges and in a vast area of North America mostly in Canada known as the Canadian Shield. Granitic rocks are used for buildings, tombstones, pedestals for statues, and as crushed stone for aggregate in cement. Several varieties of cut and polished granitic rocks are attractive and used as facing stones on buildings, or for decorative stones in tabletops and mantelpieces. In addition, the rocks adjacent to granitic rocks in some areas contain important minerals such as copper. Some of these mineral resources are discussed later in this chapter.

Granitic rocks form from molten rock material far below the surface, but uplift and deep erosion have yielded vast exposures of these rocks in many areas of

North America and other parts of the world. On this continent, one of the most extensive areas of granitic rocks is in British Columbia, Canada; smaller yet still vast areas of similar rocks are present in Alaska, California, the Rocky Mountain states, Texas, the Great Lakes region, some of the Appalachian states, and the New England states. In Acadia National Park, Maine, for instance, magma cooled within the crust, forming granitic rocks that now constitute much of Mount Desert Island, which comprises most of the park. Later, magma of a different composition was injected into the granitic rocks, where it formed dark-colored bands (Figure 3.1a). Since then other geologic processes have shaped the area, especially deformation and erosion by glaciers during the last 1.6 million years.

Widespread exposures of granitic rocks are also present in the Black Hills of South Dakota. No doubt the most commonly visited site in this area is Mount Rushmore National Memorial, where the stone images of four presidents have been carved into rocks of the 1.7-billion-year-old Harney Peak Granite (Figure 3.1b). A memorial to the Native American Crazy Horse is now in progress in these granitic rocks near Custer, South Dakota. These granitic rocks form the core of the Black Hills, an oval area of crust that was uplifted beginning about 80 million years ago. The Black Hills have been modified by erosion too, but in this case changes were brought about by chemical and mechanical alteration of the rocks and erosion by running water.

Between 210 and 70 million years ago, during the Jurassic and Cretaceous Periods, huge bodies of granitic rock formed in what are now the Rocky Mountain states, especially in Idaho, Colorado, and Montana, and in California and adjacent parts of Nevada and Oregon. Visitors to Yosemite, Sequoia, and Kings Canyon National Parks in California can see rock exposures consisting of little but granitic rocks (Figure 3.1c), in addition to the other features the parks have to offer, such as giant trees, wildlife, waterfalls, and broad, grass-covered meadows. Some of these rock exposures, such as El Capitan in Yosemite National Park, are well known to rock climbers. The most important geologic processes now modifying these areas are chemical and mechanical breakdown

(a)

(b)

(c)

(d)

Figure 3.1
Exposures of granitic rocks. (a) Rocks along the shoreline in Acadia National Park, Maine. (b) The presidents' images at Mount Rushmore in the Black Hills of South Dakota were carved in the Harney Peak Granite. (c) Denali, formerly known as Mount McKinley, in Alaska's Denali National Park, is composed largely of granitic rocks. (d) Vast exposures of granite and related rocks are present in Yosemite National Park, California.

of the rocks and erosion by running water, but glaciers were important during the past.

At more than 6100 m above sea level, Denali (formerly known as Mount McKinley) in Alaska, the highest peak in North America, is composed predominately of granitic rocks (Figure 3.1d). Native Alaskans named the mountain Denali, meaning "The High One." Snow, ice, and glaciers cover most of the mountain, and glaciers are responsible for most of its present scenery. Denali and nearby North Peak (5935 m high) are the most conspicuous features in Denali National Park.

Granite and related rocks figure importantly in some of our following discussions. In this chapter, for example, the types of granitic rocks and the igneous bodies they form are considered, as well as resources in adjacent rocks. And in Chapter 5, the Mount Airy Granite of North Carolina is important in our discussion of a particular kind of mechanical alteration of rocks. Knowledge about the origin of granitic rocks will also help you understand the nature of rock alteration by heat and fluids (Chapter 7) and mountain-building episodes (Chapter 13).

INTRODUCTION

Rocks that originate from lava flows or particulate matter erupted from volcanoes are common, but represent only a tiny fraction of the rocks formed by the cooling and crystallization of molten rock material known as *magma*. Most magma cools below Earth's surface, where it forms bodies of rock known as *plutons,* which vary in size, shape, and relationships to the previously existing rocks adjacent to them. Many of the rocks forming in plutons are granitic (see the Prologue), but a variety of other types are known as well. In this chapter, we are concerned (1) with the origin, composition, textures, and classification of igneous rocks resulting from either volcanic activity or the formation of plutons, and (2) with the origin, significance, and types of plutons. In the following chapter, we consider volcanism, volcanoes, and associated phenomena, which are produced by magma reaching the surface.

Even though plutons and volcanism are discussed in separate chapters, they are nevertheless related. The same types of magmas are involved in both processes, but some magmas are more mobile than others and more commonly reach the surface. Plutons typically lie beneath areas of volcanism and, in fact, serve as the sources for the overlying lavas and fragmental materials ejected during explosive volcanic eruptions. Furthermore, present-day plutons and volcanoes are found mostly at or near plate boundaries, indicating that they share a common origin. Thus ancient plutons and rocks resulting from volcanic eruptions serve as some of the evidence geologists use to identify plate boundaries that existed during the past.

THE PROPERTIES AND BEHAVIOR OF MAGMA AND LAVA

In Chapter 2 we noted that one process leading to the formation of minerals, and thus of rocks composed of minerals, was the cooling and crystallization of the molten rock material known as *magma* and *lava.* **Magma** is simply molten rock material below the surface; the same material at the surface is called **lava.** No other distinction is necessary, so the two terms simply tell us the location of the molten rock material. Any magma is less dense than the rock from which it was derived, so it tends to move up toward the surface. However, most of it cools and solidifies far below the surface, thus accounting for the origin of various plutons. Magma that does reach the surface is erupted as either **lava flows** or is forcefully ejected into the atmosphere as particles known as **pyroclastic materials** (from the Greek *pyro,* "fire" and *klastos,* "broken"). Lava flows and eruptions of pyroclastic materials are the most awe-inspiring manifestations of all processes related to magma, but as noted, result from only a small percentage of all magma that forms.

All **igneous rocks** derive from magma, but two separate processes account for their origin. They form when (1) magma or lava cools and crystallizes to form minerals, or (2) pyroclastic materials such as volcanic ash are consolidated, forming solid masses from the previously loose particles. Igneous rocks resulting from cooling lava flows and consolidation of pyroclastic materials are further characterized as **volcanic rocks** or **extrusive**

igneous rocks, whereas those forming when magma cools and crystallizes below the surface are known as **plutonic rocks** or **intrusive igneous rocks.**

This brief introduction to magma and lava is enough to clarify what these substances are and what kinds of rocks are derived from them. However, let's explore magma, the source of all igneous rocks, a bit further and consider its composition, temperature, and resistance to flow (what is called its *viscosity*).

The Composition of Magma

In Chapter 2, we made the point that by far the most abundant minerals in Earth's crust are silicates such as quartz, feldspars, and several ferromagnesian silicates, all composed mostly of silicon, oxygen, and other elements listed in Table 2.4. As a result, melting of crustal rocks yields magmas most of which are silica rich but that also contain considerable aluminum, calcium, sodium, iron, magnesium, potassium, and several other elements in lesser quantities. Another source of magma is the melting of rocks in Earth's upper mantle, which are composed largely of ferromagnesian silicates. Thus a magma derived from this source contains comparatively less silica and more iron and magnesium.

Silica is the primary constituent of most magmas, but it varies enough to distinguish magmas characterized as felsic, intermediate, and mafic (Table 3.1). **Felsic magma,** with more than 65% silica, is silica rich and contains considerable sodium, potassium, and aluminum, but little calcium, iron, and magnesium. **Mafic magma,** in contrast, is silica poor, but contains proportionately more calcium, iron, and magnesium. As its name implies, **intermediate magma** has a composition intermediate between felsic and mafic magma (Table 3.1).

How Hot Are Magma and Lava?

No direct measurements of magma temperatures below Earth's surface have been made. Erupting lavas generally have temperatures in the range of 1000°C to 1200°C, although temperatures of 1350°C have been recorded above Hawaiian lava lakes where volcanic gases reacted with the atmosphere.

Most direct temperature measurements of lavas have been taken at volcanoes characterized by little or no explosive activity where geologists can safely approach the lava (Figure 3.2). Therefore, little is known of the temperatures of felsic lavas because eruptions of such lavas are rare, and when they do occur, they tend to be explosive. The temperatures of some lava domes, most of which are bulbous masses of felsic magma, have been measured at a distance by using an instrument called an *optical pyrometer*. The surfaces of these domes have temperatures up to 900°C, but the exterior of a dome is probably much cooler than its interior.

When Mount St. Helens erupted in 1980, it ejected felsic magma as particulate matter in pyroclastic flows. Two weeks later, these flows still had temperatures between 300°C and 420°C, and a steam explosion took place when water encountered some of the still-hot deposits more than a year later. The reason why magma or lava retains heat so well is that rock is such a poor conductor of heat. Accordingly, the interiors of thick lava flows may remain hot for months or years, whereas plutons, depending on their size and depth, may not completely cool for thousands or millions of years.

Viscosity—Resistance to Flow

All liquids possess the property of **viscosity,** or simply their resistance to flow. For many liquids, such as water, the viscosity is very low so they are highly fluid and flow readily. For other liquids, though, viscosity is so high that they flow much more slowly. Good examples are cold motor oil and syrup, both of which are quite viscous and thus flow only with difficulty. But when these same liquids are heated, their viscosity is much lower and they flow more easily. That is, they become more fluid with increasing temperature. Accordingly, you might suspect the temperature controls the viscosity of magma and lava, and this inference is partly correct. We can generalize and say that hot magma or lava moves more readily than cooler magma or lava, but we must qualify this statement by noting that temperature is not the only control of viscosity.

Silica content strongly controls magma and lava viscosity. With increasing silica content, numerous networks of silica tetrahedra form and retard flow because for flow to take place, the strong bonds of the networks must be ruptured. Mafic magma and lava with 45–52% silica have fewer silica tetrahedra networks and as a result are more mobile than felsic magma and lava flows. One mafic flow in 1783 in Iceland flowed about 80 km, and some ancient flows in Washington State can be traced for more than 500 km. Felsic magma, in contrast, because of its higher viscosity, does not reach the

Table 3.1

The Most Common Types of Magmas and Their Characteristics			
Type of Magma	Silica Content (%)	Sodium, Potassium, & Aluminum	Calcium, Iron, & Magnesium
Mafic	45–52		Increase
Intermediate	53–65	↓	↑
Felsic	>65	Increase	

[handwritten margin note: High Viscosity]

Figure 3.2
A geologist uses a thermocouple to determine the temperature of a lava flow.

surface as commonly as mafic magma. And when felsic lava flows do occur, they tend to be slow moving and thick and to move only short distances. A thick, pasty lava flow that erupted in 1915 from Lassen Peak in California moved only about 300 m before it ceased flowing.

HOW DOES MAGMA ORIGINATE AND CHANGE?

Most people are familiar with magma that reaches Earth's surface as either lava flows or pyroclastic materials ejected during explosive volcanic eruptions. The lava flows issuing from Kilauea in Hawaii fascinate many observers, and the activity at Mount St. Helens, Washington, in 1980 and Mount Pinatubo in the Philippines in 1991 remind us of the violence of some eruptions. Yet most people have little understanding of how magma originates in the first place, how it can change or evolve, or how it reaches the surface. Indeed, many believe the misconceptions that lava comes from a continuous layer of molten material beneath the crust or that it comes from Earth's molten core.

Some magma rises from depths of as much as 100 to 300 km, but most of it forms at much shallower depths in the upper mantle or lower crust and accumulates in reservoirs known as **magma chambers.** Beneath spreading ridges in the ocean basins, where the crust is thin, magma chambers exist at a depth of only a few kilometers. Along convergent plate boundaries where an oceanic plate is subducted beneath another oceanic plate or beneath a continental plate, magma chambers are only a few tens of kilometers deep. The volume of a magma chamber may be several cubic kilometers of molten rock within the otherwise solid lithosphere. This magma might simply cool and crystallize in place, thus forming various intrusive igneous rock bodies, except that some migrates to the surface.

Bowen's Reaction Series

During the early part of this century, N. L. Bowen hypothesized that mafic, intermediate, and felsic magmas could all derive from a parent mafic magma. He knew that minerals do not all crystallize simultaneously from a cooling magma, but rather crystallize in a predictable sequence. Based on his observations and laboratory experiments, Bowen proposed a mechanism, now called **Bowen's reaction series,** to account for the derivation of intermediate and felsic magmas from a mafic magma. Bowen's reaction series consists of two branches: a *discontinuous branch* and a *continuous branch* (Figure 3.3). As the temperature of a magma decreases, crystallization of minerals occurs along both branches simultaneously, but for convenience we will discuss them separately.

In the discontinuous branch, which contains only ferromagnesian silicates, one mineral changes to another over specific temperature ranges (Figure 3.3). As the temperature decreases, a temperature range is reached in which a given mineral begins to crystallize. A previously formed mineral reacts with the remaining liquid magma (the melt) so that it forms the next mineral in the sequence. For instance, olivine [$(Mg,Fe)_2SiO_3$] is the first ferromagnesian silicate to crystallize. As the magma continues to cool, it reaches the temperature range at

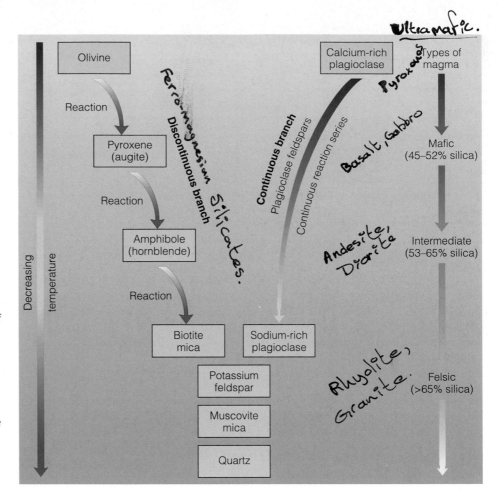

Figure 3.3
Bowen's reaction series. It consists of a discontinuous branch along which a succession of ferromagnesian silicates crystallize as the magma's temperature decreases, and a continuous branch along which plagioclase feldspars with increasing amounts of sodium crystallize. Notice also that the composition of the initial mafic magma changes as crystallization takes place along the two branches.

which pyroxene is stable; a reaction occurs between the olivine and the remaining melt, and pyroxene forms.

With continued cooling, a similar reaction takes place between pyroxene and the melt, and the pyroxene structure is rearranged to form amphibole. Further cooling causes a reaction between the amphibole and the melt, and its structure is rearranged so that the sheet structure typical of biotite mica forms. Although the reactions just described tend to convert one mineral to the next in the series, the reactions are not always complete. Olivine, for example, might have a rim of pyroxene, indicating an incomplete reaction. If a magma cools rapidly enough, the early-formed minerals do not have time to react with the melt, and thus all the ferromagnesian silicates in the discontinuous branch can be in one rock. In any case, by the time biotite has crystallized, essentially all magnesium and iron present in the original magma have been used up.

Plagioclase feldspars, which are nonferromagnesian silicates, are the only minerals in the continuous branch of Bowen's reaction series (Figure 3.3). Calcium-rich plagioclase crystallizes first. As cooling of the magma proceeds, calcium-rich plagioclase reacts with the melt, and plagioclase containing proportionately more sodium crystallizes until all of the calcium and sodium are used up. In many cases cooling is too rapid for a complete

transformation from calcium-rich to sodium-rich plagioclase to take place. Plagioclase forming under these conditions is *zoned,* meaning that it has a calcium-rich core surrounded by zones progressively richer in sodium.

As minerals crystallize along the two branches of Bowen's reaction series, iron and magnesium are depleted, because they are used in ferromagnesian silicates, whereas calcium and sodium are used up in plagioclase feldspars. At this point, any leftover magma will be enriched in potassium, aluminum, and silicon, which combine to form orthoclase ($KalSi_3O_8$), a potassium feldspar, and if water pressure is high the sheet silicate muscovite forms. Any remaining magma is enriched in silicon and oxygen (silica) and forms the mineral quartz (SiO_2). The crystallization of orthoclase and quartz is not a true reaction series as is the crystallization of ferromagnesian silicates and plagioclase feldspars, because they form independently rather than by a reaction of orthoclase with the melt.

The Origin of Magma at Spreading Ridges

One fundamental observation we can make regarding the origin of magma is that Earth's temperature increases with depth. This temperature increase, known as the *geothermal gradient,* averages about 25°C/km. Accordingly, rocks at depth are hot but remain solid because their melting temperature rises with increasing pressure (Figure 3.4a). However, beneath spreading ridges the temperature locally exceeds the melting temperature, at least in part, because pressure decreases. That is, plate separation at ridges probably causes a decrease in pressure on the already hot rocks at depth, thus initiating melting (Figure 3.4a). In addition, the presence of water can also decrease the melting temperature beneath spreading ridges because water aids thermal energy in breaking the chemical bonds in minerals (Figure 3.4b).

Another explanation for the origin of spreading-ridge magmas is that localized, cylindrical plumes of hot mantle material, called *mantle plumes,* rise beneath the ridges and spread out in all directions. Perhaps localized concentrations of radioactive minerals within the crust and upper mantle decay and provide the heat necessary to melt rocks and thus generate magma.

The magmas formed beneath spreading ridges are invariably mafic (45–52% silica). But the upper mantle rocks from which these magmas are derived are characterized as ultramafic (<45% silica), consisting largely of ferromagnesian silicates and lesser amounts of nonferromagnesian silicates. To explain how mafic magma originates from ultramafic rock, geologists propose that the magma is formed from source rock that only partially melts. This phenomenon of partial melting takes place because various minerals have different melting temperatures.

Recall the sequence of minerals in Bowen's reaction series (Figure 3.3). The order in which these minerals melt is the opposite of their order of crystallization. Accordingly, quartz, potassium feldspar, and sodium-rich plagioclase melt before most of the ferromagnesian silicates and the calcic varieties of plagioclase. So when ultramafic rock begins to melt, the minerals richest in silica melt first, followed by those containing less silica. Therefore, if melting is not complete, a mafic magma containing proportionately more silica than the source rock results. Once this mafic magma forms, some of it rises to the surface, where it forms lava flows, and some simply cools beneath the surface to form various intrusive igneous bodies.

Subduction Zones and the Origin of Magma

Another basic observation we can make regarding magma is that, where an oceanic plate is subducted beneath either a continental plate or another oceanic plate, a belt of volcanoes and plutons is found near the leading edge of the overriding plate (Figure 3.5). It would seem, then, that subduction and the origin of magma must be related in some way, and indeed they are. Furthermore, magma at these convergent plate boundaries is mostly intermediate (53–65% silica) or felsic (>65% silica).

Once again, geologists invoke the phenomenon of partial melting to explain the origin and composition of magma at subduction zones. As a subducted plate descends toward the asthenosphere, it eventually reaches the depth where the temperature is high enough to initiate partial melting. In addition, the wet oceanic crust descends to a depth at which dewatering takes place, and as the water rises into the overlying mantle, it enhances melting and magma forms (Figure 3.4b).

Recall that partial melting of ultramafic rock at spreading ridges yields mafic magma. Similarly, partial melting of mafic rocks of the oceanic crust yields intermediate (53–65% silica) and felsic (>65% silica) magmas, both of which are richer in silica than the source rock. Moreover, some of the silica-rich sediments and sedimentary rocks of continental margins are probably carried downward with the subducted plate and contribute their silica to the magma. Also, mafic magma

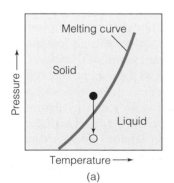
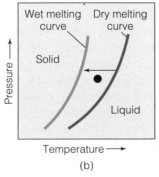

Figure 3.4
The effects of pressure and water on melting. (a) Melting temperature rises with increasing pressure, so a pressure decrease on already hot rocks can initiate melting. (b) When water is present the melting curve shifts to the left, because water provides an additional agent to break chemical bonds. Accordingly, rocks melt at a lower temperature if water is present.

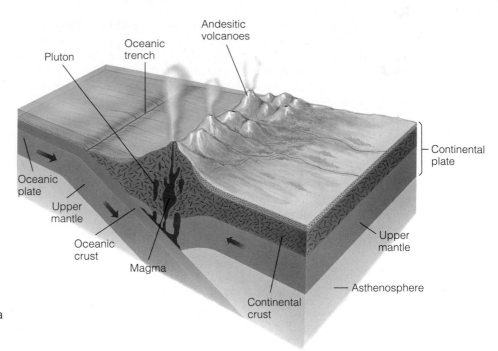

Andesitic
volcanoes

Pluton

Oceanic
trench

Continental
plate

Oceanic
plate

Upper
mantle

Oceanic
crust

Upper
mantle

Magma

Asthenosphere

Continental
crust

Figure 3.5
Subduction of an oceanic plate
beneath either another oceanic plate
or a continental plate, as shown here,
produces magma. Some of the magma
forms intrusive igneous bodies, and
some is erupted to form volcanoes.

rising through the lower continental crust must be contaminated with silica-rich materials, which changes its composition.

Processes Resulting in Chemical Changes of Magma

Once a magma forms, its composition may change by **crystal settling,** which involves the physical separation of minerals by crystallization and gravitational settling (Figure 3.6). Olivine, the first ferromagnesian silicate to form in the discontinuous branch of Bowen's reaction series, has a specific gravity greater than that of the remaining magma and tends to sink down in the melt. Accordingly, the remaining melt becomes comparatively rich in silica, sodium, and potassium, because much of the iron and magnesium were removed when minerals containing these elements crystallized.

Although crystal settling does take place in magmas, it does not do so on the scale envisioned by Bowen. In some thick, sheetlike, intrusive igneous bodies called *sills,* the first-formed minerals in the reaction series are indeed concentrated. The lower parts of these bodies contain more olivine and pyroxene than the upper parts, which are less mafic. But even in these bodies, crystal settling has yielded little felsic magma from an original mafic magma.

If felsic magma could be derived on a large scale from mafic magma as Bowen thought, there should be far more mafic magma than felsic magma. To yield a particular volume of granite (a felsic igneous rock), about 10 times as much mafic magma would have to be present initially for crystal settling to yield the volume of granite in question. If this were so, then mafic intrusive igneous rocks should be much more common than felsic ones. However, just the opposite is the case, so it appears that mechanisms other than crystal settling must account for the large volume of felsic magma. Partial melting of mafic oceanic crust and silica-rich sediments of continental margins during subduction yields magma richer in silica than the source rock. Furthermore, magma rising through the continental crust can absorb some felsic materials and become more enriched in silica.

The composition of a magma can also change by **assimilation,** a process whereby a magma reacts with preexisting rock, called **country rock,** with which it comes in contact (Figure 3.6). The walls of a volcanic conduit or magma chamber are, of course, heated by the adjacent magma, which may reach temperatures of 1300°C. Some of these rocks can be partly or completely melted, provided their melting temperature is less than that of the magma. Because the assimilated rocks seldom have the same composition as the magma, the composition of the magma is changed.

The fact that assimilation occurs can be demonstrated by *inclusions,* incompletely melted pieces of rock that are fairly common within igneous rocks. Many inclusions were simply wedged loose from the country rock as the magma forced its way into preexisting fractures (Figure 3.6).

No one doubts that assimilation takes place, but its effect on the bulk composition of most magmas must be slight. The reason is that the heat for melting must come from the magma itself, and this would have the effect of cooling the magma. Only a limited amount of rock can be assimilated by a magma, and that amount is usually insufficient to bring about a major compositional change.

Magma chamber

Crystal
settling

Assimilated pieces
of country rock

Country rock

Inclusion

(a)

(b)

Figure 3.6
(a) Early-formed ferromagnesian silicates have a high specific gravity and settle and accumulate in the magma chamber. Fragments of rock dislodged by upward-moving magma might melt and be incorporated into the magma, or they might remain as inclusions.(b) Dark-colored inclusions in granitic rock.

Neither crystal settling nor assimilation can produce a significant amount of felsic magma from a mafic one. But both processes, if operating concurrently, can change the composition of a mafic magma much more than either process acting alone. Some geologists think that this is one way many intermediate magmas form where oceanic lithosphere is subducted beneath continental lithosphere.

The fact that a single volcano can erupt lavas of different composition indicates that magmas of differing composition must be present. It seems likely that some of these magmas would come into contact and mix with one another. If this is the case, we would expect that the composition of the magma resulting from **magma mixing** would be a modified version of the parent magmas. Suppose a rising mafic magma mixes with a felsic magma of about the same volume (Figure 3.7). The resulting "new" magma would have a more intermediate composition.

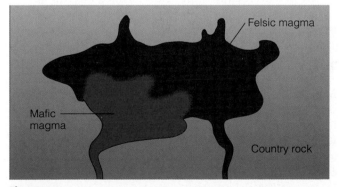

Felsic magma

Mafic
magma

Country rock

Figure 3.7
Magma mixing. Two magmas mix and produce a magma with a composition different from either of the parent magmas.

IGNEOUS ROCKS—WHAT ARE THEY, AND WHAT ARE THEIR CHARACTERISTICS?

In the Introduction we briefly defined *plutonic* or *intrusive igneous rocks* and *volcanic* or *extrusive igneous rocks*. Here we will have considerably more to say about the texture, composition, and classification of these rocks, which constitute one of the three major rock families depicted in the rock cycle (see Figure 1.17).

Igneous Rock Textures

The term *texture* refers to the size, shape, and arrangement of mineral grains composing igneous rocks. Size is the most important because grain size is related to the cooling history of a magma or lava, and generally indicates whether an igneous rock is intrusive or extrusive. The atoms in magma and lava are in constant motion, but when cooling begins, some atoms bond to form small nuclei. As other atoms in the liquid chemically bond to these nuclei, they do so in an orderly geometric arrangement and the nuclei grow into crystalline *mineral grains,* the individual particles that compose igneous rocks.

During rapid cooling, as takes place in lava flows and some shallow intrusive bodies of magma, the rate at which mineral nuclei form exceeds the rate of growth and an aggregate of many small mineral grains results. The result is a fine-grained or **aphanitic texture,** in which individual minerals are too small to be seen without magnification (Figure 3.8a). With slow cooling, the rate of growth exceeds the rate of nuclei formation, and relatively large

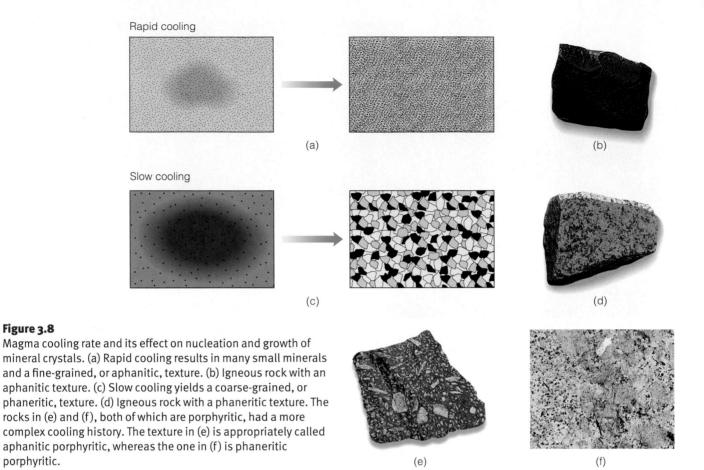

Rapid cooling

(a)

(b)

Slow cooling

(c)

(d)

(e)

(f)

Figure 3.8
Magma cooling rate and its effect on nucleation and growth of mineral crystals. (a) Rapid cooling results in many small minerals and a fine-grained, or aphanitic, texture. (b) Igneous rock with an aphanitic texture. (c) Slow cooling yields a coarse-grained, or phaneritic, texture. (d) Igneous rock with a phaneritic texture. The rocks in (e) and (f), both of which are porphyritic, had a more complex cooling history. The texture in (e) is appropriately called aphanitic porphyritic, whereas the one in (f) is phaneritic porphyritic.

mineral grains form, thus yielding a coarse-grained or **phaneritic texture,** in which minerals are clearly visible (Figure 3.8c). Aphanitic textures generally indicate an extrusive origin, whereas rocks with phaneritic textures are usually intrusive. However, shallow intrusions might have an aphanitic texture, and the rocks that form in the interiors of some thick lava flows might be phaneritic.

Another common texture in igneous rocks is one termed **porphyritic,** in which minerals of markedly different size are present in the same rock. The larger minerals are *phenocrysts* and the smaller ones are referred to as the *groundmass,* which is simply the grains between phenocrysts (Figure 3.8e, f). The groundmass can be either aphanitic or phaneritic; the only requirement for a porphyritic texture is that the phenocrysts be considerably larger than the minerals in the groundmass. Igneous rocks with porphyritic textures are designated as *porphyry* as in basalt porphyry. These rocks have more complex cooling histories than those with aphanitic or phaneritic textures that might involve, for example, magma partly cooling beneath the surface followed by its eruption and rapid cooling at the surface.

A lava may cool so rapidly that its constituent atoms do not have time to become arranged in the ordered, three-dimensional frameworks typical of minerals. As a consequence of such rapid cooling, a *natural glass* such as *obsidian* forms (Figure 3.9a). Even though obsidian with its glassy texture is not composed of minerals, it is still considered to be an igneous rock.

Some magmas contain large amounts of water vapor and other gases. These gases may be trapped in cooling lava where they form numerous small holes or cavities known as **vesicles;** rocks that have many vesicles are termed *vesicular,* as in vesicular basalt (Figure 3.9b).

A **pyroclastic** or **fragmental texture** characterizes igneous rocks formed by explosive volcanic activity. For example, ash may be discharged high into the atmosphere and eventually settle to the surface where it accumulates; if it is turned into solid rock, it is considered a pyroclastic igneous rock.

The Composition of Igneous Rocks

Most igneous rocks, just as the magmas from which they originate, are characterized as mafic (45–52% silica), intermediate (53–65% silica), or felsic (>65% silica). A few are referred to as ultramafic (<45% silica), but these are probably derived from mafic magma by a process discussed later. The parent magma plays an important role in determining the mineral composition of igneous rocks, yet it is possible for the same magma to yield a variety of igneous rocks, because its composition can change as a

David P. Hill

Monitoring Volcanic Activity

David P. Hill graduated from San Jose State University and earned an M.S. in geophysics from the Colorado School of Mines and a Ph.D. from the California Institute of Technology. He has been a geophysicist with the U.S. Geological Survey since 1961. His work in the Mammoth Lakes area has been widely cited in newspapers and magazines, including *Science News,* as a good example of how scientific research can have an impact on the general public.

Much of my career with the U.S. Geological Survey has involved the study of earthquakes and the clues they provide on deformation of Earth's crust (seismotectonics) and volcanic processes. Since 1983 I have been in charge of the U.S. Geological Survey's efforts to monitor and better understand the recurring episodes of earthquake swarms and ground uplift in Long Valley caldera in eastern California. Long Valley caldera is a large oval depression (15 by 30 km) at the base of the eastern escarpment of the Sierra Nevada that was formed by a massive volcanic eruption 730,000 years ago. Volcanic activity has continued in the area with the most recent eruptions having occurred just 500 years ago. The current earthquake activity and ground deformation is symptomatic of the movement of magma in the upper 5 to 10 km of the crust and is typical of the sort of geologic unrest that often precedes volcanic eruptions.

This activity raises a host of fascinating scientific questions, but it also raises important issues regarding the interaction of science and society. The resort town of Mammoth Lakes lies at the southwestern margin of Long Valley caldera, and the initial news in 1982 that the activity might be related to volcanic activity was met with disbelief and anger by the residents of Mammoth Lakes and Mono County. Responding to the continuing activity in the caldera over subsequent years has proved to be an educational process for both local residents and scientists studying and monitoring the activity. The local residents have come to better appreciate the geologic processes that have sculpted the spectacular setting where they live, and the scientists have come to appreciate the challenge involved in effectively communicating the results of their research to the public in a useful way. The latter is a challenge that faces science in general as taxpayers and politicians increasingly demand to know what they are getting for their money. It is a particularly acute challenge for those of us pursuing research on geologic hazards (earthquakes, volcanoes, landslides, etc.) because the results of our research and the manner in which we present them can have an immediate impact on the economy and social well-being of the communities involved, as well as on public safety.

Over the last couple of decades, Earth scientists have learned a great deal about the nature of volcanic unrest from detailed studies of volcanoes around the world. A number of these volcanoes have erupted, and in several cases, the eruptions were successfully predicted. Notable examples of successful predictions include five eruptions of Mount Redoubt, Alaska, from December 14, 1989, through April 15, 1990; eruptions of Mount Spurr, Alaska, on June 27 and September 17, 1992; and the climactic eruption of Mount Pinatubo in the Philippines on June 15, 1991 (the largest eruption in the world this century after the 1915 eruption of Mount Katmi in Alaska). The high-stakes drama associated with the successful prediction of the Mount Pinatubo eruption, which provided time for the evacuation of over 60,000 people and the removal of several hundred million dollars worth of aircraft equipment from nearby Clark Air Force Base, was documented in PBS's *Nova* program on Mount Pinatubo.

The problem of reliably predicting volcanic eruptions is far from solved, however. Volcanoes show a wide range of behaviors. The precursory unrest that led to the successful prediction of the eruptions from andesitic, central-vent volcanoes mentioned above lasted from a few months to less than a week. In contrast, volcanic unrest at large calderas such as Long Valley may persist for decades without culminating in an eruption. The Campi Flegri caldera on the outskirts of Naples, Italy, for example, is currently quiet following episodes of ground deformation and earthquake-swarm activity that began in 1969 and elevated the caldera floor by over 3 m in 1984. The floor of Yellowstone caldera (in Yellowstone National Park) rose nearly 1 m between 1932 and 1985 and has since been gradually subsiding. At Rabaul caldera in Papua New Guinea, however, two decades of unrest, which included an episode of rapid ground uplift and intense earthquake-swarm activity in 1984, culminated in a moderate but locally destructive eruption in September 1994. Although the short-term precursor to this eruption provided only 27 hours of warning, it was enough for the safe evacuation of nearly 30,000 people from the town of Rabaul. This wide range in behavior in volcanic systems places a premium on establishing and maintaining clear communication between scientists, civil authorities, the media, and the public in areas exposed to volcanic hazards.

The Earth sciences are central to our understanding the risks posed by geologic hazards and a growing number of environmental issues that face modern society. They will continue to offer a range of scientifically exciting and socially significant opportunities in the future. ∎

(a)

(b)

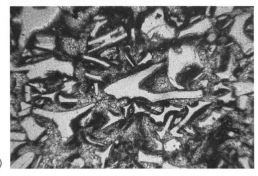

(c)

Figure 3.9
(a) The glassy texture of obsidian forms when lava cools too quickly for mineral crystals to form. (b) Vesicles form when gases expand in some volcanic rocks giving them a vesicular texture. (c) Enlarged view of an igneous rock with a pyroclastic or fragmental texture. The colorless, angular objects are pieces of volcanic glass measuring up to 2 mm.

result of crystal settling, assimilation, magma mixing, and the sequence in which minerals crystallize.

Classifying Igneous Rocks

With few exceptions, igneous rocks are classified on the basis of textural features and composition. Notice in Figure 3.10 that all the rocks, except peridotite, constitute pairs; the members of a pair have the same composition but different textures. Basalt and gabbro, andesite and diorite, and rhyolite and granite are compositional (mineralogical) equivalents, but basalt, andesite, and rhyolite are aphanitic and most commonly extrusive, whereas gabbro, diorite, and granite have phaneritic textures that generally indicate an intrusive origin. All these pairs exist in a textural continuum. The extrusive and intrusive members of each pair can usually be differentiated by texture, but many shallow intrusive rocks have textures that cannot be readily distinguished from those of extrusive igneous rocks.

The igneous rocks shown in Figure 3.10 are also differentiated by composition. Reading across the chart from rhyolite to andesite to basalt, for example, the relative proportions of nonferromagnesian and ferromagnesian silicates differ. The differences in composition are gradual, however, so that a compositional continuum

exists. In other words, rocks exist with compositions intermediate between rhyolite and andesite, and so on.

Ultramafic Rocks   Ultramafic rocks (<45% silica) are composed largely of ferromagnesian silicates. The ultramafic rock *peridotite* contains mostly olivine, lesser amounts of pyroxene, and generally a little plagioclase feldspar (Figures 3.10 and 3.11). Another ultramafic rock (pyroxenite) is composed predominately of pyroxene. Because these minerals are dark colored, the rocks are generally black or dark green. Peridotite is thought to be the rock type composing the upper mantle (see Chapter 10). Ultramafic rocks are generally thought to have originated by concentration of the early-formed ferromagnesian minerals that separated from mafic magmas.

Ultramafic lava flows are known in rocks older than 2.5 billion years, but younger ones are rare or absent. The reason is that to erupt an ultramafic lava must have a near-surface temperature of about 1600°C; the surface temperatures of present-day mafic lava flows are generally between 1000°C and 1200°C. During early Earth history, though, more radioactive decay heated the mantle to as much as 300°C hotter than now and ultramafic lavas could be erupted onto the surface. Because the amount of heat has decreased through time, Earth has cooled, and eruptions of ultramafic lava flows ceased.

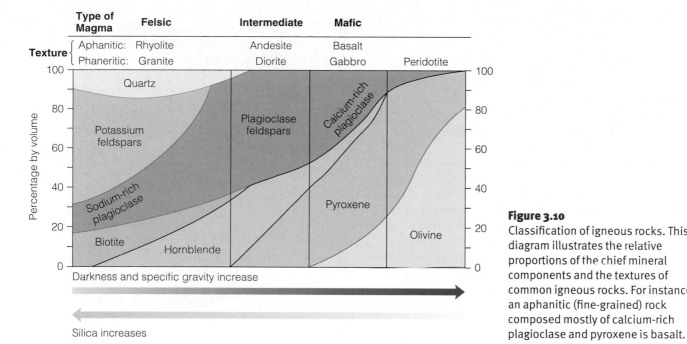

Type of Magma	Felsic		Intermediate	Mafic	
Texture { Aphanitic:	Rhyolite		Andesite	Basalt	
Phaneritic:	Granite		Diorite	Gabbro	Peridotite

Percentage by volume

Quartz
Potassium feldspars
Sodium-rich plagioclase
Biotite
Hornblende
Plagioclase feldspars
Calcium-rich plagioclase
Pyroxene
Olivine

Darkness and specific gravity increase

Silica increases

Figure 3.10
Classification of igneous rocks. This diagram illustrates the relative proportions of the chief mineral components and the textures of common igneous rocks. For instance, an aphanitic (fine-grained) rock composed mostly of calcium-rich plagioclase and pyroxene is basalt.

Figure 3.11
The ultramafic rock peridotite. Notice in Figure 3.10 that peridotite is the only phaneritic (coarse-grained) igneous rock not having an aphanitic (fine-grained) equivalent. Peridotite is rare at the surface but is thought to be the rock making up Earth's mantle.

Basalt-Gabbro *Basalt* and *gabbro* are the fine-grained and coarse-grained rocks, respectively, that crystallize from mafic magmas (45–52% silica) (Figure 3.12a and b). Thus, both have the same composition—mostly calcium-rich plagioclase and pyroxene, with smaller amounts of olivine and amphibole (Figure 3.10). Because they contain a large proportion of ferromagnesian silicates, basalt and gabbro are dark colored; those that are porphyritic typically contain calcium plagioclase or olivine phenocrysts.

Basalt is generally considered to be the most common extrusive igneous rock. Extensive basalt lava flows

were erupted in vast areas in Washington, Oregon, Idaho, and northern California (see Chapter 4). Oceanic islands such as Iceland, the Galapagos, the Azores, and the Hawaiian Islands are composed mostly of basalt, and basalt makes up the upper part of the oceanic crust.

Gabbro is much less common than basalt, at least in the continental crust or where it can be easily observed. Small intrusive bodies of gabbro are present in the continental crust, but intermediate to felsic intrusive rocks such as diorite and granite are much more common. The lower part of the oceanic crust is composed of gabbro, however.

(a) Basalt
(b) Gabbro

Figure 3.12
Mafic igneous rocks. (a) Basalt has an aphanitic (fine-grained) texture. Notice the small vesicles. (b) Gabbro is phaneritic (coarase-grained). Shiny crystal faces are visible in this specimen.

Andesite-Diorite Magmas intermediate in composition (53–65% silica) crystallize to form *andesite* and *diorite,* which are compositionally equivalent fine- and coarse-grained igneous rocks (Figure 3.13). Andesite and diorite are composed predominately of plagioclase feldspar, with the typical ferromagnesian component being amphibole or biotite (Figure 3.10). Andesite is generally medium to dark gray, but diorite has a salt-and-pepper appearance because of its white to light gray plagioclase and dark ferromagnesian silicates (Figure 3.13).

Andesite is a common extrusive igneous rock formed from lavas erupted in volcanic chains at convergent plate margins. The volcanoes of the Andes Mountains of South America and the Cascade Range in western North America are composed in part of andesite. Intrusive bodies composed of diorite are fairly common in the continental crust but are not nearly as abundant as granitic rocks.

Rhyolite-Granite *Rhyolite* and *granite* (>65% silica) crystallize from felsic magmas and are therefore silica-rich rocks (Figure 3.14). They consist largely of potassium feldspar, sodium-rich plagioclase, and quartz, with perhaps some biotite and rarely amphibole (Figure 3.10). Because nonferromagnesian silicates predominate, rhyolite and granite are generally light colored. Phyolite is fine grained, although most often it contains phenocrysts of potassium feldspar or quartz, and granite is coarse grained. Grainite porphyry is also fairly common.

Rhyolite lava flows are much less common than andesite and basalt flows. Recall that the greatest control of viscosity in a magma is silica content. Thus, if a felsic magma rises to the surface, it begins to cool, the pressure on it decreases, and gases are released explosively, usually yielding rhyolitic pyroclastic materials. The rhyolitic lava flows that do occur are thick and highly viscous and move only short distances.

Among geologists, granite has come to mean any coarsely crystalline igneous rock with a composition corresponding to that of the field shown in Figure 3.10. Strictly speaking, not all rocks in this field are granites. For example, a rock with a composition close to the line separating granite and diorite is usually called *granodiorite*. To avoid the confusion that might result from introducing more rock names, we will follow the practice of referring to rocks to the left of the granite-diorite line in Figure 3.10 as *granitic*.

(a) Andesite

(b) Diorite

Figure 3.13
Intermediate igneous rocks. (a) Andesite. This specimen has hornblende phenocrysts and is thus an andesite porphyry. (b) Diorite has a salt-and-pepper appearance because it contains both light- and dark-colored minerals.

Granitic rocks are by far the most common intrusive igneous rocks, although they are restricted to the continents. Most granitic rocks were intruded at or near convergent plate margins during episodes of mountain building. When these mountainous regions are uplifted and eroded, the vast bodies of granitic rocks forming their cores are exposed. The granitic rocks of the Sierra Nevada of California form a composite body measuring about 640 km long and 110 km wide, and the granitic rocks of the Coast Ranges of British Columbia, Canada, are even more voluminous.

(a) Rhyolite

(b) Granite

Figure 3.14
Felsic igneous rocks. (a) Rhyolite and (b) granite are typically light-colored because they contain mostly nonferromagnesian silicates. The dark spots in the granite specimen are biotite mica.

Pegmatite The term *pegmatite* refers to a particular texture rather than a specific composition, but most pegmatites are composed largely of quartz, potassium feldspar, and sodium-rich plagioclase, thus corresponding closely to granite. Their most remarkable feature is the size of their minerals, which measure at least 1 cm across and in some pegmatites are measured in tens of centimeters or meters (Figure 3.15). In fact, some minerals in pegmatites are truly gigantic. A few pegmatites are mafic or intermediate in composition and are appropriately called *gabbro* and *diorite pegmatites*. Many pegmatites are associated with large granite intrusive bodies and are composed of minerals that formed from the fluid and vapor phases that remained after most of the granite crystallized.

The water-rich vapor phase that exists after most of a magma has crystallized as granite has properties that differ from the magma from which it separated. It has a lower density and viscosity and commonly invades the adjacent rocks where it crystallizes. This water-rich vapor phase ordinarily contains a number of elements that rarely enter into the common minerals that form granite. Pegmatites crystallizing to form very coarsely crystalline granite are simple pegmatites, whereas those with minerals containing elements such as lithium, beryllium, cesium, tin, and several others are complex pegmatites. Some complex pegmatites contain 300 different mineral species, a few of which are important economically. In addition, several gem minerals such as emerald and aquamarine, both of which are varieties of the silicate mineral beryl, and tourmaline are found in some pegmatites (see Perspective 3.1). Many rare minerals of lesser value and well-formed crystals of common minerals, such as quartz, are also mined and sold to collectors and museums.

The formation and growth of mineral-crystal nuclei in pegmatites are similar to those processes in magma, but with one critical difference: The water-rich vapor phase from which pegmatites crystallize inhibits the formation of nuclei. However, some nuclei do form, and because the appropriate ions in the liquid can move easily and attach themselves to a growing crystal, individual mineral grains have the opportunity to grow to very large sizes.

Other Igneous Rocks A few igneous rocks, including tuff, volcanic breccia, obsidian, pumice, and scoria, are identified by their textures (Figure 3.16). Much of the fragmental material erupted by volcanoes is *ash,* a designation for pyroclastic materials measuring less than 2.0 mm in diameter, most of which consists of broken pieces or shards of volcanic glass. The consolidation of ash forms the pyroclastic rock *tuff* (Figure 3.17). Most tuff is

Figure 3.15
Pegmatite is a textural term for very coarse-grained igneous rocks. Most pegmatites are compositionally similar to granite. The minerals in this specimen measure 2 to 3 cm, and minerals several meters long have been found in some pegmatites.

Composition	Felsic ← → Mafic	
Vesicular	Pumice	Scoria
Glassy	Obsidian	
Pyroclastic or Fragmental	← Volcanic Breccia →	
	Tuff/welded tuff	

(Texture)

Figure 3.16
Classification of igneous rocks for which texture is the main consideration.

Figure 3.17
Exposure of tuff in Colorado.

Perspective 3.1

Complex Pegmatites

As noted in the text, simple pegmatites consist of coarse crystalline rocks similar to granite, but complex pegmatites possess minerals having lithium, cesium, tin, and several other elements that are not normally found in minerals elsewhere. Although complex pegmatites vary considerably, they share some features in common. They all possess concentric zones that differ in composition and texture. Ideally four zones are present, which are usually designated as the *border, wall, intermediate,* and *core* zones, the last of which is usually composed entirely of quartz.

The origin of these zones is the subject of continuing debate between proponents of two hypotheses. One group holds that crystallization took place from the border zone inward, with the fluid phase changing as successive layers of minerals crystallized. According to the opposing group, a simple pegmatite formed first and was later partially or completely replaced as hydrothermal fluids circulated through it. No compelling evidence exists that supports one hypothesis to the exclusion of the other. Another feature of complex pegmatites is the occurrence of giant mineral crystals in the inner zones. Crystals of muscovite measuring 2.44 m across have been recovered from pegmatites in Canada, and huge feldspar and quartz crystals are known from pegmatites in many areas.

Pegmatites have been the source of a variety of mineral commodities. In the United States, pegmatites were mined in New England soon after colonists settled there. In 1803, large transparent sheets of muscovite were recovered from the Ruggles Mine in New Hampshire for use in windows. And by about 1825, feldspar minerals were mined from pegmatites in Connecticut. Gemstones have also been recovered from a number of these pegmatites, especially the Dunton pegmatite in Maine that has yielded hundreds of gem-quality tourmaline crystals (Figure 1).

Anyone familiar with gemstones knows that many gem-quality minerals come from pegmatites, but pegmatites are also the source of many minerals used in various industrial applications. Feldspars, quartz, and micas as well as

Figure 1
Tourmaline from the Dunton pegmatite mine in Maine.

minerals containing elements such as tin, cesium, lithium, rubidium, and beryllium are mined from pegmatites. Pegmatites of the tin-spodumeme belt of North Carolina contain vast deposits of lithium. And pegmatites are the only known sources of some of the rare earth elements.

Pegmatites are particularly common in the Black Hills of South Dakota, where more than 20,000 have been identified in the country rock adjacent to the Harney Peak Granite. The stone images of Presidents Washington, Lincoln, Jefferson, and Theodore Roosevelt on Mount Rushmore were carved into rocks of the Harney Peak Granite, which was designated Mount Rushmore National Memorial in 1927 (see Figure 3.1b). These pegmatites formed about 1.7 billion years ago when the granite was emplaced as a composite pluton consisting of numerous sills and dikes. More recent uplift and ero-

sion of the area, during the Cretaceous Period, exposed the granite and its associated pegmatites. Complex pegmatites account for only about 1% of those identified in the Black Hills or elsewhere. One of the best known in the Black Hills is the Etta pegmatite, with spodumene crystals more than 12 m long (Figure 2). Lithium from spodumene was mined from the Etta pegmatite until 1960, but since then lithium has been derived from more economical sources such as dry lake beds in arid regions.

Complex pegmatites worldwide have yielded numerous gem minerals; one of the most famous in North America is the Himalaya Mine in the Mesa Grande pegmatite district of San Diego County, California. It is one of the most productive gem pegmatites anywhere in the world; among other minerals it has yielded some of the best specimens of gem-quality tourmaline

found anywhere (see Chapter 2 opening photo). And this is only one pegmatite in one of about 20 pegmatite districts in the Peninsular Ranges batholith in southern California; several others in this area have also yielded gem-quality mineral specimens and minerals of economic value. Most major mineral collections in North American museums include specimens from the Himalaya Mine.

A particularly well exposed and accessible pegmatite is the Harding Pegmatite Mine in Taos County, New Mexico, which is now owned by the Earth and Planetary Sciences Department of the University of New Mexico. It was mined from 1919 to 1930, but now serves as a textbook example for geology students as well as rock hounds (amateur mineral and rock collectors). Visitors to the mine are allowed to collect and keep small mineral specimens.

Spodumene crystals

Figure 2
Giant spodumene crystals in the Etta pegmatite in the Black Hills of South Dakota. The crystal above the miner's head measures more than 12 m long.

silica rich and light colored and is appropriately called *rhyolite tuff*. Some ash flows are so hot that as they come to rest, the ash particles fuse together and form a *welded tuff*. Consolidated deposits of larger pyroclastic materials, such as cinders, blocks, and bombs, are *volcanic breccia*.

Both *obsidian* and *pumice* are varieties of volcanic glass (Figure 3.18). Obsidian may be black, dark gray, red, or brown, depending on the presence of tiny particles of iron minerals. Obsidian breaks with the conchoidal fracture that is typical of glass. Analyses of many samples indicate that most obsidian has a high silica content and is compositionally similar to rhyolite.

Pumice is a variety of volcanic glass containing numerous bubble-shaped vesicles that develop when gas escapes through lava and forms a froth (Figure 3.18b). Some pumice forms as crusts on lava flows, and some forms as particles erupted from explosive volcanoes. If pumice falls into water, it can be carried great distances because it is so porous and light that it floats. Another vesicular rock is *scoria*, which has more vesicles than solid rock.

INTRUSIVE IGNEOUS BODIES: PLUTONS— THEIR CHARACTERISTICS AND ORIGINS

Unlike volcanism and the origin of volcanic rocks, both of which can be observed, intrusive igneous activity can be studied only indirectly. Intrusive igneous bodies known as **plutons** form when magma cools and crystallizes within the crust (Figure 3.19). Plutons can be observed after erosion has exposed them at the surface, but geologists cannot duplicate the conditions under which they form except in small laboratory experiments. Accordingly, geologists face a much greater challenge in interpreting the mechanisms whereby plutons form as opposed to studying extrusive igneous processes. The magma that cools to form plutons is emplaced mostly at divergent and convergent plate boundaries, which are also areas of active volcanism.

Several types of plutons are recognized, all of which are defined by their geometry (three-dimensional shape) and their relationship to the country rock (Figure 3.19). Geometrically, plutons may be characterized as massive or irregular, tabular, cylindrical, or mushroom shaped. Plutons are also described as concordant or discordant. A **concordant pluton**, such as a sill, has boundaries parallel to the layering in the country rock. A **discordant**

(b) Pumice

(a) Obsidian

Figure 3.18
(a) Obsidian is a natural glass. It forms when lava cools so quickly that minerals fail to develop. (b) Pumice is also glassy and extremely vesicular. Because it has so many vesicles, it has a low density and will float in water.

pluton, such as a dike, has boundaries that cut across the layering of the country rock (Figure 3.19).

Dikes and Sills

Both dikes and sills are tabular or sheetlike plutons, but dikes are discordant whereas sills are concordant (Figure 3.19). **Dikes** are common intrusive features (Figure 3.20), most of which are small bodies measuring 1 or 2 m across, but they range from a few centimeters to more than 100 m thick. Dikes are emplaced within preexisting fractures or where the fluid pressure is great enough for them to form their own fractures during emplacement.

Erosion of the Hawaiian volcanoes exposes dikes in rift zones, the large fractures that cut across these volcanoes. The Columbia River basalts in Washington (discussed in Chapter 4) issued from long fissures, and the magma that cooled in the fissures formed dikes. Some of the large historic fissure eruptions are underlain by dikes; for example, dikes underlie both the Laki fissure eruption of 1783 in Iceland and the Eldgja fissure, also in Iceland, where eruptions occurred in A.D. 950 from a fissure 300 km long.

Concordant sheetlike plutons are **sills,** many of which are a meter or less thick, although some are much thicker (Figures 3.19 and 3.20). A well-known sill in the United States is the Palisades sill that forms the Palisades along the west side of the Hudson River in New York and New Jersey. It is exposed for 60 km along the river and is up to 300 m thick. Most sills have been intruded into sedimentary rocks, but eroded volcanoes also reveal that sills are commonly injected into piles of volcanic rocks. In fact, some inflation of volcanoes preceding eruptions may be caused by the injection of sills (see Chapter 4).

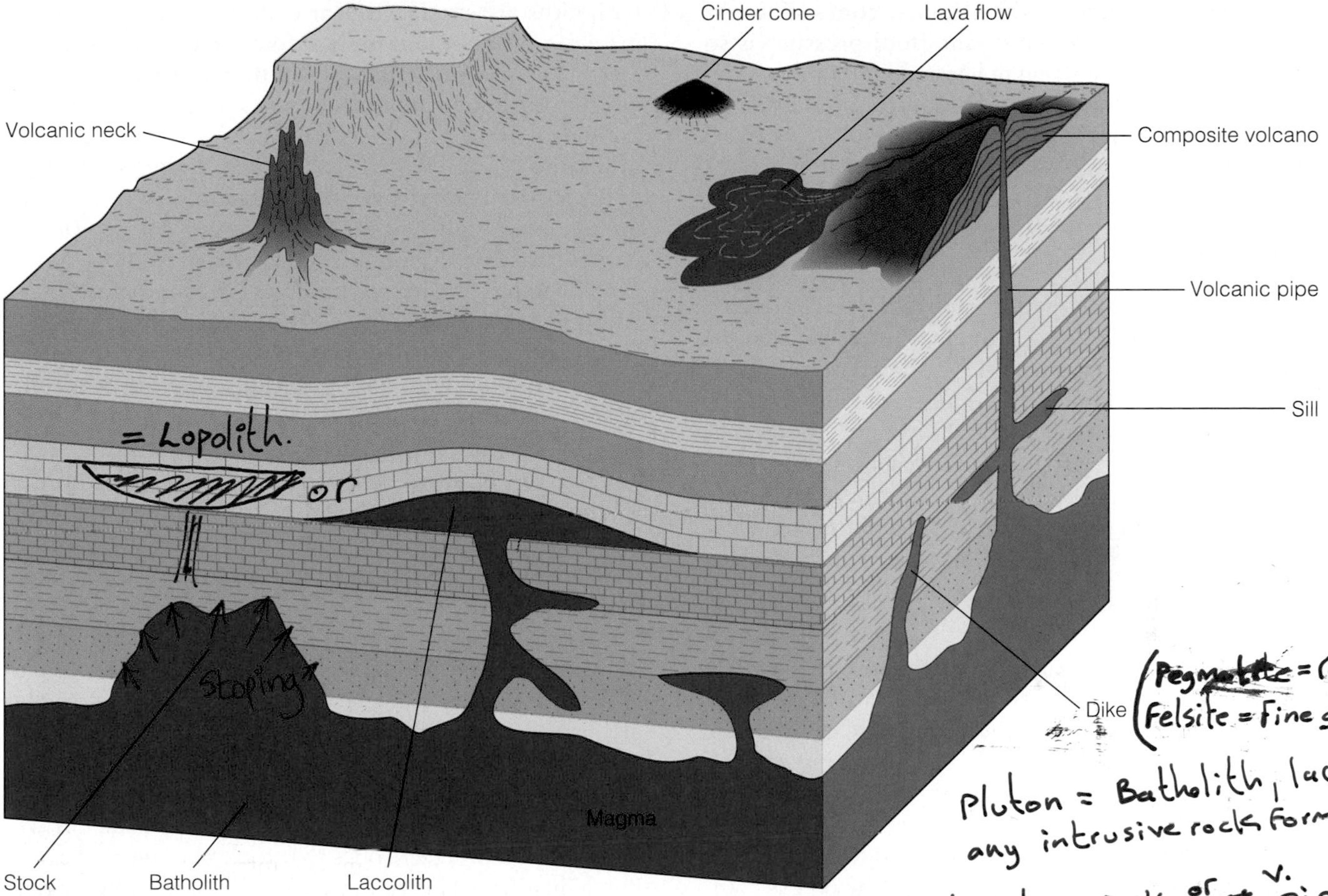

Cinder cone
Lava flow
Volcanic neck
Composite volcano
Volcanic pipe
Sill
= Lopolith.
or
Stoping
Dike
Magma
Stock
Batholith
Laccolith
(Pegmatite = course..
Felsite = Fine grained)
Pluton = Batholith, laccolith..
any intrusive rock formation.
Discordant pluton = Dike, or v. pipe

Figure 3.19
Block diagram showing the various types of plutons. Notice that some plutons cut across the layering in the country rock and are thus discordant, whereas others parallel the layering and are concordant.

Concordant pluton = Sill

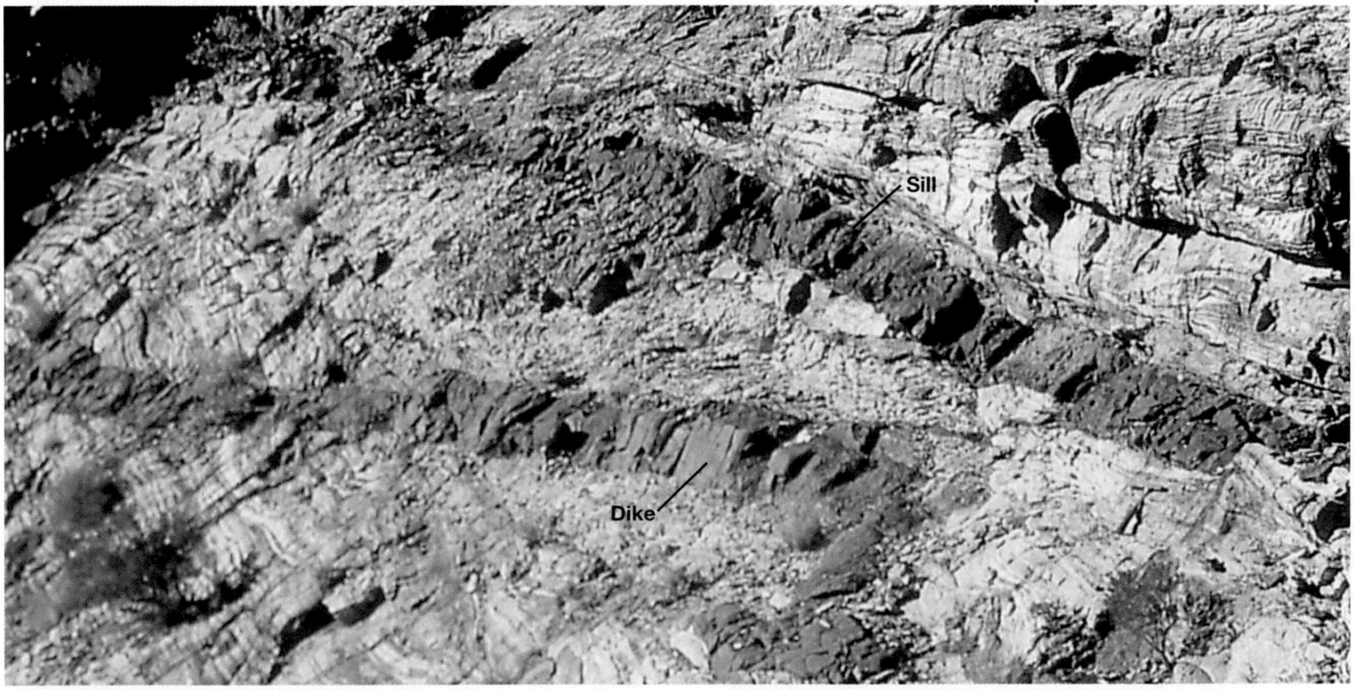

Sill
Dike

Figure 3.20
The dark layer cutting diagonally across the rock layers is a dike. The other dark layer is a sill because it parallels the layering.

In contrast to dikes, which follow zones of weakness, sills are emplaced when the fluid pressure is so great that the intruding magma actually lifts the overlying rocks. Because emplacement requires fluid pressure exceeding the force exerted by the weight of the overlying rocks, sills are typically shallow intrusive bodies.

Laccoliths

Laccoliths are similar to sills in that they are concordant, but instead of being tabular, they have a mushroomlike geometry (Figure 3.19). They tend to have a flat floor and are domed up in their central part. Like sills, laccoliths are rather shallow intrusive bodies that actually lift up the overlying strata when the magma is intruded. In this case, however, the strata are arched up over the pluton (Figure 3.19). Most laccoliths are rather small bodies. The best-known laccoliths in the United States are in the Henry Mountains of southeastern Utah.

Volcanic Pipes and Necks

A volcano has a cylindrical conduit known as a **volcanic pipe** connecting its crater with an underlying magma chamber (Figure 3.19). Through this structure magma rises to the surface. When a volcano ceases to erupt, its slopes are attacked by water, gases, and acids and it erodes, but the magma that solidified in the pipe is commonly more resistant to alteration and erosion. Consequently, much of the volcano is eroded but the pipe remains as a remnant called a **volcanic neck** (Figure 3.19). Several volcanic necks are found in the southwestern United States, especially in Arizona and New Mexico, and others are recognized elsewhere (see Perspective 3.2).

Batholiths and Stocks

By definition a **batholith,** the largest of all plutons, must have at least 100 km^2 of surface area, and most are far larger (Figure 3.19). A **stock,** in contrast, is similar but smaller. Some stocks, however, are simply parts of large plutons that once exposed by erosion are batholiths. Both batholiths and stocks are generally discordant, although locally they may be concordant, and batholiths, especially, consist of multiple intrusions. In other words, a batholith is a large composite body produced by repeated, voluminous intrusions of magma in the same region. The coastal batholith of Peru, for instance, was emplaced during a period of 60 to 70 million years and is made up of as many as 800 individual plutons.

The igneous rocks composing batholiths are mostly granitic, although diorite may also be present. Most batholiths are emplaced near continental margins during episodes of mountain building. One example is the Sierra Nevada batholith of California (Figure 3.21), which formed over millions of years during a mountain-building episode known as the Nevadan orogeny. Later uplift and erosion exposed this huge composite pluton at the surface. Other large batholiths in North America include the Idaho batholith and the Coast Range batholith in British Columbia, Canada.

A number of mineral resources are found in rocks of batholiths and stocks and in the adjacent country rocks. Granitic rocks are the primary source of gold, which forms from mineral-rich solutions moving through cracks and fractures of the igneous body. The copper deposits at Butte, Montana, are in rocks near the margins of the granitic rocks of the Boulder batholith (Figure 3.22). Near Salt Lake City, Utah, copper is mined from the mineralized rocks adjacent to the Bingham stock, a composite pluton composed of granite and granite porphyry.

As noted earlier, batholiths appear to be emplaced in the cores of mountain ranges that resulted from plate collisions. However, large exposures of granitic rocks are also present within the interiors of continents where mountains are absent. A large area in Canada is underlain by extensive granitic rocks as well as by other rock types. These granites were apparently emplaced during mountain-building episodes that took place during early Earth history. The mountains have long since been eroded away, the remaining rocks representing the eroded "roots" of these ancient mountains.

HOW ARE BATHOLITHS EMPLACED IN EARTH'S CRUST?

Geologists realized long ago that the emplacement of batholiths posed a space problem; that is, what happened to the rock that formerly occupied the space now occupied by a batholith? One proposed answer was that no displacement had occurred, but rather that batholiths had been formed in place by alteration of the country rock through a process called *granitization*. According to this view, granite did not originate as a magma but rather from hot, ion-rich solutions that simply altered the country rock and transformed it into granite. Granitization is a solid-state phenomenon, so it is essentially an extreme type of metamorphism (see Chapter 7).

Granitization is no doubt a real phenomenon, but most granitic rocks show clear evidence of an igneous origin. For one thing, if granitization had taken place one would expect the change from country rock to granite would take place gradually over some distance.

(a)

(b)

Figure 3.21
(a) View of granitic rocks of the Sierra Nevada batholith in Yosemite National Park, California. The near-vertical cliff is El Capitan, meaning "The Chief." It rises more than 900 m above the valley floor, making it the highest unbroken cliff in the world. (b) Granitic rocks in a small stock at Castle Crags State Park, California.

Figure 3.22
A copper mine at Butte, Montana. The copper deposits are in rocks at the margin of the Boulder batholith.

Perspective 3.2

Some Remarkable Volcanic Necks

L ava flows and pyroclastic materials composing the flanks of volcanoes are not usually as resistant to chemical and mechanical alteration (weathering) as are the rocks that solidify in a volcanic pipe. Accordingly, when extinct volcanoes are weathered and eroded, the volcanic pipe commonly remains as an erosional remnant—a volcanic neck. The geologic origin of volcanic necks is well known, but the necks themselves, rising as monolithic structures above otherwise rather flat land, are awe-inspiring and the subject of legends.

Volcanic necks are found in many areas of recently active volcanism. A rather small one in the town of Le Puy, France, rises only 79 m above the surrounding countryside. Even though small compared to many other volcanic necks, it is not only scenic but also the site on which the 11th-century chapel of Saint Michel d'Aiguilhe was built (Figure 1). This monolith is so steep that materials and tools used in the chapel's construction had to be hauled up in baskets.

Many volcanic necks are found in the United States, especially in Arizona and New Mexico. The most impressive one is Shiprock in northwestern New Mexico, which rises nearly 550 m above the surrounding plain and is visible from 160 km away. Radiating outward from this conical structure are three dikes that stand like walls above the adjacent plain (Figure 2). The Navajo call Shiprock "Tsae-bidahi" which means Winged Rock or Rock with Wings. According to Navajo legend it represents a giant bird that brought the Navajo people from the north, and the dikes are snakes that have turned to stone.

Figure 1
A volcanic neck rising 79 m above the surface in Le Puy, France. Workers on the chapel had to haul building materials and their tools up in baskets.

(a)

Figure 2
(a) Shiprock, a volcanic neck in north-western New Mexico, rises nearly 550 m above the surrounding plain. (b) View of one of the dikes radiating from Shiprock.

(b)

Absolute dating of one of the dikes indicates that Shiprock is about 27 million years old. When Shiprock formed, apparently during explosive eruptions, the magma penetrated metamorphic, igneous, and sedimentary rocks, including the Mancos Shale, the rock unit now exposed at the surface adjacent to Shiprock. The material composing Shiprock itself is known as tuff-breccia, consisting of fragmented volcanic debris along with pieces of various sedimentary rocks, some granite, and metamorphic rocks. Shiprock was a favorite with rock climbers for many years until the Navajos put an end to all climbing on their reservation.

Devil's Tower in northeastern Wyoming (Figure 3) is likely another volcanic neck, but another interpretation is also possible. Geologists agree that it cooled from a small body of magma and that erosion exposed it in its present form. However, opinion is divided on whether it is truly a volcanic neck or an eroded laccolith. In either case,

Devil's Tower along with other similar features in the area resulted from magma intruded into the crust about 45 to 50 million years ago. President Theodore Roosevelt established Devil's Tower as our first national monument in 1906.

At 260 m high, Devil's Tower is visible from 48 km away, and served as a landmark for early travelers in the area. Some of you might recall that Devil's Tower was featured in the 1977 movie *Close Encounters of the Third Kind.* The Cheyenne and Sioux Indians call Devil's Tower Mateo Tepee, meaning "Grizzly Bear Lodge." It was also called the "Bad God's Tower," and reportedly "Devil's Tower" is a translation of this phrase. The tower's most conspicuous features are the near-vertical lines that according to Cheyenne legend are scratch marks made by a gigantic grizzly bear. One legend holds that the bear made the marks while pursuing a group of children. Another tells of six brothers and a woman who were also pursued by

a grizzly bear. One brother carried a small rock, and when he sang a song it grew into the present size of Devil's Tower.

The "scratch marks" are actually lines formed by the intersections of *columnar joints,* fractures that form in response to cooling and contraction in some igneous bodies and lava flows (see Chapter 4). Many of the columns formed by this kind of fracturing are six sided, but columns with four, five, and seven sides are present as well. The larger columns are 2.5 m across, and the pile of rubble at the tower's base is simply an accumulation of collapsed columns.

Every year, Devil's Tower is visited by numerous tourists who marvel at its size and scenic appeal. A number of other similar structures are also found in the nearby parts of Wyoming and South Dakota, and in the Southwest volcanic necks can be seen in many areas. Certainly the most famous is Shiprock, which is easily seen from the highway. However, because it is sacred to the local Native Americans, visits to the rock itself are discouraged.

Figure 3
Devil's Tower in northeastern Wyoming. It rises about 260 m above its base and can be seen from 48 km away. It might be a volcanic neck or an eroded laccolith. The vertical lines result from the intersections of fractures known as *columnar joints.* According to Cheyenne legend, however, the lines are deep scratches made by a gigantic grizzly bear.

However, in almost all cases no such gradual change can be detected. Thus, granitic rocks most commonly have what geologists refer to as a sharp contact with adjacent rocks. Another feature indicating an igneous origin for granitic rocks is the alignment of elongate minerals parallel with their contacts, which must have occurred when magma was injected. A few granitic rocks lack sharp contacts, and gradually change in character until they resemble the adjacent country rock. These probably did originate by granitization. In the opinion of most geologists, however, only small quantities of granitic rock could form by this process, so it cannot account for the huge volume of granitic rocks of batholiths. These geologists think that an igneous origin for almost all granite is clear, but they still must deal with the space problem.

One solution is that these large igneous bodies melted their way into the crust. In other words, they simply assimilated the country rock as they moved upward (Figure 3.6). The presence of inclusions, especially near the tops of such intrusive bodies, indicates that assimilation does occur. Nevertheless, as we noted previously, assimilation is a limited process because magma is cooled as country rock is assimilated; calculations indicate that far too little heat is available in a magma to assimilate the huge quantities of country rock necessary to make room for a batholith.

Geologists now generally agree that batholiths were emplaced by *forceful injection* as magma moved up toward the surface. Recall that granite is derived from viscous felsic magma and therefore rises slowly. It appears that the magma deforms and shoulders aside the country rock, and as it rises further, some of the country rock fills the space beneath the magma (Figure 3.23). A somewhat analogous situation was discovered in which large masses of sedimentary rock known as *rock salt* rise through the overlying rocks to form *salt domes*.

Salt domes are recognized in several areas of the world, including the Gulf Coast of the United States. Layers of rock salt exist at some depth, but salt is less dense than most other types of rock materials. When under pressure, it rises toward the surface even though it remains solid, and as it moves up, it pushes aside and deforms the country rock (Figure 3.24). Natural

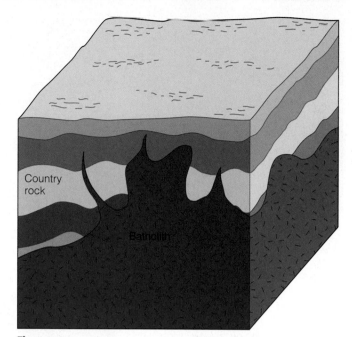

Figure 3.23
Emplacement of a hypothetical batholith. As the magma rises, it shoulders aside and deforms the country rock.

examples of rock salt flowage are known, and it can easily be demonstrated experimentally. In the arid Middle East, for example, salt moving up in the manner described actually flows out at the surface.

Some batholiths do indeed show evidence of having been emplaced forcefully by shouldering aside and deforming the country rock. This mechanism probably occurs in the deeper parts of the crust where tempera-

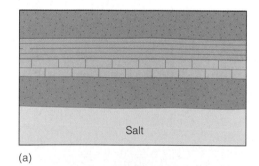

(a)

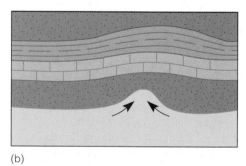

(b)

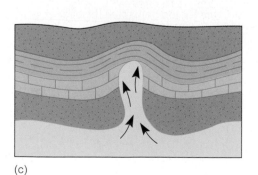

(c)

Figure 3.24
Three stages in the origin of a salt dome. Rock salt is a low-density sedimentary rock that (a) when deeply buried (b) tends to rise toward the surface, (c) pushing aside and deforming the country rock and forming a dome. Salt domes are thought to rise in much the same manner as batholiths are intruded into the crust.

(a)

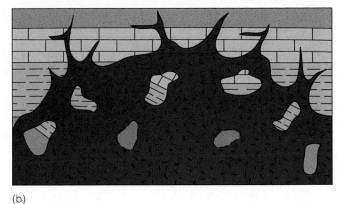
(b)

Figure 3.25
Emplacement of a batholith by stoping. (a) Magma is injected into fractures and planes between layers in the country rock. (b) Blocks of country rock are detached and engulfed in the magma, thus making room for the magma to rise further. Some of the engulfed blocks might be assimilated, and some might remain as inclusions (see Figure 3.6).

ture and pressure are high and the country rocks are easily deformed in the manner described. At shallower depths, the crust is more rigid and tends to deform by fracturing. In this environment, batholiths may be emplaced by **stoping,** a process in which rising magma detaches and engulfs pieces of country rock (Figure 3.25).

According to this concept, magma moves up along fractures and the planes separating layers of country rock. Eventually, pieces of country rock are detached and settle into the magma. No new room is created during stoping; the magma simply fills the space formerly occupied by country rock (Figure 3.25).

Chapter Summary

1. Magma is molten rock material below Earth's surface, whereas lava is magma that flows on the surface. The silica content of magmas varies and serves to differentiate felsic, intermediate, and mafic magmas.

2. The viscosity of magma and lava depends mostly on temperature and composition. Silica-rich (felsic) lava is more viscous than silica-poor (mafic) lava.

3. Minerals crystallize from magma and lava when small crystal nuclei form and grow.

4. Volcanic rocks generally have aphanitic textures because of their rapid cooling, whereas slow cooling and phaneritic textures characterize plutonic rocks. Igneous rocks with a porphyritic texture have mineral crystals of markedly different sizes. Other igneous rock textures include vesicular, glassy, and pyroclastic.

5. The composition of igneous rocks is determined largely by the composi-

tion of the parent magma. It is possible, though, for an individual magma to yield igneous rocks of different compositions.

6. Under ideal cooling conditions, a mafic magma yields a sequence of different minerals that are stable within specific temperature ranges. This sequence, called Bowen's reaction series, consists of a discontinuous branch and a continuous branch.

   a. The discontinuous branch contains only ferromagnesian silicates, each of which reacts with the melt to form the next mineral in the sequence.

   b. The continuous branch involves changes only in plagioclase feldspar as sodium replaces calcium in the crystal structure.

7. The ferromagnesian silicates that form first in Bowen's reaction series can settle and become concentrated near the base of a magma chamber

or intrusive body. Such settling of iron- and magnesium-rich minerals causes a chemical change in the remaining melt.

8. A magma can be changed compositionally when it assimilates country rock, but this process usually has only a limited effect. Magma mixing may also bring about compositional changes in magmas.

9. Most igneous rocks are classified on the basis of their textures and composition. Two fundamental groups of igneous rocks are recognized: volcanic or extrusive rocks, most of which are aphanitic, and plutonic or intrusive rocks, most of which are phaneritic.

   a. Common volcanic rocks include rhyolite, andesite, basalt, and tuff.

   b. Common plutonic rocks include granite, diorite, and gabbro.

10. Pegmatites are coarse-grained igneous rocks, most of which have

an overall composition similar to that of granite. Crystallization from a water-rich vapor phase left over after the crystallization of granite accounts for the very large mineral crystals in pegmatites.

11. Plutons are igneous bodies that formed in place or were intruded into Earth's crust. Various types of plutons are classified by their geometry and whether they are concordant or discordant.

12. Common plutons include dikes (tabular geometry, discordant); sills (tabular geometry, concordant); volcanic necks (cylindrical geometry, discordant); laccoliths (mushroom-shaped, concordant); and batholiths and stocks (irregular geometry, mostly discordant).

13. By definition, batholiths must have at least 100 km² of surface area; stocks are similar but smaller. Batholiths are large composite bodies consisting of many plutons emplaced over a long period of time.

14. Most batholiths appear to have formed in the cores of mountain ranges during episodes of mountain building.

15. Some geologists think that granite batholiths are emplaced when felsic magma moves up and shoulders aside and deforms the country rock. The upward movement of rock salt and the formation of salt domes provide a somewhat analogous situation.

Important Terms

aphanitic texture
assimilation
batholith
Bowen's reaction series
concordant pluton
country rock
crystal settling
dike
discordant pluton
felsic magma
igneous rock

intermediate magma
laccolith
lava
lava flow
mafic magma
magma
magma chamber
magma mixing
phaneritic texture
pluton
plutonic (intrusive igneous) rock

porphyritic texture
pyroclastic materials
pyroclastic (fragmental) texture
sill
stock
stoping
vesicle
viscosity
volcanic neck
volcanic pipe
volcanic (extrusive igneous) rock

Review Questions

1. An igneous rock having the composition of granite and very large mineral crystals is:

   a. _____ basalt;
   b. _____ pegmatite;
   c. _____ obsidian;
   d. _____ tuff;
   e. _____ rhyolite.

2. The first mineral to crystallize in the continuous branch of Bowen's reaction series is:

   a. _____ iron-deficient pyroxene;
   b. _____ calcium-rich plagioclase;
   c. _____ sodium-rich amphibole;
   d. _____ muscovite and biotite;
   e. _____ silica-rich quartz.

3. The size of the mineral grains composing an igneous rock is a useful criterion for determining whether the rock is _____ or _____:

   a. _____ discordant/concordant;
   b. _____ vesicular/fragmental;
   c. _____ porphyritic/felsic;
   d. _____ assimilated/ultramafic;
   e. _____ plutonic/volcanic.

4. Which of the following pairs of igneous rocks have the same mineral composition?

   a. _____ granite-scoria;
   b. _____ andesite-rhyolite;
   c. _____ basalt-gabbro;
   d. _____ pumice-diorite;
   e. _____ peridotite-rhyolite.

5. Which one of the following is a discordant pluton?

   a. _____ sill;
   b. _____ laccolith;
   c. _____ lava flow;
   d. _____ dike;
   e. _____ porphyritic andesite.

6. An extrusive igneous rock composed mostly of pyroxene and calcium plagioclase is:

   a. _____ basalt;
   b. _____ pumice;
   c. _____ gabbro;
   d. _____ diorite;
   e. _____ granite.

7. Any igneous rock having minerals large enough to be seen without magnification is said to have a(an) _____ texture and is probably _____:
   a. _____ phaneritic/intrusive;
   b. _____ laccolith/a batholith;
   c. _____ assimilated/volcanic;
   d. _____ aphanitic/a lava flow;
   e. _____ fragmental/obsidian.

8. The process whereby magma reacts with and incorporates country rock is:
   a. _____ crystal differentiation;
   b. _____ granitization;
   c. _____ plutonism;
   d. _____ magma mixing;
   e. _____ assimilation.

9. An intermediate magma is one:
   a. _____ with more magnesium than calcium;
   b. _____ having 53% to 65% silica;
   c. _____ that crystallizes to form gabbro and basalt;
   d. _____ in which early-formed quartz crystals settle;
   e. _____ that formed only during early Earth history.

10. Which of the following regarding batholiths is true?
    a. _____ They form as a series of overlapping lava flows;
    b. _____ Most are composed of granitic rock and consist of multiple intrusions;
    c. _____ The only ones known are small cylindrical bodies that were emplaced explosively;
    d. _____ The Cascade Range volcanoes in the Pacific Northwest are eroded batholiths;
    e. _____ They are found in areas where salt domes are common.

11. The phenomenon whereby a rising body of magma detaches and engulfs pieces of country rock thereby making room for itself is known as:
    a. _____ stoping;
    b. _____ granitization;
    c. _____ magma mixing;
    d. _____ crystal settling;
    e. _____ magmatic doming.

12. An igneous rock possessing a combination of minerals and markedly different sizes is:
    a. _____ a natural glass;
    b. _____ formed by explosive volcanism;
    c. _____ a porphyry;
    d. _____ muscovite;
    e. _____ produced by very rapid cooling.

13. Describe the process whereby minerals form and grow in a cooling magma.

14. Explain why volcanic rocks are generally aphanitic, whereas most plutonic rocks are phaneritic. Give an example of an igneous rock with each texture.

15. Why are felsic lava flows so much more viscous than mafic lava flows? Also, how does viscosity control the shape of a lava flow?

16. Describe the sequence of events leading to the origin of a volcanic neck.

17. How do crystal settling and assimilation bring about changes in the composition of magma? Cite evidence indicating that both of these processes actually take place.

18. Compare the continuous and discontinuous branches of Bowen's reaction series. Why are potassium feldspar and quartz not considered part of either branch?

19. Sills and dikes have the same three-dimensional shape or geometry, so how is it possible to distinguish one from the other? Explain fully.

20. Briefly describe what batholiths are and where and how they form.

21. What are pyroclastic materials and what kinds of igneous rock are formed from them?

22. Explain why ultramafic lava flows occurred during early Earth history but very rarely since then.

23. Use Figure 3.10 to classify an aphanitic igneous rock composed mostly of potassium feldspars, quartz, plagioclase feldspars, and biotite.

24. What does the term *pegmatite* mean, and why are such large mineral crystals found in pegmatites?

Points to Ponder

1. In the discontinuous branch of Bowen's reaction series, olivine forms in a specific temperature range, but as the magma continues to cool, it reacts with the remaining melt and changes to pyroxene. Pyroxene in turn changes to amphibole, and amphibole changes to biotite with continued cooling. How is it possible to have any of these minerals other than biotite in an igneous rock?

2. What kinds of evidence would indicate that a batholith was emplaced as magma rather than by granitization?

3. How can color and the size of mineral grains in igneous rock give clues about the composition and cooling history of a magma?

4. Two rock specimens have the following compositions:

   Specimen 1: 15% biotite, 15% sodium-rich plagioclase, 60% potassium feldspar, and 10% quartz.

   Specimen 2: 10% olivine, 55% pyroxene, 5% hornblende, and 30% calcium-rich plagioclase.

   How would these two rocks differ in color and specific gravity? Also, what was the viscosity of the magmas from which these rocks crystallized?

World Wide Web Activities

For these Web site addresses, along with current updates and exercises, log on to
http://www.brookscole.com/geo/

➤ ROB'S GRANITE PAGE

This site is the work of Robert M. Reed of the Department of Geological Sciences, University of Texas at Austin. As you might guess, it is concerned mainly with granite. It has many links to other Web sites with information about granite, as well as information about Reed's own research. Check out the various links from his home page.

1. Click on the *Introduction to Granite* heading. It will take you to a page that, among other things, has pictures of granite. Test your knowledge of what the minerals that compose granite look like by pointing your cursor at the various minerals in this image and seeing if you can identify them. Click your mouse to see if you are correct.
2. Click on the *General Stuff* heading. It will take you to a page with links to many other sites. Click on the diamond for *A nice discussion of some California geology (Pt. Reyes), including granite rocks* heading. This site contains a discussion of the geology of the Point Reyes Peninsula, including links to specific references about the granitic rocks of this region. What is the age of these granites? What is their origin?

➤ IGNEOUS ROCKS

This Web site contains images and information on 10 common igneous rocks. It is part of the Soil Science 223 Rocks and Minerals Reference Web site. Click on any of the *rock names*, and compare the information and images to the information in this chapter.

➤ UNIVERSITY OF TULSA DEPARTMENT OF GEOSCIENCES IGNEOUS ROCKS AND PROCESSES

This Web site contains much information about igneous rocks and processes in general, as well as a few links to other sites.

1. Compare the classifications for igneous rocks presented here with those in this chapter.
2. Click on the *Henry Mountains* site. The Henry Mountains are an excellent example of a laccolith. Compare the images of the Henry Mountains in this site to the diagram of the laccolith shown in Figure 3.19.

➤ PETE'S BASALT PAGE

As its name implies, this site is devoted to the igneous rock basalt and the magma from which it crystallizes. Click on *Introductory Basaltology* and learn about the composition and abundance of basaltic magma and see images of the rocks formed from this magma. Also, take the *Virtual Geologic Tour of Kaua'i* and see images and learn about the geology of this Hawaiian Island. While taking the tour, click on *mantle plume* and see how an oceanic hot spot works.

➤ IGNEOUS ROCKS

This site is part of the Georgia Science On-Line project. It was created by Dr. Pamela J. W. Gore for Geology 101 at DeKalb College, Clarkston, Georgia. It has good images of all common igneous rocks, along with descriptions of their textures and compositions. It also has a classification chart similar to the one in the text. Scroll down and read about the various igneous rocks shown.

1. How do scoria and vesicular basalt differ?
2. What are the common phenocrysts in andesite porphyry?
3. What is the composition of pumice? What are its uses?

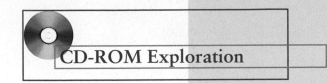

CD-ROM Exploration

➤ Exploring your *Earth Systems Today* CD-ROM will add to your understanding of the material in this chapter.

TOPIC: EARTH'S MATERIALS

MODULE: ROCKS AND ROCK CYCLE

Explore activities in this module to see if you can discover the following for yourself:

Using the "rock laboratory" portion of this module, observe the formation of both a phaneritic and a porphyritic igneous rock through the process of crystallization. Based on what you see in the "rock laboratory" write a short description of how a phaneritic rock and a porphyritic rock crystallize.

What is the origin of "interlocking texture" among crystals in an igneous rock? How does forming from a liquid contribute to development of this texture?

Chapter 4

On May 18, 1980, Mount St. Helens in Washington erupted violently. A huge explosion resulted in 63 fatalities and devastation of 600 km² of forest. Shortly after the explosion, this 19-km-high steam and ash cloud was erupted.

OBJECTIVES

At the end of this chapter, you will have learned that

- In addition to lava, volcanoes emit gases and eject pyroclastic materials such as ash.
- Although each volcano is unique, most can nevertheless be classified as one of several major types.

- The shapes of shield volcanoes, cinder cones, and composite volcanoes differ because of different eruptive styles.
- Volcanoes characterized as lava domes tend to erupt explosively and are thus dangerous.
- Some eruptions yield vast sheets of lava or pyroclastic materials rather than volcanoes.

- The volcanic explosivity index is a semiquantitative measure of an eruption's size.
- Some volcanoes are carefully monitored to help geologists predict eruptions.
- Most volcanoes are located at or close to divergent and convergent plate boundaries.

Volcanism

Prologue

housands of volcanoes have erupted during historic time, but few are as well known as the eruptions of August 24–25, A.D. 79, that destroyed the cities of Pompeii, Herculaneum, and Stabiae. All were thriving Roman communities along the shore of the Bay of Naples in what is now Italy (Figure 4.1a). Fortunately for us, the event was recorded in some detail by Pliny the Younger, whose uncle, Pliny the Elder, died while trying to investigate the eruption.

When the eruption began, both Plinys were about 30 km away in the town of Misenum across the Bay of Naples. At first they were not particularly alarmed when a large cloud appeared over Mount Vesuvius, but soon it became apparent that an eruption was beginning. Pliny the Elder, an admiral in the Roman Navy and naturalist, set sail to investigate and perhaps to evacuate friends from the communities closest to the volcano. When he arrived, pyroclastic materials ejected by the volcano had already accumulated along the shore, preventing him from landing, so he changed course and made landfall at Stabiae. The next morning while fleeing from noxious fumes emitted by the volcano, the elder Pliny died, probably of a heart attack.

Pliny the Younger's account of Mount Vesuvius's eruption is so vivid that similar eruptions during which large quantities of pumice are blasted into the air are still referred to as *plinian*. Pompeii, a city of about 20,000 people and about 9 km downwind from the volcano, was buried in nearly 3 m of pyroclastic materials, which covered all but the tallest buildings. Pompeii was so utterly destroyed that, even though its location was known, it was largely forgotten until 1595 when some of the city was rediscovered during the construction of an aqueduct. During the 17th and 18th centuries, the city was ravaged for artifacts to grace the homes of wealthy Europeans.

Systematic excavations, which began in the 1800s, have exposed much of Pompeii, now a popular tourist attraction (Figure 4.1b). Probably the most famous attractions are the molds of human bodies that formed when the volcanic debris hardened before the bodies decayed. About 2000 victims have been discovered in the city, including a dog still chained to a post, but what happened to the city's other residents is not known. Some probably escaped by sea or over land, but many may remain buried in the debris beyond the city.

Although Pompeii was the largest city destroyed by Mount Vesuvius, it was not the only one to suffer such a fate. A number of others, such as Stabiae, still remain buried beneath a vast blanket of debris. Herculaneum was just as close to the volcano as Pompeii, but—as opposed to Pompeii, which was buried rather gradually—it was overwhelmed in minutes by surges of incandescent pyroclastic materials in what are known as *glowing avalanches* (*nuée ardentes*). The debris covered the town to a depth of about 20 m. Prior to 1982, only about a dozen skeletons had been discovered, so it was thought that many of Herculaneum's citizens had escaped; excavations of the city's waterfront, however, revealed hundreds of human skeletons. Many of these were discovered in chambers that probably housed fishing boats, and many more skeletons have since been discovered along the ancient beach.

Before the A.D. 79 eruptions began, Mount Vesuvius was probably not considered a threat, although residents of that area were certainly aware of volcanoes. However, this particular volcano had been quiet for at least 300 years, so it no doubt was of little concern. But it is only one in a chain of volcanoes along Italy's south coast, where the African plate is subducted beneath the European plate. Since A.D. 79, Mount Vesuvius has erupted 80 times, most violently in 1631 and 1906; it last erupted in 1944.

Mount Vesuvius lies just to the south of a 13-km-diameter volcanic depression, or caldera, known as Campi Phlegraei (Figure 4.1a). The last eruption within the caldera took place in 1538, but movement

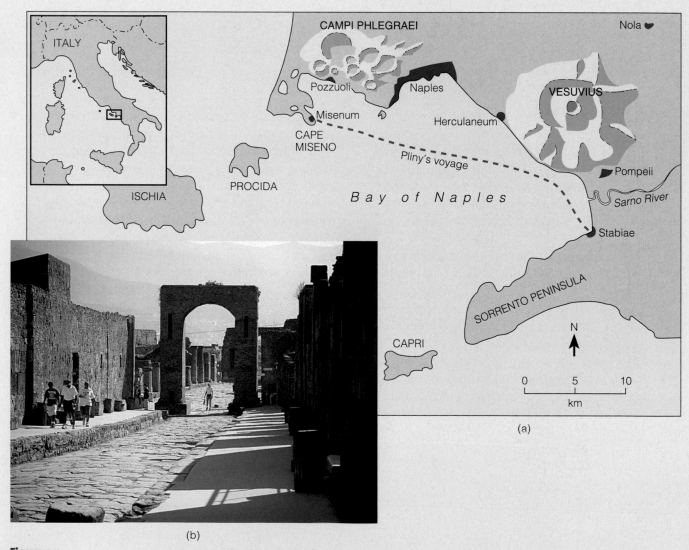

(a)

(b)

Figure 4.1
(a) The Mount Vesuvius region on the shores of the Bay of Naples, Italy. Vesuvius erupted in A.D. 79 and destroyed the cities of Pompeii, Herculaneum, and Stabiae. Campi Phlegraei northwest of Vesuvius is another area of active volcanism. (b) The excavated ruins of Pompeii are now a popular tourist attraction.

of magma beneath the surface has generated so many earthquakes that about half the population of Pozzuoli, a city within the caldera, has moved away. Furthermore, the city has experienced vertical ground movements also related to moving masses of magma. In Roman times, Pozzuoli stood high, but by A.D. 1000 it had sunk to about 11 m below sea level, only to rise 12 m between 1000 and 1538. In fact, during a 1538 volcanic outburst from nearby Monte Nuovo, Pozzuoli rose 4 m in just two days! Another episode of uplift began in 1982 when the city rose 1.8 m in about a year and a half.

Numerous communities on the shores of the Bay of Naples are within easy reach of eruptions from Mount Vesuvius or Campi Phlegraei caldera. Indeed, the city of Naples is situated on the eastern margin of Campi Phlegraei and only a short distance west of Mount Vesuvius (Figure 4.1a). Probably no other major population center in the world is in more danger from devastation by volcanoes. The fact that the region remains geologically active was tragically demonstrated in 1980 when an earthquake killed 3000 people in the Naples area.

INTRODUCTION

The complex interactions among Earth's major systems are vividly illustrated by volcanic eruptions. Volcanism, especially the emission of gases and pyroclastic materials, has an immediate and profound impact on the atmosphere, hydrosphere, and biosphere, at least in the vicinity of an eruption. And in some cases, the effects are felt worldwide as in the eruptions of Tambora in 1815, Krakatau in 1883, and Pinatubo in 1991. Lava flows, in contrast, might destroy homes and croplands, but otherwise have little impact on humans. Depictions of lava flows in movies notwithstanding, humans have little to fear from these incandescent streams of molten rock. Probably no other geologic phenomenon has captured the public imagination more than volcanism.

Most magma cools and crystallizes to form plutonic rocks (Chapter 3), but some rises to the surface where it is erupted as lava flows or pyroclastic materials. Truly, erupting volcanoes are the most impressive manifestation of Earth's dynamic internal processes. Incandescent streams of lava and lava fountains are impressive, especially at night, and explosive eruptions of pyroclastic materials remind us of the violence of some eruptions—the 1980 eruption of Mount St. Helens in Washington, for instance (Figure 4.2).

Although most of us are unlikely to ever experience an eruption, such events are commonplace in some parts of the world. Residents of the island of Hawaii, the Philippines, Japan, and Iceland are well aware of volcanic eruptions, but eruptions in the United States are not particularly common except in Hawaii and Alaska. In the mainland United States, only California and Washington have had eruptions during this century, both in the Cascade Range, which stretches from northern California through Oregon, Washington, and into southern British Columbia, Canada. Canada has experienced no eruptions during historic time.

The fact that lava flows and explosive eruptions can cause property damage, injuries, and fatalities, and at least short-term changes in the atmosphere indicates that some eruptions are catastrophic events. And indeed some are, from the human perspective, but, ironically, when considered in the context of Earth history, volcanism is actually a constructive process. Oceanic islands such as the Azores, the Galapagos, Iceland, and the Hawaiian Islands owe their existence to volcanism; oceanic crust is continually produced by volcanism at spreading ridges, and gases released at volcanoes during Earth's early history probably formed the atmosphere and surface waters. Furthermore, chemical and mechanical alteration of lava flows and pyroclastic materials in tropical areas such as Indonesia convert them to productive soils.

Volcanism has yielded some interesting features and scenery in several of our western national parks and monuments. For instance, Craters of the Moon National Monument, Idaho; Lassen Volcanic National Park, California; Crater Lake National Park, Oregon; and Mount Rainier National Park, Washington, provide visitors with unique opportunities to see the effects of several types of volcanic activity. In fact, many good examples of interesting volcanic features are found in North America, several of which are featured in this chapter.

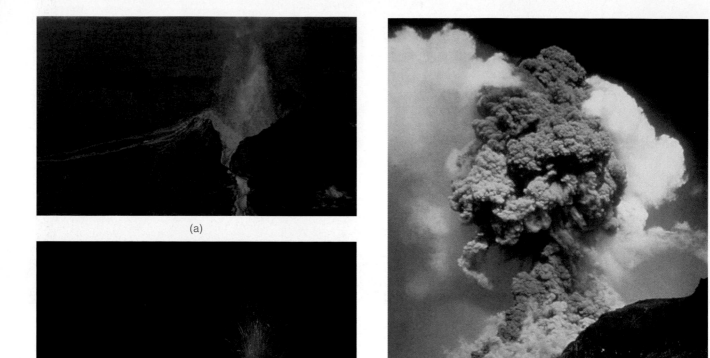

Figure 4.2
Volcanism. (a) A lava fountain and lava flow issuing from Kilauea on the island of Hawaii. (b) Lava fountains, such as these in Hawaii, are particularly impressive at night. (c) Volcanic ash erupted from Mount Ngauruhoe, New Zealand, during January 1974.

VOLCANISM

Volcanism includes those processes whereby lava and its contained gases and pyroclastic materials are extruded onto the surface or into the atmosphere, as well as being the origin of volcanoes and extrusive igneous rocks. At present, about 550 volcanoes are *active*—that is, they have erupted during historic time. At any one time about 12 volcanoes are erupting somewhere in the world; although much of this activity is minor, large eruptions are not uncommon. Mauna Loa and Kilauea on the island of Hawaii; Mount Fuji in Japan; Mount Vesuvius, Italy; Mount St. Helens, Washington; and Mount Pinatubo in the Philippines are active volcanoes. So far, only two other bodies in the solar system appear to have active volcanoes: Io, a moon of Jupiter, and perhaps Triton, one of Neptune's moons (see Chapter 20). In fact, Io is by far the most volcanically active body in the solar system. It is about the size and density of Earth's moon, and many of its 100 or so volcanoes are erupting at any given time.

Besides active volcanoes, numerous *dormant* volcanoes exist that have not erupted recently but may do so again. Mount Vesuvius in Italy showed no signs of activity in human memory until A.D. 79, when it erupted violently and destroyed the cities of Herculaneum, Pompeii, and Stabiae (see the Prologue). Mount Pinatubo in the Philippines lay dormant for 600 years until 1991 when it produced what was probably the largest volcanic outburst in 50 years. Some volcanoes have not erupted during historic time and show no evidence of doing so again. Thousands of these *extinct*, or *inactive*, volcanoes are known.

Volcanic Gases

When magma rises toward the surface, the pressure is reduced, and the contained gases begin to expand. In felsic magmas, which are highly viscous, expansion is inhibited, and gas pressure increases. Eventually the pressure may become great enough to cause an explosion and produce pyroclastic materials such as ash. In contrast, low-viscosity mafic magmas allow gases to expand and escape easily. Accordingly, mafic magmas generally erupt rather quietly.

Samples taken from present-day volcanoes indicate that 50 to 80% of all volcanic gases are water vapor.

(b)

(a)

(c)

Figure 4.3
Volcanic gases. (a) Gases emitted at the Sulfur Works in Lassen Volcanic National Park, California. (b) Trees killed by carbon dioxide rising from beneath Mammoth Mountain volcano, California. (c) The CO_2 gas at Mammoth Mountain has not affected humans, but this sign warns of the potential danger.

Lesser amounts of carbon dioxide, nitrogen, sulfur gases—especially sulfur dioxide and hydrogen sulfide—and very small amounts of carbon monoxide, hydrogen, and chlorine are also commonly emitted. In areas of recent volcanism, such as Lassen Volcanic National Park in California, gases continue to be emitted, and one cannot help but notice the rotten egg odor of hydrogen sulfide gas (Figure 4.3a).

Most gases released during eruptions quickly dissipate in the atmosphere and pose little danger to humans, but on several occasions gases have caused numerous fatalities. In 1783, toxic gases, probably sulfur dioxide, erupted from Laki fissure in Iceland had devastating effects. About 75% of the nation's livestock died, and the haze resulting from the gas caused lower temperatures and crop failures; about 24% of Iceland's population died as a result of the ensuing Blue Haze Famine. The country suffered its coldest winter in 225 years in 1783–1784, with temperatures 4.8°C below the long-term average. The eruption also produced what Benjamin Franklin called a "dry fog" that was responsible for dimming the

intensity of sunlight in Europe. The severe winter of 1783–1784 in Europe and eastern North America is attributed to the presence of this "dry fog" in the upper atmosphere.

The particularly cold spring and summer of 1816 are attributed to the 1815 eruption of Tambora in Indonesia, the largest and most deadly eruption during historic time. The eruption of Mayon volcano in the Philippines during the previous year may have contributed to the cool spring and summer of 1816 as well. Another large historic eruption that had widespread climatic effects was the eruption of Krakatau in 1883.

More recently, in 1986, in the African nation of Cameroon 1746 people died when a cloud of carbon dioxide engulfed them. The gas accumulated in the waters of Lake Nyos, which occupies a volcanic crater. No agreement exists on what caused the gas to suddenly burst forth from the lake, but once it did, it flowed downhill along the surface because it was denser than air. In fact, the density and velocity of the gas cloud were great enough to flatten vegetation, including trees, a few kilometers from the lake. Unfortunately, thousands of animals and many people, some as far as 23 km from the lake, were asphyxiated.

In 1990, forest rangers noticed that trees in an area of about 170 acres on the south side of Mammoth Mountain volcano in eastern California began dying (Figure 4.3b). Previously the trees had been thriving, but a swarm of earthquakes in 1989 apparently triggered the release of carbon dioxide (CO_2) gas from beneath the volcano. Investigations showed that (CO_2) gas accounted for 20 to 95% of the gas content of the soil. During photosynthesis, plant leaves produce oxygen (O_2) from CO_2, but their roots must absorb O_2. Diminished quantities of O_2 in the soil were responsible for the tree deaths.

Residents of the island of Hawaii have coined the term *vog* for volcanic smog. Kilauea volcano has been erupting continuously since 1983, releasing small amounts of lava, copious quantities of carbon dioxide, and about 1000 tons of sulfur dioxide per day. Carbon dioxide has been no problem, but sulfur dioxide produces a haze and the unpleasant odor of sulfur. Vog probably poses little or no problem for tourists, but a long-term threat exists for residents of the west side of the island, where vog is most common.

WHAT WOULD YOU DO?

A debate is raging in your community about whether global warming is a real phenomenon or not. One critic of the global warming theory claims that volcanoes contribute more greenhouse gases to the atmosphere than all the gases resulting from burning coal since the Industrial Revolution began. What could you do to shed some light on this aspect of the debate?

Lava Flows and Pyroclastic Materials

Lava flows are frequently portrayed in movies and on television as fiery streams of incandescent rock material posing a great danger to humans. Actually, lava flows are the least dangerous manifestation of volcanism, although they may destroy buildings and cover agricultural land. Most lava flows do not move particularly fast, and because they are fluid, they follow existing low areas. Thus, once a flow erupts from a volcano, determining the path it will take is fairly easy, and anyone in areas likely to be affected can be evacuated.

The geometry of lava flows differs considerably, depending on their viscosity and the preexisting topography. Unless they are confined to a valley, comparatively fluid flows are thin and widespread, whereas more viscous flows tend to be lobate and to have distinct margins.

Two types of lava flows, both of which were named for Hawaiian flows, are generally recognized. A **pahoehoe** (pronounced pah-hoy-hoy) flow has a ropy surface almost like taffy (Figure 4.4a). The surface of an **aa** (pronounced ah-ah) flow is characterized by rough, jagged angular blocks and fragments (Figure 4.4b). Some flows solidify as pahoehoe or aa throughout, but some pahoe-

(a)

(b)

Figure 4.4
(a) Pahoehoe erupted from the Pu'u O'o vent on Kilauea volcano's southeastern flank. (b) An aa flow in the east rift zone of Kilauea volcano in 1983. The flow front is about 2.5 m high.

(a)

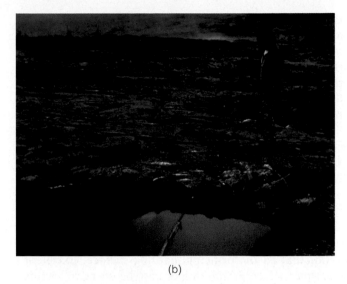

(b)

Figure 4.5
(a) This hollow beneath a lava flow is a lava tube. (b) Part of the tube's roof has collapsed so the still active flow can be seen.

hoe flows change to aa in the downflow direction; an aa flow will not change to pahoehoe in a downflow direction. Pahoehoe flows are less viscous than aa flows; indeed, the latter are viscous enough to break up into blocks and move forward as a wall of rubble.

Even low-viscosity lava flows generally do not move very quickly. One of the fastest flows ever measured in Hawaii moved at an average speed of about 9.5 km/hr. Flows can move faster, though, when insulated on all sides as in a *lava tube,* where a speed of more than 50 km/hr has been recorded. A conduit known as a **lava tube** beneath the surface of a lava flow forms when the margins and upper surface of the flow solidify. Thus confined and insulated, the flow can move quite rapidly

and over great distances. As an eruption ceases, the tube drains, leaving an empty conduit (Figure 4.5a). Part of the roof of a lava tube may collapse, forming a *skylight* through which an active flow can be observed, or access can be gained to an inactive lava tube (Figure 4.5b). Lava tubes are common in many areas of active or recently active volcanism. Good examples can be found in several western states, and in Hawaii lava flows through tubes many kilometers long and in some cases discharges into the sea.

Such features as pressure ridges and spatter cones may mark the surface of lava flows. Pressure on the partly solidified crust of a still-moving flow causes the surface to buckle into **pressure ridges** (Figure 4.6a).

(a)

(b)

Figure 4.6
(a) Pressure ridge on a 1982 lava flow in Hawaii. (b) A spatter cone on a lava flow in Hawaii.

Gases escaping from a flow hurl globs of lava into the air. As these globs fall back to the surface and adhere to one another, they form small, steep-sided **spatter cones** (Figure 4.6b). These cones may rise several meters above the surface of a flow. Spatter cones are common on lava flows in Hawaii, and ancient ones can be seen in Craters of the Moon National Monument, Idaho.

Columnar joints are common in many lava flows, especially mafic flows, but they also occur in other kinds of flows and in some intrusive igneous rocks (Figure 4.7). A lava flow contracts as it cools and produces forces that cause fractures called *joints* to open. On the surface of a flow, these joints commonly form polygonal (often six-sided) cracks. These cracks extend downward into the flow, forming parallel columns with their long axes perpendicular to the principal cooling surface. Excellent examples of columnar joints can be seen at Devil's Postpile National Monument in California, Devil's Tower National Monument in Wyoming (see Perspective 3.2), the Giant's Causeway in Ireland, and many other areas (see Perspective 4.1).

Much of the igneous rock in the upper part of the oceanic crust is of a distinctive type; it consists of bulbous masses of basalt resembling pillows, hence the name **pillow lava** (Figure 4.8). It was long recognized that pillow lava forms when lava is rapidly chilled beneath water, but its formation was not observed until 1971. Divers near Hawaii saw pillows form when a blob of lava broke through the crust of an underwater lava flow and cooled almost instantly, forming a pillow-shaped structure with a glassy exterior. Remaining fluid inside then broke through the crust of the pillow, resulting in an accumulation of interconnected pillows (Figure 4.8a).

Figure 4.7
Columnar jointing in a lava flow at John Day Fossil Beds National Monument, Oregon.

(a)

(b)

Figure 4.8
(a) These bulbous masses of pillow lava form when magma is erupted underwater. (b) Ancient pillow lava now on land in Marin County, California.

Much pyroclastic material is erupted as **ash,** a designation for pyroclastic particles measuring less than 2.0 mm. Ash may be erupted in two ways: an ash fall or an ash flow. During an ash fall, ash is ejected into the atmosphere and settles to the surface over a wide area. In 1947, ash erupted from Mount Hekla in Iceland fell 3800 km away on Helsinki, Finland. About 10 million years ago, in what is now northeastern Nebraska, numerous rhinoceroses, horses, camels, and other mammals were buried by volcanic ash that was apparently erupted in New Mexico, more than 1000 km away. Ash is also erupted in ash flows, which are coherent clouds of ash and gas that commonly flow along or close to the land surface. These flows can move at more than 100 km per hour, and some of them cover vast areas.

In populated areas adjacent to volcanoes, ash falls and ash flows can pose some serious problems. Furthermore, volcanic ash in the atmosphere is a serious hazard to aviation. Since 1980, about 80 aircraft have been damaged when they encountered clouds of volcanic ash. The most serious incident took place in 1989 when volcanic ash from Redoubt volcano in Alaska caused all four jet engines to fail on KLM Flight 867. The plane, carrying 231 passengers, nearly crashed when it fell more than 3 km before the crew could restart the engines. Although the plane landed safely in Anchorage, Alaska, it required $80 million in repairs.

Pyroclastic materials measuring from 2 to 64 mm are known as *lapilli,* and any particle larger than 64 mm is called a *bomb* or *block* depending on its shape. Bombs have twisted, streamlined shapes indicating they were erupted as globs of magma that cooled and solidified during their flight through the air (Figure 4.9). Blocks

are angular pieces of rock ripped from a volcanic conduit or pieces of a solidified crust of magma. Because of their large size, volcanic bomb and block accumulations are not nearly as widespread as ash deposits; instead, they are confined to the immediate area of eruption.

WHAT ARE VOLCANOES?

According to one definition, a **volcano** is a conical mountain formed around a vent where lava, pyroclastic materials, and gases are erupted. Even though many volcanoes are conical, some are simply bulbous masses of magma, and others are dome-shaped or resemble an inverted shield lying on the ground. In all cases, they have a conduit or conduits leading to a magma chamber beneath the surface. Vulcan, the Roman deity of fire, was the inspiration for calling these mountains *volcanoes,* and because of their danger and obvious connection to Earth's interior they have been held in awe by many cultures.

Probably no other geologic process, except perhaps earthquakes, has more lore associated with it. Native Americans of the Northwest tell of a titanic battle between the volcano gods Skel and Llao to account for huge eruptions that took place about 6600 years ago in Oregon, and in Hawaiian legends the volcano goddess Pele resides in the crater of Kilauea on Hawaii. In one of her frequent rages, Pele causes earthquakes and lava flows, and she may hurl flaming boulders at those who offend her.

Volcanoes come in many shapes and sizes, but geologists recognize several major categories, each of which has a distinctive eruptive style. One must realize, however, that each volcano is unique in terms of its overall history of eruptions and development. The frequency of eruptions, for example, varies considerably; the Hawaiian volcanoes have erupted repeatedly during historic time, whereas others, such as Mount St. Helens, have erupted periodically after long periods of inactivity. One of the duties of the U.S. Geological Survey is monitoring active volcanoes and developing methods of forecasting eruptions.

Most volcanoes have a circular depression, or **crater,** at their summit. Craters form as a result of the extrusion of gases and lava from a volcano and are connected via a conduit to a magma chamber below the surface. It is not unusual, though, for magma to erupt from vents on the flanks of large volcanoes where smaller, parasitic cones develop. For instance, Shastina is a large parasitic cone on the flank of Mount Shasta in California, and Mount Etna on Sicily has some 200 smaller vents on its flanks.

Some volcanoes are characterized by a **caldera** rather than a crater. Craters are generally less than 1 km

Figure 4.9
Pyroclastic materials. The large object on the left is a volcanic bomb; it is about 20 cm long. The streamlined shape of bombs indicates they were erupted as globs of magma that cooled and solidified as they descended. The granular objects in the upper right are pyroclastic materials known as *lapilli.* The pile of gray-white material on the lower right is ash.

Perspective 4.1

Columnar Jointing

Peculiar marks, shapes patterns, and structures in rocks are not unusual. Some are merely oddities of little or no importance although they might be attractive or interesting. Among some of the more interesting structures are columns yielded by the phenomenon known as *columnar jointing*. They are found in numerous areas of current or past volcanism (Figure 4.7) and in some intrusive igneous rocks as well, such as in Devil's Tower, Wyoming.

Columnar jointing occurs most commonly in basalt and andesite lava flows, but it is also known in other volcanic rocks, including some ash flow tuffs, as well as shallow intrusive bodies such as dikes, sills, and volcanic necks. As noted in the text, columnar joints form in response to cooling and contraction, and the columns so formed tend to have their long axes perpendicular to the cooling surface—the surface of a lava flow, for example. In the upper part of a cooling flow or igneous body, the columns tend to be irregular and poorly developed, whereas in the lower part of the same flow they are much better developed.

Many columns are remarkably vertical and straight. However, most igneous bodies are irregularly shaped and not homoge-

neous throughout, so complex cooling may yield curved columns, such as those at Devil's Tower, Wyoming. In addition, the degree of development of columns is variable. In some igneous rocks they are difficult to discern, whereas in others they are remarkably prominent features. The columns are typically six-sided but those with three, four, five, and seven sides are also found. At Devil's Postpile National Monument, California, for instance, six-sided columns account for 55% of those present, and most of the rest (37%) are five-sided.

In his 1968 book *Chariots of the Gods,* Erich von Daniken claimed that the colossal structures (Figure 1) on the island of Pohnpei in the Federated States of Micronesia were evidence for his thesis that astronauts from elsewhere had visited Earth long ago. According to him, the hexagonal stones

used in the structures were so regularly shaped that they could not be natural, and thus must have been shaped and moved by beings with a more advanced technology than possessed by the natives of the area. Although the structures represent a phenomenal engineering feat, the source of the columns on the island is known and there is no reason to suppose that the native islanders could not move the stones and pile them into the remarkable structures. However, nothing is known of why the 500-year-old stone structures were erected.

Certainly one of the most impressive examples of columnar jointing is at the Giant's Causeway, a popular tourist attraction in Northern Ireland (Figure 2a). The columns are particularly straight and vertical and in several locations have been given fanciful names such as the Honeycomb and the

Figure 1
These structures on the island of Pohnpei in the Federated States of Micronesia are made up of long, six-sided (hexagonal) stones. The stones were derived from a lava flow in which columnar joints formed as it cooled.

Wishing Chair. According to one legend, a Scottish giant named Benendoner built the causeway from Scotland to Ireland. But when threatened by the Irish giant Finn McCool, he fled back to Scotland and destroyed the causeway except at both ends, one in Ireland and the other on the island of Staffa at Fingal's Cave, another place to see well-developed columnar joints. Another legend holds that Finn McCool fell in love with a giant lady on the island of Staffa and built the causeway to bring her to Ireland.

Among the many places in North America where columnar joints can be observed, those at Devil's Postpile National Monument, California, are probably the best developed. Here, columns formed in a lava flow no more than 100,000 years old show the typi-cal hexagonal shapes in both side and top views. However, in some parts of the flow the columns are strongly curved (Figure 2b), indicating complex cooling. Long after the lava flow cooled and solidified, glaciers covered the area that scratched and polished the surface of the columns. Not far from Devil's Postpile are vast exposures of the 760,000-year-old Bishop Tuff with its radial or rosette columnar joints. These unusual structures appear to have formed around gas vents in the cooling tuff deposit.

Hundreds of localities are known where columns in lava flows and shallow plutons are present. Visitors to Yellowstone National Park, Wyoming; North Cascades National Park, Washington; and Devil's Tower National Monument, Wyoming, can see these structures there as well as at many other places not encompassed by our national park and monument system. Less well known are the columns in the Palisades sill along the Hudson River in New Jersey.

(a)

Figure 2
Excellent examples of columnar jointing. (a) The Giant's Causeway in Northern Ireland. (b) Devil's Postpile National Monument, California.

(b)

in diameter, whereas calderas greatly exceed this dimension and commonly have steep sides; the Toba caldera in Sumatra measures 100 km long and 30 km wide. One of the best-known calderas in the United States is misnamed Crater Lake in Oregon—Crater Lake is actually a caldera (Figure 4.10). It formed about 6600 years ago after voluminous eruptions partially drained the magma chamber. This drainage left the summit of the mountain, Mount Mazama, unsupported, and it collapsed into the magma chamber, forming a caldera more than 1200 m deep and measuring 9.7 × 6.5 km. Most calderas probably formed when a summit collapsed during particularly large, explosive eruptions, as in the case of Crater Lake.

Shield Volcanoes

Shield volcanoes resemble the outer surface of a shield lying on the ground, with the convex side up (Figure 4.11). They have low, rounded profiles with gentle slopes ranging from about 2 to 10 degrees. Their low slopes reflect the fact that they are composed mostly of low-viscosity mafic lava flows, so the flows spread out and form thin layers. Eruptions from shield volcanoes, sometimes called *Hawaiian-type volcanoes,* are quiet compared with those of volcanoes such as Mount St. Helens; lavas most commonly rise to the surface with little explosive activity, so they usually pose little danger to humans. Lava fountains, some up to 400 m high, contribute some pyroclastic materials to shield volcanoes (Figure 4.2b), but otherwise they are composed largely of basalt lava flows; flows comprise more than 99% of the Hawaiian volcanoes above sea level.

Although eruptions of shield volcanoes tend to be rather quiet, some of the Hawaiian volcanoes have, on occasion, produced sizable explosions. Explosions take place when magma comes in contact with groundwater, causing it to vaporize instantly. One such explosion occurred in 1790 while Chief Keoua was leading about 250 warriors across the summit of Kilauea volcano to engage a rival chief in battle. About 80 of Keoua's warriors were killed by a cloud of hot volcanic gases. The current activity at Kilauea is impressive for another reason as well; it has been going on continuously since January 3, 1983, making it the longest recorded eruption.

Shield volcanoes are most common in the ocean basins, such as the Hawaiian Islands and Iceland, but some are also present on the continents—for example, in east Africa. The island of Hawaii consists of five huge shield volcanoes, two of which, Kilauea and Mauna Loa, are active much of the time. Mauna Loa at nearly 100 km across its base and more than 9.5 km above the surrounding seafloor is the largest volcano in the world (Figure 4.11b). Its volume is estimated at about 50,000 km^3. By contrast, the largest volcano in the continental United States, Mount Shasta in northern California, has a volume of only about 350 km^3.

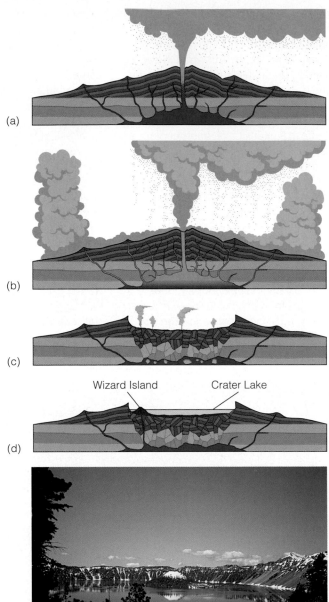

(a)

(b)

(c)

Wizard Island   Crater Lake

(d)

(e)

Figure 4.10
The sequence of events leading to the origin of Crater Lake, Oregon. (a–b) Ash clouds and ash flows partly drain the magma chamber beneath Mount Mazama. (c) The collapse of the summit and formation of the caldera. (d) Postcaldera eruptions partly cover the caldera floor, and the small volcano known as Wizard Island forms. (e) View from the rim of Crater Lake showing Wizard Island.

Shield volcanoes have a summit crater or caldera and a number of smaller cones on their flanks through which lava is erupted (Figure 4.11a). A vent opened on the flank of Kilauea and grew to more than 250 m high between June 1983 and September 1986.

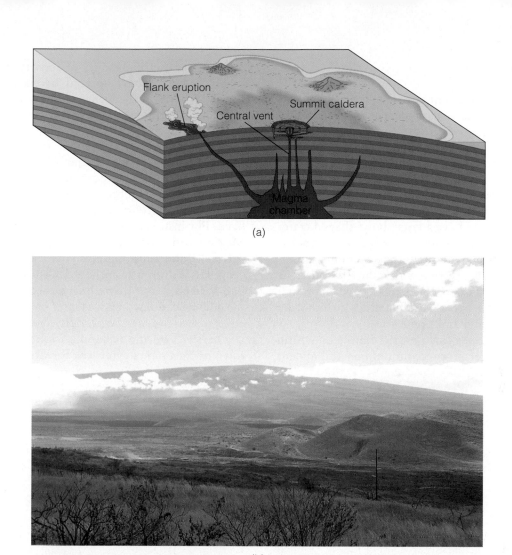

(a)

(b)

(c)

Figure 4.11

(a) A shield volcano. Each layer shown consists of numerous, thin basalt lava flows. (b) View of Mauna Loa, a large, active shield volcano on the island of Hawaii. (c) An extinct shield volcano in Lassen County, California. Note in the illustration and the images that the flanks of shield volcanoes are not steep.

Cinder Cones

Volcanic peaks composed of pyroclastic materials that resemble cinders are known as **cinder cones** (Figure 4.12). They form when pyroclastic materials are ejected into the atmosphere and fall back to the surface to accumulate around the vent, forming small, steep-sided cones. The slope angle may be as much as 33 degrees, depending on the angle that can be maintained by the irregularly shaped pyroclastic materials. Cinder cones are rarely more than 400 m high, and many have a large, bowl-shaped crater (Fig. 4.12b). Some cinder cones are very nearly symmetric; that is, the pyroclastic materials accumulate uniformly around the vent, forming a symmetric cone. The symmetry may be less than perfect, though, when prevailing winds cause the pyroclastic materials to build up higher in the downwind direction.

Many cinder cones form on the flanks or within the calderas of larger volcanic mountains and appear to represent the final stages of activity, particularly in areas formerly characterized by basalt lava flows. Wizard Island in Crater Lake, Oregon, is a small cinder cone that formed after the summit of Mount Mazama collapsed to form a caldera (Figure 4.10). Cinder cones are common in the southern Rocky Mountain states, particularly New Mexico and Arizona, and many others are found in California, Oregon, and Washington.

Eruptive activity at cinder cones is rather short-lived. For instance, on February 20, 1943, a farmer in Mexico noticed fumes emanating from a crack in his cornfield, and a few minutes later ash and cinders were erupted. Within a month, a 300-m-high cinder cone had formed. Lava flows broke through the flanks of the new volcano, which was later named Paricutín, and covered two nearby towns. Activity ceased in 1952.

In 1973 on the Icelandic island of Heimaey, the town of Vestmannaeyjar was threatened by a new cin-

(a)

(b)

Figure 4.12
(a) A 230-m-high cinder cone in Lassen Volcanic National Park, California. It was last active during the 1650s.
(b) The large, bowl-shaped crater at the summit of the cinder cone shown in (a).
(c) Vestmannaeyjar, Iceland, was threatened by a lava flow from Eldfell, a cinder cone that formed in 1973. Within two days the new volcano had grown to 100 m high. Another cinder cone known as Helgafel is also visible.

(c)

der cone. The initial eruption began on January 23, and within two days a cinder cone, later named Eldfell, rose to about 100 m above the surrounding area (Figure 4.12b). Pyroclastic materials from the volcano buried parts of the town, and by February a massive aa lava flow was advancing toward the town. The flow's leading edge ranged from 10 to 20 m thick, and its central part was as much as 100 m thick. By spraying the leading edge of the flow with seawater, which caused it to cool and solidify, the residents of Vestmannaeyjar successfully diverted the flow before it did much damage to the town.

Composite Volcanoes

Composite volcanoes, also called *stratovolcanoes,* are composed of both pyroclastic layers and lava flows (Figure 4.13). Typically, both materials have an intermediate composition, and the flows cool to form andesite. Recall that lava of intermediate composition is more viscous than mafic lava. Besides lava flows and pyroclastic layers, a significant proportion of a composite volcano is made up of **lahars** (volcanic mudflows). Some lahars form when rain falls on layers of loose pyroclastic materials and creates a muddy slurry that

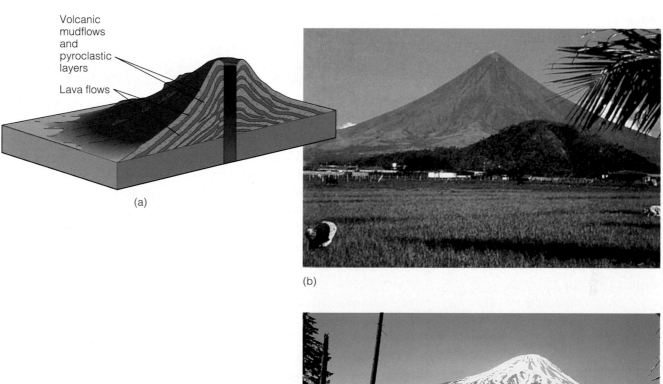

Volcanic
mudflows
and
pyroclastic
layers

Lava flows

(a)

(b)

(c)

Figure 4.13
(a) Composite volcanoes, also known as *stratovolcanoes,* are composed of lava flows, layers of pyroclastic materials, and volcanic mudflows or lahars. Note that the slope is steep near the summit but decreases toward the base. (b) Mayon volcano in the Philippines is one of the most nearly symmetrical composite volcanoes in the world. It erupted during 1999 for the 13th time this century. (c) Mount St. Helens, Washington, from the southwest in 1978.

moves downslope. On November 13, 1985, mudflows resulting from a rather minor eruption of Nevado del Ruiz in Columbia killed about 23,000 people (see Chapter 12 Prologue) (Table 4.1).

Composite volcanoes are steep-sided near their summits, perhaps as much as 30 degrees, but the slope decreases toward the base where it is generally less than 5 degrees (Figure 4.13a). Mount St. Helens, Washington, is an excellent example. Its concave slopes rise ever-steeper to the summit with its central vent through which lava and pyroclastic materials are periodically erupted (Figure 4.13c).

Composite volcanoes are the typical large volcanoes of the continents and island arcs. Familiar examples include Fujiyama in Japan and Mount Vesuvius in Italy as well as many of the volcanic peaks in the Cascade Range of western North America (see Perspective 4.2).

On June 15, 1991, Mount Pinatubo in the Philippines erupted violently, discharging huge quantities of ash and gases into the atmosphere. An estimated 3 to 5 km^3 of pyroclastic materials, mostly ash, were erupted, making this the largest volcanic eruption in more than half a century (Figure 4.14). Fortunately, warnings of an impending eruption were heeded, and

Table 4.1

Some Notable Volcanic Eruptions

Date	Volcano	Deaths
Aug. 24, 79	Mt. Vesuvius, Italy	At least 3360 killed in Pompeii and Herculaneum.
1586	Kelut, Java	Mudflows kill 10,000.
Dec. 16, 1631	Mt. Vesuvius, Italy	3500 killed.
Aug. 4, 1672	Merapi, Java	3000 killed by mudflows and pyroclastic flows.
Dec. 10, 1711	Awu, Indonesia	3000 killed by pyroclastic flows.
Sept. 22, 1760	Makian, Indonesia	Eruption kills 2000; island evacuated for seven years.
May 21, 1782	Unzen, Japan	14,500 die in debris avalanche and tsunami.
June 8, 1783	Lakagigar, Iceland	Largest historic lava flows: 12 km^3; 9350 die.
July 26, 1783	Asama, Japan	Pyroclastic flows and floods kill 1200+.
Apr. 10, 1815	Tambora, Indonesia	92,000 killed; another 80,000 reported to have died from famine and disease.
Oct. 8, 1822	Galunggung, Java	4011 die in pyroclastic flows and mudflows.
Mar. 2, 1856	Awu, Indonesia	Pyroclastic flows kill 2806.
Aug. 27, 1883	Krakatau, Indonesia	36,417 die; most killed by tsunami.
June 7, 1892	Awu, Indonesia	1532 die in pyroclastic flows.
May 8, 1902	Mt. Pelée, Martinique	St. Pierre destroyed by pyroclastic flow; 28,000 killed.
Oct. 24, 1902	Santa María, Guatemala	5000 killed.
June 6, 1912	Novarupta, Alaska	Largest 20th-century eruption: about 33 km^3 of pyroclastic materials erupted; no fatalities.
May 19, 1919	Kelut, Java	Mudflows kill 5110, devastate 104 villages.
Jan. 21, 1951	Lamington, New Guinea	2942 killed by pyroclastic flows.
Mar. 17, 1963	Agung, Indonesia	1148 killed.
Aug. 12, 1976	Soufrière, Guadeloupe	74,000 residents evacuated.
May 18, 1980	Mount St. Helens, Washington	63 killed; 600 km^2 of forest devastated.
Mar. 28, 1982	El Chichón, Mexico	Pyroclastic flows kill 1877.
Nov. 13, 1985	Nevado del Ruiz, Colombia	Mudflows kill 23,000.
Aug. 21, 1986	Oku volcanic field, Cameroon	1746 asphyxiated by cloud of CO_2 released from Lake Nyos.
June 1991	Unzen, Japan	43 killed; at least 8500 fled.
June 1991	Mt. Pinatubo, Philippines	~281 killed during initial eruption; 83 killed by later mudflows; 358 died of illness; 200,000 evacuated.
Feb. 2, 1993	Mayon, Philippines	At least 70 killed; 60,000 evacuated.
Nov. 22, 1994	Mt. Morapis, Indonesia	Pyroclastic flows kill 60; more than 6000 evacuated.
July 1999	Soufriére Hills, Montserrat	19 killed; 12,000 evacuated.

Source: American Geological Institute Data Sheets, except for last five entries.

(a)

(b)

Figure 4.14
(a) Mount Pinatubo in the Philippines is one of many volcanoes in a belt nearly encircling the Pacific Ocean basin. It is shown here erupting on June 12, 1991. A huge, thick cloud of ash and steam rises above Clark Air Force Base, from which about 15,000 people had already been evacuated to Subic Bay Naval Base. Following this eruption, the remaining 900 people at the base were also evacuated. (b) Homes partly buried by a volcanic mudflow (lahar) on June 15, 1991. Note that the roof at the far right is still partly covered by pyroclastic materials.

200,000 people were evacuated from areas around the volcano, yet the eruption still caused 722 deaths (Table 4.1). Another composite volcano in the Philippines, Mayon volcano, erupted for the 13th time this century beginning in June 1999.

Most volcanoes encountered by casual observers can be classified as shield volcanoes, cinder cones, or composite volcanoes, but some types do not fit easily into any of these categories. Lava domes, discussed in the next section, are one of these, but even more unusual are tuff rings and maar craters (see Perspective 4.3).

Lava Domes

If the upward pressure in a volcanic conduit is great enough, the most viscous magmas move up and form bulbous, steep-sided **lava domes** (Figure 4.15). Lava domes are generally composed of felsic lavas although some are of intermediate composition. Because felsic magma is so viscous, it moves up very slowly; the lava dome that formed in Santa María volcano in Guatemala in 1922 took two years to grow to 500 m high and

Figure 4.15
Lava domes form when viscous magma, generally of felsic composition, is forced up through a volcanic conduit. The 2 km³ of material composing Lassen Peak in northern California makes it the world's largest lava dome. It formed about 27,000 years ago, but erupted most recently from 1914 to 1917.

1200 m across. Lava domes contribute significantly to many composite volcanoes. Beginning in 1980, a number of lava domes were emplaced in the crater of

Eruptions of Cascade Range Volcanoes

During the summer of 1914, Lassen Peak in northern California began erupting without warning and culminated with the "Great Hot Blast," a huge steam explosion on May 22, 1915 (Figure 1). Fortunately, the area was sparsely settled, and little property damage and no deaths resulted, even though a large area of forest on the volcano's eastern and northeastern flanks were devastated. Activity largely ceased by 1917,

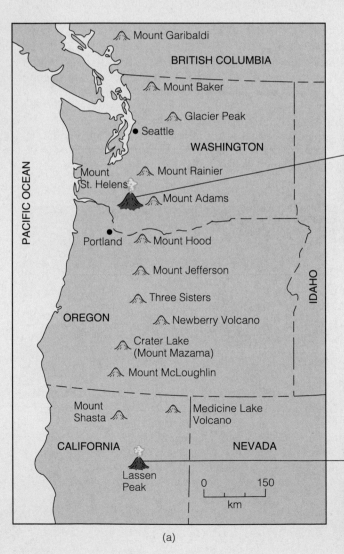

(b)

(c)

(a)

Figure 1
(a) The major volcanoes of the Cascade Range, several of which have been active during the last 200 years. (b) The eruption of Mount St. Helens, Washington, on May 18, 1980. A lateral blast occurred when a bulge on the mountain's north face collapsed and reduced the pressure on the gas-charged magma within the mountain. (c) Lassen Peak in northern California erupted numerous times from 1914 to 1917. This eruption took place in 1915.

but hot springs, boiling mud pots, and gas vents known as *fumaroles* remind us that a source of heat is present beneath the surface (Figure 4.3a).

Lassen Peak is the world's largest lava dome. It formed about 27,000 years ago as a bulbous mass of viscous lava injected into the flank of an older, eroded composite volcano known as Mount Tehama.

It is one of 15 large volcanoes in the Cascade Range of northern California, Oregon, Washington, and southern British Columbia, Canada (Figure 1, Table 1). After Lassen Peak's 1914–1917 eruptions, the Cascade volcanoes remained quiet for 63 years. Then, on March 16, 1980, Mount St. Helens in southern Washington showed signs of renewed activity, and on May 18 it erupted violently, causing the worst volcanic disaster in U.S. history (Figure 1b).

The awakening of Mount St. Helens, its first eruption in 123 years, came as no surprise to geologists of the U.S. Geological Survey (USGS), who warned in 1978 that Mount St. Helens is "an especially dangerous volcano because [of] its past behavior and [its] relatively high frequency of eruptions during the last 4,500 years."* Although no one could predict precisely when Mount St. Helens would erupt, the USGS report included maps showing areas in which damage from an eruption could be expected. Forewarned with such data, local officials were better prepared to formulate policies when the eruption did occur.

On March 27, 1980, Mount St. Helens began erupting steam and ash and continued to do so during the rest of March and most of April. By late March, a visible bulge had developed on its north face as molten rock was injected into the mountain, and the bulge continued to expand at about

---

*D. R. Crandell and D. R. Mullineaux, "Potential Hazards from Future Eruptions of Mt. St. Helens Volcano, Washington," *United States Geological Survey Bulletin 1383-C* (1978): C1.

Table 1

History of Eruptions of Some Cascade Range Volcanoes

Location	Volcano	Last Erupted/Comments
British Columbia	Mount Garibaldi (composite volcano)	10,000 years ago. Might be extinct.
Washington	Mount Baker (composite volcano)	Erupted 5 or 6 times between 1820 and 1870. Has erupted extensive pyroclastic flows. Fumeroles on flanks and in crater.
	Glacier Peak (composite volcano)	240 to 300 years ago. Typically erupts large quantities of ash and pumice.
	Mount Rainier (composite volcano)	About 5 times between 1820 and 1882. Highest Cascade volcano. History of massive debris flows and mudflows.
	Mount St. Helens (composite volcano)	1980 to present. Most active volcano in Cascade Range. Erupted 14 or 15 times during last 4000 years.
Oregon	Mount Hood (composite volcano)	Minor explosive eruptions 1859, 1865, and possibly 1907. Fumeroles in crater.
	Newberry Volcano* (shield volcano)	600 years ago. Large summit caldera. Obsidian flow about 1600 years ago.
	Crater Lake (Mount Mazama) (composite volcano)	Explosive eruptions and origin of caldera 6600–7000 years ago. Cinder cone formed in caldera 800 to 900 years ago.
California	Medicine Lake Volcano* (shield volcano)	A.D. 1065. Largest volcano in Cascade Range (600 km^3).
	Mount Shasta (composite volcano)	Probably erupted in 1786. Has erupted about once every 600 years during last 4500 years. Largest composite volcano in Cascade Range (350 km^3).
	Lassen Peak (lava dome)	1914–1917 (some activity continued until 1921). Hydrothermal activity in several areas. Eruptions at Cinder Cone near Lassen Peak during 1600s.

*Newberry Volcano, Oregon, and Medicine Lake Volcano, California, are just east of the main Cascade Range but are usually considered in discussions of the range.

1.5 m per day. On May 18, an earthquake shook the area, the unstable bulge collapsed, and the pent-up volcanic gases below expanded rapidly, creating a tremendous northward-directed lateral blast that blew out the north side of the mountain. The lateral blast accelerated from 350 to 1080 km/hr, obliterating virtually everything in its path. Some 600 km^2 of forest was completely destroyed; trees were snapped off at their bases and strewn about the countryside, and trees as far as 30 km from the bulge were seared by the intense heat. Tens of thousands of animals were killed; roads, bridges, and buildings were destroyed; and 63 people perished.

Shortly after the lateral blast, volcanic ash and steam erupted, forming a 19-km-high cloud above the volcano (see chapter opening photo). The ash cloud drifted east-northeast, and the resulting ash fall at Yakima, Washington, 130 km to the east, caused almost total darkness at midday. Detectable amounts of ash were deposited over a huge area. Flows of hot gases and volcanic ash raced down the north flank of the mountain, causing steam explosions when they encountered bodies of water or moist ground. Steam explosions continued for weeks, and at least one occurred a year later.

Snow and glacial ice on the upper slopes of Mount St. Helens melted and mixed with ash and other surface debris to form thick, pasty volcanic mudflows. The largest and most destructive mudflow surged down the valley of the North Fork of the Toutle River. Ash and mudflows displaced water in lakes and streams and flooded downstream areas. Ash and other particles carried by the floodwaters were deposited in stream channels; many kilometers from Mount St. Helens, the navigation channel of the Columbia River was reduced from 12 m to less than 4 m as a result of such deposition.

Although the damage resulting from the eruption of Mount St. Helens was significant and the deaths were tragic, it was not a particularly large or deadly eruption compared with some historic eruptions (see Table 4.1). The eruption of Tambora in 1815 is the greatest volcanic eruption in recorded history in terms of both casualties and the amount of material erupted; it produced at least 80 times more ash than the 0.9 km^3 that spewed forth from Mount St. Helens.

Several other currently dormant Cascade Range volcanoes also pose threats to populated areas. Eruptions of Mount Shasta in northern California would cause damage and perhaps fatalities in several nearby communities, and Mount Hood, Oregon, lies less than 65 km from the densely populated Portland area. But the most dangerous is probably Mount Rainier, Washington (Figure 2).

Rather than lava flows, ash falls, or even a colossal explosion as in the case of Mount St. Helens, the greatest threat from Mount Rainier is volcanic mudflows or lahars similar to the one in Figure 4.14b. During the last 10,000 years, there have been at least 60 such flows. The largest consisted of nearly 4 km^3 of debris, and it covered an area now occupied by more than 120,000 people. No one can predict when another mudflow will take place, but at least one community has taken the threat seriously enough to formulate an emergency evacuation plan. Unfortunately, they would have only 1 or 2 hours to carry out an evacuation.

Figure 2
Mount Rainier as seen from the waterfront of Tacoma, Washington. Its summit elevation of 4392 m makes Mount Rainier the highest peak in the Cascade Range. The summit is only about 80 km from where this picture was taken.

Mount St. Helens; most of these were destroyed during subsequent eruptions. Since 1983, Mount St. Helens has been characterized by sporadic dome growth. In June 1991, a dome in Japan's Unzen volcano collapsed, causing a flow of debris and hot ash that killed 43 people in a nearby town (Table 4.1).

Lava domes are often responsible for extremely explosive eruptions. In 1902 viscous magma accumulated beneath the summit of Mount Pelée on the island of Martinique. Eventually, the pressure within the mountain increased until it could no longer be contained, and the side of the mountain blew out in a tremendous explosion. When this occurred, a mobile, dense cloud of pyroclastic materials and gases called a **nuée ardente** (French for "glowing cloud") was ejected and raced downhill at about 100 km/hr, engulfing the city of St. Pierre (Figure 4.16).

A tremendous blast hit St. Pierre that leveled buildings; hurled boulders, trees, and pieces of masonry down the streets; and moved a 3-ton statue 16 m. Accompanying the blast was a swirling cloud of incandescent ash and gases with an internal temperature of 700°C that incinerated everything in its path. The nuée ardente passed through St. Pierre in 2 or 3 minutes, only to be followed by a firestorm as combustible materials burned and casks of rum exploded. But by then most of the 28,000 residents of the city were already dead. In fact, in the area covered by the nuée ardente, only 2 survived!* One survivor was on the outer edge of the nuée ardente, but even there he was terribly burned and his family and neighbors were all killed. The other survivor, a stevedore incarcerated the night before for disorderly conduct, was in a windowless cell partly below ground level. He remained in his cell badly burned for four days after the eruption until rescue workers heard his cries for help. He later became an attraction in the Barnum & Bailey Circus, where he was advertised as "The only living object that survived in the 'Silent City of Death' where 40,000 beings were suffocated, burned or buried by one belching blast of Mont Pelée's terrible volcanic eruption."†

---

*Although commonly reported that only two people survived the eruption, at least 69 and possibly as many as 111 people survived beyond the extreme margins of the nuée ardente and on ships in the harbor. Many, however, were badly injured.

†Quoted from A. Scarth (1999), *Vulcan's Fury: Man Against the Volcano* (New Haven, CT: Yale University Press), p. 177.

(a)

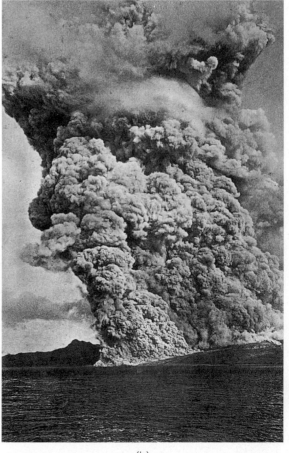

(b)

Figure 4.16
(a) St. Pierre, Martinique, after it was destroyed by a nuée ardente erupted from Mount Pelée in 1902. Only 2 of the city's 28,000 inhabitants survived. (b) A large nuée ardente from Mount Pelée a few months after the one that destroyed St. Pierre.

Perspective 4.3

Tuff Rings and Maar Craters

Everyone is familiar with the classic image of a volcano, with its slopes becoming steeper toward a snow-capped summit. Mount Fuji in Japan, Pinatubo in the Philippines, and, before its 1980 eruption, Mount St. Helens, Washington, all closely resembled this ideal volcano. Views of these large composite volcanoes are striking and to most people represent volcanoes in general. However, a number of other, less impressive shapes are common. Earlier we mentioned cinder cones and shield volcanoes, both of which differ markedly from composite volcanoes. And even more variations of volcano features are known from many areas.

One of the most familiar views on Oahu in the Hawaiian Islands is Diamond Head (Figure 1). It has a broad, low profile and has what appears to be an exceptionally large crater. Diamond Head is actually a volcano known as a *tuff ring*. It formed about 300,000 years ago when pyroclastic materials and pieces of country rock excavated during explosive eruptions accumulated around a volcanic vent. By volcano standards, it is a rather small structure. It got its name from sparkling calcite crystals that fooled some early explorers into thinking they were diamonds. A short climb to the summit of the 232-m-high structure, now a state monument, offers excellent 360-degree views of Honolulu and the nearby shoreline.

Perhaps Fort Rock, Oregon, is not as scenic as Diamond Head, but it is just as interesting. This peculiar ring-shaped structure is composed of vol-

Figure 1
Diamond Head, the mountain on the skyline, is a tuff ring on the island of Oahu in the Hawaiian Islands.

canic ash of basaltic composition. It rises about 100 m above the surrounding rather flat countryside and resembles a ruined fortress, hence the name Fort Rock (Figure 2). It came into existence during the Pleistocene Epoch, when a broad, shallow lake was present in the area. As basaltic magma rose beneath the lake, steam explosions hurled fragments into the air that accumulated to form a complete ring. Since it formed, erosion has breached the outer part of the ring in one area, and wave erosion around its outer margin has yielded steep cliffs, giving the structure its fortresslike appearance. The lake has long since vanished and now the structure rises above an otherwise rather featureless plain.

In many areas of active or recently active volcanism, shallow, flat-floored depressions known as *maar craters* are present (Figure 3). One of the best known in the United States is Ubeehe Crater in Death Valley National Park, California (see Figure 2, Perspective 18.1). Maar craters are also rather small volcanic features with low rims composed of volcanic rock fragments and fragmented country rock. They vary from 60 to 1980 m wide and 10 to 200 m deep and are commonly the sites of small lakes. Maar craters apparently lie above volcanic vents through which gas-charged magma was explosively ejected. Zuni Salt Lake in New Mexico at 1980 m across and 122 m deep is one of the largest maar craters known (Figure 3).

Figure 2
Fort Rock, Oregon. Fort Rock originated when basaltic magma rose beneath a lake causing steam explosions and the formation of a ring of debris. Since it formed, the 100-m-high structure has been eroded by waves.

(a)

Figure 3
(a) A maar crater in Alaska. (b) At 1980 m across and 122 m deep, Zuni Salt Lake in New Mexico is among the largest known maar craters.

(b)

DO ALL ERUPTIONS BUILD UP VOLCANOES?

Nearly everyone is aware of magnificent composite volcanoes such as those in Figure 4.13. Indeed, when the term *volcano* is mentioned, mountains such as these immediately come to mind. However, as noted earlier, even though some volcanoes are the typical conical mountains commonly envisioned, numerous volcanoes with other shapes are found in many areas (see Figures 4.11 and 4.15). In fact, in some areas of volcanism, volcanoes fail to develop at all. For instance, during *fissure eruptions* fluid lava pours out and simply builds up rather flat-lying areas, whereas huge explosive eruptions might yield *pyroclastic sheet deposits,* which, as their name implies, have a sheetlike geometry.

Fissure Eruptions

Some 164,000 km² of eastern Washington and parts of Oregon and Idaho were covered by overlapping basalt lava flows during the Miocene and Pliocene Epochs (between about 17 and 5 million years ago). Now known as the Columbia River basalts, they are well exposed in canyons eroded by the Snake and Columbia Rivers (Figure 4.17). Rather than being erupted from a central vent, these flows issued from long cracks or fissures and are thus known as **fissure eruptions.** Lava erupted from these fissures was so fluid (had such low viscosity) that it simply spread out, covering vast areas and building up a **basalt plateau,** which is simply a broad, flat, elevated area underlain by lava flows.

The Columbia River basalt flows have an aggregate thickness of about 1000 m, and some individual flows cover huge areas—for example, the Roza flow, which is 30 m thick, advanced along a front about 100 km wide and covered 40,000 km².

Fissure eruptions and basalt plateaus are not common, although several large areas with these features are known. Currently, this kind of activity is occurring only

in Iceland. A number of volcanic mountains are present in Iceland, but the bulk of the island is composed of basalt flows erupted from fissures. Two large fissure eruptions, one in A.D. 930 and the other in 1783, account for about half of the magma erupted in Iceland during historic time. The 1783 eruption issued from the Laki fissure, which is more than 30 km long; about 12.5 km³ of lava flowed several tens of kilometers from the fissure; the lava covered an area of more than 560 km² and in one place filled a valley to a depth of about 200 m.

Pyroclastic Sheet Deposits

More than 100 years ago, geologists were aware of vast areas covered by felsic volcanic rocks a few meters to hundreds of meters thick. It seemed improbable that these could have formed as vast lava flows, but it also

(a)

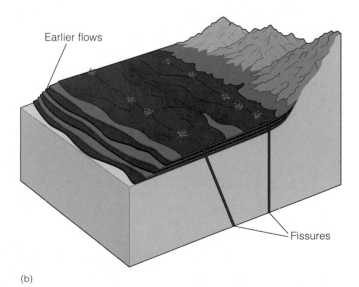

(b)

Figure 4.17
(a) The Columbia River basalts. (b) A block diagram showing fissure eruptions and the origin of a basalt plateau.

seemed equally unlikely that they were ash fall deposits. Based on observations of historic pyroclastic flows, such as the nuée ardente erupted by Mount Pelée in 1902, it now seems probable that these ancient rocks originated as pyroclastic flows, hence the name **pyroclastic sheet deposits**. They cover far greater areas than any observed during historic time, and apparently erupted from long fissures rather than from a central vent. The pyroclastic materials of many of these flows were so hot that they fused together to form *welded tuff*.

It now appears that major pyroclastic flows issue from fissures formed during the origin of calderas (Figure 4.18). For instance, pyroclastic flows were erupted during the formation of a large caldera in the area of present-day Crater Lake, Oregon. Similarly, the Bishop Tuff of eastern California appears to have been erupted shortly before the formation of the Long Valley caldera. Interestingly, earthquake activity in the Long Valley caldera and nearby areas beginning in 1978 may indicate that magma is moving up beneath part of the caldera. Thus, the possibility of future eruptions in that area cannot be discounted.

Figure 4.18

Pyroclastic flow deposits in the Marella River valley in the Philippines. The flows that filled this valley to depths of 50 to 200 m deep were erupted from Mount Pinatubo on June 15, 1991. Some of the flows moved up to 16 km from the volcano.

IS IT POSSIBLE TO FORECAST ERUPTIONS?

Numerous volcanoes have erupted explosively during historic time and have the potential to do so again. Most dangerous or potentially dangerous volcanoes are at or near the margins of tectonic plates, especially at convergent plate boundaries, but some, such as the Hawaiian volcanoes, are far from plate boundaries. At any one time about a dozen volcanoes are erupting, but most eruptions cause little or no property damage, injuries, or fatalities. Unfortunately, some do. The 1815 eruption of Tambora in Indonesia resulted in more than 100,000 deaths, and a rather minor eruption of Nevado del Ruiz in Colombia in 1985 triggered volcanic mudflows that killed 23,000 people (Table 4.1).

Only a few of Earth's potentially dangerous volcanoes are monitored, including some in Japan, Italy, Russia, New Zealand, and the United States. Two facilities in the United States are devoted to volcano monitoring: Hawaiian Volcano Observatory on Kilauea volcano and the David A. Johnston Cascades Volcano Observatory in Vancouver, Washington. Many of the methods now commonly used to monitor volcanoes were developed at the Hawaiian Volcano Observatory.

Volcano monitoring involves recording and analyzing various changes in the physical and chemical attributes of volcanoes. Tiltmeters are used to detect changes in the slopes of a volcano as it inflates as magma rises beneath it, whereas a Geodimeter uses a laser beam to measure horizontal distances, which also change as a volcano inflates (Figure 4.19). Geologists also monitor gas emissions, changes in groundwater level and temperature, and changes in the local magnetic and electrical fields. Even the accumulation of snow and ice is evaluated to anticipate hazards from flooding should an eruption take place. Of critical importance in volcano monitoring and warning of an imminent eruption is the detection of harmonic tremor. **Harmonic tremor** is continuous ground motion that may last for minutes or hours as opposed to the sudden, sharp jolts produced by most earthquakes. Such activity indicates that magma is moving beneath the surface, and it precedes the eruptions of Hawaiian volcanoes and it preceded the eruption of Mount St. Helens in 1980.

To more fully anticipate the future activity of a particular volcano, the eruptive history of the volcano must also be known. Accordingly, geologists study the record of past eruptions preserved in rocks. Prior to 1980, Mount St. Helens, Washington, was considered one of the most likely Cascade volcanoes to erupt because detailed studies indicated that it has a record of 14 or 15 explosive eruptions during the past 4500 years. In fact, maps prepared prior to 1980 showing areas in which damage from an eruption could be expected were helpful in planning and evacuating once an eruption did take place.

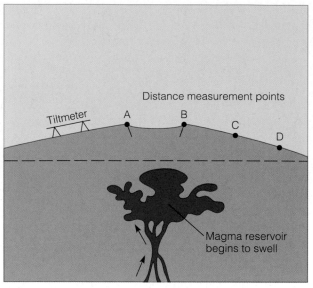

(a) Stage 1

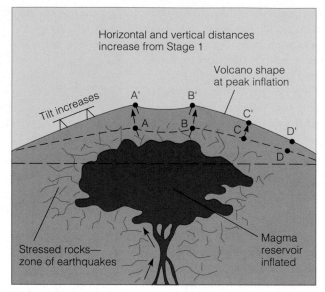

(b) Stage 2

Figure 4.19
Volcano monitoring. These illustrations show three stages in the eruption of a Hawaiian volcano, and how tilt and distance measurements are made. (a) The volcano begins to inflate as the magma chamber grows larger. (b) Inflation reaches its peak. (c) The volcano erupts and then deflates, returning to its original shape.

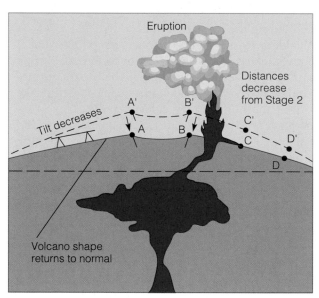

(c) Stage 3

Geologists successfully gave timely warnings of impending eruptions of Mount St. Helens, Washington, and Mount Pinatubo in the Philippines, but in both cases the climactic eruptions were preceded by eruptive activity of lesser intensity. In some cases, however, the warning signs are much more subtle and difficult to interpret. A good example comes from studies in the Long Valley caldera in eastern California. Because numerous small earthquakes and other warning signs indicated to geologists of the U.S. Geological Survey (USGS) that magma was moving beneath the surface, they issued a low-level warning in 1987, but nothing happened. Volcanic activity in the Long Valley caldera area had occurred as recently as 250 years ago, and there is every reason to think that it might occur again; however, when remains an open question. Unfortunately, because the local populace was largely unaware of the geologic history of the region, the USGS did a poor job in communicating their concerns, and premature news releases of the warning caused more concern than was justified. In any case, local residents were outraged because the warning caused a decrease in tourism (Mammoth Mountain on the margins of the caldera is the second largest ski area in the county) and property values plummeted. Monitoring continues in the Long Valley caldera, and the signs of renewed volcanism, including earthquake swarms, trees killed by carbon dioxide gas apparently emanating from magma (see Figure 4.3b), and hot springs, cannot be ignored.

For the better-monitored volcanoes, such as those in Hawaii, it is now possible to make accurate short-term forecasts of eruptions. But for many volcanoes, little or no information is available for making predictions. For instance, on January 14, 1993, Colombia's Galeras volcano erupted without warning, killing 6 of 10 volcanologists on a field trip and 3 Colombian tourists (see Guest Essay, page 120). Ironically, the volcanologists were attending a conference on improving methods for predicting volcanic eruptions.

WHAT WOULD YOU DO?

Suppose that in the next 20 years it is possible to predict that a volcanic eruption will occur within a two-week period. Your responsibility is to decide whether or not to evacuate a large metropolitan area should an eruption actually take place. What kinds of specific information would you request from geologists, civil defense personal, police, transportation officials, and the health-care community?

HOW LARGE IS AN ERUPTION?

Most people are aware of magnitude as a measure expressing the size of an earthquake. The August 1999 earthquake in Turkey was 7.4 on the magnitude scale, and because it was centered in a densely populated region, thousands of fatalities and injuries resulted (see Chapter 9). Geologists have also devised a **volcanic explosivity index (VEI)** to express the size of

volcanic eruptions (Table 4.2). Unlike the earthquake magnitude scale, though, it is only semiquantitative, being based partly on subjective criteria.

The volcanic explosivity index ranges from 0 (nonexplosive) to 8 (megacolossal) and is based on several aspects of a particular eruption such as the volume of material explosively ejected and the height of the eruption plume. However, the volume of lava, fatalities, and property damage are not considered when making a VEI determination. For instance, the 1985 eruption of Nevado del Ruiz in Colombia killed 23,000 people, yet had a VEI of only 3. In contrast, the colossal eruption (VEI = 6) of Katmai in Alaska in 1911 caused no fatalities or injuries. Since A.D. 1500, only the 1815 eruption of Tambora had a value of 7; it was both large and deadly (see Table 4.1). Nearly 5700 eruptions during the last 10,000 years have been assigned VEI numbers, but none has exceeded 7, and most (62%) were assigned a value of 2.

DISTRIBUTION OF VOLCANOES

Most of the world's active volcanoes are found in well-defined zones or belts rather than being randomly distributed. The **circum-Pacific belt** has more than 60% of all active volcanoes, including those in the Andes of South America; the volcanoes of Central America, Mexico, and the Cascade Range of North America; as well as the Alaskan volcanoes and those in Japan, the Philippines, Indonesia, and New Zealand (Figure 4.20). Also included in the circum-Pacific belt are the southernmost active volcanoes at Mount Erebus in Antarctica, and a large caldera at Deception Island

Table 4.2

The Volcanic Explosivity Index (VEI)

VEI	Description	Plume Height	Volume of Tephra[1]	Duration of Continuous Blast (hours)	Classification[2]	Frequency
0	Nonexplosive	0–100 m	1000s m³	< 1	Hawaiian	Daily
1	Gentle	100–1000s m	10,000s m³	< 1	Hawaiian/Strombolian	Daily
2	Explosive	1–5 km	1,000,000s m³	1–6	Strombolian/Vulcanian	Weekly
3	Severe	3.15 km	10,000,000s m³	1–12	Vulcanian	Yearly
4	Cataclysmic	10–25 km	100,000,000s m³	1–12	Vulcanian/Plinian	10s of years
5	Paroxysmal	> 25 km	1 km³	6–12	Plinian	100s of years
6	Colossal	> 25 km	10s km³	> 12	Plinian/Ultra-Plinian	100s of years
7	Supercolossal	> 25 km	100s km³	> 12	Ultra-Plinian	1000s of years
8	Megacolossal	> 25 km	1000s km³	> 12	Ultra-Plinian	10,000s of years

[1]Tephra is a collective term for all pyroclastic materials erupted by a volcano.

[2]The terms in this column refer to types of eruptions: Hawaiian—lava flows but little explosive activity; Strombolian—eruption in which fluid basaltic lava is "jetted" from a central vent; Vulcanian—an eruption of explosively ejected incandescent lava; Plinian—high-velocity eruption of streams of fragmented magma and gases from a vent.

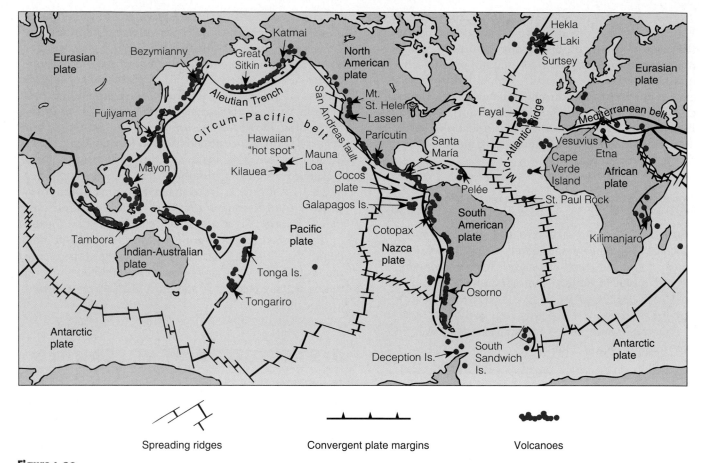

Figure 4.20
Most volcanoes are at or near plate boundaries. Two major volcano belts are recognized: The circum-Pacific belt contains about 60% of all active volcanoes, about 20% are in the Mediterranean belt, and most of the rest are located along mid-oceanic ridges.

that erupted most recently during 1970. In fact, this belt nearly encircling the Pacific Ocean basin is popularly referred to as the Ring of Fire, alluding to the fact that so many active volcanoes are present there. Several volcanoes in the circum-Pacific belt are erupting at any particular time.

The second most common area of active volcanism is in the **Mediterranean belt** (Figure 4.20). About 20% of all active volcanism takes place in this belt where the famous Italian volcanoes such as Mount Etna and Vesuvius and the Greek volcano Santorini are found. Mount Etna has issued lava flows more than 150 times since 1500 B.C., when activity was first recorded. A particularly violent eruption of Santorini in 1390 B.C. might be the basis for the myth regarding the lost continent of Atlantis (see Chapter 11), and in A.D. 79 an eruption of Vesuvius destroyed Pompeii and other nearby cities (see the Prologue).

Nearly all the remaining 20% of active volcanoes are at or near mid-oceanic ridges or the extensions of these ridges onto land (Figure 4.20). These include the East Pacific Rise and the longest of all mid-oceanic ridges, the Mid-Atlantic Ridge. The latter is situated

near the center of the Atlantic Ocean basin, accounting for the volcanism in Iceland and elsewhere. It continues around the southern tip of Africa, where it connects with the Indian Ridge. Branches of the Indian Ridge extend into the Red Sea and East Africa, where such volcanoes as Kilamanjaro in Tanzania and Erta Ale in Ethiopia with its continuously active lava lake are found.

Anyone familiar with volcanoes will have noticed that no mention has been made so far of the Hawaiian volcanoes. This is not an oversight. Mauna Loa and Kilauea on the island of Hawaii, and Loihi seamount just to the south, are the notable exceptions to the distribution of volcanoes in well-defined belts. Their location and significance are discussed in the following section.

PLATE TECTONICS, VOLCANOES, AND PLUTONS

In Chapter 3 we discussed the origin and evolution of magma and concluded that (1) mafic magmas are generated beneath spreading ridges and (2) intermediate and felsic magmas form where an oceanic plate is subducted beneath another oceanic plate or a continental plate. Accordingly, most of Earth's volcanism and emplacement of plutons takes place at divergent and convergent plate boundaries.

Much of the mafic magma that forms beneath spreading ridges is simply emplaced at depth as vertical dikes and gabbro plutons (Figure 4.21). Some of this magma rises to the surface where it usually forms submarine lava flows and pillow lavas (Figure 4.8). Indeed, the oceanic crust is composed largely of gabbro and basalt. Much of this submarine volcanism goes undetected, but researchers in submersible craft have observed the results of these eruptions.

Pyroclastic materials are not common in this environment because mafic lava is very fluid, allowing gases to easily escape; and at great depth, water pressure prevents gases from expanding. Accordingly, the explosive eruptions that yield pyroclastic materials are not common. Should an eruptive center along a ridge build above sea level, however, pyroclastic materials might be erupted at lava fountains, but most of the magma issues forth as fluid lava flows that form shield volcanoes.

Excellent examples of divergent plate boundary volcanism are found along the Mid-Atlantic Ridge, particularly where it is taking place above sea level as in Iceland. In November 1963, a new volcanic island, later named Surtsey, rose from the sea just south of Iceland. The East Pacific Rise and the Indian Ridge are also areas of similar volcanism. Not all divergent plate boundaries are beneath sea level as in the previous examples. For instance, divergence and igneous activity is taking place in Africa at the East African Rift system (see Chapter 12).

The circum-Pacific and Mediterranean volcanic belts are made up of composite volcanoes near the leading edges of overriding plates at convergent plate boundaries (Figure 4.21). The overriding plate may be oceanic as in the case of the Aleutian Islands, or it may be continental as, for instance, the South American plate with its chain of volcanoes along its western margin. As already noted, these volcanoes consist largely of pyroclastic materials and lava flows of intermediate to felsic composition. Recall that partial melting of mafic oceanic crust of a subducted oceanic plate generates the magmas, some of which are emplaced along the plate margins as plutons, especially batholiths, and some are erupted to build up composite volcanoes. Some of the more viscous magmas, generally of felsic composition, are emplaced as lava domes within existing volcanoes, thus accounting for the explosive eruptions that characterize convergent plate boundaries.

In previous sections of this chapter, we have alluded to several eruptions at convergent plate boundaries. Good examples are the explosive eruptions of Mount Pinatubo and Mayon volcano in the Philippines, both of which are situated near a plate boundary beneath which an oceanic plate is subducted. Mount St. Helens, Washington (see Perspective 4.2), is similarly situated, but it is on a continental rather than an oceanic plate. Mount Vesuvius in Italy, one of several active volcanoes in that region, lies on a plate that the northern margin of the African plate is subducted beneath (see the Prologue).

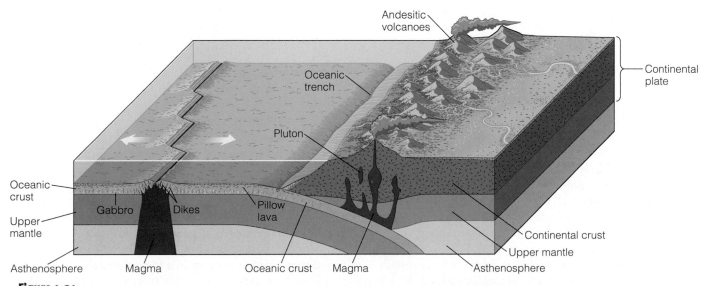

Figure 4.21
Both intrusive and extrusive igneous activity take place at divergent plate boundaries (spreading ridges) and where plates are subducted at convergent plate boundaries. Oceanic crust is composed largely of vertical dikes and gabbro, and its upper part consists of submarine lavas, especially pillow lavas. Magma forms where an oceanic plate is subducted beneath another oceanic plate or beneath a continental plate as shown here. Some of the magma forms plutons, especially batholiths, but some is erupted to form volcanoes.

Stanley N. Williams

Volcanology: The Challenge of Volcanoes

Stanley N. Williams was a geology major at Beloit College and then went to Dartmouth College for his M.S. study with Dick Stoiber on the 1902 eruption of Santa María volcano in Guatemala, which was one of the largest of the twentieth century. He stayed and worked on his Ph.D. in Nicaragua where the volcano Masaya has frequently erupted during historic times. Studying active volcanoes has meant working on more than 100 volcanoes in 21 countries. It has also meant being the survivor in 1993 when six colleagues died, when they tried to understand the danger of a volcano in Colombia. He is returning to field studies of active volcanoes because he is caught by curiosity. He has been a professor at Arizona State University since 1991.

As a volcanologist, I have been on the scene of many volcanoes showing threatening activity and at several disasters—the aftermath of eruptions where thousands of people died. Each volcano has produced very important progress in our ability to understand how volcanoes erupt.

Mount St. Helens was recognized by two U.S. Geological Survey geologists as the volcano of the Cascade Mountains most likely to erupt in the twentieth century. Just about one year after they published a paper, an earthquake produced a crack in the volcano's summit and caused a small explosion. My mentor, Dick Stoiber, and I were there within 18 hours to measure the gases emitted by the eruptions. We used a remote sensor to quantify the amount of sulfur dioxide (SO_2) released. We were amazed that the eruptions were releasing the smallest amounts of SO_2 we had ever measured at an erupting volcano. We had learned something important—groundwater can enter cracks on active volcanoes and become so hot that it flashes to steam and produces an eruption. Two months later came the famous lateral blast and subsequent partial collapse of Mount St. Helens.

Volcanologists learn a great deal from each tragedy but avoiding the next one is the really serious challenge. There are about 1500 active volcanoes around the world, and 500 million people live very close to them. To understand what is happening when a volcano becomes active, we need much information: What has this volcano done historically and prehistorically? What areas are most likely to be affected by a future eruption? How can the public be trained to mitigate the hazards? How does what is happening now compare to the "normal" activity for this volcano? Unfortunately, very little is usually known about the recent eruptions, and almost nothing is known about the "normal" activity. Nevertheless, we must talk with the authorities about what we do know and how we interpret it. In doing so, we need to avoid jargon and try to help them understand the natural processes. Many different people—medical doctors, engineers, architects, civil defense personnel, economists, political leaders, and others—must be involved. Together, we try to make good plans to inform people so that they are not terrified by volcanoes and are able to help themselves if there is an eruption.

As I write this, I am still struggling with information about one of the biggest volcanoes in the world (Popocatépetl), which stands close to the world's largest city (Mexico City). It has been releasing much more gas than before for three years, and the number and energy of earthquakes do seem to be much different. What are we going to do that might help minimize the chance of another crisis or disaster? The search for answers is what makes volcanology so exciting and challenging. ∎

Mauna Loa and Kilauea on the island of Hawaii and Loihi just 32 km to the south are within the interior of a rigid plate far from any spreading ridge or subduction zone. Thus, they are unrelated to divergence or convergence. It is postulated that a **mantle plume** creates a local "hot spot" beneath Hawaii. However, the magma is derived from the upper mantle as it is at spreading ridges and accordingly is mafic, so it builds up shield volcanoes.

Loihi is particularly interesting because it represents a stage in the origin of a new Hawaiian island. It is a submarine volcano that rises more than 3000 m above the adjacent seafloor, but its summit is still about 940 m below sea level.

Even though the Hawaiian volcanoes are unrelated to spreading ridges or subduction zones, their evolution is related to plate movements. Notice in Figure 4.22 that the ages of the rocks composing the various Hawaiian Islands increase toward the northwest; Kauai formed 3.8 to 5.6 million years ago, whereas Hawaii began forming less than 1 million years ago, and Loihi began forming even more recently. Continuous movement of the Pacific plate over the hot spot, now beneath Hawaii and Loihi, has formed the islands in succession.

Mantle plumes and hot spots have also been proposed to explain volcanism in a few other areas. A mantle plume may exist beneath Yellowstone National Park in Wyoming, for instance. Some source of heat at depth is responsible for the present-day hot springs and geysers such as Old Faithful, but many geologists think the heat source is a body of intruded magma that has not yet completely cooled, rather than a mantle plume.

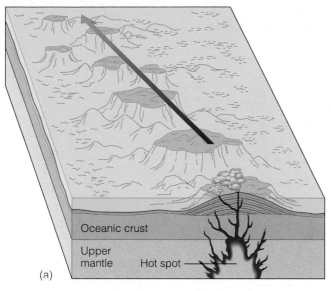

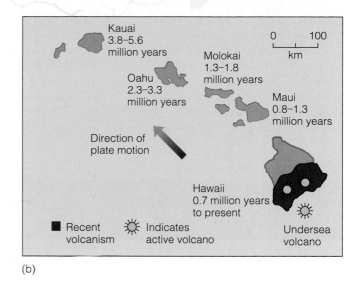

Figure 4.22
(a) Generalized diagram showing the origin of the Hawaiian Islands. As a lithospheric plate moves over the hot spot beneath Hawaii, a succession of volcanoes form. Present-day volcanism occurs only on Hawaii and beneath the sea just to the south. (b) Map showing the age of the islands in the Hawaiian chain.

Chapter Summary

1. Volcanism is the process whereby magma and its associated gases erupt at the surface. Magma erupts as lava flows, or is ejected explosively as pyroclastic materials.

2. Only a few percent by weight of a magma consists of gases, most of which is water vapor. Sulfur gases emitted during large eruptions can have far-reaching climatic effects.

3. The surfaces of aa lava flows consist of rough, jagged, angular blocks, whereas pahoehoe flows have smoothly wrinkled surfaces.

4. Many lava flows are characterized by pressure ridges, spatter cones, and lava tubes. Columnar joints form in some lava flows when they cool. Pillow lavas are erupted under water and consist of interconnected bulbous masses.

5. Volcanoes are mountains built up around a vent where lava flows and/or pyroclastic materials are erupted.

6. Shield volcanoes have low, rounded profiles and are composed mostly of mafic flows that cooled to form basalt. Small, steep-sided cinder cones form where pyroclastic materials resembling cinders are erupted and accumulate. Composite volcanoes are composed of lava flows of intermediate composition, layers of pyroclastic materials, and volcanic mudflows known as *lahars*.

7. Viscous masses of lava, generally of felsic composition, are forced up through the conduits of some volcanoes and form bulbous lava domes. Volcanoes with lava domes are dangerous because they erupt explosively and frequently eject nuée ardentes.

8. The summits of volcanoes are characterized by a circular or oval crater or a much larger caldera. Many calderas form by summit collapse when an underlying magma chamber is partly drained.

9. Fluid mafic lava erupted from long fissures (fissure eruptions) spreads over large areas to form basalt plateaus.

10. Pyroclastic flows erupted from fissures formed during the origin of calderas cover vast areas. Such eruptions of pyroclastic materials form pyroclastic sheet deposits.

11. An eruption's size is indicated by a number derived from the volcano explosivity index. The quantity of material ejected during an eruption as well as the height of the eruption cloud are criteria evaluated to determine a number from 0 to 8. The largest eruption during historic time measured 7 on this scale.

12. To effectively monitor volcanoes, geologists evaluate changes in a volcano's slope and shape, changes in gas emissions, changes in groundwater temperature and level, as well as changes in the local electrical and magnetic properties of rocks. Some of the most useful criteria used to make predictions are the detection of harmonic tremor and evaluation of a particular volcano's past history.

13. Most active volcanoes are distributed in linear belts. The circum-Pacific belt and Mediterranean belt contain about 80% of all active volcanoes.

14. Volcanism in the circum-Pacific and Mediterranean belts is at convergent plate margins where subduction occurs. Partial melting of the subducted plate generates intermediate and felsic magmas, most of which form plutons.

15. Magma derived by partial melting of the upper mantle beneath spreading ridges accounts for the mafic plutons and lavas of ocean basins.

16. The two active volcanoes on the island of Hawaii and one just to the south are thought to lie above a hot mantle plume. The Hawaiian Islands developed as a series of volcanoes formed on the Pacific plate as it moved over this mantle plume.

Important Terms

aa
ash
basalt plateau
caldera
cinder cone
circum-Pacific belt
columnar joint
composite volcano (stratovolcano)
crater

fissure eruption
harmonic tremor
lahar
lava dome
lava tube
mantle plume
Mediterranean belt
nuée ardente
pahoehoe

pillow lava
pressure ridge
pyroclastic sheet deposit
shield volcano
spatter cone
volcanic explosivity index (VEI)
volcanism
volcano

Review Questions

1. Much of the upper part of the oceanic crust is made up of interconnected bulbous masses of igneous rock known as:

   a. _____ pyroclastic materials;
   b. _____ volcanic bombs;
   c. _____ parasitic cones;
   d. _____ lapilli;
   e. _____ pillow lava.

2. Basalt plateaus form as a result of:

   a. _____ repeated eruptions of cinder cones;
   b. _____ accumulation of thick layers of pyroclastic materials;
   c. _____ eruptions of fluid lava from long fissures;
   d. _____ deposition by volcanic mudflows;
   e. _____ erosion of composite volcanoes.

3. Two Cascade Range volcanos have erupted in the United States during this century. They are _____ and _____.

   a. _____ Mount Rainier, Washington/Crater Lake, Oregon;
   b. _____ Mount St. Helens, Washington/Lassen Peak, California;
   c. _____ Newberry Volcano, Oregon/Mount Garibaldi, British Columbia;
   d. _____ Mount Shasta, California/Mount Hood, Oregon;
   e. _____ Glacier Peak, Washington/Mount Mazama, Oregon.

4. The volcanoes on the island of Hawaii and at Loihi seamount just to the south:

   a. _____ are composite volcanoes;
   b. _____ have not erupted during historic time;
   c. _____ are particularly dangerous;
   d. _____ erupt mostly pyroclastic materials;
   e. _____ are not located at a plate boundary.

5. The fact that _____ have slopes rarely steeper than 10 degrees is because they are composed mostly of low-viscosity lava flows:

   a. _____ shield volcanoes;
   b. _____ lava tubes;
   c. _____ basalt plateaus;
   d. _____ pyroclastic sheet deposits;
   e. _____ cinder cones.

6. One of the warning signs of an impending volcanic eruption is harmonic tremor, which is:

   a. _____ changes in groundwater levels;
   b. _____ inflation of a volcano as magma rises;
   c. _____ ground shaking lasting

d. _____ emission of large quantities of gases;

e. _____ a marked temperature increase.

7. Which volcanic gas was responsible for the deaths at Lake Nyos in Africa and the dead trees at Mammoth Lakes, California?

a. _____ methane;

b. _____ carbon dioxide;

c. _____ hydrogen sulfide;

d. _____ water vapor;

e. _____ chlorine.

8. A lava flow with a smooth, ropy surface is termed:

a. _____ a spatter cone;

b. _____ lapilli;

c. _____ pahoehoe;

d. _____ pillow lava;

e. _____ vesicular.

9. A nuée ardente is a(an):

a. _____ lava flow of especially high viscosity;

b. _____ thick accumulation of volcanic bombs;

c. _____ type of eruption likely to produce a basalt plateau;

d. _____ underwater eruption that yields bulbous masses of lava;

e. _____ incandescent cloud of gases and particles.

10. Most active volcanoes are found in the:

a. _____ mid-oceanic ridge volcanic zone;

b. _____ Sierra Nevada–Cascade volcanic province;

c. _____ Mediterranean divergence boundary;

d. _____ circum-Pacific belt;

e. _____ eastern Atlantic subduction zone.

11. The summits of some volcanoes are characterized by very wide, steep-sided depressions known as _____, most of which form by _____:

a. _____ explosion pits/voluminous eruptions;

b. _____ calderas/summit collapse;

c. _____ basalt plateaus/fissure eruptions;

d. _____ lava domes/forceful injection;

e. _____ parasitic cones/lava flows.

12. In addition to lava flows and pyroclastic materials, composite volcanoes are also made up of a significant amount of:

a. _____ lahars;

b. _____ pressure ridges;

c. _____ spatter cones;

d. _____ columnar joints;

e. _____ pahoehoe.

13. Why are most composite volcanoes at convergent plate boundaries, whereas most shield volcanoes are along divergent plate boundaries?

14. How do aa and pahoehoe lava flows differ, and what accounts for the difference(s)?

15. Why kind(s) of materials erupted by volcanoes are a hazard to commercial aviation? Explain fully.

16. Draw cross sections showing the internal structure of a shield volcano and a cinder cone. What is each type of volcano composed of?

17. Explain how a nuée ardente originates and why nuée ardentes are particularly dangerous.

18. What kinds of information do geologists evaluate to warn of potential volcanic eruptions?

19. Explain why the Hawaiian Islands are progressively older toward the northwest, and why active volcanoes are present only on the island of Hawaii and just to the south?

20. How does a crater differ from a caldera, and how does the latter form? Give an example of a well-known caldera.

21. Why are most volcanoes found in distinct belts rather than being randomly distributed around the world? Where are these belts?

22. What are lava domes, of what kind(s) of magma are they composed, and why are they especially dangerous?

23. Explain why eruptions of mafic magma are rather quiet, whereas eruptions of felsic magma are commonly explosive.

24. How do columnar joints and spatter cones form? Where are good places to see examples of each?

Points to Ponder

1. During this century, two Cascade Range volcanoes have erupted. What kinds of evidence would indicate that some of the other volcanoes in this range might erupt in the future?

2. Suppose you find rocks on land consisting of layers of pillow lava overlain by deep-sea sedimentary rocks. Where did the pillow lava layers originally form, and what type of rock would you expect to lie beneath them?

3. If several eruptions the size of Mount Pinatubo's occurred in one year, what types of widespread effects could be expected?

4. What geologic events would have to occur in order for a chain of volcanoes to form along the east coasts of Canada and the United States?

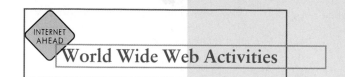
➤ MICHIGAN TECHNOLOGICAL UNIVERSITY VOLCANOES PAGE

This site provides scientific and educational information about volcanoes to the public. It contains information about current global volcanic activity, research about volcanoes, as well as links to governmental agencies and research institutions.

1. Click on the *Worldwide Volcanic Reference Map* site. Which regions are experiencing the greatest volcanic activity? Click on the *Recent and Ongoing Volcanic Activity* site. Check out one of the active volcanic sites. What type of volcano is erupting? What is the history of this volcano? Has it been monitored in the past, and is it currently being monitored?

2. Click on the *Remote Sensing of Volcanoes* site. What type of research is being conducted in the area of remote sensing of volcanoes? How is this beneficial to humans?

➤ U.S. GEOLOGICAL SURVEY CASCADES VOLCANO OBSERVATORY

This site contains information about volcanoes and other natural hazards in the western United States and elsewhere in the world. It is an excellent source of information on the hazards of volcanic eruptions and contains specific information about Mount St. Helens, as well as other volcanoes in the western United States and elsewhere.

1. Click on the *Mt. St. Helens* site under the *Volcanic Information* section. What is the current activity level of Mount St. Helens? What are some of the current research projects on Mount St. Helens?

2. Click on the *Visit a Volcano* site under the *Educational Outreach* section. What are some of the volcanoes you can visit? Report on one of the volcanoes listed on this site.

➤ VOLCANOES AND GLOBAL CLIMATE CHANGE

This is a NASA Facts site and contains information on the relationship between volcanoes and global cooling and ozone depletion. Read about how volcanic eruptions affect global temperatures and how scientists monitor the various components erupted from a volcano.

➤ VOLCANO WORLD

This excellent volcano site is maintained and supported by NASA's Public Use of Earth and Space Science Data Over the Internet program. It contains a wealth of information about volcanoes and volcanic parks and monuments.

1. Click on the *What's Erupting Now* site. It will take you to a map showing all current volcanic activity and a listing of volcanoes. Click on one of the red triangles, which will take you to the volcano shown on the map where you can read about the volcano and see images and videos of its eruption. In what region of the world is current volcanism most active?

2. Click on the *Exploring Earth's Volcanoes* site. It will take you to a map of the world divided into regions. By clicking on a particular region, you can view active volcanoes from that region and learn more about the volcanoes. Compare the volcanoes from the different regions of the world.

3. Click on the *Volcanoes of Other Worlds* site. Check out volcanic activity on the Moon, Mars, and Venus.

➤ THE ELECTRONIC VOLCANO

This site is maintained by personnel at the library and Department of Earth Sciences, Dartmouth College, Hanover, New Hampshire. It contains information on active volcanoes, volcanic hazards, catalogs and maps of active volcanoes, current events and research, and a list of journals with articles on active volcanoes. Scroll down to Paricutin volcano and read the account. When and where did Paricutin erupt? Scroll to Benjamin Franklin's "1784 paper." What did he conclude regarding eruptions and climate?

> ## VOLCANO WATCH

This is a weekly newsletter mostly about the volcanoes on the island of Hawaii. It is written for the general public by scientists at the USGS's Hawaiian Volcano Observatory. The site contains information regarding the current status of erupting volcanoes in Hawaii.

> ## TERRESTRIAL VOLCANOES

A site with information on many aspects of volcanism, it includes maps and photos of notable eruptions and descriptions of eruptions and their effects.
1. How much lava was erupted during the 1783 eruption at Laki, Iceland?
2. What are the relationships between lava domes and composite volcanoes?
3. Why does Mount Pinatubo in the Philippines continue to be a major hazard?

> ## GLOBAL VOLCANISM PROGRAM

This database at the Smithsonian Institution contains information on eruptions that have occurred during the last 10,000 years. The site features *Volcanoes of the World, Volcano Net Links,* and the *Bulletin of the Global Volcanism Network,* a weekly newsletter compiled from data submitted by more than 1000 correspondents.

> ## OTHER SITES

Web sites about volcanoes and volcanism are numerous. Ones to visit include:

USGS Photoglossary of Volcano Terms. One can read definitions and see images of many volcanic features.
Cenozoic-Mesozoic Volcanism . . . Eastern Sierra Nevada. Take a Field Trip to Owens Valley and Mammoth Lakes in the east-central part of the Sierra Nevada and see images and descriptions of volcanic features. Also, this is an area of potential future volcanism.
USGS Volcano & Hydrologic Monitoring. Great site to see what methods are used to monitor active volcanoes.
Hawaii Center for Volcanology. Devoted to Loihi volcano, the newest volcano in the Hawaiian chain.
Soufrière Hills, Montserrat, West Indies. This site has images and updates on the continuing volcanism on this Caribbean island.

CD-ROM Exploration

> Exploring your *Earth Systems Today* CD-ROM will add to your understanding of the material in this chapter.

TOPIC: EARTH'S DISASTERS

MODULE: VOLCANOES

Explore activities in this module to see if you can discover the following for yourself:

Using aspects of this activity, examine for yourself the relationship of magma chemistry and eruptive style.

Using aspects of this activity, examine the global distribution of volcanoes and the nature of volcanic landforms.

Select a volcano from among those available in this activity and study its geologic setting, geologic hazards, geologic history, and current eruptive activity.

Chapter 5

Weathering and erosion are responsible for this spectacular scenery at Bryce Canyon National Park, Utah.

OBJECTIVES

At the end of this chapter, you will have learned that

- Weathering causes physical and chemical changes in Earth materials, whereas erosion removes materials from the weathering site.
- Several physical processes disaggregate Earth materials but do not change their composition.
- Several factors account for the rate at which chemical processes change the composition of Earth materials.

- Weathering yields the raw materials for both soils and sedimentary rocks.
- Different types of soil form in response to weathering under specific circumstances.
- A variety of factors are important in determining each type of soil and its fertility.
- Soil erosion and physical and chemical deterioration of soils cause many problems, including decreases in productivity.

- Phyical and chemical weathering processes are important in the origin or concentration of some mineral resources.

Weathering, Erosion, and Soil

Prologue

Several geologic processes including volcanism, glaciation, and shoreline erosion have yielded many areas of exceptional scenery in North America. Some of the most striking examples of scenery produced by the combined effects of weathering (mechanical and chemical alteration) and erosion of Earth materials are found in several of the semiarid western states. All rocks and minerals at or near the surface are constantly attacked by physical forces such as the wedging action of water freezing in cracks and a variety of chemical processes as well as the activities of organisms. However, the effects of these processes are not the same everywhere, because rocks vary in their resistance to change. As a result, weathering and erosion yield a number of oddly shaped features going by such fanciful names as *spires, splinters, monuments, arches,* and *hoodoos.*

A particularly striking example of the effects of weathering and erosion is found in Bryce Canyon National Park, Utah, where badlands topography is well represented. Brilliantly colored rocks have been intricately sculpted to form mazes of interconnected gullies and pillars and spires of rock (see the chapter opening photo). Scattered localities of similar badlands topography are present in many other areas from Alberta, Canada, to Arizona and New Mexico.

Badlands develop in dry areas with sparse vegetation and nearly impermeable yet easily eroded rocks. Rain falling on such unprotected rocks rapidly runs off and intricately dissects the surface by forming numerous closely spaced, small gullies and steep ravines separated by sharp angular slopes, fluted ridges, arches, and pinnacles. The rocks at Bryce Canyon National Park are commonly described as limy siltstones, meaning they are composed of silt-sized (1/256- to 1/16-mm) particles with calcite ($CaCO_3$) acting as a cement to hold the particles together. These rocks formed 40 to 50 million years ago in a lake and have subsequently been uplifted along a large fracture and now form what are known as the Pink Cliffs. Incidentally, Bryce Canyon is not a canyon—rather, it is the eroded eastern margin of a high, fairly flat area known as the Paunsaugunt Plateau. The Paiute referred to the area as "red rocks standing like men in a bend-shaped canyon."

Another area to see the remarkable effects of weathering and erosion is in Arches National Park, Utah. Here the rocks, mostly sandstone, have yielded to weathering and erosion to produce a number of isolated spires and balanced rocks (Figure 5.1), as well as the famous arches for which the park was named. Unfortunately, the term *arch* has been used to describe a variety of features with different origins, but here we restrict it to mean an opening through a wall of rock that has formed in response to weathering and erosion.

Figure 5.1
This isolated spire and the rounded remnants of rocks on the right in Arches National Park, Utah, are called *hoodoos* and *goblins*. They resulted from differential weathering and erosion of sandstone.

The origin of the arches in the park is well known, and, in fact, they are forming at present. Probably the most famous of the arches is Delicate Arch (Figure 5.2a), which represents an advanced stage of arch formation and evolution. Arches form when weathering and erosion along parallel fractures leave a slender fin of rock. Some parts of these fins are more susceptible to weathering and erosion than others, and as the sides are altered and eroded, a recess may form. If it does, eventually pieces of the unsupported rock at the top of the recess fall away, forming an arch as the original recess is enlarged (Figure 5.2b). Thus arches are simply the remnants of rock fins formed by weathering and erosion along fractures.

The fact that arches continue to evolve is apparent from historical observations. In 1940, for instance, Skyline Arch was enlarged when a large block fell from its underside. The park also has a number of examples of collapsed arches, which commonly leave pinnacles and spires when they fall. Both Arches and Bryce Canyon National Parks as well as several others with similar features are well worth visiting.

(a)

(b)

Figure 5.2
(a) Delicate Arch in Arches National Park, Utah, formed by differential weathering and erosion of fractured sedimentary rocks, is 9.7 m wide and 14 m high. (b) Hole-in-the-Wall or Baby Arch shows the early development of an arch in a fin of rock. It measures 7.6 m wide and 4.5 m high.

INTRODUCTION

Weathering, defined as the physical breakdown (disintegration) and chemical alteration (decomposition) of rocks and minerals at or near Earth's surface, is such a pervasive phenomenon that many people take it for granted or completely overlook it. Nevertheless, it takes place continuously on all Earth materials and on rocklike materials such as the concrete in sidewalks, foundations, and bridges. The phenomenon of weathering includes all those physical and chemical processes whereby minerals and rocks alter so that they are more nearly in equilibrium with a new set of environmental conditions. For instance, many rocks form within the crust where temperatures and pressures are high and little or no water or oxygen is present. At or near the surface these same rocks are exposed to low temperatures and pressures, atmospheric gases, water, acids, and the activities of organisms. In short, weathering provides another example of the interactions among Earth's systems.

Geologists are interested in weathering because it is an essential part of the rock cycle (see Figure 1.17). The **parent material,** or rock being weathered, is disaggregated to form smaller pieces, and some of its constituent minerals are altered or dissolved and removed from the weathering site. This removal of weathered materials, known as **erosion,** is accomplished by running water, wind, moving masses of ice called *glaciers,* or waves and marine currents that **transport** the materials elsewhere to be deposited as *sediment,* which may become *sedimentary rock.* Transport, deposition, and the conversion of sediment to sedimentary rock is discussed more fully in Chapter 6. Some weathered materials, though, are further altered to form *soil.* Thus weathering provides the raw materials for both sedimentary rocks and soils, and is also responsible for the origin and enrichment of some mineral resources. For instance, the ore of aluminum forms by intense weathering in the tropics.

Even though weathering takes place continuously, its effects commonly vary from area to area and even within the same body of rock. Rocks are not structurally and chemically homogeneous throughout, so **differential weathering** takes place, which yields uneven surfaces. And coupled with *differential erosion,* that is, variable rates of erosion, some unusual and even bizarre features may develop (Figure 5.3, and see the Prologue). Indeed, some of the picturesque and peculiar features seen in such areas as Bryce Canyon National Park, Utah; Badlands National Park, South Dakota; and Dinosaur National Park, Alberta, Canada, resulted from differential weathering and erosion.

Two types of weathering are recognized, *mechanical* and *chemical.* Both proceed simultaneously at the original weathering site, during erosion and transport,

(a)

(b)

Figure 5.3
(a) These knobs and spires of rock in Austria resulted from differential weathering and erosion. (b) Honeycomb weathering of rocks at Pebble Beach, California. In coastal areas this kind of weathering results from physical disaggregation of mineral grains. The partitions between cavities are protected by coatings of microscopic algae.

and even in the areas where weathered materials are deposited as sediment. In short, all surface materials are subjected to weathering, although one type of weathering might predominate over the other. Mechanical and chemical weathering are discussed separately in the following sections simply for convenience.

HOW DO PHYSICAL PROCESSES DISAGGREGATE EARTH MATERIALS?

Mechanical weathering takes place when physical forces break Earth materials into smaller pieces that retain the chemical composition of the parent material. Granite, for instance, might be mechanically weathered to produce smaller pieces of granite, or disintegration might liberate individual minerals from it (Figure 5.4). Thus mechanically weathered granite yields granite rock fragments and mineral fragments of quartz, potassium feldspars, plagioclase feldspars, and several other minerals in lesser quantities, all of which have the same chemical composition as they did in the parent material. The physical processes responsible for mechanical weathering include frost action, pressure release, thermal expansion and contraction, salt crystal growth, and the activities of organisms.

Frost Action

Frost action involving repeated freezing and thawing of water in cracks and pores in rocks is particularly effective where temperatures commonly fluctuate above and below freezing. In the high mountains of the western United States and Canada, frost action is effective even during summer months. As one would expect, it is of little or no importance in the tropics or where water is permanently frozen.

When water seeps into a crack and freezes, it expands by about 9% and exerts great force on the walls of the crack, thereby widening and extending it by **frost wedging.** As a result of repeated freezing and thawing, pieces of rock are eventually detached from the parent material (Figure 5.5). Frost wedging is particularly effective if the crack is convoluted. If the crack is a simple wedge-shaped opening, much of the force of expansion is released up toward the surface.

The debris produced by frost wedging and other weathering processes in mountains commonly accumulates as large cones of **talus** lying at the bases of slopes (Figure 5.6). The debris forming the talus is simply angular pieces of rock from a larger body that has been mechanically and to a lesser degree chemically weathered. Most rocks have a system of fractures called *joints* along which frost action is particularly effective. Water seeps along the joint surfaces and eventually wedges pieces of rock loose; these then tumble downslope to accumulate with other loosened rocks.

In the phenomenon known as **frost heaving,** a mass of sediment or soil undergoes freezing, expansion, and actual lifting, followed by thawing, contraction, and lowering of the mass. Frost heaving is particularly evident where water freezes beneath roadways and sidewalks.

Pressure Release

The mechanical weathering process called **pressure release** is especially evident in rocks that formed as deeply buried intrusive bodies such as batholiths, but it

Figure 5.4
Exposure of granitic rock altered mostly by mechanical weathering. The cones of material in the foreground consist of individual mineral grains and granitic rock fragments. In fact, most of the exposed rocks have been so thoroughly weathered that only a few rounded masses of rock appear unaltered.

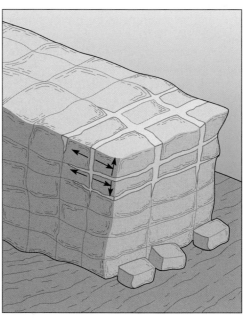

Figure 5.5
Frost wedging occurs when water seeps into cracks and expands as it freezes. Angular pieces of rock are pried loose by repeated freezing and thawing.

occurs in other types of rocks as well. When a batholith forms, the magma crystallizes under tremendous pressure (the weight of the overlying rock) and is stable under these pressure conditions. If the batholith is uplifted and the overlying rock is eroded, the pressure is reduced. But the rock contains energy that is released by outward expansion and the formation of **sheet joints,** large fractures that more or less parallel the rock surface (Figure 5.7). Slabs of rock bounded by sheet joints may slip, slide, or spall (break) off of the host rock—a process known as **exfoliation**—and accumulate as talus. The large rounded domes of rock resulting from this

Figure 5.6
Accumulation of talus at the base of a slope in Lassen Volcanic National Park, California. Notice that the source rocks for the talus are highly fractured.

Figure 5.7
Slabs of granitic rock bounded by sheet joints, in the Sierra Nevada of California.

(a)

(b)

Figure 5.8
Good examples of exfoliation domes. (a) North Dome (left) and Basket Dome in Yosemite National Park, California. (b) Stone Mountain, Georgia.

process are **exfoliation domes;** good examples are found in Yosemite National Park, California, and at Stone Mountain, Georgia (Figure 5.8).

The fact that solid rock can expand and produce fractures is counterintuitive, but is nevertheless a well-known phenomenon. In deep mines, masses of rock sud-denly detach from the sides of the excavation, often with explosive violence. Spectacular examples of these *rock bursts* have been recorded in deep mines, where they and the related but less violent phenomenon called *popping* pose a danger to mine workers. In South Africa, about 20 miners are killed by rock bursts every year.

In some quarrying operations,* the removal of surface materials to a depth of only 7 or 8 m has led to the formation of sheet joints in the underlying rock (Figure 5.9). At quarries in Vermont and Tennessee, the excavation of marble exposed rocks that were formerly buried and under great pressure. When the overlying rock was removed, the marble expanded and sheet joints formed. Some slabs of rock bounded by sheet joints burst so violently that quarrying machines weighing more than a ton were thrown from their tracks, and some quarries had to be abandoned because fracturing rendered the stone useless.

Expansion and Contraction Caused by Heating and Cooling

During **thermal expansion and contraction,** the volume of rocks changes in response to heating and cooling. In a desert, where the temperature may vary as much as 30°C in one day, rocks expand when heated and contract as they cool. Rock is a poor conductor of heat, so its outside heats up more than its inside; the surface expands more than the interior, producing stresses that may cause fracturing. Furthermore, dark minerals absorb heat faster than light-colored minerals, so differential expansion occurs even between the mineral grains of some rocks.

Repeated thermal expansion and contraction is a common phenomenon, but are the forces generated sufficient to overcome the internal strength of a rock? Experiments in which rocks are heated and cooled repeatedly to simulate years of such activity indicate that thermal expansion and contraction is not an important agent of mechanical weathering.† Despite these experimental results, some rocks in deserts do indeed appear to show the effects of this process.

Daily temperature variation is the most common cause of alternate expansion and contraction, but these changes take place over periods of hours. In contrast,

fire can cause very rapid expansion. During a forest fire, rocks may heat very rapidly, especially near the surface because they conduct heat so poorly. The heated surface layer expands more rapidly than the interior, and thin sheets paralleling the rock surface become detached.

Growth of Salt Crystals

Under some circumstances, salt crystals forming from solution can cause disaggregation of rocks. Growing crystals exert enough force to widen cracks and crevices or dislodge particles in porous, granular rocks such as sandstone. Even in crystalline rocks such as granite, **salt crystal growth** may pry loose individual mineral grains. To the extent that salt crystal growth produces forces that expand openings in rocks, it is similar to frost wedging. Most salt crystal growth occurs in hot arid areas, although it probably affects rocks in some coastal regions as well.

Organisms and Mechanical Weathering

Animals, plants, and bacteria all participate in the mechanical and chemical alteration of rocks. Burrowing animals, such as worms, termites, reptiles, rodents, and many others, constantly mix soil and sediment particles and bring material from depth to the surface where further weathering may occur. Even materials ingested by worms are further reduced in size, and animal burrows give gases and water easier access to greater depths. The roots of plants, especially large bushes and trees, wedge themselves into cracks in rocks and further widen them (Figure 5.10). Tree roots growing under or through sidewalks and foundations may do considerable damage.

---

*A *quarry* is a surface excavation, generally for the extraction of building stone.

†Thermal expansion and contraction may be a significant mechanical weathering process on the moon, where extreme temperature changes occur quickly.

Figure 5.10
This small tree is growing in the cracks between columns in a lava flow. Its roots will enlarge the crack, thus contributing to mechanical weathering of the rock.

Figure 5.11
The yellowish and black irregular masses on these rocks in Switzerland are lichens (composite organisms consisting of fungi and algae). Lichens derive their nutrients from the rock, so lichen-covered rocks are chemically weathered more rapidly than lichen-free rocks.

CHEMICAL WEATHERING— DECOMPOSITION OF EARTH MATERIALS

Chemical weathering includes those processes whereby rock materials are decomposed by alteration of parent material. Several clay minerals (sheet silicates) form by the chemical and structural alteration of other minerals such as potassium feldspars and plagioclase feldspars, which are framework silicates (see Figure 2.15). Other minerals are completely decomposed during chemical weathering as their ions are taken in solution, but chemically stable ones might simply be liberated from the parent material. Important agents of chemical weathering include atmospheric gases, especially oxygen, water, and acids. Organisms also play an important role. Rocks with lichens (composite organisms consisting of fungi and algae) on their surfaces undergo more extensive chemical alteration than lichen-free rocks (Figure 5.11). In addition, plants remove ions from soil water and reduce the chemical stability of soil minerals, and plant roots release organic acids. Other chemical weathering processes include solution, oxidation, and hydrolysis.

Solution

During **solution** the ions of a substance separate in a liquid, and the solid substance dissolves. Water is a remarkable solvent because its molecules have an asymmetric shape; they consist of one oxygen atom with two hydrogen atoms arranged so that the angle between the two hydrogens is about 104 degrees (Figure 5.12a). Because of this asymmetry, the oxygen end of the molecule retains a slight negative electrical charge, whereas the hydrogen end retains a slight positive charge. When a soluble substance such as the mineral halite (NaCl) comes in contact with a water molecule, the positively charged sodium ions are attracted to the negative end of the water molecule, and the negatively charged chloride ions are attracted to the positively charged end of the water molecule (Figure 5.12b). Thus, ions are liberated from the crystal structure, and the solid dissolves.

Most minerals are not very soluble in pure water because the attractive forces of water molecules are not sufficient to overcome the forces between particles in minerals. The mineral calcite ($CaCO_3$), the major constituent of the sedimentary rock limestone and the metamorphic rock marble, is practically insoluble in pure water but rapidly dissolves if a small amount of acid is present. An easy way to make water acidic is by dissociating the ions of carbonic acid. That is, the ions become separated as carbonic acid breaks down into other substances.

$$H_2O + CO_2 \rightleftharpoons H_2CO_3 \rightleftharpoons H^+ + HCO_3^-$$

WATER CARBON CARBONIC HYDROGEN BICARBONATE
        DIOXIDE   ACID   ION   ION

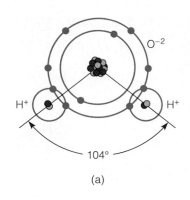

(a)

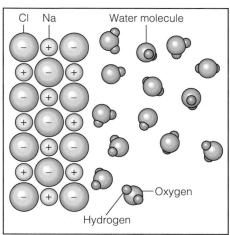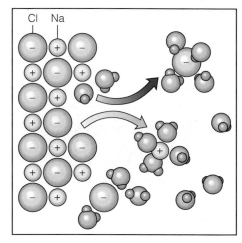

Figure 5.12
(a) The structure of a water molecule. The asymmetric arrangement of hydrogen atoms causes the molecule to have a slight positive electrical charge at its hydrogen end and a slight negative charge at its oxygen end. (b) Solution of sodium chloride (NaCl), the mineral halite, in water. Note that the sodium atoms are attracted to the oxygen end of a water molecule, whereas chloride ions are attracted to the hydrogen end of the molecule.

(b)

According to this chemical equation, water and carbon dioxide combine to form *carbonic acid,* a small amount of which dissociates to yield hydrogen and bicarbonate ions. The concentration of hydrogen ions determines the acidity of a solution; the more hydrogen ions present, the stronger the acid.

Carbon dioxide from several sources may combine with water and react to form acid solutions. The atmosphere is mostly nitrogen and oxygen, but about 0.03% is carbon dioxide, causing rain to be slightly acidic. Human activities have added gases to the atmosphere that contribute to the problem of acid rain (see Perspective 5.1). Decaying organic matter and the respiration of organisms produce carbon dioxide in soils, so groundwater is also generally slightly acidic. Arid regions have sparse vegetation and few soil organisms, so groundwater has a limited supply of carbon dioxide and tends to be alkaline rather than acidic—that is, it has a low concentration of hydrogen ions.

Whatever the source of carbon dioxide, once an acidic solution is present, calcite rapidly dissolves according to the following reaction:

$$CaCO_3 + H_2O + CO_2 \rightleftharpoons Ca^{+2} + 2HCO_3^-$$

CALCITE   WATER   CARBON   CALCIUM   BICARBONATE
                    DIOXIDE   ION   ION

Because of the dissociation of the ions in carbonic acid, this reaction may also be written as

$$CaCO_3 + H^+ + HCO_3^- \rightleftharpoons Ca^{+2} + 2HCO_3^-$$

CALCITE   HYDROGEN   BICARBONATE   CALCIUM   BICARBONATE
           ION   ION   ION   ION

The dissolution of the calcite in limestone and marble has had dramatic effects in many places, ranging from small cavities to large caverns such as Mammoth Cave in Kentucky and Carlsbad Caverns in New Mexico (see Chapter 16). The varying solubility of limestone and other rocks with the carbonate radical (CO_3) such as dolostone is well demonstrated by the bold cliffs they form in semiarid and arid regions where groundwater is alkaline (Figure 5.13). Subdued exposures of these rocks are more common in humid regions with their acidic groundwater.

Oxidation

The term **oxidation** has a variety of meanings for chemists, but in chemical weathering it refers to reactions with oxygen to form an oxide (one or more metallic elements combined with oxygen) or, if water is present, a hydroxide (a metallic element or radical com-

Figure 5.13
Exposure of the Bighorn Dolomite in Wyoming. Limestone and dolostone are resistant to chemical weathering in the semiarid to arid West where groundwater is alkaline. The same rocks in more humid climates tend to have more subdued exposures.

bined with OH). For example, iron rusts when it combines with oxygen to form the iron oxide hematite:

$$4Fe + 3O_2 \rightarrow 2Fe_2O_3$$

IRON   OXYGEN   IRON OXIDE
(HEMATITE)

Of course, atmospheric oxygen is abundantly available for oxidation reactions, but oxidation is generally a slow process unless water is present. Most oxidation is carried out by oxygen dissolved in water.

Oxidation is very important in the alteration of ferromagnesian silicates such as olivine, pyroxenes, amphiboles, and biotite. Iron in these minerals combines with oxygen to form the reddish iron oxide hematite (Fe_2O_3) or the yellowish or brown hydroxide limonite ($FeO(OH) \cdot nH_2O$). The yellow, brown, and red colors of many soils and sedimentary rocks are caused by the presence of small amounts of hematite or limonite (see Chapter 6).

An oxidation reaction of particular concern in some areas is the oxidation of iron sulfides such as the mineral pyrite (FeS_2). Pyrite is commonly associated with coal, so in mine tailings* pyrite oxidizes to form sulfuric acid (H_2SO_4) and iron oxide. Acid soils and waters in coal-mining areas are produced in this manner and present a serious environmental hazard (Figure 5.14).

Hydrolysis

Hydrolysis is the chemical reaction between the hydrogen (H^+) ions and hydroxyl (OH^-) ions of water and a

---

Tailings are the rock debris of mining; they are considered too poor for further processing and are left as heaps on the surface.

Figure 5.14
The oxidation of pyrite in mine tailings forms acid water, as in this small stream. More than 11,000 km of U.S. streams, mostly in the Appalachian region, are contaminated by abandoned coal mines that leak sulfuric acid.

mineral's ions. In hydrolysis, hydrogen ions actually replace positive ions in minerals. Such replacement changes the composition of minerals by liberating soluble compounds and iron that then may be oxidized.

As an illustration of hydrolysis, consider the chemical alteration of feldspars. Potassium feldspars such as orthoclase ($KAlSi_3O_8$) are common in many rock types, as are the plagioclase feldspars (which vary in composition from $CaAl_2Si_2O_8$ to $NaAlSi_3O_8$). All feldspars are framework silicates, but when altered, they yield materials in solution and clay minerals, such as kaolinite, which are sheet silicates.

The chemical weathering of potassium feldspar by hydrolysis occurs as follows:

$$2KAlSi_3O_8 + 2H^+ + 2HCO_3^- + H_2O \rightarrow$$

ORTHOCLASE   HYDROGEN   BICARBONATE   WATER
ION   ION

$$Al_2Si_2O_5(OH)_4 + 2K^+ + 2HCO_3^- + 4SiO_2$$

CLAY (KAOLINITE)   POTASSIUM   BICARBONATE   SILICA
ION   ION

Perspective 5.1

Industrialization and Acid Rain

One consequence of industrialization is atmospheric pollution. Several of the most industralized nations, including the United States, Canada, and the former Soviet Union, have reduced their emissions into the atmosphere, but many developing nations continue to increase theirs. Some of the results of atmospheric pollution include smog, possible disruption of the ozone layer, global warming, and acid rain.

Acidity is an indication of hydrogen ion concentration and is measured on the pH scale (Figure 1). A pH of 7 is neutral, whereas values less than 7 indicate acidic conditions and values greater than 7 indicate alkaline, or basic, conditions. Normal rainfall has a pH of about 5.6, making it slightly acidic. Acid rain is defined as rainfall with a pH value of less than 5.0. Besides acid rain, one may experience acid snow in colder regions and acid fog with pH values as low as 1.7 in some industrialized areas.

Recall that water and carbon dioxide in the atmosphere react to form carbonic acid that dissociates and yields hydrogen ions and bicarbonate ions. The effect of this reaction is that all rainfall is slightly acidic. Thus, acid rain is the direct result of the self-cleansing nature of the atmosphere; that is, many suspended particles of gases in the atmosphere are soluble in water and are removed from the atmosphere during precipitation.

Several natural processes, including volcanism and the metabolism of soil bacteria, introduce gases into the atmosphere that cause acid rain. Human activities, however, produce added atmospheric stress. The burning of fossil fuels, for instance, has added excess carbon dioxide to the atmosphere. Nitrogen oxide from internal combustion engines and nitrogen dioxide, which is formed in the atmosphere from nitrogen oxide, react to form nitric acid. Although carbon dioxide and nitrogen gases contribute to acid rain, the greatest culprit is sulfur dioxide, which is primarily released by burning coal that contains sulfur. When coal is burned, sulfur is oxidized to form sulfur dioxide (SO_2):

$$S \text{ (in coal)} + O_2 \text{ (gas)} \rightarrow SO_2 \text{ (gas)}$$

Sulfur dioxide is released into the atmosphere from coal-burning facilities, where it reacts with oxygen to form sulfur trioxide (SO_3):

$$2SO_2 \text{ (gas)} + O_2 \text{ (gas)} \rightarrow 2SO_3 \text{ (gas)}$$

And finally, sulfur trioxide reacts with water droplets in the atmosphere to form sulfuric acid (H_2SO_4), the main component of acid rain:

$$SO_3 \text{ (gas)} + H_2O \text{ (liquid)} \rightarrow H_2SO_4$$

The phenomenon of acid raid was first recognized by Robert Angus Smith in England in 1872, about a century after the beginning of the Industrial Revolution. Not until 1961, however, did acid rain become a public environmental concern. At that time, it was realized that acid rain is corrosive and irritating, kills vegetation, and has a detrimental effect on surface waters. Since then, the effects of acid rain have been recognized in Europe (especially in eastern Europe where so much coal is burned), the eastern United States, and eastern Canada. During the past three decades, acid rain has been getting worse, particularly in Europe and eastern North America (Figure 2). During the last decade, the developed countries have made efforts to reduce the impact of acid rain; in the United States the Clean Air Act of 1990 outlined specific steps to reduce the emissions of pollutants that cause acid rain.

The areas most affected by acid rain invariably lie downwind from coal-burning power plants or other industries that emit sulfur gases. Chemical plants and smelters (plants where metal ores are refined) discharge large quantities of sulfur gases and other substances such as heavy metals. The effect of acid rain in these areas may be modified by the local geology. If an area

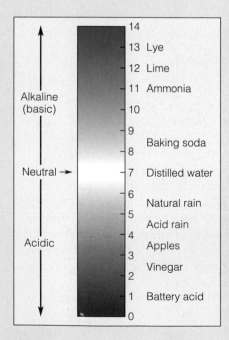

Figure 1
The pH scale. Values less than 7 are acidic, whereas those greater than 7 are alkaline. This is a logarithmic scale, so a decrease of one unit is a tenfold increase in acidity.

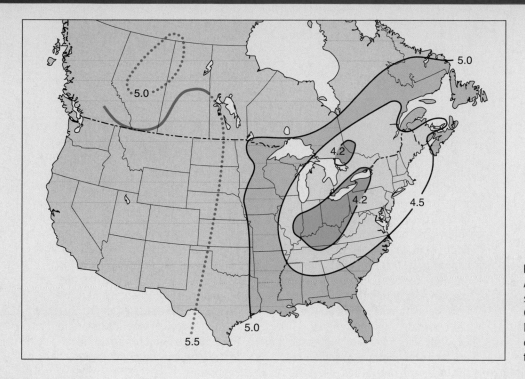

Figure 2
Average pH values for rain and snow in the United States and Canada. The area encompassed by a line has a pH value equal to or less than the value shown on the line.

is underlain by limestone or alkaline soils, acid rain tends to be neutralized by the limestone or soil. Areas underlain by granite, on the other hand, are acidic to begin with and have little or no effect on the rain.

The effects of acid rain vary. Small lakes become more acid as they lose the ability to neutralize acid rainfall. As the lakes increase in acidity, various types of organisms disappear, and, in some cases, all life-forms eventually die. Acid rain also causes increased weathering of limestone and marble (recall that both are soluble in weak acids) and, to a lesser degree, sandstone. Such effects are particularly visible on buildings, monuments, and tombstones; a notable example is Gettysburg National Military Park in Pennsylvania, which lies in an area that receives some of the most acidic rain in the country.

Although industries that emit sulfur gases have clear effects on vegetation as much as 80 km downwind, some people have questioned whether acid rain has much effect on forests and crops distant from these sources. But many forests in the eastern United States show signs of stress that cannot be attributed to other causes. In Germany's Black Forest, the needles of

firs, spruce, and pines are turning yellow and falling off.

Millions of tons of sulfur dioxide are released yearly into the atmosphere in the United States, mostly from coal-burning power plants. Power plants built before 1975 have no emission controls, but the problems they pose must be addressed if emissions are to be reduced to an acceptable level. The most effective way to reduce emissions from these older plants is with flue-gas desulfurization, a process that removes up to 90% of the sulfur dioxide from exhaust gases. Flue-gas desulfurization has some drawbacks. One is that some plants are simply too old to be profitably upgraded. Other problems with flue-gas desulfurization include disposal of sulfur wastes, the lack of control on nitrogen gas emissions, and reduced efficiency of the power plant, which must burn more coal to make up the difference.

Other ways to control emissions include burning low-sulfur coal, fluidized bed combustion, and the conservation of electricity; the less electricity used, the lower the emissions of pollutants. Natural gas contains practically no sulfur, but converting to this alternate energy source would require the installation of expensive new furnaces in existing plants.

Acid rain, like global warming, is a worldwide problem that knows no national boundaries. Wind may blow pollutants from the source in one country to another where the effects are felt. For instance, much of the acid rain in eastern Canada actually comes from sources in the United States.

Developed nations have the economic resources to reduce emissions, but many underdeveloped nations cannot afford to do so. Furthermore, many nations have access to only high-sulfur coal and cannot afford to install flue-gas desulfurization devices. Nevertheless, acid rain can be controlled only by the cooperation of all nations contributing to the problem.

Using Geologic Analyses to Suggest a New Date for the Great Sphinx

Robert M. Schoch received undergraduate degrees in geology and anthropology at George Washington University. He subsequently earned a Ph.D. in geology and geophysics from Yale University. Since 1984 he has been a faculty member at the College of General Studies, Boston University, where, besides his teaching and work related to the Great Sphinx, he has continued to do research and write about various aspects of stratigraphy, paleontology, and environmental science.

The Great Sphinx appears to stand guard over the three Fourth Dynasty (circa 2500 B.C.) pyramids, the only survivors among the seven wonders of the ancient world, on the Giza Plateau west of Cairo, Egypt. For thousands of years, this huge enigmatic statue carved from solid bedrock limestone has inspired awe in observers and raised more questions than answers. What does the Sphinx represent? Why was it carved? How old is it? There are no inscriptions that unambiguously identify when the Great Sphinx was carved. Indeed, the most important inscriptions discussing the Sphinx, which themselves are subject to a considerable range of interpretations, were composed at least 1000 years after the latest possible date for the carving of the Sphinx. By that time, circa 1400 B.C., the Sphinx was already considered an "ancient" structure whose origins were obscure.

Egyptologists have had a range of opinions concerning the age of the Sphinx. Some late nineteenth- and early twentieth-century researchers suggested that the Sphinx may well antedate the pyramids, but most Egyptologists since the 1930s have accepted the opinion of the late Selim Hassan that the Sphinx was probably erected during the reign of the pharaoh Khafre (also known as Chephren), builder of the second pyramid, around 2500 B.C. However, Hassan warned that his attribution of the Sphinx was based solely on circumstantial evidence and deduction. Hassan placed great significance on the fact that the Sphinx is situated due east of Khafre's pyramid. Hassan speculated that the Sphinx may have been built as part of Khafre's original pyramid/funerary complex. Egyptologists have also suggested that the face of the Sphinx may resemble the face seen on statues of Khafre. Many later Egyptologists accepted Hassan's attribution unquestioningly and dogmatically.

It was with this background that I entered the controversy over the dating of the Sphinx. I have attempted to apply geologic techniques to arrive at a possible temporal context for the Sphinx. The nature of the surface weathering and erosion patterns and subsurface patterns of weathering (elucidated through a series of low-energy seismic surveys carried out around the base of the statue) suggest that the Sphinx was exposed to long, heavy rains at some point in its existence. I have therefore concluded that the earliest portions of the body of the Sphinx were carved out well prior to the dry, arid conditions prevailing on the Giza Plateau, which sits on the edge of the Sahara Desert, since the third millennium B.C. Correlating the known paleoclimatic history of the region with the nature of the weathering and erosional features seen on the Sphinx, I believe that the statue may have its origins around 5000 B.C. or earlier. This would make it roughly contemporaneous with parts of the oldest towers, walls, and moats of the biblical city of Jericho in nearby Palestine.

Recent research has seriously questioned much of the evidence that had been used to date the Sphinx to 2500 B.C. For instance, based on careful analyses by forensic experts, the battered face of the monument does not match the face seen on statues of Khafre after all. Considerable evidence indicates that parts of the Sphinx were repaired and even recarved numerous times, including during Old Kingdom times (circa 2500–2200 B.C.), as well as over the next several thousands of years. The date of 2500 B.C. seems not to pertain to the oldest parts of the Sphinx but to a major restoration effort when a much older structure (which may not have even looked like the Sphinx) was refurbished and reused by the Fourth Dynasty Egyptians. My redating of the Sphinx extends its origins back in time but does not deny the use of the statue by later generations of Egyptians.

Still, my research and hypotheses have aroused a considerable amount of controversy among Egyptologists and archaeologists. Many have refused to accept my analyses and have attempted to hypothesize ways to explain my data that would maintain the age of the Sphinx at about 2500 B.C. This is the nature of science: Disagreements occur, more data must be sought, further analyses performed, and ideas modified. I have suggested that new dating methods, based on measuring the concentration of certain isotopes that are produced on the surface of a carved rock by the bombardment of cosmic rays, might be applied to the Sphinx. To carry out such studies would entail collecting minute samples of rock from the statue, something that the Egyptian authorities are hesitant to allow. Definitively pinning down the date and constructional history of the Great Sphinx will ultimately require the collection of more data and further analyses. Thus far, however, I personally continue to believe that, taken in total, the simplest hypothesis explaining all current lines of evidence bearing on the problem is that the earliest portions of the Sphinx date back to at least circa 5000 B.C. ∎

In this reaction, hydrogen ions attack the ions in the orthoclase structure, and some liberated ions are incorporated in a developing clay mineral, while others simply go into solution. On the right side of the equation is excess silica that would not fit into the crystal structure of the clay mineral. This dissolved silica is an important source of cement that binds together the particles in some sedimentary rocks (see Chapter 6).

Plagioclase feldspars are also altered by hydrolysis, but in this case soluble calcium and sodium compounds form rather than potassium compounds. Otherwise the reaction is the same. In fact, these dissolved compounds, particularly calcium compounds, are what make water hard. Calcium in water is a problem because it inhibits the reaction of detergents with dirt and precipitates as scaly mineral matter in water pipes and water heaters (see Chapter 16).

Figure 5.15
Fractured rocks are more susceptible to chemical weathering than rocks with fewer fractures, if other factors are the same. Notice that the fractures in these granitic rocks have been accentuated by weathering.

HOW FAST DOES CHEMICAL WEATHERING TAKE PLACE?

Chemical weathering processes operate on the surface of particles, so they alter rocks and minerals from the outside inward. In fact, it is not uncommon to see rocks with a rind of weathered material near the surface but completely unaltered inside. The rate at which chemical weathering proceeds, however, depends on several factors. One is simply the presence or absence of fractures, because fluids seep along fractures, accounting for more intense chemical weathering along these surfaces (Figure 5.15). Thus, given the same rock type under similar conditions, the more fractures, the more rapid the chemical weathering. Of course other factors also control the rate of chemical weathering, including particle size, climate, and parent material.

Particle Size and the Rate of Chemical Weathering

Because chemical weathering affects particle surfaces, the greater the surface area, the more effective the weathering. It is important to realize that small particles have larger surface areas compared to their volume than do large particles. Notice in Figure 5.16 that a block measuring 1 m on a side has a total surface area of 6 m², but when the block is broken into particles measuring 0.5 m on a side, the total surface area increases to 12 m². And if these particles are all reduced to 0.25 m on a side, the total surface area increases to 24 m². Note that, although the surface area in this example increases, the total volume remains the same at 1 m³.

We can make two important statements regarding the block in Figure 5.16. First, as it is split into a number of smaller blocks, its total surface area increases. Second, the smaller any single block is, the more surface area it has compared to its volume. Because chemical weathering is a surface process, the fact that small objects have proportionately more surface area compared to volume than do large objects has profound implications. We can conclude that mechanical weathering, which reduces the size of particles, contributes to chemical weathering by exposing more surface area.

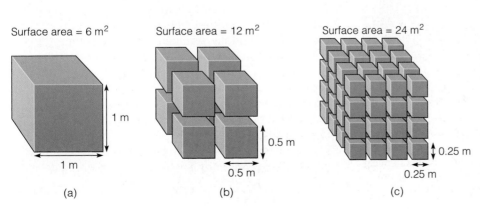

Surface area = 6 m²   Surface area = 12 m²   Surface area = 24 m²

1 m

1 m

0.5 m

0.5 m

0.25 m

0.25 m

(a)   (b)   (c)

Figure 5.16
Particle size and chemical weathering. As a rock is reduced into smaller and smaller particles, its surface area increases but its volume remains the same. In (a) the surface area is 6 m², in (b) it is 12 m², and in (c) 24 m², but the volume remains the same at 1 m³. Small particles have more surface area in proportion to their volume than do large particles.

Climate and Chemical Weathering

Chemical processes proceed more rapidly at high temperatures and in the presence of liquids. Accordingly, it is not surprising that chemical weathering is more effective in the tropics than in arid and arctic regions because temperatures and rainfall are high and evaporation rates are low (Figure 5.17). In addition, vegetation and animal life are much more abundant in the tropics. Consequently, the effects of weathering extend to depths of several tens of meters, but commonly extend only centimeters to a few meters deep in arid and arctic regions. You should realize, though, that chemical weathering goes on everywhere, except perhaps where earth materials are permanently frozen.

The Importance of Parent Material

Some rocks are more resistant to chemical alteration than others and are not altered as rapidly by chemical processes. The metamorphic rock quartzite, composed of quartz (SiO_2), is an extremely stable substance that alters very slowly compared with most other rock types. In contrast, rocks such as basalt, which contain large amounts of calcium-rich plagioclase and pyroxene minerals, decompose rapidly because these minerals are chemically unstable. In fact, the stability of common minerals is just the opposite of their order of crystallization in Bowen's reaction series (Table 5.1). The minerals that form last in this series are chemically stable, whereas

Table 5.1

Stability of Silicate Minerals

	FERROMAGNESIAN SILICATES	NONFERROMAGNESIAN SILICATES
Increasing Stability ↓	Olivine	Calcium plagioclase
	Pyroxene	
	Amphibole	Sodium plagioclase
	Biotite	Potassium feldspar
		Muscovite
		Quartz

those that form early are more easily altered by chemical processes.

One manifestation of chemical weathering is **spheroidal weathering** (Figure 5.18). In spheroidal weathering, a stone, even one that is rectangular to begin with, weathers to form a more spherical shape because that is the most stable shape it can assume. The reason? On a rectangular stone, the corners are attacked by weathering processes from three sides, and the edges are attacked from two sides, but the flat surfaces are weathered more or less uniformly (Figure 5.18). Consequently, the corners and edges are altered more rapidly, the material sloughs off them, and a more spherical shape develops. Once a spherical shape is present, all surfaces are weathered at the same rate.

The effects of spheroidal weathering can be observed in many rock bodies, particularly those having a rectangular pattern of fractures (joints) similar to those in Figure 5.15. Fluids seep along the joint surfaces, resulting in more intense weathering at the edges and corners of the rectangular blocks, thus yielding more nearly spherical objects. Fractured granitic rocks are especially susceptible to spheroidal weathering, but good examples can be found in all rock types (Figure 5.18).

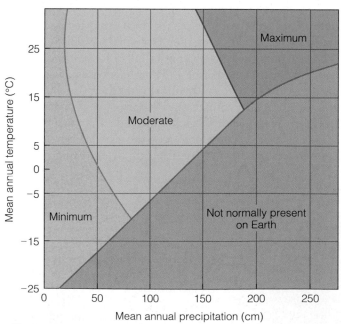

Figure 5.17
Relationships of chemical weathering rates and climate. Chemical weathering is at a maximum where temperature and rainfall are high, and at a minimum in arid environments whether hot or cold.

WHAT WOULD YOU DO?

Acid rain is one consequence of industrialization, but it is a problem that can be solved. In your community, a local copper smelter is obviously pouring out sulfur gases, thereby contributing to the problem both locally and regionally. As chairperson of a committee of concerned citizens, you make some recommendations to the smelter owners to clean up their emissions. What specific recommendations would you make? Suppose further that the smelter owners say that the remedies are too expensive and if they are forced to implement them, they will close the smelter. Would you still press forward even though your efforts might mean economic disaster for your community?

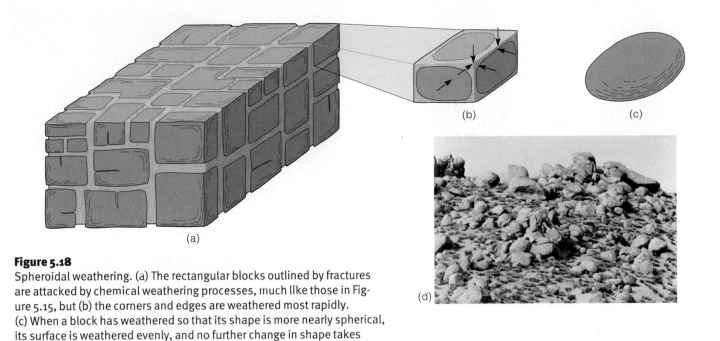

Figure 5.18
Spheroidal weathering. (a) The rectangular blocks outlined by fractures are attacked by chemical weathering processes, much like those in Figure 5.15, but (b) the corners and edges are weathered most rapidly. (c) When a block has weathered so that its shape is more nearly spherical, its surface is weathered evenly, and no further change in shape takes place. (d) Exposure of granitic rocks reduced to spherical boulders.

SOIL—ONE PRODUCT OF MECHANICAL AND CHEMICAL WEATHERING

A layer of **regolith** consisting of unconsolidated rock and mineral fragments covers most of Earth's land surface. *Regolith* is a collective term for sediment regardless of how it was deposited, as well as layers of pyroclastic materials and the residue formed in place by weathering. Some regolith that we call **soil** consists of weathered materials, air, water, and organic matter and can support vegetation. Soil is an essential link between the parent material below and life above. Almost all land-dwelling organisms depend on soil for their existence. Plants grow in soil from which they derive their nutrients and most of their water, whereas land-dwelling animals depend directly or indirectly on plants for nutrients.

Only about 45% of a good soil for farming or gardening is composed of weathered material, mostly sand, silt, and clay. Much of the remaining 55% is simply void spaces filled with either air or water, and a small but important amount consists of humus. **Humus** consists of carbon derived by bacterial decay of organic matter and is highly resistant to further decay. Even a fertile soil might contain as little as 5% humus, but it is nevertheless important as a source of plant nutrients and it enhances moisture retention. Furthermore, it gives the upper layers of many soils their dark color; soils with little humus are lighter colored and not very productive.

Some weathered materials in soils are simply sand- and silt-sized mineral grains, especially quartz, but other weathered materials may be present as well. These solid particles are important because they hold soil particles apart, allowing oxygen and water to circulate more freely. Clay minerals are also important constituents of soils and aid in the retention of water as well as supplying nutrients to plants. Soils with excess clay minerals, however, drain poorly and are sticky when wet and hard when dry.

Soils are commonly characterized as *residual* or *transported*. If a body of rock weathers and the weathering residue accumulates over it, the soil so formed is residual, meaning that it formed in place (Figure 5.19a). In contrast, transported soil develops on weathered material that has been eroded and transported from the weathering site and deposited elsewhere, such as on a stream's floodplain (Figure 5.19b). Many fertile transported soils of the Mississippi River valley and the Pacific Northwest developed on deposits of wind-blown dust called *loess*.

THE SOIL PROFILE

Observed in vertical cross section, a soil consists of distinct layers, or **soil horizons,** that differ from one an-other in texture, structure, composition, and color (Figure 5.19c). Starting from the top, the horizons are designated O, A, B, and C, but the boundaries between horizons are transitional rather than sharp. Because soil-forming processes begin at the surface and work downward, the upper layer of soil is more altered from the parent material than are the layers below.

Horizon O, which is generally only a few centimeters thick, consists of organic matter. The remains of plant materials are clearly recognizable in the upper part of horizon O, but its lower part consists of humus.

Horizon A, called *topsoil,* contains more organic matter than horizons B and C, and it is also characterized by intense biological activity because plant roots, bacteria, fungi, and animals such as worms are abundant. Threadlike soil bacteria give freshly plowed soil its earthy aroma. In soils developed over a long period of time, the A horizon consists mostly of clays and chemically stable minerals such as quartz. Water percolating down through horizon A dissolves soluble minerals and carries them away or down to lower levels in the soil by a process called *leaching,* so horizon A is also known as the **zone of leaching** (Figure 5.19).

Horizon B, or *subsoil,* contains fewer organisms and less organic matter than horizon A. Horizon B is known as the **zone of accumulation** because soluble minerals leached from horizon A accumulate as irregular masses. If horizon A is stripped away by erosion, leaving horizon B exposed, plants do not grow as well; and if horizon B is clayey, it is harder when dry and stickier when wet than other soil horizons.

Horizon C, the lowest soil layer, consists of partially altered parent material grading down into unaltered parent material (Figure 5.19c). In horizons A and B, the composition and texture of the parent material have been so thoroughly altered that it is no longer recognizable. In contrast, rock fragments and mineral grains of the parent material retain their identity in horizon C. Horizon C contains little organic matter.

WHAT FACTORS ARE IMPORTANT IN SOIL FORMATION?

All soils owe their existence to mechanical and chemical weathering, but soils differ in texture, color, thickness, and fertility. Accordingly, we are interested in the factors controlling the attributes and locations of various soils as well as the rate at which they form. Climate, parent material, organic activity, relief and slope, and time are the critical factors in soil formation. Complex interactions among these factors determine soil type, thickness, and fertility (Figure 5.20).

Climate and Soil

Soil scientists acknowledge that climate is the single most important factor influencing soil type and depth. Intense chemical weathering in the tropics generally yields deep soils from which most of the soluble mineral matter has been removed by leaching. In arctic and desert climates, soils tend to be thin, contain significant quantities of soluble minerals, and are composed mostly of materials derived by mechanical weathering (Figure 5.20).

A very general classification recognizes three major soil types characteristic of different climatic settings.

(a)

(b)

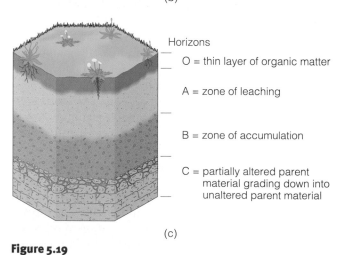

Horizons

O = thin layer of organic matter

A = zone of leaching

B = zone of accumulation

C = partially altered parent material grading down into unaltered parent material

(c)

Figure 5.19
(a) Residual soil developed on bedrock in California. (b) Transported soil exposed in a small gully in Nevada. (c) The soil horizons in a fully developed soil. Notice that in (a) horizons A, B, and C are easily seen, but in (b) horizon C is covered by debris.

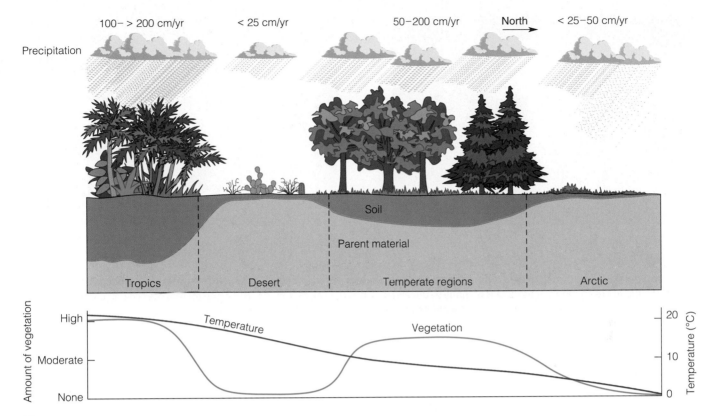

Figure 5.20
Schematic representation showing soil formation as a function of the relationships between climate and vegetation, which alter parent material over time. Soil-forming processes operate most vigorously where precipitation and temperatures are high.

Soils that develop in humid regions such as the eastern United States and much of Canada are **pedalfers,** a name derived from the Greek word *pedon,* meaning "soil," and from the chemical symbols for aluminum (Al) and iron (Fe). Because these soils form where abundant moisture is present, most of the soluble minerals have been leached from horizon A. Although it may be gray, horizon A is generally dark colored because of abundant organic matter, and aluminum-rich clays and iron oxides tend to accumulate in horizon B.

Soils found in much of the arid and semiarid western United States, especially the southwest, are **pedocals.** Pedocal derives its name in part from the first three letters of *calcite.* These soils contain less organic matter than pedalfers, so horizon A is generally lighter colored and contains more unstable minerals because of less intense chemical weathering. As soil water evaporates, calcium carbonate leached from above commonly precipitates in horizon B, where it forms irregular masses of *caliche* (Figure 5.21a). Precipitation of sodium salts in some desert areas where soil water evaporation is intense yields *alkali soils* that are so alkaline that they support few plants (Figure 5.21b).

Laterite is a soil formed in the tropics where chemical weathering is intense and leaching of soluble minerals is complete. These soils are red, commonly extend to depths of several tens of meters, and are composed largely of aluminum hydroxides, iron oxides, and clay minerals; even quartz, a chemically stable mineral, is generally leached out (Figure 5.22).

Although laterites support lush vegetation, as in tropical rain forests, they are not very fertile. The native vegetation is sustained by nutrients derived mostly from the surface layer of organic matter. When these soils are cleared of their native vegetation, the surface accumulation of organic matter is rapidly oxidized, and there is little to replace it. Consequently, when societies practicing slash-and-burn agriculture clear these soils, they can raise crops for only a few years at best. Then the soil is depleted of plant nutrients, the clay-rich laterite bakes brick hard in the tropical sun, and the farmers move on to another area, where the process is repeated.

Parent Material

The same rock type can yield different soils in different climatic regimes, and in the same climatic regime the same soils can develop on different rock types. Thus it seems climate is more important than parent material in determining the type of soil that develops. Nevertheless, rock type does exert some control. For example, the metamorphic rock quartzite will have a thin soil over it because it is chemically stable, whereas an adjacent body of granite will have a much deeper soil (Figure 5.23).

(a)

(b)

Figure 5.21
(a) This boulder has been turned over in order to show the scaly white material known as *caliche* that formed on its underside. Irregular masses of caliche are common in horizon B of many pedocals. (b) An alkali soil in Nevada. The white material is sodium carbonate or potassium carbonate. Notice that only a few hardy plants grow in this area.

Figure 5.22
Laterite, shown here in Madagascar, is a deep, red soil that forms in response to intense chemical weathering in the tropics.

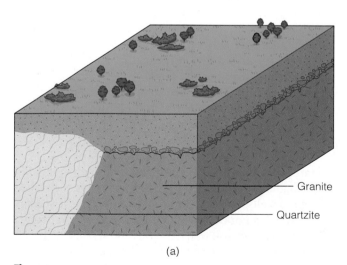
(a)

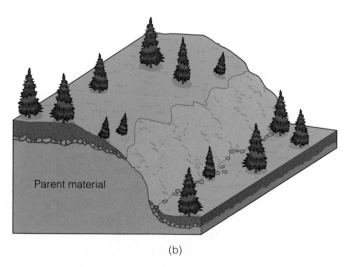
(b)

Figure 5.23
(a) The influence of parent material on soil development. Quartzite is resistant to chemical weathering, whereas granite alters more quickly. (b) The effect of slope on soil formation. Where slopes are steep, erosion occurs faster than soil can form.

Soil that develops on basalt will be rich in iron oxides because basalt contains abundant ferromagnesian silicates, but rocks lacking these minerals will not yield an iron oxide-rich soil no matter how thoroughly they are weathered. Also, weathering of a pure quartz sandstone will yield no clay, whereas weathering of clay will yield no sand.

Activities of Organisms

Soils not only depend on organisms for their fertility but also provide a suitable habitat for organisms ranging from microscopic, single-celled bacteria to burrowing animals such as ground squirrels and gophers. Earthworms—as many as 1 million per acre—ants, sowbugs, termites, centipedes, millipedes, and nematodes, along with various types of fungi, algae, and single-celled animals, make their homes in soil. All these contribute to the formation of soils and provide humus when they die and are decomposed by bacterial action.

Much of the humus in soils is provided by grasses or leaf litter that microorganisms decompose to obtain food. In so doing, they break down organic compounds within plants and release nutrients back into the soil. In addition, organic acids produced by decaying soil organisms are important in further weathering of parent materials and soil particles.

Burrowing animals constantly churn and mix soils, and their burrows provide avenues for gases and water. Soil organisms, especially some types of bacteria, are extremely important in changing atmospheric nitrogen into a form of soil nitrogen suitable for use by plants.

The Lay of the Land—Relief and Slope

Relief is the difference in elevation between high and low points in a region. In some mountainous areas relief is measured in hundreds of meters, whereas it rarely exceeds a few meters in, for instance, the Great Plains of the United States and Canada. Because climate is such an important factor in soil formation and climate changes with elevation, areas with considerable relief have different soils in mountains and adjacent lowlands.

Slope influences soil formation in two ways. One is simply *slope angle:* Steep slopes have little or no soil, because weathered materials erode faster than soil-forming processes can operate (Figure 5.23b). The other slope factor is *slope direction,* that is, the direction a slope faces. In the Northern Hemisphere, north-facing slopes receive less sunlight than south-facing slopes. In fact, if north-facing slopes are steep they may not receive any sunlight at all. Accordingly, north-facing slopes have cooler internal temperatures, support different vegetation, and, if in a cold climate, remain snow covered or frozen longer.

Time

Soil-forming processes begin at the surface and work downward, so horizon A has been altered longer than the other horizons, and thus parent material is no longer recognizable. Even in horizon B parent material is usually not discernible, but it is in horizon C. In fact, soil properties are determined by climate and organisms altering parent material through time (Figure 5.20), so the longer the processes have operated the more fully developed a soil will be. If weathering takes place for an extended period, especially in humid climates, soil fertility decreases as plant nutrients are leached out, unless new materials are delivered. For instance, agricultural lands adjacent to rivers such as the Nile in Egypt have their soils replenished yearly during floods. In areas of active tectonism, erosion of uplifted areas provide fresh materials that are transported to nearby lowlands, where they contribute to soils.

How much time is needed to develop a centimeter of soil or a fully developed soil a meter or so deep? No definitive answer can be given because weathering proceeds at vastly different rates depending on climate and parent material, but an overall average might be about 2.5 cm per century. However, a lava flow a few centuries old in Hawaii may have a well-developed soil on it, whereas a flow the same age in Iceland will have considerably less soil. Given the same climatic conditions, soil develops faster on unconsolidated sediment than it does on bedrock.*

Under optimum conditions, soil-forming processes operate rapidly in the context of geologic time. From the human perspective, though, soil formation is a slow process; consequently, soil is regarded as a nonrenewable resource.

EXPANSIVE SOILS

News reports commonly cover geologic events that cause fatalities, injuries, and property damage such as floods, earthquakes, volcanic eruptions, and landslides. These more sensational events overshadow the fact that processes that rarely make the news cause more property damage. One of these, soil creep, is considered in Chapter 14, but here we are concerned with **expansive soils,** soils containing clay minerals that increase in volume when wet and shrink when they dry out. Soils with clays that expand 6% are considered highly expansive, and some are even more expansive. About $6 billion in damage to foundations,

---

*Bedrock is a general term for the rock underlying soil or unconsolidated sediment.

roadways, sidewalks, and other structures takes place each year in the United States, mostly in the Rocky Mountain states, the Southwest, and in some of the states along the Gulf of Mexico (Figure 5.24).

When soils expand and contract, overlying structures are first uplifted, then subside, thus experiencing forces that usually are not equally applied. Part of a house, for instance, might be uplifted more than an adjacent part of the same structure. Avoiding areas of expansive soils is the best way to prevent damage, but what can be done to minimize the damage to existing

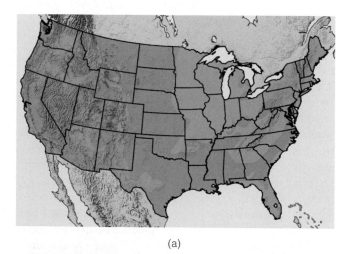

(a)

(b)

Figure 5.24
(a) The brown pattern shows the distribution of expansive soils in the United States. Many areas of expansive soils are also present in the blue areas, but are too small to show at this scale. (b) The undulations in this sidewalk near Dallas, Texas, were caused by expansive soil. *Source: (b) Ed Nuhfer, Director, Teaching Effectiveness & Faculty Development, CU–Denver, Campus Box 137, P.O. Box 173364, Denver, CO 80217-3364.*

structures? In some cases the soils can be removed, mixed with chemicals that change the way they react with water, or covered by a layer of nonexpansive fill, all expensive but perhaps necessary remedies. Also, soils should be kept as dry as possible to inhibit expansion, and specialized building methods can be employed such as placing structures on piers or reinforced foundations designed to minimize the effects of expansion.

EROSION AND PHYSICAL AND CHEMICAL DETERIORATION OF SOILS

Given that a fully mature soil takes centuries to thousands of years to form, soil losses that exceed the rate of formation are viewed with alarm. Likewise, any reduction of soil fertility and productivity is cause for concern, especially in those parts of the world where soils already provide only a marginal existence. Soil erosion and decreases in productivity are considered forms of **soil degradation,** both of which are problems in many parts of the world. Erosion is, of course, an ongoing process that occurs quite naturally, but it is usually slow enough for soil formation to keep pace. Unfortunately, human activities commonly add to the problem and introduce elements into the natural system that would not otherwise be present.

According to studies by the World Resources Institute, 17% of the world's soils were degraded to some extent by human activities between 1945 and 1990. In North America 5.3% of the soil is estimated to be degraded, whereas the figure is much higher for the other continents. Three general types of soil degradation are recognized, each of which consists of several separate processes: *erosion, chemical deterioration,* and *physical deterioration.*

Wind and running water are responsible for most soil erosion. Human activities that contribute to erosion include removing natural vegetation by agricultural practices such as plowing, overgrazing, overexploitation for firewood, and deforestation. When a soil's natural vegetation is removed and soil is pulverized by plowing, the fine particles are easily blown away. The Dust Bowl that developed in several Great Plains states during the 1930s is a poignant example of just how effective wind erosion is on exposed soil (see Perspective 5.2). Rainfall also disrupts soil particles and carries soil with it as it runs off at the surface. Steep slopes from which the vegetation has been removed by overgrazing, deforestation, or construction are especially susceptible to erosion.

Although wind is an effective agent of erosion in some areas, running water is much more effective at eroding soil. Some soil is removed by **sheet erosion,** which is water more or less evenly distributed over the

(a)

(b)

Figure 5.25
(a) Rill erosion in a field in Michigan during a rainstorm. The rill was later plowed over. (b) A large gully in the upper basin of the Rio Reventado in Costa Rica.

surface and which removes thin layers of soil. **Rill erosion,** in contrast, takes place when running water scours small, troughlike channels. If these channels are shallow enough to be eliminated by plowing, they are called *rills,* but channels too deep (about 30 cm) to be plowed over are *gullies* (Figure 5.25). Running water in both rills and gullies carry eroded soil, but because rills can be plowed over, the remaining soil can still be used for agriculture. However, where **gullying** is extensive, croplands can no longer be tilled and must be abandoned.

Clearing tropical rain forests leaves the surface unprotected and generally results in increased soil erosion (Figure 5.26). More rainwater runs off at the surface, gullying becomes more common, and flooding is more frequent because trees with their huge water-holding capacity are no longer present. Furthermore, the soil becomes more compact and less absorbent, inhibiting infiltration and increasing runoff. Clearing woodlands in less humid regions such as much of the eastern United States also results in accelerated erosion, at least initially. Studies of lake deposits in several areas indicate that clearing woodlands is followed by several years of rapid erosion that then decreases markedly when the land is covered by crops. But even then erosion rates may be 10 times greater than before clearing.

A soil undergoes chemical deterioration when its nutrients are depleted and its productivity decreases. Loss of soil nutrients is most notable in many of the populous developing nations where soils are overused to maintain high levels of agricultural productivity. Chemical deterioration is also caused by insufficient use of fertilizers and by clearing soils of their natural vegetation. Examples of chemical deterioration can be found everywhere soil is used, but it is most prevalent in South America, where it accounts for 29% of all soil degradation.

Figure 5.26
Accelerated soil erosion on a bare surface in Madagascar that was once covered by lush forest.

Perspective 5.2

The Dust Bowl—
An American Tragedy

The stock market crash of 1929 ushered in the Great Depression, a time when millions of people were unemployed and many had no means to provide food and shelter. Urban areas were affected most severely by the depression, but rural areas suffered as well, especially during the great drought of the 1930s. Prior to the 1930s, farmers had enjoyed a degree of success unparalleled in U.S. history. During World War I, the price of wheat soared, and after the war when Europe was recovering, the government subsidized wheat prices. High prices and mechanized farming resulted in more and more land being tilled. Even the weather cooperated, and land in the western United States that would otherwise have been marginally productive was plowed. Deep-rooted prairie grasses that held the soil in place were replaced by shallow-rooted wheat.

Beginning in about 1930, drought conditions prevailed throughout the country; only two states—Maine and Vermont—were not drought-stricken. Drought conditions varied from moderate to severe, but its consequences were particularly severe in the southern Great Plains. Some rain fell, but not enough to maintain agricultural production. And since the land, even marginal land, had been tilled, the native vegetation was no longer available to keep the topsoil from blowing away. And blow away it did—in huge quantities.

A large region in the southern Great Plains that was particularly hard hit by the drought, dust storms, and soil erosion came to be known as the Dust Bowl. Although its boundaries were not well defined, it included parts of Kansas, Colorado, and New Mexico, as well as the panhandles of Oklahoma and Texas (Figure 1a); the Dust Bowl and its less affected fringe area covered more than 400,000 km²!

Dust storms were common during the 1930s, and some reached phenomenal sizes (Figure 1b). One of the largest storms occurred in 1934 and covered more than 3.5 million km². It lifted dust nearly 5 km into the air, obscured the sky over large parts of six states, and blew hundreds of millions of tons of soil eastward where it settled on New York City, Washington, D.C., and other eastern cities, as well as on ships as far as 480 km out in the Atlantic Ocean. The Soil Conservation Service reported dust storms of regional extent on 140 occasions during 1936 and 1937. Dust was everywhere. It seeped into houses, suffocated wild animals and livestock, and adversely affected human health.

The dust was, of course, the material derived from the tilled lands; in other words, much of the topsoil in many regions was simply blown away. Blowing dust was not the only problem; sand piled up along fences, drifted against houses and farm machinery, and covered what otherwise might have been productive soils. Agricultural production fell precipitously in the Dust Bowl, farmers could not meet their mortgage payments, and by 1935 tens of thousands were homeless, on relief, or leaving (Figure 2). Many of these people went west to California and became the migrant farm workers immortalized in John Steinbeck's novel *The Grapes of Wrath*.

The Dust Bowl was an economic disaster of great magnitude. Droughts had stricken the southern Great Plains before, and have done so since—from August 1995 well into the summer of 1996, for instance—but the drought of the 1930s was especially severe. Political and economic factors also contributed to the disaster. Due in part to the artificially inflated wheat prices, many farmers were deeply in debt—mostly because they had purchased farm machinery in order to produce more and benefit from the high prices. Feeling economic pressure because of their huge debts, they tilled marginal land and employed few, if any, soil conservation measures.

If the Dust Bowl has a bright side, it is that the government, farmers, and the public in general no longer take soil for granted or regard it as a substance that needs no nurturing. In addition, a number of soil conservation methods developed then have now become standard practices.

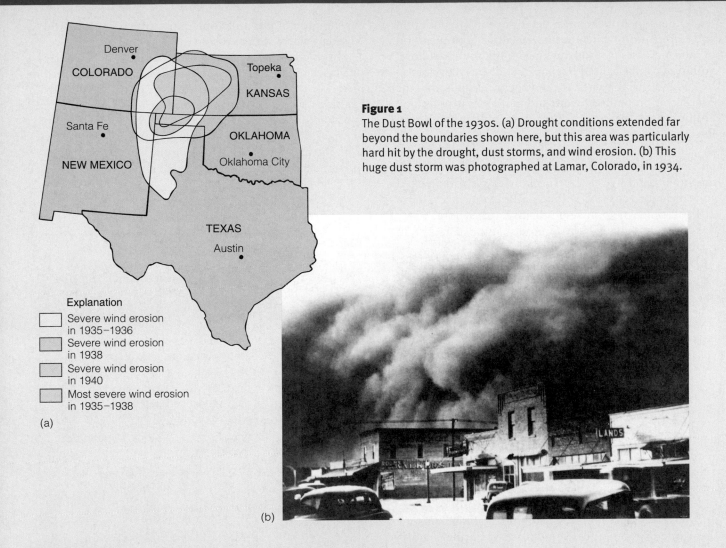

Figure 1
The Dust Bowl of the 1930s. (a) Drought conditions extended far beyond the boundaries shown here, but this area was particularly hard hit by the drought, dust storms, and wind erosion. (b) This huge dust storm was photographed at Lamar, Colorado, in 1934.

Explanation
- Severe wind erosion in 1935–1936
- Severe wind erosion in 1938
- Severe wind erosion in 1940
- Most severe wind erosion in 1935–1938

(a)

(b)

Figure 2
By the mid-1930s, tens of thousands of people were on relief, homeless, or had left the Dust Bowl. In 1939, Dorthea Lange photographed this family of seven in Pittsburgh County, Oklahoma. Many of these people became the migrant farm workers immortalized in John Steinbeck's novel *The Grapes of Wrath*.

Other types of chemical deterioration are pollution and *salinization,* which occurs when the concentration of salts increases in a soil, making it unfit for agriculture. Pollution can be caused by improper disposal of domestic and industrial wastes, oil and chemical spills, and the concentration of insecticides and pesticides in soils. Soil pollution is a particularly severe problem in Eastern Europe.

Physical deterioration of soils results when soil particles are compacted under the weight of heavy machinery and livestock, especially cattle. Compacted soils are more costly to plow, and plants have a more difficult time emerging from them. Furthermore, water does not readily infiltrate, so more runoff occurs; this in turn accelerates the rate of water erosion.

If soil losses to erosion are minimal, soil-forming processes can keep pace, and the soil remains productive. Should the loss rate exceed the formation rate, however, the most productive upper layer of soil, horizon A, is removed, exposing horizon B, which is much less productive. The Soil Conservation Service of the Department of Agriculture estimates that 25% of U.S. cropland is eroding faster than soil-forming processes can replace it. Such losses are problems, of course, but there are additional consequences. For one thing, the eroded soil is transported elsewhere, perhaps onto neighboring cropland, onto roads, or into channels. Sediment accumulates in canals and irrigation ditches, and agricultural fertilizers and insecticides are carried into streams and lakes.

In North America, the rich prairie soils of the midwestern United States and the Great Plains of the United States and Canada are suffering significant soil degradation. Nevertheless, this degradation, which is characterized as moderate, is less serious than in many other parts of the world where it is severe or extreme. These soils remain productive, although their overall productivity has decreased somewhat during the last several decades and more fertilizers are needed to maintain productivity. Other areas of concern are the central valleys of California, an area in Washington State, and some parts of Mississippi and Missouri where water erosion rates are high.

Problems experienced during the past, particularly during the 1930s, have stimulated the development of methods to minimize soil erosion on agricultural lands (Table 5.2). Crop rotation, contour plowing, and the construction of terraces have all proved helpful (Figure 5.27). So has no-till planting, in which the residue from the harvested crop is left on the ground to protect the surface from the ravages of wind and water.

Table 5.2

Soil Conservation Practices	
Terracing	Creating flat areas on sloping ground. One of the oldest and most effective ways of preserving soil and water.
Strip-cropping	Growing of different crops on alternate, parallel strips of ground to minimize wind and water erosion. Alternating strips of corn and alfalfa, for instance.
Crop rotation	Yearly alternation of crops on the same land. Significant reduction in soil erosion when soil-depleting crops are alternated with soil-enriching crops.
Contour plowing	Plowing along a slope's contours so that furrows and ridges are perpendicular to the slope.
No-till planting	Planting seeds through the residue of a previously harvested crop.
Windbreaks	Planting trees or large shrubs along the margins of a field. Especially effective in reducing wind erosion.

Figure 5.27
One soil conservation practice is contour plowing, which involves plowing parallel to the contours of the land. The furrows and ridges are perpendicular to the direction that water would otherwise flow downhill, and thus inhibit erosion.

WEATHERING AND MINERAL RESOURCES

In a preceding section, we discussed intense chemical weathering in the tropics and the origin of laterite, which is composed largely of aluminum hydroxides, iron oxides, and clay minerals. And even though these soils are not very productive for agriculture, one aspect of them is of great economic importance. If the parent material is rich in aluminum, *bauxite*, the ore of aluminum, might accumulate in horizon B. Because such intense chemical weathering does not presently occur in North America, the United States and Canada depend on foreign sources for aluminum ore. Some aluminum ore is present in Arkansas, Alabama, and Georgia, which had tropical climates about 50 million years ago, but it is cheaper to import aluminum ore than to mine these deposits.

Accumulations of valuable minerals such as bauxite formed by the selective removal of soluble substances by chemical weathering are *residual concentrations*. In addition to bauxite, a number of other residual concentrations are important, including ore deposits of iron, manganese, clays, nickel, phosphate, tin, diamonds, and gold.

Some limestones contain small amounts of iron carbonate minerals. When the limestone dissolves during chemical weathering, a residual concentration of insoluble iron oxides accumulates. Residual concentrations of insoluble manganese oxides form in a similar fashion from manganese-rich source rocks. Some of the sedimentary iron deposits (see Chapter 6) of the Lake Superior region were enriched by chemical weathering when original soluble constituents were carried away.

Most commercial clay deposits formed by hydrothermal alteration of granitic rocks or by sedimentary processes, but some formed as residual concentrations. A number of kaolinite deposits in the southern United States were formed by the chemical weathering of feldspars in pegmatites and of clay-bearing limestones and dolostones. Kaolinite is a type of clay mineral used in the manufacture of paper and ceramics.

Gossans, oxidized ores, and supergene enrichment of ores are interrelated, and all result from chemical weathering. A *gossan* is a yellow to reddish deposit composed largely of hydrated iron oxides that formed by the oxidation and leaching of sulfide minerals such as pyrite (FeS_2). The dissolution of sulfide minerals forms sulfuric acid, which causes other metallic minerals to dissolve, and these tend to be carried down toward the groundwater table (Figure 5.28). *Oxidized ores* form just above the groundwater table as a result of chemical reactions with these descending solutions. Some of the minerals formed in this zone contain copper, zinc, and lead.

Supergene enrichment of ores occurs where metal-bearing solutions penetrate below the water table (Figure 5.28). Such deposits are characterized by the replacement of sulfide minerals of the primary deposit with sulfide minerals introduced by the descending solutions. The iron in iron sulfides may be replaced by other metals such as lead, zinc, nickel, and copper that have a greater affinity for sulfur. Indeed, supergene chalcocite (Cu_2S), an important copper ore, forms as a replacement of primary pyrite (FeS_2) and chalcopyrite ($CuFeS_2$). Notice that both chalcocite and chalcopyrite are copper-bearing minerals, but the former is a richer source of copper than the latter.

Gossans have been used occasionally as sources of iron, but they are far more important as indicators of underlying ore deposits. One of the oldest known underground mines exploited such ores about 3400 years ago in what is now southern Israel. Supergene-enriched ore bodies are generally small but extremely rich sources of various metals. The largest copper mine in the world, at Bingham, Utah, was originally mined for supergene ores, but currently only primary ores are being mined.

WHAT WOULD YOU DO?

You have inherited a piece of property ideally located for everything you consider important. Unfortunately, as you prepare to have a house built, your contractor tells you that the soil is rich in clay that expands when wet and contracts when dry. You nevertheless go ahead with construction, but must now decide on what measures should be taken to prevent damage to the structure. Make several proposals that might solve the problem. Which one or ones do you think would be the most cost effective?

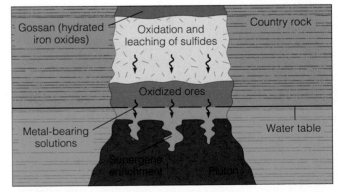

Figure 5.28
A diagrammatic representation of a gossan and the origin of oxidized ores and supergene enrichment of ores. Gossans are not usually used for their ores but as indicators of ore deposits beneath.

Chapter Summary

1. Mechanical and chemical weathering are processes whereby parent material is disintegrated and decomposed so that it is more nearly in equilibrium with new physical and chemical conditions. The products of weathering include solid particles, soluble compounds, and ions in solution.

2. The residue of weathering can be further modified to form soil, or it can be deposited as sediment, which might become sedimentary rock.

3. Mechanical weathering processes include frost action, pressure release, thermal expansion and contraction, salt crystal growth, and the activities of organisms. Particles liberated by mechanical weathering retain the chemical composition of the parent material.

4. Solution, oxidation, and hydrolysis are chemical weathering processes; they result in a chemical change of the weathered products. Clay minerals, various ions in solution, and soluble compounds are formed during chemical weathering.

5. Chemical weathering proceeds most rapidly in hot, wet environments, but it occurs in all areas, except perhaps where water is permanently frozen.

6. Mechanical weathering aids chemical weathering by breaking parent material into smaller pieces, thereby exposing more surface area.

7. Mechanical and chemical weathering produce regolith, some of which is soil if it consists of solids, air, water, and humus and supports plants.

8. Soils are characterized by horizons that are designated, in descending order, as O, A, B, and C; soil horizons differ from one another in texture, structure, composition, and color.

9. The factors controlling soil formation include climate, parent material, organic activity, relief and slope, and time.

10. Soils called *pedalfers* develop in humid regions such as the eastern United States and much of Canada.

11. Laterite is a soil resulting from intense chemical weathering, as in the tropics. Such soils are deep and red and are sources of aluminum ores if derived from aluminum-rich parent material.

12. Soil degradation caused by erosion (mostly sheet and rill erosion and gullying), chemical deterioration, and physical deterioration is a problem in some areas. Human activities such as construction, agriculture, and deforestation can accelerate soil degradation.

13. Intense chemical weathering is responsible for the origin of residual concentrations, many of which contain valuable minerals such as iron, lead, copper, and clay.

14. Gossans, oxidized ores, and supergene enrichment of ores all result from chemical weathering.

Arid and semiarid region soils are *pedocals,* many of which contain irregular masses of caliche in horizon B.

Important Terms

chemical weathering
differential weathering
erosion
exfoliation
exfoliation dome
expansive soil
frost action
frost heaving
frost wedging
gullying
humus
hydrolysis

laterite
mechanical weathering
oxidation
parent material
pedalfer
pedocal
pressure release
regolith
rill erosion
salt crystal growth
sheet erosion
sheet joint

soil
soil degradation
soil horizon
solution
spheroidal weathering
talus
thermal expansion and contraction
transport
weathering
zone of accumulation
zone of leaching

Review Questions

1. Most of Earth's land surface is covered by soil and unconsolidated rock material called:
   a. _____ regolith;
   b. _____ horizon A;
   c. _____ talus;
   d. _____ parent material;
   e. _____ humus.

2. Limestone, composed of calcite ($CaCO_3$), is nearly insoluble in pure water but dissolves rapidly if _____ is present.
   a. _____ silicon dioxide;
   b. _____ sodium sulfate;
   c. _____ residual manganese;
   d. _____ carbonic acid;
   e. _____ clay.

3. The process whereby hydrogen and hydroxyl ions of water replace ions in minerals is:
   a. _____ carbonization;
   b. _____ supergene enrichment;
   c. _____ hydrolysis;
   d. _____ exfoliation;
   e. _____ solution.

4. Which one of the following is *not* a mechanical weathering process?
   a. _____ pressure release;
   b. _____ oxidation;
   c. _____ frost wedging;
   d. _____ thermal expansion and contraction;
   e. _____ salt crystal growth.

5. Laterite is:
   a. _____ a deep, red tropical soil;
   b. _____ a type of erosion common in arid regions;
   c. _____ the most common mechanical weathering process;
   d. _____ responsible for the origin of exfoliation domes;
   e. _____ produced mostly by frost wedging and pressure release.

6. Horizon C differs from the other soil horizons in that it:
   a. _____ is more fertile;
   b. _____ contains the most humus;
   c. _____ has pieces of partly altered parent material;
   d. _____ is made up of caliche;
   e. _____ has been weathered the longest.

7. The accumulation of angular blocks of rock at the base of a slope is known as:
   a. _____ pedocal;
   b. _____ soil;
   c. _____ parent material;
   d. _____ humus;
   e. _____ talus.

8. Horizon B of a soil is also known as the:
   a. _____ alkali zone;
   b. _____ topsoil;
   c. _____ zone of accumulation;
   d. _____ talus layer;
   e. _____ organic-rich bed.

9. The primary weathering process responsible for the origin of exfoliation domes is:
   a. _____ pressure release;
   b. _____ oxidation-solution;
   c. _____ frost heaving;
   d. _____ soil degradation;
   e. _____ differential weathering.

10. The kind of soil typical of arid regions is _____ whereas _____ is (are) much more common in humid areas:
    a. _____ pedocal/pedalfer;
    b. _____ regolith/laterite;
    c. _____ exfoliation/humus;
    d. _____ hydrolysis/caliche;
    e. _____ talus/residual oxides.

11. When the ions in a mineral dissociate in a fluid, it has been:
    a. _____ mechanically weathered;
    b. _____ converted to clay;
    c. _____ oxidized;
    d. _____ changed to soil;
    e. _____ dissolved.

12. Spheroidal weathering takes place because:
    a. _____ the corners and edges of stones weather more rapidly than flat surfaces;
    b. _____ most naturally occurring rocks are spherical to begin with;
    c. _____ oxidation changes limestone to clay;
    d. _____ pedalfers are composed mostly of iron, aluminum, and clay;
    e. _____ thermal expansion and contraction is such an efficient chemical weathering process.

13. Explain why particle size is an important factor in chemical weathering.

14. How does mechanical weathering differ from and contribute to chemical weathering?

15. Why are most minerals not very soluble in pure water? Give an example of how a soluble mineral goes into solution.

16. Explain how the factors of climate, parent material, and time determine the depth and fertility of soil.

17. Describe the types of soil degradation. What practices can be used to prevent or at least decrease soil erosion?

18. How does weathering differ from erosion and transport?

19. Draw a soil profile, and list the characteristics of each profile for arid- and humid-area soils.

20. Explain why groundwater in humid regions tends to be acidic, whereas in arid regions it is generally alkaline.

21. What is differential weathering and why does it occur? Give some examples of features produced by differential weathering.

22. Explain fully how an exfoliation dome forms. In what kinds of rocks do these features commonly develop, and where would you go to see examples?

23. What are residual concentrations, how do they form, and why are they important?

24. Show the chemical reaction or explain what happens to potassium feldspars during hydrolysis.

1. Consider the following: A soil is 1.5 m thick, new soil forms at the rate of 2.5 cm per century, and the erosion rate is 4 mm per year. How much soil will be left after 100 years?

2. You have two identical samples of granite, sample A and sample B. Discuss the products that would result from only mechanical weathering of sample A. Also, describe the products that would result from only chemical weathering of sample B.

3. What kinds of mechanical and chemical weathering would you expect to take place on Earth's moon, Mars, and Venus? Explain fully.

4. How do human practices contribute to soil degradation? What can be done to minimize the impact of such practices on soils?

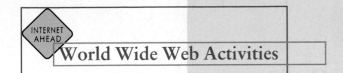

For these Web site addresses, along with current updates and exercises, log on to
http://www.brookscole.com/geo/

➤ DEVILS MARBLES

This site is maintained by Patrick Jennings and contains various images of Devils Marbles from Australia. Devils Marbles are examples of spheroidal weathering. Click on any of the images on this page. They will take you to larger images and some information about the origin and location of those "marbles."

➤ ROB'S GRANITE PAGE

This site, which is obviously mostly about granite, is maintained by Robert M. Reed of the Department of Geological Sciences, University of Texas at Austin. Click on *Llano Uplift* and then click *Enchanted Rock State Park*. Here you will see images of the Town Mountain Granite. Note especially the well-developed exfoliation domes, which are some of the finest examples seen anywhere. Also click on *Turkey Peak* and see an excellent example of differential weathering and erosion.

➤ WORLD OF CHEMISTRY: THE HOME PAGE OF RALPH LOGAN

This site is maintained by Ralph H. Logan, chemistry instructor at North Lake College in Dallas, Texas. It is mostly devoted to chemistry but also has information of geologic interest. Scroll down to *Places to Visit* and click *Frequently Asked Questions on Chemical Concepts by Category,* and then click *Ecology & Energy Source Questions*. In this section, there are answers to such questions as "What can you tell me about acid rain?" "Why should we be concerned about carbon dioxide in the atmosphere?" and "What is the greenhouse effect?" Answers to questions regarding energy resources in this section will be addressed in the next chapter.

➤ JEFF'S HOME PAGE & LINK TO THE SUSTAINABLE AGRICULTURE EDUCATION PAGE

This site is maintained by Jeffery M. Dunn, who is majoring in environmental and natural resource policy at Michigan State University in Lansing, Michigan. Click on *Sustainable Agriculture Educational Project Homepage* and see what is meant by sustainable agriculture. Next, click on *Soil Erosion in Agricultural Systems*. What are the positive and negative aspects of no-till planting and leaving crop residue on the field after harvesting?

Chapter 6

Sedimentary rocks, mostly siltstone and sandstone, and volcanic ash layers exposed at Scott's Bluff National Monument, Nebraska.

OBJECTIVES

After reading this chapter, you will have learned that

■ Sediment comes from weathering of rock and is transported and deposited by a variety of processes.

■ Compaction and cementation convert unconsolidated sediment into sedimentary rocks.

■ Texture and composition are the criteria used to classify sedimentary rocks.

■ Individual layers of sediment differ in texture, composition, or both.

■ Many widespread layers of sedimentary rocks were deposited during marine transgressions and regressions, both of which can be recognized by the evidence they leave in the geologic record.

■ Sedimentary rocks possess a variety of structures that formed when they were deposited, and some also contain fossils.

■ Several geologic processes lead to the preservation of fossils.

■ Structures and fossils in sedimentary rocks can be used to determine depositional environments; that is, to read the story told by sedimentary rocks.

■ Some sedimentary rocks contain resources or may be resources themselves.

Sediment and Sedimentary Rocks

Prologue

One feature of the rock cycle (see Figure 1.17) is that all rock types are related. That is, any one of the major rock families (igneous, sedimentary, or metamorphic) can be derived from rocks of the other families. This being the case, we might expect some confusion about rocks showing features of more than one rock-forming process. A good example is provided by the rocks at John Day Fossil Beds National Monument, which is made up of three widely separated units in central Oregon. Many of the rocks are characterized as *volcanoclastic,* meaning they contain large amounts of volcanic rock debris and/or pyroclastic materials. This, coupled with the fact that they have few of the internal structures typical of sedimentary rocks, resulted in their being ignored for many years by geologists interested in sedimentary rocks. Similarly, geologists who study igneous rocks and processes also ignored them, because they thought the rocks were sedimentary.

Most geologists would probably now agree that many of the rocks are in fact sedimentary, although there are also undisputed igneous rocks in the monument as well. In any case, several rock units known as *formations* are now recognized, all have been well studied, some contain the remains of ancient plants and mammals, and the numerous ash beds present make the rocks easily dated by absolute dating techniques (see Chapter 8) (Figure 6.1a). The rocks are between 55

and 6 million years old, and along with their interest as repositories of ancient life, their colors and erosional patterns provide some truly inspiring scenery (Figure 6.1b and c).

The oldest rock unit, the 39- to 55-million-year-old Clarno Formation (Figure 6.1b), is composed of volcanic ash and volcanic mudflows that buried plants and animals living in a semitropical rain forest. Today the area is semiarid, but when the Clarno Formation formed it was covered by forests of palms, avocados, and ferns and was occupied by ancient horses, rhinoceroses, carnivorous mammals, as well as the now extinct mammals known as oreodonts and titanotheres (Figure 6.2).

Overlying the Clarno Formation are rocks of the 20- to 39-million-year-old John Day Formation, consisting of a complex of ash flow tuffs, welded tuffs, basalt and rhyolite flows, and ash-rich layers of claystone, sandstone, and conglomerate (Figure 6.1c). Fossil plants indicate a drier climate, deciduous trees having replaced those of the previous subtropical forest, and grasslands were more common. The remains of ancient horses, camels, cats, dogs, rhinoceroses, and rodents have been recovered from several locations within the formation.

Beginning about 18 million years ago, following a time of 2 million years for which no rocks were present, lava flows of the Picture Gorge Basalt covered the area. Weathering and erosion of these lava flows provided much of the debris that makes up the Mascall Formation

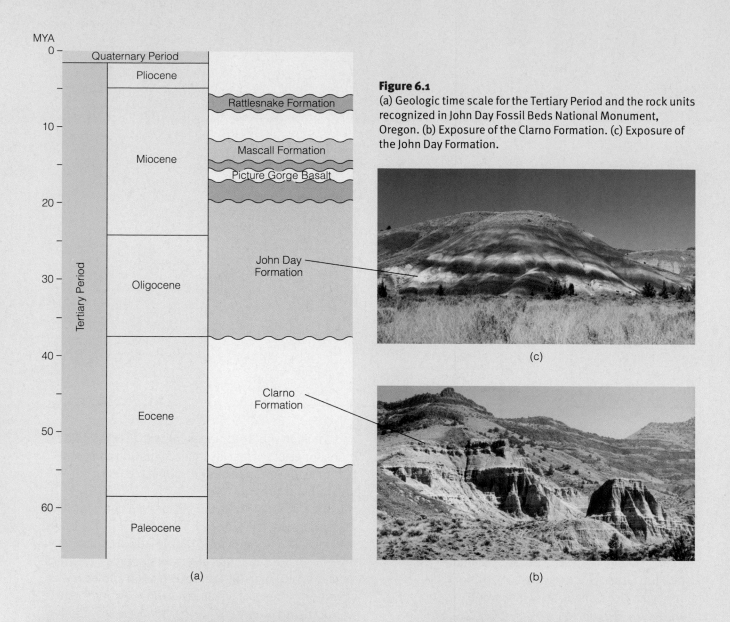

Figure 6.1
(a) Geologic time scale for the Tertiary Period and the rock units recognized in John Day Fossil Beds National Monument, Oregon. (b) Exposure of the Clarno Formation. (c) Exposure of the John Day Formation.

Figure 6.2
Fossils from the Clarno Formation in John Day Fossil Beds National Monument, Oregon, indicate a subtropical climate. Lush forests were occupied by a variety of mammals including (1) titanotheres, (2) a carnivore, (3) ancient horses, (4) tapirs, and (5) early rhinoceroses.

(Figure 6.1a). These rocks also contain the remains of numerous mammals, many of them descendants of those found in the older formations, and giants such as early elephants. Finally, the Rattlesnake Formation originated when debris eroded from nearby highlands covered a large area. These rocks, composed mostly of sand and gravel, contain far fewer animal and plant remains than the older formations, but those present indicate a cooler, dryer climate. Since the deposition of the Rattlesnake Formation, all rocks in the John Day Fossil Beds area have been slightly deformed, and, more importantly, they have been eroded to yield the present landscape.

INTRODUCTION—WHAT IS SEDIMENT, AND HOW DOES IT ORIGINATE?

In Chapter 5 we emphasized the fact that weathering is an important part of the rock cycle because it yields particles and dissolved substances, both of which might be raw materials for *sedimentary rocks,* the second major family of rocks (see Figure 1.17). All **sedi**mentary rocks are composed of **sediment**—solid particles derived by mechanical and chemical weathering, minerals precipitated from solution by chemical processes, or minerals secreted by organisms when they build their skeletons. Thus all sediment is derived by one weathering process or another, eroded from the weathering site, transported elsewhere, and deposited as a loose aggregate of particles. Sand and gravel in stream channels and on beaches as well as mud on the seafloor are examples of sedimentary deposits (Figure 6.3).

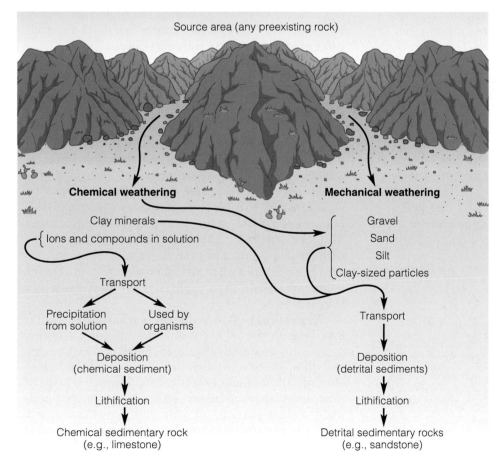

Figure 6.3
The derivation of sediment from preexisting rocks. Whether yielded by chemical or mechanical weathering, solid particles and materials in solution are transported and deposited as sediment, which if lithified becomes sedimentary rock. This illustration simply shows part of the rock cycle in more detail (see Figure 1.17).

Most sedimentary rocks form when sediment is transformed into solid rock, but a few skip the unconsolidated sediment stage and form directly as solids. Some marine organisms extract dissolved mineral matter from seawater to construct their skeletons. In doing so they form wave-resistant structures commonly called *coral reefs* that are, in fact, made up of the skeletons of corals, mollusks, and a variety of other animal skeletons and encrusting algae. In any case, the reef structure is solid when formed and constitutes one form of the sedimentary rock known as *limestone*. However, if reef material is ripped up and deposited on the seafloor—during a storm, for example—the pieces of reef material are sediment.

One important criterion for classifying sedimentary particles is their size (Table 6.1), particularly for solid particles, or *detrital sediment*, derived by weathering as opposed to *chemical sediment*, minerals extracted from solution by inorganic chemical processes, or used in the skeletons of organisms. Particles described as *gravel* measure more than 2 mm, whereas sand measures l/16–2 mm and silt is any particle between l/256 and l/16 mm. Composition is not a consideration in naming these particles. Most gravel is made up of rock fragments, that is, small pieces of granite, basalt, or any other rock type, and quartz, whereas the most common mineral in sand is quartz, but a number of others might be present as well. Particles smaller than l/256 mm are termed *clay*, but *clay* has two meanings. One is simply a size designation, but it also refers to certain types of sheet silicates known as *clay minerals*. However, most clay minerals are also clay sized.

Earth's crust is composed mostly of *crystalline rocks,* a term referring loosely to metamorphic rocks and most igneous rocks. Nevertheless, sediment and sedimentary rocks, comprising perhaps 5% of the crust, are by far the most common at or near the surface where they cover much of the seafloor and about two-thirds of the continents. All types of rocks are important in deciphering Earth history, but sedimentary rocks have a special significance in this endeavor because they preserve evidence of the surface processes responsible for their origin. Accordingly, geologists can make inferences about the past distribution of streams, lakes, deserts, glaciers, and shorelines and determine where resources might be found in these rocks. And many sedimentary rocks also contain the remains or traces of organisms, or what are called *fossils,* thus providing details about ancient climates and most of our record of prehistoric life. Studies of sedimentary and igneous rocks and fossils in John Day Fossil Beds National Monument in Oregon, for instance, reveal much about the region's physical and biological history (see the Prologue).

SEDIMENT TRANSPORT AND DEPOSITION

Weathering and erosion are fundamental processes in the origin of sediment and sedimentary rocks, but so is *sediment transport,* that is, the movement of detrital and chemical sediment by natural processes. Because they are moving solids, glaciers can carry sedimentary particles of any size, whereas wind transports only sand and smaller sediment. Waves and marine currents transport sediment along shorelines, but by far the most common way to transport sediment from its weathering site to other locations is by running water. Even the weakest currents can move ions in solution, which constitute an unseen but important part of all the material carried in streams, but more vigorous currents are necessary to transport sand and gravel.

During sediment transport, *abrasion* reduces the size of particles and the sharp corners and edges are worn smooth, a process known as **rounding,** as pieces of sand and gravel collide with one another (Figure 6.4a). Transport and processes operating where sediment is deposited also result in **sorting,** which refers to the size distribution in an aggregate of sediment. For instance, geologists characterize sediment as *well sorted* if all the particles are about the same size, and *poorly sorted* if a wide range of particle sizes are present (Figure 6.4b). Both rounding and sorting have important implications for other aspects of sediments and sedimentary rocks, such as how fluids move through them. They are also used to determine what processes were responsible for sediment deposition, a topic covered more fully in a later section.

Regardless of how sediment is transported, and it may be transported a considerable distance from its source, it is eventually deposited. Small particles of clay and silt might be transported to a lake where the lack of currents allows them to settle, thus forming a layer of mud; likewise, mud might settle on a stream's floodplain. Larger particles of sand and gravel commonly accumulate in stream channels and on beaches, whereas chemical sediment might be deposited in shallow sea-

Table 6.1

Classification of Sedimentary Particles	
Size (mm)	Sediment Name
>2	Gravel
1/16–2	Sand
1/256–1/16	Silt ⎫ Mud*
<1/256	Clay ⎭

*Mixtures of silt and clay are generally referred to as mud.

(a)

(b)

Figure 6.4
Rounding and sorting in sediments. (a) A deposit of well-sorted and well-rounded gravel. These particles measure about 5 cm across. (b) Angular, poorly sorted gravel.

water, where organisms use dissolved substances to make their skeletons. Any of these geographic areas in which sediment is deposited is a **depositional environment** where physical, chemical, and biological processes impart various characteristics to sedimentary deposits.

No completely satisfactory classification of depositional environments exists, but geologists recognize three major depositional settings: continental, transitional, and marine, each with several specific depositional environments (Figure 6.5). Deserts, lakes, streams and their adjacent floodplains, and areas affected by glaciation are the main continental depositional environments. Deltas and beaches are considered transitional environments because processes operating both

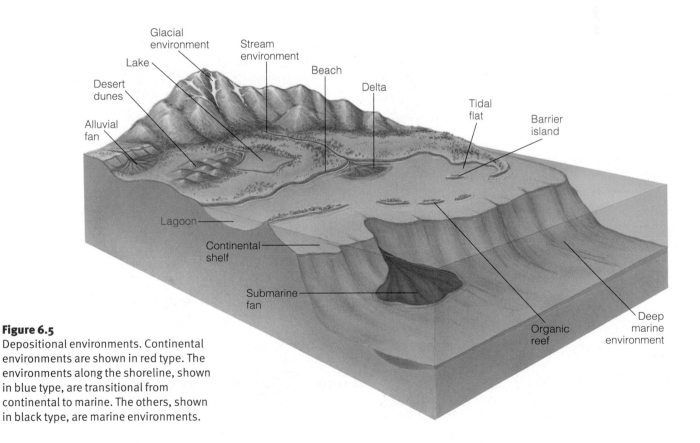

Figure 6.5
Depositional environments. Continental environments are shown in red type. The environments along the shoreline, shown in blue type, are transitional from continental to marine. The others, shown in black type, are marine environments.

on continents and in the marine realm affect the deposits. For instance, a delta might form where a stream enters the sea but waves, tides, and marine currents also affect the deposit. Seaward of the transitional environments are barrier islands, the continental shelf, reefs, and submarine fans that are influenced only by processes operating in the oceans (Figure 6.5).

HOW IS SEDIMENT TRANSFORMED INTO SEDIMENTARY ROCK?

Chemical sediments such as calcium carbonate mud are presently accumulating in the shallow waters of Florida Bay and the Great Bahama Bank, and detrital gravel, sand, and mud are being deposited in a variety of environments. In all cases, these sediments consist of loose aggregates of mineral grains and rock fragments. To make sedimentary rocks from these sediments requires **lithification,** the process whereby sediment is converted to solid rock by compaction and/or cementation (Figure 6.6).

Although compaction and cementation are responsible for lithification, the importance of one or the other depends on the type of sediment. For purposes of illustration, let us consider two deposits of detrital sediment,

one consisting of mud and the other of sand. In both cases, the newly deposited sediment consists of particles and *pore spaces,* the voids between particles. As more sediment accumulates on top of these layers they are subjected to **compaction** from the weight of the overlying sediments, and the amount of pore space decreases, thus reducing the volume of the deposit. Our hypothetical deposit of mud might have as much as 80% water-filled pore space, but during compaction the water is squeezed out and the volume of the deposit can be reduced by up to 40% (Figure 6.6). Our sand deposit might initially have 50% pore space, although it is usually somewhat less, and it too can be compacted so that the sand grains fit more tightly together.

Compaction alone is generally sufficient for lithification of mud, but for larger particles such as sand and gravel, **cementation** is necessary to convert the sediment to sedimentary rock. *Cement* consists of chemically precipitated minerals in the pore spaces of sediment that have the effect of binding the particles together into a solid mass (Figure 6.6). Recall from Chapter 5 that calcium carbonate ($CaCO_3$) readily dissolves in water containing a small amount of carbonic acid, and that chemical weathering of feldspars and other minerals yields silica (SiO_2) in solution. Circulating groundwater containing these compounds precipitates calcite ($CaCO_3$) and quartz (SiO_2) cement in the pore spaces of sediment.

Calcium carbonate and silicon dioxide are by far the most common cements in sedimentary rocks, but

Figure 6.6
Lithification of detrital sediments by compaction and cementation. Notice that little compaction takes place in sand and gravel.

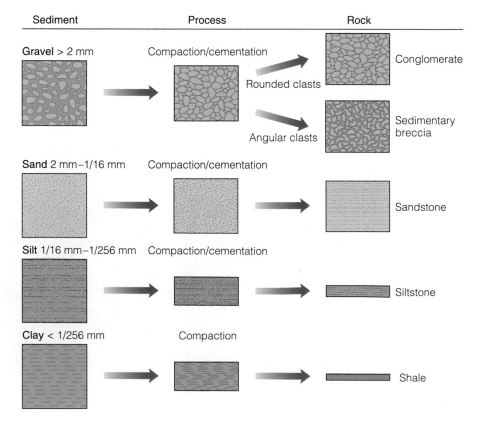

iron oxides and hydroxides, such as hematite (Fe_2O_3) and limonite ($FeO(OH) \cdot nH_2O$), respectively, are the chemical cements in some rocks. Much of the iron cement in sedimentary rocks is derived from oxidation of ferromagnesian silicates present in the original deposit, but circulating groundwater carries some in. The vast exposures of red, yellow, and brown sedimentary rocks in the southwestern United States and elsewhere owe their color to small amounts of iron oxide or hydroxide cement (Figure 6.7).

The preceding discussion has been adequate to explain lithification in detrital sediments but has not fully addressed this process in chemical sediments. By far the most common sediment in this category is calcium carbonate mud and various larger particles such as seashell fragments that when lithified form limestone. The same process of compaction and cementation occur in these sediments, but compaction is generally less effective because cementation commonly takes place soon after deposition. In any case, cement binding these sediments together is calcium carbonate derived by partial solution of some of the particles in the deposit.

Figure 6.7
These sedimentary rocks in Wyoming are red because they contain iron oxide cement.

TYPES OF SEDIMENTARY ROCKS

In the Introduction we noted that sedimentary rocks constitute only a small part of Earth's crust, but along with sediments they cover most of the seafloor and about 75% of the continents. Accordingly, sedimentary rocks and sediments are the most commonly encountered Earth materials. We have discussed the origin of sediment, its transport, deposition, and lithification, so we now turn to a consideration of sedimentary rocks and their classification. Geologists recognize two broad classes or types of sedimentary rocks, *detrital* and *chemical*, the latter including a subcategory known as *biochemical* (Table 6.2).

Detrital Sedimentary Rocks

Detrital sedimentary rocks are made up of *detritus*, the solid particles such as gravel and sand derived from preexisting rocks by mechanical and chemical weathering. They have a *clastic texture*, meaning they are composed of particles or fragments known as *clasts* (Figure 6.8a). The several varieties of detrital sedimentary rocks are classified primarily by the size of their constituent particles, although composition is used to modify some of the rock names (Table 6.2).

(a)

(b)

Figure 6.8
(a) Photomicrograph of a sandstone showing a clastic texture consisting of fragments of minerals, mostly quartz averaging about 0.5 mm in diameter. (b) Photomicrograph of the crystalline texture of a limestone showing a mosaic of calcite crystals, which measure about 1 mm across.

Table 6.2

Classification of Sedimentary Rocks

DETRITAL SEDIMENTARY ROCKS		
SEDIMENT NAME AND SIZE	DESCRIPTION	ROCK NAME
Gravel (>2 mm)	Rounded gravel particles	Conglomerate
	Angular gravel particles	Sedimentary breccia
Sand (1/16–2 mm)	Mostly quartz sand	Quartz sandstone ⎫ Sandstones
	Quartz with >25% feldspar	Arkose ⎭
Mud (<1/16 mm)	Mostly silt	Siltstone
	Silt and clay	Mudstone* ⎫ Mudrocks
	Mostly clay	Claystone* ⎭

CHEMICAL SEDIMENTARY ROCKS		
TEXTURE	COMPOSITION	ROCK NAME
Varies	Calcite ($CaCO_3$)	Limestone ⎫ Carbonates
Varies	Dolomite [$CaMg(CO_3)_2$]	Dolostone ⎭
Crystalline	Gypsum ($CaSO_4 \cdot 2H_2O$)	Rock gypsum ⎫ Evaporites
Crystalline	Halite (NaCl)	Rock salt ⎭

BIOCHEMICAL SEDIMENTARY ROCKS		
TEXTURE	COMPOSITION	ROCK NAME
Clastic	Calcium carbonate ($CaCO_3$) shells	Limestone (various types such as chalk and coquina)
Usually crystalline	Altered microscopic shells of silicon dioxide (SiO_2)	Chert
—	Mostly carbon from altered plant remains	Coal

*Mudrocks possessing the property of fissility, *meaning they break along closely spaced planes, are commonly called* shale.

Conglomerate and Sedimentary Breccia   Both *conglomerate* and *sedimentary breccia* are composed of gravel-sized particles, that is, detrital particles measuring more than 2 mm (Table 6.2; Figure 6.9). Composition is not a consideration in naming these rocks; both are made up mostly of rock fragments of various sizes but some large, individual mineral grains may also be present. The only difference between the two rocks is the shape of their gravel-sized particles; conglomerate has rounded gravel, whereas sedimentary breccia has angular gravel called *rubble.*

Conglomerate is a fairly common rock, but sedimentary breccia is rare because gravel-sized particles become rounded very quickly during sediment transport or by processes operating in environments where gravel is deposited. For instance, gravel transported for a few kilometers in a stream or moved to and fro by waves on a beach is invariably rounded. Thus, if you encounter sedimentary breccia, you can conclude that the angular gravel composing it experienced little or no transport. Considerable energy is needed to transport gravel, so it tends to be deposited in high-energy environments such as stream channels and on beaches.

(a)   (b)

Figure 6.9
Detrital sedimentary rocks composed of gravel-sized particles. Rounded gravel is present in conglomerate (a), whereas sedimentary breccia (b) consists of angular gravel.

Sandstone   *Sand* is simply a designation for detrital particles measuring between l/16 and 2 mm regardless of composition, so any mineral or rock fragment can be in *sandstone.* Although composition is not a consideration for sandstone classification, geologists recognize varieties based on composition (Table 6.2; Figure 6.10). *Quartz sandstone* is the most common

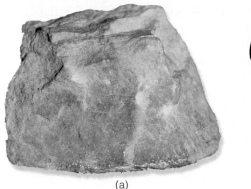

(a)

(b)

Figure 6.10
Quartz sandstone (a) is the most common variety of sandstone, but arkose (b), a feldspar-rich sandstone, is also fairly abundant. Iron oxides are responsible for the red parts of the rock.

and, as the name implies, is composed mostly of quartz grains. Another variety of sandstone called *arkose* contains at least 25% feldspar minerals.

Given that Earth's crust is made up of an estimated 51% feldspars (39% plagioclase and 12% potassium feldspars), 24% ferromagnesian silicates (olivine, pyroxene, amphiboles, and biotite), 12% quartz, and 13% other minerals, it might seem odd that quartz is by far the most common constituent of sandstones. However, the chance that any specific mineral will end up in a sedimentary rock depends on its availability, mechanical durability, and chemical stability. Feldspars and quartz are both abundant, but quartz is hard, lacks

cleavage, and is chemically stable, whereas feldspars are not quite as hard, possess two directions of cleavage along which they break more easily, and are more susceptible to chemical weathering. Other mechanically durable and chemically stable minerals such as zircon and tourmaline are rare in sedimentary rocks simply because they are uncommon in source rocks.

The only other particles of any consequence in sandstones are grains of chert derived from rock composed of microscopic quartz crystals and the micas (muscovite and biotite), although these rarely comprise more than a few percent of any sandstone. But what about the other ferromagnesian silicates, such as olivine, pyroxenes, and amphiboles, which make up about 24% of Earth's crust? Although common, with the exception of biotite they are scarce in sandstones because of their chemical instability (see Table 5.1).

Sandstone forms in a variety of depositional environments, such as stream channels, beaches and barrier islands, deltas, and the continental shelf. It is a particularly common sedimentary rock, and easily recognized by its gritty appearance.

Mudrocks *Mudrock* is a general term that encompasses all detrital rocks composed of silt- and clay-sized particles (Table 6.2; Figure 6.11). Varieties of mudrocks include *siltstone*, composed mostly of silt-sized particles,

(a)

(b)

Figure 6.11
Mudrocks. Exposures of mudstone (a) and shale (b). Notice that the mudstone lacks any discernible layering so it is characterized as *massive*. Even a close examination of these rocks reveals little in the way of layering. In contrast, the shale in (b) has a platy look and splits along closely spaced planes so it is fissile.

mudstone, a mixture of silt- and clay-sized particles, and *claystone,* composed primarily of clay-sized particles. Recall that the term *clay* has two meanings; it refers to sheet silicates known as *clay minerals,* and it is a size designation for particles smaller than 1/256 mm. And even though mudstones and claystones contain significant amounts of clay minerals, their identity is not important in classifying these rocks. Some mudstones and claystones are designated as *shale* if they are fissile, meaning that they break along closely spaced parallel planes (Figure 6.11b).

About 40% of all detrital sedimentary rocks are mudrocks, making them the most common of these rocks. Because silt and clay particles are so small, they can be transported by weak currents and kept suspended in water by minor turbulence. Consequently deposition takes place where turbulence is at a minimum, as in the quiet offshore waters of lakes or in lagoons.

Chemical and Biochemical Sedimentary Rocks

Various compounds and ions taken into solution during chemical weathering are the raw materials for **chemical sedimentary rocks.** They are so named because chemical processes are responsible for their origin, as when minerals form either as the result of inorganic chemical processes or the chemical activities of organisms. Some of these rocks have a *crystalline texture,* meaning they are composed of an interlocking mosaic of mineral crystals (Figure 6.8b). Others, however, have a clastic texture; some limestones, for instance, are composed of fragments of seashells. Organisms play an important role in the origin of those chemical sedimentary rocks designated as **biochemical sedimentary rocks.**

Limestone and Dolostone   Limestone and dolostone are known as **carbonate rocks** because each is made up mostly of minerals containing the carbonate radical (CO_3) (see Figure 2.9); limestone consists of calcite ($CaCO_3$) and dolostone is made up of dolomite [$CaMg(CO_3)_2$], which are carbonate minerals (see Chapter 2). In Chapter 5 we noted that calcite rapidly dissolves in acidic water, but the chemical reaction leading to dissolution is reversible, so calcite can be precipitated from solution under some circumstances. So some limestone, although probably not very much, is formed by inorganic chemical precipitation. Travertine, a finely crystalline type of limestone precipitated around hot springs, is one example. And some limestones consist of small spherical grains known as *ooids,* which form as layers of calcite chemically precipitate around some kind of nucleus such as a sand grain or shell fragment. Lithified deposits of ooids are known as *oolitic limestones* (Figure 6.12a).

(a)

(b)

(c)

(d)

Figure 6.12
Four types of limestone, two of which, (b) and (c), are biochemical sedimentary rocks. (a) Present-day ooids from the Bahamas. The largest of these ooids measures about 2 mm. A rock composed of ooids is oolitic limestome. (b) Coquina is composed of the broken shells of organisms. (c) Chalk cliffs in Denmark. Chalk is made up of microscopic shells. (d) Limestone containing numerous fossil shells.

Limestone is a particularly common sedimentary rock, most of which has a large component of calcite that was extracted from seawater by organisms. Corals, clams, oysters, snails, algae, and a number of other marine organisms construct their shells of aragonite, an unstable form of calcium carbonate that alters to calcite. When the organisms die, their skeletons or fragments of skeletons from clay- to gravel-size particles accumulate as sediment, which if lithified forms limestone. The limestone known as *coquina* consists entirely of shell fragments cemented by calcium carbonate, and *chalk* is a variety of limestone made up of microscopic shells of organisms (Figure 6.12b and c). Because organisms play such an important role in their origin, most limestones are conveniently classified as *biochemical sedimentary rocks* (Figure 6.12d).

Dolostone is similar to limestone but most or all of it probably formed secondarily by the alteration of limestone. The consensus among geologists is that dolostone forms when magnesium replaces some of the calcium in calcite, thereby converting $CaCO_3$ to dolomite $CaMg(CO_3)_2$. One way this might take place is in an environment such as a lagoon where evaporation rates are high, and much of the calcium in seawater is used in calcite ($CaCO_3$) and gypsum ($CaSO_4 \cdot 2H_2O$). Magnesium (Mg) becomes concentrated in the water, which then becomes denser, sinks, and permeates preexisting limestone converting it to dolostone by adding magnesium to calcite.

Evaporites   In Chapter 5 we discussed the fact that during chemical weathering some minerals are taken into solution. Some of these dissolved substances are used by organisms and account for the origin of several varieties of limestone, but others are precipitated from evaporating water by chemical processes, forming rocks known as **evaporites.** *Rock salt,* composed of halite (NaCl), and *rock gypsum,* composed of gypsum ($CaSO_4 \cdot 2H_2O$) are the most common rocks characterized as evaporites (Figure 6.13). Both have a crystalline texture, and their origin is well understood

All groundwater, water in lakes and streams, and seawater contains dissolved substances. If evaporation occurs, as it does in some lakes and restricted, marginal parts of seas, the volume of water decreases, thereby increasing the amount of dissolved mineral matter in relation to the volume of the fluid. If evaporation continues, the fluid eventually reaches the saturation point, the point at which it can no longer contain all the dissolved material and precipitation of minerals occurs. Rock salt is simply sodium chloride, and rock gypsum is calcium sulfate; both are formed by the process just outlined.

Compared to mudrocks, sandstone, and limestone, evaporites are not very common but nevertheless are significant deposits in some areas. For instance, vast deposits of rock salt and rock gypsum underlie parts of Michigan, Ohio, and New York; the Louann Salt of the Gulf Coast region underlies a large area; and similar deposits are present in Canada and on other continents. In addition to rock salt and rock gypsum, a number of other evaporite minerals and rocks are known, but most are rare. Some are important, though, as sources of chemical compounds, and sylvite, a potassium chloride (KCl), is used in manufacturing fertilizers, dyes, and soaps.

Chert   *Chert* is a hard rock composed of microscopic crystals of quartz (SiO_2) (Table 6.2; Figure 6.14). Several color varieties of chert are recognized, including *flint,* which is black because of inclusions of organic matter, and *jasper,* which is colored red or brown by iron oxides. Because chert is hard and lacks cleavage, it can be shaped to form sharp cutting edges. Many cultures have used chert to manufacture tools, spear points, and arrowheads.

Figure 6.13
Evaporites. (a) Core of rock salt from an oil well in Michigan. (b) Rock Gypsum.

(a)

Figure 6.14
(a) Chert is a dense, hard rock composed of microscopic crystals of quartz. (b) Thin layer of bedded chert.

(b)

Some chert is found as irregular masses or *nodules* within other rocks, especially limestones, and as distinct layers of *bedded chert* (Figure 6.14b). Chert nodules in limestone are clearly secondary; that is, they have replaced part of the host rock, apparently being precipitated from solution. Some bedded chert may precipitate from seawater, but because so little silica is dissolved in seawater, a biochemical origin is more likely. It seems that bedded chert forms from deposits of shells of silica-secreting, single-celled organisms such as radiolarians and diatoms. Unfortunately, these shells are easily altered, so the evidence for a biochemical origin of bedded cherts is obscured.

Coal Coal is one of the few exceptions to the definition of a rock as an aggregate of one or more minerals. It consists of compressed, altered remains of land plants, and despite not conforming to the rock definition, is nevertheless a biochemical sedimentary rock. It forms in swamps and bogs where the water is oxygen deficient or where organic matter accumulates faster than it decomposes. In oxygen-deficient swamps and bogs, the bacteria that decompose vegetation can

live without oxygen, but their wastes must be oxidized, and because little or no oxygen is present, wastes accumulate and kill the bacteria. Thus bacterial decay ceases and the vegetation is not completely decomposed. These partly altered plant remains form organic muck, which commonly smells of hydrogen sulfide (the rotten-egg odor of swamps). When buried and compressed, the muck becomes *peat*, which looks rather like coarse pipe tobacco (Figure 6.15a). Where peat is abundant, as in Ireland and Scotland, it is used for fuel.

The formation of peat is the first step in the origin of coal. If peat is more deeply buried and compressed, and especially if it is heated too, it is converted to

(a)

(b)

(c)

Figure 6.15
(a) Peat is partly decomposed plant material. It represents the first stage in the origin of coal. (b) Lignite is a dull variety of coal in which plant remains are still visible. (c) Bituminous coal is shinier and darker than lignite, and only rarely are plant remains visible.

dull black coal called *lignite*, in which plant remains are still clearly visible (Figure 6.15b). During the change from organic muck to lignite, the easily vaporized or volatile elements of the vegetation such as oxygen, nitrogen, and hydrogen are driven off, enriching the residue in carbon; lignite contains about 70% carbon, whereas only about 50% is present in peat.

Bituminous coal, which contains about 80% carbon, is dense, black, and so thoroughly altered that plant remains can only rarely be seen (Figure 6.15c). It is a higher-grade coal than lignite because it burns more efficiently, but the highest-grade coal is *anthracite*, a metamorphic type of coal (see Chapter 7). It contains up to 98% carbon and, when burned, yields more heat per unit volume than other types of coal.

SEDIMENTARY FACIES— RECOGNIZABLE DIFFERENCES IN ROCK LAYERS

If a layer of sediment or sedimentary rock is traced laterally, it generally shows changes in composition, texture, or both. These lateral changes result from the simultaneous operation of different processes in adjacent depositional environments. For example, sand may be deposited in a high-energy nearshore marine environment, whereas mud and carbonate sediments accumulate simultaneously in the laterally adjacent low-energy offshore environments (Figure 6.16). Deposition in each of these environments produces **sedimentary facies,** bodies of sediment possessing distinctive physical, chemical, and biological attributes.

Any aspect of sedimentary rocks that makes them recognizably different from adjacent rocks of the same age, or approximately the same age, can be used to establish a sedimentary facies. Figure 6.16 illustrates three sedimentary facies: a sand facies, a mud facies, and a carbonate

facies. If lithified, these sediments become sandstone, mudstone (or shale), and limestone facies, respectively.

Marine Transgressions and Regressions

Many sedimentary rocks in the interiors of continents show clear evidence of having been deposited in marine environments. The rocks in Figure 6.17d, for instance, consist of a sandstone facies that was deposited in a nearshore marine environment overlain by shale and limestone facies deposited in offshore environments. This vertical sequence of facies can be explained by deposition taking place during a time when sea level rose with respect to the continent and the shoreline moved inland, thereby giving rise to a **marine transgression** (Figure 6.17). As the shoreline advances inland, the depositional environments parallel to the shoreline do likewise. Remember that each laterally adjacent environment in Figure 6.17 is the depositional site of a different sedimentary facies. As a result of a marine transgression, the facies that formed in the offshore environments are superposed over the facies deposited in the nearshore environment, thus accounting for the vertical succession of sedimentary facies in Figure 6.17d.

Another important aspect of marine transgressions is that an individual facies can be deposited over a huge geographic area. Even though the nearshore environment is long and narrow at any particular time, deposition takes place continuously as the environment migrates landward during a marine transgression. The sand deposited under these conditions may be tens to hundreds of meters thick but has horizontal dimensions of length and width measured in hundreds of kilometers.

The opposite of a marine transgression is a **marine regression.** If sea level falls with respect to a continent, the shoreline and environments that parallel the shoreline move in a seaward direction. The vertical sequence produced by a marine transgression has facies of the

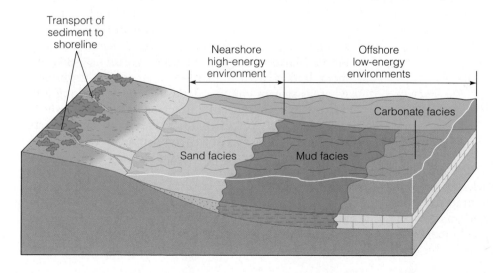

Transport of sediment to shoreline

Nearshore high-energy environment

Offshore low-energy environments

Carbonate facies

Sand facies

Mud facies

Figure 6.16
Deposition in adjacent environments yields distinct bodies of sediment, each of which is designated a sedimentary facies.

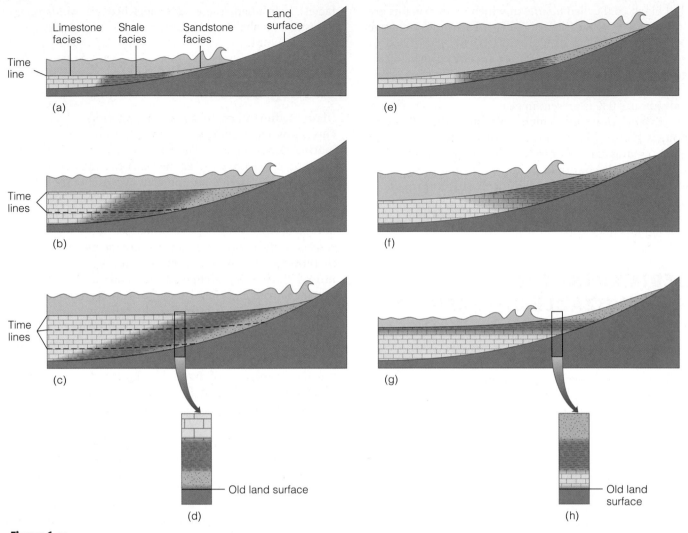

Figure 6.17

(a), (b), and (c) Three stages of a marine transgression. (d) Diagrammatic representation of the vertical sequence of facies resulting from the transgression. (e), (f), and (g) Three stages of a marine regression. (h) Diagrammatic representation of the vertical sequence of facies resulting from the regression.

nearshore environment superposed over facies of off-shore environments (Figure 6.17h). Marine regressions can also account for the deposition of a facies over a large geographic area.

READING THE STORY IN SEDIMENTARY ROCKS

When geologists investigate sedimentary rocks in the field, they are observing the products of events that took place during the past. The only record of those events is preserved in the rocks, so geologists must evaluate those aspects of sedimentary rocks that allow inferences to be made about the original processes and the environment of deposition. Sedimentary textures such as sorting and rounding can give clues to the depositional process. Wind-blown dune sands, for example, tend to be well sorted and well rounded, but poor sorting is characteristic of glacial deposits. The geometry or three-dimensional shape of rock bodies is another important criterion in environmental interpretation. Marine transgressions and regressions yield sediment bodies with a blanket or sheetlike geometry, whereas stream channel deposits tend to be long and narrow and are therefore described as having a shoe-string geometry (Figure 6.18). Other important aspects of sedimentary rocks in environmental analysis include sedimentary structures and fossils.

Sedimentary Structures

When sediment is deposited, it contains a variety of features known as **sedimentary structures** that formed as a

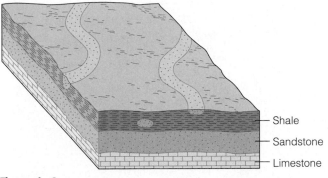

Figure 6.18
Two different geometries of sedimentary rock bodies. The shale, sandstone, and limestone all have blanket geometries. The elongate sandstones within the shale have a shoestring geometry.

Shale
Sandstone
Limestone

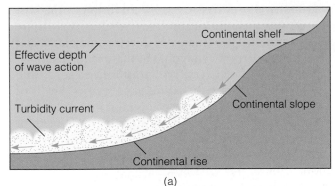

result of physical and biological processes operating in the depositional environment. One of the most common of these sedimentary structures is distinct layers known as **strata,** or **beds** (Figure 6.19). Beds vary in thickness from less than a millimeter up to many meters. Many sedimentary rock layers are separated from one another by a surface known as a *bedding plane,* above and below which the rocks differ markedly in composition, grain size, color, or a combination of features. Such abrupt changes indicate rapid changes in the type of sediment that accumulated or perhaps a period of nondeposition or erosion followed by renewed deposition. In contrast, a layer may grade upward from one rock type into another, thus indicating a gradual change in deposition. Almost all sedimentary rocks show some kind of stratification or bedding; a few, such as limestones that formed as coral reefs, lack this feature.

In **graded bedding,** grain size decreases upward within a single bed (Figure 6.20). Most graded bedding appears to have formed from turbidity current

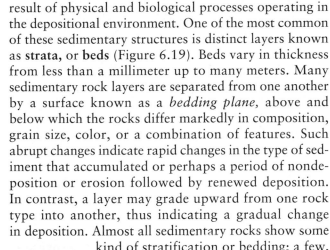

Figure 6.19
Bedding or stratification is shown in these rocks as alternating layers of mudrock (shale in this case) and sandstone.

Continental shelf

Effective depth of wave action

Turbidity current

Continental slope

Continental rise

(a)

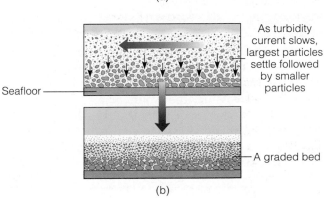

As turbidity current slows, largest particles settle followed by smaller particles

Seafloor

A graded bed

(b)

(c)

Figure 6.20
Graded bedding. (a) Turbidity currents flow downslope along the seafloor (or lake bottom) because of their density. (b) Graded bedding forms as deposition of progressively smaller particles takes place. (c) Graded bedding in a stream channel where flow velocity decreases rapidly.

deposition. *Turbidity currents* are underwater flows of sediment–water mixtures that are denser than sediment-free water. These flows move downslope along the bottom of the sea or a lake until they reach the relatively level seafloor or lake floor. There, they rapidly slow down and begin depositing transported sediment, the coarsest first followed by progressively smaller particles (Figure 6.20a and b). Although turbidity currents account for most graded bedding, some forms in stream channels during the waning stages of floods (Figure 6.20c).

Many sedimentary rocks are characterized by **cross-bedding,** in which layers are arranged at an angle to the surface on which they are deposited (Figure 6.21). Cross-bedding is common in desert dunes and in sediments in stream channels and shallow marine environments. Invariably, cross-beds result from transport by wind or water currents and deposition on the downcurrent side of dunelike structures. Cross-beds are inclined downward, or dip, in the direction of flow, so are good indicators of ancient current directions, or **paleocurrents** (Figure 6.21).

In sand deposits, one can commonly observe small-scale, ridgelike **ripple marks** on the surfaces of beds. One type of ripple mark is asymmetric in cross section with a gentle upstream slope and a steep downstream slope. Known as *current ripple marks* (Figure 6.22b), they form as a result of currents that move in one direction as in a stream channel. Like cross-bedding, current ripple marks are good paleocurrent indicators. In contrast, the to-and-fro motion of waves produces ripples that tend to be symmetric in cross section; known as *wave-formed ripple marks,* they form mostly in the shallow, nearshore waters of oceans and lakes (Figure 6.22d).

When clay-rich sediment dries, it shrinks and forms intersecting fractures known as **mud cracks** (Figure 6.23). These features in ancient sedimentary rocks indicate that the sediment was deposited where periodic drying was possible, as on a river floodplain, near a lake shore, or where muddy deposits are exposed on marine shorelines at low tide. Many other sedimentary structures are known, including some that form long after deposition (see Perspective 6.1).

Fossils

Fossils, the remains or traces of ancient organisms, are mostly the hard skeletal parts such as shells, bones, and teeth, but under exceptional conditions even the soft-part anatomy may be preserved. For example, several frozen woolly mammoths have been discovered in Alaska and Siberia with hair, flesh, and internal organs preserved. The remains of organisms are known as *body fossils* to distinguish them from *trace fossils* such as

Figure 6.21
The inclined layers in the center of this image are cross-beds. Notice that the cross-bedding is bounded above and below by nearly horizontal layers. Because the cross-beds are inclined downward toward the left, we know that the currents responsible for them flowed in that direction.

tracks, trails, nests, and burrows, which are indications of ancient organic activity (Figure 6.24).

For any potential fossil to be preserved, it must escape the ravages of destructive processes such as running water, waves, scavengers, exposure to the atmosphere, and bacterial decay. Obviously, the soft parts of organisms are devoured or decomposed most rapidly, but even the hard skeletal parts will be destroyed unless they are buried and protected in mud, sand, or volcanic ash. Even if buried, bones and shells may be dissolved by groundwater or destroyed by alteration of the host rock during metamorphism. Nevertheless, fossils are quite common. The remains of microscopic plants and animals are the most common, but these require specialized methods of recovery, preparation, and study and are not sought out by casual fossil collectors. Shells of marine animals are also very common and easily collected in many areas, and even the bones and teeth of dinosaurs are much more common than most people realize (see Perspective 6.2).

WHAT WOULD YOU DO?

No one was present millions of years ago to record data about the climate, geography, and geologic processes. So how is it possible to decipher unobserved past events? In other words, what features in rocks, and especially sedimentary rocks, would you look for to determine what happened in the far distant past? Also, explain why your interpretations might have relevance to predicting future events.

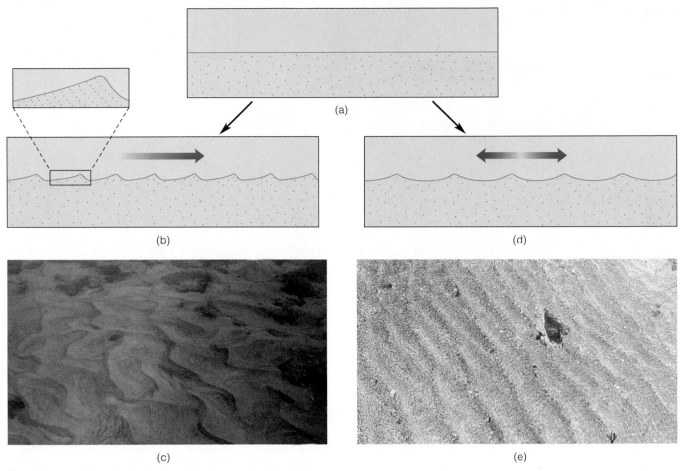

Figure 6.22

Ripple marks. (a) Undisturbed layer of sand. (b) Current ripple marks form in response to flow in one direction as in a stream channel. The enlargement of one ripple shows its internal structure. Note that individual layers within the ripple are inclined, showing an example of cross-bedding. (c) Current ripples that formed in a small stream channel; flow was from right to left. (d) The to-and-fro currents of waves in shallow water deform the surface of the sand layer into wave-formed ripple marks. (e) Wave-formed ripple marks on ancient rocks.

Figure 6.23

(a) Mud cracks form in clay-rich sediments when they dry and shrink. (b) Mud cracks in ancient rocks in Glacier National Park, Montana. Notice that the cracks have been filled by sediment.

Perspective 6.1

Concretions and Geodes

R ipple marks, cross-bedding, and mud cracks are designated *primary sedimentary structures* because they formed at the time of sediment deposition or shortly thereafter. In contrast, *secondary sedimentary structures* form in sediment long after deposition, and unlike primary structures they tell nothing of how sediment was deposited in the first place. They are, however, interesting in their own right, and some are attractive. Many varieties of secondary sedimentary structures are known but only concretions and geodes are discussed here.

*Concretion*s consist of any mass of material that can easily be separated from the enclosing rock (Figure 1). They are found in shale, limestone, and coal beds, but probably the most common types are sandstone concretions result-

ing from more thorough cementation in restricted parts of sand deposits. They vary from l cm up to 9 m across, and although commonly spherical, their shapes can be quite irregular, giving rise to oddly shaped objects. Because they are harder than the enclosing host rock, concretions commonly accumulate at the surface during weathering.

Imatra stones, or *marlekor,* are particularly interesting concretions (Figure 1b). Typically disk-shaped and composed of calcium carbonate ($CaCO_3$), imatra stones are found in clay layers of glacial lake deposits. Most are only a few centimeters across and have simple shapes, but some unusual shapes result when simple forms grow together or outgrowths form (Figure 1b). Some of these concretions have such regular geometric shapes that amateur rock and fossil collectors sometimes mis-

take these and similar ones for objects manufactured by humans.

Perhaps the most attractive concretions are those called *septaria,* or *septarian nodules.* These are large spherical concretions with a series of cracks that widen toward the concretion's center. Smaller cracks more or less parallel the margin of the concretion and intersect the larger cracks (Figure 2). Apparently, the cracks formed when the original concretion shrank during dehydration; later the cracks filled with mineral crystals, mostly calcite. Some septaria are released from the host rock during weathering and eroded so that the interior cracks are visible, giving the concretion the appearance of a turtle shell (Figure 2b).

Some oddly shaped concretions in Cretaceous rocks in Oklahoma, and

(a)

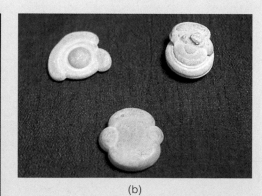

(b)

Figure 1
(a) These spherical concretions each measuring 6 to 8 cm across have grown together to form a composite concretion.
(b) These disk-shaped concretions called *imatra stones,* or *marlekor,* are from glacial lake deposits in Connecticut.

(a)

(b)

Figure 2
(a) Cross section showing the internal structure of a septarian nodule.
(b) Surface view of septarian nodule resembling a turtle shell.

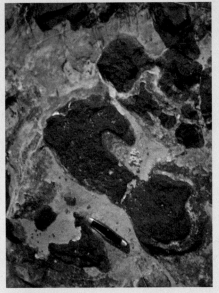

Figure 3
Concretions eroded from Cretaceous rocks in Oklahoma. These were mistakenly identified as human shoe prints. The specimens measure about 20 to 25 cm long.

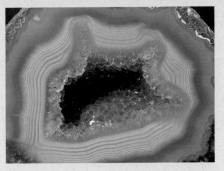
(a)

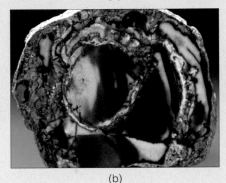
(b)

Figure 4
(a) A geode formed by the partial filling of a cavity by color-banded agate, which conforms to the walls of the cavity, and by inward-pointing quartz crystals. (b) A thunder egg from Oregon.

elsewhere, have been mistakenly identified as human shoe prints in ancient rocks (Figure 3). Some people claim these constitute evidence that humans and dinosaurs existed at the same time. However, even a superficial examination of the rocks reveals the "shoe prints" are nothing more than concretions exposed by differential weathering and erosion of the host rock. Their slight resemblance to shoe prints is purely coincidental.

Geodes are hollow, globular, attractive objects commonly used as decorative stones. Sawed specimens mounted on some kind of base are available at just about any rock and mineral shop. Most geodes are small, measuring 30 cm or less in diameter, although some are much larger. They form when minerals grow along the margins of a cavity, partially filling it (Figure 4a). The outermost layer of minerals is a thin, sometimes discontinuous layer of agate, a variety of

color-banded, compact, microcrystalline quartz (SiO_2). The banding in this agate conforms to the walls of the cavity in which the minerals grew. Next inward toward the cavity's center are inward-pointing crystals of quartz or calcite ($CaCO_3$), or crystals of various sulfates such as barite ($BaSO_4$) and celestite ($SrSO_4$), all of which were precipitated from solution.

Geodes, or at least geodelike objects, are not restricted to sedimentary rocks. *Thunder eggs* resembling geodes are found in some igneous rock, especially welded tuffs and rhyolite (Figure 4b). These formed when a gas cavity filled with minerals. According to one legend, the spirits occupying neighboring volcanoes in Oregon hurled spherical rocks at one another which they had stolen from the thunder bird, hence the name *thunder eggs*. The knobby, ribbed exterior of a thunder egg is not particularly impressive, but inside they are filled with

agate, chalcedony, jasper, or opal, all of which are varieties of silicon dioxide (SiO_2).

Concretions and geodes are interesting but provide no information on how their host rocks formed in the first place. Nevertheless, they might reveal something about changes that took place following deposition and lithification. And of course their interest as curiosities and decorative objects makes their study worthwhile.

Fossils—Much More Common Than Most People Realize

Only a small proportion of animals and plants that ever lived become fossils, yet fossils are quite common, much more so than most people realize. The reason for this abundance? So many billions of organisms have existed over so many millions of years that if only one or part of one in 10,000 were fossilized, the total number of fossils is truly astonishing. Of course, many fossils are destroyed by dissolution, by the conversion of limestone to dolostone, and during metamorphism, and many remain deeply buried and as yet undiscovered. But even with these several qualifications, the number of fossils found so far and those that are readily accessible is still phenomenal.

By far the most common fossils are microscopic remains of plants and animals, but collecting and analyzing them requires specialized techniques not normally used by casual fossil collectors. These microfossils are very useful, in the oil industry, for instance, because large numbers can be recovered from very small rock samples and used to determine the age of their host rock. The most common, easily collected fossils are those of marine invertebrate animals, sea-dwelling creatures lacking a segmented vertebral column, such as clams, oysters, corals, sea lilies, and brachiopods (see Figure 6.24b). In some areas these fossils are so common that they litter the surface as their host rocks weather.

Fossils of vertebrate animals, those possessing a segmented vertebral column, are not nearly as common as those of invertebrates, but even the remains of fish, amphibians, reptiles, birds, and mammals are more abundant than most people think. Thousands of fossil fish are present on single surfaces within 50-million-year-old sedimentary rocks in Wyoming, and pieces of dinosaur bones and teeth are easily collected in some places. It is true that complete skeletons of dinosaurs and other land-dwelling animals are not common, but even so some remarkable concentrations of their bones have been discovered in many areas.

At Howe Quarry in Wyoming, more than 4000 dinosaur bones have been recovered from a deposit measuring only 18 × 16 meters (Figure 1). But the

Figure 1
Howe Quarry in Wyoming, where more than 4000 dinosaur bones have been recovered.

most remarkable aspect of this discovery was the 12 legs found preserved in an upright position penetrating the mud deposit, indicating that several large dinosaurs became mired in mud and perished. The parts of the dinosaurs above the mud layer decomposed and formed a heap of bones, whereas the legs were preserved in the position in which the dinosaurs were trapped. In northwestern Montana, a bone bed was discovered in 1981 that contains an estimated 10,000 juvenile to adult duck-billed dinosaurs measuring between 3.6 and 7.6 m long. Apparently, a vast herd was overcome by volcanic ash and gases and subsequently buried in ash. A bone bed in Canada contains the remains of hundreds of horned dinosaurs that evidently drowned while attempting to cross a river.

Numerous fossils of Triassic-aged marine reptiles known as ichthyosaurs are present at Berlin-Ichthyosaur State Park in Nevada (Figure 2). Ichthyosaurs were fast-swimming predators that somewhat resemble present-day dolphins and porpoises. The remains of dozens of these animals have been found in various layers in the park, including as many as nine that are now housed in a special viewing area. Many of these animals died when they were stranded along the shore of a vast inland sea at low tide, thus providing a record of the largest ichthyosaurs known; they were more than 15 m long and must have weighed several tons.

About 10 million years ago in what is now northeastern Nebraska, a vast grassland was inhabited by short-legged rhinoceroses, camels, three-

(a)

(b)

Figure 2
(a) Flipper bones of an ichthyosaur at Berlin-Ichthyosaur State Park, Nevada. These and other bones of perhaps nine ichthyosaurs are partly excavated and protected in a structure. (b) Frieze of an ichthyosaur about 18 m long, near the entrance to the viewing area in the park.

toed horses, saber-toothed deer, land turtles, and a variety of other animals. Thousands of these animals died when a huge cloud of volcanic ash covered the area. Many animals were killed in the initial ashfall, and their bodies lay at the surface where they were partly decomposed and scavenged before burial. In contrast, later redistribution of the ash by wind killed hundreds of rhinoceroses and buried them quickly, as indicated by the large number of complete skeletons with even delicate middle ear bones preserved (Figure 3).

Scientists known as *paleontologists*, who study life history as revealed by fossils, find, recover, and study numerous fossils, but the remarkable concentrations of bones in these examples provide answers to some otherwise difficult questions. For instance, some aspects of social behavior and the makeup of a dinosaur herd might be revealed. The ancient calamities accounting for such remarkable concentrations of bones provide paleontologists with a glimpse of what life was like millions of years ago.

Figure 3
Paleontologists excavating rhinoceros (foreground) and horse (background) skeletons from volcanic ash near Orchard, Nebraska.

Figure 6.24
Body fossils (a) and (b) and trace fossils (c). (a) Bones of a 2.3-m-long, Mesozoic-aged marine reptile in the museum at the Glacier Garden in Lucerne, Switzerland. (b) Shells of extinct Mesozoic organisms known as ammonites in the Comstock Rock Shop, Virginia City, Nevada. (c) Trace fossils consisting of bird tracks in the 50-million-year-old Green River Formation of Wyoming. This is actually the layer of rock that was deposited on the layer containing the tracks, so it is a cast of the tracks.

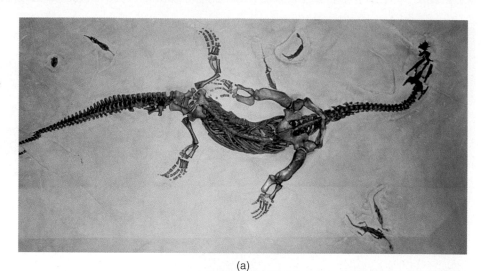

(a)

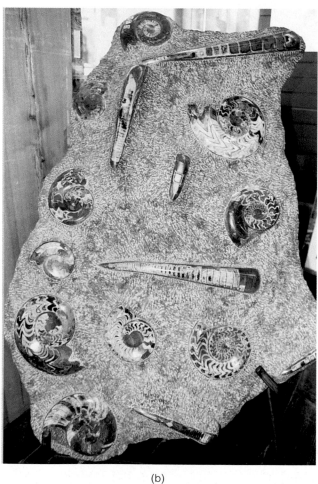

(b)

(c)

Some fossils retain their original composition and structure and are preserved as unaltered remains, but many have been altered in some way. Dissolved minerals can be precipitated in the pores of bones, teeth, and shells or can fill the spaces within cells of wood. Wood may be preserved by silica replacing the woody tissues; it then is referred to as *petrified,* a term that means "to become stone" (Figure 6.25a). Silicon dioxide (SiO_2) or iron sulfide (FeS_2) can completely replace the calcium carbonate ($CaCO_3$) shells of marine animals (Figure 6.25b). Insects and the leaves, stems, and roots of plants are commonly preserved as thin carbon films that show the details of the original organism (Figure 6.25c, d).

Figure 6.25
Various types of fossil preservation. (a) Petrified tree stump in Florissant Fossil Beds National Monument, Colorado. Volcanic mudflows 3 to 6 m deep buried the lower parts of many trees at this site. (b) Shells replaced by iron sulfide (FeS$_2$). Carbon film of a palm frond (c) and an insect (d).

Shells and bones in sediment (Figure 6.26) may be dissolved leaving a cavity called a *mold* shaped like the shell or bone. If a mold is filled in, it becomes a *cast* (Figure 6.26).

If it were not for fossils, we would have no knowledge of trilobites, dinosaurs, and other extinct organisms. Thus, fossils constitute our only record of ancient life. They are not simply curiosities but have several practical uses. In many geologic studies, it is necessary to correlate or determine age equivalence of sedimentary rocks in different areas. Such correlations are most commonly demonstrated with fossils; we will discuss correlation more fully in Chapter 8. Fossils are also useful in determining environments of deposition.

Determining the Environment of Deposition

Ancient sedimentary rocks acquired many of their properties as a result of the physical, chemical, and biological processes that operated in the depositional environment. To determine what the depositional environment was, geologists investigate these rock properties, reasoning that the processes responsible for them are the same as those going on at present. For instance, we have every reason to think that wave-formed ripple marks originated by the to-and-fro motion of waves throughout geologic time; thus, we are justified in stating that these structures in an ancient sandstone formed

Figure 6.26
Molds and casts. (a) Burial of a shell in sediment. (b) The shell dissolves leaving a cavity, or mold. (c) The mold is filled by sediment, thus forming a cast. (d) This fossil turtle shell shows how both a mold and a cast form. Note that some of the original turtle shell remains as the outermost part of the fossil. The hollow interior of the shell served as a mold that was filled with sediment, thus forming a cast of the shell's interior. Later, much of the shell was lost, revealing the cast showing details of the inner part of the shell.

(a)

(b)

(c)

(d)

just as they do now. In short, we are simply applying the principle of uniformitarianism (see Chapter 1). Accordingly, geologists with their knowledge of various present-day processes such as sediment transport, wave action, and deposition by streams can make inferences regarding the depositional environment of ancient sedimentary rocks.

While conducting field studies, geologists commonly make some preliminary interpretations. Some sedimentary particles such as ooids in limestones (see Figure 6.12a) form in shallow marine environments where currents are vigorous. Large-scale cross-bedding is typical of but not restricted to desert dunes. Fossils of land plants and animals can be washed into transitional environments, but most of them are preserved in deposits of continental environments. Fossil shells of such marine-dwelling animals as corals obviously indicate marine depositional environments.

Much environmental interpretation is done in the laboratory where the data and rock samples collected during fieldwork can be more fully analyzed. The analysis may include microscopic and chemical examination of rock samples, identification of fossils, and graphic representations showing the three-dimensional shapes of rock units and their relationships to other rock units. In addition, the features of sedimentary rocks are compared with those of sediments from present-day depositional environments; once again, the contention is that features in ancient rocks, such as cross-beds and mud cracks, formed during the past in response to the same

processes responsible for them now. Finally, when all data have been analyzed, an environmental interpretation is made.

The following examples illustrate how environmental interpretations are made. The Navajo Sandstone of the southwestern United States covers a vast area, perhaps as much as 500,000 km^2. It has an irregular sheet geometry, reaches about 300 m thick in the area of Zion National Park, Utah, and consists mostly of well-sorted, well-rounded sand grains measuring about 0.2 to 0.5 mm in diameter (Figure 6.27). Some of the sandstone beds also possess tracks of dinosaurs and other land-dwelling animals, ruling out the possibility of a marine origin for the rock unit. These features, and the fact that the Navajo Sandstone has cross-beds up to 30 m high (Figure 6.27) and current ripple marks, both of which appear to have formed in sand dunes, lead to the conclusion that the sandstone represents an ancient desert dune deposit. The cross-beds are inclined downward, or dip generally to the southwest, indicating that the wind blew mostly from the northeast.

In the Grand Canyon of Arizona several types of rocks are exposed and recognized as *formations*, which are simply widespread units of rock recognizably different from those above and below. The term *formation* can be applied to any rock type, but it is especially applicable to sedimentary rocks such as the Tapeats Sandstone, Bright Angel Shale, and Muav Limestone. A vertical sequence of these three formations is well exposed in the lower part of the Grand Canyon (Fig-

Figure 6.27
These sedimentary rocks making up Checkerboard Mesa in Zion National Park, Utah, belong to the Jurassic-aged Navajo Sandstone, which represents an ancient windblown dune deposit. Vertical fractures intersect inclined layers known as *cross-beds,* giving this cliff its checkerboard appearance.

Figure 6.28
View of the Tapeats Sandstone, Bright Angel Shale, and Muav Limestone in the Grand Canyon in Arizona. These formations were deposited during a widespread marine transgression. Compare this vertical sequence of rocks with the one shown in Figure 6.17d.

ure 6.28), and all contain features, including fossils, clearly indicating they were deposited in transitional and marine environments. In fact, all three were forming simultaneously in different adjacent environments, but during a marine transgression they came to be superposed in the order now seen. That is, offshore facies came to overlie nearshore facies (see Figure 6.17). Similar sequences of rocks of approximately the same age in Utah, Colorado, Wyoming, Montana, and South Dakota indicate that this was indeed a widespread marine transgression.

WHAT IMPORTANT RESOURCES ARE FOUND IN SEDIMENTS AND SEDIMENTARY ROCKS?

Sediments and sedimentary rocks or the materials they contain have a variety of uses. Sand and gravel are essential to the construction industry, pure clay deposits are used for ceramics, and limestone is used in the manufacture of cement and in blast furnaces where iron ore is refined to make steel. Evaporites are the source of table salt as well as a number of chemical compounds, and rock gypsum is used to manufacture wallboard. Sand composed mostly of quartz, or what is called *silica sand,* has a variety of uses, including the manufacture of glass, refractory bricks for blast furnaces, and molds for casting iron, aluminum, and copper alloys.

Some valuable sedimentary deposits form when minerals become separated from other transported sediment because of their greater density. These *placer deposits,* as they are called, are surface accumulations resulting from the separation and concentration of material of greater density from those of less density in streams and on beaches. Much of the gold recovered during the California gold rush (1848–1853) was mined from placer deposits, and placers of a variety of other minerals such as diamonds and tin are important.

The tiny island nation of Nauru, with one of the highest per capita incomes in the world, has an economy based almost entirely on mining and exporting phosphate-bearing sedimentary rock used in fertilizers. More than half of Florida's mineral value comes from mining phosphate rock (see Chapter 11).

Dolostones in Missouri are the host rocks for ores of lead and zinc. Diatomite is a sedimentary rock composed of the microscopic shells of single-celled plants that have a skeleton of silica (SiO_2). This lightweight, porous rock is used in gas purification and to filter a number of fluids such as molasses, fruit juices, water, and sewage. The United States is the world leader in diatomite production, mostly from mines in California, Oregon, and Washington.

Petroleum and Natural Gas

Both petroleum and natural gas are *hydrocarbons,* meaning they are composed solely of hydrogen and carbon. Hydrocarbons form from the remains of microscopic organisms that exist in the seas and in some large lakes. When these organisms die, their remains settle to the seafloor or lake floor where little oxygen is available to decompose them. They are then buried under layers of sediment. As the depth of burial increases, they are heated and transformed into petroleum and natural gas.

The rock in which the hydrocarbons formed is the *source rock,* but for them to accumulate in economic quantities, they must migrate from the source rock into some kind of *reservoir rock* where they are trapped beneath some kind of cap rock. Otherwise, both would migrate upward and eventually seep out at the surface (Figure 6.29). Indeed, a number of oil seeps are known; one of the most famous is the La Brea Tar Pits in Los Angeles, California. Effective reservoir rocks contain considerable pore space where appreciable quantities of hydrocarbons can accumulate. Furthermore, reservoir rocks must possess high *permeability,* or the capacity to transmit fluids; otherwise, hydrocarbons cannot be extracted in reasonable quantities.

Many hydrocarbon reservoirs consist of nearshore marine sandstones in proximity to fine-grained, organic-rich source rocks. Such oil and gas traps are called *stratigraphic traps* (Figure 6.29a), because they owe their existence to variations in the rock layers or strata. Ancient coral reefs are also good stratigraphic traps. Indeed, some of the oil in the Persian Gulf region is trapped in ancient reefs. *Structural traps* result when

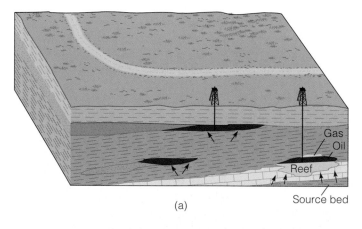

(a)

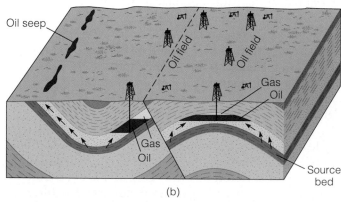

(b)

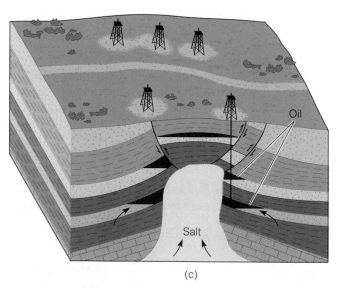

(c)

Figure 6.29

Oil and natural gas traps. The arrows in the diagrams indicate the migration of hydrocarbons. (a) Two examples of stratigraphic traps. (b) Two examples of structural traps, one formed by folding, the other by faulting. (c) An example of structures adjacent to a salt dome where oil and natural gas might be trapped.

rocks are deformed by folding, fracturing, or both (Figure 6.29b). In areas where sedimentary rocks have been deformed into a series of folds, hydrocarbons migrate to the high parts of these structures. Displacement of rocks along faults (fractures along which movement has occurred) also yields situations conducive to trapping hydrocarbons (Figure 6.29b).

In the Gulf Coast region, hydrocarbons are commonly found in structures adjacent to salt domes. A vast layer of rock salt was deposited in this region during the Jurassic Period as the ancestral Gulf of Mexico formed when North America separated from North Africa. Rock salt is a low-density sedimentary rock, and when deeply buried beneath more dense sediments such as sand and mud, it rises toward the surface in pillars known as *salt domes*. As the rock salt rises, it penetrates and deforms the overlying rock layers, forming structures along its margins that may trap petroleum and gas (Figure 6.29c).

Although large concentrations of petroleum are present in many areas of the world, more than 50% of all proven reserves are in the Persian Gulf region! Furthermore, some of the oil fields are gigantic; at least 20 are expected to yield more than 5 billion barrels of oil each, and several have already surpassed this figure.

Many nations including the United States are heavily dependent on imports of Persian Gulf oil, a dependence that will increase in the future. Within a few decades, however, the world's petroleum resources will likely be nearly exhausted. Most geologists think that all the truly gigantic oil fields have already been found but concede that some significant discoveries are yet to be made. One must view these potential discoveries in the proper perspective, however. For example, the discovery of an oil field comparable to that of the North Slope of Alaska (about 10 billion barrels) constitutes about a two-year supply for the United States at the current consumption rate.

Oil Shale and Tar Sands

Other sources of petroleum that will probably become increasingly important in the future are *oil shale* and *tar sands*. Oil shale consists of rock composed of small particles and an organic substance known as *kerogen*. When the appropriate processes are used, liquid oil and combustible gases can be extracted from the kerogen of oil shale. The use of oil shale as a fuel is not new. During the Middle Ages (from about A.D. 600 to 1350) people in Europe used oil shale as solid fuel for domestic heating, and during the 1850s, small oil shale industries existed in the eastern United States, but were discontinued when drilling and pumping of oil began in 1858.

The United States has about two-thirds of all known oil shale, most of it in the Green River Formation of Wyoming, Colorado, and Utah, but large deposits are also present elsewhere, especially in South America. According to one estimate, 80 billion barrels of oil could be recovered from the Green River Formation with present technology. Currently, though, no oil is produced from oil shale in the United States because the process is more expensive than conventional drilling and pumping. And even though oil shale represents a huge untapped resource, at current and expected rates of oil consumption extracting oil from oil shale will not solve all our energy needs. Furthermore, large-scale mining would be necessary, as would large quantities of water, both of which would have profound impacts on the environment.

Tar sand is a type of sandstone in which viscous, asphaltlike hydrocarbons fill the pore spaces. This substance is the sticky residue of once-liquid petroleum from which the volatile constituents have been lost. Liquid petroleum can be extracted from tar sands, but to do so requires mining and processing large amounts of rock. Because the United States has few tar sand deposits, it cannot look to this source as a significant future energy resource. The Athabaska tar sands in Alberta, Canada, however, are one of the largest deposits of this type. These deposits are currently being mined, and may contain several hundred billion barrels of recoverable petroleum.

Coal

Historically, most coal mined in the United States has been bituminous coal from the Appalachian coal basin (Figure 6.30). These coal deposits formed in coastal swamps during the Pennsylvanian Period between 286 and 320 million years ago. Huge lignite and subbituminous coal deposits in the western United States are becoming increasingly important resources (Figure 6.30). During 1995, more than a billion tons of coal was mined in the United States, more than half of it coming from mines in Wyoming, West Virginia, and Kentucky.

Anthracite coal is an especially desirable resource, because it burns hot with a smokeless flame. Unfortunately, it is the least common type of coal, so most coal used for heating buildings and for generating electrical energy is bituminous (see Figure 6.15c). Bituminous coal is also used to make *coke,* a hard gray substance consisting of the fused ash of bituminous coal. Coke is prepared by heating the coal and driving off the volatile matter, and is used to fire blast furnaces for steel production. Synthetic oil and gas and a number of other products are also made from bituminous coal and lignite.

Uranium

Most uranium used in nuclear reactors in North America comes from the complex potassium-, uranium-,

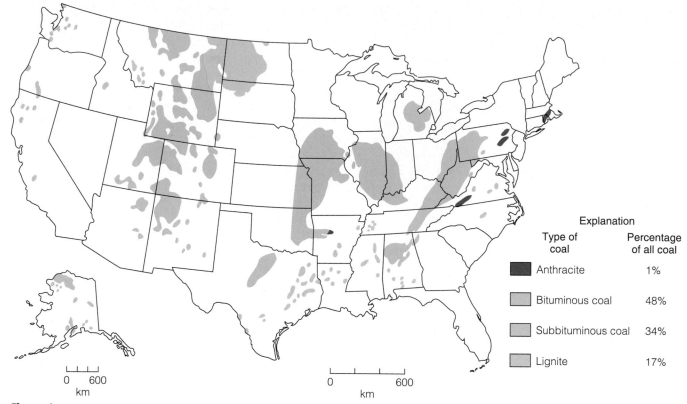

Figure 6.30
Distribution of coal deposits in the United States.

Explanation

Type of coal	Percentage of all coal
Anthracite	1%
Bituminous coal	48%
Subbituminous coal	34%
Lignite	17%

vanadium-bearing mineral *carnotite* found in some sedimentary rocks. Some uranium is also derived from *uraninite* (UO_2), a uranium oxide found in granitic rocks and hydrothermal veins. Uraninite is easily oxidized and dissolved in groundwater, transported elsewhere, and chemically reduced and precipitated in the presence of organic matter.

The richest uranium ores in the United States are widespread in the Colorado Plateau area of Colorado and adjoining parts of Wyoming, Utah, Arizona, and New Mexico. These ores, consisting of fairly pure masses and encrustations of carnotite, are associated with plant remains in sandstones that formed in ancient stream channels. Although most of these ores are associated with fragmentary plant remains, some petrified trees also contain large quantities of uranium.

Large reserves of low-grade uranium ore also are found in the Chattanooga Shale. The uranium is finely disseminated in this black, organic-rich mudrock that underlies large parts of several states including Illinois, Indiana, Ohio, Kentucky, and Tennessee. Canada is the world's largest producer and exporter of uranium.

Banded Iron Formation

Banded iron formation is a chemical sedimentary rock of great economic importance. These rocks consist of alternating thin layers of chert and iron minerals, mostly the iron oxides hematite and magnetite (Figure 6.31).

Figure 6.31
Banded iron formation near Ishpeming, Michigan. At this location the rocks are brilliantly colored alternating layers of red chert and silver iron minerals.

Susan M. Landon

Exploring for Oil and Natural Gas

Susan M. Landon began her career in 1974 with Amoco Production Company and, in 1989, opened her own consulting office in Denver, Colorado. She is currently a partner in the exploration group, Thomasson Partner Associates. In 1990 she was elected president of the American Institute of Professional Geologists and, in 1992, treasurer of the American Association of Petroleum Geologists.

I am an independent petroleum geologist. I specialize in applying geologic principles to oil and gas exploration in frontier areas—places where little or no exploration has occurred and few or no hydrocarbons have been discovered. It is very much like solving a mystery. The Earth provides a variety of clues—rock type, organic content, stratigraphic relationships, structure, and the like—that geologists must piece together to determine the potential for the presence of hydrocarbons.

For example, as part of my work for Amoco, I explored the Precambrian Midcontinent Rift frontier in the north central portion of the United States. Some rifts, like the Gulf of Suez and the North Sea, are characterized by significant hydrocarbon reserves, and the presence of an unexplored rift basin in the center of North America is intriguing. A copper mine in the Upper Peninsula of Michigan, the White Pine Mine, has historically been plagued by oil bleeding out of fractures in the shale. For many years, this had been documented as academically interesting because the rocks are much older than those that typically have been associated with hydrocarbon production.

Field and laboratory work documented that the copper-bearing shale at the White Pine Mine contained adequate organic material to be the source of the oil. The thermal history of the basin was modeled to determine the timing of hydrocarbon generation. If hydrocarbons had been generated prior to deposition of an effective seal and formation of a trap, the hydrocarbons would have leaked out naturally into the atmosphere.

Further work identified sandstones with enough porosity to serve as reservoirs for hydrocarbons. Seismic data were acquired and interpreted to identify specific traps. We then had to convince Amoco management that this prospect had high enough potential to contain hydrocarbon reserves to offset the significant risks and costs. In this case, management agreed that the risk was offset by the potential for a very large accumulation of hydrocarbons, and a well was authorized. Amoco drilled a 5441-m well in Iowa to test the prospect, at a cost of nearly $5 million. The well was dry (economically unsuccessful), but the geologic information obtained as a result of drilling the well will be used to continue to define prospective drilling sites in the Midcontinent Rift.

My career in the petroleum industry began with Amoco Production Company, and after 15 years, I made the decision to leave the company to work independently. For several years, I consulted for a variety of companies, assisting them in exploration projects. I am now involved in a partnership with several other exploration geologists and geophysicists, and we are actively exploring for oil and natural gas in the United States. We develop ideas, defining areas that we believe may be prospective, and attract other companies as partners to assist us in further exploration and drilling. Currently, I am working on projects located in Wyoming, southern Illinois, southern Michigan, Iowa, and Minnesota. I was Manager of Exploration Training when I left Amoco, and as a result, I currently teach a few courses (such as Petroleum Geology for Engineers), which allows me to travel to places like Cairo, Egypt, Quito, Ecuador, and Houston, Texas. ∎

Banded iron formations are present on all the continents and account for most of the iron ore mined in the world today. Vast banded iron formations are present in the Lake Superior region of the United States and Canada and in the Labrador trough of eastern Canada.

The origin of banded iron formations is not fully understood, and none are currently forming. Fully 92% of all banded iron formations were deposited in shallow seas during the Proterozoic Eon between 2.5 and 2.0 billion years ago. A highly reactive element, iron in the presence of oxygen combines to form rustlike oxides that are not readily soluble in water. During early Earth history, little oxygen was present in the atmosphere, so little was dissolved in seawater. However, soluble reduced iron (Fe^{+2}) and silica were present in seawater.

Geologic evidence indicates that abundant photosynthesizing organisms were present about 2.5 billion years ago. These organisms, such as bacteria, release oxygen as a by-product of respiration; thus, they released oxygen into seawater and caused large-scale precipitation of iron oxides and silica as banded iron formations.

Chapter Summary

1. Detrital sediment consists of weathered solid particles, whereas chemical sediment consists of minerals extracted from solution by inorganic chemical processes and the activities of organisms.

2. Sedimentary particles are designated in order of decreasing size as gravel, sand, silt, and clay.

3. During transport, sedimentary particles are rounded and sorted, although the degree of rounding and sorting depends on particle size, transport distance, and depositional process.

4. Any area where sediment is deposited is a depositional environment. Major depositional settings are continental, transitional, and marine, each of which includes several specific depositional environments.

5. Lithification takes place when sediments are compacted and cemented, and thus converted into sedimentary rock. Silica and calcium carbonate are the most common chemical cements, but iron oxide and iron hydroxide cements are important in some rocks.

6. Sedimentary rocks are generally classified as detrital or chemical:

   a. Detrital sedimentary rocks consist of solid particles derived from preexisting rocks.

   b. Chemical sedimentary rocks are derived from substances in solution by inorganic chemical processes or the activities of organisms. A subcategory called *biochemical sedimentary rocks* is recognized.

7. Carbonate rocks contain minerals with the carbonate radical $(CO_3)^{-2}$ as in limestone and dolostone. Dolostone forms when magnesium partly replaces the calcium in limestone.

8. Evaporites include rock salt and rock gypsum, both of which form by inorganic precipitation of minerals from evaporating water.

9. Coal is a type of biochemical sedimentary rock composed of the altered remains of land plants.

10. Sedimentary facies are bodies of sediment or sedimentary rock that are recognizably different from adjacent sediments or rocks.

11. Vertical sequences of rocks with offshore facies overlying nearshore facies form when sea level rises with respect to the land, causing a marine transgression. A rise in the land relative to sea level causes a marine regression, which results in nearshore facies overlying offshore facies.

12. Sedimentary structures such as bedding, cross-bedding, and ripple marks commonly form in sediments during, or shortly after deposition. These features help geologists determine ancient current directions and depositional environments of sedimentary rocks.

13. Sediments and sedimentary rocks are the host materials for most fossils. Fossils provide the only record of prehistoric life and are useful for correlation and environmental interpretations.

14. Depositional environments of ancient sedimentary rocks are determined by studying sedimentary textures and structures, examining fossils, and making comparisons with present-day sediments deposited by known processes.

15. Many sediments and sedimentary rocks including sand, gravel, evaporites, coal, and banded iron formations are important natural resources. Most oil and natural gas are found in sedimentary rocks.

Important Terms

bed (bedding)
biochemical sedimentary rock
carbonate rock
cementation
chemical sedimentary rock
compaction
cross-bedding
depositional environment
detrital sedimentary rock

evaporite
fossil
graded bedding
lithification
marine regression
marine transgression
mud crack
paleocurrent
ripple mark

rounding
sediment
sedimentary facies
sedimentary rock
sedimentary structure
sorting
strata (stratification)

Review Questions

1. Sedimentary breccia is a rare type of rock because:
   a. _____ graded bedding forms in the deep sea;
   b. _____ gravel is rounded quickly during transport;
   c. _____ clay is less common than other detrital particles;
   d. _____ sand deposits are typically well sorted;
   e. _____ magnesium replaces some of the calcium in calcite.

2. The process of precipitating minerals in pore spaces of sediment, thus binding it together, is:
   a. _____ dolomitization;
   b. _____ carbonization;
   c. _____ cementation;
   d. _____ sorting;
   e. _____ rounding.

3. Graded bedding is characterized by:
   a. _____ beds deposited at an angle to the surface on which they accumulate;
   b. _____ biochemical precipitation of minerals;
   c. _____ angular sand and poorly sorted gravel;
   d. _____ carbonate rocks and evaporites;
   e. _____ an upward decrease in grain size.

4. A vertical sequence of facies in which offshore deposits are overlain by nearshore deposits result from:
   a. _____ marine regression;
   b. _____ deposition in a stream channel and on its adjacent floodplain;
   c. _____ deposition of evaporites in a desert;
   d. _____ drying out and cracking of clay-rich sediments;
   e. _____ compaction of mud.

5. If a deposit of detrital sediment consists of particles of markedly different sizes, it is said to be:
   a. _____ lithified;
   b. _____ cross-bedded;
   c. _____ biochemical;
   d. _____ poorly sorted;
   e. _____ conglomerate.

6. When magnesium replaces some of the calcium in calcite, limestone is converted to:
   a. _____ dolostone;
   b. _____ sedimentary breccia;
   c. _____ arkose;
   d. _____ rock gypsum;
   e. _____ bituminous coal.

7. Assuming that all of the following are in ancient rocks, which one is *not* a trace fossil?
   a. _____ reptile footprint;
   b. _____ worm burrow;
   c. _____ clam shell;
   d. _____ dinosaur nest;
   e. _____ elephant tracks.

8. Traps for petroleum and natural gas formed by deformation such as folding and fracturing of rocks are known as _____ traps:
   a. _____ lithification;
   b. _____ reservoir;
   c. _____ structural;
   d. _____ compaction;
   e. _____ stratigraphic.

9. Which one of the following can be used to determine ancient current directions?
   a. _____ mud cracks;
   b. _____ worm burrows;
   c. _____ current ripple marks;
   d. _____ graded bedding;
   e. _____ evaporites.

10. Rock salt and rock gypsum:
    a. _____ form when silica replaces limestone;
    b. _____ are the most common evaporite rocks;
    c. _____ are an alternate source of oil and natural gas;
    d. _____ are the main ores of iron;
    e. _____ are detrital sedimentary rocks.

11. Detrital particles measuring between l/16 and 2 mm are referred to as:
    a. _____ calcium carbonate;
    b. _____ clay;
    c. _____ arkose;
    d. _____ sand;
    e. _____ iron oxide.

12. Arkose is a type of sandstone containing:
    a. _____ ooids and shell fragments;
    b. _____ mostly ferromagnesian silicates;
    c. _____ a sticky residue from evaporated oil;
    d. _____ wave-formed ripple marks;
    e. _____ more than 25% feldspars.

13. Explain why the sediment in wind-blown dunes is better sorted than that in glacier deposits.

14. What are marine transgressions and regressions? Illustrate or explain the vertical sequence of facies that might develop during each.

15. How does coal originate, and what are the varieties of coal? Which variety makes the best fuel?

16. In what fundamental way(s) do chemical sedimentary rocks differ from detrital sedimentary rocks?

17. Illustrate and describe two sedimentary structures that can be used to determine paleocurrent direction.

18. What are body fossils and trace fossils? Also, explain how molds and casts form.

19. Describe the processes responsible for lithification of detrital sedimentary rocks. Are these processes equally important in all detrital sediments? Explain.

20. How does dolostone differ from limestone, and what is one possible way dolostone forms?

21. Why are most limestones considered to be biochemical sedimentary rocks?

22. How is it possible to use knowledge of sorting, rounding, and cross-bedding to determine the depositional environment of an ancient sandstone?

23. What are banded iron formations, and why are they important?

24. Explain how evaporites form. What are the most common evaporite rocks?

1. As a field geologist, you encounter a rock unit consisting of well-sorted, well-rounded sandstone. In addition, the unit has large cross-beds and contains reptile footprints. What can you infer about the depositional environment?

2. While visiting one of our national parks, you notice a vertical sequence of rocks consisting of sandstone at the base followed upward by shale and limestone. All these rocks contain fossil sea lilies, corals, and clams. Explain how these rocks were deposited and how they came to be superposed in the vertical sequence observed.

3. The United States has a total coal reserve of 243 billion metric tons and uses about 860 million metric tons of coal annually. Assuming that all of this coal can be mined, how long will it last at the current rate of consumption? Why is it improbable that all the reserve can be mined?

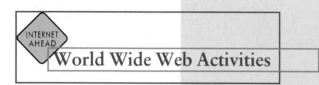

INTERNET AHEAD

World Wide Web Activities

For these Web site addresses, along with current updates and exercises, log on to
http://www.brookscole.com/geo/

➤ SEDIMENTARY ROCKS

This Web site contains images and information on 11 common sedimentary rocks. It is part of the Soil Science 223 Rocks and Minerals Reference Web site. Click on any of the *rock names*, and compare the information and images to the information in this chapter.

➤ GEOLOGY 110 WEB PAGES OF SEDIMENTARY STRUCTURES

This Web site contains links to two slide collections of sedimentary structures that were used in the Geology 110 class at Duke University. Click on either the *Sedimentary Structures, Part 1* or *Sedimentary Structures, Part 2* heading to view slides of different sedimentary structures. Use this collection to review the different sedimentary structures discussed in this chapter.

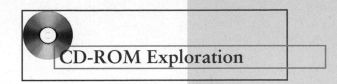

CD-ROM Exploration

➤ Exploring your *Earth Systems Today* CD-ROM will add to your understanding of the material in this chapter.

TOPIC: EARTH'S MATERIALS

MODULE: ROCKS AND ROCK CYCLE

Explore activities in this module to see if you can discover the following for yourself:

Using the "rock laboratory" portion of this module, observe the formation of a clastic sedimentary rock (sandstone) through the process of lithification. What role does water play in the lithification process? Based on what you see in the "rock laboratory," write a short description of how a sandstone is lithified.

Compaction and cementation must work together to form sedimentary rock. Observe how this happens using the animation and make notes on what you see.

How are igneous rocks and sedimentary rocks different in their texture and mode of formation?

Chapter 7

Open-pit asbestos mine, Quebec, Canada.

OBJECTIVES

At the end of this chapter, you will have learned that:

- Metamorphic rocks result from the transformation of other rocks by various processes occurring beneath Earth's surface.
- Heat, pressure, and fluid activity are the three agents of metamorphism.
- Contact, dynamic, and regional metamorphism are the three types of metamorphism.

- Metamorphic rocks are typically divided into two groups, primarily on the basis of texture.
- Metamorphic rocks with a foliated texture include slate, phyllite, schist, gneiss, and amphibolite.
- Metamorphic rocks with a nonfoliated texture include marble, quartzite, greenstone, and hornfels.
- Metamorphic rocks can be grouped into metamorphic zones based on the presence of index minerals that form

under specific temperature and pressure conditions.
- The appearance of particular index minerals results from increasing metamorphic intensity.
- Metamorphism is associated with all three types of plate boundaries, but is most common along convergent plate boundaries.
- Many metamorphic minerals and rocks are valuable metallic ores, building materials, and gemstones.

Metamorphism and Metamorphic Rocks

Prologue

Its homogeneity, softness, and various textures have made marble, a metamorphic rock formed from limestone or dolostone, a favorite rock of sculptors throughout history. As the value of authentic marble sculptures has increased through the years, the number of forgeries has also increased. With the price of some marble sculptures in the millions of dollars, private collectors and museums need some means of ensuring the authenticity of the work they are buying. Aside from the monetary considerations, it is important that forgeries not become part of the historical and artistic legacy of human endeavor.

Experts have traditionally relied on the artistic style and weathering characteristics to determine whether a marble sculpture is authentic or a forgery. Because marble is not very resistant to weathering, forgers have resorted to a variety of methods to produce the weathered appearance of an authentic ancient work. Now, however, using new techniques, geologists can distinguish a naturally weathered marble surface from one that has been artificially altered.

Marbles result when the agents of metamorphism (heat, pressure, and fluid activity) are applied to carbonate rocks. The type of marble formed depends, in part, on the original composition of the parent carbonate rock as well as the type and intensity of metamorphism. Therefore, one way to authenticate a marble sculpture is to determine the origin of the marble itself. During the Preclassical, Greek, and Roman periods, the islands of Naxos, Thasos, and Paros in the Aegean Sea—as well as the Greek mainland, Turkey, and Italy—were all sites of major marble quarries.

To identify the source of the marble in various sculptures, geologists employ a wide variety of analytical techniques. These include hand-specimen and thin-section analysis of the marble, trace-element analysis by X-ray fluorescence, stable isotopic ratio analysis for carbon and oxygen, and other more esoteric techniques. Carbon and oxygen isotopic analysis, however, has proved the most powerful and reliable method for source-area determination, because each quarry yields marble with a distinctive set of carbon and oxygen isotope values.

The J. Paul Getty Museum in Malibu, California, employed some of these techniques to help authenticate an ancient Greek kouros (a sculptured figure of a Greek youth) thought to have been carved around 530 B.C. (Figure 7.1). The kouros was offered to the Getty Museum in 1984 for a reported price of $7 million. Some of its stylistic features, however, caused some experts to question its authenticity.

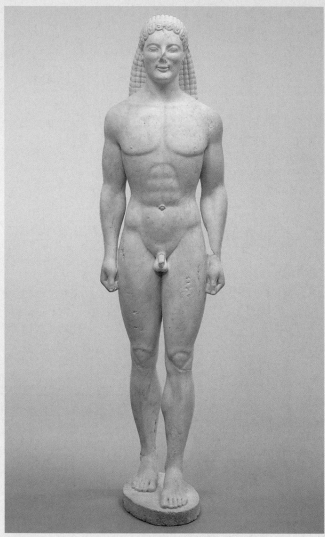

Figure 7.1
This Greek kouros, which stands 206 cm tall, has been the object of an intensive authentication study by the Getty Museum. Using a variety of geologic tests, it has been determined that the kouros was carved from dolomitic marble that probably came from the Cape Vathy quarries on the island of Thasos.

Consequently, the museum had a variety of geochemical and mineralogical tests performed in an effort to determine the authenticity of the kouros.

Isotopic analysis of the weathered surface and fresh interior of the kouros confirmed that the marble probably came from the Cape Vathy quarries on the island of Thasos in the Aegean Sea. But these results did not prove the age of the kouros—it might still have been a forgery carved from marble taken from an archaeological site on the island.

The kouros was carved from dolomite marble, and its surface is covered with a complex thin crust (0.01–0.05 mm thick) consisting of whewellite, a calcium oxalate monohydrate mineral. To ascertain that the crust is the result of long-term weathering and not a modern forgery, dolomitic marble samples were subjected to a variety of forgery techniques that tried to replicate the surface of the kouros. To try to match the appearance of the weathered surface of the kouros, samples were soaked or boiled in various mixtures for periods of time ranging from hours to months, and their surfaces were treated and re-treated. Such tests yielded only a few examples that appeared similar to the surface of the kouros. Even those samples, however, appeared different when examined under high magnification or subjected to geochemical analysis. In fact, all samples clearly showed that they were the result of recent alteration and not of long-term weathering processes.

Although scientific tests have not unequivocally proved authenticity, they have shown that the weathered surface layer of the kouros bears more similarities to naturally occurring weathered surfaces than to known artifically produced surfaces. Furthermore, no evidence indicates that the surface alteration of the kouros is of modern origin.

Despite intensive study by scientists, archaeologists, and art historians, opinion is still divided as to the authenticity of the Getty kouros. Most scientists accept that the kouros was carved sometime around 530 B.C., but most art historians are doubtful. Pointing to inconsistencies in its style of sculpture for that period, they think it is a modern forgery. Because of the continuing doubts about the statue's authenticity, the recently opened J. Paul Getty Center has mounted it in an exhibition listing the evidence for and against its authenticity.

Regardless of the ultimate conclusion on the Getty kouros, geologic testing to authenticate marble sculptures is now an important part of many museums' curatorial functions. In addition, a large body of data about the characteristics and origin of marble is being amassed as more sculptures and quarries are analyzed.

INTRODUCTION

Metamorphic rocks (from the Greek *meta* meaning "change" and *morpho* meaning "shape") constitute the third major group of rocks. They result from the transformation of other rocks by metamorphic processes that usually occur beneath Earth's surface (see Figure 1.17). During metamorphism, rocks are subjected to sufficient heat, pressure, and fluid activity to change their mineral composition and/or texture, thus forming new rocks. These transformations take place in the solid state, and the type of metamorphic rock formed depends on the original composition and texture of the parent rock, the agents of metamorphism, and the amount of time the parent rock was subjected to the effects of metamorphism.

A useful analogy for metamorphism is the process of baking. The resulting cake, just like a metamorphic rock, depends on the ingredients, their proportions, how they are mixed together, how much water or milk is added, and the temperature and length of time the cake is baked.

A large portion of Earth's continental crust is composed of metamorphic and igneous rocks. Together, they form the crystalline basement rocks that underlie the sedimentary rocks of a continent's surface. This basement rock is widely exposed in regions of the continents known as *shields,* which have been very stable

during the past 600 million years (Figure 7.2). Metamorphic rocks also constitute a sizable portion of the crystalline core of large mountain ranges. Some of the oldest known rocks, dated at 3.96 billion years from the Canadian Shield, are metamorphic, indicating they formed from even older rocks.

Why is it important to study metamorphic rocks? For one thing, they provide information about geologic processes operating within Earth and about the way these processes have varied through time. From the

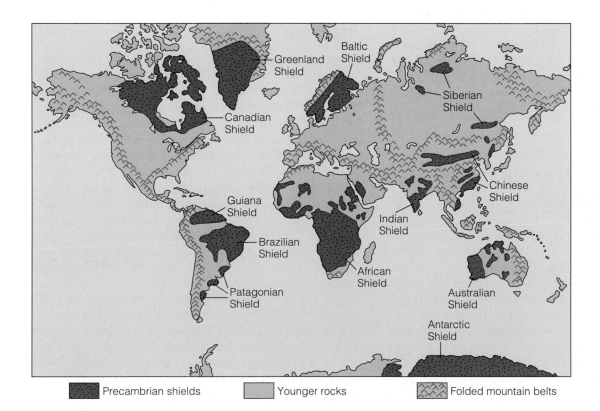

Figure 7.2
Metamorphic rock occurrences. Shields are the exposed portion of the crystalline basement rocks that underlie each continent; these areas have been very stable during the past 600 million years. Metamorphic rocks also constitute the crystalline core of major mountain belts.

Precambrian shields   Younger rocks   Folded mountain belts presence of certain minerals in metamorphic rocks, geologists can determine the approximate temperatures and pressures that parent rocks were subjected to during metamorphism and thus gain insights into the physical and chemical changes that occur at different depths within the crust. Furthermore, metamorphic rocks such as marble and slate are used as building materials, and certain metamorphic minerals are economically important. For example, garnets are used as gemstones or abrasives; talc is used in cosmetics, in the manufacture of paint, and as a lubricant; asbestos is used for insulation and fireproofing (see Perspective 7.1); and kyanite is used in the production of heat-resistant materials used in spark plugs.

WHAT WOULD YOU DO?

The problem of asbestos removal from public buildings is an important national health and political issue. The current policy of the Environmental Protection Agency (EPA) mandates that all forms of asbestos be treated as identical hazards. Yet studies indicate only one form of asbestos is a known health hazard. Because the cost of asbestos removal from all U.S. public buildings has been estimated to be as much as $100 billion, many people are questioning whether it is cost effective to remove asbestos from all such public buildings where it has been installed.

As a leading researcher on the health hazards of asbestos, you have been asked to testify before a congressional committee on whether it is worthwhile to spend so much money for asbestos removal. How would you address this issue in terms of formulating a policy that balances the risks versus the benefits of removing asbestos from public buildings? What role could geologists play in formulating this policy?

WHAT ARE THE AGENTS OF METAMORPHISM?

The three agents of metamorphism are heat, pressure, and fluid activity. During metamorphism, the original rock undergoes change to achieve equilibrium with its new environment. The changes may result in the formation of new minerals and/or a change in the texture of the rock, brought about by the reorientation of the original minerals. In some instances, the change is minor, and features of the parent rock can still be recognized. In other cases the rock changes so much that the identity of the parent rock can be determined only with great difficulty, if at all.

Besides heat, pressure, and fluid activity, time is also important to the metamorphic process. Chemical reactions proceed at different rates and thus require different amounts of time to complete. Reactions involving silicate compounds are particularly slow, and because most metamorphic rocks are composed of silicate minerals, it is thought that metamorphism is a slow geologic process.

Heat

Heat is an important agent of metamorphism because it increases the rate of chemical reactions that may produce minerals different from those in the original rock. The heat may come from intrusive magmas or result from deep burial in the crust such as occurs during subduction along a convergent plate boundary.

When rocks are intruded by bodies of magma, they are subjected to intense heat that affects the surrounding rock; the most intense heating usually occurs adjacent to the magma body and gradually decreases with distance from the intrusion. The zone of metamorphosed rocks that forms in the country rock adjacent to an intrusive igneous body is usually rather distinct and easy to recognize.

Recall that temperature increases with depth and that Earth's geothermal gradient averages about 25°C/km. Rocks forming at the surface may be transported to great depths by subduction along a convergent plate boundary and thus subjected to increasing temperature and pressure. During subduction, some minerals may be transformed into other minerals that are more stable under the higher temperature and pressure conditions.

Pressure

When rocks are buried, they are subjected to increasingly greater **lithostatic pressure**; this pressure, which results from the weight of the overlying rocks, is applied equally in all directions (Figure 7.3a). A similar situation occurs when an object is immersed in water. For example, the deeper a Styrofoam cup is submerged in the ocean, the smaller it gets because water pressure increases with depth and is exerted on the cup equally in all directions, thereby compressing the Styrofoam (Figure 7.3b).

Just as in the Styrofoam cup example, rocks are subjected to increasing lithostatic pressure with depth such that the mineral grains within a rock become more closely packed. Under these conditions, the minerals may *recrystallize*; that is, they become smaller, denser minerals.

Along with lithostatic pressure resulting from burial, rocks may also experience **differential pressures** (Figure 7.4). In this case, the pressures are not equal on all sides, and the rock is consequently distorted. Differential pressures typically occur during deformation associated with mountain building and can produce distinctive metamorphic textures and features.

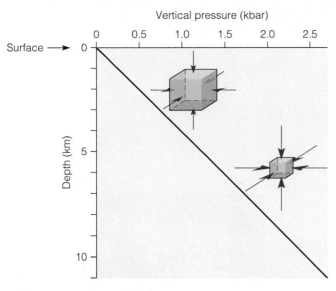

Vertical pressure (kbar)

1 kilobar (kbar) = 1,000 bars
Atmospheric pressure at sea level = 1 bar

(a)

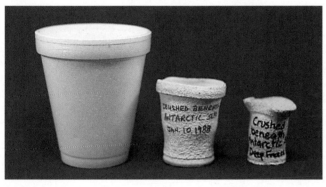

(b)

Figure 7.3
(a) Lithostatic pressure is applied equally in all directions in Earth's crust due to the weight of overlying rocks. Thus pressure increases with depth, as indicated by the sloping black line.
(b) A similar situation occurs when 200-ml Styrofoam cups are lowered to ocean depths of approximately 750 m and 1500 m. Increased water pressure is exerted equally in all directions on the cups, and they consequently decrease in volume, while still maintaining their general shape.

Fluid Activity

In almost every region of metamorphism, water and carbon dioxide (CO_2) are present in varying amounts along mineral grain boundaries or in the pore spaces of rocks. These fluids, which may contain ions in solution, enhance metamorphism by increasing the rate of chemical reactions. Under dry conditions, most minerals react very slowly, but when even small amounts of fluid are introduced, reaction rates increase, mainly because ions move readily through the fluid and thus enhance chemical reactions and the formation of new minerals.

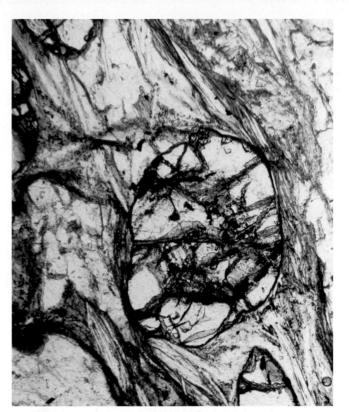

Figure 7.4
Differential pressure is pressure that is unequally applied to an object. Rotated garnets are a good example of the effects of differential pressure applied to a rock during metamorphism. These rotated garnets come from a schist in northeast Sardinia.

The following reaction provides a good example of how new minerals can be formed by **fluid activity**. Here, seawater moving through hot basaltic rock of the oceanic crust transforms olivine into the metamorphic mineral serpentine:

$$2Mg_2SiO_4 + 2H_2O \rightarrow Mg_3Si_2O_5(OH)_4 + MgO$$

OLIVINE   WATER   SERPENTINE   CARRIED AWAY IN SOLUTION

The chemically active fluids that are part of the metamorphic process come primarily from three sources. The first is water trapped in the pore spaces of sedimentary rocks as they form; as these rocks are subjected to heat and pressure, the water is heated, thus accelerating the various chemical reaction rates. A second source is the volatile fluid within magma; as these hot fluids disperse through the surrounding rock, they frequently react with and alter the minerals of the country rock by adding or removing ions. The third source is the dehydration of water-bearing minerals such as gypsum ($CaSO_4 \cdot 2H_2O$) and some clays. When these minerals, which contain water as part of their crystal chemistry, are subjected to heat and pressure, the water may be driven off and enhance metamorphism.

Perspective 7.1

Asbestos

A sbestos (from the Latin, meaning "unquenchable") is a general term applied to any silicate mineral that easily separates into flexible fibers (Figure 1). The combination of such features as noncombustibility and flexibility makes asbestos an important industrial material of considerable value. In fact, asbestos has more than 3000 known uses, including brake linings, fireproof fabrics, and heat insulators.

Asbestos can be divided into two broad groups, serpentine and amphibole asbestos. *Chrysotile,* a hydrous magnesium silicate with the chemical formula $Mg_3Si_2O_5(OH)_4$, is the fibrous form of serpentine asbestos; it is the most valuable type and constitutes the bulk of all commercial asbestos. Chrysotile's strong, silky fibers are eas-

Figure 1
Specimen of chrysotile from Thetford, Quebec, Canada. Chrysotile is the fibrous form of serpentine asbestos.

ily spun and can withstand temperatures of up to 2750°C.

The vast majority of chrysotile asbestos is in serpentine, a type of rock formed by the alteration of ultramafic igneous rocks such as peridotite under low- and medium-grade metamorphic conditions. Serpentine is thought to form from the alteration of olivine by hot, chemically active, residual fluids emanating from cooling magma. The chrysotile

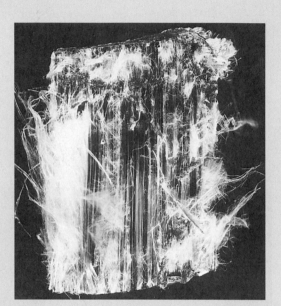

WHAT ARE THE THREE TYPES OF METAMORPHISM?

Three major types of metamorphism are recognized: contact metamorphism in which magmatic heat and fluids act to produce change; dynamic metamorphism, which is principally the result of high differential pressures associated with intense deformation; and regional metamorphism, which occurs within a large area and is caused primarily by mountain-building forces. Even though we will discuss each type of metamorphism separately, the boundary between them is not always distinct and depends largely on which of the three metamorphic agents was dominant.

Contact Metamorphism

Contact metamorphism takes place when a body of magma alters the surrounding country rock. At shallow depths, an intruding magma raises the temperature of the surrounding rock, causing thermal alteration. Furthermore, the release of hot fluids into the country rock by the cooling intrusion can also aid in the formation of new minerals.

Important factors in contact metamorphism are the initial temperature and size of the intrusion as well as the fluid content of the magma and/or country rock. The initial temperature of an intrusion is controlled, in part, by its composition: Mafic magmas are hotter than felsic magmas (see Chapter 3) and hence have a greater thermal effect on the rocks directly surrounding them. The size of the intrusion is also important. In the case of small intrusions, such as dikes and sills, usually only those rocks in immediate contact with the intrusion are affected. Because large intrusions, such as batholiths, take a long time to cool, the increased temperature in the surrounding rock may last long enough for a larger area to be affected.

Fluids also play an important role in contact metamorphism. Many magmas are wet and contain hot,

asbestos forms veinlets of fiber within the serpentine and may comprise up to 20% of the rock. Other chrysotile results when the metamorphism of magnesium limestone or dolostone produces discontinuous serpentine bands that develop within the carbonate beds.

At least five varieties of amphibole asbestos are known, but *crocidolite,* a sodium-iron amphibole with the chemical formula $Na_2(Fe^{+3})_2(Fe^{+2})_3Si_8O_{22}(OH)_2$ is the most common. Crocidolite, which is also known as blue asbestos, is a long, coarse, spinning fiber that is stronger but more brittle than chrysotile and also less resistant to heat. The other varieties of amphibole asbestos have fewer uses and are used primarily for insulation.

Crocidolite is found in such metamorphic rocks as slates and schists. It is thought that crocidolite forms by the solid-state alteration of other minerals within the high-temperature and high-pressure environment that results from deep burial. Unlike chrysotile, crocidolite is rarely found associated with igneous intrusions.

Despite the widespread use of asbestos, the federal Environmental Protection Agency (EPA) has instituted a gradual ban on all new asbestos products. The ban was imposed because some forms of asbestos can cause lung cancer and scarring of the lungs if its fibers are inhaled. Because the EPA apparently paid little attention to the issue of risks versus benefits when it enacted this rule, the U.S. Fifth Circuit Court of Appeals overturned the EPA ban on asbestos in 1991.

The threat of lung cancer has also resulted in legislation mandating the removal of asbestos already in place in all public buildings, including all public and private schools. Recently, however, important questions have been raised concerning the threat posed by asbestos and the additional potential hazards that may arise from its improper removal.

The current policy (1993) of the EPA mandates that all forms of asbestos be treated as identical hazards. Yet studies indicate that only the amphibole forms constitute a known health hazard. Chrysotile, whose fibers tend to be curly, does not become lodged in the lungs. Furthermore, its fibers are generally soluble and disappear in tissue. In contrast, crocidolite has long, straight, thin fibers that penetrate the lungs and stay there. These fibers irritate the lung tissue and over a long period of time can lead to lung cancer. Thus, crocidolite, and not chrysotile, is overwhelmingly responsible for asbestos-related lung cancer. Because about 95% of the asbestos in place in the United States is chrysotile, many people are questioning whether the dangers from asbestos have been somewhat exaggerated.

Removing asbestos from buildings where it has been installed might cost as much as $100 billion, and some recent studies have indicated that the air in buildings containing asbestos has essentially the same amount of airborne asbestos fibers as the air outdoors. In fact, unless the material containing the asbestos is disturbed, asbestos does not shed fibers. Furthermore, improper removal of asbestos can lead to contamination. In most cases of improper removal, the concentration of airborne asbestos fibers is far higher than if the asbestos had been left in place.

The problem of asbestos contamination is a good example of how geology affects our lives and why a basic knowledge of science is important.

chemically active fluids that may emanate into the surrounding rock. These fluids can react with the rock and aid in the formation of new minerals. In addition, the country rock may contain pore fluids that, when heated by the magma, also increase reaction rates.

Temperatures can reach nearly 900°C adjacent to an intrusion, but they gradually decrease with distance. The effects of such heat and the resulting chemical reactions usually occur in concentric zones known as **aureoles** (Figure 7.5). The boundary between an intrusion and its aureole may be either sharp or transitional (Figure 7.6).

Metamorphic aureoles vary in width depending on the size, temperature, and composition of the intrusion, as well as on the composition of the surrounding country rock. Typically, large intrusive bodies have several metamorphic zones, each characterized by distinctive mineral assemblages indicating the decrease in temperature with distance from the intrusion (Figure 7.5). The zone closest to the intrusion, and hence subject to the highest temperatures, may contain high-temperature metamorphic minerals (that is, minerals in equilibrium with the higher-temperature environment) such as sillimanite. The outer zones may be characterized by lower-temperature metamorphic minerals such as chlorite, talc, and epidote.

The formation of new minerals by contact metamorphism depends not only on proximity to the intrusion but also on the composition of the country rock. Shales, mudstones, impure limestones, and impure dolostones are particularly susceptible to the formation of new minerals by contact metamorphism, whereas pure sandstones or pure limestones typically are not.

Two types of contact metamorphic rocks are generally recognized: those resulting from baking of country rock and those altered by hot solutions. Many of the rocks resulting from contact metamorphism have the texture of porcelain; that is, they are hard and fine grained. This is particularly true for rocks with a high clay content, such as shale. Such texture results because the clay minerals in the rock are baked, just as a clay pot is baked when fired in a kiln.

Figure 7.5
A metamorphic aureole typically surrounds many igneous intrusions. The metamorphic aureole associated with this idealized granite batholith contains three zones of mineral assemblages reflecting the decreases in temperature with distance from the intrusion. An andalusite–cordierite hornfels forms the inner zone adjacent to the batholith. This is followed by an intermediate zone of extensive recrystallization in which some biotite develops, and farthest from the intrusion is the outer zone, which is characterized by spotted slates.

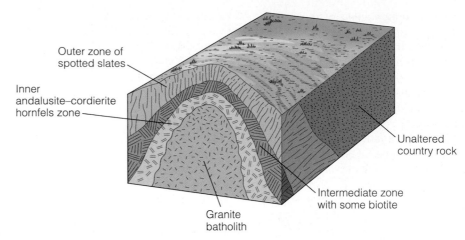

Outer zone of spotted slates

Inner andalusite–cordierite hornfels zone

Unaltered country rock

Intermediate zone with some biotite

Granite batholith

Igneous rock

Metamorphic rock

Figure 7.6
A sharp and clearly defined boundary (red line) occurs between the intruding light-colored igneous rock on the left and the dark-colored metamorphosed country rock on the right. The intrusion is part of the Peninsular Ranges Batholith, east of San Diego, California.

During the final stages of cooling, when an intruding magma begins to crystallize, large amounts of hot, watery solutions are often released. These solutions may react with the country rock and produce new metamorphic minerals. This process, which usually occurs near Earth's surface, is called *hydrothermal alteration* (from the Greek *hydor,* meaning "water," and *therme,* meaning "heat") and may result in valuable mineral deposits. Geologists think that many of the world's ore deposits result from the migration of metallic ions in hydrothermal solutions. Examples include copper, gold, iron ores, tin, and zinc in various localities including Australia, Canada, China, Cyprus, Finland, Russia, and the western United States.

Dynamic Metamorphism

Most **dynamic metamorphism** is associated with fault (fractures along which movement has occurred) zones where rocks are subjected to high differential pressures. The metamorphic rocks resulting from pure dynamic metamorphism are called *mylonites* and are typically restricted to narrow zones adjacent to faults. Mylonites are hard, dense, fine-grained rocks, many of which are characterized by thin laminations (Figure 7.7). Tectonic settings where mylonites occur include the Moine Thrust Zone in northwest Scotland, the Adirondack Highlands in New York, and portions of the San Andreas fault in California (see Chapter 13).

Regional Metamorphism

Most metamorphic rocks result from **regional metamorphism,** which occurs over a large area and is usually caused by tremendous temperatures, pressures, and deformation within the deeper portions of the crust. Regional metamorphism is most obvious along convergent plate margins where rocks are intensely deformed and recrystallized during convergence and subduction. Metamorphic rocks usually show a gradation of metamorphic intensity from areas subjected to the most intense pressures and/or highest temperatures to areas of lower pressures and temperatures. Such a gradation in metamorphism can be recognized by the metamorphic minerals that are present.

Regional metamorphism is not just confined to convergent margins. It also occurs in areas where plates

Figure 7.7
Mylonite from the Adirondack Highlands, New York. Note the thin laminations.

diverge, although usually at much shallower depths because of the high geothermal gradient associated with these areas.

From field studies and laboratory experiments, certain minerals are known to form only within specific temperature and pressure ranges. Such minerals are known as **index minerals,** because their presence allows geologists to recognize low-, intermediate-, and high-grade metamorphic zones (Figure 7.8).

When a clay-rich rock such as shale is metamorphosed, new minerals form as a result of metamorphic processes. The mineral chlorite, for example, is produced

Figure 7.8
Change in mineral assemblage and rock type with increasing metamorphism of shale. When a clay-rich rock such as shale is subjected to increasing metamorphism, new minerals form as shown by the various colored bars. The progressive appearance of particular minerals allows geologists to recognize low-, intermediate-, and high-grade metamorphic zones.

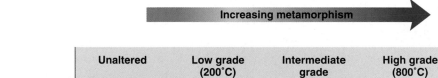

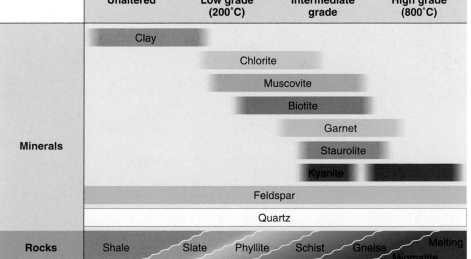

under relatively low temperatures of about 200°C, so its presence indicates low-grade metamorphism. As temperatures and pressures continue to increase, new minerals form that are stable under those conditions. Thus, there is a progression in the appearance of new minerals from chlorite, whose presence indicates low-grade metamorphism, to sillimanite, whose presence indicates high-grade metamorphism and temperatures exceeding 500°C.

Different rock compositions develop different index minerals. When sandy dolomites are metamorphosed, for example, they produce an entirely different set of index minerals. Thus, a specific set of index minerals commonly forms in specific rock types as metamorphism progresses.

Although such common minerals as mica, quartz, and feldspar can occur in both igneous and metamorphic rocks, other minerals such as andalusite, sillimanite, and kyanite generally occur only in metamorphic rocks derived from clay-rich sediments. Although these three minerals all have the same chemical formula (Al_2SiO_5), they differ in crystal structure and other physical properties because each forms under a different range of pressures and temperatures. Thus they are sometimes used as index minerals for metamorphic rocks formed from clay-rich sediments.

HOW ARE METAMORPHIC ROCKS CLASSIFIED?

For purposes of classification, metamorphic rocks are commonly divided into two groups: those exhibiting a foliated (from the Latin *folium*, meaning "leaf") texture and those with a nonfoliated texture (Table 7.1).

Foliated Metamorphic Rocks

Rocks subjected to heat and differential pressure during metamorphism typically have minerals arranged in a parallel fashion giving them a **foliated texture** (Figure 7.9). The size and shape of the mineral grains deter-

Table 7.1

Classification of Common Metamorphic Rocks

Texture	Metamorphic Rock	Typical Minerals	Metamorphic Grade	Characteristics of Rocks	Parent Rock
Foliated	Slate	Clays, micas, chlorite	Low	Fine-grained, splits easily into flat pieces	Mudrocks, volcanic ash
	Phyllite	Fine-grained quartz, micas, chlorite	Low to medium	Fine-grained, glossy or lustrous sheen	Mudrocks
	Schist	Micas, chlorite, quartz, talc, hornblende, garnet, staurolite, graphite	Low to high	Distinct foliation, minerals visible	Mudrocks, carbonates, mafic igneous rocks
	Gneiss	Quartz, feldspars, hornblende, micas	High	Segregated light and dark bands visible	Mudrocks, sandstones, felsic igneous rocks
	Amphibolite	Hornblende, plagioclase	Medium to high	Dark-colored, weakly foliated	Mafic igneous rocks
	Migmatite	Quartz, feldspars, hornblende, micas	High	Streaks or lenses of granite intermixed with gneiss	Felsic igneous rocks mixed with sedimentary rocks
Nonfoliated	Marble	Calcite, dolomite	Low to high	Interlocking grains of calcite or dolomite, reacts with HCl	Limestone or dolostone
	Quartzite	Quartz	Medium to high	Interlocking quartz grains, hard, dense	Quartz sandstone
	Greenstone	Chlorite, epidote, hornblende	Low to high	Fine-grained, green	Mafic igneous rocks
	Hornfels	Micas, garnets, andalusite, cordierite, quartz	Low to medium	Fine-grained, equidimensional grains, hard, dense	Mudrocks
	Anthracite	Carbon	High	Black, lustrous, subconcoidal fracture	Coal

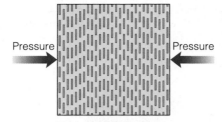
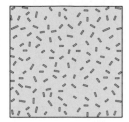

Random arrangement of elongated minerals before pressure is applied to two sides

Elongated minerals arranged in a parallel fashion as a result of pressure applied to two sides

(a)

(a)

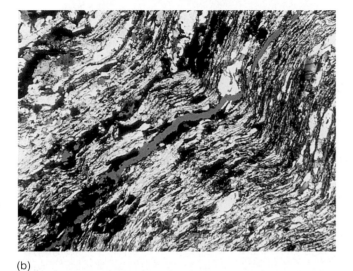

(b)

Figure 7.9
(a) When rocks are subjected to differential pressure, the mineral grains are typically arranged in a parallel fashion, producing a foliated texture. (b) Photomicrograph of a metamorphic rock with a foliated texture showing the parallel arrangement of mineral grains.

mine whether the foliation is fine or coarse. If the foliation is such that the individual grains cannot be recognized without magnification, the rock is said to be slate (Figure 7.10a). A coarse foliation results when granular minerals such as quartz and feldspar are segregated into roughly parallel and streaky zones that differ in composition and color as in a gneiss. Foliated metamorphic rocks can be arranged in order of increasingly coarse grain size and perfection of foliation.

Slate is a very fine-grained metamorphic rock that commonly exhibits *slaty cleavage* (Figure 7.10b). Slate is the result of low-grade regional metamorphism of shale or, more rarely, volcanic ash. Because it can easily be split along cleavage planes into flat pieces, slate is an excellent rock for roofing and floor tiles, billiard and pool table tops, and blackboards (Figure 7.10c). The different colors of most slates are caused by minute amounts of graphite (black), iron oxide (red and purple), and/or chlorite (green).

(b)

(c)

Figure 7.10
(a) Hand specimen of slate. (b) This panel of Arvonia Slate from Albemarne Slate Quarry, Virginia, shows bedding (upper right to lower left) at an angle to the slaty cleavage. (c) Slate roof of Chalet Enzian, Switzerland.

Phyllite is similar in composition to slate but is coarser grained. The minerals, however, are still too small to be identified without magnification. Phyllite can be distinguished from slate by its glossy or lustrous sheen (Figure 7.11). It represents an intermediate grain size between slate and schist.

Schist is most commonly produced by regional metamorphism. The type of schist formed depends on the intensity of metamorphism and the character of the parent rock (Figure 7.12). Metamorphism of many rock types can yield schist, but most schist appears to have formed from clay-rich sedimentary rocks (Table 7.1).

All schists contain more than 50% platy and elongated minerals, all of which are large enough to be clearly visible. Their mineral composition imparts a *schistosity* or *schistose foliation* to the rock that commonly produces a wavy type of parting when split. Schistosity is common in low- to high-grade metamorphic environments, and each type of schist is known by its most conspicuous mineral or minerals, such as mica schist, chlorite schist, and talc schist.

Gneiss is a metamorphic rock that is streaked or has segregated bands of light and dark minerals. Gneisses are composed mostly of granular minerals such as quartz and/or feldspar, with lesser percentages of platy or elongated minerals such as micas or amphiboles (Figure 7.13). Quartz and feldspar are the principal light-colored minerals, whereas biotite and hornblende are the typical dark-colored minerals. Gneiss typically breaks in an irregular manner, much like coarsely crystalline nonfoliated rocks.

Most gneiss probably results from recrystallization of clay-rich sedimentary rocks during regional metamorphism (Table 7.1). Gneiss also can form from igneous rocks such as granite or older metamorphic rocks.

Another fairly common foliated metamorphic rock is *amphibolite*. A dark-colored rock, it is composed mainly of hornblende and plagioclase. The alignment of the hornblende crystals produces a slightly foliated texture. Many amphibolites result from medium- to high-grade metamorphism of igneous rocks rich in ferromagnesian minerals, such as basalt.

In some areas of regional metamorphism, exposures of "mixed rocks" having both igneous and high-grade metamorphic characteristics are present. In these rocks, called *migmatites,* streaks or lenses of granite are usually

Figure 7.11
Specimen of phyllite.

(a)

(b)

Figure 7.12
Schist. (a) Garnet–mica schist.
(b) Hornblende–mica–garnet schist.

Figure 7.13
Gneiss is characterized by segregated bands of light and dark minerals. This folded gneiss is exposed at Wawa, Ontario, Canada.

intermixed with high-grade ferromagnesian-rich metamorphic rocks, thereby imparting a wavy appearance to the rocks (Figure 7.14).

Most migmatites are thought to be the product of extremely high-grade metamorphism, and several models for their origin have been proposed. Part of the problem in determining the origin of migmatites is explaining how the granitic component formed. According to one

Figure 7.14
Migmatites consist of high-grade metamorphic rock intermixed with streaks or lenses of granite. This migmatite is exposed at Thirty Thousand Islands of Georgian Bay, Lake Huron, Ontario, Canada.

model, the granitic magma formed in place by the partial melting of rock during intense metamorphism. Such an origin is possible providing that the host rocks contained quartz and feldspars and that water was present. Another possibility is that the granitic components formed by the redistribution of minerals by recrystallization in the solid state, that is, by pure metamorphism.

Nonfoliated Metamorphic Rocks

In some metamorphic rocks, the mineral grains do not show a discernible preferred orientation. Instead, these rocks consist of a mosaic of roughly equidimensional minerals and are characterized as having a **nonfoliated texture** (Figure 7.15). Most nonfoliated metamorphic rocks result from contact or regional metamorphism of rocks in which no platy or elongate minerals are present. Frequently, the only indication that a granular rock has been metamorphosed is the large grain size resulting from recrystallization. Nonfoliated metamorphic rocks are generally of two types: those composed mainly of only one mineral (for example, marble or quartzite) and

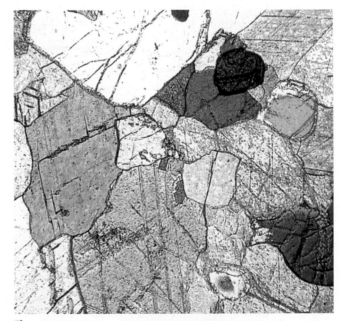

Figure 7.15
Nonfoliated textures are characterized by a mosaic of roughly equidimensional minerals as in this photomicrograph of marble.

those in which the different mineral grains are too small to be seen without magnification, such as greenstone and hornfels.

Marble is a well-known metamorphic rock composed predominantly of calcite or dolomite; its grain size ranges from fine to coarsely granular (Figures 7.1 and 7.16). Marble results from either contact or regional metamorphism of limestones or dolostones (Table 7.1). Pure marble is snowy white or bluish, but varieties of all colors exist because of the presence of mineral impurities in the parent sedimentary rock. The softness of marble, its uniform texture, and its various colors have made it the favorite rock of builders and sculptors throughout history (see the Prologue).

Quartzite is a hard, compact rock formed from quartz sandstone under medium- to high-grade metamorphic conditions during contact or regional metamorphism (Figure 7.17). Because recrystallization is so complete, metamorphic quartzite is of uniform strength and therefore usually breaks across the component quartz grains rather than around them when it is struck. Pure quartzite is white, but iron and other impurities commonly impart a reddish or other color to it. Quartzite is commonly used as foundation material for road and railway beds.

The name *greenstone* is applied to any compact, dark green, altered, mafic igneous rock that formed under low- to high-grade metamorphic conditions. The green color results from the presence of chlorite, epidote, and hornblende.

Hornfels, a fine-grained, nonfoliated metamorphic rock resulting from contact metamorphism, is composed of various equidimensional mineral grains. The composition of hornfels directly depends on the composition of the original rock, and many compositional varieties are known. The majority of hornfels, however, are apparently derived from contact metamorphism of clay-rich sedimentary rocks or impure dolostones.

Anthracite is a black, lustrous, hard coal that contains a high percentage of fixed carbon and a low percentage of volatile matter. It usually forms from the metamorphism of lower-grade coals by heat and pressure and is thus considered by many geologists to be a metamorphic rock.

WHAT ARE METAMORPHIC ZONES AND FACIES?

The first systematic study of metamorphic zones was conducted during the late 1800s by George Barrow and other British geologists working in the Dalradian schists of the southwestern Scottish Highlands. Here clay-rich sedimentary rocks have been subjected to regional metamorphism, and the resulting

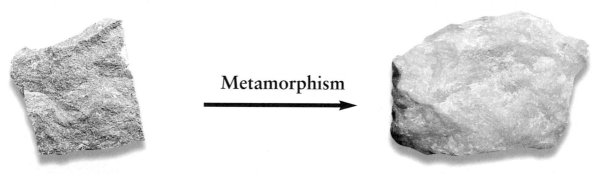

Figure 7.16
Marble results from the metamorphism of the sedimentary rock limestone or dolostone.

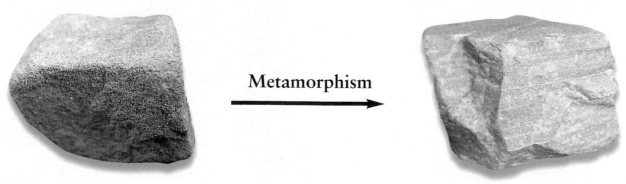

Figure 7.17
Quartzite results from the metamorphism of quartz sandstone.

Conservation and Museum Science: Geology in the Preservation and Understanding of Cultural Heritage

Eric Doehne, a conservation scientist at the Getty Conservation Institute in Marina del Rey, California, works as part of an interdisciplinary group to further the conservation of art, architecture, and monuments. He received his Ph.D. in geology from the University of California, Davis. His research interests include studies of stone-weathering mechanisms, the origin of complex patinas on ancient Greek and Roman stone surfaces, and the microdynamics of deterioration phenomena using the environmental scanning electron microscope (ESEM).

W hen I first arrived at Haverford College for my freshman year, I went through the catalog and promptly crossed out all the majors I knew did not interest me. That left me with history, political science, and geology. I took courses in all three subjects, but it was geology field trips and the realization that so much in geology was yet to be discovered that caught my imagination. In graduate school, I found I had a talent for instrumental analysis and shared many diverse interests with my adviser, including applying geochemical analytic techniques to global change, environmental, and art historical research questions. I also developed a strong interest in understanding how rocks weather.

I was fortunate to find a position that required such skills at an institution that fills an unusual niche—the Getty Conservation Institute (GCI), part of the J. Paul Getty Trust. The Institute has an ambitious mission: to further the conservation of the world's cultural heritage. The GCI focuses its resources on scientific research, the training of conservators, documentation, and international collaborative projects such as the conservation of Queen Nefertari's tomb in Egypt, Roman mosaics in Cyprus, 3.5-million-year-old hominid footprints in Tanzania, and elaborate carvings in Buddhist caves along the silk route in China. Many of these endeavors have a geologic component.

The conservation of our cultural heritage has received much attention in recent years, due to public concern over accelerated deterioration and loss caused by increased pollution, development, and tourism. The field of conservation science is relatively new and brings together collaborators from governments, universities, museums, industry, and consultants from many fields. Conservation scientists who study monuments and architecture often have a geologic background. Others have training as chemists, engineers, archaeologists, or architects. Working with other professionals from diverse disciplines and sharing insights across traditional academic boundaries makes for an ever-changing and challenging career.

It is sometimes surprising how little is known about important questions in conservation science. For example, in Venice and many other locations, a serious problem in the conservation of monuments is the deterioration of stone and brick caused by the crystallization of salts. Part of my research has been dedicated to better understanding the microdynamics of phenomena such as salt crystallization by using a special microscope that allows high-magnification observation of what actually happens to different materials as they deteriorate. The information from these experiments is used to improve conservation treatments and better control deleterious environmental conditions.

As an allied field of study, museum science has a longer history than conservation science and is concerned with improving our knowledge of the techniques and materials used to create the objects in our museums, as well as assessing their state of preservation. The related specialty of archaeometry is centered on questions of age, authenticity, and provenance (the source of an object). Many of the professionals in these fields have a background in geology because the analytic techniques and materials (such as pigments and stone) are familiar.

As part of my research, I have studied some of the oldest human cultural artifacts, using the latest high-tech analytic methods. For example, I recently helped identify human blood cells in a black pigment sample from a Chumash Indian rock-art site near Santa Barbara, California. I also contributed to a research project at the J. Paul Getty Museum on the kouros statue, a free-standing sculpture of a naked youth (see the Prologue). The Getty kouros is thought to be a rare example of Archaic Greek sculpture carved from dolomite marble in the 6th century B.C., although its authenticity has been questioned on stylistic grounds. All the data from the kouros investigation was found to be consistent with authenticity. However, we were unable to rule out the admittedly remote possibility of forgery.

Scientific advances in conservation and museum science often have the potential for wide application because creative, low-cost solutions to common problems are needed in every country. These two fields operate in an arena where (1) politics, cultural values, and history all play important roles and (2) communicating with nonscientists is as important as keeping in touch with colleagues. The future of conservation and museum science will depend on increasing public support for museums and conservation projects, as well as involving more scientists from universities and industry in collaborative programs. ■

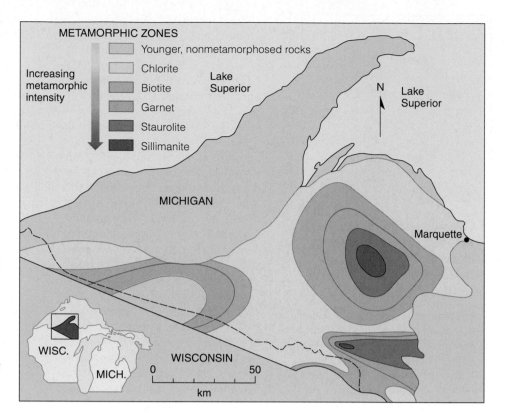

Figure 7.18
Metamorphic zones in the Upper Peninsula of Michigan. The zones in this region are based on the presence of distinctive silicate mineral assemblages resulting from the metamorphism of sedimentary rocks during an interval of mountain building and minor granitic intrusion during the Proterozoic Eon, about 1.5 billion years ago. The lines separating the different metamorphic zones are isograds.

metamorphic rocks can be divided into different zones based on the presence of distinctive silicate mineral assemblages. These mineral assemblages, each recognized by the presence of one or more index minerals, indicate different degrees of metamorphism. The index minerals Barrow and his associates chose to represent increasing metamorphic intensity were chlorite, biotite, garnet, staurolite, kyanite, and sillimanite (Figure 7.8). Note that these are the metamorphic minerals produced from clay-rich sedimentary rocks. Other mineral assemblages and index minerals are produced from rocks with different original compositions.

The successive appearance of metamorphic index minerals indicates gradually increasing or decreasing intensity of metamorphism. Going from lower- to higher-grade zones, the first appearance of a particular index mineral indicates the location of the minimum temperature and pressure conditions needed for the formation of that mineral. When the locations of the first appearances of that index mineral are connected on a map, the result is a line of equal metamorphic intensity, or an *isograd*. The region between isograds is known as a **metamorphic zone**. By noting the occurrence of metamorphic index minerals, geologists can construct a map showing the metamorphic zones of an entire area (Figure 7.18).

Numerous studies of different metamorphic rocks have demonstrated that although the texture and composition of any rock may be altered by metamorphism, the overall chemical composition may be little changed. Thus, the different mineral assemblages found in in-creasingly higher-grade metamorphic rocks derived from the same parent rock result from changes in temperature and pressure.

A **metamorphic facies** is a group of metamorphic rocks characterized by particular mineral assemblages formed under the same broad temperature–pressure conditions (Figure 7.19). Each facies is named after its most characteristic rock or mineral. For example, the green

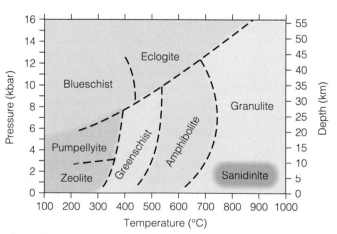

Figure 7.19
A pressure–temperature diagram showing where various metamorphic facies occur. A facies is characterized by a particular mineral assemblage that formed under the same broad temperature–pressure conditions. Each facies is named after its most characteristic rock or mineral.

metamorphic mineral chlorite, which forms under relatively low temperatures and pressures, yields rocks belonging to the *greenschist facies.* Under increasingly higher temperatures and pressures, other metamorphic facies, such as the *amphibolite* and *granulite facies,* develop.

Although usually applied to areas where the original rocks were clay rich, the concept of metamorphic facies can be used with modification in other situations. It cannot, however, be used in areas where the original rocks were pure quartz sandstones or pure limestones or dolostones. Such rocks would yield only quartzites and marbles, respectively.

HOW DOES METAMORPHISM RELATE TO PLATE TECTONICS?

Although metamorphism is associated with all three types of plate boundaries (see Figure 1.16), it is most common along convergent plate margins. Metamorphic rocks form at convergent plate boundaries because temperature and pressure increase as a result of plate collisions.

Figure 7.20 illustrates the various temperature–pressure regimes produced along an oceanic–continental convergent plate boundary and the type of metamorphic facies and rocks that can result. When an oceanic plate collides with a continental plate, tremendous pressure is generated as the oceanic plate is subducted. Because rock is a poor heat conductor, the cold descending oceanic plate heats slowly, and metamorphism is caused mostly by increasing pressure with depth. Metamorphism in such an environment produces rocks typical of the *blueschist facies* (low temperature, high pressure), which is characterized by the blue-colored amphibole mineral glaucophane (Figure 7.19). Geologists use the occurrence of blueschist facies rocks as evidence of ancient subduction zones. An excellent example of blueschist metamorphism can be found in the California Coast Ranges. Here rocks of the Franciscan Complex were metamorphosed under low-temperature, high-pressure conditions that clearly indicate the presence of a former subduction zone (Figure 7.21).

As subduction along the oceanic–continental plate boundary continues, both temperature and pressure increase with depth and can result in high-grade metamorphic rocks. Eventually, the descending plate begins to melt and generates magma that moves upward. This rising magma may alter the surrounding rock by contact metamorphism, producing migmatites in the deeper portions of the crust and hornfels at shallower depths. Such an environment is characterized by high temperatures and low to medium pressures.

Whereas metamorphism is most common along convergent plate margins, many divergent plate boundaries are characterized by contact metamorphism. Rising magma at mid-oceanic ridges heats the adjacent rocks, producing contact metamorphic minerals and textures. Besides contact metamorphism, fluids emanating from the rising magma—and its reaction with seawater—very commonly produce metal-bearing hydrothermal solutions rich in copper, iron, lead, zinc, and other metals, which are precipitated along mid-oceanic ridges. These deposits may eventually be brought to Earth's surface by later tectonic activity. The copper ores of Cyprus are a good example of such hydrothermal activity (see Chapter 12).

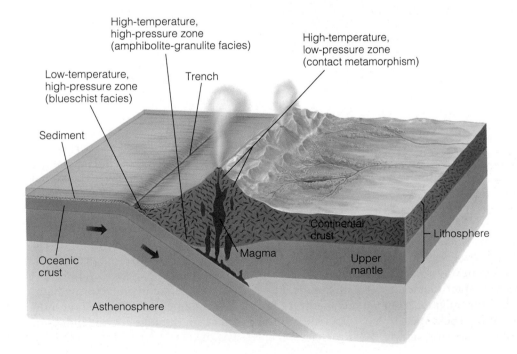

Figure 7.20
Metamorphic facies resulting from various temperature–pressure conditions produced along an oceanic–continental convergent plate boundary.

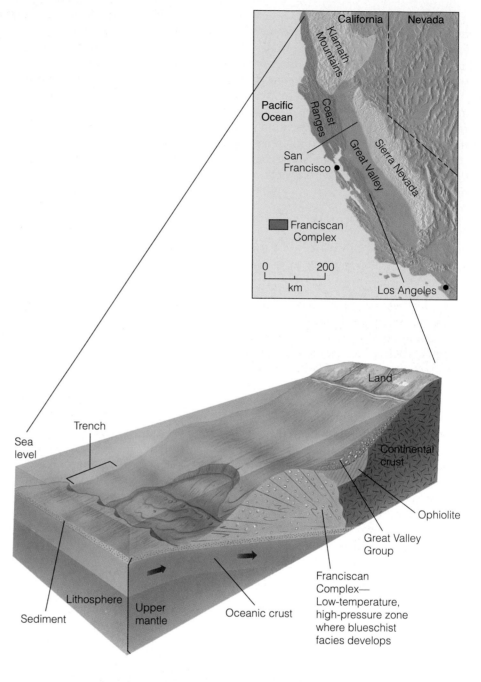

Figure 7.21
Index map of California showing the location of the Franciscan Complex and a diagrammatic reconstruction of the environment in which it was regionally metamorphosed under low-temperature, high-pressure subduction conditions approximately 150 million years ago. The red line on the index map shows the orientation of the reconstruction to the current geography.

METAMORPHISM AND NATURAL RESOURCES

Many metamorphic rocks and minerals are valuable natural resources. While these resources include various types of ore deposits, the two most familiar and widely used metamorphic rocks, as such, are marble and slate, which, as previously discussed, have been used for centuries in a variety of ways.

Many ore deposits result from contact metamorphism during which hot, iron-rich fluids migrate from igneous intrusions into the surrounding rock, thereby producing rich ore deposits. The most common sulfide ore minerals associated with contact metamorphism are bornite, chalcopyrite, galena, pyrite, and sphalerite; two common oxide ore minerals are hematite and magnetite. Tin and tungsten are also important ores associated with contact metamorphism (Table 7.2).

Other economically important metamorphic minerals include talc for talcum powder; graphite for pencils and dry lubricants (see Perspective 7.2); garnets and corundum, which are used as abrasives or gemstones, depending on their quality; and andalusite, kyanite, and sillimanite, all of which are used in manufacturing high-temperature porcelains and temperature-resistant minerals for products such as spark plugs and the linings of furnaces.

Perspective 7.2

Graphite

Graphite (from the Greek *grapho,* meaning "write"), a soft, gray-to-black mineral that has a greasy feel, is composed of the element carbon. Graphite occurs in two varieties: crystalline, which consists of thin, flat, nearly pure black flakes; and massive, an impure variety found in compact masses.

Graphite has the same composition as diamond, but its carbon atoms are strongly bonded together in sheets, with the sheets weakly held together by van der Waals bonds (see Figure 2.10). Because the sheets are loosely held together, they easily slide over one another, giving graphite its ability to mark paper and serve as a dry lubricant.

Graphite occurs mainly in metamorphic rocks produced by contact and regional metamorphism. It is found in marble, quartzite, schist, gneiss, and anthracite. Contact metamorphism of impure limestones by igneous intrusions produces some of the graphite found in marbles. The graphite resulting from regional metamorphism of sedimentary rocks probably came from organic matter present in the sediments. Some evidence, however, indicates that the graphite in Precambrian-age rocks (> 545 million years) may be the result of the reduction of calcium carbonate ($CaCO_3$) by an inorganic process. Graphite is also found in igneous rocks, pegmatite dikes, and veins; it is thought to have formed in these environments from the primary constituents of the magma or from the hot fluids and vapors released by the cooling magma.

Major producers of graphite are Mexico, Russia, and South Korea. In the United States, graphite has been mined in 27 states, but production is now generally limited to Alabama and New York.

Graphite is used for many purposes. The oldest use is in pencil leads, where it is finely ground, mixed with clay, and baked. The amount of clay and the baking time give pencil leads their desired hardness. Other important uses include batteries, brake linings, carbon brushes, crucibles, foundry facings, lubricants, refractories, and steelmaking.

Synthetic graphite can be produced from anthracite coal or petroleum coke and now accounts for most graphite production. Its extreme purity (99–99.5% pure) makes it especially valuable where high purity is required, as in the rods that slow the reaction rates in nuclear reactors.

Table 7.2

The Main Ore Deposits Resulting from Contact Metamorphism

Ore Deposit	Major Mineral	Formula	Use
Copper	Bornite Chalcopyrite	Cu_5FeS_4 $CuFeS_2$	Important sources of copper, which is used in various aspects of manufacturing, transportation, communications, and construction
Iron	Hematite Magnetite	Fe_2O_3 Fe_3O_4	Major sources of iron for manufacture of steel, which is used in nearly every form of construction, manufacturing, transportation, and communications
Lead	Galena	PbS	Chief source of lead, which is used in batteries, pipes, solder, and elsewhere where resistance to corrosion is required
Tin	Cassiterite	SnO_2	Principal source of tin, which is used for tin plating, solder, alloys, and chemicals
Tungsten	Scheelite Wolframite	$CaWO_4$ $(Fe,Mn)WO_4$	Chief sources of tungsten, which is used in hardening metals and manufacturing carbides
Zinc	Sphalerite	$(Zn,Fe)S$	Major source of zinc, which is used in batteries and in galvanizing iron and making brass

1. Metamorphic rocks result from the transformation of other rocks, usually beneath Earth's surface, as a consequence of one or a combination of three agents: heat, pressure, and fluid activity.

2. Heat for metamorphism comes from intrusive magmas or deep burial. Pressure is either lithostatic or differential. Fluids trapped in sedimentary rocks or emanating from intruding magmas can enhance chemical changes and the formation of new minerals.

3. The three major types of metamorphism are contact, dynamic, and regional.

4. Index minerals—minerals that form only within specific temperature and pressure ranges—allow geologists to recognize low-, intermediate-, and high-grade metamorphic zones.

5. Metamorphic rocks are classified primarily according to their texture. In a foliated texture, platy minerals have a preferred orientation. A nonfoliated texture does not exhibit any discernible preferred orientation of the mineral grains.

6. Foliated metamorphic rocks can be arranged in order of grain size and/or perfection of their foliation. Slate is fine grained, followed by (in coarser-grained order) phyllite and schist; gneiss displays segregated bands of minerals. Amphibolite is another fairly common foliated metamorphic rock.

7. Common nonfoliated metamorphic rocks are marble, quartzite, greenstone, and hornfels.

8. Metamorphic zones are based on index minerals and are areas of equal metamorphic intensity. Metamorphic facies are characterized by particular assemblages of minerals that formed under specific metamorphic conditions. These facies are named for a characteristic constituent mineral or rock type.

9. Metamorphism can occur along all three kinds of plate boundaries but most commonly occurs at convergent plate margins.

10. Metamorphic rocks formed near Earth's surface along an oceanic–continental plate boundary result from low-temperature, high-pressure conditions. As a subducted oceanic plate descends, it is subjected to increasingly higher temperatures and pressures that result in higher-grade metamorphism.

11. Many metamorphic rocks and minerals, such as marble, slate, graphite, talc, and asbestos, are valuable natural resources.

aureole
contact metamorphism
differential pressure
dynamic metamorphism
fluid activity

foliated texture
heat
index minerals
lithostatic pressure
metamorphic facies

metamorphic rock
metamorphic zone
nonfoliated texture
regional metamorphism

1. The metamorphic rocks resulting from pure dynamic metamorphism are:

    a. _____ aureoles;
    b. _____ migmatites;
    c. _____ mylonites;
    d. _____ amphibolites;
    e. _____ anthracites.

2. The metamorphic rock that is streaked or has segregated bands of light and dark minerals is called:

    a. _____ schist;
    b. _____ gneiss;
    c. _____ quartzite;
    d. _____ marble;
    e. _____ phyllite.

3. The metamorphic rock formed from clay-rich sedimentary rocks is:

    a. _____ marble;
    b. _____ quartzite;
    c. _____ anthracite;
    d. _____ hornfels;
    e. _____ none of these.

4. Which type of metamorphism has high differential pressures as the main agent of change?

    a. _____ contact;
    b. _____ regional;
    c. _____ dynamic;
    d. _____ hydrothermal;
    e. _____ none of these.

5. Pressure applied equally owing to the weight of overlying rock is called:

    a. _____ lithostatic;
    b. _____ differential;
    c. _____ hydrothermal;
    d. _____ aureole;
    e. _____ none of these.

6. Metamorphic rocks that occur in broad zones and facies over large areas are called:

    a. _____ contact;
    b. _____ dynamic;
    c. _____ regional;
    d. _____ hydrothermal;
    e. _____ none of these.

7. Heat and fluid activity are the main agents of change in which type of metamorphism?

    a. _____ regional;
    b. _____ hydrothermal;
    c. _____ dynamic;
    d. _____ contact;
    e. _____ none of these.

8. Most metamorphic rocks are of what type?

    a. _____ dynamic;
    b. _____ regional;
    c. _____ contact;
    d. _____ hydrothermal;
    e. _____ none of these.

9. Which of the following rock(s) have a nonfoliated texture?

    a. _____ slate;
    b. _____ marble;
    c. _____ schist;
    d. _____ phyllite;
    e. _____ gneiss.

10. A line of equal metamorphic intensity on a map is called a(an):

    a. _____ facies;
    b. _____ zone;
    c. _____ isograd;
    d. _____ index;
    e. _____ none of these.

11. Which of the following rocks has slightly coarser grain size than slate?

    a. _____ schist;
    b. _____ phyllite;
    c. _____ gneiss;
    d. _____ migmatite;
    e. _____ none of these.

12. What grade of metamorphism is indicated by greenstone?

    a. _____ high;
    b. _____ intermediate;
    c. _____ low;
    d. _____ very low;
    e. _____ none of these.

13. Metamorphic facies are *not*:

    a. _____ characterized by an assemblage of minerals;
    b. _____ reflective of temperature and pressure conditions;
    c. _____ equal to a metamorphic zone;
    d. _____ named after a most characteristic rock or mineral;
    e. _____ none of these.

14. Regional metamorphism is most closely associated with what type of plate boundary?

    a. _____ divergent;
    b. _____ convergent;
    c. _____ transform;
    d. _____ isograd;
    e. _____ none of these.

15. Which of the following sites will likely have the least metamorphic rock?

    a. _____ core of a mountain range;
    b. _____ oceanic crust;
    c. _____ shields;
    d. _____ answers (a) and (c);
    e. _____ none of these.

16. Name the three types of metamorphism, and briefly explain each.

17. What are the two main textural groups of metamorphic rocks? Briefly describe the key characteristics of each.

18. Where and how does regional metamorphism take place (in relation to plate tectonics)? What regional metamorphic facies are produced, and how are they arranged?

19. What are mylonites? Explain how they may form in dynamic metamorphism.

20. What is contact metamorphism? How do the agents of metamorphism yield an aureole during contact metamorphism?

21. In metamorphic processes, what are the two important types of pressure? What different effects are produced by each?

22. Describe progressively decreasing regional metamorphic grades, including rock types and index materials, beginning with rocks nearest the source of heat and pressure.

23. Name the more common foliated metamorphic rocks, and describe their characteristics. How would you distinguish each?

24. What are the differences among metamorphic grade, facies, and zone?

25. List four economically important metamorphic minerals, and comment on their usefulness.

Points to Ponder

1. If there were no plate tectonics on Earth, do you think Earth would have any metamorphic rocks? Explain.

2. If any of the metamorphic rocks possessed planes of weakness that could make them vulnerable to possible failure as bedrock (such as sudden breaking under the weight of a dam or other construction project), what rock type(s) would you predict might fail? Is one of the two main textural groups more susceptible to failure than the other? Why?

3. Using Figure 7.8, compare the following mineral assemblage: intermediate-versus high-temperature. Describe what minerals exist at intermediate temperatures of metamorphism, but not at high temperatures. What happens to these minerals in metamorphism?

4. Using Figure 7.18, briefly describe the origin of metamorphic zones in the "bull's eye"-appearing feature west of Marquette, Michigan. Speculate on why the zones are likely arranged in this fashion.

5. Using Figure 7.19, go to a point that is represented by 450°C and 6 kilobars of pressure. What metamorphic facies is represented by those conditions? If pressure is raised to 10 kilobars, what facies is represented by the new conditions? What change in depth of burial is required to effect the pressure change of 6 to 10 kilobars?

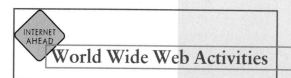

INTERNET AHEAD

World Wide Web Activities

For these Web site addresses, along with current updates and exercises, log on to
http://www.brookscole.com/geo/

➤ **METAMORPHIC ROCKS**

This Web site contains images and information on 11 common metamorphic rocks. It is part of the Soil Science 223 Rocks and Minerals Reference Web site. Click on any of the *rock names,* and compare the information and images to the information in this chapter.

➤ **UNIVERSITY OF TULSA—DEPARTMENT OF GEOSCIENCES METAMORPHIC ROCKS AND PROCESSES**

This Web site contains much information about metamorphic rocks and processes in general. Read over the material about metamorphic rocks and processes at this site.

➤ **THE GEOLOGY OF THE POINT REYES PENINSULA**

This Web site contains information about the geology of the Point Reyes Peninsula, California, including information about the metamorphic rocks in the area. Click on the *Pre-Cretaceous metamorphic rocks (pKm)* site. What are the different types of metamorphic rocks found in this area?

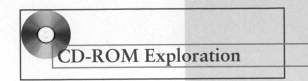

CD-ROM Exploration

➤ Exploring your *Earth Systems Today* CD-ROM will add to your understanding of the material in this chapter.

TOPIC: EARTH'S MATERIALS

MODULE: ROCK TYPES

Explore activities in this module to see if you can discover the following for yourself:

Go to the "rock laboratory" in this activity and observe how shale is transformed into slate. What progressive changes do you see in the process of metamorphism? Based on what you see in the "rock laboratory," write a short description of all the changes that clay in mudstone must experience to become mica in slate.

How are metamorphic rocks different from the other two types of rocks in the way that they form? (What would you look for to tell metamorphic rocks from other types of rocks?)

Chapter 8

The White Cliffs of Dover, shown here at Eastbourne, England, consist of the sedimentary rock chalk. The chalk was deposited in warm, shallow seas during the Cretaceous Period (144 to 66 million years ago).

OBJECTIVES

At the end of this chapter, you will have learned that

- The concept of geologic time and its measurement have changed throughout human history.
- The principle of uniformitarianism is fundamental to geology.
- Lord Kelvin seemingly invalidated the uniformitarian foundation of geology but his basic premise was false.

- The fundamental principles of relative dating provide a means to interpret geologic history.
- The three types of unconformities— disconformities, angular unconformities, and nonconformities— are erosional surfaces separating younger from older rocks and represent significant intervals of geologic time for which we have no record.

- Time equivalency of rock units can be demonstrated by various correlation techniques.
- Different absolute-dating methods can be used to date geologic events in terms of years before the present.
- The most accurate radiometric dates are obtained from igneous rocks.
- The geologic time scale evolved primarily during the 19th century through the efforts of many people.

Geologic Time

Prologue

What is time? We seem obsessed with it and organize our lives around it with the help of clocks, calendars, and appointment books. Yet most of us feel we don't have enough of it—we are always running "behind" or "out of time." According to biologists and psychologists, animals and children less than 2 years old exist in a "timeless present." Some scientists think that our early ancestors may also have lived in a state of timelessness with little or no perception of a past or future. According to Buddhist, Taoist, and Mayan beliefs, time is circular, and like a circle, all things are destined to return to where they once were. Thus, in these belief systems, there is no beginning or end, but rather a cyclicity to everything.

In some respects, time is defined by the methods used to measure it. Many prehistoric monuments are oriented to detect the summer solstice, and sundials were used to divide the day into measurable units. As civilization advanced, mechanical devices were invented to measure time, the earliest being the water clock, first used by the ancient Egyptians and further developed by the Greeks and Romans. The pendulum clock was invented in the seventeenth century and provided the most accurate timekeeping for the next two and one-half centuries.

Today the quartz watch is the most popular timepiece. Powered by a battery, a quartz crystal vibrates approximately 100,000 times per second. An integrated circuit counts these vibrations and converts them into a digital or dial reading on your watch face. An inexpensive quartz watch today is more accurate than the best mechanical watch, and precision-manufactured quartz clocks are accurate to within 1 second per 10 years.

Precise timekeeping is important in our technological world. Ships and aircraft plot their locations by satellite, relying on an extremely accurate time signal. Deep-space and planetary probes, such as the *Voyagers* and the *Mars Pathfinder* and robotic rover *Sojourner* (see Chapter 20), require radio commands timed to billionths of a second, and physicists exploring the motion inside the nucleus of an atom deal in trillionths of a second as easily as we talk about minutes.

To achieve such accuracy, scientists use atomic clocks. First developed in the 1940s, these clocks rely on an atom's oscillating electrons, a rhythm so regular that they are accurate to within a few thousandths of a second per day. An atomic clock accurate to within 1 second per 3 million years was recently installed at the National Institute of Standards and Technology.

Whereas physicists deal with incredibly short intervals of time, astronomers and geologists are concerned with geologic time measured in millions or billions of years. When astronomers look at a distant galaxy, they are seeing what it looked like billions of years ago. When geologists investigate rocks in the walls of the Grand Canyon, they are deciphering events that occurred over an interval of 2 billion years. Geologists can measure decay rates of such radioactive elements as uranium, thorium, and rubidium to determine how long ago an igneous rock formed. Furthermore, geologists know that Earth's rotational velocity has been slowing a few thousandths of a second per century as a result of the frictional effects of tides, ocean currents, and varying thicknesses of polar ice. Five hundred million years ago, a day was only 20 hours long, and at the current rate of slowing, 200 million years from now a day will be 25 hours long.

Time is a fascinating topic that has been the subject of numerous essays and books. And although we can comprehend concepts such as milliseconds and understand how a quartz watch works, deep time, or geologic time, is still difficult for most people to comprehend. In fact, geology could not have advanced as a science until the foundation of geologic time was firmly established and accepted.

INTRODUCTION

Time sets geology apart from most of the other sciences, and an appreciation of the immensity of geologic time is fundamental to understanding the physical and biological history of our planet. In fact, the understanding and acceptance of the enormity of geologic time is one of the major contributions geology has made to the sciences.

Most people have difficulty comprehending geologic time because of its immensity and because we tend to view time from the perspective of our own existence. Ancient history is what occurred among humans hundreds or even thousands of years ago, but when geologists talk of ancient geologic history, they are referring to events that happened millions or even billions of years ago!

One of the many ways to try and gain a perspective of geologic time is to imagine counting pennies (or any other coins or objects) at the rate of one penny per second, 24 hours per day, and 365 days a year. At that rate it would take 11.6 days to count one million pennies, 11,574 days (31.7 years) to count one billion pennies, and 53,240 days (145.9 years) to count 4.6 billion pennies—which is the age of Earth. Another interesting way to visualize the immensity of geologic time and when important geologic and biologic events occurred is to compress the age of Earth (4.6 billion years) into one year (365 days). When that is done, one finds that humans don't even appear until late in the evening on December 31!

Besides gaining an appreciation for the immensity of geologic time, why is the study of geologic time important? One of the most important lessons to learn

from this chapter is how to reason and apply the fundamental geologic principles to solve geologic problems. Most students will not become professional geologists, but the logic used in applying the principles of relative dating to reconstruct the geologic history of an area involves basic reasoning skills that you can transfer and use in almost any profession.

The study of geologic time is also important for a number of other reasons. The discovery of radioactivity at the end of the 19th century gave geologists a way to determine the age of minerals and rocks. Advances and refinements in the various radiometric dating techniques during the twentieth century have changed the way we view Earth in terms of when events occurred in the past and the rates of geologic change through time. The ability to accurately determine past climatic changes and their causes has important implications for the current debate on global warming and its effects on humans (see Perspective 8.1).

When discussing geologic time, geologists use two different frames of reference. **Relative dating** involves placing geologic events in a sequential order as determined from their position in the geologic record. Relative dating will not tell us how long ago a particular event occurred, only that one event preceded another. A useful analogy for relative dating is a television guide without the times when programs are shown. In this example, you cannot tell what time a particular program will be shown, but by watching a few shows and checking the guide, you can determine if you have missed the show or how many more shows are scheduled before the one you want to see.

The various principles used to determine relative dating were discovered hundreds of years ago, and since

Perspective 8.1

Geologic Time and Climate Change

With all the debate concerning global warming and its possible implications, it is extremely important to be able to reconstruct past climatic regimes as accurately as possible. If we are to model how Earth's climate system has responded to changes in the past and use that information for simulations of future climatic scenarios, it is essential that we have as precise a geologic calendar as possible.

One way to study climatic changes is to examine lake sediment or ice cores that have organic matter in them. By taking closely spaced samples and dating the organic matter in the cores using the carbon 14 dating technique (see "Radiocarbon and Tree-Ring Dating Methods" section), a detailed chronology for each core examined can be constructed. Changes in isotope ratios, pollen, and plant and invertebrate fossil assemblages can then be accurately dated, and the time and duration of climate changes correlated over increasingly larger areas. Without a means of precise dating, there would be no way to accurately model past climatic changes with the precision needed to predict possible future climate changes on a human time scale.

An interesting method that is becoming more common in reconstructing past climates is the analysis of stalagmites from caves. Stalagmites are icicle-shaped structures rising from a cave floor and formed of calcium carbonate precipitated from evaporating water (see Chapter 16). A stalagmite therefore records a layered stratigraphy because each newly precipitated layer of calcium carbonate is younger than the previously precipitated layer. Thus a stalagmite's layers are oldest in the center at its base and get progressively younger outward. Using techniques developed during the past 10 years, geologists can achieve very precise radiometric dates on individual layers of a stalagmite using high-precision ratios of uranium 234 to thorium 230. This technique lets geologists determine the age of materials much older than can be obtained by carbon 14 dating and is reliable back to about 500,000 years.

An interesting recent study of stalagmites from Crevice Cave, in Missouri, revealed a history of climatic and vegetation change in the midcontinent region of the United States during the interval between 75,000 and 25,000 years ago. By analyzing carbon 13 and oxygen 18 isotope profiles during this interval, geologists were able to deduce that average temperature fluctuations of 4°C correlated with major changes in vegetation. During the interval between 75,000 and 55,000 years ago, the climate oscillated between warm and cold, and vegetation alternated among forest, savannah, and prairie. Fifty-five thousand years ago the climate cooled and there was a sudden change from grasslands to forest, which persisted until 25,000 years ago. This corresponds to the time when global ice sheets began building and advancing.

High-precision uranium 234–thorium 230 dating techniques in stalagmite studies provide an accurate chronology allowing geologists to model climate systems of the past and perhaps to determine what causes global climatic changes and their durations. Without these sophisticated dating techniques and others like them, geologists would not be able to make precise correlations and accurately reconstruct past environments and climates. By analyzing past environmental and climate changes and their durations, geologists hope they can use these data, sometime in the near future, to predict and possibly modify regional climatic changes.

then they have been used to construct the *relative geologic time scale* (Figure 8.1). These principles are still widely used today.

Absolute dating results in specific dates for rock units or events expressed in years before the present. In our analogy of the television guide, the time when the programs were actually shown would be the absolute dates. In this way, you not only could determine whether you have missed a show (relative dating), but also would know how long it would be until a show you want to see would be shown (absolute dating).

Radiometric dating is the most common method of obtaining absolute ages. Such dates are calculated from the natural rates of decay of various radioactive elements present in trace amounts in some rocks. Not until the discovery of radioactivity near the end of the

19th century could absolute ages be accurately applied to the relative geologic time scale. Today, the geologic time scale is really a dual scale: a relative scale based on rock and fossil sequences, with radiometric dates expressed as years before the present (Figure 8.1).

HOW HAS THE CONCEPT OF GEOLOGIC TIME AND EARTH'S AGE CHANGED THROUGHOUT HUMAN HISTORY?

The concept of geologic time and its measurement has changed through human history. Many early Christian scholars and clerics tried to establish the date of creation by analyzing historical records and the genealogies found in Scripture. Based on their analyses, they generally believed that Earth and all its features were no more than about 6000 years old. The idea of a very young Earth provided the basis for most Western chronologies of Earth history prior to the eighteenth century.

During the eighteenth and nineteenth centuries, several attempts were made to determine Earth's age on the basis of scientific evidence rather than revelation. For example, the French zoologist Georges Louis de Buffon (1707–1788) assumed Earth gradually cooled to its present condition from a molten beginning. To simulate this history, he melted iron balls of various diameters and allowed them to cool to the surrounding temperature. By extrapolating their cooling rate to a ball the size of Earth, he determined that Earth was at least 75,000 years old. Although this age was much older than that derived from Scripture, it was vastly younger than we now know the planet to be.

Other scholars were equally ingenious in attempting to calculate Earth's age. For example, if deposition rates could be determined for various sediments, geologists reasoned that they could calculate how long it would take to deposit any rock layer. They could then extrapolate how old Earth was from the total thickness of sedimentary rock in its crust. Rates of deposition vary, however, even for the same type of rock. Furthermore, it is impossible to estimate how much rock has been removed by erosion, or how much a rock sequence has been reduced by compaction. As a result of these variables, estimates ranged from less than 1 million years to more than 1 billion years.

Another attempt to determine Earth's age involved calculating the age of the oceans. If the ocean basins were filled very soon after the origin of the planet, then they would be only slightly younger than Earth itself. The best-known calculations for the oceans' age were made by the Irish geologist John Joly in 1899. He reasoned that Earth's ocean waters were originally fresh and their present salinity was the result of dissolved salt being carried into the ocean basins by rivers. By measuring the present amount of salt in the world's rivers, and knowing the volume of ocean water and its salinity, Joly calculated that it would have taken at least 90 million years for the oceans to reach their present salinity level. This was still much younger than the now accepted age of 4.6 billion years for Earth, mainly because Joly had no way of calculating how much salt had been recycled or the amount of salt stored in continental salt deposits and seafloor clay deposits.

Besides trying to determine Earth's age, the naturalists of the eighteenth and nineteenth centuries were also formulating some of the fundamental geologic principles that are used in deciphering Earth history. From the evidence preserved in the geologic record, it was clear to them that Earth is very old and that geologic processes have operated over long periods of time.

WHO IS JAMES HUTTON, AND WHY ARE HIS CONTRIBUTIONS TO GEOLOGY IMPORTANT?

The Scottish geologist James Hutton (1726–1797) is considered by many to be the father of modern geology. His detailed studies and observations of rock exposures and present-day geologic processes were instrumental in establishing the **principle of uniformitarianism** (see Chapter 1), the concept that the same processes have operated over vast amounts of time. Because Hutton relied on known processes to account for Earth history, he concluded that Earth must be very old and wrote that "we find no vestige of a beginning, and no prospect of an end."

Unfortunately, Hutton was not a particularly good writer, so his ideas were not widely disseminated or accepted. In 1830 Charles Lyell published a landmark book, *Principles of Geology*, in which he championed Hutton's concept of uniformitarianism. Instead of relying on catastrophic events to explain various Earth features, Lyell recognized that imperceptible changes brought about by present-day processes could, over long periods of time, have tremendous cumulative effects. Through his writings, Lyell firmly established uniformitarianism as the guiding philosophy of geology. Furthermore, the recognition of virtually limitless amounts of time was also necessary for and instrumental in the acceptance of Darwin's 1859 theory of evolution.

After finally establishing that present-day processes have operated over vast periods of time, geologists were nevertheless nearly forced to accept a very young age for

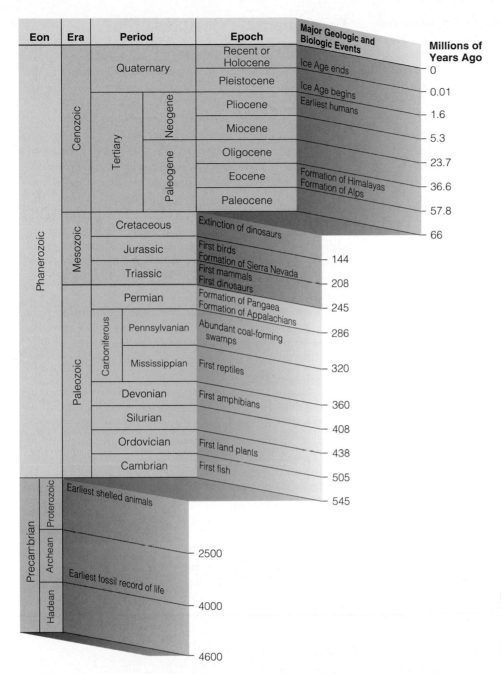

Eon	Era	Period		Epoch	Major Geologic and Biologic Events	Millions of Years Ago
Phanerozoic	Cenozoic	Quaternary		Recent or Holocene	Ice Age ends	0
				Pleistocene	Ice Age begins	0.01
		Tertiary	Neogene	Pliocene	Earliest humans	1.6
				Miocene		5.3
			Paleogene	Oligocene		23.7
				Eocene	Formation of Himalayas Formation of Alps	36.6
				Paleocene		57.8
	Mesozoic	Cretaceous			Extinction of dinosaurs	66
		Jurassic			First birds Formation of Sierra Nevada	144
		Triassic			First mammals First dinosaurs	208
	Paleozoic	Permian			Formation of Pangaea Formation of Appalachians	245
		Carboniferous	Pennsylvanian		Abundant coal-forming swamps	286
			Mississippian		First reptiles	320
		Devonian			First amphibians	360
		Silurian				408
		Ordovician			First land plants	438
		Cambrian			First fish	505
Precambrian	Proterozoic				Earliest shelled animals	545
	Archean					2500
	Hadean				Earliest fossil record of life	4000
						4600

Figure 8.1
The geologic time scale. Some of the major geologic and biologic events are indicated along the right-hand margin.

Earth when a highly respected English physicist, Lord Kelvin (1824–1907), claimed, in a paper written in 1866, to have destroyed the uniformitarian foundation of geology. Starting with the generally accepted belief that Earth was originally molten, Kelvin assumed that it has gradually been losing heat and that, by measuring this heat loss, he could determine its age.

Kelvin knew from deep mines in Europe that Earth's temperature increases with depth, and he reasoned that Earth is losing heat from its interior. By knowing the melting temperatures of rocks, the size of Earth, and the rate of heat loss, Kelvin was able to calculate the age at which Earth was entirely molten. From these calculations, he concluded that Earth could

not be older than 400 million years or younger than 20 million years. This wide range in age reflected uncertainties in average temperature increases with depth and the various melting points of Earth's constituent materials.

After finally establishing that Earth was very old and showing how present-day processes can be extended back over long periods of time to explain geologic features, geologists were in a quandary. Either they had to accept Kelvin's dates and squeeze events into a shorter time frame, or they had to abandon the concept of seemingly limitless time that was the underpinning of uniformitarian geology and one of the foundations of Darwinian evolution.

Kelvin's reasoning and calculations were sound, but his basic premises were false, thereby invalidating his conclusions. Kelvin was unaware that Earth has an internal heat source, radioactivity, that has allowed it to maintain a fairly constant temperature through time.* His 40-year campaign for a young Earth ended with the discovery of radioactivity near the end of the nineteenth century. His calculations were therefore no longer valid, and his proof for a geologically young Earth collapsed. Moreover, although the discovery of radioactivity destroyed Kelvin's arguments, it provided geologists with a clock that could measure Earth's age and validate what geologists had been saying all along; namely, that Earth was indeed very old!

WHAT ARE RELATIVE-DATING METHODS, AND WHY ARE THEY IMPORTANT?

Before the development of radiometric dating techniques, geologists had no reliable means of absolute dating and therefore depended solely on relative-dating methods. These methods allow events to be placed in sequential order only and do not tell us how long ago an event took place. Although the principles of relative dating may now seem self-evident, their discovery was an important scientific achievement because they provided geologists with a means to interpret geologic history and develop a relative geologic time scale.

Six fundamental geologic principles are used in relative dating: superposition, original horizontality, lateral continuity, cross-cutting relationships, inclusions, and fossil succession.

Fundamental Principles of Relative Dating

The seventeenth century was an important time in the development of geology as a science because of the widely circulated writings of the Danish anatomist Nicolas Steno (1638–1686). Steno observed that when streams flood, they spread out across their floodplains and deposit layers of sediment that bury organisms dwelling on the floodplain. Subsequent floods produce new layers of sediments that are deposited or superposed over previous deposits. When lithified, these layers of sediment become sedimentary rock. Thus, in an undisturbed succession of sedimentary rock layers, the oldest layer is at the bottom and the youngest layer is at the top. This **principle of superposition** is the basis for relative-age determinations of strata and their contained fossils (Figure 8.2).

Steno also observed that, because sedimentary particles settle from water under the influence of gravity, sediment is deposited in essentially horizontal layers, thus illustrating the **principle of original horizontality** (Figure 8.2). Therefore, a sequence of sedimentary rock layers that is steeply inclined from the horizontal must have been tilted after deposition and lithification.

Steno's third principle, the **principle of lateral continuity**, states that sediment extends laterally in all directions until it thins and pinches out or terminates against the edge of the depositional basin (Figure 8.2).

James Hutton is credited with discovering the **principle of cross-cutting relationships.** Based on his detailed studies and observations of rock exposures in Scotland,

Figure 8.2
The Grand Canyon of Arizona illustrates three of the six fundamental principles of relative dating. The sedimentary rocks of the Grand Canyon were originally deposited horizontally in a variety of marine and continental environments (principle of original horizontality). The oldest rocks are at the bottom of the canyon, and the youngest rocks are at the top, forming the rim (principle of superposition). The exposed rock layers extend laterally for some distance (principle of lateral continuity).

---

*Actually, Earth's temperature has decreased through time, because the original amount of radioactive materials has been decreasing and thus is not supplying as much heat. However, the temperature is decreasing at a rate considerably slower than would be required to lend any credence to Kelvin's calculations.

(a)   (b)

Figure 8.3
The principle of cross-cutting relationships. (a) A dark-colored dike has been intruded into older light-colored granite along the north shore of Lake Superior, Ontario, Canada. (b) A fault displacing tilted beds along Templin Highway, Castaic, California.

Hutton recognized that an igneous intrusion or fault must be younger than the rocks it intrudes or displaces (Figure 8.3).

Although this principle illustrates that an intrusive igneous structure is younger than the rocks it intrudes, the association of sedimentary and igneous rocks may cause problems in relative dating. Buried lava flows and intrusive igneous bodies such as sills look very similar in a sequence of strata (Figure 8.4). A buried lava flow, however, is older than the rocks above it (principle of superposition), while a sill is younger than all the beds below it and younger than the bed immediately above it.

To resolve such relative-age problems as these, geologists look to see if the sedimentary rocks in contact with the igneous rocks show signs of baking or alteration by heat (see the section on contact metamorphism

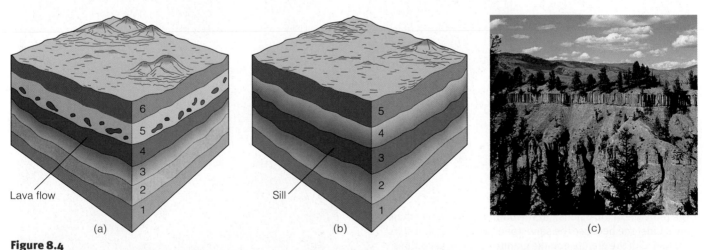

Lava flow   Sill

(a)   (b)   (c)

Figure 8.4
Relative ages of lava flows, sills, and associated sedimentary rocks may be difficult to determine. (a) A buried lava flow in bed 4 baked the underlying bed, and bed 5 contains inclusions of the lava flow. The lava flow is younger than bed 3 and older than beds 5 and 6. (b) The rock units above and below the sill in bed 3 have been baked, indicating that the sill is younger than beds 2 and 4, but its age relative to bed 5 cannot be determined. (c) Buried lava flow, Yellowstone National Park, Wyoming. Lava flow displays columnar jointing.

in Chapter 7 page 196). A sedimentary rock showing such effects must be older than the igneous rock with which it is in contact. In Figure 8.4, for example, a sill produces a zone of baking immediately above and below it because it intruded into previously existing sedimentary rocks. A lava flow, in contrast, bakes only those rocks below it.

Another way to determine relative ages is by using the **principle of inclusions.** This principle holds that inclusions, or fragments of one rock contained within a layer of another, are older than the rock layer itself. The batholith shown in Figure 8.5a contains sandstone inclusions, and the sandstone unit shows the effects of baking. Accordingly, we conclude that the sandstone is older than the batholith. In Figure 8.5b, however, the sandstone contains granite rock fragments, indicating that the batholith was the source rock for the inclusions and is therefore older than the sandstone.

Fossils have been known for centuries (see Chapter 6), yet their utility in relative dating and geologic mapping was not fully appreciated until the early nineteenth century. William Smith (1769–1839), an English civil engineer involved in surveying and building canals in southern England, independently recognized the principle of superposition by reasoning that the fossils at the bottom of a sequence of strata are older than those at the top of the sequence. This recognition served as the basis for the **principle of fossil succession,** or the *principle of faunal and floral succession,* as it is sometimes called (Figure 8.6).

According to this principle, fossil assemblages succeed one another through time in a regular and predictable order. The validity and successful use of this principle depends on three points: (1) Life has varied through time, (2) fossil assemblages are recognizably different from one another, and (3) the relative ages of the fossil assemblages can be determined. Observations of fossils in older versus younger strata clearly demonstrate that life-forms have changed. Because this is true, fossil assemblages (point 2) are recognizably different. Furthermore, superposition can also be used to demonstrate the relative ages of the fossil assemblages.

Unconformities

Our discussion so far has been concerned with conformable sequences of strata, sequences in which no depositional breaks of any consequence occur. A sharp bedding plane (see Figure 6.19) separating strata may represent a depositional break of minutes, hours, years, or even tens of years, but it is inconsequential when considered in the context of geologic time.

Surfaces of discontinuity that encompass significant amounts of geologic time are **unconformities,** and any interval of geologic time not represented by strata in a particular area is a *hiatus* (Figure 8.7). Thus, an unconformity is a surface of nondeposition or erosion that separates younger strata from older rocks. As such,

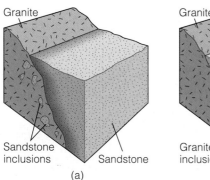
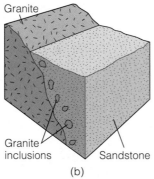

(a) (b)

(c)

Figure 8.5
The principle of inclusions. (a) The batholith is younger than the sandstone because the sandstone has been baked at its contact with the granite and the granite contains sandstone inclusions. (b) Granite inclusions in the sandstone indicate that the batholith was the source of the sandstone and therefore is older. (c) Outcrop in northern Wisconsin showing basalt inclusions (black) in granite (white). Accordingly, the basalt inclusions are older than the granite.

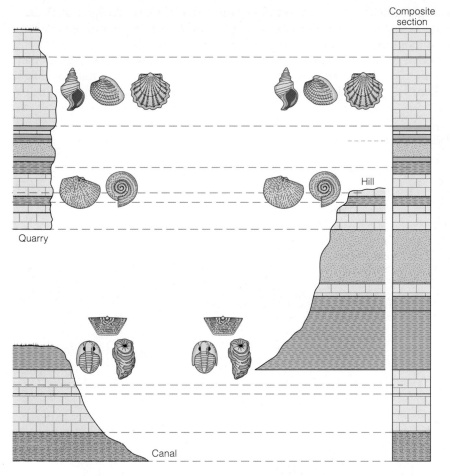

Figure 8.6
This generalized diagram shows how William Smith used fossils to identify strata of the same age in different areas (principle of fossil succession). The composite section on the right shows the relative ages of all strata in this area.

Composite section

Hill

Quarry

Canal

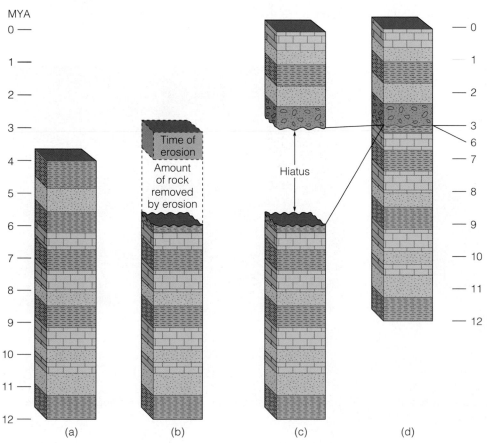

MYA

Time of erosion

Amount of rock removed by erosion

Hiatus

(a) (b) (c) (d)

Figure 8.7
A simplified diagram showing the development of an unconformity and a hiatus.
(a) Deposition began 12 million years ago (MYA) and continued more or less uninterrupted until 4 MYA. (b) A 1-million-year episode of erosion occurred, and during that time strata representing 2 million years of geologic time were eroded. (c) A hiatus of 3 million years exists between the older strata and the strata that formed during a renewed episode of deposition that began 3 MYA. (d) The actual stratigraphic record. The unconformity is the surface separating the strata and represents a major break in our record of geologic time.

it represents a break in our record of geologic time. The famous 12-minute gap in the Watergate tapes of Richard Nixon's presidency is somewhat analogous. Just as we have no record of the conversations that were occurring during this period of time, we have no record of the events that occurred during a geologic hiatus.

Three types of unconformities are recognized. A **disconformity** is a surface of erosion or nondeposition between younger and older beds that are parallel with one another (Figure 8.8). Unless a well-defined erosional surface separates the older from the younger parallel beds, the disconformity frequently resembles an ordi-

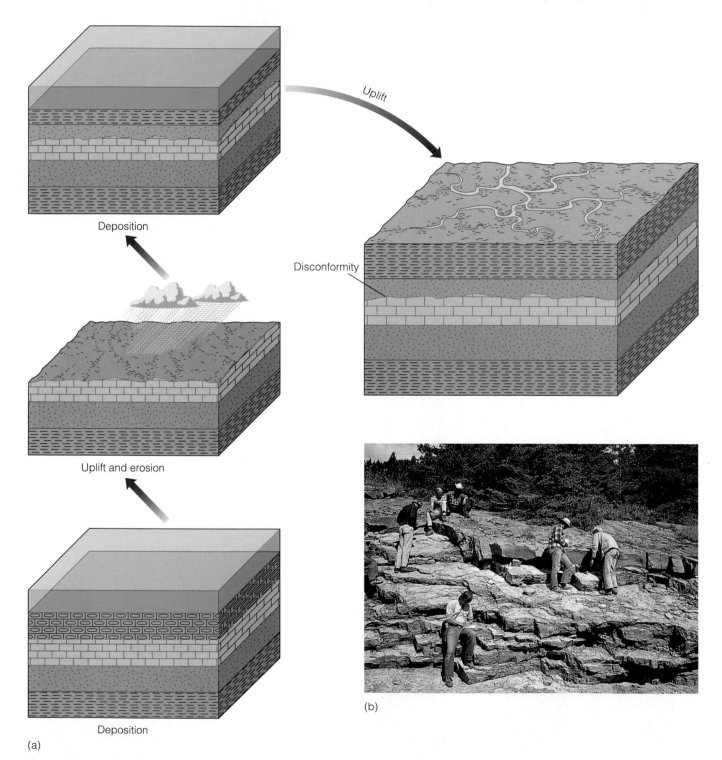

Deposition

Uplift

Disconformity

Uplift and erosion

Deposition

(a)

(b)

Figure 8.8

(a) Formation of a disconformity. (b) Disconformity between Mississippian and Jurassic strata in Montana. The geologist at the upper left is sitting on Jurassic strata, and his right foot is resting on Mississippian rocks.

nary bedding plane. Accordingly, many disconformities are difficult to recognize and must be identified on the basis of fossil assemblages.

An **angular unconformity** is an erosional surface on tilted or folded strata over which younger strata have been deposited (Figure 8.9). Both younger and older strata may dip, but if their dip angles are different (generally the older strata dip more steeply), an angular unconformity is present.

The angular unconformity illustrated in Figure 8.9b is probably the most famous in the world. It was here at Siccar Point, Scotland, that James Hutton realized that

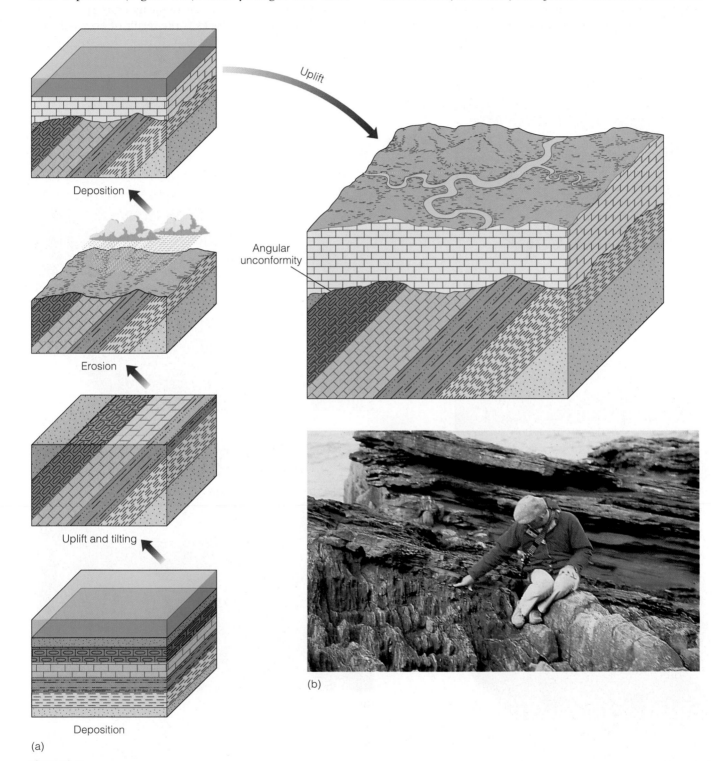

Figure 8.9
(a) Formation of an angular unconformity. (b) Angular unconformity at Siccar Point, Scotland. James Hutton first realized the significance of unconformities at this site in 1788.

severe upheavals had tilted the lower rocks and formed mountains that were then worn away and covered by younger, flat-lying rocks. The erosional surface between the older tilted rocks and the younger flat-lying strata meant that a significant gap existed in the geologic record. Although Hutton did not use the term *unconformity*, he was the first to understand and explain the significance of such discontinuities in the geologic record.

The third type of unconformity is a **nonconformity.** Here an erosion surface cut into metamorphic or igneous rocks is covered by sedimentary rocks (Figure 8.10). This type of unconformity closely resembles an intrusive igneous contact with sedimentary rocks. The principle of inclusions is helpful in determining whether the relationship between the underlying igneous rocks and overlying sedimentary rocks is the result of an intrusion

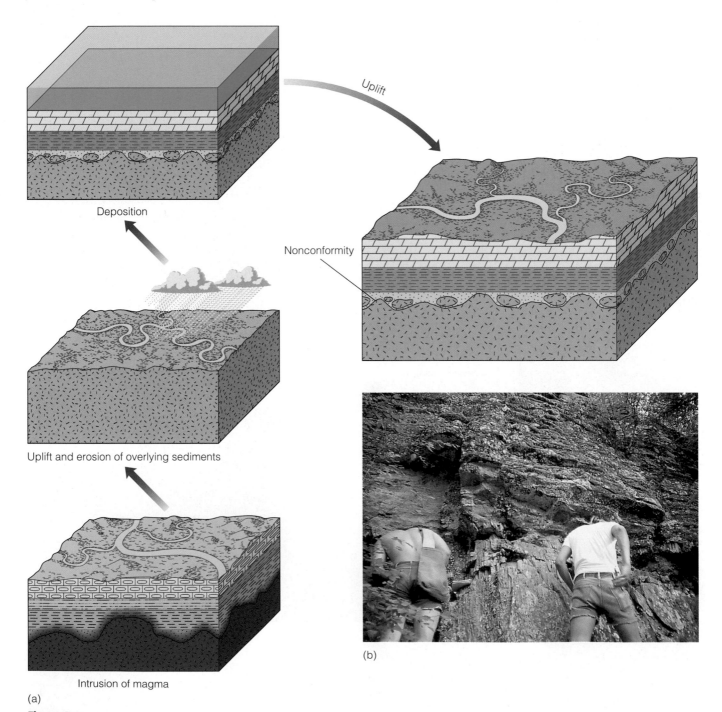

Deposition

Uplift

Nonconformity

Uplift and erosion of overlying sediments

Intrusion of magma

(a)

(b)

Figure 8.10
(a) Formation of a nonconformity. (b) Nonconformity between Precambrian metamorphic rocks and the overlying Cambrian-age Deadwood Formation, Wyoming.

or erosion. In the case of an intrusion, the igneous rocks are younger, but in the case of erosion, the sedimentary rocks are younger. Being able to distinguish between a nonconformity and an intrusive contact is very important because they represent different sequences of events.

Applying the Principles of Relative Dating

We can decipher the geologic history of the area represented by the block diagram in Figure 8.11 by applying the various relative-dating principles just discussed. The methods and logic used in this example are the same as those applied by nineteenth-century geologists in constructing the geologic time scale.

According to the principles of superposition and original horizontality, beds A–G were deposited horizontally; then they were either tilted, faulted (H), and eroded, or after deposition, they were faulted (H), tilted, and then eroded (Figure 8.12a–c). Because the fault cuts beds A–G, it must be younger than the beds according to the principle of cross-cutting relationships.

Beds J–L were then deposited horizontally over this erosional surface, producing an angular unconformity (I) (Figure 8.12d). Following deposition of these three beds, the entire sequence was intruded by a dike (M), which, according to the principle of cross-cutting rela-

tionships, must be younger than all the rocks it intrudes (Figure 8.12e).

The entire area was then uplifted and eroded; next beds P and Q were deposited, producing a disconformity (N) between beds L and P and a nonconformity (O) between the igneous intrusion M and the sedimentary bed P (Figure 8.12f and g). We know that the relationship between igneous intrusion M and the overlying sedimentary bed P is a nonconformity because of the presence of inclusions of M in P (principle of inclusions).

At this point, there are several possibilities for reconstructing the geologic history of this area. According to the principle of cross-cutting relationships, dike R must be younger than bed Q because it intrudes into it. It can have intruded anytime *after* bed Q was deposited; however, we cannot determine whether R was formed right after Q, right after S, or after T was formed. For purposes of this history, we will say that it intruded after the deposition of bed Q (Figure 8.12g and h).

Following the intrusion of dike R, lava S flowed over bed Q, followed by the deposition of bed T (Figure 8.12i and j). Although the lava flow (S) is not a sedimentary unit, the principle of superposition still applies because it flowed onto the surface, just as sediments are deposited on Earth's surface.

We have established a relative chronology for the rocks and events of this area by using the principles of relative dating. Remember, however, that we have no way of knowing how many years ago these events

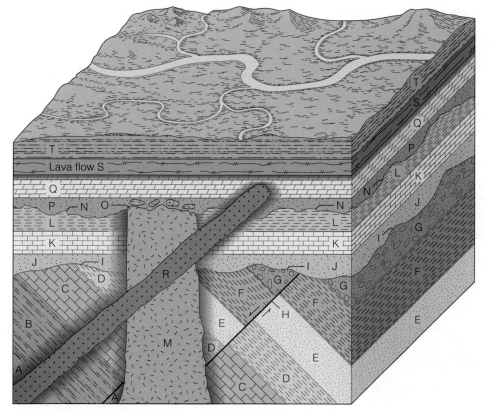

Figure 8.11
A block diagram of a hypothetical area in which the various relative-dating principles can be applied to determine its geologic history.

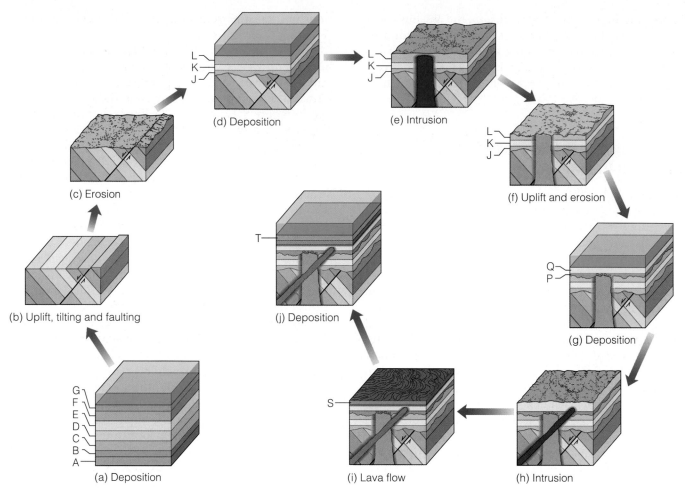

Figure 8.12

(a) Beds A–G are deposited. (b) The preceding beds are tilted and faulted. (c) Erosion. (d) Beds J–L are deposited, producing an angular unconformity. (e) The entire sequence is intruded by a dike. (f) The entire sequence is uplifted and eroded. (g) Beds P and Q are deposited, producing a disconformity (N) and a nonconformity (O). (h) Dike R intrudes. (i) Lava (S) flows over bed Q, baking it. (j) Bed T is deposited.

occurred unless we can obtain radiometric dates for the igneous rocks. With these dates, we can establish the range of absolute ages between which the different sedimentary units were deposited and also determine how much time is represented by the unconformities.

WHAT WOULD YOU DO?

You have been chosen to be part of the first astronaut crew to land on Mars. You were selected because you are a geologist, and therefore your primary responsibility is to map the geology of the landing site. An important goal of the mission will be to work out the geologic history of the area. How will you go about doing this? Will you be able to use the principles of relative dating? How will you go about correlating the various rock units? Will you be able to determine absolute ages? How would you do this?

HOW DO GEOLOGISTS CORRELATE ROCK UNITS?

To decipher Earth history, geologists must demonstrate the time equivalency of rock units in different areas. This process is known as **correlation.**

If surface exposures are adequate, units may simply be traced laterally (principle of lateral continuity), even if occasional gaps exist (Figure 8.13). Other criteria used to correlate units are similarity of rock type, position in a sequence, and key beds. *Key beds* are units, such as coal beds or volcanic ash layers, that are sufficiently distinctive to allow identification of the same unit in different areas (Figure 8.13).

Generally, no single location in a region has a geologic record of all events that occurred during its history; therefore, geologists must correlate from one area to another in order to determine the complete geologic history of the region. An excellent example is provided

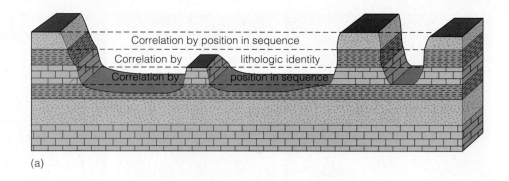

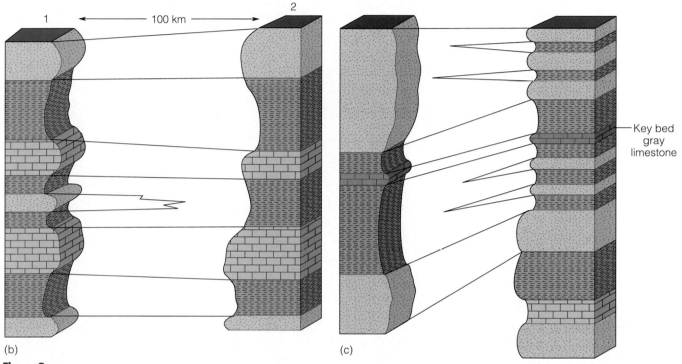

Figure 8.13
Correlation of rock units. (a) In areas of adequate exposures, rock units can be traced laterally even if occasional gaps exist.
(b) Correlation by similarities in rock type and position in a sequence. The sandstone in section 1 is assumed to intertongue or grade
laterally into the shale at section 2. (c) Correlation using a key bed, a distinctive gray limestone.

by the history of the Colorado Plateau (Figure 8.14). A
record of events occurring over approximately two billion years is present in this region. Because of the forces
of erosion, the entire record is not preserved at any single location. Within the walls of the Grand Canyon are
rocks of the Precambrian and Paleozoic Era, whereas
Paleozoic and Mesozoic Era rocks are found in Zion
National Park, and Mesozoic and Cenozoic Era rocks
are exposed in Bryce Canyon (Figure 8.14). By correlating the uppermost rocks at one location with the lowermost equivalent rocks of another area, the history of the
entire region can be deciphered.

Although geologists can match up rocks on the
basis of similar rock type and stratigraphic position,
correlation of this type can only be done in a limited
area where beds can be traced from one site to another.
To correlate rock units over a large area or to correlate

age-equivalent units of different composition, fossils
and the principle of fossil succession must be used.

Fossils are useful as time indicators because they
are the remains of organisms that lived for a certain
length of time during the geologic past. Fossils that are
easily identified, are geographically widespread, and
existed for a rather short interval of geologic time are
particularly useful. Such fossils are **guide fossils** or
index fossils (Figure 8.15). The trilobite *Isotelus* and
the clam *Inoceramus* meet these criteria and are therefore good guide fossils. In contrast, the brachiopod
Lingula is easily identified and widespread, but its geologic range of Ordovician to Recent makes it of little
use in correlation.

Because most fossils have fairly long geologic
ranges, geologists construct *assemblage range zones* to
determine the age of the sedimentary rocks containing

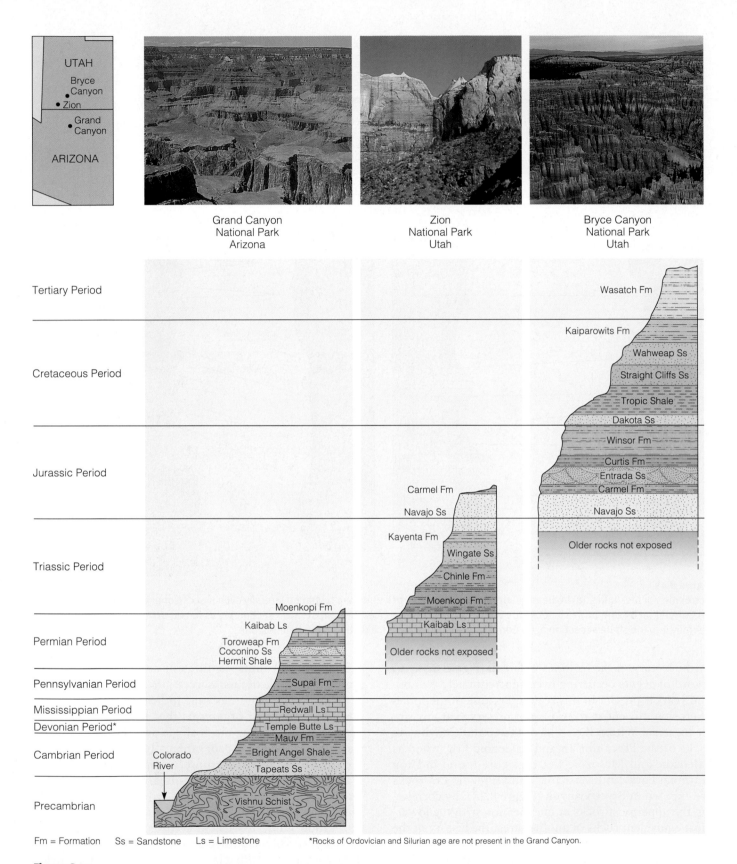

Grand Canyon
National Park
Arizona

Zion
National Park
Utah

Bryce Canyon
National Park
Utah

	Grand Canyon	Zion	Bryce Canyon
Tertiary Period			Wasatch Fm
Cretaceous Period			Kaiparowits Fm / Wahweap Ss / Straight Cliffs Ss / Tropic Shale / Dakota Ss
Jurassic Period		Carmel Fm / Navajo Ss	Winsor Fm / Curtis Fm / Entrada Ss / Carmel Fm / Navajo Ss
		Kayenta Fm	Older rocks not exposed
Triassic Period		Wingate Ss / Chinle Fm / Moenkopi Fm	
	Moenkopi Fm		
Permian Period	Kaibab Ls / Toroweap Fm / Coconino Ss / Hermit Shale	Kaibab Ls / Older rocks not exposed	
Pennsylvanian Period	Supai Fm		
Mississippian Period	Redwall Ls		
Devonian Period*	Temple Butte Ls		
Cambrian Period	Mauv Fm / Bright Angel Shale / Tapeats Ss		
	Colorado River		
Precambrian	Vishnu Schist		

Fm = Formation   Ss = Sandstone   Ls = Limestone   *Rocks of Ordovician and Silurian age are not present in the Grand Canyon.

Figure 8.14
Correlation of rocks within the Colorado Plateau. By correlating the rocks from various locations, the history of the entire region can be deciphered.

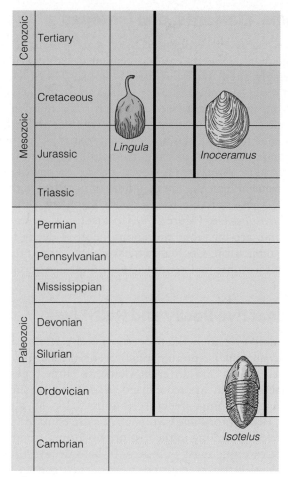

Figure 8.15
The geologic ranges of three marine invertebrates. The brachiopod *Lingula* is of little use in correlation because of its long geologic range. The trilobite *Isotelus* and the bivalve *Inoceramus* are good guide fossils because they are geographically widespread, are easily identified, and have short geologic ranges.

the fossils. Assemblage range zones are established by plotting the overlapping geologic ranges of different species of fossils. The first and last occurrences of two species are used to establish an assemblage zone's boundaries (Figure 8.16). Correlation of assemblage zones generally yields correlation lines that are considered time equivalent. In other words, the strata encompassed by the correlation lines are thought to be the same age.

Subsurface Correlation

In addition to the surface geology, geologists are also interested in subsurface geology, because it provides additional information about geologic features beneath Earth's surface. A variety of techniques and methods are used to acquire and interpret data about the subsurface geology of an area.

When drilling for oil or natural gas, cores or rock chips called *well cuttings* are commonly recovered from the drill hole. These samples are studied under the microscope and reveal such important information as rock type, porosity (the amount of pore space), permeability (the ability to transmit fluids), and the presence of oil stains. In addition, the samples can also be processed for a variety of microfossils that aid in determining the geologic age of the rock and the environment of deposition.

Geophysical instruments may be lowered down the drill hole to record such rock properties as electrical resistivity and radioactivity, thus providing a record or *well log* of the rocks penetrated. Cores, well cuttings,

Figure 8.16
Correlation of two sections by using assemblage range zones. These zones are established by the overlapping ranges of fossils A–E.

and well logs are all extremely useful in making subsurface correlations (Figure 8.17).

Subsurface rock units may also be detected and traced by the study of seismic profiles. Energy pulses, such as those from explosions, travel through rocks at a velocity determined by rock density, and some of this energy is reflected from various horizons (contacts between contrasting layers) back to the surface, where it is recorded (see Figure 11.8). Seismic stratigraphy is particularly useful in tracing units in areas such as the continental shelves, where it is very expensive to drill holes and other techniques have limited use.

WHAT ARE ABSOLUTE-DATING METHODS, AND WHY ARE THEY IMPORTANT?

Although most of the isotopes of the 92 naturally occurring elements are stable, some are radioactive and spontaneously decay to other more stable isotopes of elements, releasing energy in the process. The discovery, in 1903 by Pierre and Marie Curie, that radioactive decay produces heat meant that geologists finally had a mechanism for explaining Earth's internal heat that did not rely on residual cooling from a molten origin. Furthermore, geologists had a powerful tool to date geologic events accurately and to verify the long time periods postulated by Hutton, Lyell, and Darwin.

Atoms, Elements, and Isotopes

As we discussed in Chapter 2, all matter is made up of chemical elements, each of which is composed of extremely small particles called *atoms*. The nucleus of an atom is composed of *protons* and *neutrons* with *electrons* encircling it (see Figure 2.6). The number of protons defines an element's *atomic number* and helps determine its properties and characteristics. The combined number of protons and neutrons in an atom is its *atomic mass number*. However, not all atoms of the same element have the same number of neutrons in their nuclei. These variable forms of the same element are called *isotopes* (see Figure 2.7). Most isotopes are stable, but some are unstable and spontaneously decay to a more stable form. Geologists measure the decay rate of unstable isotopes to determine the absolute age of rocks.

Radioactive Decay and Half-Lives

Radioactive decay is the process whereby an unstable atomic nucleus is spontaneously transformed into an atomic nucleus of a different element. Three types of radioactive decay are recognized, all of which result in a change of atomic structure (Figure 8.18). In **alpha decay,** two protons and two neutrons are emitted from the nucleus, resulting in the loss of two atomic numbers and four atomic mass numbers. In **beta decay,** a fast-moving electron is emitted from a neutron in the nucleus, changing that neutron to a proton and consequently increasing the atomic number by one, with no resultant atomic mass number change. **Electron capture** results when a proton captures an electron from an elec-

Figure 8.17
A schematic diagram showing how well logs are made. As the logging tool is withdrawn from the drill hole, data are transmitted to the surface where they are recorded and printed as a well log. The curve labeled SP in this diagrammatic electric log is a plot of self-potential (electrical potential caused by different conductors in a solution that conducts electricity) with depth. The curve labeled R is a plot of electrical resistivity with depth. Electric logs yield information about the rock type and fluid content of subsurface formations. Electric logs are also used to correlate from well to well.

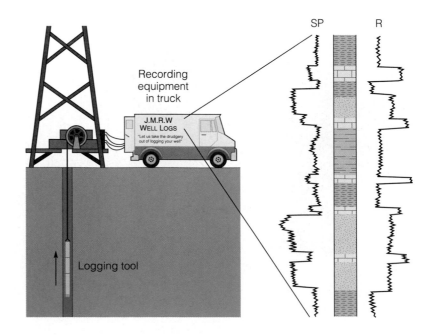

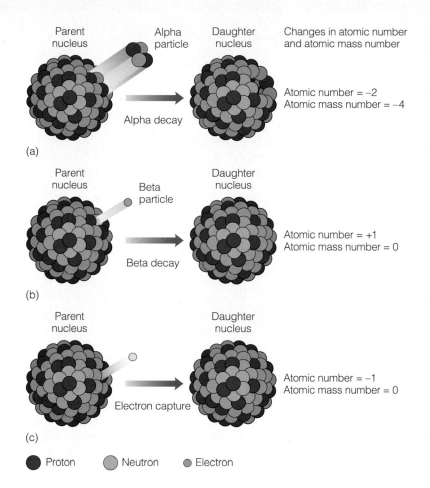

Changes in atomic number and atomic mass number

(a) Parent nucleus — Alpha particle — Daughter nucleus

Alpha decay

Atomic number = –2
Atomic mass number = –4

(b) Parent nucleus — Beta particle — Daughter nucleus

Beta decay

Atomic number = +1
Atomic mass number = 0

(c) Parent nucleus — Daughter nucleus

Electron capture

Atomic number = –1
Atomic mass number = 0

● Proton   ◐ Neutron   ● Electron

Figure 8.18
Three types of radioactive decay. (a) Alpha decay, in which an unstable parent nucleus emits two protons and two neutrons. (b) Beta decay, in which an electron is emitted from the nucleus. (c) Electron capture, in which a proton captures an electron and is thereby converted to a neutron.

tron shell and thereby converts to a neutron, resulting in the loss of one atomic number but not changing the atomic mass number.

Some elements undergo only one decay step in the conversion from an unstable form to a stable form. For example, rubidium 87 decays to strontium 87 by a sin-

gle beta emission, and potassium 40 decays to argon 40 by a single electron capture. Other radioactive elements undergo several decay steps (see Perspective 8.2). Uranium 235 decays to lead 207 by seven alpha and six beta steps, while uranium 238 decays to lead 206 by eight alpha and six beta steps (Figure 8.19).

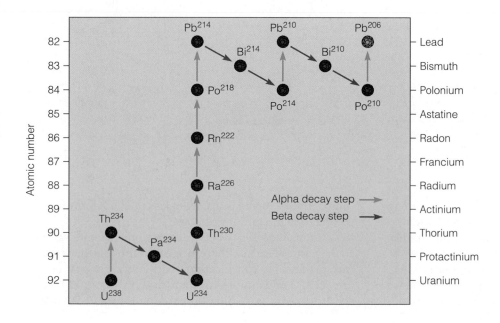

Figure 8.19
Radioactive decay series for uranium 238 to lead 206. Radioactive uranium 238 decays to its stable end product, lead 206, by eight alpha and six beta decay steps. A number of different isotopes are produced as intermediate steps in the decay series.

Perspective 8.2

Radon: The Silent Killer

Radon is a colorless, odorless, naturally occurring radioactive gas with a 3.8-day half-life. It is part of the uranium 238/lead 206 radioactive decay series (Figure 8.19) and occurs in any rock or soil that contains uranium 238. Radon concentrations are reported in picocuries per liter (pCi/L) of air (a curie is the standard measure of radiation, and a picocurie is one-trillionth of a curie, or the equivalent of the decay of about two radioactive atoms per minute). Outdoors, radon escapes into the atmosphere, where it is diluted and dissipates to harmless levels (0.2 pCi/L is the ambient outdoor level of radon). Radon levels for indoor air range from less than 1 pCi/L to about 3000 pCi/L, but average about 1.5 pCi/L. The Environmental Protection Agency (EPA) considers radon levels exceeding 4 pCi/L to be unhealthy and recommends remedial action be taken to lower them. Continued exposure to elevated levels of radon over an extended period of time is thought by many researchers to increase the risk of lung cancer.

Radon is one of the natural decay products of uranium 238. It rapidly decays by the emission of an alpha particle, producing two short-lived radioactive isotopes—polonium 218 and polonium 214 (Figure 8.19). Both isotopes are solid and can become trapped in your lungs every time you breathe. When polonium decays, it emits alpha and beta particles, which can damage lung cells and cause lung cancer.

Your chances of being adversely affected by radon depend on numerous interrelated factors such as geographic location, the geology of the area, the climate, how buildings are constructed, and the amount of time spent in your house. Because radon is a naturally occurring gas, contact with it is unavoidable, but atmospheric concentrations of it are probably harmless. Only when concentrations of radon build up in poorly ventilated structures does it become a potential health risk.

Concern about the health risks posed by radon first arose during the 1960s when the news media revealed that some homes in the West had been built with uranium mine tailings. Since then, geologists have found that high indoor radon levels can be caused by natural uranium in minerals of the rock and soil on which buildings are constructed. In response to the high cost of energy during the 1970s and 1980s, old buildings were insulated, and new buildings were constructed to be as energy efficient and air-tight as possible. Ironically, these energy-saving measures also sealed in radon.

Radon enters buildings through dirt floors, cracks in the floor or walls, joints between floors and walls, floor drains, sumps, and utility pipes as well as any cracks or pores in hollow block walls (Figure 1). Radon can also be released into a building whenever the water is

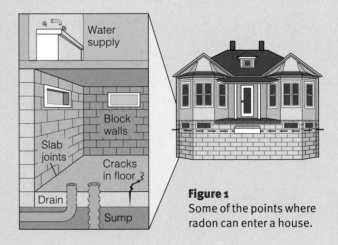

Figure 1
Some of the points where radon can enter a house.

When discussing decay rates, it is convenient to refer to them in terms of half-lives. The **half-life** of a radioactive element is the time it takes for one-half of the atoms of the original unstable *parent element* to decay to atoms of a new, more stable *daughter element*. The half-life of a given radioactive element is constant and can be precisely measured. Half-lives of various radioactive elements range from less than a billionth of a second to many billions of years.

Radioactive decay occurs at a geometric rate rather than a linear rate. Therefore, a graph of the decay rate produces a curve rather than a straight line (Figure 8.20). For example, an element with 1,000,000 parent atoms will have 500,000 parent atoms and 500,000

turned on, particularly if the water comes from a private well. Municipal water is generally safe because it has usually been aerated before it gets to your home.

To find out if your home has a radon problem, you must test for it with commercially available, relatively inexpensive, simple home-testing devices. If radon readings are above the recommended EPA levels of 4 pCi/L, several remedial measures can be taken to reduce your risk. These include sealing up all cracks in the foundation, pouring a concrete slab over a dirt floor, increasing the circulation of air throughout the house, especially in the basement and crawl space, providing filters for drains and other utility openings, and limiting time spent in areas with higher concentrations of radon.

It is important to remember that, although the radon hazard covers most of the country, some areas are more likely to have higher natural concentrations of radon than others (Figure 2).

Rocks such as uranium-bearing granites, metamorphic rocks of granitic composition, and black shales (high carbon content) are quite likely to cause indoor radon problems. Other rocks such as marine quartz sandstone, noncarbonaceous shales and siltstones, most volcanic rocks, and igneous and metamorphic rocks rich in iron and magnesium typically do not cause radon problems. The permeability of the soil overlying the rock can also affect the indoor levels of radon gas. Some soils are more permeable than others and allow more radon to escape into the overlying structures.

Climate and the type of construction affect not only how much radon enters a structure but also how much escapes. Concentrations of radon are highest during the winter in northern climates because buildings are sealed as tightly as possible. Homes with basements are more likely to have higher radon levels than those built on concrete slabs. While research continues into the

sources of indoor radon and ways of controlling it, the most important thing people can do is to test their home, school, or business for radon.

Currently a heated debate is ongoing among scientists about the large-scale health hazards resulting from radon exposure and how much money should be spent for its remediation. On the one hand, the EPA and the Surgeon General estimate that exposure to high levels of indoor radon cause between 5000 and 20,000 lung cancer deaths each year. On the other hand, some scientists dispute these figures because of the difficulty of attributing mortality rates for lung cancer directly to radon, particularly when so many other factors, such as smoking, are involved. Central to this debate are two questions: What concentration levels of indoor radon are acceptable, and exactly how serious is the risk from radon at those levels? Unfortunately, the data for making these determinations are simply not available at this time.

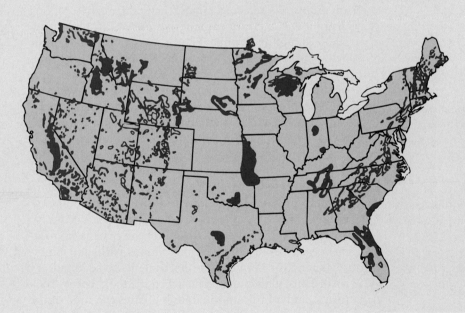

Figure 2
Areas in the United States where granite, phosphate-bearing rocks, carbonaceous shales, and uranium occur. These rocks are all potential sources of radon gas.

daughter atoms after one half-life. After two half-lives, it will have 250,000 parent atoms (one-half of the previous parent atoms, which is equivalent to one-fourth of the original parent atoms) and 750,000 daughter atoms. After three half-lives, it will have 125,000 parent atoms (one-half of the previous parent atoms or one-eighth of the original parent atoms) and 875,000 daughter atoms,

and so on until the number of parent atoms remaining is so few that they cannot be accurately measured by present-day instruments.

By measuring the parent–daughter ratio and knowing the half-life of the parent (which has been determined in the laboratory), geologists can calculate the age of a sample containing the radioactive element. The

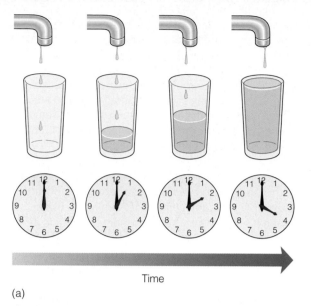

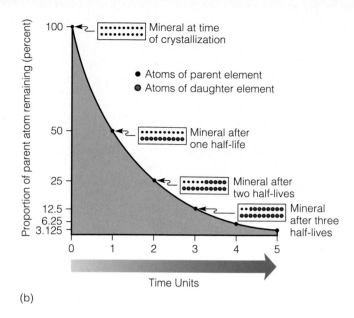

Figure 8.20

(a) Uniform, linear depletion is characteristic of many familiar processes. (b) Geometric radioactive decay curve, in which each time unit represents one half-life, and each half-life is the time it takes for one-half of the parent element to decay to the daughter element.

parent–daughter ratio is usually determined by a *mass spectrometer,* an instrument that measures the proportions of atoms of different masses.

Sources of Uncertainty

The most accurate radiometric dates are obtained from igneous rocks. As a magma cools and begins to crystallize, radioactive parent atoms are separated from previously formed daughter atoms. Because they are the right size, some radioactive parent atoms are incorporated into the crystal structure of certain minerals. The stable daughter atoms, though, are a different size than the radioactive parent atoms and consequently cannot fit into the crystal structure of the same mineral as the parent atoms. Therefore, a mineral crystallizing in a cooling magma will contain radioactive parent atoms but no stable daughter atoms (Figure 8.21). Thus, the time being measured is the time of crystallization of the mineral containing the radioactive atoms, and not the time of formation of the radioactive atoms.

Except in unusual circumstances, sedimentary rocks cannot be radiometrically dated because one would be measuring the age of a particular mineral rather than the time that it was deposited as a sedimentary particle. One of the few instances in which radiometric dates can be obtained on sedimentary rocks is when the mineral glauconite is present. Glauconite is a greenish mineral containing radioactive potassium 40, which decays to argon 40 (Table 8.1). It forms in certain marine environments as a result of chemical reactions with clay minerals during the conversion of sediments to sedimentary rock. Thus, glauconite forms when the sedimentary rock forms, and a radiometric date indicates the time of the sedimentary rock's origin. Being a gas, however, the daughter product argon can easily escape from a mineral. Therefore, any date obtained from glauconite, or any other mineral containing the potassium 40/argon 40 pair, must be considered a minimum age.

To obtain accurate radiometric dates, geologists must be sure that they are dealing with a *closed system,* meaning that neither parent nor daughter atoms have been added or removed from the system since crystallization and that the ratio between them results only from radioactive decay. Otherwise, an inaccurate date

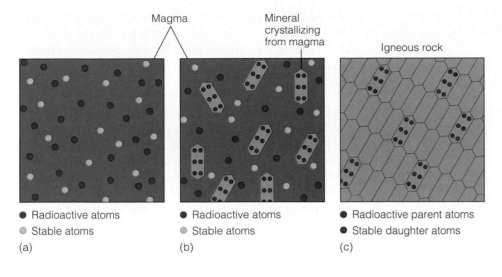

Figure 8.21
(a) A magma contains both radioactive and stable atoms. (b) As the magma cools and begins to crystallize, some radioactive atoms are incorporated into certain minerals because they are the right size and can fit into the crystal structure. Therefore, at the time of crystallization, the mineral will contain 100% radioactive parent atoms and 0% stable daughter atoms. (c) After one half-life, 50% of the radioactive parent atoms will have decayed to stable daughter atoms.

Magma

Mineral crystallizing from magma

Igneous rock

● Radioactive atoms
● Stable atoms
(a)

● Radioactive atoms
● Stable atoms
(b)

● Radioactive parent atoms
● Stable daughter atoms
(c)

will result. If daughter atoms have leaked out of the mineral being analyzed, the calculated age will be too young; if parent atoms have been removed, the calculated age will be too great.

Leakage may take place if the rock is heated or subjected to intense pressure as can sometimes occur during metamorphism. If this happens, some of the parent or daughter atoms may be driven from the mineral being analyzed, resulting in an inaccurate age determination. If the daughter product was completely removed, then one would be measuring the time since metamorphism (a useful measurement itself), and not the time since crystallization of the mineral (Figure 8.22). Because heat affects the parent–daughter ratio, metamorphic rocks are difficult to date accurately. Remember that while the parent–daughter ratio may be affected by heat, the decay rate of the parent element remains constant, regardless of any physical or chemical changes.

To obtain an accurate radiometric date, geologists must make sure that the sample is fresh and unweathered and that it has not been subjected to high temperatures or intense pressures after crystallization. Furthermore, it is sometimes possible to cross-check the radiometric date obtained by measuring the parent–daughter ratio of two different radioactive elements in the same mineral. For example, naturally occurring uranium consists of both uranium 235 and uranium 238 isotopes. Through various decay steps, uranium 235 decays to lead 207, whereas uranium 238 decays to lead 206 (Figure 8.19). If the minerals containing both uranium isotopes have remained closed systems, the ages obtained from each parent–daughter ratio should agree closely and therefore should indicate the time of crystallization of the magma. If the ages do not closely agree, other samples must be taken and ratios measured to see which, if either, date is correct.

Table 8.1

Five of the Principal Long-Lived Radioactive Isotope Pairs Used in Radiometric Dating					
ISOTOPES		HALF-LIFE OF PARENT (YEARS)	EFFECTIVE DATING RANGE (YEARS)	MINERALS AND ROCKS THAT CAN BE DATED	
PARENT	DAUGHTER				
Uranium 238	Lead 206	4.5 billion	10 million to 4.6 billion	Zircon	
				Uraninite	
Uranium 235	Lead 207	704 million			
Thorium 232	Lead 208	14 billion			
Rubidium 87	Strontium 87	48.8 billion	10 million to 4.6 billion	Muscovite	
				Biotite	
				Potassium feldspar	
				Whole metamorphic or igneous rock	
Potassium 40	Argon 40	1.3 billion	100,000 to 4.6 billion	Glauconite	Hornblende
				Muscovite	Whole volcanic rock
				Biotite	

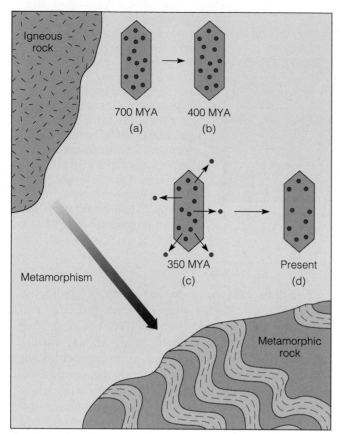

Figure 8.22
The effect of metamorphism in driving out daughter atoms from a mineral that crystallized 700 million years ago (MYA). (a) The mineral is shown immediately after crystallization, (b) then at 400 million years, when some of the parent atoms had decayed to daughter atoms. (c) Metamorphism at 350 MYA drives the daughter atoms out of the mineral into the surrounding rock. (d) Assuming the rock has remained a closed chemical system throughout its history, dating the mineral today yields the time of metamorphism, while dating the whole rock provides the time of its crystallization, 700 MYA.

Recent advances and the development of new techniques and instruments for measuring various isotope ratios have enabled geologists to analyze not only increasingly smaller samples, but with a greater precision than ever before. Presently the measurement error for many radiometric dates is typically less than 0.5% of the age, and in some cases it is even better than 0.1%. Thus, for a rock 540 million years old (near the beginning of the Cambrian Period), the possible error could range from nearly 2.7 million years to as low as less than 540,000 years.

Long-Lived Radioactive Isotope Pairs

Table 8.1 shows the five common, long-lived parent–daughter isotope pairs used in radiometric dating. Long-lived pairs have half-lives of millions or billions of years.

All of these were present when Earth formed and are still present in measurable quantities. Other shorter-lived radioactive isotope pairs have decayed to the point that only small quantities near the limit of detection remain.

The most commonly used isotope pairs are the uranium–lead and thorium–lead series, which are used principally to date ancient igneous intrusives, lunar samples, and some meteorites. The rubidium–strontium pair is also used for very old samples and has been effective in dating the oldest rocks on Earth as well as meteorites. The potassium–argon method is typically used for dating fine-grained volcanic rocks from which individual crystals cannot be separated; hence, the whole rock is analyzed. Because argon is a gas, great care must be taken to ensure that the sample has not been subjected to heat, which would allow argon to escape; such a sample would yield an age that is too young. Other long-lived radioactive isotope pairs exist, but they are rather rare and are used only in special situations.

Fission Track Dating

The emission of atomic particles resulting from the spontaneous decay of uranium within a mineral damages its crystal structure. The damage appears as microscopic linear tracks that are visible only after etching the mineral with hydrofluoric acid. The age of the sample is determined on the basis of the number of fission tracks present and the amount of uranium the sample contains: the older the sample, the greater the number of tracks (Figure 8.23).

Fission track dating is of particular interest to geologists because the technique can be used to date samples ranging from only a few hundred to hundreds of millions of years in age. It is most useful for dating samples between about 40,000 and 1.5 million years

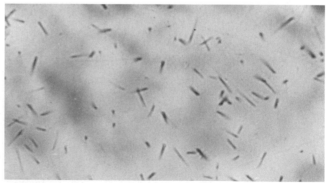

Figure 8.23
Each fission track (about 16 μm in length) in this apatite crystal is the result of the radioactive decay of a uranium atom. The apatite crystal, which has been etched with hydrofluoric acid to make the fission tracks visible, comes from one of the dikes at Shiprock, New Mexico, and has a calculated age of 27 million years.
Source: Photo courtesy of Charles W. Naeser, USGS.

Molly Miller

Paleontology: The Perspective of Time

Dr. Miller earned her B.S. at the College of Worchester and her Ph.D. at the University of California–Los Angeles. Dr. Miller has been a professor of geology at Vanderbilt University for many years. Currently she is working on the evolution of animals living in ancient lakes and streams in the Late Paleozoic (250 million years ago).

When I was ten my family went camping at a state park that had an abandoned rock quarry. Meeting my first rattlesnake there was the highlight of that trip until I discovered that the fossil I had found in the quarry was an unchanged part of an animal that was ALIVE over 350 million years ago. When I went to college, I wanted to prepare for a career where I could improve people's lives rather than to concentrate on something esoteric like the (very!) ancient history of the earth and its inhabitants. However, I was struck by the phrase in our textbook that said "In studying the history of life, we see ourselves in the perspective of time." That convinced me that it is worthwhile to study how life and Earth have changed so that we understand how we fit into the grand scheme, the overall picture.

Rather than analyze the preserved parts of ancient animals, I concentrate on the traces and burrows they made as they moved in the sand or mud on the bottom of the ocean (or river or lake). It's pretty amazing that we can tell what animals were doing millions of years ago, even if there is no record of their shells or bones. We do this in part by studying the burrowing habits of modern animals so that we can recognize their activity where it is recorded in rocks. For instance, coastal shrimp that burrow a meter or more into the sandy bottoms of lagoons worldwide construct distinctive complex tunnels. Because they occur in high densities (over 100 per square meter) and because they constantly toss nutrient-rich sand onto the sediment surface, they facilitate exchange of nutrients and other chemicals between the ocean water and the sediments on the ocean bottom. The presence of these animals' burrows in rocks that are millions of years old suggest that they have been playing a similar role in coastal ecosystems for hundreds of millions of years.

We can use the record of animal activity to identify major climate and environmental changes that occurred in the distant past. In late 1995, colleague John Isbell and I were studying sedimentary rocks high in the Transantarctic Mountains (Antarctica). We discovered the deposits of a small lake that formed by the temporary melting of the huge ancient (300-million-year-old [Ma]) ice sheet that covered the combined southern continents during the Late Paleozoic. Although the lake was near the South Pole at the time, it lasted long enough to be colonized by bottom-dwelling animals that left traces of their activity in the sand and mud at the bottom of the lake. The ice sheet advanced again, depositing unsorted sediment on top of the lake deposits. However, the burrows in the rocks give clear evidence that significant climate amelioration occurred even at very high latitudes during the period of major glaciation about 300 Ma.

Despite all the history of change that we can piece together using information from the preserved record of ancient animal activity, unsolved mysteries remain. One of the most intriguing is the peculiar *Zoophycos,* a complicated burrow system that broadly resembles a conifer. *Zoophycos* is common in rocks and sediments deposited in diverse environments for the past 500 million years. Unknown to motorists, thousands of these fossil burrow systems in rocks are exposed along interstate highways from Kentucky to New York, and animals probably are making them in ocean-bottom sediment at this minute. Yet we have little understanding of what type (or types) of animals produced this, or why they engage in the complex behavior that this burrow system represents. We may never figure out *Zoophycos.* However, we do know that its producers predated humans by hundreds of millions of years, and we can be quite sure that they will carry on long after we are extinct. Although we can't answer all the questions, the study of ancient life and environments gives us unique insight about where we fit in. ∎

---

ago, a period for which other dating techniques are not particularly suitable. One of the problems in fission track dating occurs when the rocks have later been subjected to high temperatures. If this happens, the damaged crystal structures are repaired by annealing, and consequently the tracks disappear. In such instances, the calculated age will be younger than the actual age.

Radiocarbon and Tree-Ring Dating Methods

Carbon is an important element in nature and is one of the basic elements found in all forms of life. It has three isotopes; two of these, carbon 12 and 13, are stable, whereas carbon 14 is radioactive (see Figure 2.7). Carbon 14 has a half-life of 5730 years plus or minus

30 years. The **carbon 14 dating technique** is based on the ratio of carbon 14 to carbon 12 and is generally used to date once-living material.

The short half-life of carbon 14 makes this dating technique practical only for specimens typically younger than about 50,000 years. With isotopic enrichment, some investigators have dated samples as old as 75,000 years. Consequently, the carbon 14 dating method is especially useful in archaeology and has greatly helped unravel the events of the latter portion of the Pleistocene Epoch. For example, carbon 14 dates of maize from the Tehuacan Valley of Mexico have forced archeologists to rethink their ideas of where the first center for maize domestication in Mesoamerica arose. Carbon 14 dating is also helping to answer the question of when humans began populating North America.

Carbon 14 is constantly formed in the upper atmosphere when cosmic rays, which are high-energy particles (mostly protons), strike the atoms of upper-atmospheric gases, splitting their nuclei into protons and neutrons. When a neutron strikes the nucleus of a nitrogen atom (atomic number 7, atomic mass number 14), it may be absorbed into the nucleus and a proton emitted. Thus, the atomic number of the atom decreases by one, while the atomic mass number stays the same. Because the atomic number has changed, a new element, carbon 14 (atomic number 6, atomic mass number 14), is formed. The newly formed carbon 14 is rapidly assimilated into the carbon cycle and, along with carbon 12 and 13, is absorbed in a nearly constant ratio by all living organisms (Figure 8.24). When an organism dies, however, carbon 14 is not replenished, and the ratio of carbon 14 to carbon 12 decreases as carbon 14 decays back to nitrogen by a single beta decay step (Figure 8.24).

Currently the ratio of carbon 14 to carbon 12 is remarkably constant in both the atmosphere and living organisms. There is good evidence, however, that the production of carbon 14, and thus the ratio of carbon 14 to carbon 12, has varied somewhat over the past several thousand years. This was determined by comparing ages established by carbon 14 dating of wood samples against those established by counting annual tree rings in the same samples (Figure 8.25). As a result, carbon 14 ages have been corrected to reflect such variations in the past.

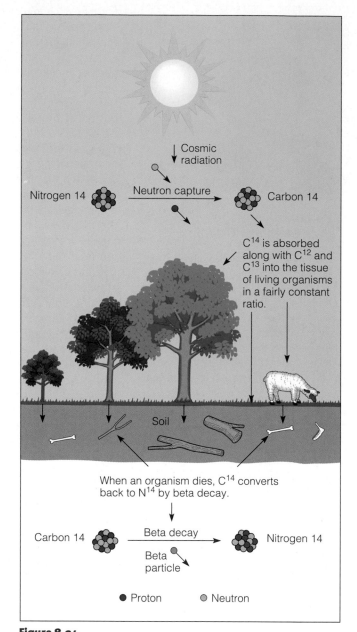

Figure 8.24
The carbon cycle showing the formation, dispersal, and decay of carbon 14.

Figure 8.25
Discrepancies exist between carbon 14 dates and those obtained by counting annual tree rings. Back to about 600 B.C., carbon 14 dates are too old, and those from about 600 B.C. to about 5000 B.C. are too young. Consequently, corrections must be made in the carbon 14 dates for this time period.

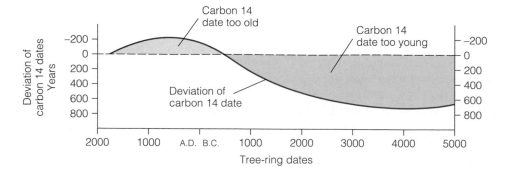

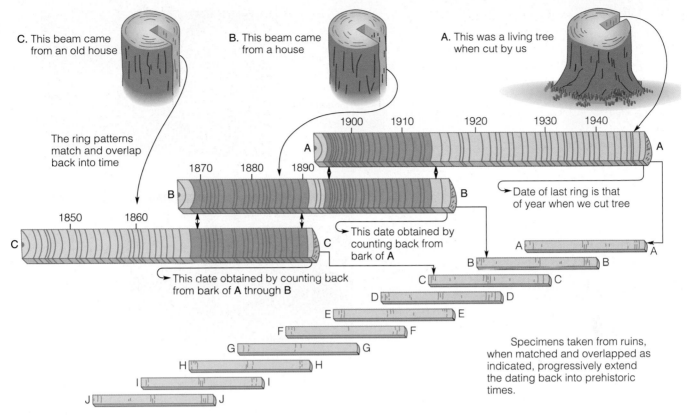

Figure 8.26
In the cross-dating method, tree-ring patterns from different woods are matched against each other to establish a ring-width chronology backward in time.

Tree-ring dating is another useful method for dating geologically recent events. The age of a tree can be determined by counting the growth rings in the lower part of the stem. Each ring represents one year's growth, and the pattern of wide and narrow rings can be compared among trees to establish the exact year in which the rings were formed. The procedure of matching ring patterns from numerous trees and wood fragments in a given area is referred to as *cross-dating*. By correlating distinctive tree-ring sequences from living to nearby dead trees, a time scale can be constructed that extends back to about 14,000 years ago (Figure 8.26). By matching ring patterns to the composite ring scale, wood samples whose ages are not known can be accurately dated.

The applicability of tree-ring dating is somewhat limited because it can only be used where continuous tree records are found. It is therefore most useful in arid regions, particularly the southwestern United States.

HOW WAS THE GEOLOGIC TIME SCALE DEVELOPED?

The geologic time scale is a hierarchical scale in which the 4.6-billion-year history of Earth is divided into time units of varying duration (Figure 8.1). It was not developed by any one individual but rather evolved, primarily during the nineteenth century, through the efforts of many people. By applying relative-dating methods to rock outcrops, geologists in England and western Europe defined the major geologic time units without the benefit of radiometric-dating techniques. Using the principles of superposition and fossil succession, they could correlate various rock exposures and piece together a composite geologic section. This composite section is in effect a relative time scale because the rocks are arranged in their correct sequential order.

By the beginning of the twentieth century, geologists had developed a relative geologic time scale but did not yet have any absolute dates for the various time-unit boundaries. Following the discovery of radioactivity near the end of the nineteenth century, radiometric dates were added to the relative geologic time scale (Figure 8.1).

Because sedimentary rocks, with rare exceptions, cannot be radiometrically dated, geologists have had to rely on interbedded volcanic rocks and igneous intrusions to apply absolute dates to the boundaries of the various subdivisions of the geologic time scale (Figure 8.27). An ashfall or lava flow provides an excellent

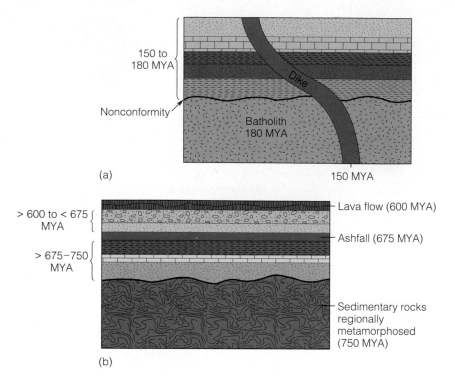

(a)

(b)

Figure 8.27
Absolute ages of sedimentary rocks can be determined by dating associated igneous rocks. In (a) and (b), sedimentary rocks are bracketed by rock bodies for which absolute ages have been determined.

marker bed that is a time-equivalent surface, supplying a minimum age for the sedimentary rocks below and a maximum age for the rocks above. Ashfalls are particularly useful because they may fall over both marine and nonmarine sedimentary environments and can provide a connection between these different environments.

Thousands of absolute ages are now known for sedimentary rocks of known relative ages, and these absolute dates have been added to the relative time scale. In this way, geologists have been able to determine the absolute ages of the various geologic periods and to determine their durations (Figure 8.1).

Chapter Summary

1. Relative dating involves placing geologic events in a sequential order as determined from their position in the geologic record. Absolute dating results in specific dates for events, expressed in years before the present.

2. During the eighteenth and nineteenth centuries, attempts were made to determine Earth's age based on scientific evidence rather than revelation. Although some attempts were quite ingenious, they yielded a variety of ages that now are known to be much too young.

3. James Hutton thought that present-day processes operating over long periods of time could explain all the geologic features of Earth. His observations were instrumental in establishing the principle of uniformitarianism.

4. Uniformitarianism, as articulated by Charles Lyell, soon became the guiding principle of geology. It holds that the laws of nature have been constant through time and that the same processes operating today have operated in the past, although not necessarily at the same rates.

5. Besides uniformitarianism, the principles of superposition, original horizontality, lateral continuity, cross-cutting relationships, inclusions, and fossil succession are basic for determining relative geologic ages and for interpreting Earth history.

6. Surfaces of discontinuity encompassing significant amounts of geologic time are common in the geologic record. Such surfaces are unconformities and result from times of nondeposition, erosion, or both.

7. Correlation is the practice of demonstrating equivalency of units in different areas. Time equivalence is most commonly demonstrated by correlating strata containing similar fossils.

8. Radioactivity was discovered during the late nineteenth century, and soon thereafter radiometric-dating techniques allowed geologists to determine absolute ages for geologic events.

9. Absolute-age dates for rock samples are usually obtained by determining how many half-lives of a radioactive parent element have elapsed since the sample originally crystallized. A half-life is the time it takes for one-half of the radioactive parent element to decay to a stable daughter element.

10. The most accurate radiometric dates are obtained from long-lived radioactive isotope pairs in igneous rocks. The most reliable dates are those obtained by using at least two different radioactive decay series in the same rock.

11. Carbon 14 dating can be used only for organic matter such as wood, bones, and shells and is effective back to about 50,000 years ago. Unlike the long-lived isotope pairs, the carbon 14 dating technique determines age by the ratio of radioactive carbon 14 to stable carbon 12.

12. Through the efforts of many geologists applying the principles of relative dating, a relative geologic time scale was established.

13. Most absolute ages of sedimentary rocks and their contained fossils are obtained indirectly by dating associated metamorphic or igneous rocks.

Important Terms

absolute dating
alpha decay
angular unconformity
beta decay
carbon 14 dating technique
correlation
disconformity
electron capture

fission track dating
guide fossil
half-life
nonconformity
principle of cross-cutting relationships
principle of fossil succession
principle of inclusions
principle of lateral continuity

principle of original horizontality
principle of superposition
principle of uniformitarianism
radioactive decay
relative dating
tree-ring dating
unconformity

Review Questions

1. Demonstrating time equivalency of rock units in different areas is called:

   a. _____ absolute dating;
   b. _____ correlation;
   c. _____ relative dating;
   d. _____ unconformity;
   e. _____ none of these.

2. Which of the following is *not* one of Steno's original principles?

   a. _____ lateral continuity;
   b. _____ fossil succession;
   c. _____ original horizontality;
   d. _____ superposition;
   e. _____ none of these.

3. Hutton's locality at Siccar Point, Scotland, is a good example of an unconformity called a(an):

   a. _____ disconformity;
   b. _____ angular unconformity;
   c. _____ nonconformity;
   d. _____ hiatus;
   e. _____ none of these.

4. The geological principle demonstrated first by William Smith is today called:

   a. _____ cross-cutting relationships;
   b. _____ superposition;

   c. _____ original horizontality;
   d. _____ fossil succession;
   e. _____ none of these.

5. What step below is essential to development of an angular unconformity, but not to development of a disconformity?

   a. _____ initial deposition of layers;
   b. _____ uplift and tilting;
   c. _____ erosion;
   d. _____ deposition after erosion;
   e. _____ none of these.

6. In which type of radioactive decay is a neutron changed to a proton in the nucleus owing to emission of an electron?

   a. _____ alpha decay;
   b. _____ beta decay;
   c. _____ electron capture;
   d. _____ fission track;
   e. _____ none of these.

7. Which of the following long-lived radioactive isotopes used in radiometric dating has a half-life less than 100,000 years?

   a. _____ uranium 238;
   b. _____ uranium 235;
   c. _____ thorium 232;
   d. _____ rubidium 87;
   e. _____ none of these.

8. If a feldspar grain within a sedimentary rock (e.g., a sandstone) is radiometrically dated, the date obtained will indicate when:

   a. _____ the feldspar crystal formed;
   b. _____ the sedimentary rock formed;
   c. _____ the parent radioactive isotope formed;
   d. _____ the daughter radioactive isotope(s) formed;
   e. _____ none of these.

9. Considering the half-life of potassium 40, which is 1.3 billion years, what fraction of the original potassium 40 can be expected within a given mineral crystal after 3.9 billion years?

   a. _____ 1/2;
   b. _____ 1/4;
   c. _____ 1/8;
   d. _____ 1/16;
   e. _____ 1/32.

10. In radiocarbon dating, what isotopic ratio decreases as carbon 14 decays back to nitrogen?

   a. _____ nitrogen 14 to carbon 14;
   b. _____ carbon 14 to carbon 12;
   c. _____ carbon 13 to carbon 12;
   d. _____ nitrogen 14 to carbon 12;
   e. _____ none of these.

11. How does metamorphism affect the potential for accurate radiometric dating using any and all techniques discussed in this chapter? How would such radiometric dates be affected by metamorphism, and why?

12. List and briefly define the six principles of relative dating. Which is the most fundamental or basic? Explain why.

13. What is the significance of an unconformity in correlation and relative dating? Define the types of unconformities and, in doing so, note their key features.

14. What is a mass spectrometer, and why would it be important in determining radiometric dates?

15. How many steps does a parent uranium 238 atom go through before it finally becomes a stable daughter atom of lead 206? How many are alpha decay steps, and how many are beta? What type of decay step is associated with a decrease in atomic mass number?

16. What is radon, and how is radon related to decay within a geologically important parent–daughter radioactive isotope pair (see Figure 8.19)? Why is radon dangerous to living things, and in what concentrations is it judged dangerous?

17. Describe the principle of uniformitarianism according to Hutton and Lyell. What was the significance of this principle?

18. If you wanted to complete a radiometric dating project on a metamorphic rock using its constituent potassium feldspars for such an analysis, what parent–daughter radioactive isotope pair would you choose to work with? Why? What is the effective dating range of this isotope pair?

19. Describe the uncertainties associated with trying to radiometrically date any sedimentary rock.

20. Describe the reasons why radiocarbon dating works, its main uses, and the limitations and corrections of this method.

21. Explain the concept of an assemblage range zone and how that relates to index or guide fossils.

22. Explain briefly how well logs are made and their usefulness in the search for oil and gas. What is meant by SP and R on a well log?

23. Describe how the principle of inclusions would be important in recognizing a nonconformity.

24. Explain briefly what is meant by intertonguing (or lateral gradation) in correlation. Also, explain briefly what a key bed is and its use in correlation.

25. Where did Lord Kelvin go wrong? Explain his rationale about age-of-Earth calculations and what discovery subsequently showed that Kelvin's calculations were, in fact, in error.

Points to Ponder

1. In some places, where disconformities are particularly difficult to discern from a physical point of view, use of the principle of fossil succession helps us delineate such unconformities. How do you suppose using fossils could help us find such hard to see disconformities?

2. Both radon 222 and carbon 14 are radioactive gasses. Consider what it is about the decay of radon 222 that makes it a human carcinogen, whereas carbon 14 is not.

3. Using Figure 8.14, determine the differences between correlation by position in sequence versus correlation by lithologic identity.

4. If a rock or mineral were radiometrically dated using two or more radioactive isotope pairs (e.g., uranium 238 to lead 206 and rubidium 87 to thorium 87) and the analyses for those isotope pairs yielded distinctly different results, what possible explanations could be offered to explain how this happened? How can one rock have two "correct" ages?

5. Of the six principles of relative dating, which would be most effective in studying the geological history of Mars using recently returned high-resolution imagery of the planet's surface? Think of some possible examples of Martian geology that illustrate these key principles.

6. Suppose that you serve on a regional planning commission that is currently considering a plan for constructing what is said to be a much-needed river dam that will create a recreational lake. Opponents of the dam project have come to you with a geological report and map showing that a fault underlies the area of the proposed dam and the fault trace can be clearly seen at the surface. Opponents say that the fault may be active, and thus some day it will move suddenly, bursting the dam and sending a wall of water downstream. You seek the advice of a local geologist who has worked in the area of the dam and she tells you that she found a lava flow covering the fault less than a mile from the proposed dam project site. Do you think that you might be able to use one of our basic relative-dating principles along with a radiometric date from the lava flow to help convince the opponents that the fault has not moved in any direction (vertically or laterally) anytime in the recent past? How would you do this, and what type of reasoning would you use?

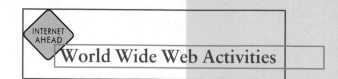

➤ UNIVERSITY OF CALIFORNIA MUSEUM OF PALEONTOLOGY

Visit this site for an excellent description of any aspect of geologic time, paleontology, and evolution. Click on the *On-line Exhibits* to go to the *Paleontology without Walls* home page, which is an introduction to the UCMP Virtual Exhibits. Click on the *Geologic Time* site. It will take you to the *Geology and Geologic Time* home page, which discusses geologic time. You can then click on one of the eras to learn more about what occurred during that time interval. Click on the *Cenozoic* site to learn more about how geologic time scales are constructed.

➤ UNITED STATES GEOLOGICAL SURVEY: RADON IN EARTH, AIR, AND WATER

Maintained by the U.S. Geological Survey (USGS), this site contains information about the geology of radon, radon potential in the United States, and USGS radon publications, as well as links to other sites with information about radon.
1. Click on *The Geology of Radon* site. This site contains the publication "The Geology of Radon" by James K. Otton, Linda C. S. Gundersen, and R. Randall Schumann, one of a series of general-interest publications prepared by the USGS. Click on the various sites listed to learn about what radon is, the geology of radon, and the potential dangers of radon.
2. Click on *Radon Potential of the USA* site. From the map shown, where in the United States is the geologic radon potential greatest? The least? From your study of "The Geology of Radon" in *The Geology of Radon* site, what factors are causing high radon potential in the various areas of the United States?

➤ RADIOCARBON WEB-INFO

This site is maintained by the radiocarbon labs of Waikato and Oxford universities. The site contains information about the basis of carbon 14 dating, applications, measurement methods, and other carbon 14 Web sites.
1. Click on the *Basis of the Method* site. What is the carbon 14 method? Who developed it? How does it work?
2. Click on the *Measurement Methods* site. What are the three principal methods of measuring residual carbon 14 activity?
3. Click on the *Applications* site. What are the various ways carbon 14 can be used to date objects and events? Click on some of the archaeology sites to see how carbon 14 dating is being used in current projects such as the radiocarbon dating of the Dead Sea Scrolls.
4. Click on the *Corrections to C14 Dates* site. What types of corrections need to be made to carbon 14 dates?

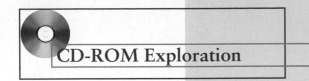

CD-ROM Exploration

➤ Exploring your *Earth Systems Today* CD-ROM will add to your understanding of the material in this chapter.

TOPIC: EARTH'S PROCESSES

MODULE: GEOLOGIC TIME

Explore activities in this module to see if you can discover the following for yourself:

Using the hour-glass animation, explore the decay rate of a radioactive system and how the slope of the line plotted corresponds to the age through the half-life relationship.

Using the parent–daughter cross-plot graph, examine how the amount of daughter changes as the amount of parent changes. In addition, note how all samples of the same element have the same decay rate regardless of the element's age.

Chapter 9

Damage from a January 1995 earthquake in Kobe, Japan. The shaking lasted only 20 seconds but caused more than 5000 fatalities, 25,000 injuries, and left 300,000 people homeless. The 50,000 destroyed buildings accounted for more than $30 billion in property damage.

OBJECTIVES

At the end of this chapter, you will have learned that

- Most earthquakes take place in well-defined zones at transform, divergent, and convergent plate boundaries.
- Energy is stored in rocks and is released when they fracture, thus producing various types of waves that travel outward in all directions from their source.
- An earthquake's epicenter is found by analyzing earthquake waves at no fewer than three seismic stations.

- The amount of damage and peoples' reactions to an earthquake are used to determine an earthquake's intensity according to the Modified Mercalli Intensity Scale.
- The Richter Magnitude Scale and Moment Magnitude Scale are used to express the amount of energy released during an earthquake.
- Great hazards are associated with earthquakes, such as ground shaking, fire, tsunami, and ground failure.

- Adequate preparation, building practices, and monitoring in earthquake-prone areas can minimize the destructive effects of earthquakes.
- The fact that earthquakes will take place in given areas is well established.
- Efforts by scientists to make accurate, short-term earthquake predictions have met with only limited success so far.

Earthquakes

Prologue

A t 3:02 A.M., on August 17, 1999, violent shaking from an earthquake awakened millions of people in Turkey. Unfortunately for many, their houses or apartment buildings collapsed causing an estimated 17,000 deaths, at least 50,000 injuries, and tens of thousands of survivors were left homeless. The amount of damage was staggering. More than 150,000 buildings were moderately to heavily damaged, and another 90,000 suffered slight damage (Figure 9.1). Collapsed buildings were everywhere, streets were strewn with rubble, and all communications were knocked out. All in all, a disaster of truly epic proportions, but hardly the first nor will it be the last in this part of the world. And as if this

Figure 9.1
The Izmit, Turkey, earthquake of August 17, 1999. (a) Map showing the plate tectonic relationships in the region. The earthquake took place on a large fracture known as the North Anatolian fault. (b) and (c) Some of the earthquake damage.

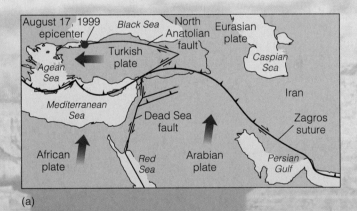

(a)

(b)

(c)

were not enough, the same area was struck only three months later, on November 12, 1999, by an aftershock nearly as large as the original earthquake, that killed another 374 people and injured about 3000!

Most residents of North America have heard of the San Andreas fault, a huge fracture in Earth's crust, that cuts through coastal California and that occasionally spawns large earthquakes. Here, two large lithospheric plates slide horizontally past one another, generating the energy that causes the earthquakes. The situation is very similar in northern Turkey where the North Anatolian fault cuts east–west across the country (Figure 9.1). Movements on this fault have caused several large earthquakes during this century, the most recent being the 7.4 magnitude one centered near Izmit, Turkey, and its 7.1 magnitude aftershock centered only a few tens of kilometers to the east.

What makes Turkey such an earthquake-prone area, and why are there so many deaths and so much destruction when earthquakes of similar size along the San Andreas fault usually result in far fewer casualties and much less damage? Looking at Figure 9.1 will help explain the first part of this question in terms of the plate tectonic regime of the region. Notice in this figure that a divergent plate boundary exists in the Red Sea and that the Arabian plate is moving northward against the Eurasian plate. One consequence of these moving plates is that the smaller Turkish plate is being forced west along the North Anatolian fault. Periodic movements on this fault result in earthquakes; eight of nine earthquakes during the last 60 years have had magnitudes greater than 7.

Of course, Turkey does not exist in isolation. The plate motions just described also affect a much larger region extending east–west through the Mediterranean Sea, through Turkey and northern Iran, Iraq, and India. The details of plate interactions differ in different parts of this extensive earthquake belt. In the case of Turkey, horizontal movement takes place between plates along a transform plate boundary, whereas farther west in Italy and Greece, earthquakes are generated at a convergent plate boundary where oceanic crust is subducted beneath continental crust. And to the east, as in India, a collision between two continental plates causes earthquakes.

Now to the second part of the question—why so many deaths and so much destruction? Since 1988, more than 120,000 people have died in earthquakes in the belt just mentioned—40,000 in Iran in 1990 and 30,000 in India in 1993, for instance. Of course, one factor in determining the number of fatalities in a large earthquake is population density. Several of these earthquakes, including the most recent ones in Turkey, have indeed struck large population centers. An even more important factor, however, is the types of structures. Many buildings in this region are made of brick or unreinforced concrete or have stone roofs, all of which are especially unstable when shaken. Unfortunately, if an earthquake occurs when most people are inside, the death toll is enormous. We will have much more to say about types of structures, building practices, and other factors important in determining fatalities and damage by earthquakes later in this chapter.

INTRODUCTION

Moving plates and volcanism as well as interactions among the hydrosphere, biosphere, atmosphere, and solid Earth are all manifestations of Earth's dynamic nature. Earthquakes also indicate that Earth is an internally active planet. As one of nature's most frightening and destructive phenomena, earthquakes have always aroused a sense of fear and are thus the subject of myths and legends. Even when an earthquake begins, no one can tell how long it will last or how violent it will be. About 13 million people have died in earthquakes during the last 4000 years, with about 2.7 million of these deaths during this century alone (Table 9.1). A case in point is the August 1999 earthquake in Turkey, in which thousands died and millions of dollars in property damage occurred (see the Prologue).

Geologists define an **earthquake** as the shaking or trembling caused by the sudden release of energy, usually as a result of faulting, which involves displacement of rocks along fractures. (The various types of faults are discussed in Chapter 13.) Following an earthquake,

Table 9.1

Some Significant Earthquakes

Year	Location	Magnitude (estimated before 1935)	Deaths (estimated)
1556	China (Shanxi Province)	8.0	1,000,000
1755	Portugal (Lisbon)	8.6	70,000
1811–1812	USA (New Madrid, Missouri)	7.5	20
1886	USA (Charleston, South Carolina)	7.0	60
1906	USA (San Francisco, California)	8.3	700
1923	Japan (Tokyo)	8.3	143,000
1964	USA (Alaska)	8.6	131
1970	China (Yunnan Province)	7.7	15,621
1971	USA (San Fernando, California)	6.6	65
1976	China (Tangshan)	8.0	242,000
1985	Mexico (Mexico City)	8.1	9,500
1988	Armenia	7.0	25,000
1989	USA (Loma Prieta, California)	7.1	63
1990	Iran	7.3	40,000
1992	Turkey	6.8	570
1992	Egypt (Cairo)	5.9	550
1993	India	6.4	30,000
1994	USA (Northridge, California)	6.7	61
1995	Japan (Kobe)	7.2	5,000+
1995	Russia	7.6	2,000+
1996	China (Lijiang)	6.5	304
1997	Iran	5.5	554
1997	Iran	7.3	2,400+
1998	Afghanistan	6.1	5,000+
1999	Taiwan	7.6	2,400
1999	Turkey	7.4	17,000
1999	Turkey	7.1	374

continuing adjustments along a fault may generate a series of earthquakes known as **aftershocks.** Most of these are smaller than the main shock, but they can cause considerable damage to already weakened structures. Indeed, much of the damage and many of the fatalities caused by the 1755 Lisbon, Portugal, earthquake resulted from aftershocks. After a small earthquake, aftershocks usually cease within a few days, but after a large earthquake they may continue for months.

The geologic definition of an earthquake is accurate but not nearly as imaginative or colorful as explanations held by many people during the past. In many cultures the cause of earthquakes was attributed to movements of some kind of animal on which Earth rested. In Japan, it was a giant catfish; in Mongolia, a giant frog; in China, an ox; in parts of South America, a whale; and to the Algonquin of North America, an immense tortoise.

According to a story from India, Earth rests on the backs of four elephants standing on the back of a turtle, which in turn is balanced on a cobra; movement by any of these animals causes earthquakes (Figure 9.2). And a legend from Mexico holds that earthquakes occur when the devil, El Diablo, rips open the crust so he and his friends can reach the surface.

Even in more recent times many people believed that earthquakes were divine retribution or warnings to the unrepentant. This view was strongly reinforced by the November 1, 1755 (All Saints' Day), Lisbon, Portugal, earthquake when the churches were crowded with worshipers. So strong was the shaking that it was felt all over Europe, and chandeliers rattled as far away as North America. A combination of collapsing buildings, huge waves that devastated the waterfront, and a fire that swept through the city resulted in 30,000 to

Figure 9.2
A legend from India holds that Earth rests on the backs of four elephants standing on the back of a turtle standing on a cobra. Movements of any of these creatures result in earthquakes.

40,000 deaths, and another 20,000 died in the following months as a direct result of the earthquake.

The Greek philosopher Aristotle (384–322 B.C.) was the first to offer what he thought was a natural explanation for earthquakes. According to him, atmospheric winds drawn into Earth's interior caused fires and swept through subterranean cavities as they tried to escape. This moving underground air was allegedly the cause of earthquakes and occasional volcanic eruptions. Today, geologists know that most earthquakes result from faulting at divergent, convergent, and transform plate boundaries. It is true that earthquakes occur preceding and during volcanic eruptions as well, but even though these can and do cause damage, at least locally, volcanism is not responsible for most large earthquakes.

ELASTIC REBOUND THEORY

Based on studies conducted after the 1906 San Francisco earthquake, H. F. Reid of Johns Hopkins University proposed the **elastic rebound theory** to explain how energy is released during earthquakes. Reid studied three sets of measurements taken across a portion of the San Andreas fault that had broken during the 1906 earthquake. The measurements revealed that points on opposite sides of the fault had moved 3.2 m during the 50-year period prior to breakage in 1906, with the west side moving northward (Figure 9.3).

According to Reid, rocks on opposite sides of the San Andreas fault had been storing energy and bending slightly for at least 50 years before the 1906 earthquake. Any straight line such as a fence or road that crossed the San Andreas fault would gradually be bent, as rocks on one side of the fault moved relative to rocks on the other side (Figure 9.3). Eventually, the strength of the rocks was exceeded, and rupture occurred. When this happened, the rocks on opposite sides of the fault rebounded or "snapped back" to their former undeformed shape, and the energy stored was released as earthquake waves radiating out from the break.

Additional field and laboratory studies conducted by Reid and others have confirmed that elastic rebound is the mechanism by which energy is released during earthquakes. In laboratory studies, rocks subjected to forces equivalent to those occurring in the crust initially change their shape. As more force is applied, however, they resist further deformation until their internal strength is exceeded. At that point, they break and snap back to their original undeformed shape, releasing internally stored energy.

The energy stored in rocks undergoing elastic deformation is analogous to the energy stored in a tightly wound watch spring. The tighter the spring is wound, the more energy is stored, thus making more energy available for release. If the spring is wound so tightly that it breaks, then the stored energy is released as the spring rapidly unwinds and partially regains its original shape. Perhaps an even more meaningful analogy is simply bending a long, straight stick over one's knee. As the stick bends, it deforms and eventually reaches the point at which it fractures. When this happens, the two pieces of the original stick snap back into their original straight position. Likewise, rocks subjected to intense forces bend until they break and then return to their original position, releasing energy in the process.

SEISMOLOGY

Seismology, the study of earthquakes, emerged as a true science during the 1880s when instruments were developed to effectively record earthquake waves. As early as A.D. 132, the Chinese scholar Chang Heng invented the first earthquake detector, but it only revealed that an earthquake took place in one of two directions from the device (Figure 9.4). By the 1880s, scientists tried using pendulums to record earthquake waves, but with little success until 1875, when Filippo Cecchi in Italy made the first successful **seismograph,** an instrument that detects, records, and measures the vibrations produced by an earthquake. By 1880 im-

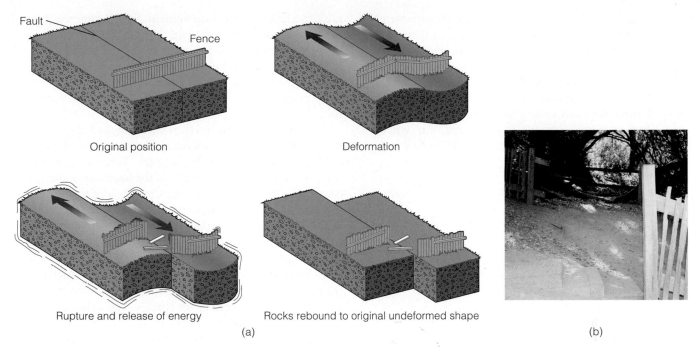

Figure 9.3

(a) According to the elastic rebound theory, when rocks are deformed, they store energy and bend. When the internal strength of the rocks is exceeded, they fracture, releasing energy as they rebound to their former undeformed shape. This sudden release of energy causes an earthquake. (b) During the 1906 San Francisco earthquake, this fence in Marin County was displaced nearly 5 m.

provements in seismograph sensitivity began yielding meaningful data about earthquakes. Although some seismographs today still use suspended masses to detect earthquake waves (Figure 9.5), most now employ electronic sensors, and, of course, computer printouts have largely replaced the strip charts of earlier seismographs.

Measuring ground motion during an earthquake is a problem because any instrument on Earth moves as the ground moves. Notice in Figure 9.5a that the seismograph employs a weight suspended on a cable, which is essentially a pendulum. Contrary to intuition, ground motion does not make the pendulum swing. Instead, the pendulum, because of its inertia, tends to remain stationary while the ground and instrument attached to it move. A seismograph of this type responds to ground motions at right angles to the pendulum, so a complete seismic station must have at least two such instruments: one sensitive to north–south motions, and one to east–west motions. In addition, a seismograph with a spring-supported weight detects vertical movements (Figure 9.5b).

When an earthquake occurs, energy in the form of *seismic waves* radiates out from the point of release; the record of these seismic waves detected by a seismograph is a *seismogram* (Figure 9.5). These waves are somewhat analogous to the ripples that move out concentrically from the point where a stone is thrown into a pond, but unlike waves on a pond, seismic waves move outward in all directions from their source.

Figure 9.4

The first earthquake detector invented in about A.D. 132 by the Chinese scholar Chang Heng. Movement of the vase dislodged a ball from a dragon's mouth into the mouth of a frog below. For instance, if a ball from the dragons on either the east or west sides of the vase were dislodged, then the earthquake must have come from one of those two directions.

Earthquakes take place because rocks are capable of storing energy but their strength is limited, so if enough force is present, they rupture and thus release their stored energy. That is, most earthquakes result when movement occurs along fractures (faults), most of which are related to plate movements. Once a fracture begins, it moves along the fault at several kilometers per second for as long as conditions for failure exist. Anywhere from a few meters to several hundred kilometers of a fault might experience movement. The longer the fracture along which movement occurs, the more time it takes for the stored energy to be released, and the longer the ground will shake. During some very large earthquakes, the ground might shake for 3 minutes, a seemingly brief period but interminable when the otherwise solid Earth shakes violently.

The point within Earth where fracturing begins—that is, the point at which energy is first released—is an earthquake's **focus,** or *hypocenter.* What one usually hears in news reports, though, is the location of the **epicenter,** the point at the surface directly above the focus (Figure 9.6). For instance, according to a report by the U.S. Geological Survey the August 1999 earthquake in Turkey had an epicenter about 11 km southeast of the city of Izmit, and its focal depth was about 17 km. The closer structures are to an earthquake's epicenter, the more likely they are to be damaged or destroyed.

Seismologists recognize three categories of earthquakes based on focal depth. *Shallow-focus* earthquakes have focal depths of less than 70 km from the surface, whereas those with foci between 70 and 300 km are *intermediate focus,* and the foci of those more than 300 km deep are characterized as *deep focus.* However, earthquakes are not evenly distributed among these three categories. Approximately 90% of all earthquake foci are at depths of less than 100 km, whereas only about 3% of all earthquakes are deep. Shallow-focus earthquakes are, with few exceptions, the most destructive.

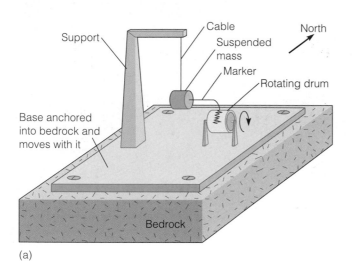

(a)

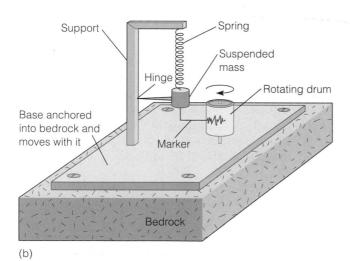

(b)

Figure 9.5
(a) A horizontal-motion seismograph. Because of its inertia, the heavy mass that contains the marker remains stationary while the rest of the structure moves along with the ground during an earthquake. As long as the length of the arm is not parallel to the direction of ground movement, the marker will record the earthquake waves on the rotating drum. This seismograph would record waves from north or south, but to record waves from the east or west another seismograph at right angles to this one is needed. (b) A vertical-motion seismograph. This seismograph operates on the same principle as a horizontal-motion instrument and records vertical ground movement.

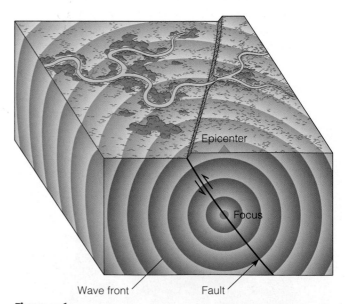

Figure 9.6
The focus of an earthquake is the location where rupture begins and energy is released. The place on the surface vertically above the focus is the epicenter. Seismic wave fronts move out in all directions from their source, the focus of an earthquake.

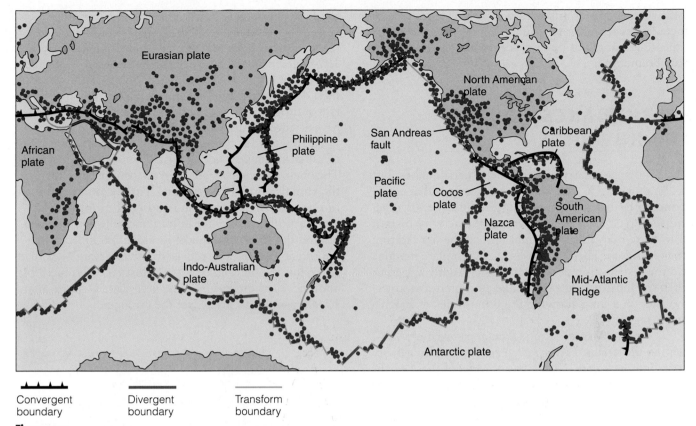

Figure 9.7
The relationship between the distribution of earthquake epicenters and plate boundaries. Approximately 80% of earthquakes occur within the circum-Pacific belt, 15% within the Mediterranean-Asiatic belt, and the remaining 5% within plate interiors or along oceanic spreading ridges. Each dot represents a single earthquake epicenter.

An interesting relationship exists between earthquake foci and plate boundaries. Earthquakes generated along divergent or transform plate boundaries are invariably shallow focus, while many shallow- and nearly all intermediate- and deep-focus earthquakes occur along convergent margins (Figure 9.7). Furthermore, a pattern emerges when the focal depths of earthquakes near island arcs and their adjacent ocean trenches are plotted. Notice in Figure 9.8 that the focal depth increases beneath the Tonga Trench in a narrow, well-defined zone that dips approximately 45 degrees. Dipping seismic zones, called *Benioff zones*, are common to

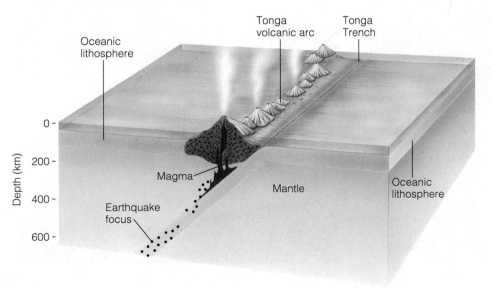

Figure 9.8
Focal depth increases in a well-defined zone that dips approximately 45 degrees beneath the Tonga volcanic arc in the South Pacific. Dipping seismic zones are called Benioff zones.

convergent plate boundaries where one plate is subducted beneath another. Such dipping seismic zones indicate the angle of plate descent along a convergent plate boundary.

EARTHQUAKES—WHERE AND HOW OFTEN?

No place on Earth is immune to earthquakes, but almost 95% take place in seismic belts that correspond to boundaries where plates converge, diverge, and slide past each other. Earthquake activity distant from plate margins is minimal but can be devastating when it occurs. The relationship between plate margins and the distribution of earthquakes is readily apparent when the locations of earthquake epicenters are superimposed on a map showing the boundaries of Earth's plates (Figure 9.7).

The majority of all earthquakes (approximately 80%) occur in the *circum-Pacific belt,* a zone of seismic activity encircling the Pacific Ocean basin. Most of these earthquakes result from convergence along plate margins, as in the case of the 1995 Kobe, Japan, earthquake (Figure 9.9a). The earthquakes along the North American Pacific Coast, especially in California, are also in this belt,

but here plates slide past one another rather than converge. The October 17, 1989, Loma Prieta earthquake in the San Francisco area (Figure 9.9b) and the January 17, 1994, Northridge earthquake (see Perspective 9.1) happened along this plate boundary. Some of the world's most devastating earthquakes, resulting in billions of dollars of property damage and more than 500,000 deaths, have occurred within the circum-Pacific belt (Table 9.1).

The second major seismic belt, accounting for 15% of all earthquakes, is the *Mediterranean-Asiatic belt.* This belt extends westerly from Indonesia through the Himalayas, across Iran and Turkey, and continues through the Mediterranean region of Europe. The devastating 1990 earthquake in Iran that killed 40,000 people and the 1993 earthquake in India that killed 30,000 are recent examples of the destructive earthquakes that strike this region (Figure 9.10). Most recently, in 1999, another earthquake in this belt killed more than 17,000 people in Turkey (see the Prologue).

(a)

(b)

Figure 9.9
Earthquake damage in the circum-Pacific belt. (a) Damage in Oakland, California, resulting from the October 1989 Loma Prieta earthquake. The columns supporting the upper deck of Interstate 880 failed, causing the upper deck to collapse onto the lower one. (b) Some of the damage in Kobe, Japan, caused by the January 1995 earthquake in which more than 5000 people died. The earthquake in Japan took place at a convergent plate boundary, whereas the Loma Prieta earthquake was generated along a transform plate boundary.

Figure 9.10
Earthquake damage in the Mediterranean-Asiatic belt. In 1993 India experienced its worst earthquake in more than 50 years. Thousands of brick and stone houses collapsed, killing at least 30,000 people.

The remaining 5% of earthquakes take place mostly in the interiors of plates and along oceanic spreading-ridge systems, that is, at divergent plate boundaries. Most of these earthquakes are not strong, although several major intraplate earthquakes are worthy of mention. For example, the 1811 and 1812 earthquakes near New Madrid, Missouri, killed approximately 20 people and nearly destroyed the town of New Madrid. So strong were these earthquakes that they were felt from the Rocky Mountains to the Atlantic Ocean and from the Canadian border to the Gulf of Mexico. In addition, the earthquakes caused church bells to ring as far away as Boston, Massachusetts (1600 km). Within the immediate area, numerous buildings were destroyed and forests were flattened; the land sank several meters in some areas, causing flooding; and reportedly the Mississippi River reversed its flow during the shaking and changed its course slightly. Eyewitnesses described the scene at New Madrid as follows:

> The earth was observed to roll in waves a few feet high with visible depressions between. By and by these swells burst throwing up large volumes of water, sand, and coal.
> . . . Undulations of the earth resembling waves, increasing in elevations as they advanced, and when they attained a certain fearful height the earth would burst.
> The shocks were clearly distinguishable into two classes, those in which the motion was horizontal and those in which it was perpendicular. The latter was attended with explosions and the terrible mixture of noises, . . . but they were by no means as destructive as the former.

Cpt. Sarpy tied up at this [small] island on the evening of the 15th of December, 1811. In looking around they found that a party of river pirates occupied part of the island and were expecting Sarpy with the intention of robbing him. As soon as Sarpy found that out he quietly dropped lower down the river. In the night the earthquake came and next morning when the accompanying haziness disappeared the island could no longer be seen. It had been utterly destroyed as well as its pirate inhabitants.*

Another major intraplate earthquake struck Charleston, South Carolina, on August 31, 1886, killing 60 people and causing $23 million in property damage (Figure 9.11). In December 1988, a large intraplate earthquake struck near Tennant Creek in Australia's Northern Territory.

The cause of intraplate earthquakes is not well understood, but geologists think they arise from localized stresses caused by the compression that most plates experience along their margins. The release of these stresses and hence the resulting intraplate earthquakes are due to local factors. A useful analogy might be that of moving a house. Regardless of how careful the movers are, it is impossible to move something so large without its internal parts shifting slightly. Similarly, plates are not likely to move without some internal stresses that occasionally cause earthquakes. Interestingly, many intraplate earthquakes are associated with very ancient and presumed inactive faults that are reactivated at various intervals.

*C. Officer and J. Page, *Tales of the Earth* (New York: Oxford University Press, 1993), pp. 49–50.

Perspective 9.1

The San Andreas Fault—One Segment of the Circum-Pacific Belt

The circum-Pacific belt is well known for both its volcanic activity and its earthquakes. Indeed, about 60% of all volcanic eruptions and 80% of all energy released by earthquakes take place in this belt that nearly encircles the Pacific Ocean basin (see Figure 9.7). And although it is commonly called the Ring of Fire, alluding to its large number of active volcanoes, you should not conclude that volcanoes cause earthquakes or the other way around. The fact that these two geologic phenomena are common in the circum-Pacific belt is accounted for by the interactions of moving plates. Where plates diverge, as at the East Pacific Rise, volcanism and shallow-focus earthquakes occur, whereas convergent plate boundaries, as along the west coast of South America, are characterized by volcanism and shallow-, intermediate-, and deep-focus earthquakes.

One well-known and well-studied segment of the circum-Pacific belt is the 1300-km-long San Andreas fault extending from the Gulf of California north through coastal California until it terminates at the Mendocino fracture zone off California's north coast (see Figure 9.7). In plate tectonic terminology, it marks a transform plate boundary with the North American plate sliding horizontally past the Pacific plate (see Chapter 12). In some places the two plates slide more or less continuously, releasing energy in the form of numerous small earthquakes (many of which can be detected only by sensitive instruments).

However, some parts of the fault are locked—that is, not moving or continuously releasing energy. When they fail, these locked segments have the potential to cause large earthquakes.

Many people think of a fault as a single crack in the ground, but most faults, and the San Andreas is no exception, consist of a complex zone of fractured rocks anywhere from a few meters to kilometers wide. The San Andreas fault zone, in many areas 1 to 2 km wide, is easily seen on the surface and in aerial photographs (Figure 1). Furthermore, it has many subsidiary faults branching from it or paralleling it.

Rocks on opposite sides of the San Andreas fault periodically lurch past one another, generating large earthquakes. Probably the most famous one destroyed San Francisco on April 18, 1906. No other earthquake has received such intense study or yielded more scientifically useful information. It resulted when 465 km of the fault ruptured, causing about 6 m of horizontal displacement in some areas (see Figure 9.3b). Initial reports put the death toll at 700 to 800, but it was probably closer to 3000, and more than half of the city's approximately 400,000 residents were made homeless. The shaking lasted nearly 1 minute and caused property damage estimated at $400 million in 1906 dollars!

Some 28,000 buildings were destroyed, many of them by the three-day fire that raged out of control and

(a)

Figure 1
(a) View across the San Andreas fault at Tomales Bay north of San Francisco. The low area occupied by the bay is underlain by shattered rocks of the San Andreas fault zone. Rocks underlying the hills in the distance are on the North American plate, whereas those at the point where the photograph was taken are on the Pacific plate. (b) This shop in Olema, California, is rather whimsically called The Epicenter, alluding to the fact that it is in the San Andreas fault zone.

(b)

devastated about 12 km² of the city (Figure 2). In fact, the San Francisco fire caused more damage than the earthquake, although the earthquake was responsible for the fire, and also ruptured water lines so that the fires could not be effectively brought under control. Water was finally pumped from San Francisco Bay to fight the fires, but by then most of the city was in ruins.

Since 1906 the San Andreas fault and its subsidiary faults have spawned many more earthquakes; one of the most tragic was centered at Northridge, California, a small community north of Los Angeles. This earthquake actually took place on a buried fault along which rocks were displaced vertically rather than horizontally. During the early morning hours of January 17, 1994, Northridge and surrounding areas were shaken for 40 seconds. When it was over, 61 people were dead and thousands injured; an oil main and at least 250 gas lines had ruptured, igniting numerous fires; nine freeways were destroyed; and thousands of homes and other buildings were damaged or destroyed (Figure 3). So many power lines were knocked down and circuits blown that 3.1 million people were without electricity, and at least 40,000 had no water because of broken water mains.

More than 1000 aftershocks followed the main Northridge earthquake, many of them strong enough to contribute to the already considerable damage. The destruction left in the earthquake's wake was enormous; it amounted to $15 to $30 billion in property damage. But much had been learned in other large earthquakes, and many of the newer structures in this area had been built to stricter standards and generally escaped unscathed or with only minor damage. However, many older unreinforced buildings

and more recent wood-frame apartments built over ground-floor garages were destroyed or heavily damaged.

Caltrans, the state transportation department, began a program of reinforcing bridges and freeway overpasses soon after the Sylmar 1971 earthquake in the Los Angeles area, and began a second round of reinforcing structures following the 1989 Loma Prieta earthquake in northern California. Most of the reinforced structures suffered little or no damage during the Northridge earthquake, but several awaiting reinforcement collapsed, including the vital east–west Santa Monica Freeway.

Figure 2
San Francisco following the 1906 earthquake. View along Sacramento Street showing damaged buildings and the approaching fire.

Earthquakes along the San Andreas and related faults will continue to occur. But other segments of the circum-Pacific belt as well as the Mediterranean-Asiatic belt are also quite active, and these areas too will continue to experience earthquakes. Keep in mind as we discuss this topic further that even though the two earthquakes featured in this Perspective were tragic, similar and even smaller earthquakes in China, Japan, Iran, Afghanistan, Turkey, and India have killed more than 150 times as many people and brought about unimaginable human suffering. In no way do we mean to minimize the death and destruction wrought by earthquakes along the San Andreas fault, but we must be aware of and consider why even greater earthquake disasters occur elsewhere.

(a)

(c)

(b)

Figure 3
Damage resulting from the 1994 Northridge earthquake. (a) Severe damage to the Northridge Meadows apartments. Sixteen died in this building. (b) Damage done to Interstate 5 Golden State Freeway. (c) Fire caused by a gas main explosion.

Figure 9.11
Damage done to Charleston, South Carolina, by the earthquake of August 31, 1886. This earthquake is the largest reported in the eastern United States.

More than 150,000 earthquakes strong enough to be felt are recorded every year by the worldwide network of seismograph stations. In addition, an estimated 900,000 earthquakes are recorded annually by seismographs but are too small to be individually cataloged. These small earthquakes result from the energy released as continual adjustments between the various plates occur.

SEISMIC WAVES

Many people have experienced an earthquake but are probably unaware that the shaking they experience and the damage to structures are caused by the arrival of various *seismic waves*, a general term encompassing all waves generated by an earthquake. Recall that when movement on a fault takes place, energy is released in the form of two kinds of waves that radiate outward in all directions from an earthquake's focus. **Body waves**, so called because they travel through the solid body of Earth, are somewhat like sound waves, and **surface waves**, which travel along the ground surface, are analogous to undulations or waves on water surfaces.

Body Waves

An earthquake generates two types of body waves: P-waves and S-waves (Figure 9.12). **P-waves**, or *primary waves*, are the fastest seismic waves and can travel through solids, liquids, and gases. P-waves are compressional, or push–pull, waves and are similar to sound waves in that they move material forward and backward along a line in the same direction that the waves themselves are moving (Figure 9.12b). Thus, the material through which P-waves travel is expanded and compressed as the wave moves through it and returns to its original size and shape after the wave passes by. In fact, some P-waves emerging from within Earth are transmitted into the atmosphere as sound waves that at certain frequencies can be heard by humans and animals.

S-waves, or *secondary waves*, are somewhat slower than P-waves and can only travel through solids. S-waves are *shear waves* because they move the material perpendicular to the direction of travel, thereby producing shear stresses in the material they move through (Figure 9.12c). Because liquids (as well as gases) are not rigid, they have no shear strength, and S-waves cannot be transmitted through them.

The velocities of P- and S-waves are determined by the density and elasticity of the materials through which they travel. For example, seismic waves travel more slowly through rocks of greater density but more rapidly through rocks with greater elasticity. *Elasticity* is a property of solids, such as rocks, and means that once they have been deformed by an applied force, they return to their original shape when the force is no longer present. Because P-wave velocity is greater than S-wave velocity in all materials, P-waves always arrive at seismic stations first.

Surface Waves

Surface waves travel along the surface of the ground, or just below it, and are slower than body waves. Unlike

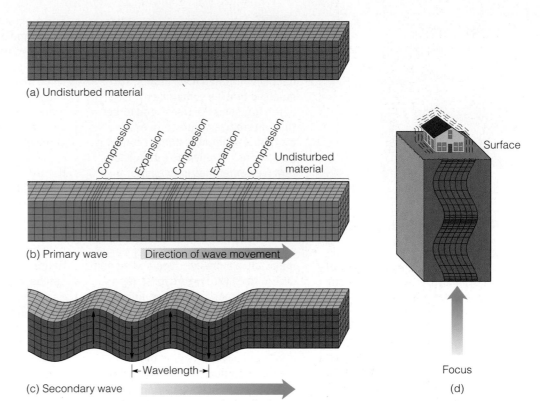

Figure 9.12
Seismic waves.
(a) Undisturbed material for reference. (b) and (c) show how body waves travel through Earth. (b) Primary waves (P-waves) compress and expand material in the same direction they travel. (c) Secondary waves (S-waves) move material perpendicular to the direction of wave movement. (d) P- and S-waves and their effect on surface structures.

(a) Undisturbed material

Compression Expansion Compression Expansion Compression

Undisturbed material

(b) Primary wave   Direction of wave movement

Wavelength

(c) Secondary wave

Surface

Focus

(d)

the sharp jolting and shaking that body waves cause, surface waves generally produce a rolling or swaying motion, much like the experience of being in a boat.

Several types of surface waves are recognized. The two most important are **Rayleigh waves (R-waves)** and **Love waves (L-waves),** named after the British scientists who discovered them, Lord Rayleigh and A. E. H. Love. Rayleigh waves are generally the slower of the two and behave like water waves in that they move forward while the individual particles of material move in an elliptic path within a vertical plane oriented in the direction of wave movement (Figure 9.13a).

The motion of a Love wave is similar to that of an S-wave, but the individual particles of the material only move back and forth in a horizontal plane perpendicular to the direction of wave travel (Figure 9.13b). This type of lateral motion can be particularly damaging to building foundations.

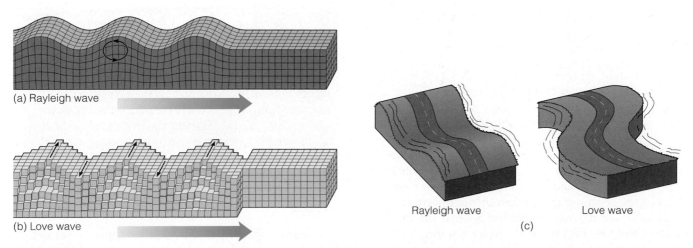

(a) Rayleigh wave

(b) Love wave

Rayleigh wave

Love wave

(c)

Figure 9.13
Surface waves. (a) Rayleigh waves (R-waves) move material in an elliptical path in a plane oriented parallel to the direction of wave movement. (b) Love waves (L-waves) move material back and forth in a horizontal plane perpendicular to the direction of wave movement. (c) The arrival of R- and L-waves causes the surface to undulate and shake from side to side.

HOW IS AN EARTHQUAKE'S EPICENTER LOCATED?

Previously we mentioned that news articles commonly report an earthquake's epicenter, but just how is the location of an epicenter determined? Once again, geologists rely on the study of seismic waves. We already know that P-waves travel faster than S-waves, nearly twice as fast in all substances, so P-waves arrive at a seismograph station first, followed some time later by S-waves. Both P- and S-waves travel directly from the focus to the seismograph station through Earth's interior, but L- and R-waves arrive last because they are the slowest, and they also travel the longest route along the surface (Figure 9.14). L- and R-waves cause much of the damage during earthquakes, but only P- and S-waves need concern us here because they are the ones important in finding an epicenter.

WHAT WOULD YOU DO?

Your child's high school science teacher is aware that you have had several courses in geology and asks for your assistance in demonstrating how earthquake waves behave in Earth and how they are recorded on a seismograph. What kinds of demonstrations can you devise using easily obtained materials to show how P- and S-waves travel through solids? Also, what could you devise to detect and record Earth vibrations? Hint—think of a pendulum and how it operates.

Seismologists, geologists who study seismology, have accumulated a tremendous amount of data over the years and now know the average speeds of P- and S-waves for any specific distance from their source. These P- and S-wave travel times are published in **time–**

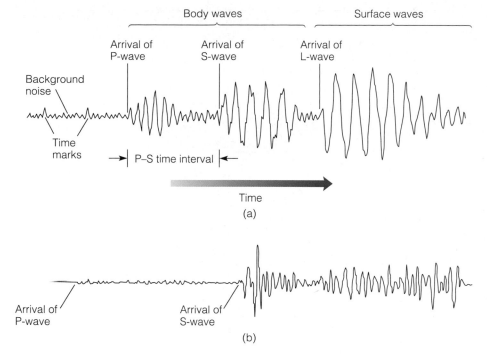

Body waves | Surface waves

Arrival of P-wave | Arrival of S-wave | Arrival of L-wave

Background noise

Time marks

P–S time interval

Time

(a)

Arrival of P-wave | Arrival of S-wave

(b)

Figure 9.14

(a) A schematic seismogram showing the arrival order and pattern produced by P-, S-, and L-waves. When an earthquake occurs, body and surface waves radiate out from the focus at the same time. Because P-waves are the fastest, they arrive at a seismograph first, followed by S-waves, and then by surface waves, which are the slowest waves. The difference between the arrival times of the P- and S-waves is the P–S time interval; it is a function of the distance of the seismograph station from the focus. (b) Seismogram for the 1906 San Francisco earthquake, recorded 14,668 km away in Göttingen, Germany. The total record represents about 26 minutes, so considerable time passed between the arrival of the P-waves and the slower moving S-waves. The arrival of surface waves, not shown here, caused the instrument to go off the scale.

distance graphs illustrating that the difference between the arrival times of the two waves is a function of distance between a seismograph and an earthquake's focus. That is, the farther the waves travel, the greater the *P–S time interval* or simply the time difference between the arrivals of P- and S-waves (Figure 9.15).

If the P–S time intervals are known from at least three seismograph stations, then the epicenter of any earthquake can be determined (Figure 9.16). Here is how it works. Subtracting the arrival time of the first P-wave from the arrival time of the first S-wave gives the P–S time interval for each seismic station. Each of

Figure 9.15

A time–distance graph showing the average travel times for P- and S-waves. The farther away a seismograph station is from the focus of an earthquake, the longer the interval between the arrivals of the P- and S-waves, and hence the greater the distance between the curves on the time–distance graph as indicated by the P–S time interval.

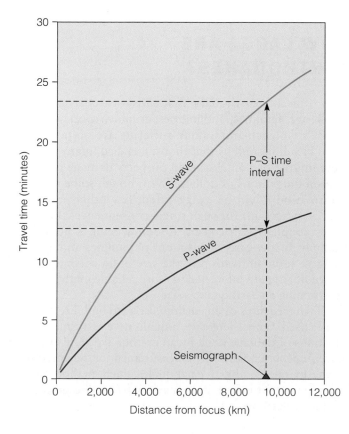

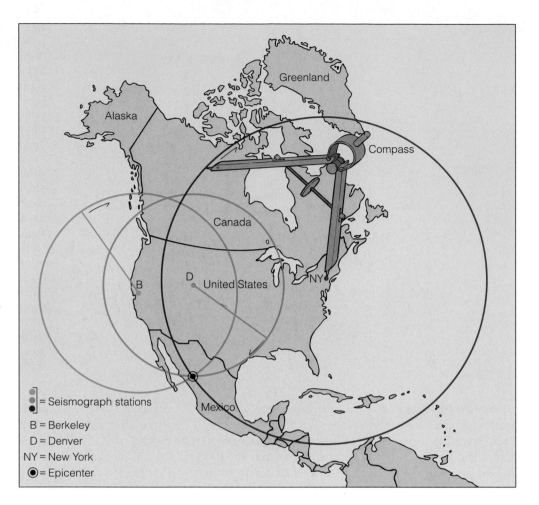

Figure 9.16
Three seismograph stations are needed to locate the epicenter of an earthquake. The P–S time interval is plotted on a time–distance graph for each seismograph station to determine the distance that station is from the epicenter. A circle with that radius is drawn from each station, and the intersection of the three circles is the epicenter of the earthquake.

● = Seismograph stations
B = Berkeley
D = Denver
NY = New York
◉ = Epicenter

these time intervals is then plotted on a time–distance graph, and a line is drawn straight down to the distance axis of the graph, thus giving the distance from the focus to each seismic station. Next, a circle whose radius equals the distance shown on the time–distance graph from each of the seismic stations is drawn on a map (Figure 9.16). The intersection of the three circles is the location of the earthquake's epicenter. It should be obvious from Figure 9.16 that P–S time intervals from at least three seismic stations are needed. If only one were used, the epicenter could be at any location on the circle drawn around that station, and two stations would give two possible locations for the epicenter.

Determining the focal depth of an earthquake is much more difficult and considerably less precise than finding its epicenter. It is usually found by making computations based on several assumptions, comparisons with the results obtained at other seismic stations, and recalculating and approximating the depth as closely as possible. Even so, the results are not highly accurate, but they do tell us that most earthquakes, probably 75%, have foci no deeper than 10 to 15 km and that a few are as much as 680 km deep.

HOW LARGE ARE EARTHQUAKES?

After any earthquake that causes extensive damage, fatalities, and injuries, graphic reports of the quake's violence and human suffering are quite common. Headlines tell us that thousands died, many more were injured or homeless, and property damage is perhaps in the billions of dollars. Even though some of the information might be exaggerated, few other natural processes account for such tragic consequences, hurricanes coupled with coastal flooding being a notable example. In any case, descriptions of fatalities and damage give some indication of the size of an earthquake, but geologists are interested in more reliable methods of determining an earthquake's size.

Two measures of an earthquake's strength are commonly used. One is *intensity,* a qualitative assessment of the kinds of damage done by an earthquake. The other, *magnitude,* is a quantitative measurement of the amount of energy released by an earthquake. Each method provides important information that can be used to prepare for future earthquakes.

Intensity

Intensity is a subjective measure of the kind of damage done by an earthquake, as well as people's reaction to it. Since the mid-nineteenth century, intensity has been used as a rough approximation of the size and strength of an earthquake. The most common intensity scale used in the United States is the **Modified Mercalli Intensity Scale**, which has values ranging from I to XII (Table 9.2).

After an assessment of the earthquake damage is made, *isoseismal lines* (lines of equal intensity) are drawn on a map, dividing the affected region into various intensity zones. The intensity value given for each zone is the maximum intensity that the earthquake produced for that zone. Even though intensity maps are not precise because of the subjective nature of the measurements, they do provide geologists with a rough approximation of the location of the earthquake, the kind and extent of the damage done, and the influence of local geology and types of building construction (Figure 9.17). Because intensity is a measure of the kind of damage done by an earthquake, insurance companies still classify earthquakes on the basis of intensity.

Generally, a large earthquake will produce greater intensity values than a small earthquake, but many other factors besides the amount of energy released by an earthquake affect its intensity. These include the distance from the epicenter, the focal depth of the earthquake, the population density and local geology of the area, the type of building construction employed, and the duration of shaking.

A comparison of the intensity map for the 1906 San Francisco earthquake and a geologic map of the area shows a strong correlation between the amount of damage done and the underlying rock and soil conditions (Figure 9.18). Damage was greatest in those areas underlain by poorly consolidated material or artificial fill because the effects of shaking are amplified in these materials, whereas damage was rather low in areas of solid bedrock. The correlation between the geology and the amount of damage done by an earthquake was further reinforced by the 1989 Loma Prieta earthquake when many of the same areas that were extensively damaged in the 1906 earthquake were once again heavily damaged.

Magnitude

If earthquakes are to be compared quantitatively, we must use a scale that measures the amount of energy released and is independent of intensity. Such a scale was developed in 1935 by Charles F. Richter, a seismologist at the California Institute of Technology. The **Richter Magnitude Scale** measures earthquake

Table 9.2

Modified Mercalli Intensity Scale

I	Not felt except by a very few under especially favorable circumstances.
II	Felt only by a few people at rest, especially on upper floors of buildings.
III	Felt quite noticeably indoors, especially on upper floors of buildings, but many people do not recognize it as an earthquake. Standing automobiles may rock slightly.
IV	During the day felt indoors by many, outdoors by few. At night some awakened. Sensation like heavy truck striking building, standing automobiles rocked noticeably.
V	Felt by nearly everyone, many awakened. Some dishes, windows, etc. broken, a few instances of cracked plaster. Disturbance of trees, poles, and other tall objects sometimes noticed.
VI	Felt by all, many frightened and run outdoors. Some heavy furniture moved, a few instances of fallen plaster or damaged chimneys. Damage slight.
VII	Everybody runs outdoors. Damage negligible in buildings of good design and construction; slight to moderate in well-built ordinary structures; considerable in poorly built or badly designed structures; some chimneys broken. Noticed by people driving automobiles.
VIII	Damage slight in specially designed structures; considerable in normally constructed buildings with possible partial collapse; great in poorly built structures. Fall of chimneys, monuments, walls. Heavy furniture overturned. Sand and mud ejected in small amounts.
IX	Damage considerable in specially designed structures. Buildings shifted off foundations. Ground noticeably cracked. Underground pipes broken.
X	Some well-built wooden structures destroyed; most masonry and frame structures with foundations destroyed; ground badly cracked. Rails bent. Landslides considerable from river banks and steep slopes. Water splashed over river banks.
XI	Few, if any (masonry) structures remain standing. Bridges destroyed. Broad fissures in ground. Underground pipelines completely out of service.
XII	Damage total. Waves seen on ground surfaces. Objects thrown upward into the air.

Source: U.S. Geological Survey.

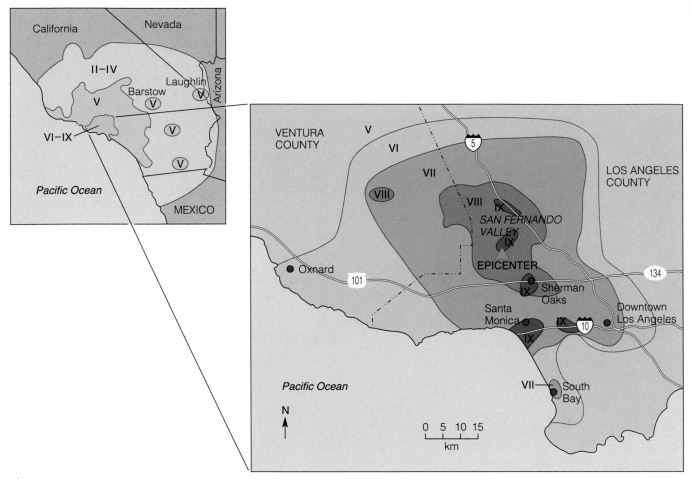

Figure 9.17
Preliminary Modified Mercalli Intensity map for the 1994 Northridge, California, earthquake, showing the region divided into intensity zones based on the kind of damage done. This earthquake had a magnitude of 6.7.

magnitude, which is the total amount of energy released by an earthquake at its source. It is an open-ended scale with values beginning at 1. The largest magnitude recorded has been 8.6, and although values greater than 9 are theoretically possible, they are highly improbable because rocks are not able to store the energy necessary to generate earthquakes of this magnitude.

The magnitude of an earthquake is determined by measuring the amplitude of the largest seismic wave as recorded on a seismogram (Figure 9.19). To avoid large numbers, Richter used a conventional base-10 logarithmic scale to convert the amplitude of the largest recorded seismic wave to a numeric magnitude value (Figure 9.19). Therefore, each whole-number increase in magnitude represents a 10-fold increase in wave amplitude. For example, the amplitude of the largest seismic wave for an earthquake of magnitude 6 is 10 times that produced by an earthquake of magnitude 5, 100 times as large as a magnitude 4 earthquake, and 1000 times that of an earthquake of magnitude 3 ($10 \times 10 \times 10 = 1000$).

A common misconception about the size of earthquakes is that an increase of one unit on the Richter Magnitude Scale—a 7 versus a 6, for instance—means a 10-fold increase in size. It is true that each whole-number increase in magnitude represents a 10-fold increase in the wave amplitude, but each magnitude increase corresponds to a roughly 30-fold increase in the amount of energy released (actually it is 31.5, but 30 is close enough for our purposes). Thus it would take about 30 earthquakes of magnitude 6 to equal the energy released in 1 earthquake with a magnitude of 7. The 1964 Alaska earthquake with a magnitude of 8.6 released almost 900 times more energy than the 1994 Northridge, California, earthquake of magnitude 6.7! And if we compare the Alaska earthquake to one with a magnitude of 5.6, it released more than 27,000 times as much energy.

We have mentioned that more than 900,000 earthquakes are recorded worldwide each year. These figures can be placed in better perspective by reference to Table 9.3, which shows that the vast majority of earthquakes

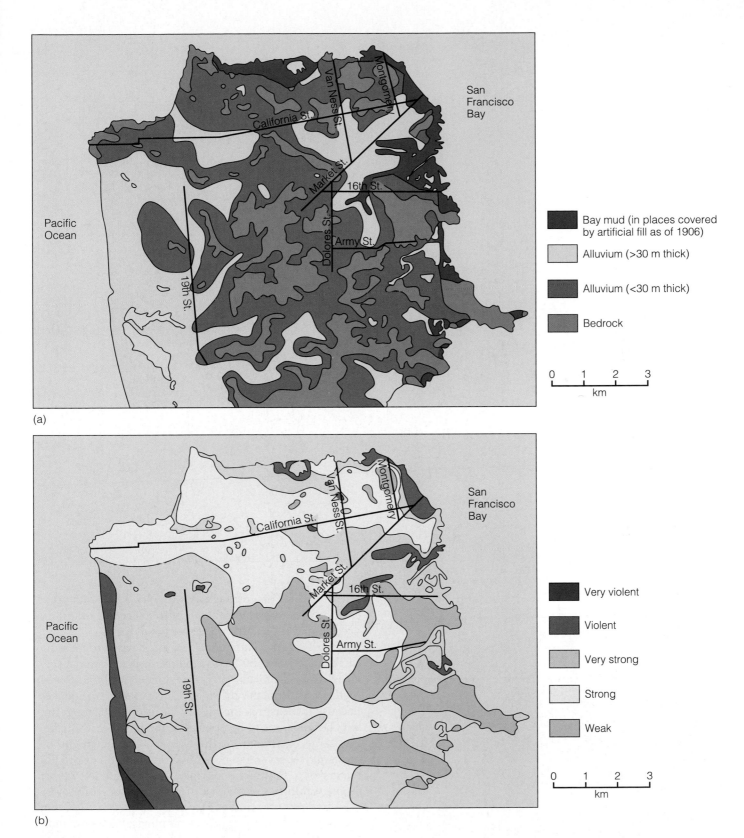

Figure 9.18
Comparison between (a) the general geology of the San Francisco area and (b) a Modified Mercalli Intensity map of the same area for the 1906 earthquake. A close correlation exists between the geology and intensity. Areas underlain by bedrock correspond to the lowest intensity values, followed by areas underlain by thin alluvium (sediment) and thick alluvium. Bay mud and/or artificial fill lie beneath the areas shaken most violently.

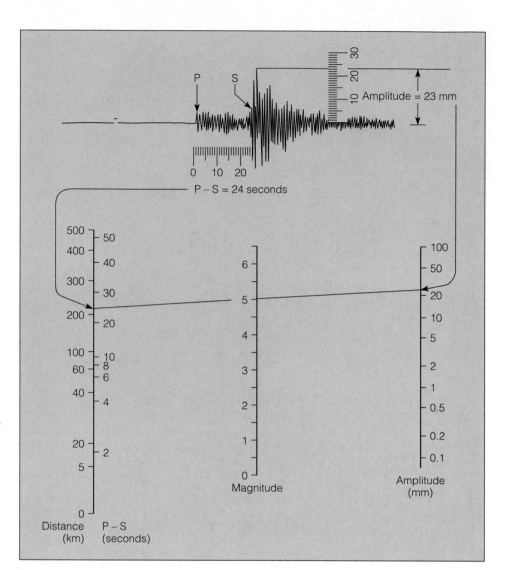

Figure 9.19
The Richter Magnitude Scale measures the total amount of energy released by an earthquake at its source. The magnitude is determined by measuring the maximum amplitude of the largest seismic wave and marking it on the right-hand scale. The difference between the arrival times of the P- and S-waves (recorded in seconds) is marked on the left-hand scale. When a line is drawn between the two points, the magnitude of the earthquake is the point at which the line crosses the center scale.

Table 9.3

Average Number of Earthquakes of Various Magnitudes per Year Worldwide

Magnitude	Effects	Average Number per Year
<2.5	Typically not felt but recorded	900,000
2.5–6.0	Usually felt; minor to moderate damage to structures	31,000
6.1–6.9	Potentially destructive, especially in populated areas	100
7.0–7.9	Major earthquakes; serious damage results	20
>8.0	Great earthquakes; usually result in total destruction	1 every 5 years

Source: Modified from Earthquake Information Bulletin and Gutenberg and Richter (1949).

have a Richter magnitude of less than 2.5 and that great earthquakes (those with a magnitude greater than 8.0) occur, on average, only once every five years.

The Richter Magnitude Scale was devised to measure earthquake waves on a particular seismograph and a specific distance from an earthquake. One of its limitations is that it underestimates the energy of very large earthquakes because it measures the highest peak on a seismogram, which represents only an instant during an earthquake. For large earthquakes, though, the energy might be released over several minutes and along hundreds of kilometers of a fault. For instance, during the 1857 Fort Tejon, California, earthquake, the ground shook for more than two minutes and energy was released for 360 km along the fault. Despite its shortcomings, Richter magnitudes still commonly appear in news releases. More recently, seismologists developed a **Moment Magnitude Scale** that considers the area of a fault along which the rupture occurred and the amount of movement of rocks adjacent to the fault. Seismologists are confident they now have a scale with which they not only can compare different-sized earthquakes more effectively, but also can evaluate the sizes of earthquakes that occurred before instruments were available to record them.

THE DESTRUCTIVE EFFECTS OF EARTHQUAKES

Certainly earthquakes are one of nature's most destructive phenomena. Little or no warning precedes an earthquake, and once they begin, little or nothing can be done to minimize their effects. Only planning before an earthquake can be very effective, but earthquake prediction may become a reality in the future (discussed in a later section). The destructive effects of earthquakes include ground shaking, fire, seismic sea waves, and landslides, as well as panic, disruption of vital services, and psychological shock. In some cases, rescue attempts are hampered by inadequate resources or planning, existing conditions of civil unrest, or simply the magnitude of the disaster.

The number of deaths and injuries as well as the amount of property damage depend on several factors. Generally speaking, earthquakes during working hours and school hours in densely populated urban areas are the most destructive and cause most fatalities and injuries. However, magnitude, duration of shaking, distance from the epicenter, geology of the affected region, and type of structures are also important considerations. Given these variables, it should not be surprising that a comparatively small earthquake can have disastrous effects, whereas a much larger one might go largely unnoticed, except perhaps by seismologists.

Ground Shaking

Ground shaking is the most obvious and immediate effect of an earthquake, but it varies depending on magnitude, distance from the epicenter, and the type of underlying materials in the area—unconsolidated sediment or fill versus bedrock, for instance. Certainly ground shaking is terrifying, and it might be violent enough for fissures to open in the ground. Nevertheless, contrary to popular myth, fissures do not swallow up people and buildings and then close on them. And although California will no doubt have big earthquakes in the future, rocks cannot store enough energy to displace a landmass as large as California into the Pacific Ocean, as one sometimes hears suggested in the tabloids.

The effects of ground shaking, collapsing buildings, falling building facades and window glass, and toppling monuments and statues cause more damage and result in more loss of life and injuries than any other earthquake hazard.

Structures built on solid bedrock generally suffer less damage than those built on poorly consolidated material such as water-saturated sediments or artificial fill (see Perspective 9.2). Structures on poorly consolidated or water-saturated material are subjected to ground shaking of longer duration and greater S-wave amplitude than those on bedrock (Figure 9.20).

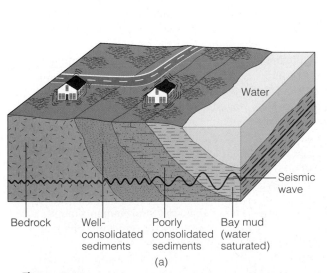

(a)

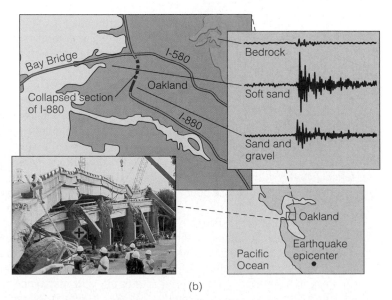

(b)

Figure 9.20
(a) The amplitude and duration of seismic waves generally increase as they pass from bedrock into poorly consolidated or water-saturated materials, resulting in greater damage to structures built on the latter. (b) During the 1989 Loma Prieta earthquake, collapse of the upper deck of the Interstate 880 freeway in Oakland, California, occurred where it rested on artificial fill over mud, whereas adjacent parts of the same freeway on firmer materials withstood the 7.1 magnitude shock. Seismograms (upper right) of aftershocks show that shaking is more violent in mud than in sand and gravel or bedrock.

Perspective 9.2

Designing Earthquake-Resistant Structures

One way to reduce property damage, injuries, and loss of life is to design and build structures as earthquake resistant as possible. Many things can be done to improve the safety of current structures and of new buildings.

California has a Uniform Building Code that sets minimum standards for building earthquake-resistant structures that is used as a model around the world. The California code is far more stringent than federal earthquake building codes and requires that structures be able to withstand a 25-second main shock. Unfortunately, many earthquakes are of far longer duration. For example, the main shock of the 1964 Alaskan earthquake lasted approximately three minutes and was followed by numerous aftershocks. Although many of the extensively damaged buildings in this earthquake had been built according to the California code, they were not designed to withstand shaking of such long duration. Nevertheless, in California and elsewhere, structures built since the California code went into effect have fared much better during moderate to major earthquakes than those built before its implementation.

To design earthquake-resistant structures, engineers must understand the dynamics and mechanics of earthquakes, including the type and duration of the ground motion and how rapidly the ground accelerates during an earthquake. An understanding of the area's geology is also important because certain ground materials such as water-saturated sediments or land-fill can lose their strength and cohesiveness during an earthquake (see Figure 9.20). Finally, engineers must be aware of how different structures behave under different earthquake conditions.

With the level of technology currently available, a well-designed, properly constructed building should be able to withstand small, short-duration earthquakes of less than 5.5 magnitude with little or no damage. In moderate earthquakes (5.5–7.0 magnitude), the damage suffered should not be serious and should be repairable. In a major earthquake of greater than 7.0 magnitude, the building should not collapse, although it may later have to be demolished.

Many factors enter into the design of an earthquake-resistant structure, but the most important is that the building be tied together; that is, the foundation, walls, floors, and roof should all be joined together to create a structure that can withstand both horizontal and vertical shaking (Figure 1). Almost all the structural failures resulting from earthquake ground movement occur at weak connections, where the various parts of a structure are not securely tied together. Buildings with open or unsupported first stories are particularly susceptible to damage. Some reinforcement must be done or collapse is a distinct possibility (Figure 1).

The size and shape of a building can also affect its resistance to earthquakes (Figure 2). Rectangular, box-shaped buildings are inherently stronger than those of irregular size or shape because different parts of an irregular building may sway at different rates, increasing the stress and likelihood of structural failure (Figure 2b).

Tall buildings, such as skyscrapers, must be designed so that a certain amount of swaying or flexing can occur, but not so much that they touch neighboring buildings during swaying (Figure 2d). If a building is brittle and does not give, it will crack and fail. Besides designed flexibility, engineers must ensure that a building does not vibrate at the same frequency as the ground does during an earthquake. When that happens, the force applied by the seismic waves at ground level is multiplied several times by the time they reach the top of the building (Figure 2c). This condition is particularly troublesome in areas of poorly consolidated sediment (Figure 3). Fortunately, buildings can be designed so that they will sway at a different frequency than the ground.

During the Mexico City earthquake of 1985, many buildings between 6 and 15 stories high were either badly damaged or collapsed, even if constructed of reinforced concrete. In contrast, the same kinds of buildings with fewer or more stories suffered comparatively little. The reason? The 6- to 15-story buildings vibrated at the same frequency as the ground and accordingly swayed so much that structural elements failed, and in some cases one floor collapsed on another (Figure 3).

Damage to high-rise structures can also be minimized or prevented by using diagonal steel beams to help prevent swaying. In addition, tall buildings in earthquake-prone areas are now commonly placed on layered steel and

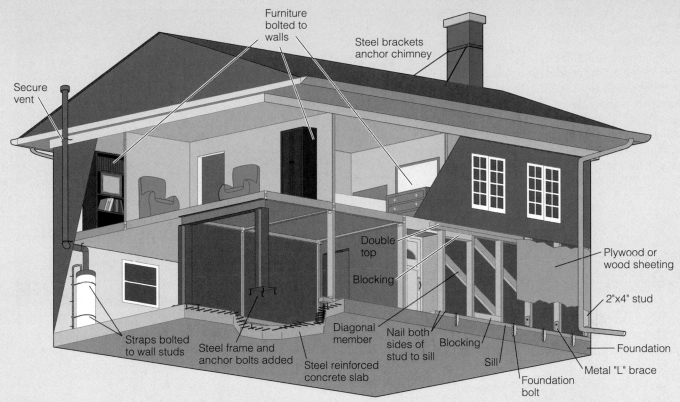

Figure 1

This illustration shows some of the things a homeowner can do to reduce damage to a building because of ground shaking during an earthquake. Notice that the structure must be solidly attached to its foundation, and bracing the walls helps prevent damage from horizontal motion.

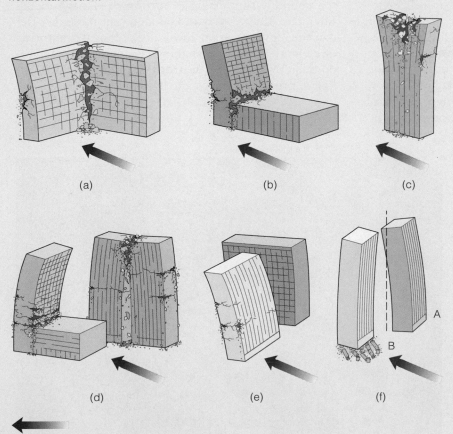

(a)　(b)　(c)

(d)　(e)　(f)

Direction of seismic wave

Figure 2

The effects of ground shaking on various tall buildings of differing shapes. (a) Damage will occur if two wings of a building are joined at right angles and experience different motions. (b) Buildings of different heights will sway differently, leading to damage at the point of connection. (c) Shaking increases with height and is greatest at the top of a building. (d) Closely spaced buildings may crash into each other due to swaying. (e) A building whose long axis is parallel to the direction of the seismic waves will sway less than a building whose axis is perpendicular. (f) Two buildings of different design will behave differently even when subjected to the same shaking conditions. Building A sways as a unit and remains standing, while building B, whose first story is composed of only tall columns, collapses because most of the swaying takes place in the "soft" first story.

rubber structures and devices similar to shock absorbers that help decrease the amount of sway.

What about structures built many years ago? Almost every city and town has older single and multistory structures, constructed of unreinforced brick masonry, poor-quality concrete, and rotting or decaying wood. Just as in new buildings, the most important thing that can be done to increase the stability and safety of older structures is to tie together the different components of each building. This can be done by adding a steel frame to unreinforced parts of a building such as a garage, bolting the walls to the foundation, adding reinforced beams to the exterior, and using beam and joist connectors whenever possible. Although such modifications are expensive, they are usually cheaper than having to replace a building that was destroyed by an earthquake.

Figure 3

This 15-story reinforced-concrete building collapsed due to the ground shaking that occurred during the 1985 Mexico City earthquake. The soft lakebed sediments on which Mexico City is built amplified the seismic waves as they passed through.

In addition, fill and water-saturated sediments tend to liquefy, or behave as a fluid, a process known as *liquefaction*. When shaken, the individual grains lose cohesion and the ground flows. Dramatic examples of damage resulting from liquefaction include Niigata, Japan, where large apartment buildings were tipped on their sides after the water-saturated soil of the hillside collapsed (Figure 9.21), and Turnagain Heights, Alaska, where many homes were destroyed when the Bootlegger Cove Clay lost all of its strength when it was shaken by the 1964 earthquake (see Figure 14.21).

During the Loma Prieta earthquake that caused the third game of the 1989 World Series to be postponed, those districts in the San Francisco–Oakland Bay area built on artificial fill or reclaimed bay mud suffered the most damage. In the Marina district of San Francisco,

many buildings were destroyed, and a fire, fed by broken gas lines, lit up the night sky. The failure of the columns supporting a portion of the two-tiered Interstate 880 freeway in Oakland sent the upper tier crashing down onto the lower one, killing 42 motorists (see Figure 9.9b). The shaking lasted less than 15 seconds but resulted in 63 deaths, 3800 injuries, and $6 billion in property damage and left at least 12,000 people homeless.

Figure 9.21

The effects of ground shaking on water-saturated soil are dramatically illustrated by the collapse of these buildings in Niigata, Japan, during a 1964 earthquake. The buildings were designed to be earthquake resistant and fell over on their sides intact.

Besides the magnitude of an earthquake and the underlying geology, the material used and the type of construction also affect the amount of damage done (see Perspective 9.2). Adobe and mud-walled structures are the weakest of all and almost always collapse during an earthquake. Unreinforced brick structures and poorly built concrete structures are also particularly susceptible to collapse. For example, the 1976 earthquake in Tangshan, China, completely leveled the city because hardly any structures were built to resist seismic forces. In fact, most had unreinforced brick walls, which have no flexibility, and consequently they collapsed during the shaking (Figure 9.22). The 6.4 magnitude earthquake that struck India in 1993 killed about 30,000 whereas the 6.7 magnitude Northridge earthquake resulted in only 61 deaths. Both earthquakes occurred in densely populated regions, but in India the brick and stone buildings could not withstand ground shaking; most collapsed, entombing their occupants (see Figure 9.10).

Fire

In many earthquakes, particularly in urban areas, fire is a major hazard. Almost 90% of the damage done in the 1906 San Francisco earthquake was caused by fire. The shaking severed many of the electrical and gas lines, which touched off flames and started numerous fires all over the city. Because water mains were ruptured by the earthquake, there was no effective way to fight the fires that raged out of control for three days, destroying much of the city. During the 1989 Loma Prieta earthquake, a fire broke out in the Marina district of San Francisco (Figure 9.23) but was contained within a small area. In contrast to 1906, in 1989 San Francisco had a system of valves throughout its water and gas pipeline system so that lines could be isolated from breaks.

During the September 1, 1923, earthquake in Japan, fires destroyed 71% of the houses in Tokyo and practically all the houses in Yokohama. In all, a total of 576,262 houses were destroyed by fire, and 143,000 people died, many as a result of the fire. A horrible example occurred in Tokyo where thousands of people gathered along the banks of the Sumida River to escape the raging fires. Suddenly, a firestorm swept over the area, killing more than 38,000 people. The fires from this earthquake were particularly devastating because most of the buildings were constructed of wood; many fires were started by chemicals and fanned by 20 km/hr winds.

Killer Waves—Tsunami

On April 1, 1946, the residents of Hilo, Hawaii, were completely unaware of an earthquake more than

Figure 9.22
Many of the approximately 242,000 people who died in the 1976 earthquake in Tangshan, China, were killed by collapsing structures. Many of the buildings were constructed from unreinforced brick, which has no flexibility, and quickly fell down during the earthquake.

3500 km away in the Aleutian Islands, but this earthquake killed 159 people in that city and caused $25 million in property damage (Figures 9.24 and 9.25). Thus, earthquakes can kill at a considerable distance, in some cases even much farther away than in this example. This earthquake generated what is popularly called a *tidal wave* but more correctly termed a *seismic sea wave* or **tsunami,** the latter is a Japanese term meaning "harbor wave." The term *tidal wave* nevertheless persists in popular literature and some news accounts, but these waves are not caused by or related to tides. Indeed, tsunami are destructive sea waves generated when large amounts of energy are rapidly released into a body of water. Many result from submarine earthquakes, but volcanoes at sea or submarine landslides can also cause them. Recall from Chapter 4 that the 1883 eruption of Krakatau between Java and Sumatra generated a large sea wave that killed 36,000 on nearby islands.

Once a tsunami is generated, it can travel across an entire ocean and cause devastation far from its source. In the open sea, tsunami travel at several hundred kilometers per hour, and commonly go unnoticed as they pass beneath ships because they are usually less than 1 m high and the distance between wave crests is typically hundreds of kilometers. When they enter shallow water, however, the wave slows down and water piles up to heights anywhere from a meter or two to many meters high. The 1946 tsunami that struck Hilo, Hawaii, was 16.5 m high. In any case, the tremendous energy possessed by a tsunami is concentrated on a

Figure 9.23
San Francisco Marina district fire caused by broken gas lines during the 1989 Loma Prieta earthquake.

Figure 9.24
As a tsunami crashes into the street behind them, residents of Hilo, Hawaii, run for their lives. This tsunami was generated by an earthquake in the Aleutian Islands and resulted in massive property damage to Hilo and the deaths of 159 people.

Figure 9.25
This painting is an artist's concept of the tsunami that hit the Anchorage, Alaska, waterfront following the 1964 earthquake. It is based on eyewitness accounts.

after the first wave hits, more waves follow at 20- to 60-minute intervals. About 80 minutes after the 1755 Lisbon, Portugal, earthquake, the first of three tsunami, the largest more than 12 m high, destroyed the waterfront area and killed numerous people. Following the arrival of a 2 m high tsunami in Crescent City, California, in 1964, curious people went to the waterfront to inspect the damage. Unfortunately, 10 were killed by a following 4 m high wave.

One of nature's warning signs of an approaching tsunami is that some are preceded by a sudden withdrawal of the sea from a coastal region. In fact, the sea might withdraw so far that it cannot be seen and the seafloor is laid bare over a huge area. On more than one occasion, people have rushed out to inspect exposed reefs or collect fish and shells, only to be swept away when the tsunami arrives. Obviously, tsunami can cause many fatalities and considerable destruction in coastal areas, and a

shoreline when it hits either as a large breaking wave or, in some cases, as what appears to be a very rapidly rising tide.

The Hawaiian Islands are especially vulnerable to tsunami, but other areas around the Pacific Rim have also been devastated by these waves. Japan and parts of the coast of South America have experienced a number of tsunami, and following the 1964 Alaska earthquake the waterfronts at Anchorage, Alaska (Figure 9.25), and Crescent City, California, were heavily damaged and a number of people were swept away.

A common popular belief is that a tsunami is a single large wave that crashes onto a shoreline. Any tsunami consists of a series of waves that pour onshore for as much as 30 minutes followed by an equal time during which water rushes back to sea. Furthermore,

tsunami in about 1390 B.C., coupled with the volcanic eruption that caused it, may have contributed to the demise of the Minoan civilization on Crete (see Chapter 11, Prologue).

Following the tragic 1946 tsunami that hit Hilo, Hawaii, the U.S. Coast and Geodetic Survey established a Tsunami Early Warning System in Honolulu, Hawaii, (Figure 9.26). This system combines seismographs and instruments that can detect earthquake-generated waves. Whenever a strong earthquake occurs anywhere within the Pacific basin, its location is determined, and instruments are checked to see if a tsunami has been generated. If it has, a warning is sent out to evacuate people from low-lying areas that may be affected. Nevertheless, tsunami remain a threat to people in coastal areas, especially around the Pacific Ocean (Table 9.4).

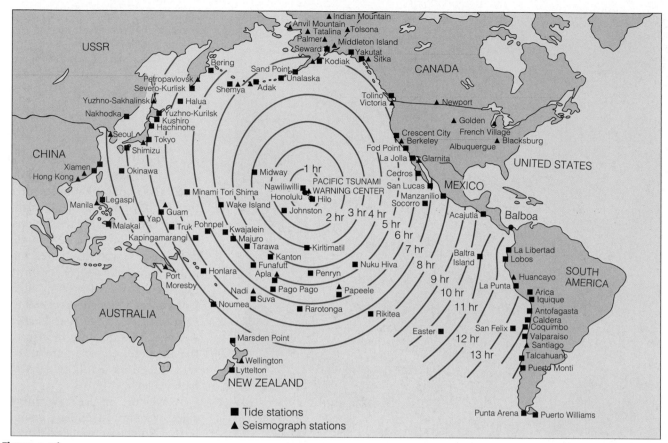

Figure 9.26

The Pacific Tsunami Warning System. Reporting stations and tsunami travel times to Honolulu, Hawaii.

Table 9.4

Tsunami Fatalities in the Pacific Since 1990			
DATE	LOCATION	MAXIMUM WAVE HEIGHT	FATALITIES
September 2, 1992	Nicaragua	10 m	170
December 12, 1992	Flores Island	26 m	>1000
July 12, 1993	Okushiri, Japan	31 m	239
June 2, 1994	East Java	14 m	238
November 14, 1994	Mindoro Island	7 m	49
October 9, 1995	Jalisco, Mexico	11 m	1
January 1, 1996	Sulawesi Island	3.4 m	9
February 17, 1996	Irian, Java	7.7 m	161
February 21, 1996	North coast of Peru	5 m	12
July 17, 1998	Papua New Guinea	15 m	>2200

Source: F. I. Gonzales. 1999. Tsunami! Scientific American, 280, no. 5, p. 59.

Ground Failure

Earthquake-triggered landslides are particularly dangerous in mountainous regions and have been responsible for tremendous amounts of damage and many deaths. The 1959 earthquake in Madison Canyon, Montana, for example, caused a huge rock slide (Figure 9.27), while the 1970 Peru earthquake caused an avalanche that destroyed the town of Yungay (see the Prologue to Chapter 14). Most of the 100,000 deaths from the 1920 earthquake in Gansu, China, resulted when cliffs composed of loess (wind-deposited silt) collapsed. More than 20,000 people were killed when two-thirds of the town of Port Royal, Jamaica, slid into the sea following an earthquake on June 7, 1692.

(a)

Source of landslide

Landslide deposit

(b)

Figure 9.27
On August 17, 1959, an earthquake with a Richter magnitude of 7.3 shook southwestern Montana and a large area in adjacent states.
(a) The fault scarp in this image was produced when the block on the right moved up several meters compared to the one on the left.
(b) The earthquake triggered a landslide (visible in the distance) that blocked the Madison River in Montana and created Earthquake Lake (foreground). The slide entombed about 26 people in a campground at the valley bottom.

EARTHQUAKE PREDICTION

Can earthquakes be predicted? A successful prediction must include a time frame for the occurrence of an earthquake, its location, and its strength. Despite the tremendous amount of information geologists have gathered about the cause of earthquakes, successful predictions are still rare. Nevertheless, if reliable predictions can be made, they can greatly reduce the number of deaths and injuries.

From an analysis of historic records and the distribution of known faults, **seismic risk maps** can be constructed that indicate the likelihood and potential severity of future earthquakes based on the intensity of past earthquakes. An international effort by scientists from several countries recently resulted in the publication of the first Global Seismic Hazard Assessment Map in December 1999 (Figure 9.28). Although such maps cannot be used to predict when an earthquake will take place in any particular area, they are useful in anticipating future earthquakes and helping people prepare for them.

WHAT WOULD YOU DO?

You and several other citizens are asked to make recommendations for your community, which has experienced moderate to large earthquakes during the past. Consider zoning regulations, building codes for private dwellings, hospitals, public buildings, and high-rise structures, as well as emergency contingency plans. What kinds of recommendations would you make, and what and who would you ask for professional guidance?

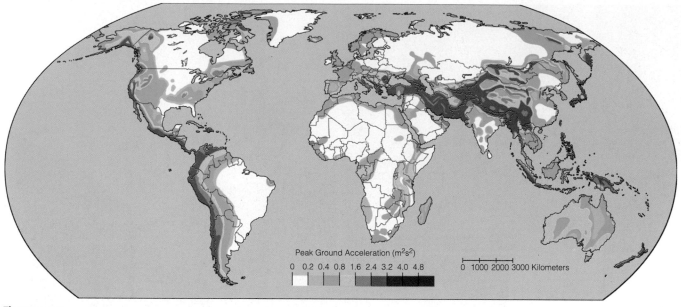

Figure 9.28

The Global Seismic Hazard Assessment Program recently published this seismic hazard map showing peak ground accelerations. The values are based on a 90% probability that the indicated horizontal ground acceleration during an earthquake is not likely to be exceeded in 50 years. The higher the number, the greater the hazard. As expected, the greatest seismic risks are in the circum-Pacific belt and the Mediterranean-Asiatic belt.

Earthquake Precursors

Studies conducted over the past several decades indicate that most earthquakes are preceded by both short-term and long-term changes within Earth. Such changes are called **precursors.**

One long-range prediction technique used in seismically active areas involves plotting the location of major earthquakes and their aftershocks to detect areas that have had major earthquakes in the past but are currently inactive. Such regions are locked and not releasing energy. Nevertheless, pressure is continuing to accumulate in these regions due to plate motions, making these *seismic gaps* prime locations for future earthquakes. Several seismic gaps along the San Andreas fault have the potential for future major earthquakes (Figure 9.29). A major earthquake that damaged Mexico City in 1985 occurred along a seismic gap in the convergence zone along the west coast of Mexico (see the Prologue to Chapter 12).

Changes in elevation and tilting of the land surface have frequently preceded earthquakes and may be warnings of impending quakes. Extremely slight changes in the angle of the ground surface can be measured by *tiltmeters*. Tiltmeters have been placed on both sides of the San Andreas fault to measure tilting of the ground surface that is thought to result from increasing pressure in the rocks. Data from measurements in central California indicate significant tilting occurred immediately preceding small earthquakes. Furthermore, extensive tiltmeter work performed in Japan prior to the 1964 Niigata earthquake clearly showed a relationship between increased tilting and the main shock. Although more research is needed, such changes appear to be useful in making short-term earthquake predictions.

Other earthquake precursors include fluctuations in the water level of wells and local changes in Earth's magnetic field and the electrical resistance of the ground. These fluctuations are thought to result from changes in the amount of pore space in rocks due to increasing pressure. A change in animal behavior prior to an earthquake also is frequently mentioned. It may be that animals are sensing small and subtle changes in Earth prior to a quake that humans simply do not sense.

The Chinese used all the precursors just mentioned, except seismic gaps, to successfully predict a large earthquake in Haicheng on February 4, 1975. The earthquake had a magnitude of 7.5 and destroyed hundreds of buildings but claimed very few lives because most people had been evacuated from the buildings and were outdoors when it occurred. Unfortunately, visits by U.S. scientists revealed that the prediction resulted from unique circumstances that really could not be applied to earthquake prediction elsewhere. And for that matter, the Chinese failed to predict a 1976 earthquake that killed 242,000 people.

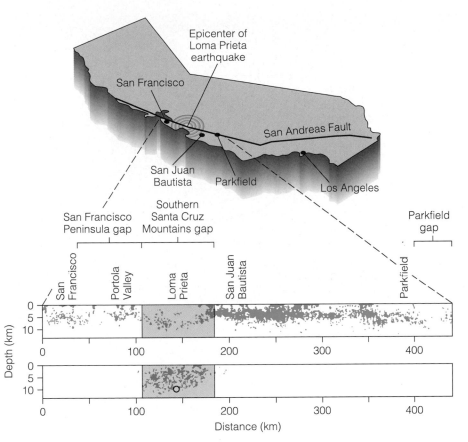

Figure 9.29
Three seismic gaps are evident in this cross section along the San Andreas fault from north of San Francisco to south of Parkfield. The first is between San Francisco and Portola Valley, the second near Loma Prieta Mountain, and the third is southeast of Parkfield. The top section shows the epicenters of earthquakes between January 1969 and July 1989. The bottom section shows the southern Santa Cruz Mountains gap after it was filled by the October 17, 1989, Loma Prieta earthquake (open circle) and its aftershocks.

Dilatancy Model Many of the precursors just discussed can be related to the **dilatancy model,** which is based on changes occurring in rocks subjected to very high pressures. Laboratory experiments have shown that rocks undergo an increase in volume, known as dilatancy, just before rupturing. As pressure builds in rocks along faults, numerous small cracks are produced that alter the physical properties of the rocks. Water enters the cracks and increases the fluid pressure; this further increases the volume of the rocks and decreases their inherent strength until failure eventually occurs, producing an earthquake.

The dilatancy model is consistent with many earthquake precursors (Figure 9.30). Although additional research is needed, this model has the potential for predicting earthquakes under certain circumstances.

Earthquake Prediction Programs

Currently, only four nations—the United States, Japan, Russia, and China—have government-sponsored earthquake prediction programs. These programs include laboratory and field studies of the behavior of rocks before, during, and after large earthquakes, as well as monitoring activity along major active faults.

Most earthquake prediction work in the United States is done by the U.S. Geological Survey (USGS) and involves research into all aspects of earthquake-related phenomena. One of the more ambitious programs undertaken by the USGS was the Parkfield earthquake prediction experiment. Over the past 130 years, moderate-sized earthquakes have occurred on an average of every 21 to 22 years along a 24 km segment of the San Andreas fault at Parkfield, California. Based on the regularity of these earthquakes and the fact they have all been very similar, the USGS forecasted that another moderate-sized earthquake would occur in this region between 1986 and 1993. To study earthquake precursors and assess the possibility of short-term predictions of moderate-sized earthquakes, the USGS installed a variety of instruments to monitor conditions along the San Andreas fault in the Parkfield area. When the predicted deadline passed without the promised earthquake occurring, some geologists questioned whether earthquakes could ever be predicted with any certainty. What the Parkfield failure showed, however, was that Earth's interior is far more complex than had previously been assumed and that accurate predictions of earthquakes are going to be more difficult than geologists originally thought.

The Chinese have perhaps one of the most ambitious earthquake prediction programs in the world,

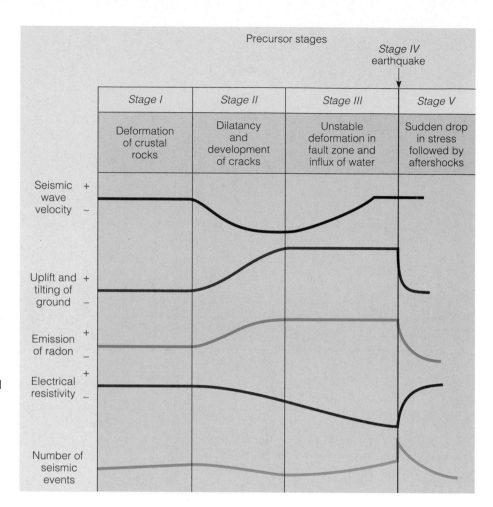

Figure 9.30
The relationship between dilatancy and various other earthquake precursors. The onset of dilatancy matches a change in each of the precursors illustrated. For example, a drop in seismic wave velocity corresponds to the onset of dilatancy and the development of cracks.

which is understandable considering their long history of destructive earthquakes. Their earthquake prediction program was initiated soon after two large earthquakes occurred at Xingtai (300 km southwest of Beijing) in 1966. The program includes extensive study and monitoring of all possible earthquake precursors. In addition, the Chinese also emphasize changes in phenomena that can be observed and heard without the use of sophisticated instruments. They successfully predicted the 1975 Haicheng earthquake, as we already noted, but failed to predict the devastating 1976 Tangshan earthquake that killed at least 242,000 people (Figure 9.22).

Progress is being made toward dependable, accurate earthquake predictions, and studies are underway to assess public reactions to long-, medium-, and short-term earthquake warnings. However, unless short-term warnings are actually followed by an earthquake, most people will probably ignore the warnings, as they frequently do now for hurricanes, tornadoes, and tsunami. Perhaps the best we can hope for is that people in seismically active areas will take measures to minimize their risk from the next major earthquake (Table 9.5).

EARTHQUAKE CONTROL

Reliable earthquake prediction is still in the future, but can anything be done to control or at least partly control earthquakes? Because of the tremendous energy involved, it seems unlikely that humans will ever be able to prevent earthquakes. However, it might be possible to gradually release the energy stored in rocks, thus decreasing the probability of a large earthquake and extensive damage.

During the early to mid-1960s, Denver, Colorado, experienced many small earthquakes. This was surprising because Denver had not been prone to earthquakes in the past. In 1962 David M. Evans, a geologist, suggested that the earthquakes were directly related to the injection of contaminated waste water into a disposal well 3674 m deep at the Rocky Mountain Arsenal, northeast of Denver. The U.S. Army initially denied that there was any connection, but a USGS study concluded that the pumping of waste fluids into the disposal well was the cause of the earthquakes.

Figure 9.31 shows the relationship between the average number of earthquakes in Denver per month and the average amount of contaminated fluids injected

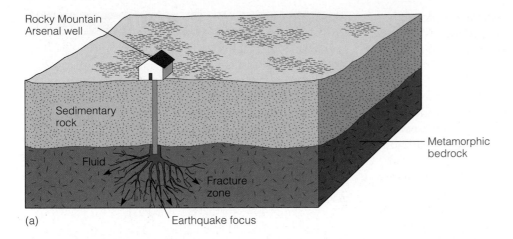

(a)

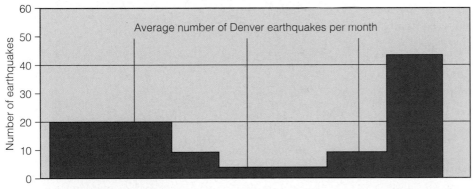

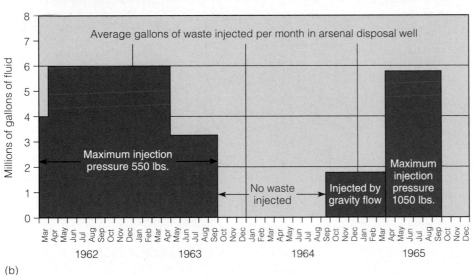

(b)

Figure 9.31
(a) A block diagram of the Rocky Mountain Arsenal well and the underlying geology. (b) A graph showing the relationship between the amount of waste injected into the well per month and the average number of Denver earthquakes per month. There have not been any significant earthquakes in Denver since injection of waste water into the disposal well ceased in 1965.

into the disposal well per month. Obviously, a high degree of correlation between the two exists, and the correlation is particularly convincing considering that during the time when no waste fluids were injected, earthquake activity decreased dramatically. The area beneath the Rocky Mountain Arsenal consists of highly fractured gneiss overlain by sedimentary rocks. When water was pumped into these fractures, it decreased the friction on opposite sides of the fractures and, in

essence, lubricated them so that movement occurred, causing the earthquakes that Denver experienced.

Experiments conducted in 1969 at an abandoned oil field near Rangely, Colorado, confirmed the arsenal hypothesis. Water was pumped in and out of abandoned oil wells, the pore-water pressure in these wells was measured, and seismographs were installed in the area to measure any seismic activity. Monitoring showed that small earthquakes were occurring in the area when fluid was injected and that earthquake activity declined when the fluids were pumped out. What the geologists were doing was starting and stopping earthquakes at will, and the relationship between pore-water pressures and earthquakes was established.

Based on these results, some geologists have proposed that fluids be pumped into the locked segments of active faults to cause small- to moderate-sized earthquakes. They think this would relieve the pressure on the fault and prevent a major earthquake from occurring. Although this plan is intriguing, it also has many potential problems. For instance, there is no guarantee that only a small earthquake might result. Instead, a major earthquake might occur, causing tremendous property damage and loss of life. Who would be responsible? Certainly, a great deal more research is needed before such an experiment is performed, even in an area of low population density.

It appears that until such time as earthquakes can be accurately predicted or controlled, the best means of defense is careful planning and preparation (Table 9.5).

Table 9.5

What You Can Do to Prepare for an Earthquake

Anyone who lives in an area that is subject to earthquakes or who will be visiting or moving to such an area can take certain precautions to reduce the risks and losses resulting from an earthquake.

Before an earthquake:

1. Become familiar with the geologic hazards of the area where you live and work.

2. Make sure your house is securely attached to the foundation by anchor bolts and that the walls, floors, and roof are all firmly connected together.

3. Heavy furniture such as bookcases should be bolted to the walls; semiflexible natural gas lines should be used so that they can give without breaking; water heaters and furnaces should be strapped and the straps bolted to wall studs to prevent gas-line rupture and fire. Brick chimneys should have a bracket or brace that can be anchored to the roof.

4. Maintain a several-day supply of freshwater and canned foods, and keep a fresh supply of flashlight and radio batteries as well as a fire extinguisher.

5. Maintain a basic first-aid kit, and have a working knowledge of first-aid procedures.

6. Learn how to turn off the various utilities at your house.

7. Above all, have a planned course of action for when an earthquake strikes.

During an earthquake:

1. Remain calm and avoid panic.

2. If you are indoors, get under a desk or table if possible, or stand in an interior doorway or room corner as these are the structurally strongest parts of a room; avoid windows and falling debris.

3. In a tall building, do not rush for the stairwells or elevators.

4. In an unreinforced or other hazardous building, it may be better to get out of the building rather than to stay in it. Be on the alert for fallen power lines and the possibility of falling debris.

5. If you are outside, get to an open area away from buildings if possible.

6. If you are in an automobile, stay in the car, and avoid tall buildings, overpasses, and bridges if possible.

After an earthquake:

1. If you are uninjured, remain calm and assess the situation.

2. Help anyone who is injured.

3. Make sure there are no fires or fire hazards.

4. Check for damage to utilities and turn off gas valves if you smell gas.

5. Use your telephone only for emergencies.

6. Do not go sightseeing or move around the streets unnecessarily.

7. Avoid landslide and beach areas.

8. Be prepared for aftershocks.

Chapter Summary

1. Earthquakes are vibrations of Earth, caused by the sudden release of energy usually along a fault.

2. The elastic rebound theory holds that pressure builds in rocks on opposite sides of a fault until the strength of the rocks is exceeded and rupture occurs. When the rocks rupture, stored energy is released as they snap back to their original position.

3. Seismology is the study of earthquakes. Earthquakes are recorded on seismographs, and the record of an earthquake is a seismogram.

4. The point where energy is released is an earthquake's focus, whereas its epicenter is vertically above the focus on the surface.

5. Approximately 80% of all earthquakes occur in the circum-Pacific belt, 15% within the Mediterranean-Asiatic belt, and the remaining 5% mostly in the interior of plates or along oceanic spreading-ridge systems.

6. The two types of body waves are P-waves and S-waves. Both travel through Earth, although S-waves do not travel through liquids. P-waves are the fastest and are compressional, whereas S-waves are shear. Surface waves travel along or just below the surface. The two types of surface waves are Rayleigh and Love waves.

7. The epicenter of an earthquake is located by the use of a time–distance graph of the P- and S-waves from any given distance. Three seismographs are needed to locate the epicenter.

8. Intensity is a measure of the kind of damage done by an earthquake and is expressed by values from I to XII in the Modified Mercalli Intensity Scale.

9. Magnitude measures the amount of energy released by an earthquake and is expressed in the Richter Magnitude Scale. Each increase in the magnitude number represents about a 30-fold increase in energy released. The Seismic-Moment Magnitude Scale more accurately estimates the energy released during very large earthquakes.

10. Ground shaking is the most destructive of all earthquake hazards. The amount of damage done by an earthquake depends on the geology of the area, the type of building construction, the magnitude of the earthquake, and the duration of shaking.

11. Tsunami are seismic sea waves produced by earthquakes, submarine landslides, and eruptions of volcanoes at sea. They can do tremendous damage to coastlines.

12. Seismic risk maps are helpful in making long-term predictions about the severity of earthquakes based on past occurrences.

13. Earthquake precursors include changes preceding an earthquake that can be used to predict earthquakes. Precursors include seismic gaps, changes in surface elevation, tilting, fluctuations in water-well levels, and anomalous animal behavior.

14. A variety of earthquake research programs are underway in the United States, Japan, Russia, and China. Studies indicate that most people would probably not heed a short-term earthquake warning.

15. Fluid injection into locked segments of an active fault holds promise as a possible means of earthquake control.

Important Terms

aftershock
body wave
dilatancy model
earthquake
elastic rebound theory
epicenter
focus
intensity

Love wave (L-wave)
magnitude
Modified Mercalli Intensity Scale
Moment Magnitude Scale
precursor
P-wave
Rayleigh wave (R-wave)
Richter Magnitude Scale

seismic risk map
seismograph
seismology
surface wave
S-wave
time–distance graph
tsunami

1. An earthquake's epicenter is:
   a. _____ usually in the lower part of the mantle;
   b. _____ a point on the surface directly above the focus;
   c. _____ determined by analyzing surface wave arrival times at seismic stations;
   d. _____ a measure of energy released during an earthquake;
   e. _____ the damage corresponding to a value of IV on the Modified Mercalli Intensity Scale.

2. Earthquakes following a large earthquake are known as:
   a. _____ precursors;
   b. _____ tsunami;
   c. _____ seismic gaps;
   d. _____ aftershocks;
   e. _____ seismic moments.

3. Earthquake waves travel at different speeds, so they arrive at a seismograph in which order?
   a. _____ surface wave first followed by L-wave and R-wave;
   b. _____ S-wave first followed by R-wave and P-wave;
   c. _____ L-wave first followed by P-wave and R-wave;
   d. _____ body wave first followed by surface wave and S-wave;
   e. _____ P-wave first followed by S-wave and surface wave.

4. With few exceptions, the most damaging earthquakes are:
   a. _____ deep focus;
   b. _____ caused by volcanic eruptions;
   c. _____ those occurring along spreading ridges;
   d. _____ shallow focus;
   e. _____ those with Richter magnitudes of about 3.

5. It would take about _____ earthquakes with a Richter magnitude of 3 to equal the energy released in one earthquake with a magnitude of 6.
   a. _____ 9;
   b. _____ 2,000,000;
   c. _____ 27,000;
   d. _____ 30;
   e. _____ 250.

6. A tsunami is a:
   a. _____ part of a fault with a seismic gap;
   b. _____ precursor to an earthquake;
   c. _____ seismic sea wave;
   d. _____ particularly large and destructive earthquake;
   e. _____ an earthquake with a focal depth exceeding 300 km.

7. A qualitative assessment of the damage done by an earthquake is expressed by:
   a. _____ intensity;
   b. _____ dilatancy;
   c. _____ seismicity;
   d. _____ magnitude;
   e. _____ liquefaction.

8. Most earthquakes take place in the:
   a. _____ spreading-ridge zone;
   b. _____ Mediterranean-Asiatic belt;
   c. _____ rifts in continental interiors;
   d. _____ circum-Pacific belt;
   e. _____ Appalachian fault zone.

9. In which one of the following areas would you most likely experience an earthquake?
   a. _____ England;
   b. _____ Kansas;
   c. _____ Germany;
   d. _____ Florida;
   e. _____ Japan.

10. According to the elastic rebound theory:
    a. _____ most large earthquakes take place at divergent plate boundaries;
    b. _____ body waves cause more damage than surface waves;
    c. _____ seismic risk maps can be used to assess the probability of an earthquake occurring in a particular region;
    d. _____ a time–distance graph is needed to find an earthquake's focus;
    e. _____ rocks bend, then fracture and snap back into their original position.

11. Which one of the following statements is correct?
    a. _____ Tsunami are caused by especially high tides and earthquakes with epicenters on land.
    b. _____ Monitoring precursors allows seismologists to make accurate short-range predictions of earthquakes.
    c. _____ The Richter Magnitude Scale is the most useful measure of an earthquake's size for insurance companies.
    d. _____ The time interval between arrivals of P- and S-waves at a seismic station depends on distance from an earthquake's focus.
    e. _____ The safest kind of building to be in during an earthquake is one constructed of stones cemented together.

12. A P-wave is one in which:
    a. _____ movement is perpendicular to the direction of wave travel;
    b. _____ Earth's surface moves as a series of waves;
    c. _____ materials move forward and back along a line in the same direction that the wave moves;
    d. _____ large waves crash onto a shoreline following a submarine earthquake;
    e. _____ movement at the surface is similar to that in water waves.

13. What are precursors, and how can they be used to predict earthquakes?

14. Explain fully how an earthquake's epicenter is located.

15. Explain where and how intermediate- and deep-focus earthquakes are generated.

16. Describe how a suspended mass can be used to detect earthquake waves.

17. What are the differences between earthquake intensity and magnitude? Why was the Seismic-Moment Magnitude Scale developed?

18. Fully describe how a tsunami is generated, how it travels, and what impact it has on shorelines.

19. Describe body waves and surface waves. That is, how fast do they travel, through what kinds of materials do they move, and what kinds of each wave are recognized?

20. Fully describe the phenomenon of liquefaction.

21. What plate tectonic settings account for earthquakes along the west coasts of North America and South America? In which of these two areas would you expect deep-focus earthquakes?

22. Describe the dilatancy model and how it might help predict earthquakes.

23. How does the elastic rebound theory account for energy released during an earthquake?

24. What is the difference between the focus and epicenter of an earthquake?

Points to Ponder

1. From the following arrival times of P- and S-waves shown in the chart below and the graph in Figure 9.15, calculate how far away from each seismograph station the earthquake occurred. How would you determine the epicenter of this earthquake?

2. Using the graph in Figure 9.19, answer the following question. A seismograph in Berkeley, California, records the arrival time of an earthquake's P-waves at 6:59:54 P.M. and the S-waves at 7:00:02 P.M. The maximum amplitude of the S-waves as recorded on the seismogram was 75 mm. What was the magnitude of the earthquake, and how far away from Berkeley did it occur?

3. Discuss why insurance companies use the qualitative Modified Mercalli Intensity Scale instead of the quantitative Richter Magnitude Scale in classifying earthquakes.

4. On June 8, 1994, a deep-focus (600 km) 8.2 magnitude earthquake occurred beneath Bolivia and was felt as far away in North America as Seattle, Washington, Minneapolis, Minnesota, and Toronto, Canada. Some damage was done in the area of the epicenter, but no deaths were reported. Discuss why this particular earthquake should be felt so far away, while other equally strong earthquakes are not felt much beyond the area of the epicenter. Why do you think so little damage was done and no deaths occurred from such a strong earthquake? Finally, what do you think was the cause of this deep-focus earthquake, and what can it tell geologists about the interior of Earth?

	Arrival Time of P-Wave	Arrival Time of S-Wave
Station A:	2:59:03 P.M.	3:04:03 P.M.
Station B:	2:51:16 P.M.	3:01:16 P.M.
Station C:	2:48:25 P.M.	2:55:55 P.M.
Distance from the earthquake:		
Station A:		
Station B:		
Station C:		

➤ CALIFORNIA INSTITUTE OF TECHNOLOGY SEISMOLOGICAL LABORATORY

This site provides much information about earthquakes and some of the projects going on in the various labs at the California Institute of Technology.
1. Click on the *Record of the Day* site. Where was that earthquake? What was its magnitude?
2. Click on the *Southern California Earthquake Center Data Center* site. Check out the current activity and location of earthquakes in the southern California area. Where has most of the activity been located for the past week?

➤ U.S. GEOLOGICAL SURVEY PASADENA FIELD OFFICE

This site contains many links to other Web sites containing information about earthquake activity in the southern California region as well as elsewhere in the world. Click on the *Earthquake Information* site under the *Earthquakes, Monitoring & Publications* section. This site contains a listing on earthquake information for southern California and other areas. Click on any of the sites, and check out the latest earthquake activity in that region.

➤ U.S. GEOLOGICAL SURVEY NATIONAL EARTHQUAKE INFORMATION CENTER

This site provides current and general seismicity information as well as links to other sites concerned with earthquake activity. Click on the *Current Seismicity Information* site. Click on one of the *Current Seismicity* map sites such as *World* to see the distribution of earthquakes, their depth, and magnitude. In the *World* site, where is the greatest amount of seismic activity taking place? When and where did the largest magnitude earthquake most recently occur? Given the information in this chapter, does the current distribution of earthquakes reflect where you would expect most earthquakes to take place?

➤ ABOUT.COM

At the About.Com site index, click *G*, which brings up a menu. Click *Geology,* and then under NetLinks click *Earthquakes.* This brings up a list with the topics "Current Earthquakes," "Earthquake Hazards," "Earthquake Prediction," "Earthquakes in Your Region," "Historic Earthquakes," and "Seismology." Listed under each of these headings are several other sites of interest. For instance, under "Earthquake Hazards," one can learn about tsunami.

➤ 1906 SAN FRANCISCO EARTHQUAKE

This site, maintained by the U.S. Geological Survey, contains a variety of topics related to the devastating 1906 San Francisco earthquake, including eyewitness accounts, many photographs, and information about possible future earthquakes.

➤ THE VIRTUAL TIMES: THE NEW MADRID EARTHQUAKE

A great site for information on the 1811 and 1812 earthquakes in the New Madrid, Missouri, region. An excellent opportunity to learn about large earthquakes that occurred far from a plate boundary.

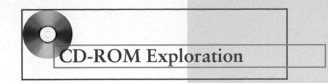

CD-ROM Exploration

➤ Exploring your *Earth Systems Today* CD-ROM will add to your understanding of the material in this chapter.

TOPIC: EARTH'S DISASTERS

MODULE: EARTHQUAKES/TSUNAMI

Explore activities in this module to see if you can discover the following for yourself:

Using activities in this module, examine the distribution of earthquakes in space (including depth) and time (i.e., 1975–present). How are earthquakes related to plate boundaries?

Using activities in this module, study those regions of Earth known for generating tsunami and those regions of Earth known to experience the effects of tsunami.

Using activities in this module, find an earthquake epicenter. In addition, observe current seismic activity in Alaska. What regions of the U.S. are at greatest seismic risk?

Chapter 10

Probing Earth's interior. This 30-story building on the Kola Peninsula in northwestern Russia houses a drill rig that has penetrated more than 12 km into Earth's crust.

OBJECTIVES

At the end of this chapter, you will have learned that

- Geologists use seismic waves to determine Earth's internal structure.
- Earth has a central core, overlain by a thick mantle, and a thin outer layer of crust.
- Seismic tomography is yielding a more refined view of Earth's internal structure.

- Geologists determine the density, composition, and structure of the core, mantle, and crust based on studies of seismic waves, Earth's overall density, meteorites, and inclusions in volcanic rocks.
- Earth possesses considerable internal heat that continuously escapes at the surface.

- The force of gravity varies depending on several factors, thus accounting for gravity anomalies.
- The principle of isostasy holds that Earth's crust is buoyed up by a denser medium below.
- Earth possesses a magnetic field that varies through time and in intensity and direction.

Earth's Interior

Prologue

During most of historic time, people perceived Earth's interior as an underground world of vast caverns, heat, and sulfur gases, populated by demons or the souls of those waiting to be judged (Figure 10.1). The Greek philosopher Aristotle (384–322 B.C.) thought that winds and fires in subterranean cavities caused earthquakes. Romans thought that Vulcan, the god of fire, had several underground workshops where his beating on anvils caused the ground to rumble and volcanoes to erupt. By the 1800s, scientists had some sketchy ideas about Earth's internal structure, but outside scientific circles some bizarre ideas were proposed, some of which are still popular.

In 1869, Cyrus Reed Teed claimed Earth was hollow and that humans lived on the inside. And in 1913, Marshall B. Gardner held that Earth is a large, hollow sphere with a 1300-km-thick outer shell surrounding a central sun. He also claimed that Eskimos and the now extinct mammoths (relatives of elephants) dwelled within this hollow Earth. Even today, one still sees claims that huge openings exist at the poles, and that Arctic and Antarctic explorers have traveled from one pole to the other through these openings.

Figure 10.1
In 1678, Athanasius Kircher (1602–1680) published *Mundus Subterraneus*, which contained this drawing showing what he described as the "ideal system of subterranean fire cells from which volcanic mountains arise, as it were, like vents."

Although making no claim to present a reliable picture of the interior, Jules Verne's 1864 novel *Journey to the Center of the Earth* described the adventures of Professor Hardwigg, his nephew, and an Icelandic guide as they descended into the interior through the crater of Mount Sneffels in Iceland. During their travels, they followed a labyrinth of passageways until they finally arrived 140 km below the surface. Here they encountered a vast cavern containing "the central sea," illuminated by some electrical phenomenon related to the northern lights. Along the margins of the sea, they saw forests of prehistoric ferns and palms and a herd of mastodons (extinct relatives of elephants) complete with a gigantic human shepherd. Living in the central sea were large Mesozoic-aged marine reptiles and huge turtles. Their adventure ended when the trio was carried upward to the surface on a raft by a rising plume of water.

When Jules Verne wrote this novel in 1864, scientists knew what Earth's average density was, and that pressure and temperature increases with depth ruled out the possibility of caverns deep within the planet. But little else was known, although humans had probed beneath the surface with mines and wells for centuries. However, even the gold mines in South Africa, the deepest in the world, extend to only about 3 km below the surface, barely penetrating the uppermost part of Earth's crust. The deepest drill holes reach about 15 km, but this is only about 0.2% of the distance to Earth's center. To put this in perspective, if Earth were the size of an apple, a 15-km-deep drill hole would be roughly equivalent to a pinprick penetrating less than halfway through the apple's skin!

In 1958, an ambitious project to drill through the oceanic crust to the mantle was launched. Known as *Mohole,* the project attracted considerable attention and support from scientists and the public, and in 1962 the U.S. Congress appropriated more than $40 million to finance it. Unfortunately, *Mohole* encountered technological difficulties, and by 1966 funding had dried up. Soon thereafter the project was quickly forgotten.

Even though Earth's interior cannot be observed directly, scientists have a reasonably good idea of its composition and structure. No vast openings or passageways exist, as in Jules Verne's story; the deepest known caverns extend to only about 1500 m deep. And even at the modest depth of 140 km that Professor Hardwigg and his companions are supposed to have descended, the pressure and temperature are so great that rock actually flows when subjected to a force, although it remains solid. In deep mines the rocks are under such great pressure that rock bursts and popping are constant problems (see Chapter 5). In short, the behavior of solids at depth where temperature and pressure are great is very different from their brittle behavior at or near the surface.

INTRODUCTION

Earth's interior is so inaccessible that most people think little about it. One can appreciate the stunning beauty of the northern lights and yet be completely unaware that they exist because of an interaction of Earth's internally generated magnetic field and the solar wind, a continuous stream of electrically charged particles emanating from the Sun. Much of Earth's geologic activity—including volcanism, earthquakes, moving plates, and the origin of mountains—is caused by internal heat. In fact, the slow release of heat from the interior is one major factor that makes Earth a dynamic planet.

Scientists have known for more than 200 years that planet Earth is not homogeneous throughout. Indeed, Sir Isaac Newton (1642–1727) noted in a study of the planets that Earth's average density— that is, its mass per unit volume—is 5.0 to 6.0 g/cm^3 (pure water has a density of 1 g/cm^3). In 1797 Henry Cavendish calculated a density value very close to the 5.5 g/cm^3 now accepted. To accurately determine Earth's density, first its size must be known. In the third century B.C., a Greek librarian performed a clever experiment and determined that Earth's circumference is about 40,000 km—a figure only slightly different from the one now accepted. From this, one can calcu-

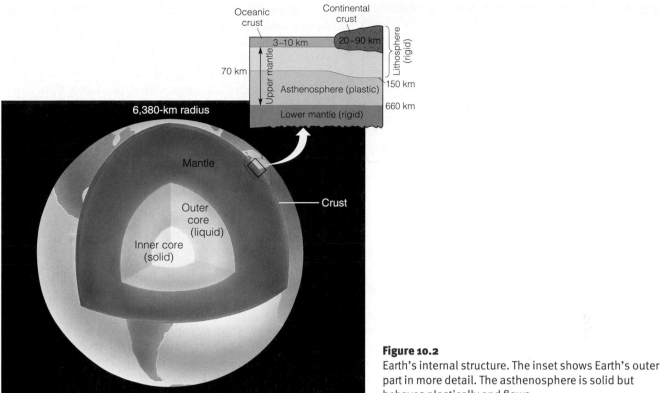

Oceanic crust
Continental crust
3–10 km
20–90 km
70 km
Upper mantle
Asthenosphere (plastic)
Lower mantle (rigid)
Lithosphere (rigid)
150 km
660 km

6,380-km radius

Mantle

Outer core (liquid)

Inner core (solid)

Crust

Figure 10.2
Earth's internal structure. The inset shows Earth's outer part in more detail. The asthenosphere is solid but behaves plastically and flows.

late the diameter and radius, and then determine volume, of the nearly spherical Earth. But density can be found only indirectly by, for instance, comparing the gravitational attraction between Earth and its Moon and between metal spheres of known mass. After carrying out these calculations, scientists derived a value of 5.52 g/cm³. Given that Earth's average density of 5.5 g/cm³ is considerably greater than that of surface rocks, most of which range from 2.5 to 3.0 g/cm³, much of the interior must consist of materials that are denser than average.

Earth is generally depicted as consisting of concentric layers differing in composition and density, with each separated from adjacent layers by rather distinct

boundaries (Figure 10.2). Recall that the outermost layer, or the **crust,** is Earth's thin skin. Below the crust and extending about halfway to Earth's center is the **mantle,** which comprises more than 80% of Earth's volume (Table 10.1). The central part is the **core,** which is divided into a solid inner core and a liquid outer part (Figure 10.2).

Because no direct observations of Earth's interior can be made, this model of its internal structure is based on indirect evidence, mostly from the study of seismic waves. Nevertheless, the model is widely accepted by scientists and is becoming increasingly refined as more sophisticated methods of probing what some call "inner space" are developed.

Table 10.1

Earth Data				
	VOLUME (THOUSANDS OF KM³)	PERCENTAGE OF TOTAL VOLUME	MASS (TRILLIONS OF METRIC TONS)	PERCENTAGE OF TOTAL MASS
Inner core	7,512,800	0.70%	19,000,000,000	31.79%
Outer core	169,490,000	15.68		
Mantle	896,990,000	83.02	40,500,000,000	67.77
Continental crust	4,760,800	0.44	250,000,000	0.42
Oceanic crust	1,747,200	0.16		
Atmosphere, water, ice			14,351,000	0.02

SEISMIC WAVES—CLUES TO EARTH'S INTERNAL STRUCTURE

The behavior and travel times of P- and S-waves provide geologists with much information about Earth's internal structure. Seismic waves travel outward as wave fronts from their source areas, although it is most convenient to depict them as *wave rays,* which are lines showing the direction of movement of small parts of wave fronts (Figure 10.3). Any disturbance such as a passing train or construction equipment can cause seismic waves, but only those generated by large earthquakes, explosive volcanism, asteroid impacts, and nuclear explosions can travel completely through Earth.

As we noted in Chapter 9, the velocities of P- and S-waves are determined by the density and elasticity of the materials they travel through, both of which increase with depth. Wave velocity is slowed by increasing density but increases in materials with greater elasticity. Because elasticity increases with depth faster than density, a general increase in the velocity of seismic waves takes place as they penetrate to greater depths. P-waves travel faster than S-waves under all circumstances, but unlike P-waves, S-waves cannot be transmitted through a liquid because liquids have no shear strength (rigidity)—liquids simply flow in response to shear stress.

If Earth were a homogeneous body, P- and S-waves would travel in straight paths as shown in Figure 10.4a. But as a seismic wave travels from one material into another of different density and elasticity, its velocity and direction of travel change. That is, the wave is bent, a phenomenon known as **refraction,** in much the same way as light waves are refracted as they pass from air into a more dense medium such as water. Because seismic waves pass through materials of differing density and elasticity, they are continually refracted so that their paths are curved; wave rays travel only in a straight line and are not refracted when their direction of travel is perpendicular to a boundary (Figure 10.5).

In addition to refraction, seismic rays are also **reflected,** much as light is reflected from a mirror. When seismic waves encounter a boundary separating materials of different density or elasticity, some of the rays' energy is *reflected* back to the surface (Figure 10.5). If

Figure 10.3
Seismic wave fronts move out in all directions from their source, the focus of an earthquake in this example. Wave rays are lines drawn perpendicular to wave fronts.

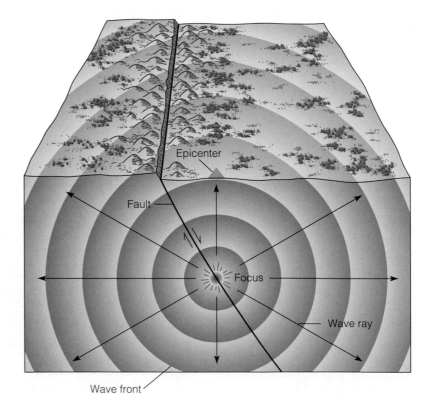

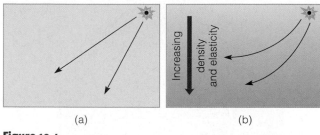

Figure 10.4
(a) If Earth had the same composition and density throughout, seismic wave rays would follow straight paths. (b) Because density and elasticity increase with depth, wave rays are continually refracted so that their paths are curved.

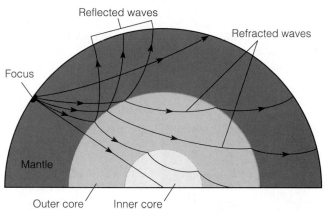

Figure 10.5
Refraction and reflection of P-waves. When seismic waves encounter a boundary separating Earth materials of different density or elasticity, they are refracted and some of their energy is reflected back to the surface. Note that the only wave ray not refracted is the one perpendicular to the mantle–core boundary and outer core–inner core boundary.

we know the wave velocity and the time required for it to travel from its source to the boundary and back to the surface, we can calculate the depth of the reflecting boundary. Such information is useful in determining not only the depths of the various layers but also the depths of sedimentary rocks that may contain petroleum. Seismic reflection is a common tool used in petroleum exploration.

Although changes in seismic wave velocity occur continuously with depth, P-wave velocity increases suddenly at the base of the crust and decreases abruptly at a depth of about 2900 km (Figure 10.6). These marked changes in seismic wave velocity indicate a boundary called a **discontinuity** across which a significant change in Earth materials or their properties occurs. Such discontinuities are the basis for subdividing Earth's interior into concentric layers.

The contribution of seismology to the study of Earth's interior cannot be overstated. Beginning in the early 1900s, scientists recognized the utility of seismic wave studies and, between 1906 and 1936, largely worked out Earth's internal structure on the basis of these studies.

HOW WAS EARTH'S CORE DISCOVERED?

In 1906 R. D. Oldham of the Geological Survey of India discovered that seismic waves arrived later than expected at seismic stations more than 130 degrees from an earthquake focus. He postulated the existence of a core that transmits seismic waves at a slower rate than shallower Earth materials. We now know that P-wave velocity decreases markedly at a depth of 2900 km, indicating a major discontinuity now recognized as the core–mantle boundary (Figure 10.6).

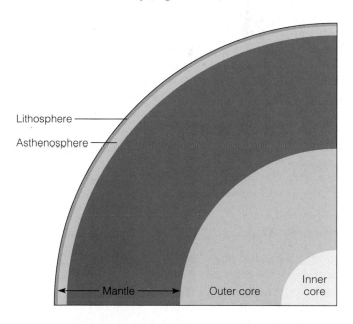

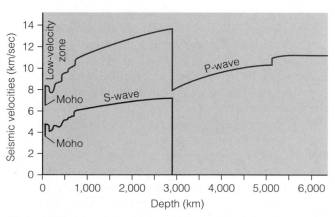

Figure 10.6
Profiles showing seismic wave velocities versus depth. Several discontinuities are shown across which seismic wave velocities change rapidly.

The sudden decrease in P-wave velocity at the core–mantle boundary causes P-waves entering the core to be refracted in such a way that very little P-wave energy reaches the surface in the area between 103 and 143 degrees from an earthquake focus (Figure 10.7). This area in which little P-wave energy is recorded by seismographs is a **P-wave shadow zone**; it was discovered by the German seismologist Beno Gutenberg in 1914.

The P-wave shadow zone is not a perfect shadow zone. That is, some weak P-wave energy reaches the surface within the zone. Several hypotheses were proposed to explain this phenomenon, but all were rejected by the Danish seismologist Inge Lehmann (Figure 10.8a), who in 1936 postulated that the core is not entirely liquid, as previously thought. She demonstrated that reflection from a solid inner core could account for the arrival of weak P-waves in the P-wave shadow zone (Figure 10.8b). Lehmann's proposal of a solid inner core was quickly accepted by seismologists.

In 1926 the British physicist Harold Jeffreys realized that S-waves were not simply slowed by the core but were completely blocked by it. So, besides a P-wave shadow zone, a much larger and more complete **S-wave shadow zone** also exists (Figure 10.9). At locations greater than 103 degrees from an earthquake focus, no S-waves are recorded, indicating that S-waves cannot be transmitted through the core. S-waves will not pass through a liquid, so it seems that the outer core must be liquid or behave as a liquid. The inner core, however, is thought to be solid because P-wave velocity increases at the base of the outer core.

Density and Composition of the Core

Below the mantle at a depth of 2900 km and extending to Earth's center lies the core. It has a diameter of 6960 km (Figure 10.2), about the same as the total diameter of Mars. It constitutes 16.4% of Earth's volume and nearly one-third of its mass (Table 10.1).

We can estimate the core's density and composition by using seismic evidence and laboratory experiments. For instance, a device known as a diamond-anvil pressure cell has been developed in which small samples are studied while being subjected to pressures and temperatures similar to those in the core. Furthermore, meteorites, which are thought to represent remnants of the material from which the solar system formed, can be used to make estimates of density and composition. For example, the irons—meteorites composed of iron and nickel alloys—may represent the differentiated interiors of large asteroids and approximate the density and composition of Earth's core. The density of the outer core varies from 9.9 to 12.2 g/cm³, and that of the inner core ranges from 12.6 to 13.0 g/cm³ (Table 10.2). At Earth's center, the pressure is equivalent to about 3.5 million times normal atmospheric pressure.

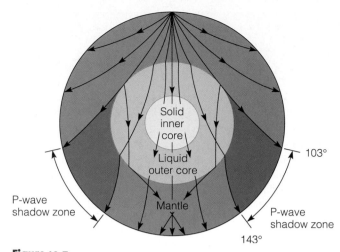

Figure 10.7
P-waves are refracted so that no direct P-wave energy reaches Earth's surface in the P-wave shadow zone.

Figure 10.8
(a) Inge Lehmann, a Danish seismologist who in 1936 postulated that Earth has a solid inner core. (b) Lehmann proposed that reflection from an inner core could explain the arrival of weak P-wave energy in the P-wave shadow zone.

(a)

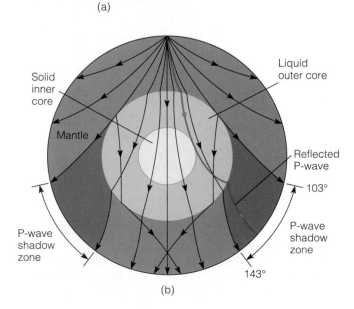

(b)

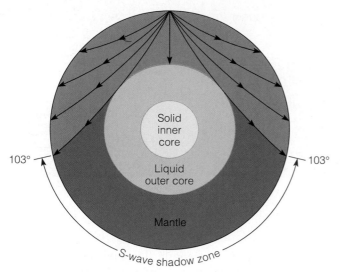

Figure 10.9
The presence of an S-wave shadow zone indicates that S-waves are being blocked within Earth.

Table 10.2

Earth's Composition and Density

	COMPOSITION	DENSITY (G/CM³)
Inner core	Iron with 10–20% nickel	12.6–13.0
Outer core	Iron with perhaps 12% sulfur, silicon, oxygen, nickel, and potassium	9.9–12.2
Mantle	Peridotite (composed mostly of ferromagnesian silicates)	3.3–5.7
Oceanic crust	Upper part basalt, lower part gabbro	~3.0
Continental crust	Average composition of granodiorite	~2.7

The core cannot be composed of minerals common at the surface because, even under the tremendous pressures at great depth, they would still not be dense enough to yield an average density of 5.5 g/cm³ for Earth. Both the outer and inner core are thought to be composed largely of iron, but pure iron is too dense to be the sole constituent of the outer core. It must be "diluted" with elements of lesser density. Laboratory experiments and comparisons with iron meteorites indicate that perhaps 12% of the outer core consists of sulfur and possibly some silicon, oxygen, nickel, and potassium (Table 10.2).

In contrast, pure iron is not dense enough to account for the estimated density of the inner core. Many geologists think that perhaps 10 to 20% of the inner core consists of nickel. These metals form an iron–nickel alloy thought to be sufficiently dense under the pressure at that depth to account for the density of the inner core.

Any model of the core's composition and physical state must explain not only the variations in density but also (1) why the outer core is liquid whereas the inner core is solid and (2) how the magnetic field is generated within the core (discussed later in this chapter). When the core formed during early Earth history, it was probably entirely molten and has since cooled to the point that its interior has crystallized. Indeed, the inner core continues to grow as Earth slowly cools, and liquid of the outer core crystallizes as iron. Recent evidence also indicates that the inner core rotates faster than the outer core, moving about 20 km/year relative to the outer core.

The temperature at the core–mantle boundary is estimated at 2500°C to 5000°C, yet the high pressure within the inner core prevents melting. In contrast, the outer core is under even less pressure, but more impor-

tant than the differences in pressure are compositional differences between the inner and outer core. The sulfur content of the outer core helps depress its melting temperature. An iron–sulfur mixture melts at a lower temperature than does pure iron, or an iron–nickel alloy, so despite the high pressure, the outer core is molten.

EARTH'S MANTLE

Another significant discovery about Earth's interior was made in 1909 when the Yugoslavian seismologist Andrija Mohorovičić detected a discontinuity at a depth of about 30 km. While studying arrival times of P-waves from Balkan earthquakes, Mohorovičić noticed that seismic stations a few hundred kilometers from an earthquake's epicenter were recording two distinct sets of P- and S-waves.

From his observations, Mohorovičić concluded that a sharp boundary separating rocks with different properties exists at a depth of about 30 km. He postulated that P-waves below this boundary travel at 8 km/sec, whereas those above the boundary travel at 6.75 km/sec. When an earthquake occurs, some waves travel directly from the focus to a seismic station, while others travel through the deeper layer and some of their energy is refracted back to the surface (Figure 10.10). Waves traveling through the deeper layer travel farther to a seismic station, but they do so more rapidly and arrive before those in the shallower layer. The boundary identified by Mohorovičić separates the crust from the mantle and is now called the **Mohorovičić discontinuity,** or simply the **Moho.** It is present everywhere except beneath spreading ridges, but its depth varies: Beneath the continents, it ranges from 20 to 90 km, with an average of 35 km; beneath the seafloor, it is 5 to 10 km deep.

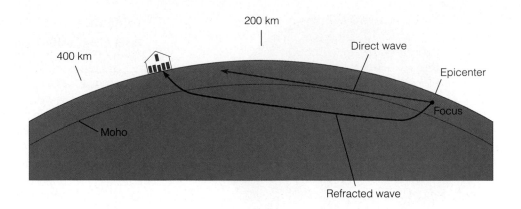

Figure 10.10
Andrija Mohorovičić studied seismic waves and detected a seismic discontinuity at a depth of about 30 km. The deeper, faster seismic waves arrive at seismic stations first, even though they travel farther. This discontinuity, now known as the Moho, is between the crust and mantle.

Structure, Density, and Composition of the Mantle

Although seismic wave velocity in the mantle generally increases with depth, several discontinuities also exist. Between depths of 100 and 250 km, both P- and S-wave velocities decrease markedly (Figure 10.11). This 100- to 250-km-deep layer is the **low-velocity zone;** it corresponds closely to the **asthenosphere,** a layer in which the rocks are close to their melting point and are less elastic, accounting for the observed decrease in seismic wave velocity. The asthenosphere is an important zone because it may be where some magmas are generated.

Furthermore, it lacks strength, flows plastically, and is thought to be the layer over which the plates of the outer, rigid **lithosphere** move.

Even though the low-velocity zone and the asthenosphere closely correspond, they are still distinct. The asthenosphere appears to be present worldwide, but the low-velocity zone is not. In fact, the low-velocity zone appears to be poorly defined or even absent beneath the ancient shields of continents.

Other discontinuities have been detected at deeper levels within the mantle. But unlike those between the crust and mantle or between the mantle and core, these probably represent structural changes in minerals rather

Figure 10.11
Variations in P-wave velocity in the upper mantle and transition zone.

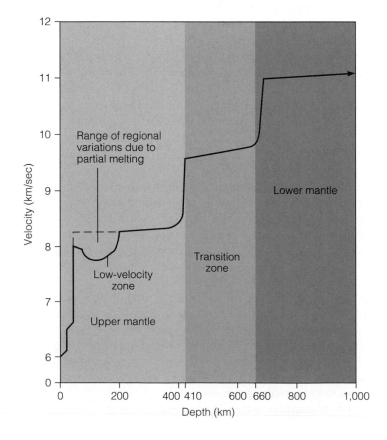

than compositional changes. In other words, geologists think the mantle is composed of the same material throughout, but the structural states of minerals such as olivine change with depth. At a depth of 410 km, seismic wave velocity increases slightly as a consequence of such changes in mineral structure (Figure 10.11). Another velocity increase occurs at about 660 km where the minerals break down into metal oxides, such as FeO (iron oxide) and MgO (magnesium oxide), and silicon dioxide (SiO_2). These two discontinuities define the top and base of a *transition zone* separating the upper mantle from the lower mantle (Figure 10.11).

A decrease in seismic wave velocity in a zone extending 200 to 300 km above the core–mantle boundary is recognized as the D″ layer. Although commonly included within the lower mantle, the D″ layer might be considerably different in composition. Experiments indicate that silicates in the mantle react with liquid from the outer core, thus forming a vertically and laterally heterogeneous layer. Some geologists think that mantle plumes originate in the D″ layer (see Chapters 3 and 12).

Although the mantle's density, which varies from 3.3 to 5.7 g/cm³, can be inferred rather accurately from seismic waves, its composition is less certain. The igneous rock *peridotite* is considered the most likely component. Peridotite contains mostly ferromagnesian silicates (60% olivine and 30% pyroxene) and about 10% feldspars (see Figure 3.11). Peridotite is considered the most likely candidate for three reasons. First, laboratory experiments indicate that it possesses physical properties that would account for the mantle's density and observed rates of seismic wave transmissions. Second, peridotite forms the lower parts of igneous rock sequences thought to be fragments of the oceanic crust and upper mantle emplaced on land (see Chapter 11). And third, peridotite is found as inclusions in volcanic rock bodies such as *kimberlite pipes* that are known to have come from depths of 100 to 300 km. These inclusions appear to be pieces of the mantle (see Perspective 10.1).

SEISMIC TOMOGRAPHY AND EARTH'S INTERIOR

The model of Earth's interior consisting of an iron-rich core and a rocky mantle is probably accurate, but it is also rather imprecise. However, geophysicists have developed a technique called *seismic tomography* that allows them to develop more accurate three-dimensional models of Earth's interior. In seismic tomography, numerous crossing seismic waves are analyzed in much the same way radiologists analyze

CAT (computerized axial tomography) scans. In CAT scans, X rays penetrate the body, and a two-dimensional image of the body's interior is formed. Repeated CAT scans, each from a slightly different angle, are computer-analyzed and stacked to produce a three-dimensional picture.

In a similar fashion, geophysicists use seismic waves to probe Earth's interior. From the time of arrival and distance traveled, the velocity of a seismic ray is computed at a seismic station. Only average velocity is determined, rather than variations in velocity. In seismic tomography, numerous wave rays are analyzed so that "slow" and "fast" areas of wave travel can be detected (Figure 10.12). Recall that seismic wave velocity is controlled partly by elasticity; cold rocks have greater elasticity and therefore transmit seismic waves faster than hot rocks.

Using this technique, geophysicists have detected areas within the mantle at a depth of about 150 km where seismic velocities are slower than expected. These anomalously hot regions lie beneath volcanic areas and beneath the mid-oceanic ridges, where convection cells of rising hot mantle rock are thought to exist. In contrast, beneath the older interior parts of continents, where tectonic activity ceased hundreds of millions or billions of years ago, anomalously cold spots are recognized. In effect, tomographic maps and three-dimensional diagrams show heat variations within Earth.

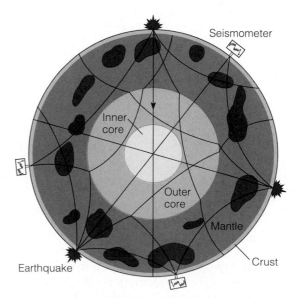

Figure 10.12
Numerous earthquake waves are analyzed to detect areas within Earth that transmit seismic waves faster or slower than adjacent areas. Areas of fast wave travel correspond to "cold" regions (blue), whereas "hot" regions (red) transmit seismic waves more slowly.

Perspective 10.1

Kimberlite Pipes—Windows to the Mantle

D iamonds have been eagerly sought after throughout history, yet until 1870 they were known only from gravel deposits in rivers, having accumulated as placer deposits (see Chapter 6). In 1870, though, the source of diamonds in South Africa was traced to cone-shaped igneous bodies known as *kimberlite pipes,* found near the town of Kimberley (Figure 1). Now we know that kimberlite pipes are the source rocks for most diamonds.

The greatest concentrations of kimberlite pipes are in southern Africa and Siberia, but they are known in many other areas as well. Most of them are found in the ancient stable cratons of continents. In North America, kimberlite pipes have been found in the Canadian Arctic, Colorado, Wyoming, Missouri, Montana, Michigan, and Virginia, and one at Murfreesboro, Arkansas, was briefly worked for diamonds. Diamonds discovered in glacial

deposits in some midwestern states indicate that kimberlite pipes are present farther north. The precise source of these diamonds has not been determined, although some kimberlite pipes have recently been identified in northern Michigan.

Kimberlite pipes are composed of dark gray or blue igneous rock known as *kimberlite,* which contains olivine, a potassium- and magnesium-rich mica, serpentine, calcite, and silica. Some of

Figure 1
Generalized cross section of a kimberlite pipe. Most kimberlite pipes measure less than 500 m in diameter at the surface.

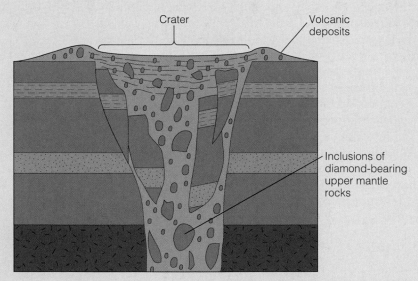

Crater

Volcanic deposits

Inclusions of diamond-bearing upper mantle rocks

Seismic tomography has also yielded additional and sometimes surprising information about the core. For example, the core–mantle boundary is not a smooth surface but has broad depressions and rises extending several kilometers into the mantle. Of course, the base of the mantle possesses the same features in reverse; geophysicists have termed these features *anticontinents* and *antimountains*. It appears that the surface of the core is

continually deformed by sinking and rising masses of mantle material.

As a result of seismic tomography, a much clearer picture of Earth's interior is emerging. It has already given us a better understanding of complex convection within the mantle, including upwelling convection currents thought to be responsible for the movement of lithospheric plates (see Chapter 12).

these rocks contain inclusions of *peridotite* that are thought to represent pieces of the mantle brought to the surface during the explosive volcanic eruptions that form kimberlite pipes.

If peridotite inclusions are in fact pieces of the mantle, they indicate that the magma in kimberlite pipes originated at a depth of at least 30 km. Indeed, the presence of diamonds and the structural form of the silica in the kimberlite can be used to establish both minimum and maximum depths for the origin of the magma. Diamond and graphite are different crystalline forms of carbon (see Figure 2.10), but diamond forms only under high-pressure, high-temperature conditions. The presence of diamond and the absence of graphite in kimberlite indicate that such conditions existed where the magma originated.

The calculated geothermal gradient and the pressure increase with depth beneath the continents are shown in Figure 2. Laboratory experiments have established a diamond–graphite inversion curve showing the pressure–temperature conditions at which graphite is favored over diamond. According to the data in Figure 2, the intersection of the diamond–graphite inversion curve with the geothermal gradient indicates that kimberlite magma came from a minimum depth of about 100 km.

Diamond can establish only a minimum depth for kimberlite because it is stable at any pressure greater than that at a depth of 100 km. The silica found in kimberlite, in contrast, is a form that indicates a maximum depth of about 300 km. Quartz is the form of silica found under low-pressure, low-temperature conditions. Under great pressure, however, the crystal structure of quartz changes to its high-pressure equivalent, coesite, and at even greater pressure it changes to stishovite.* Kimberlite pipes contain coesite but no stishovite, indicating that the kimberlite magma must have come from a depth of less than 300 km as indicated by the intersection of the coesite–stishovite inversion curve with the geothermal gradient (Figure 2).

The fact that continental crust is at most 90 km thick and averages about 35 km indicates that the magma injected to form kimberlite pipes came from the mantle. Accordingly, peridotite inclusions in kimberlites is good evidence that the mantle is composed of this rock type.

*Coesite and stishovite are also known from other high-pressure environments such as meteorite impact sites.

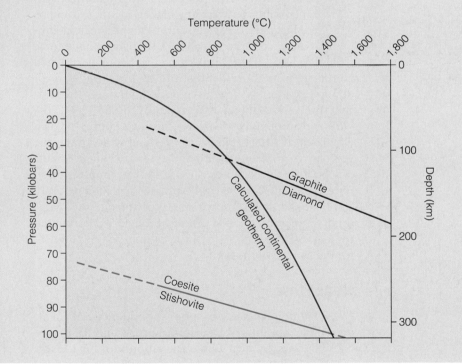

Figure 2
The forms of carbon and silica in kimberlite pipes provide information on the depth at which the magma formed. The presence of diamond and coesite in kimberlite indicates that the magma probably formed at a depth of between 100 and 300 km, as shown by the intersection of the calculated continental geotherm with the graphite–diamond and coesite–stishovite inversion curves.

EARTH'S OUTERMOST PART— CONTINENTAL AND OCEANIC CRUST

Earth's crust is the most accessible and best studied of its concentric layers, but it is also the most complex both chemically and physically. Whereas the core and mantle seem to vary mostly in a vertical dimension, the crust shows considerable vertical and lateral variation. (More lateral variation exists in the mantle than was once thought.) The crust along with that part of the upper mantle above the low-velocity zone constitutes the *lithosphere* of plate tectonic theory.

Two types of crust are recognized—continental crust and oceanic crust—both of which are less dense than the underlying mantle. **Continental crust** is the more complex, consisting of a wide variety of igneous, sedimentary, and metamorphic rocks. It is generally described as "granitic," meaning that its overall composition is similar to that of granitic rocks. Specifically, its overall composition corresponds closely to that of granodiorite, an igneous rock having a chemical composition between granite and diorite (see Figure 3.10).

Continental crust varies in density depending on rock type, but with the exception of metal-rich rocks, such as iron ore deposits, most rocks have densities of 2.0 to 3.0 g/cm³, and the overall density is about 2.70 g/cm³ (Table 10.2). P-wave velocity in the continental crust is about 6.75 km/sec; at the base of the crust, P-wave velocity abruptly increases to about 8 km/sec.

Continental crust averages about 35 km thick but is much thinner in such areas as the Rift Valleys of East Africa and a large area called the Basin and Range Province in the western United States. The crust in these areas is being stretched and thinned in what appear to be the early stages of rifting and is as little as 20 km thick. In contrast, continental crust beneath mountain ranges is much thicker and projects deep into the mantle. Beneath the Himalayas of Asia, the crust is as much as 90 km thick. Crustal thickening beneath mountain ranges is an important point that will be discussed in the section on isostasy later in the chapter.

Although variations also occur in **oceanic crust**, they are not as distinct as those for the continental crust. Oceanic crust varies from 5 to 10 km thick, being thinnest at spreading ridges. It is denser than continental crust, averaging about 3.0 g/cm³, and transmits P-waves at about 7 km/sec. Just as beneath the continental crust, P-wave velocity increases at the Moho. The P-wave velocity of oceanic crust is what one would expect if it were composed of basalt. Direct observations of oceanic crust from submersibles and deep-sea drilling confirm that its upper part is indeed composed of basalt. The lower part of the oceanic crust is composed of gabbro, the intrusive equivalent of basalt. (See Chapter 11 for a more detailed description of the oceanic crust.)

HOW HOT IS EARTH'S INTERIOR?

During the nineteenth century, scientists realized that the temperature in deep mines increases with depth. Indeed, very deep mines must be air-conditioned so that the miners can survive. More recently, the same trend has been observed in deep drill holes, but even in these we can measure temperatures directly down to a depth of only a few kilometers. This temperature increase with depth, or **geothermal gradient,** near the

surface is about 25°C/km, although it varies from area to area. In areas of active or recently active volcanism, the geothermal gradient is greater than in adjacent non-volcanic areas, and temperature rises faster beneath spreading ridges than elsewhere beneath the seafloor.

Most of Earth's internal heat is generated by radioactive decay, especially the decay of isotopes of uranium and thorium and to a lesser degree of potassium 40. When these isotopes decay, they emit energetic particles and gamma rays that heat surrounding rocks. Because rock is such a poor conductor of heat, it takes little radioactive decay to build up considerable heat, given enough time.

Unfortunately, the geothermal gradient is not useful for estimating temperatures at great depth. If we were simply to extrapolate from the surface downward, the temperature at 100 km would be so high that, despite the great pressure, all known rocks would melt. Yet except for pockets of magma, it appears that the mantle is solid rather than liquid because it transmits S-waves. Accordingly, the geothermal gradient must decrease markedly.

Current estimates of the temperature at the base of the crust are 800°C to 1200°C. The latter figure seems to be an upper limit: If it were any higher, melting would be expected. Furthermore, fragments of mantle rock in kimberlite pipes (see Perspective 10.1), thought to have come from depths of 100 to 300 km, appear to have reached equilibrium at these depths and at a temperature of about 1200°C. At the core–mantle boundary, the temperature is probably between 2500°C and 5000°C; the wide spread of values indicates the uncertainties of such estimates. If these figures are reasonably accurate, the geothermal gradient in the mantle is only about 1°C/km.

Because the core is so remote and its composition so uncertain, only very general estimates of its temperature can be made. Based on various experiments, the maximum temperature at the center of the core is estimated to be 6500°C, very close to the estimated temperature for the surface of the Sun!

Heat Flow

Even though rocks are poor conductors of heat, detectable amounts of heat from Earth's interior escape at the surface by **heat flow.** The amount of heat lost is small and can be detected only by sensitive instruments. Heavy, cylindrical probes are dropped into soft seafloor sediments, and temperatures are measured at various depths along the cylinder. On the continents, temperature measurements are made at various depths in drill holes and mines.

As one would expect, heat flow is greater in areas of active or recently active volcanism. For instance, greater heat flow occurs at spreading ridges, and lower than average values are recorded at subduction zones (Figure 10.13). Higher values are also recorded in areas of con-

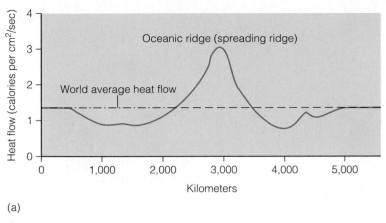

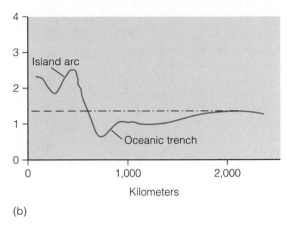

Figure 10.13

Variations in heat flow. (a) Higher than average heat flow occurs at spreading ridges and island arcs, both of which are characterized by volcanism. (b) Oceanic trenches show lower than average heat flow.

tinental volcanism, such as in Yellowstone National Park, Wyoming, Lassen National Park, California, and near Mount St. Helens, Washington. Any area possessing higher than average heat-flow values is a potential area for the development of geothermal energy (see Chapter 16).

More than 70% of the total heat lost by Earth is lost through the seafloor, but heat-flow values for both oceanic basins and continents decrease with increasing age. In the ocean basins, heat flow is higher through younger oceanic crust. This result is expected because high heat-flow values occur at spreading ridges where oceanic crust is continually formed by igneous activity. Spreading ridges are also the sites of hydrothermal vents where considerable heat is transported upward by hydrothermal convection. Heat flow through the continental crust is not as well understood, but it too shows lower values for older crust.

It should be apparent that if heat is escaping from within Earth, the interior should be cooling unless a mechanism exists to replenish it. Radioactive decay generates heat continuously, but the quantity of radioactive isotopes (except carbon 14) decreases with time as they decay to stable daughter products. Furthermore, as the core cools its heat also eventually escapes at the surface. In Chapter 3, we discussed the fact the ultramafic lava flows requiring near-surface temperatures of about 1600°C are known from Archean-aged rocks but ceased occurring about 2.5 billion years ago. The conclusion was that during its early history Earth possessed more internal heat and has been cooling continuously since then.

WHAT IS GRAVITY, AND HOW IS ITS FORCE DETERMINED?

Sir Isaac Newton formulated the law of universal gravitation in which the force of gravitational attraction (F) between two masses (m_1 and m_2) is directly proportional to the products of their mass and inversely proportional to the square of the distance (D) between their centers of mass:

$$F = G \frac{m_1 \times m_2}{D^2}$$

G in this equation is the universal gravitational constant. This law applies to any two bodies. Consequently, a gravitational attraction exists between Earth and its Moon, between the stars, and between an individual and Earth (Figure 10.14a). However, F is greater between two massive bodies, the Moon and Earth, for example, than between two bodies with smaller masses, if the distances between their centers of mass are the same. Because F is inversely proportional to the square of distance, it decreases by a factor of 4 when the distance is doubled. We generally refer to the gravitational force between an object and Earth as its *weight*.

Gravitational attraction would be the same everywhere on Earth's surface if it were perfectly spherical, homogeneous throughout, and not rotating. As a consequence of rotation, however, a centrifugal force is generated that partly counteracts the force of gravity (Figure 10.14b). An object at the equator weighs slightly less than the same object would at the poles. The force of gravity also varies with distance between the centers of masses, so an object would weigh slightly less above the surface than if it were at sea level (Figure 10.14a).

Geologists use a sensitive instrument called a *gravimeter* to measure variations in the force of gravity. A gravimeter is simple in principle; it contains a weight suspended on a spring that responds to variations in gravity (Figure 10.15). Gravimeters have been used extensively in exploration for hydrocarbons and mineral resources. Long ago, geologists realized that anomalous gravity values should exist over buried bodies of ore minerals and salt domes and that geologic structures

Figure 10.14

(a) Earth's gravitational attraction pulls all objects toward its center of mass. Objects 1 and 2 are the same distance from Earth's center of mass, but the gravitational attraction on 1 is greater because it is more massive. Objects 2 and 3 have the same mass, but the gravitational attraction on 3 is four times less than on 2 because it is twice as far from Earth's center of mass. (b) Earth's rotation generates a centrifugal force that partly counteracts the force of gravity. Centrifugal force is zero at the poles and maximum at the equator.

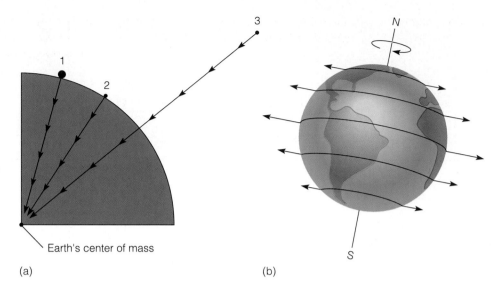

Earth's center of mass

(a)　　　　　　　　　　　　　　　(b)

such as faulted strata could be located by surface gravity surveys (Figure 10.15).

Gravity measurements are higher over an iron ore deposit than over unconsolidated sediment because of the ore's greater density (Figure 10.15a). Such departures from the expected force of gravity are **gravity anomalies**. In other words, the measurement over the body of iron ore indicates an excess of dense material, or simply a *mass excess*, between the surface and the center of Earth and is considered to be a **positive gravity anomaly**. A **negative gravity anomaly** indicating *mass deficiency* exists over low-density sediments because the force of gravity is less than the expected average (Figure 10.15b). Large negative gravity anomalies also exist over salt domes (Figure 10.15c) and at subduction zones, indicating that the crust is not in equilibrium.

Departures from Earth's expected gravitational attraction (gravity anomalies) certainly exist, but what of the tourist sites around the United States that claim gravity has somehow gone awry? All kinds of mysterious gravity-defying effects are claimed to occur in these areas, including objects rolling uphill and water running uphill, unsupported objects clinging to walls, and the famous plank illusion in which the heights of people on a level plank change when they switch positions.

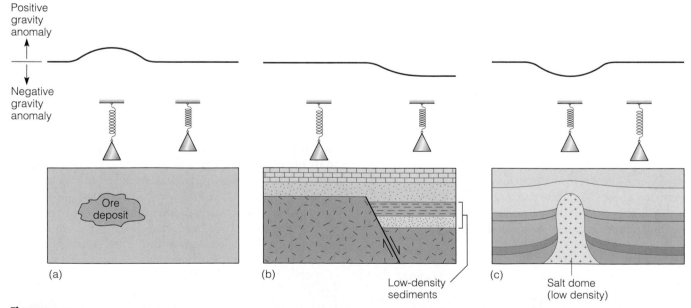

Figure 10.15

(a) The mass suspended from a spring in the gravimeter, shown diagrammatically, is pulled downward more over the dense body of ore than it is in adjacent areas, indicating a positive gravity anomaly. (b) A negative gravity anomaly over a buried structure. (c) Rock salt is less dense than most other types of rocks. A gravity survey over a salt dome shows a negative gravity anomaly.

However, all are actually clever optical illusions that can be duplicated by anyone with the interest in doing so.

FLOATING CONTINENTS?—THE PRINCIPLE OF ISOSTASY

More than 150 years ago, British surveyors in India detected a discrepancy of 177 m when they compared the results of two measurements between points 600 km apart. Even though this discrepancy was small, it was an unacceptably large error. The surveyors realized that the gravitational attraction of the nearby Himalaya Mountains probably deflected the plumb line (a cord with a suspended weight) of their surveying instruments from the vertical, thus accounting for the error. Calculations revealed, however, that if the Himalayas were simply thicker crust piled on denser material, the error should have been greater than that observed (Figure 10.16).

In 1865 George Airy proposed that, in addition to projecting high above sea level, the Himalayas—and other mountains as well—also project far below the surface and thus have a low-density root (Figure 10.16b). In effect, he was saying that mountains float on denser rock at depth. Their excess mass above sea level is compensated for by a mass deficiency at depth, which would account for the observed deflection of the plumb line during the British survey (Figure 10.16).

Another explanation was proposed by J. H. Pratt, who thought that the Himalayas were high because they were composed of rocks of lesser density than those in adjacent regions. Although Airy was correct with respect to the Himalayas, and mountains in general, Pratt was correct in that there are indeed places where the crust's elevation is related to its density. For example, (1) continental crust is thick and less dense than oceanic crust and thus stands high, and (2) the mid-oceanic ridges stand higher than adjacent areas because the crust there is hot and less dense than cooler oceanic crust elsewhere.

Gravity studies have revealed that mountains do indeed have a low-density "root" projecting deep into the mantle. If it were not for this low-density root, a gravity survey across a mountainous area would reveal a huge positive gravity anomaly. The fact that no such anomaly exists indicates that a mass excess is not present, so some of the dense mantle at depth must be displaced by lighter crustal rocks, as shown in Figure 10.16. (Seismic wave studies also confirm the existence of low-density roots beneath mountains.)

Both Airy and Pratt agreed that Earth's crust is in floating equilibrium with the more dense mantle below, and now their proposal is known as the **principle of isostasy**. This phenomenon is easy to understand by analogy to an iceberg (Figure 10.17). Ice is slightly less dense than water, and thus it floats. According to Archimedes's principle of buoyancy, an iceberg will sink in the water until it displaces a volume of water that equals its total weight. When the iceberg has sunk to an

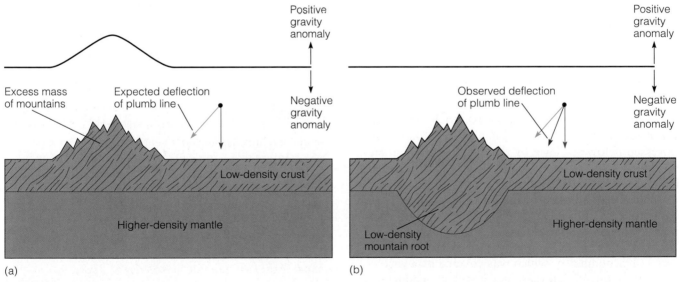

(a)

(b)

Figure 10.16

(a) A plumb line is normally vertical, pointing to the Earth's center of gravity. Near a mountain range, the plumb line should be deflected as shown if the mountains are simply thicker, low-density material resting on denser material, and a gravity survey across the mountains would indicate a positive gravity anomaly. (b) The actual deflection of the plumb line during the survey in India was less than expected. It was explained by postulating that the Himalayas have a low-density root. A gravity survey in this case would show no anomaly because the mass of the mountains above the surface is compensated for at depth by low-density material displacing denser material.

Figure 10.17
An iceberg sinks to an equilibrium position with about 10% of its mass above water level. The larger iceberg sinks farther below and rises higher above the water surface than does the smaller one. If some of the ice above water level should melt, the icebergs will rise to maintain the same proportion of ice above and below water level. Earth's crust floating in more dense material below is analogous to this example.

equilibrium position, only about 10% of its volume is above water level. If some of the ice above water level should melt, the iceberg will rise in order to maintain the same proportion of ice above and below water (Figure 10.17).

Earth's crust is similar to the iceberg, or a ship, in that it sinks into the mantle to its equilibrium level. Where the crust is thickest, as beneath mountain ranges, it sinks farther down into the mantle but also rises higher above the equilibrium surface (Figure 10.16). Continental crust being thicker and less dense than oceanic crust stands higher than the ocean basins. Should the crust be loaded, as where widespread glaciers accumulate, it responds by sinking farther into the mantle to maintain equilibrium (Figure 10.18). In Greenland and Antarctica, the surface of the crust has been depressed below sea level by the weight of glacial ice. The crust also responds isostatically to widespread erosion and sediment deposition (Figure 10.19).

Unloading of the crust causes it to respond by rising upward until equilibrium is again attained. This phenomenon, known as **isostatic rebound,** takes place in areas that are deeply eroded and in areas that were formerly glaciated. Scandinavia, which was covered by a vast ice sheet until about 10,000 years ago, is still rebounding isostatically at a rate of up to 1 m per century (Figure 10.20). Coastal cities in Scandinavia have been uplifted rapidly enough that docks constructed several centuries ago are now far from shore. Isostatic rebound has also occurred in eastern Canada where the land has risen as much as 100 m during the last 6000 years.

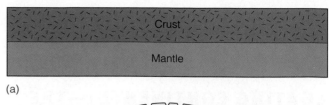
(a)

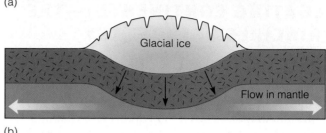
(b)

(c)

(d)

Figure 10.18
A diagrammatic representation of the response of Earth's crust to the added weight of glacial ice. (a) The crust and mantle before glaciation. (b) The weight of glacial ice depresses the crust into the mantle. (c) When the glacier melts, isostatic rebound begins, and the crust rises to its former position. (d) Isostatic rebound is complete.

If the principle of isostasy is correct, it implies that the mantle behaves as a liquid. In preceding discussions, however, we said that the mantle must be solid because

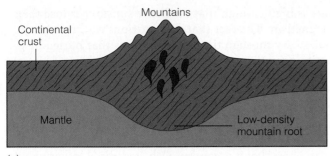

Mountains

Continental
crust

Mantle

Low-density
mountain root

(a)

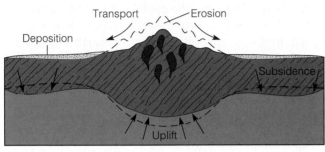

Transport   Erosion

Deposition

Subsidence

Uplift

(b)

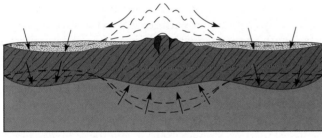

(c)

Figure 10.19
A diagrammatic representation showing the isostatic response of the crust to erosion (unloading) and widespread deposition (loading).

it transmits S-waves, which will not move through a liquid. How can this apparent paradox be resolved? When considered in terms of the short time necessary for S-waves to pass through it, the mantle is indeed solid. But when subjected to stress over long periods of time, it will yield by flowage and at these time scales can be considered a viscous liquid. A familiar substance that has the properties of a solid or a liquid depending on how rapidly deforming forces are applied is Silly Putty. It will flow under its own weight if given enough time, but shatters as a brittle solid if struck a sharp blow.

EARTH'S MAGNETIC FIELD

A simple bar magnet has a **magnetic field**, an area in which magnetic substances are affected by lines of magnetic force emanating from the magnet (Fig-

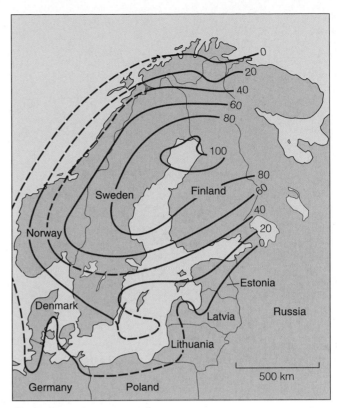

Figure 10.20
Isostatic rebound in Scandinavia. The lines show rates of uplift in centimeters per century.

ure 10.21). The magnetic field shown in Figure 10.21 is *dipolar*, meaning that it possesses two unlike magnetic poles referred to as the north and south poles. Earth possesses a dipolar magnetic field that resembles, on a large scale, that of a bar magnet (Figure 10.22).

Humans have been aware of Earth's magnetic field for centuries, but only rather recently have we had any knowledge of how it is generated even though many

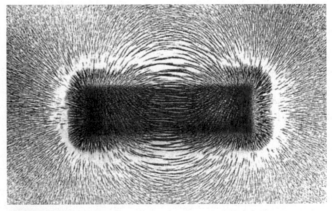

Figure 10.21
Iron filings align themselves along the lines of magnetic force radiating from a magnet.

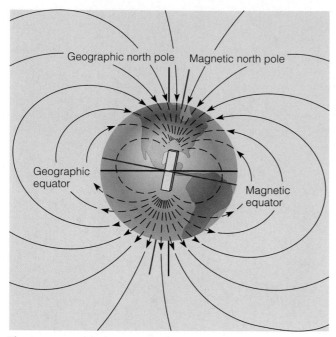

Figure 10.22
Earth's magnetic field has lines of force just like those of a bar magnet.

aspects of it are still poorly understood. We can be sure that it does not emanate from a body of deeply buried magnetic materials such as the mineral magnetite, the most magnetic of several minerals, because magnetic substances lose their magnetism when heated above a temperature known as the **Curie point.** The Curie point for magnetite is 580°C, which is far below its melting temperature. At a depth of 80 to 100 km, the tempera-

ture is high enough that magnetic substances lose their magnetism. The fact that the locations of the magnetic poles vary through time also indicates that buried magnetite is not the source of the magnetic field.

Instead, the magnetic field appears to be generated in the liquid outer core, by electrical currents (an electrical current is a flow of electrons that always generates a magnetic field). Experts on magnetism do not fully understand how the magnetic field is generated, but most agree that it is continuously generated, otherwise it would decay and Earth would have no magnetic field in as little as 20,000 years. The model most widely accepted now is that thermal and compositional convection within the liquid outer core coupled with Earth's rotation produce complex electrical currents or a *self-exciting dynamo* that in turn generates the magnetic field.

Inclination and Declination of the Magnetic Field

Notice in Figure 10.22 that the lines of magnetic force around Earth parallel its surface only near the equator. As these lines approach the poles, they are oriented at increasingly large angles with respect to the surface, and the strength of the magnetic field increases; it is weakest at the equator and strongest at the poles. Accordingly, a compass needle mounted so that it can rotate both horizontally and vertically not only points north but is also inclined with respect to the surface, except at the magnetic equator. The degree of inclination depends on the needle's location along a line of magnetic force (Figure 10.23).

Figure 10.23
Magnetic inclination. The strength of the magnetic field changes uniformly from the magnetic equator to the magnetic poles. This change in strength causes a dip needle to parallel Earth's surface only at the magnetic equator, whereas its inclination with respect to the surface increases to 90 degrees at the magnetic poles. Notice the 11 1/2-degree angle between the geographic and magnetic poles.

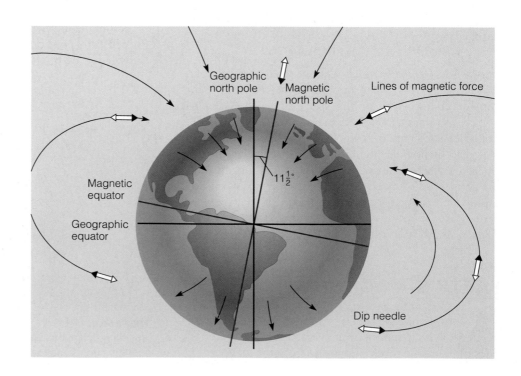

This deviation of the magnetic field from the horizontal is **magnetic inclination**. To compensate for inclination, compasses used in the Northern Hemisphere have a small weight on the south end of the needle. This property of the magnetic field is important in determining the ancient geographic positions of tectonic plates (see Chapter 12).

Another important aspect of the magnetic field is that the magnetic poles, where the lines of force leave and enter Earth, do not coincide with the geographic (rotational) poles. At present, an 11 1/2-degree angle exists between the two (Figure 10.23). Studies of the magnetic field show that the locations of the magnetic poles vary slightly over time but still correspond closely, on the average, with the locations of the geographic poles. A compass points to the north magnetic pole in the Canadian Arctic islands, some 1290 km away from the geographic pole (true north); only along the line shown in Figure 10.24 will a compass needle point to both the magnetic and geographic north poles. From any other location, an angle called **magnetic declination** exists between lines drawn from the compass position to the magnetic pole and the geographic pole (Figure 10.24). Magnetic declination, which in some locations may be as much as 30 degrees, must be taken into account during surveying and navigation because, for most places on Earth, compass needles point east or west of true north.

Magnetic Anomalies

Variations in the normal strength of the magnetic field occur on both regional and local scales, giving rise to **magnetic anomalies**. Regional variations are most likely related to the complexities of convection within the outer core where the magnetic field is generated. Local variations can be accounted for by lateral or vertical variations in rock types within the crust.

An instrument called a *magnetometer* can detect slight variations in the strength of the magnetic field, and deviations from the normal are characterized as positive or negative. A **positive magnetic anomaly** exists

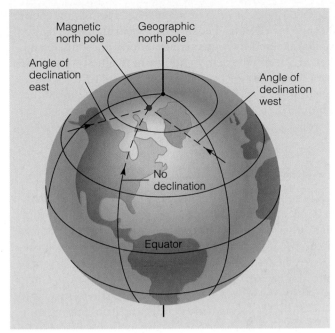

Figure 10.24
Magnetic declination. A compass needle points to the magnetic north pole rather than the geographic pole (true north). The angle formed by the lines from the compass position to the two poles is the magnetic declination.

in areas where the rocks contain more iron-bearing minerals than elsewhere. In the Great Lakes region of the United States and Canada, huge iron ore deposits containing hematite and magnetite add their magnetism to that of the magnetic field, resulting in a positive magnetic anomaly (Figure 10.25a). Positive magnetic anomalies also exist where extensive basaltic volcanism has taken place because basalt contains appreciable quantities of iron-bearing minerals (Figure 10.25b). Areas underlain by basalt lava flows, such as the Columbia River basalts of the northwestern United States (see Figure 4.17a), possess positive magnetic anomalies, whereas adjacent areas underlain by sedimentary rocks show **negative magnetic anomalies** (Figure 10.25b).

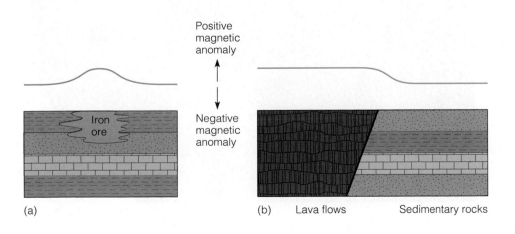

(a) (b) Lava flows Sedimentary rocks

Figure 10.25
(a) A positive magnetic anomaly over an iron ore deposit. (b) A positive magnetic anomaly over lava flows and a negative anomaly over adjacent sedimentary rocks.

Geologists have used magnetometers for magnetic surveys for decades because iron-bearing rocks can be easily detected by a positive magnetic anomaly even if they are deeply buried. In addition, magnetometers can detect a variety of buried geologic structures, such as salt domes, which show negative magnetic anomalies (Figure 10.26); these can be detected by gravity surveys as well.

Magnetic Reversals

When magma cools through the Curie point, its iron-bearing minerals gain their magnetization and align themselves with the magnetic field, recording both its direction and strength. As long as the rock is not subsequently heated above the Curie point, it will preserve that magnetism. If the rock is heated above the Curie point, however, the original magnetism is lost, and when the rock subsequently cools, the iron-bearing minerals will align with the current magnetic field.

The iron-bearing minerals of some sedimentary rocks (especially those that formed on the deep seafloor) are also oriented parallel to the magnetic field as the sediments are deposited. These rocks also preserve a record of the magnetic field at the time of their formation. Such information preserved in lava flows and some sedimentary rocks can be used to determine the directions to the magnetic poles and the latitude of the rock when it formed.

Paleomagnetism, the study of the ancient magnetic field, relies on the remanent magnetism in ancient rocks that records the direction and strength of the magnetic field at the time of their formation. Geologists refer to the present magnetic field as *normal,* that is, with the north and south magnetic poles located roughly at the

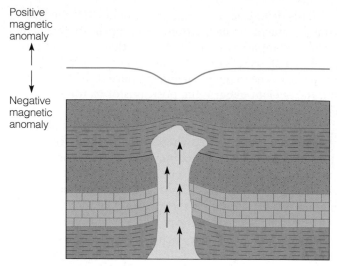

Figure 10.26
A negative magnetic anomaly over a salt dome.

north and south geographic poles. As early as 1906, though, rocks were discovered that showed reversed magnetism. Paleomagnetic studies initially conducted on continental lava flows have clearly shown that the magnetic field has completely reversed itself numerous times during the geologic past (Figure 10.27). When these **magnetic reversals** occur, the magnetic polarity is reversed so that the north arrow on a compass would point south rather than north.

Rocks that have a record of magnetism the same as the present magnetic field are described as having **normal polarity,** whereas rocks with the opposite magnetism have **reversed polarity.** These same patterns of normal and reversed polarity have been discovered in continental lava flows and in the oceanic crust (see Chapter 12).

Figure 10.27
Magnetic reversals recorded in a succession of lava flows are shown diagrammatically by red arrows, and the record of normal polarity events is shown by black arrows.

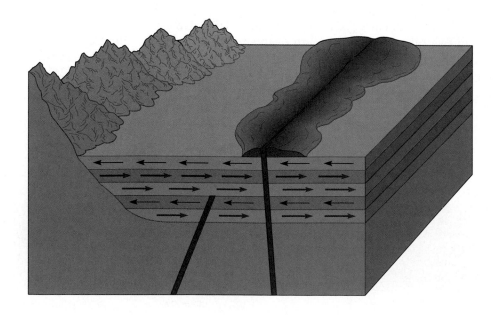

Randolph W. Bromery

Geology: An Unexpected but Rewarding Career

Randolph W. Bromery. Randolph Bromery has been serving as president of Springfield College in Springfield, Massachusetts, since 1992. He retired from the University of Massachusetts as a Commonwealth Professor Emeritus in 1992. During his 25-year career at the University of Massachusetts, he served as department chair, vice-chancellor and chancellor of the Amherst campus, executive vice-president and senior vice-president of the university system, and interim president of the Massachusetts Board of Regents for Higher Education. In 1989 he was elected president of the Geological Society of America and received the distinguished alumni award from Howard University and Johns Hopkins University. His portrait was given a place of honor in the *National Academy of Sciences Portrait Collection of African Americans in Science, Engineering and Medicine* on February 25, 1997.

Ironically, I joined the geologic profession due to my experiences in seeking other employment during the summer of 1948. I had graduated from Howard University with a major in mathematics and a minor in physics. Thus, I first attempted to find work in my field as a mathematician with the U.S. Naval Research Laboratory.

On the last of several repeat visits to the laboratory to refile my employment application, which had been "accidentally misplaced" several times, I picked up a page from the Washington, D.C., edition of an Afro-American newspaper that had been left on a trolley car seat. That page contained a very small ad stating that the U.S. Geological Survey was seeking applicants for positions as mathematicians, physicists, and chemists. Married and desperately in need of employment, I quickly transferred to a connecting trolley car and went to the U.S. Geological Survey. I filed an application and within three or four days was hired as an exploration geophysicist with the newly formed airborne magnetometer group in the Geophysics Section of the U.S. Geological Survey.

After several years with the U.S. Geological Survey conducting some of the first airborne geophysical surveys flown on the Colorado Plateau and in other parts of the United States, I returned to school, part-time and at night, completed my undergraduate and graduate requirements in geology, and received my master's degree from the American University in Washington, D.C. I then earned my Ph.D. degree in geology from Johns Hopkins University. From a study conducted by both private and federal agencies during the early 1970s, I discovered that I was one of a very small cadre of five black Americans who had doctorate degrees in geology.

My main research interests are in airborne geophysical surveys and applications of shallow-refraction seismic, gravity, electrical resistivity and conductivity surveys in hydrologic and engineering programs.

The irony associated with that earlier act of discrimination and the chance finding of that newspaper page on a trolley car in Washington, D.C., over 40 years ago, is that it deflected me into a new and unfamiliar career that has been both personally and professionally a most rewarding experience. My career in geology reached its most significant point when my professional peers elected me president of the Geological Society of America in 1989.

As I fast approach retirement from teaching, I reflect with much pride on the many undergraduate and graduate students I have taught and learned from over the last quarter of a century; the nearly two decades of geologic and geophysical consulting work I conducted in several west and central African countries; and my early career with the U.S. Geological Survey. Beginning my fifth decade as a professional geologist-geophysicist, I can say, without hesitation, that it has been a most exciting and rewarding professional career.

The economic and environmental future of our country and this planet will be shaped to a large extent by those of us in the various geologic professions. The United States now holds the leadership position in research and applications of groundwater and surface-water exploration and development techniques, Earth resources exploration and development, and waste-disposal and environmental planning and protection technology. Geology will continue to play a key role in these critically important areas, and minorities and women are now finding a place in this professional workforce. ∎

The cause of magnetic reversals is not completely known, although they appear to be related to changes in the intensity of the magnetic field. Calculations indicate that the magnetic field has weakened about 5% during the nineteenth century. If this trend continues, there will be a period during the next few thousand years when the magnetic field will be nonexistent and then will reverse. After the reversal occurs, the magnetic field will rebuild itself with opposite polarity.

1. Earth is concentrically layered into an outer oceanic crust and continental crust, below which lies a rocky mantle and an iron-rich core with a solid inner part and a liquid outer part.

2. Much of the information about Earth's interior has been derived from studies of P- and S-waves that travel through the planet. Laboratory experiments, comparisons with meteorites, and studies of inclusions in volcanic rocks provide additional information.

3. Earth's interior is subdivided into concentric layers on the basis of changes in seismic wave velocities at discontinuities.

4. Density and elasticity of Earth materials determine the velocity of seismic waves. Seismic waves are refracted when their direction of travel changes. Wave reflection occurs at boundaries across which the properties of Earth materials change.

5. The behavior of P- and S-waves within Earth and the presence of P- and S-wave shadow zones allow geologists to estimate the density and composition of Earth's interior and to estimate the size and depth of the core and mantle.

6. Earth's inner core is thought to be composed of iron and nickel, whereas the outer core is probably composed mostly of iron with 10 to 20% other substances. Peridotite is the most likely component of the mantle.

7. The oceanic and continental crusts are of basaltic and granitic composition, respectively. The boundary between the crust and the mantle is the Mohorovičić discontinuity.

8. The geothermal gradient of 25°C/km cannot continue to great depths, otherwise most of Earth would be molten. The geothermal gradient for the mantle and core is probably about 1°C/km. The temperature at the center of the core is estimated at 6500°C.

9. Detectable amounts of heat escape at the surface by heat flow. Most of Earth's internal heat is generated by radioactive decay, but some comes from the cooling core.

10. According to the principle of isostasy, the crust is floating in equilibrium with the denser mantle below. Continental crust stands higher than oceanic crust because it is thicker and less dense.

11. Positive and negative gravity anomalies can be detected where excesses and deficiencies of mass occur, respectively. Gravity surveys are useful in exploration for minerals and hydrocarbons.

12. The magnetic field is thought to be generated by electrical currents in the liquid outer core.

13. Earth is surrounded by lines of magnetic force similar to those of a bar magnet. The lines of magnetic force are inclined with respect to Earth's surface, except at the magnetic equator, accounting for the phenomenon of magnetic inclination.

14. Although the magnetic poles are close to the geographic poles, they do not coincide exactly. For most places on the planet, an angle called magnetic declination exists between lines drawn from a compass location to the magnetic and geographic north poles.

15. A magnetometer can detect departures from the normal magnetic field, which are characterized as positive or negative.

16. Although the cause of magnetic reversal is not fully understood, the polarity of the magnetic field has completely reversed itself many times during the past.

asthenosphere
continental crust
core
crust
Curie point
discontinuity
geothermal gradient
gravity anomaly (positive and negative)
heat flow
isostatic rebound

lithosphere
low-velocity zone
magnetic anomaly (positive and negative)
magnetic declination
magnetic field
magnetic inclination
magnetic reversal
mantle
Mohorovičić discontinuity (Moho)
normal polarity

oceanic crust
paleomagnetism
principle of isostasy
P-wave shadow zone
reflection
refraction
reversed polarity
S-wave shadow zone

Review Questions

1. Given that Earth's average density is 5.5 g/cm^3 and that rocks at or near the surface average 2.5 to 3.0 g/cm^3, it follows that:

   a. _____ the mantle must be liquid;
   b. _____ continental crust is denser than oceanic crust;
   c. _____ Earth materials at depth must be denser than those at the surface;
   d. _____ the liquid outer core surrounds the mantle;
   e. _____ seismic discontinuities are present within the core.

2. The Moho is:

   a. _____ a type of inclusion in volcanic rocks from great depth;
   b. _____ a seismic discontinuity at the base of the crust;
   c. _____ between 410 and 660 km below Earth's surface;
   d. _____ a layer composed of iron and nickel;
   e. _____ a zone in which rocks are plastic and flow.

3. When seismic waves travel through materials having different properties, their direction of travel changes. This phenomenon is wave_____ :

   a. _____ deflection;
   b. _____ elasticity;
   c. _____ attenuation;
   d. _____ lithification;
   e. _____ refraction.

4. Which one of the following provides evidence for the composition of the core?

   a. _____ meteorites;
   b. _____ inclusions in volcanic rock;
   c. _____ S-wave shadow zone;
   d. _____ diamonds;
   e. _____ peridotite.

5. Earth's internal heat comes from _____ and _____ :

   a. _____ radioactive decay/the core;
   b. _____ the magnetic field/gravity;
   c. _____ seismic waves/earthquakes;
   d. _____ plate tectonics/meteorite impacts;
   e. _____ magnetic reversals/gravity anomalies.

6. A decrease in the velocity of seismic waves in the lowermost 200 to 300 km of the mantle defines the:

   a. _____ asthenosphere–lithosphere boundary;
   b. _____ D″ layer;
   c. _____ Moho;
   d. _____ S-wave shadow zone;
   e. _____ geothermal gradient.

7. According to the principle of isostasy:

   a. _____ more heat escapes from oceanic crust than continental crust;
   b. _____ magnetic anomalies result from gravitational attraction between objects;
   c. _____ the crust is in floating equilibrium with the more dense mantle below;
   d. _____ the asthenosphere behaves as a brittle solid;
   e. _____ most of the lithosphere is liquid.

8. Iron-bearing minerals in magma are magnetized and align with the magnetic field when they cool through the:

   a. _____ Curie point;
   b. _____ isostasy curve;
   c. _____ declination zone;
   d. _____ magnetic polarity;
   e. _____ refraction point.

9. Continental crust is:

   a. _____ 3 to 10 km thick;
   b. _____ thinnest at spreading ridges;
   c. _____ granitic in composition;
   d. _____ the source of Earth's magnetic field;
   e. _____ more dense than oceanic crust.

10. The mantle is probably composed mostly of:

    a. _____ granitic rocks;
    b. _____ lava flows;
    c. _____ basalt and gabbro;
    d. _____ peridotite;
    e. _____ nonferromagnesian silicates.

11. Which one of the following is correct?

    a. _____ the lithosphere consists of the upper and lower mantle;
    b. _____ a buried body of magnetite generates the magnetic field;
    c. _____ isostatic rebound results in subsidence of the crust;
    d. _____ liquid iron–nickel alloy makes up most of the asthenosphere;
    e. _____ magnetic inclination is 90 degrees at the poles.

12. The temperature increase with depth is known as the:

    a. _____ isostatic curve;
    b. _____ heat-flow index;
    c. _____ geothermal gradient;
    d. _____ Curie point;
    e. _____ gravity anomaly.

13. Explain the phenomena of magnetic inclination and magnetic declination. Which can be used to determine the latitude at which ancient rocks originated? Explain.

14. Why do scientist think the inner core is solid, whereas the outer core is liquid?

15. What is the geothermal gradient, and why must it decrease with depth?

16. Explain the reasoning Mohorovičić used to determine that a discontinuity, now called the Moho, exists between the crust and mantle.

17. Explain fully why a continent stands higher than its adjacent seafloor.

18. What is meant by positive and negative gravity anomalies? How can these anomalies be used in exploring for geologic resources?

19. Why are P- and S-waves refracted and reflected within Earth?

20. What internal processes account for heat flow, and why is Earth's internal heat so important?

21. Draw a cross section showing oceanic and continental crust, the lithosphere, asthenosphere, and upper and lower mantle.

22. Explain how and where scientists think Earth's magnetic field is generated.

23. How is it possible to determine the overall mass and density of Earth?

24. Two well-known seismic discontinuities are present within the mantle. At what depths do they occur, and what is responsible for them?

Points to Ponder

1. Earth's crust is deeply eroded in one area and loaded by widespread, thick, sedimentary deposits in another. Explain fully how the crust will respond in these two areas.

2. If Earth were completely solid and compositionally homogeneous throughout, how would P- and S-waves behave as they traveled through the planet? What is their actual behavior?

3. What factors account for higher than average heat flow at spreading ridges and lower than average heat flow from ancient continental crust?

4. Use the law of universal gravitation to explain how negative and positive gravity anomalies arise.

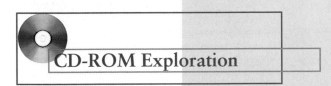

➤ EARTH'S INTERIOR AND PLATE TECTONICS

The site by Rosanna L. Hamilton has descriptions and illustrations of Earth's interior. It also has a summary of depth, volume, composition, and properties of each of Earth's internal divisions. Scroll down and click on *The Earth's Interior*.
1. What divisions of Earth's interior are recognized?
2. How much of Earth's mass is made up of the inner core?
3. Why is the inner core solid, and what is it composed of?

➤ HIGHS AND LOWS: TOPOGRAPHY AND ISOSTASY

This University of Michigan site has lecture notes along with illustrations for *Introduction to Global Change I*.
1. What two elevations dominate Earth's surface?
2. Explain why Earth is densely stratified, and why must material below the surface be denser than is the average for surface rocks?
3. How were Earth's shape, circumference, and volume determined?

➤ CAPE BRETON POST: PLANET EARTH ARTICLE

This site has three articles written by Dr. Brendan Murphy of St. Francis Xavier University for Geology 120 and 216. The articles include *Journey to the Center of the Earth, The Earth's Hidden Interior,* and *The Composition of the Earth's Core*.
1. How does the study of earthquake waves indicate that rocks within 2900 km of the surface have densities of less than 5.52?
2. What is the evidence for the probable composition of Earth's core?
3. How is heat from Earth's interior responsible for shaping our planet?

CD-ROM Exploration

➤ Exploring your *Earth Systems Today* CD-ROM will add to your understanding of the material in this chapter.

TOPIC: EARTH'S STRUCTURE

MODULE: EARTH'S LAYERS

Explore activities in this module to see if you can discover the following for yourself:

Using activities in this module, study how P- and S-waves travel through the body of Earth. Note the difference between reflection of waves and refraction of waves.

Using activities in this module, study seismic wave propagation in a theoretical homogeneous Earth, a theoretical nonlayered heterogeneous Earth, and a realistic layered Earth. Using the S-wave shadow zone, compute the depth to the core–mantle boundary.

Chapter 11

Pillow lava on the Mid-Atlantic Ridge.

OBJECTIVES

At the end of this chapter, you will have learned that

- Scientists can explore the largely hidden oceanic depths in a variety of ways.
- The margins of continents consist of several elements, but vary depending on the type of geologic activity in these marginal areas.
- Although some parts of the seafloor are flat and featureless, it also has features

such as ridges, trenches, and seamounts.
- Several seafloor features result from geologic activities at or near divergent and convergent plate boundaries.
- Seafloor sediments are derived from several sources, but most come from weathering and erosion of continents and oceanic islands and the remains of organisms.

- Reefs are wave-resistant structures formed by organisms in warm, shallow seas.
- The composition and structure of oceanic crust were first determined by studying fragments of oceanic crust now on land and then verified by direct observations on the seafloor.
- Several important resources are derived from seawater or from seafloor deposits.

The Seafloor

Prologue

Most people have heard of the mythical lost continent of Atlantis, but few are aware of the source of the Atlantis legend or the evidence cited for the former existence of this continent. Only two known sources of the Atlantis legend exist, both written in about 350 B.C. by the Greek philosopher Plato. In two of his philosophical dialogues, the *Timaeus* and the *Critias,* Plato tells of Atlantis, a large island continent that, according to him, was in the Atlantic Ocean west of the Pillars of Hercules, which we now call the Strait of Gibraltar (Figure 11.1). Plato clearly stated that Atlantis was in what we now call the Atlantic Ocean, but its location in more recent literature varies. In fact, it has been reported in just about every imaginable location on Earth. One tabloid reported it was found in the North Atlantic, and one year later the same publication gave an account of its discovery in the Pacific Ocean!

According to Plato's account, Atlantis was technologically advanced and extended its power over a huge area, including parts of southern Europe and through the Mediterranean as far east as Egypt. Yet despite its advantage of wealth, advanced technology, and a large army and navy, Atlantis was defeated in war by Athens. Following the conquest of Atlantis by Athens, Plato wrote that

> . . . there were violent earthquakes and floods and one terrible day and night came when . . . Atlantis . . . disappeared beneath the sea. And for this reason even now the sea there has become unnavigable and

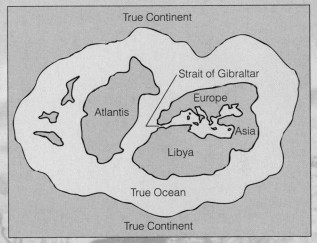

Figure 11.1
According to Plato, Atlantis was a large continent west of the Pillars of Hercules, which we now call the Strait of Gibraltar. In the *Timaeus,* Plato noted that "[Atlantis] was larger than Libya and Asia put together and was the way to other islands and from these you might pass to the whole of the opposite continent which surrounded the true ocean."

> unsearchable, blocked as it is by the mud shallows which the island produced as it sank.*

If one assumes that the destruction of Atlantis was a real event, rather than one conjured up by Plato to make a philosophical point, he nevertheless lived long after it was supposed to have occurred. According to Plato, Solon, an Athenian who lived about 200 years

---
*From the *Timaeus*. Quoted in E. W. Ramage, ed., *Atlantis: Fact or Fiction?* (Bloomington: Indiana University Press, 1978), p. 13.

before Plato, heard the story from Egyptian priests who claimed the event had taken place 9000 years before their time. Solon told the story to his grandson, Critias, who in turn told it to Plato.

Present-day proponents of the Atlantis legend generally cite two types of evidence to support their claim that Atlantis did indeed exist. First, they point to supposed cultural similarities on opposite sides of the Atlantic Ocean basin, such as the similar shapes of the pyramids of Egypt and those of Central and South America. They contend that these similarities are due to cultural diffusion from the highly developed civilization of Atlantis. According to archaeologists, however, few similarities actually exist, and those that do can be explained as the independent development of analogous features by different cultures.

Second, supporters of the legend assert that remnants of the sunken continent can be found. No "mud shallows" exist in the Atlantic as Plato claimed, but the Azores, Bermuda, the Bahamas, and the Mid-

Atlantic Ridge are alleged to be remnants of Atlantis. If a continent had actually sunk in the Atlantic, it could be easily detected by a gravity survey. Recall that continental crust has a granitic composition and a lower density than oceanic crust. So if a continent were actually present beneath the Atlantic Ocean, there would be a huge negative gravity anomaly, but no such anomaly has been detected. Furthermore, the crust beneath the Atlantic has been drilled in many places, and all samples recovered indicate that its composition is the same as that of oceanic crust elsewhere.

The popular literature also contains references to what are claimed to be actual remains of buildings and roads constructed by Atlanteans. For instance, in the Bimini Islands of the Bahamas, "pavementlike" blocks are cited as evidence of an ancient roadway, but on close inspection they are nothing more than naturally fractured pieces of limestone. Also in the Bimini Islands are cylinders of rocklike material claimed to be parts of ancient pillars. These too are not what they might seem to be. They appear to be cement manufactured after 1800 for construction that was either dumped in this area or came to rest on the seafloor as a result of shipwreck.

In short, no geological or archaeological evidence demonstrates that Atlantis actually existed. Nevertheless, some archaeologists think that the legend might be based on a real event. About 1390 B.C. a huge volcanic eruption destroyed the island of Thera

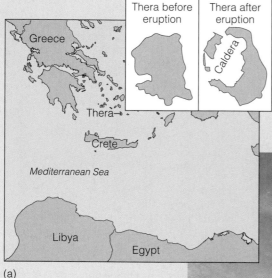

(a)

Figure 11.2
(a) The location of Thera in the Mediterranean Sea. During the huge eruption of 1390 B.C. (1628 B.C. according to some), much of the island subsided below sea level when a caldera formed (see inset). (b) An artist's rendition showing the destruction of Thera. Most of the island's inhabitants escaped.

(b)

in the Mediterranean Sea, which was an important center of early Greek civilization (Figure 11.2a). (Some researchers now think the eruption was in 1628 B.C.) The eruption was one of the most violent during historic time, and much of the island disappeared when it subsided to form a caldera. Most of the island's inhabitants escaped (Figure 11.2b), but the eruption probably contributed to the demise of the Minoan culture on Crete more than 100 km to the south. At least 10 cm of ash fell on parts of Crete, and the coastal areas of the island were probably devastated by tsunami. It is possible that Plato used an account of the destruction of Thera but fictionalized it for his own purposes, thereby giving rise to the Atlantis legend.

INTRODUCTION

One reason the Atlantis legend has persisted for so long is that until recently no one had any real knowledge of what lie beneath the oceans. Indeed, the most fundamental observation we can make regarding Earth is that much of its surface is water covered. Thus, it has vast areas largely hidden from view and continents, which at first glance might seem to be nothing more than areas *not* covered by water. However, considerable differences exist between continents and ocean basins. For one thing, ocean basins are lower than continents; recall from Chapter 10 that continental crust is thicker and less dense than oceanic crust, so according to the principle of isostasy continental crust should stand higher.

In addition to their topographic differences, oceanic crust and continental crust are also composed of different types of rocks. Oceanic crust is compositionally simple, composed of basalt and gabbro, and continually generated at spreading ridges and consumed at subduction zones. It also consists of relatively young rocks, geologically speaking. No oceanic crust is more than about 180 million years old, whereas the age of the continental crust varies from recent to nearly 4 billion years old. And even though we characterize continental crust as granitic, it actually consists of all rock types (see Chapter 10).

When Earth first formed, it was probably hot and airless and had no surface water. Volcanism was no doubt much more pervasive that at present, because Earth possessed more internal heat. Gases derived from the interior were released during volcanic eruptions and probably resulted in the origin of the atmosphere and surface waters in a process called *outgassing* (Figure 11.3). In Chapter 4 we noted that present-day volcanoes emit a variety of gases, most of which is water vapor. Erupting volcanoes surely did the same except more frequently during Earth's early history. In any case, as Earth cooled, water vapor began condensing and fell as rain, which accumulated to form the surface waters.

Now 71% of Earth's surface is covered by an interconnected body of saltwater we designate as oceans and seas. Because of this interconnection we can refer to a world ocean, but four areas are distinct enough to be

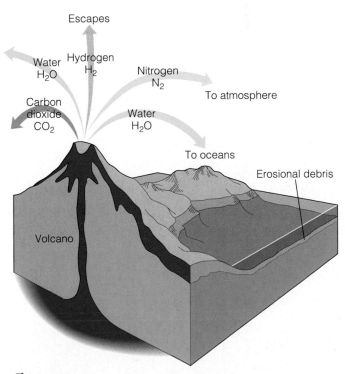

Figure 11.3
Gases derived from within Earth by outgassing formed the early atmosphere and surface waters.

recognized as the Pacific, Atlantic, Arctic, and Indian Oceans (Figure 11.4). The Pacific is by far the largest, containing almost 53% of all water on Earth (Table 11.1). The term *ocean* applies to these larger areas of the interconnected body of saltwater, and *sea* designates a smaller body of water, usually a marginal part of an ocean (Figure 11.4). Sea is also used for some bodies of water completely enclosed by land, such as the Dead Sea, Salton Sea, and the Caspian Sea, but these are actually saline lakes on continents.

In Chapter 10, we made the point that Earth's internal heat is responsible for geologic activity such as volcanism, earthquakes, the origin of mountains, and plate movements. Likewise, it is also responsible for the differentiation of the crust into two types, oceanic and continental. Earth's earliest crust is not preserved, so we can only make some reasonable speculations on what it was like. This earliest crust might have been composed of dark-colored ultramafic rock (<45% silica), which because of its density would have been subducted and

Table 11.1

Numeric Data for the Oceans

Ocean*	Surface Area (million km²)	Water Volume (million km³)	Average Depth (km)	Maximum Depth (km)
Pacific	180	700	4.0	11.0
Atlantic	93	335	3.6	9.2
Indian	77	285	3.7	7.5
Arctic	15	17	1.1	5.2

Source: P. R. Pinet, 1992. Oceanography. (St. Paul:West.)

*Excludes adjacent seas, such as the Caribbean Sea and Sea of Japan, which are marginal parts of oceans.

thus not preserved. In any case, many geologists are convinced that plate motions coupled with subduction and collisions of island arcs formed several small continents with intervening oceanic crust at least 4 billion years ago.

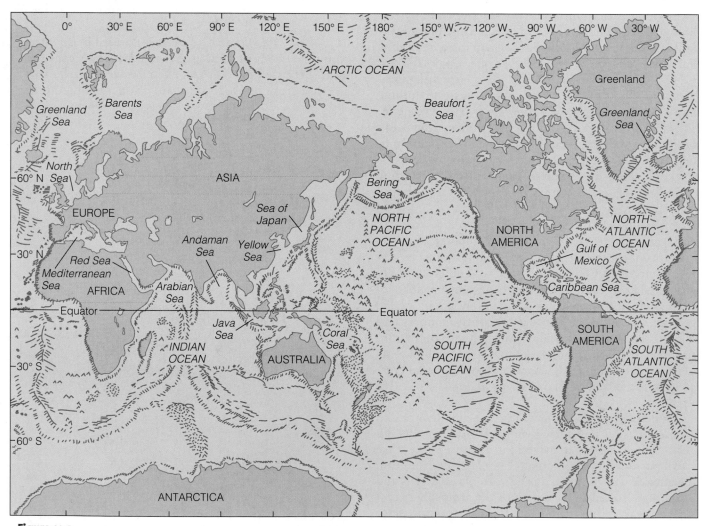

Figure 11.4

This map shows the geographic limits of the four major oceans and many of the world's seas.

EARLY EXPLORATION OF THE OCEANS

During most of historic time, people knew little of the oceans and, until fairly recently, thought that the seafloor was a vast, featureless plain. In fact, through most of this time the seafloor in one sense was more remote than the Moon's surface because it could not even be observed.

The ancient Greeks had determined Earth's size and shape rather accurately, but western Europeans were not aware of the vastness of the oceans until the 1400s and 1500s, when various explorers sought trade routes to the Indies. Even when Christopher Columbus set sail on August 3, 1492, in an effort to find a route to the Indies, he greatly underestimated the width of the Atlantic Ocean. Contrary to popular belief, he was not attempting to demonstrate Earth's spherical shape; its shape was well accepted by then. The controversy was over Earth's circumference, and the shortest route to China; on these points, Columbus's critics were correct.

These voyages added considerably to our knowledge of the oceans, but truly scientific investigations did not begin until the late 1700s. Great Britain was the dominant maritime power, and to maintain that dominance, the British sought to increase their knowledge of the oceans. The earliest British scientific voyages were led by Captain James Cook in 1768, 1772, and 1777. In 1872 the converted British warship HMS *Challenger* began a four-year voyage, during which seawater was sampled and analyzed, oceanic depths were determined at nearly 500 locations, rock and sediment samples were recovered from the seafloor, and more than 4000 new marine species were classified.

Continuing exploration of the oceans revealed that the seafloor is not flat and featureless as formerly believed. Indeed, scientists discovered that the seafloor possesses varied topography including oceanic trenches, submarine ridges, broad plateaus, hills, and vast plains (Figure 11.5). Some people have suggested that some of these features are remnants of the mythical lost continent of Atlantis (see the Prologue).

HOW ARE OCEANS EXPLORED TODAY?

The Deep Sea Drilling Project, an international program sponsored by several oceanographic institutions and funded by the National Science Foundation, began in 1968. Its first research vessel, the *Glomar*

Figure 11.5
This map showing the varied topography of the seafloor resulted from the pioneering work of Maurice Ewing, Bruce Heezen, and Marie Tharp. Their studies and maps helped confirm the existence of the Mid-Atlantic Ridge and made other important contributions to our knowledge of the deep-ocean floor.

Challenger, was equipped to drill in water more than 6000 m deep and to recover long cores of seafloor sediment and oceanic crust. During the next 15 years, the *Glomar Challenger* drilled more than 1000 holes in the seafloor.

Research vessels also sample the seafloor using *clamshell samplers* and *piston corers* (Figure 11.6). A clamshell sample is undisturbed but quite shallow, whereas a piston corer can recover samples up to 25 m long. Drilling into the seafloor is expensive but yields

considerably more information about deep-sea sediments as well as the composition of the upper oceanic crust.

The Deep Sea Drilling Project came to an end in 1983 when the *Glomar Challenger* was retired. However, an international project, the Ocean Drilling Program, continued where the Deep Sea Drilling Project left off, and a larger, more advanced research vessel, the JOIDES* *Resolution,* made its first voyage in 1985 (Figure 11.7a).

---

*JOIDES is an acronym for Joint Oceanographic Institutions for Deep Earth Sampling.

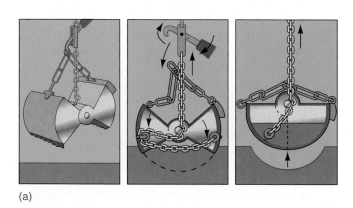

(a)

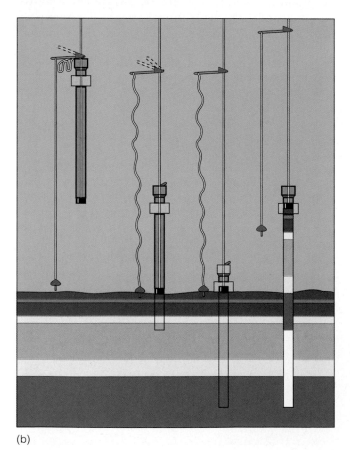

(b)

Figure 11.6
Sampling the seafloor. (a) A clamshell sampler taking a seafloor sample. (b) A piston corer falls to the seafloor, penetrates the sediment, and then is retrieved.

(a)

(b)

Figure 11.7
Oceanographic research vessels. (a) The JOIDES *Resolution* is capable of drilling the deep seafloor. (b) The submersible *Alvin* is used for observation and sampling of the deep seafloor.

In addition to surface vessels, submersibles—some remotely controlled and others carrying scientists—have been added to the research arsenal of oceanographers. In 1985 the *Argo,* towed by a surface vessel and equipped with sonar and television systems, provided the first views of the British ocean liner R.M.S. *Titanic* since it sank in 1912. The U.S. Geological Survey is also using a towed device to map the seafloor. The system uses sonar to produce images resembling aerial photographs. Researchers aboard the submersible *Alvin* (Figure 11.7b) have observed submarine hydrothermal vents and have explored parts of the oceanic ridge system.

The first measurements of the oceanic depths were made by lowering a weighted line to the seafloor and measuring the length of the line. Now an instrument called an *echo sounder* is used (Figure 11.8). Sound waves from a ship are reflected from the seafloor and detected by instruments on the ship, yielding a continuous profile of the seafloor. Depth is determined by knowing the velocity of sound waves in water and the time it takes for the waves to reach the seafloor and return to the ship.

Seismic profiling is similar to echo sounding but even more useful. Strong waves are generated at an energy source, the waves penetrate the layers beneath the seafloor, and some of the energy is reflected back to the surface from various geologic horizons. Recall from Chapter 10 that seismic waves are reflected from boundaries where the properties of Earth materials change. Seismic profiling has been particularly useful in mapping the structure of the oceanic crust beneath seafloor sediments.

Oceanographers also use gravity surveys to detect gravity anomalies. Salt domes beneath the continental margins are recognized by negative gravity anomalies, and oceanic trenches also exhibit negative gravity anomalies. Magnetic surveys have also provided important information about the seafloor (see Chapter 12).

Although scientific investigations of the oceans have been yielding important information for more than 200 years, much of our current knowledge has been acquired since World War II (1939–1945). This statement is particularly true with respect to the seafloor, because only in recent decades has instrumentation been available to study this largely hidden domain. The data collected are not only important in their own right but also have provided much of the evidence supporting plate tectonic theory (see Chapter 12).

WHAT ARE THE CONTINENTAL MARGINS?

In the Introduction we made the point that continents are not simply areas above sea level, although most people perceive of continents as land areas outlined by sea level. The true geologic margin of a continent—that is, where continental crust changes to oceanic crust—is below sea level. Accordingly, the margins of continents are submerged, so we recognize **continental margins** separating the part of a continent above sea level from the deep seafloor.

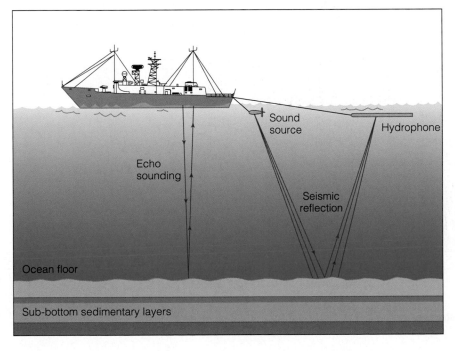

Figure 11.8
Diagram showing how echo sounding and seismic profiling are used to study the seafloor. Some of the energy generated at the energy source is reflected from various horizons back to the surface, where it is detected by hydrophones.

Sound source

Hydrophone

Echo sounding

Seismic reflection

Ocean floor

Sub-bottom sedimentary layers

A continental margin is made up of a gently sloping continental shelf, a more steeply inclined continental slope, and, in some cases, a deeper, gently sloping continental rise (Figure 11.9). Seaward of the continental margin lies the deep-ocean basin. Thus the continental margins extend to increasingly greater depths until they merge with the deep seafloor. The change from continental crust to oceanic crust generally takes place somewhere beneath the continental slope, so part of the continental slope and the continental rise actually rest on oceanic crust.

The Continental Shelf

Between the shoreline and the continental slope lies the **continental shelf,** an area where the seafloor slopes seaward at an angle of much less than 1 degree (on average, about 0.1 degree or about 2 m/km). The outer edge of the continental shelf is generally taken to correspond to the point where the inclination of the seafloor increases rather abruptly to several degrees; this *shelf–slope break* is at an average depth of 135 m (Figure 11.9). Continental shelves vary from a few tens of meters to more than 1000 km wide. The shelf along the East Coast of North America, for example, is as much as several hundred kilometers across in some places, whereas along the West Coast it is only a few kilometers wide.

Deep, steep-sided submarine canyons are most characteristic of the continental slope, but some extend well up onto the continental shelf. A number of these canyons lie offshore from the mouths of large rivers. At times during the Pleistocene Epoch (1,600,000 to 10,000 years ago), sea level was as much as 130 m lower than at present, so much of the continental shelves was above sea level. Rivers flowed across these exposed shelves and eroded deep canyons that were subsequently flooded when sea level rose. However, most submarine canyons extend to depths far greater than can be explained by river erosion during periods of lower sea level. Furthermore, many submarine canyons are not associated with rivers on land. (They are discussed more fully in a following section.)

As a result of lower sea level during the Pleistocene, much of the sediment on continental shelves accumulated in stream channels and on floodplains. In fact, in many areas such as northern Europe and parts of North America, glaciers extended onto the exposed shelves and deposited gravel, sand, and mud. Since the Pleistocene Epoch, sea level has risen, submerging the shelf sediments, which are now being reworked by marine processes. The fact that these sediments were deposited on land is indicated by evidence of human settlements and fossils of a variety of land-dwelling animals.

The Continental Slope and Rise

The seaward margin of the continental shelf is marked by the *shelf–slope break* (at an average depth of 135 m) where the relatively steep **continental slope** begins (Figure 11.9). Continental slopes average about 4 degrees, but range from 1 to 25 degrees. In many places, especially around the margins of the Atlantic, the continental slope merges with a more gently sloping **continental rise.** In other places, such as around the Pacific Ocean, slopes commonly descend directly into an oceanic trench, and a continental rise is absent (Figure 11.9).

The shelf–slope break, marking the boundary between the shelf and slope, is an important feature in terms of sediment transport and deposition. Landward of the break—that is, on the shelf—sediments are affected by waves and tidal currents, but these processes have no effect on sediments seaward of the break, where

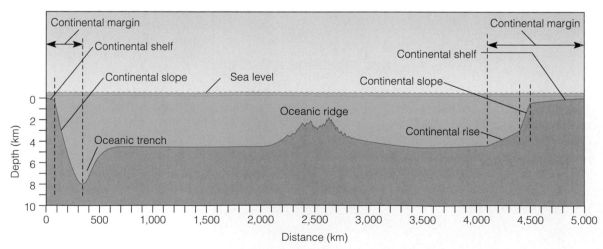

Figure 11.9
A generalized profile of the seafloor showing features of the continental margins. The vertical dimensions of the features in this profile are greatly exaggerated because the vertical and horizontal scales differ.

gravity is responsible for their transport and deposition on the slope and rise. In fact, much of the land-derived sediment crosses the shelves and is eventually deposited on the continental slopes and rises, where more than 70% of all sediments in the oceans are found. Much of this sediment is transported through submarine canyons by turbidity currents.

Turbidity Currents, Submarine Canyons, and Submarine Fans

In Chapter 6, we discussed the origin of graded bedding, most of which results from **turbidity currents,** underwater flows of sediment–water mixtures with densities greater than sediment-free water. As a turbidity current flows onto the relatively flat seafloor, it slows and begins depositing sediment, the largest particles first, followed by progressively smaller particles, thus forming graded bedding (see Figure 6.20). Deposition by turbidity currents results in the origin of a series of overlapping **submarine fans,** which constitute a large part of the continental rise (Figure 11.10). Submarine fans are distinctive features, but their outer margins are difficult to discern because they grade into deposits of the deep-ocean basins.

No one has ever observed a turbidity current in progress in the oceans, so for many years some doubted their existence. However, evidence now available removes all doubt. In 1971, for instance, abnormally turbid water was sampled just above the seafloor in the North Atlantic, indicating that a turbidity current had recently occurred. In addition, seafloor samples from many areas show a succession of layers with graded bedding and the remains of shallow-water

organisms that were apparently displaced into deeper water by turbidity currents.

Perhaps the most compelling evidence for turbidity currents is the pattern of transatlantic cable breaks that took place in the North Atlantic near Newfoundland on November 18, 1929 (Figure 11.11). Initially, it was assumed that an earthquake on that date had ruptured several transatlantic telephone and telegraph cables. However, while the breaks on the continental shelf near the epicenter occurred when the earthquake struck, cables farther seaward were broken later and in succession. The last cable to break was 720 km from the source of the earthquake, and it did not snap until 13 hours after the first break (Figure 11.11). In 1949 geologists realized that the earthquake had generated a turbidity current that moved downslope, breaking the cables in succession. The precise time each cable broke was known, so it was a simple matter to calculate the velocity of the turbidity current. It apparently moved at about 80 km/hr on the continental slope, but slowed to about 27 km/hr when it reached the continental rise.

Deep, steep-sided **submarine canyons** are present on continental shelves, but they are best developed on continental slopes (Figure 11.10). Some submarine canyons can be traced across the shelf to associated rivers on land; apparently, they formed as river valleys when sea level was lower during the Pleistocene. Many have no such association, and some extend far deeper than can be accounted for by river erosion during times of lower sea level. It is known that strong currents move through them and perhaps play some role in their origin. Furthermore, turbidity currents periodically move through these canyons and are now thought to be the primary agent responsible for their erosion.

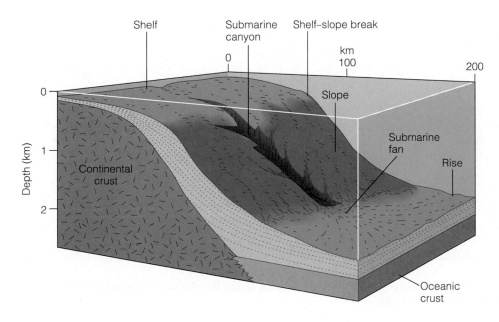

Figure 11.10
Submarine fans form by the deposition of sediments carried down submarine canyons by turbidity currents. Much of the continental rise is composed of overlapping submarine fans.

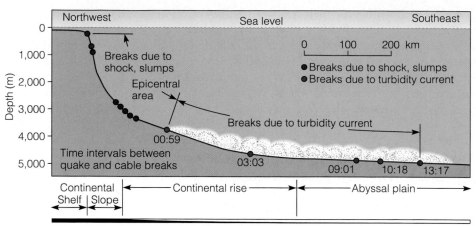

Figure 11.11
Submarine cable breaks caused by an earthquake-generated turbidity current south of Newfoundland. This profile of the seafloor shows the locations of the cables and the times at which they were severed. The vertical dimension in this profile is highly exaggerated. The profile labeled "no vertical exaggeration" shows what the seafloor actually looks like in this area.

Types of Continental Margins

In Chapter 1 we noted that plates diverge at spreading ridges, converge where two plates collide, and slide past one another along transform plate boundaries (see Figure 1.16). Continental margins are characterized as either *passive* or *active* depending on their type of plate tectonic activity. An **active continental margin** develops at the leading edge of a continental plate where oceanic lithosphere is subducted (Figure 11.12). The western margin of South America is a good example. Here an oceanic plate is subducted beneath the continent, resulting in seismicity, a geologically young mountain range, and active volcanism. In addition, the continental shelf is narrow, and the continental slope descends directly into the Peru–Chile Trench, so sediment is dumped into the trench and no continental rise develops. The western margin of North America is also considered an active

continental margin, although it is now bounded mostly by a transform fault rather than a subduction zone, as in South America. However, plate convergence and subduction still take place in the Pacific Northwest.

The configuration and geologic activity of the continental margins of eastern North and South America differ considerably from their western margins. In the east, the continental margins developed as a result of rifting of the supercontinent Pangaea. The continental crust was stretched, thinned, and fractured as rifting proceeded. As plate divergence occurred, the newly formed continental margins became the sites of deposition of land-derived sediments. These **passive continental margins** are within plates rather than at plate boundaries (Figure 11.12). They possess broad continental shelves and a continental slope and rise; vast, flat abyssal plains are commonly present adjacent to the rises (Figure 11.12). Furthermore, passive continental margins lack the intense seismic and

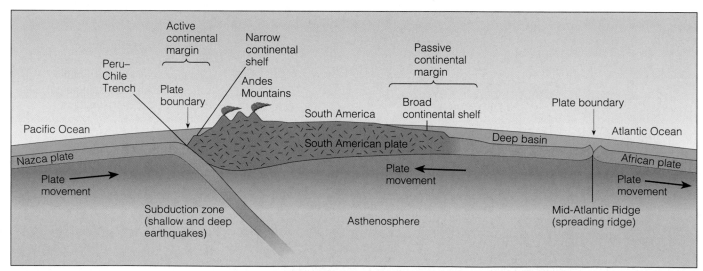

Figure 11.12
Active and passive continental margins along the west and east coasts of South America. Notice that passive continental margins are much wider than active margins. Seafloor sediment is not shown.

volcanic activity characteristic of active continental margins, and an oceanic trench is not present.

Active continental margins obviously lack a continental rise because the slope descends directly into an oceanic trench. Just as on passive continental margins, sediment is transported down the slope by turbidity currents, but it simply fills the trench rather than forming a rise. The proximity of the trench to the continent also explains why the continental shelf is so narrow. In contrast, the continental shelf of a passive continental margin is much wider because land-derived sedimentary deposits build outward into the ocean.

It should be clear from the preceding discussion that continental margins are active or passive depending on their location with respect to a plate boundary. However, just as Earth as a whole has evolved, so have continental margins, and one type can change to the other. For instance, eastern North America now has a passive continental margin, but during the Paleozoic Era it was bounded by an active continental margin. A change such as this depends on the stage of development of an ocean basin (see Perspective 11.1).

WHAT FEATURES ARE FOUND IN THE DEEP-OCEAN BASINS?

Considering that the oceans average more than 3.8 km deep, most of the seafloor lies far below the depth of sunlight penetration, which is rarely more than 100 m. Accordingly, most of the seafloor is completely dark, the temperature is generally just above 0°C, and the pressure varies from 200 to more than 1000 atmospheres depending on depth. Submersibles have carried scientists to the greatest oceanic depths, the oceanic ridges, and elsewhere so some of the seafloor has been observed directly. Nevertheless, much of the deep-ocean basin has been studied only by echo sounding, seismic profiling, sediment and oceanic crust sampling, and remote devices that have descended in excess of 11,000 m (Figure 11.13). Although oceanographers know considerably more about the deep-ocean basins than they did even a few years ago, many questions remain unanswered.

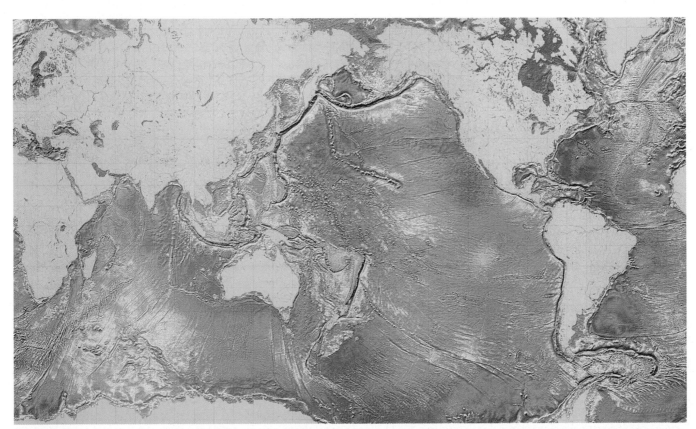

Figure 11.13
This map, based on gravity data recently declassified by the U.S. Navy, shows several features of the deep seafloor. Particularly obvious is the Mid-Atlantic Ridge and the numerous fractures oriented more or less perpendicular to the ridge. Oceanic trenches where subduction takes place show up as purple bands along the coasts of South America and Japan and elsewhere. The floor of the Pacific Ocean is dotted with submarine mountains known as *seamounts*.

Perspective 11.1

Opening and Closing Oceans

According to Canadian geologist J. Tuzo Wilson, oceans evolve through a distinct series of stages beginning with the rifting of a continent and culminating with the closure of a once expansive ocean. The generalized life cycle of ocean basins is divided into six-stages (Figure 1). This cycle of opening and closing ocean basins is now known as the *Wilson cycle.*

The onset of a Wilson cycle is marked by rifting of a continent and the origin of a complex system of linear rift valleys. As rifting proceeds, Earth's crust is stretched and thinned, and large blocks are displaced along fractures paralleling the rift margins (Figure 1). Rift valleys or basins formed in this manner are referred to as *embryonic,* because they represent the first stage in the origin of an ocean basin. Perhaps the best example of this initial stage in the Wilson cycle is the East African rift system.

Continued stretching and thinning of the continental crust mark the second stage of ocean-basin evolution. During this stage, oceanic crust is formed from basaltic magma, and the rift valley subsides below sea level, forming a long, narrow sea (Figure 1). The Red Sea and

the Gulf of Aden are in this young or juvenile stage. Stage 3 or the development of a mature ocean basin results from the continuation of the processes just described. Divergence of plates and the generation of oceanic crust at a spreading ridge continue until eventually an expansive ocean basin develops (Figure 1). As the ocean expands, sediments eroded from the continents on either side of the ocean basin are deposited in the sea, producing broad passive continental margins. The Atlantic Ocean is in this stage of development.

Following stage 3, the basin enters a declining stage as the oceanic crust becomes cooler and denser and begins to subduct at the margins of the ocean, thereby transforming passive continental margins into active ones (Figure 1). Oceanic lithosphere is consumed by subduction, and eventually even the spreading ridge itself is subducted. Parts of the East Pacific Rise are now being subducted beneath Central and South America as the Pacific Ocean basin becomes smaller.

Continuing closure of an ocean basin leads to a *terminal* stage in which island arcs and continents that were once widely separated begin to collide

(Figure 1). The Mediterranean Sea is in this stage, which, if continued, results in *suturing* as two colliding plates become welded together. An example of this stage is provided by the Himalayas of Asia, which began forming 40 to 50 million years ago when India collided with Asia. In fact, this collison is continuing, and India is currently being thrust beneath Asia at a rate of about 5 cm (2 in.) per year.

During the Late Proterozoic Eon and continuing until the Late Paleozoic Era, eastern North America experienced a sequence of events much like that illustrated in Figure 1. Beginning with the rifting, a supercontinent known as Rodina, what is now North America, separated from present-day Europe, both of which developed passive continental margins. Following the development of a large ocean basin, the plates began moving toward one another, with the consequent development of active continental margins. Finally, the two plates collided, giving rise to the supercontinent Pangaea. Rifting during the Mesozoic Era resulted in the origin of the passive continental margin now present in eastern North America.

Abyssal Plains

Beyond the continental rises of passive continental margins are **abyssal plains,** flat surfaces covering vast areas of the seafloor. In some places, they are interrupted by peaks rising more than 1 km, but in general they are the flattest, most featureless areas on Earth (Figure 11.14). The flat topography is a result of sediment deposition; where sediment accumulates in sufficient quantities, the rugged seafloor is buried beneath thick layers of sediment (Figure 11.15).

Seismic profiles and seafloor samples reveal that abyssal plains are covered with fine-grained sediment derived mostly from the continents and deposited by turbidity currents. Abyssal plains are invariably found adjacent to the continental rises, which are composed mostly of overlapping submarine fans that owe their origin to deposition by turbidity currents (Figure 11.10). Along active continental margins, sediments derived from the shelf and slope are trapped in an oceanic trench, and abyssal plains fail to develop. Accordingly, abyssal plains are common in the

Stage	Motion	Physiography	Example
Embryonic 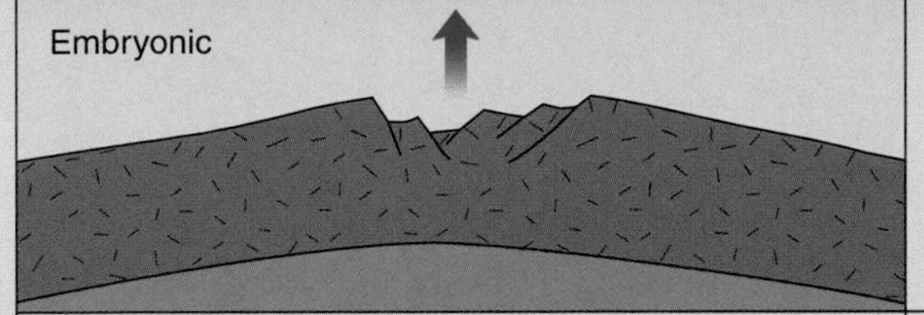	Uplift	Complex system of linear rift valleys on continent	East African rift valleys
Juvenile 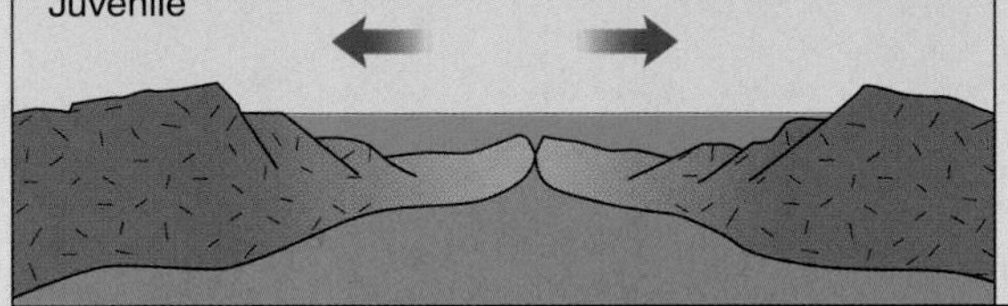	Divergence (spreading)	Narrow seas with matching coasts	Red Sea
Mature 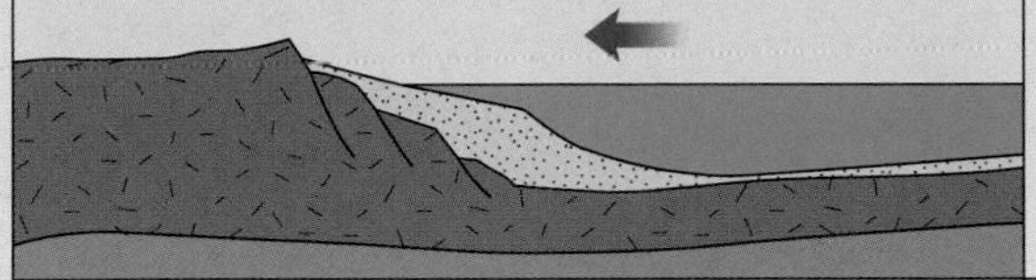	Divergence (spreading)	Ocean basin with continental margins	Atlantic, Indian, and Arctic Oceans
Declining 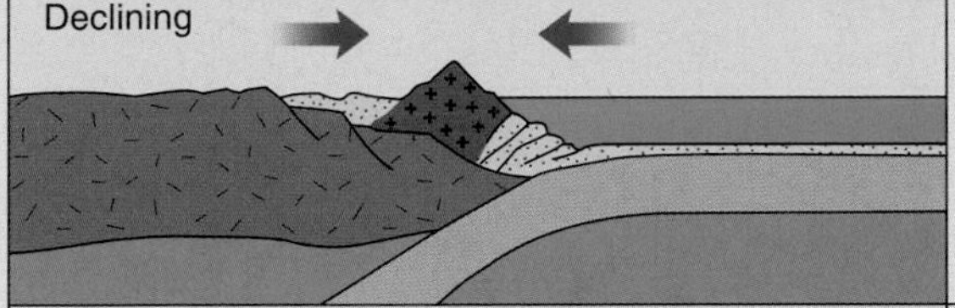	Convergence (subduction)	Island arcs and trenches around basin edge	Pacific Ocean
Terminal 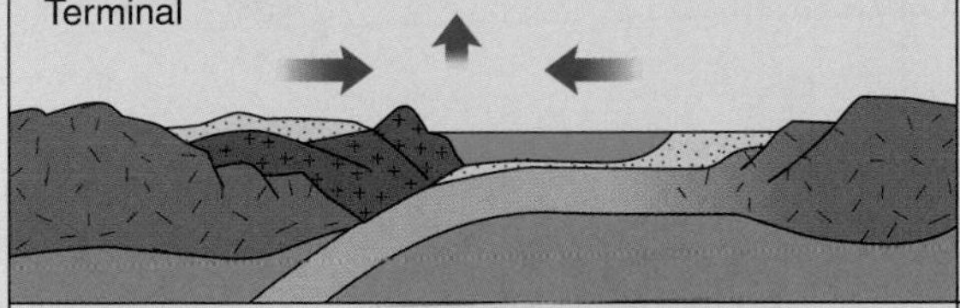	Convergence (collision) and uplift	Narrow, irregular seas with young mountains	Mediterranean Sea
Suturing 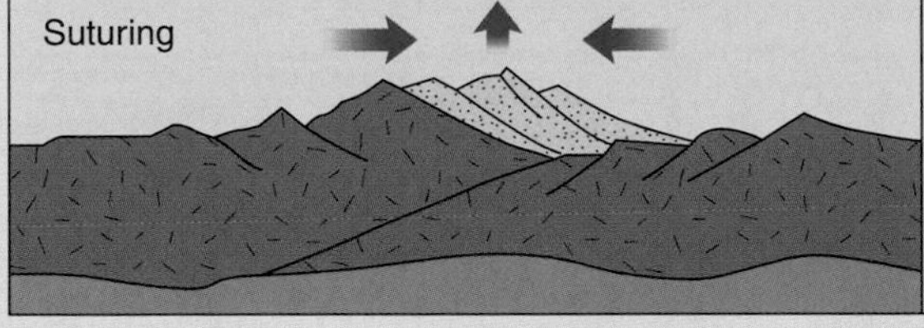	Convergence and uplift	Young to mature mountain belts	Himalayas

Figure 1
The Wilson cycle portrays ocean-basin development in six stages.

Atlantic Ocean basin but rare in the Pacific Ocean basin (Figure 11.14).

Oceanic Trenches

Oceanic trenches constitute a small percentage of the seafloor, probably less than 2%, but they are very important, for it is here that lithospheric plates are consumed by subduction (see Chapter 12). Oceanic trenches are long, narrow features restricted to active continental margins, so they are common around the margins of the Pacific Ocean basin (Figure 11.14). For instance, the Peru–Chile Trench west of South America is 5900 km long but only 100 km wide. It is more than 8000 m deep. On the landward side of oceanic trenches, the continental slope descends at angles of up to 25 degrees (Figure 11.12). Oceanic trenches are also the sites of the greatest oceanic depths; a depth of more than 11,000 m has been recorded in the Challenger Deep of the Marianas Trench.

Oceanic trenches show anomalously low heat flow compared with other areas of oceanic crust, indicating

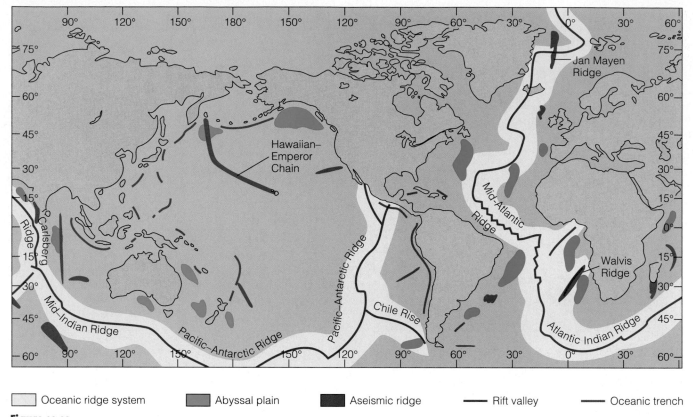

Figure 11.14
The distribution of oceanic trenches (brown), abyssal plains (green), the oceanic ridge system (yellow) and rift valleys (red), and some of the aseismic ridges (blue).

☐ Oceanic ridge system	▤ Abyssal plain	▦ Aseismic ridge	— Rift valley	— Oceanic trench

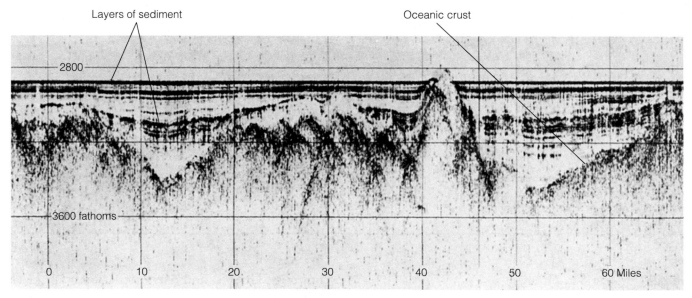

Figure 11.15
Seismic profile showing rugged seafloor topography covered by sediments of the Northern Madeira Abyssal Plain in the Atlantic Ocean.

that the crust at trenches is cooler and slightly more dense than elsewhere. Gravity surveys across trenches reveal huge negative gravity anomalies because the crust is not in isostatic equilibrium. In fact, oceanic crust at trenches is subducted, giving rise to Benioff zones in which earthquake foci become progressively deeper in the direction the subducted plate descends (see Figure 9.8). These inclined seismic zones account for most intermediate- and deep-focus earthquakes—such as the June 1994 magnitude 8.2 earthquake in Bolivia, which had a focal depth of 640 km. Finally, subduction at oceanic trenches also results in volcanism, either as an arcuate chain of volcanic islands on oceanic crust or as a chain of volcanoes along the margin of a continent, as in western South America.

Oceanic Ridges

When the first submarine cable was laid between North America and Europe during the late 1800s, a feature called the Telegraph Plateau was discovered in the North Atlantic. Using data from the 1925–1927 voyage of the German research vessel *Meteor*, scientists proposed that the plateau was actually a continuous ridge extending the length of the Atlantic Ocean basin. Subsequent investigations revealed that this proposal was correct, and we now call this feature the Mid-Atlantic Ridge (Figure 11.14).

We now know that the Mid-Atlantic Ridge is more than 2000 km wide and rises 2 to 2.5 km above the adjacent seafloor. Furthermore, it is part of a much larger **oceanic ridge** system of mostly submarine mountainous topography. It runs from the Arctic Ocean through the middle of the Atlantic, curves around South Africa where the Indian Ridge continues into the Indian Ocean, then the Pacific–Antarctic Ridge extends eastward and a branch of this, the East Pacific Rise, trends northeast until it reaches the Gulf of California (Figure 11.14). The entire system is at least 65,000 km long, far surpassing the length of any mountain range on land. Oceanic ridges are composed almost entirely of basalt and gabbro and possess features produced by tensional forces. Mountain ranges on land, in contrast, consist of granitic, metamorphic, and sedimentary rocks, and formed when rocks were folded and fractured by compressive forces (see Chapter 13).

Oceanic ridges are mostly below sea level, but they rise above the sea in some places such as Iceland, the Azores, and Easter Island. Of course, oceanic ridges are the sites where new oceanic crust is generated and plates diverge (see Chapter 12). The rate of plate divergence is important because it determines the cross-section profile of a ridge. For example, the Mid-Atlantic Ridge has a comparatively steep profile because divergence is slow, allowing the new oceanic crust to cool, shrink, and subside closer to the ridge crest than it does in areas of faster divergence such as at the East Pacific Rise. A ridge may also have a rift along its crest that opens in response to tension (Figure 11.16). A rift is particularly obvious along the Mid-Atlantic Ridge, but appears to be absent along parts of the East Pacific Rise. These rifts are commonly 1 to 2 km deep and several kilometers wide. They open as seafloor spreading takes place (discussed in Chapter 12), and are characterized by shallow-focus earthquakes, basaltic volcanism, and high heat flow.

Even though most oceanographic research is still done by echo sounding, seismic profiling, and seafloor sampling, scientists have been making direct observations of oceanic ridges and their rifts since 1974. As part of Project FAMOUS (French-American Mid-Ocean Undersea Study), submersibles have descended to the ridges and into their rifts in several areas. No active volcanism has been observed, but researchers did see pillow lavas (see Figure 4.8), lava tubes, and sheet lava flows, some of which seemed to have formed very recently. In fact, on return visits to a site they have seen the effects of volcanism that occurred since their last visit. And on January 25, 1998, a submarine volcano began erupting along the Juan de Fuca Ridge west of Oregon. Researchers aboard submersibles have also observed hot water being discharged from the seafloor

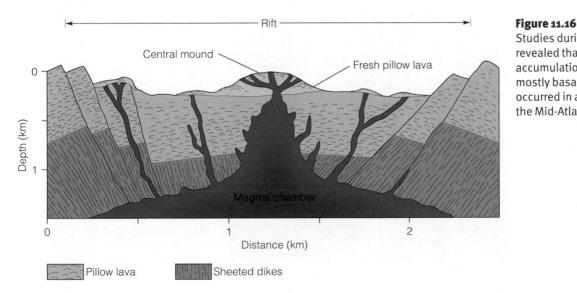

Figure 11.16
Studies during the early 1970s revealed that recent moundlike accumulations of volcanic rocks, mostly basaltic pillow lavas, occurred in a rift along the axis of the Mid-Atlantic Ridge.

at or near ridges in submarine hydrothermal vents known as *black smokers.*

Submarine Hydrothermal Vents

In 1979, researchers aboard the submersible *Alvin* descended about 2500 m to the Galapagos Rift in the eastern Pacific Ocean basin and observed hydrothermal vents on the seafloor (Figure 11.17). These vents are at or near spreading ridges where cold seawater seeps down into the oceanic crust through cracks and fissures, is heated by the hot rocks, and then rises and is discharged onto the seafloor as hot springs. Mounds of metal-rich sediments were observed, and near these mounds the researchers saw what they called black smokers (chimneylike vents) discharging plumes of hot, black water (Figure 11.17a). Since 1979 similar vents have been observed at or near spreading ridges in several other areas.

Submarine hydrothermal vents are interesting for several reasons. Near the vents live communities of organisms, including bacteria, crabs, mussels, starfish, and tube worms, many of which had never been seen before (Figure 11.17b). In most biological communities, photosynthesizing organisms form the base of the food chain and provide nutrients for the herbivores and carnivores. In vent communities, no sunlight is available for photosynthesis, and the base of the food chain consists of bacteria that oxidize sulfur compounds from the hot vent waters, thus providing their own nutrients and the nutrients for other members of the food chain.

Another interesting aspect of these submarine hydrothermal vents is their economic potential. When seawater circulates down through the oceanic crust, it is heated to as much as 400°C. The hot water reacts with the crust and is transformed into a metal-bearing solution. As the hot solution rises and discharges onto the seafloor, it cools, precipitating iron, copper, and zinc sulfides and other minerals that accumulate to form a chimneylike vent. Ap-

parently the chimneys grow rapidly. A 10-m-high chimney accidently knocked over in 1991 by the submersible *Alvin* grew to 6 m in just three months. They are, however, ephemeral; one observed in 1979 was inactive six months later. When their activity ceases, the vents eventually collapse and are incorporated into a moundlike mineral deposit.

Scientists aboard *Alvin* in April 1991 saw the results of a submarine eruption on the East Pacific Rise, which they missed by less than two weeks. Fresh lava and ash covered the area, as well as the remains of tube worms killed during the eruption. In a nearby area, a new fissure opened in the seafloor, and by December 1993 a new hydrothermal vent community had become established with tube worms fully 1.2 m long.

The economic potential of hydrothermal vent deposits is tremendous. The deposits in the Atlantis II

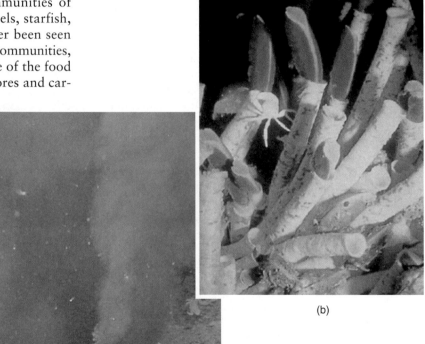

(b)

Figure 11.17
(a) A hydrothermal vent known as a *black smoker* at 2000 m on the East Pacific Rise. Seawater seeps down through the oceanic crust, becomes heated, and then rises and builds chimneys composed of anhydrite ($CaSO_4$) and sulfides of iron, copper, and zinc. The plume of "black smoke" is simply heated water saturated with dissolved minerals.
(b) Several types of organisms including these tube worms live near black smokers.

(a)

Deep of the Red Sea contain an estimated 100 million tons of metals, including iron, copper, zinc, silver, and gold. These deposits are fully as large as the major sulfide deposits mined on land—such as the Troodos Massif on Cyprus, which is thought to have formed on the seafloor by hydrothermal vent activity.

Fractures in the Seafloor

Oceanic ridges are not continuous features winding without interruption around the globe. They abruptly terminate where they are offset along fractures oriented more or less at right angles to ridge axes (Figure 11.18). These fractures run for hundreds of kilometers, although they are difficult to trace where buried beneath seafloor sediments. Many geologists are convinced that some geologic features on continents can best be accounted for by the extension of these fractures into continents.

Where oceanic ridges are offset, they are characterized by shallow seismic activity only in the area between the displaced ridge segments (Figure 11.18). Furthermore, because ridges are higher than the seafloor adjacent to them, the offset segments yield vertical relief on the seafloor. Nearly vertical escarpments 2 or 3 km high develop, as illustrated in Figure 11.18. The reason that oceanic ridges show so many fractures is that plate divergence takes place irregularly on a sphere, resulting in stresses that cause fracturing. We will have more to say about these fractures, known as *transform faults* between offset ridge segments, in Chapter 12.

Seamounts, Guyots, and Aseismic Ridges

As noted previously, the seafloor is not a flat, featureless plain, except for the abyssal plains, and even these are underlain by rugged topography (Figure 11.15). In fact, a large number of volcanic hills, seamounts, and guyots rise above the seafloor in all ocean basins, but they are particularly abundant in the Pacific. All are of volcanic origin and differ from one another mostly in size. **Seamounts** rise more than 1 km above the seafloor; if they are flat topped, they are called **guyots** rather than seamounts (Figure 11.19). Guyots are volcanoes that originally extended above sea level. But as the plate they rested on continued to move, they were carried away from a spreading ridge, and the oceanic crust cooled and descended to greater oceanic depths. So what was once an island was eroded by waves as it slowly sank beneath the sea, giving it the typical flat-topped appearance.

Many other volcanic features smaller than seamounts are also present on the seafloor, but they probably originated in the same way. These so-called *abyssal hills*, averaging only about 250 m high, are common on the seafloor and underlie thick sediments on the abyssal plains.

Other common features in the ocean basins are long, narrow ridges and broad plateaulike features rising as much as 2 to 3 km above the surrounding seafloor. These are known as **aseismic ridges** because they lack seismic activity. A few of these ridges are thought to be small fragments separated from continents during rifting. These are referred to as *microcontinents*, and are represented by such features as the Jan Mayen Ridge in the North Atlantic (Figure 11.14).

Most aseismic ridges form as a linear succession of hot spot volcanoes. These may develop at or near an oceanic ridge, but each volcano so formed is carried laterally with the plate on which it originated. The net result of this activity is a sequence of seamounts/guyots extending from an oceanic ridge (Figure 11.19); the Walvis Ridge in the South Atlantic is a good example (Figure 11.14). Aseismic ridges also form over hot spots unrelated to ridges. The Hawaiian–Emperor chain in the Pacific formed in such a manner (Figure 11.14).

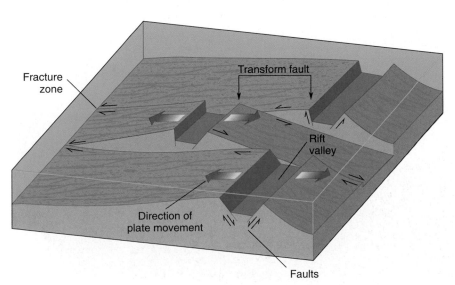

Figure 11.18
Diagrammatic view of an oceanic ridge offset along fractures. That part of a fracture between displaced segments of the ridge crest is known as a *transform fault* (see Chapter 12).

Fracture zone

Transform fault

Rift valley

Direction of plate movement

Faults

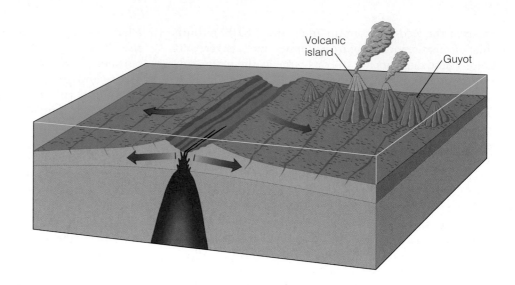

Figure 11.19
Submarine volcanism might build up above sea level, forming a volcanic island at or near a spreading ridge. As the plate the volcano rests on moves to greater depths with increasing distance from the ridge, the island is eroded, sinks beneath sea level, and becomes a guyot.

SEDIMENTATION AND SEDIMENTS ON THE DEEP SEAFLOOR

Deep-sea sediments consist mostly of fine-grained deposits because few mechanisms exist that transport coarse-grained sediment (sand and gravel) far from land. Icebergs can transport coarse sediment into the ocean basins, though, and a broad band of glacial–marine sediment is present adjacent to Antarctica and Greenland (Figure 11.20).

Most of the fine-grained sediment in the deep sea is windblown dust and volcanic ash from the continents and oceanic islands and the shells of microscopic organisms that live in the near-surface waters of the oceans. Other sources of sediment include cosmic dust and deposits resulting from chemical reactions in seawater. The manganese nodules that are fairly common in all the ocean basins are a good example of the latter (Figure 11.21). They are composed mostly of man-

ganese and iron oxides but also contain copper, nickel, and cobalt. Nodules may be an important source of some metals in the future; the United States, which imports most of its manganese and cobalt, is particularly interested in this potential resource.

The contribution of cosmic dust to deep-sea sediment is negligible. Even though some researchers estimate that as much as 40,000 metric tons of cosmic dust may fall to Earth each year, this is a trivial quantity compared to the volume of sediments derived from other sources.

Most deep seafloor sediments are *pelagic,* meaning that they settled from suspension far from land. Two categories of pelagic sediment are recognized: pelagic clay and ooze (Figure 11.22). Brown or reddish **pelagic clay,** composed of clay-sized particles derived from continents and oceanic islands, covers most of the deeper parts of the ocean basins. **Ooze,** in contrast, is composed mostly of shells of microscopic marine animals and plants. It is characterized as *calcareous ooze* if it contains mostly calcium carbonate ($CaCO_3$) skeletons of

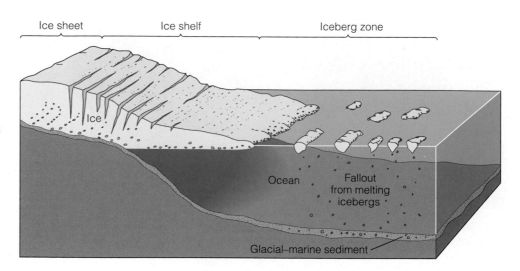

Figure 11.20
The formation of glacial–marine sediments by ice rafting, a process in which icebergs transport sediment into the ocean basins, releasing it as they melt.

Figure 11.21
(a) Manganese nodules on the seafloor. (b) Sectioned manganese nodule showing its internal structure consisting of concentric layers. This nodule measures about 6 cm across.

(a)

(b)

tiny marine organisms such as foraminifera. *Siliceous ooze* is composed of the silica (SiO_2) skeletons of such single-celled organisms as radiolarians (animals) and diatoms (plants) (Figure 11.22).

HOW ARE REEFS MADE BY ORGANISMS?

The term **reef** has a variety of meanings, such as shallowly submerged rocks that pose a hazard to navigation, but here we restrict it to mean a moundlike, wave-resistant structure composed of the skeletons of marine organisms (Figure 11.23). Although commonly called *coral reefs,* they actually have a solid framework composed of skeletons of corals and various mollusks such as clams, and encrusting organisms including sponges and algae. Reefs grow to a depth of about 45 to 50 m and are restricted to shallow, tropical seas where the water is clear and the temperature does not fall below about 20°C. The depth to which reefs can grow depends on sunlight penetration because many of the corals, for example, rely on symbiotic algae that depend on sunlight for energy.

Reefs are found in a variety of shapes and sizes, but most can be classified as one of three basic types: fringing, barrier, and atoll (Figure 11.24). *Fringing reefs* are solidly attached to the margins of an island or continent. They have a rough, tablelike surface, are as much as 1 km wide, and, on their seaward side, slope steeply down to the seafloor. *Barrier reefs* are similar to fringing reefs, except that they are separated from the mainland by a lagoon. Probably the best-known barrier reef in the world is the Great Barrier Reef of Australia, which is 2000 km long.

An *atoll* is a circular to oval reef surrounding a lagoon (Figure 11.23). They form around volcanic islands that subside below sea level as the plate they rest on is carried progressively farther from an oceanic ridge (Figure 11.19). As subsidence proceeds, the reef organisms construct the reef upward so that the living part of the reef remains in shallow water. The island

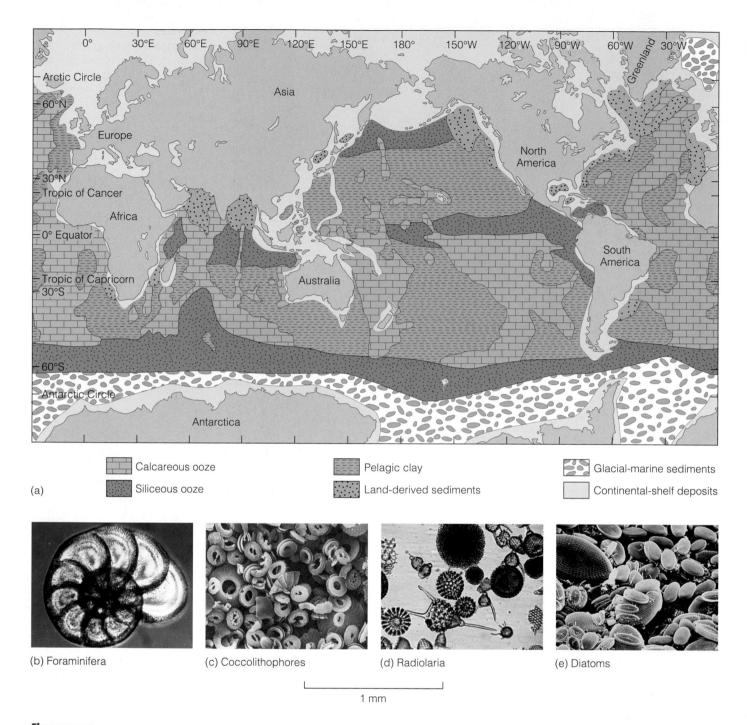

Figure 11.22

(a) A variety of sediments are present in the ocean basins, but most on the deep seafloor is pelagic clay and calcareous and siliceous ooze. Common constituents of calcareous ooze are skeltons of (b) foraminifera (floating single-celled animals) and (c) coccolithophores (floating single-celled plants), whereas siliceous ooze consists of skeletons of (d) radiolarians (single-celled floating animals) and (e) diatoms (single-celled floating plants).

(a)

Figure 11.23
Reefs are wave-resistant structures composed of the skeletons of organisms, especially corals. (a) Underwater view of a reef in the Red Sea. (b) This oval reef with a central lagoon in the Pacific Ocean is an atoll.

(b)

eventually subsides below sea level, leaving a circular lagoon surrounded by a more or less continuous reef (Figure 11.24). Atolls are particularly common in the western Pacific Ocean basin. Many of these began as fringing reefs, but as the plate they were on was carried into deeper water, they evolved first to barrier reefs and finally to atolls.

This particular scenario for the evolution of reefs from fringing to barrier to atoll was proposed more than 150 years ago by Charles Darwin while he was serving as a naturalist on the ship HMS *Beagle*. Drilling into atolls has revealed that they do indeed rest on a basement of volcanic rocks, confirming Darwin's hypothesis.

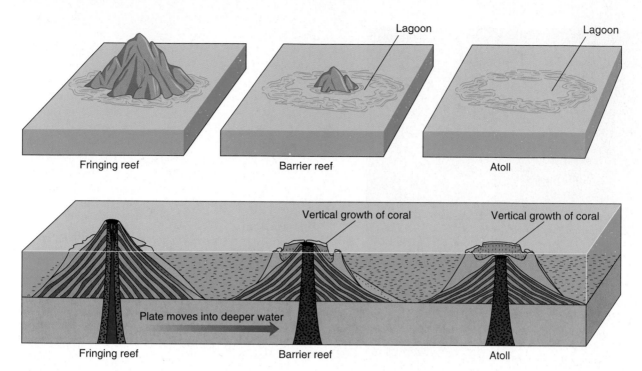

Figure 11.24

Three stages in the development of an atoll. In the first stage, a fringing reef forms around a volcanic island, but as the island is carried to greater depths on a moving plate, the reef becomes separated from the island by a lagoon. At this stage it is a barrier reef. Continued plate movement carries the island with its barrier reef into even deeper water, and as the island disappears below sea level the reef continues to grow upward, thus forming an atoll.

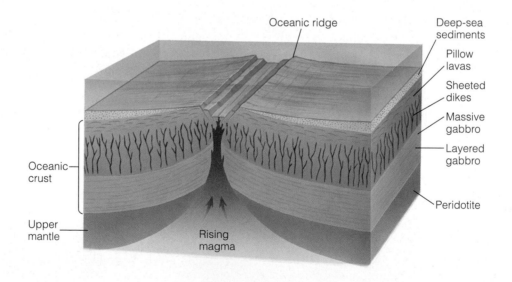

Figure 11.25

New oceanic crust consisting of the layers shown here forms as magma rises beneath oceanic ridges. The composition of the oceanic crust is known from ophiolites, sequences of rock on land consisting of deep-sea sediments, oceanic crust, and upper mantle.

The Future Beneath the Sea

Robert D. Ballard is president of the Institute for Exploration in Mystic, Connecticut, and Scientist Emeritus in the Department of Applied Physics and Engineering at Woods Hole Oceanographic Institution. He is also the founder and chairman of the Jason Foundation for Education. Ballard earned a B.S. in physical science at the University of California, Santa Barbara, and a Ph.D. in marine geology and geophysics at the University of Rhode Island, Graduate School in Oceanography.

After 30 years as an undersea explorer first at the Woods Hole Oceanographic Institution and now at the Institute for Exploration in Mystic, Connecticut—I am still fascinated by the sea. Given our exploding population, our diminished interest in the promises of the space program, and the continued development of advanced technology, I truly believe the twenty-first century will usher in an explosion in human activity in the sea. I am convinced the next generation will explore more of Earth—that is, the 71% that lies underwater—than all previous generations combined.

Just as Lewis and Clark's explorations led to the settling of the West, the exploration of the sea will lead to its colonization. The gathering and hunting of the living resources of the sea, an activity characteristic of primitive societies on land, will be replaced at sea by farming and herding. High-tech barbwire in the form of acoustic, thermal, or other barrier techniques will emerge to control and manage the living resources of the sea.

Oil and gas exploration and exploitation will continue moving into deeper and deeper depths. We have already discovered and mapped oil and gas reserves down to 12,000 feet, which represents the average depth of the ocean, and each year the oil industry brings production wells on line in waters deeper than the previous year.

In recent years, we have discovered major mineral deposits in the deep sea similar to those mined for centuries on Cyprus. They contain high concentrations of copper, lead, and sulfur, as well as silver and gold. Their formation continues today in the vast hydrothermal vent systems of the Mid-Ocean Ridge. These mineral deposits will be processed using the very geothermal energy that drives the crustal processes that lead to their formation. Some of these magnificent vent areas will also become the Yellowstone Parks of the deep sea, leading to future arguments over their commercial versus tourist value.

The unique chemosynthetic lifeforms that process the toxic material associated with the vent communities will hopefully be bioengineered to convert a portion of our waste products into less harmful or even commercially valuable by-products. These exotic creatures will also help us understand the early origin of life on our planet as well as the potential for life on other planets we once ruled out for their lack of a friendly nearby Sun.

Whether all this occurs during the next generation's time on Earth, only time will tell. But the seeds can be found in programs already under way. With this exploration will come better understanding of the ocean and the land surface beneath it, which is critical to our understanding of the planet as a whole. ■

WHAT IS THE STRUCTURE AND COMPOSITION OF THE OCEANIC CRUST?

Most of the oceanic crust is consumed at subduction zones. However, a tiny amount of it is not subducted, and it—along with seafloor sediment and pieces of the underlying upper mantle—is emplaced in mountain ranges on land, usually by moving along large fractures called *thrust faults*. These preserved slivers of upper mantle and oceanic crust are known as **ophiolites.**

Detailed studies reveal that an ideal ophiolite consists of a layer of deep-sea sedimentary rocks underlain successively by a layer of pillow lavas and sheet lava flows and a sheeted dike complex, a zone consisting mostly of vertical basaltic dikes (Figure 11.25). Farther down in an ophiolite is massive gabbro, and below that is layered gabbro that probably cooled at the top of a magma chamber. Beneath the gabbro is peridotite representing the upper mantle; this layer is sometimes altered by metamorphism to an assemblage containing serpentine. Thus, a complete ophiolite sequence consists of deep-sea sedimentary rock, oceanic crust, and upper mantle rocks (Figure 11.25).

Sampling and direct observations of the oceanic ridges and deep-sea drilling reveal that the upper part of the oceanic crust is indeed composed of a layer of pillow lavas and sheet flows underlain by a sheeted dike complex, exactly as predicted from ophiolites. It was not until 1989 that direct confirmation was made of what lay below the sheeted dike complex. In that year, scientists in a submersible descended to the walls of a seafloor fracture in the North Atlantic. There, rocks

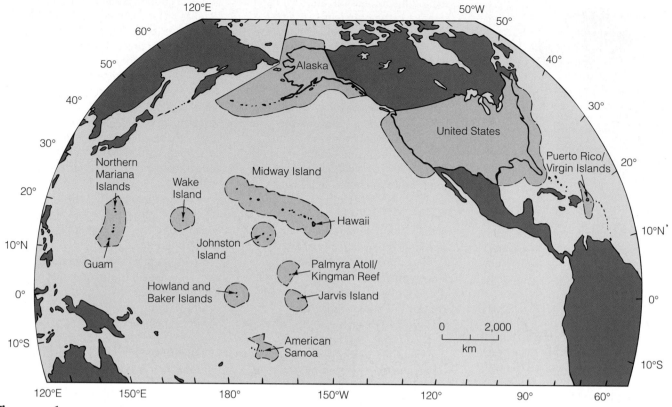

Figure 11.26
The Exclusive Economic Zone (EEZ) includes a vast area adjacent to the United States and its possessions.

of the upper mantle and lower oceanic crust were close to the seafloor and, for the first time, scientists observed the peridotite and gabbro parts of the sequence. In short, the structure and composition of oceanic crust were inferred from fragments of presumed oceanic crust on land, and these inferences were later verified by observations.

WHAT RESOURCES COME FROM SEAWATER AND THE SEAFLOOR?

Seawater contains many elements in solution, some of which are extracted for various industrial and domestic uses. In many places, sodium chloride (table salt) is produced by the evaporation of seawater, and a large proportion of the world's magnesium is produced from seawater. Numerous other elements and compounds can be extracted from seawater, but for many, such as gold, the cost is prohibitive.

In addition to substances in seawater, deposits on the seafloor or within seafloor sediments are becoming increasingly important. Many of these potential re-

sources lie well beyond continental margins, so their ownership is a political and legal problem that has not yet been resolved. Most nations bordering an ocean claim those resources within their adjacent continental margin. The United States, by a presidential proclamation issued on March 10, 1983, claims sovereign rights over an area designated as the **Exclusive Economic Zone (EEZ)**. The EEZ extends seaward 200 nautical miles (371 km) from the coast, giving the United States jurisdiction over an area about 1.7 times larger than its land area (Figure 11.26).* Also included within the EEZ are the areas adjacent to U.S. territories, such as Guam, American Samoa, Wake Island, and Puerto Rico (Figure 11.26). In short, the United States claims a huge area of the seafloor and any resources on or beneath it.

Numerous resources are found within the EEZ, some of which have been exploited for many years. Sand and gravel for construction are mined from the continental shelf in several areas. About 17% of U.S. oil and natural gas production comes from wells on the continental shelf (Figure 11.27). Some 30 sedimentary basins

---

*A number of other nations also claim sovereign rights to resources within 200 nautical miles of their coasts.

Figure 11.27
An oil-drilling platform off the coast of southern California. Although this one is in fairly shallow water, a similar platform off the Louisiana coast is anchored in water 872 m deep. About 17% of all U.S. oil production comes from wells on the continental shelves.

U.S. Geological Survey think that methane hydrates contribute to these slides, during which methane is released into the atmosphere, where along with carbon dioxide it contributes to global warming. Its contribution to global warming must be assessed because a volume of methane 3000 times greater than that in the atmosphere is present in seafloor sediments, and it is 10 times more effective than carbon dioxide as a factor in global warming.

Other resources of interest include massive sulfide deposits that form by submarine hydrothermal activity at spreading ridges. These deposits containing iron, copper, zinc, and other metals have been identified within the EEZ at the Gorda Ridge off the coasts of California and Oregon; similar deposits occur at the Juan de Fuca Ridge within the Canadian EEZ.

Other potential resources include the manganese nodules, discussed previously (Figure 11.21), and metalliferous oxide crusts found on seamounts. Manganese nodules contain manganese, cobalt, nickel, and copper; the United States is heavily dependent on imports of the first three of these elements. Within the EEZ, manganese nodules are found near Johnston Island in the Pacific Ocean and on the Blake Plateau off the east coast of South Carolina and Georgia. In addition, seamounts and seamount chains within the EEZ in the Pacific are known to have metalliferous oxide crusts several centimeters thick from which cobalt and manganese could be mined.

Another important resource forming in shallow seawater is phosphate-rich sedimentary rock known as *phosphorite* (see Perspective 11.2).

are known within the EEZ, several of which contain hydrocarbons, whereas others are areas of potential hydrocarbon production.

A potential resource within the EEZ is methane hydrate, which consists of single methane molecules bound up in networks formed by frozen water. Although methane hydrates have been known since the early part of the nineteenth century, they have only recently received much attention. Most of these deposits lie within continental margins, but so far it is not known if they can be effectively recovered and used as an energy source. According to one estimate, the amount of carbon in methane hydrates is double that in all coal, oil, and conventional natural gas reserves.

Methane hydrates are stable at water depths exceeding 500 m and at near-freezing temperatures. Along the Atlantic continental margin of the United States numerous submarine landslide scars are present where the seafloor should be stable. Geologists at the

---

WHAT WOULD YOU DO?

Hydrothermal vents on the seafloor are known sites of several metals of great importance to industrialized societies. Furthermore, it appears that these metals are being deposited even now, so if we mine one area, more of the same resources are forming elsewhere. Given these conditions, it would seem that our problems of diminishing resources are solved. So why not simply mine the seafloor? Also, many chemical elements are present in seawater. The technology exists to extract these elements such as gold, uranium, and others, so why not do so?

Perspective 11.2

Oceanic Circulation and Resources from the Sea

Earth's oceans are in constant motion as huge quantities of water circulate horizontally because surface currents such as the Gulf Stream and South Equatorial Current carry water great distances. These and other surface currents transfer heat from the equatorial regions toward the poles and thus have a modifying effect on climate. Deep-ocean waters also move horizontally as a result of temperature and salinity differences between adjacent water masses. In addition to horizontal circulation, vertical circulation takes place in the oceans when *upwelling* slowly transfers cold water from depth to the surface and *downwelling* transfers warm surface water to depth.

Upwelling is of more than academic interest. It not only transfers water from depth to the surface, but also carries nutrients, especially nitrates and phosphate, into the zone of sunlight penetration where high concentrations of floating organisms are sustained, which in turn support other organisms. In fact, other than the continental shelves and areas adjacent to hydrothermal vents on the seafloor, areas of upwelling are the only parts of the oceans where biological productivity is very high. Less than 1% of the ocean surfaces are areas of upwelling, yet they support more than 50% (by weight) of all fishes.

Figure 1
Wind from the north along the western margin of a continent, coupled with the Coriolis effect, causes surface water to move offshore, resulting in upwelling of cold, nutrient-rich deep water.

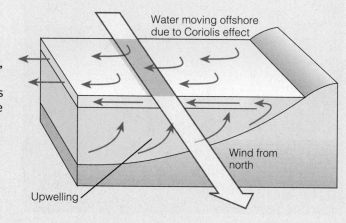

Water moving offshore due to Coriolis effect

Wind from north

Upwelling

Scientists recognize three types of upwelling, but only one need concern us here—*coastal upwelling*. Most coastal upwelling takes place along the west coasts of Africa, North America, and South America, although one notable exception is present in the Indian Ocean. Coastal upwelling involves movement of water offshore, which is replaced by water rising from depth (Figure 1). For instance, the winds blowing south along the coast of Peru, coupled with the Coriolis effect, transport surface water seaward, and cold, nutrient-rich water from depth rises to replace it. This area, too, is a major fishery, and changes in the surface-water circulation every three to seven years adjacent to South America are associated with the onset of El Niño, a

weather phenomenon with far-reaching consequences.

Our interest here lies in the fact that among the nutrients in upwelling oceanic waters is considerable phosphorous, an essential element for animal and plant nutrition. Although present in minute quantities in many sedimentary rocks, most commercial phosphorous is derived from *phosphorite,* a sedimentary rock containing such phosphate-rich minerals as fluorapatite ($Ca_5(PO_4)_3F$). Areas of upwelling along the outer margins of continental shelves are the sites of deposition of most of these so-called bedded phosphorites, which are interlayered with carbonate rocks, chert, and detrital rocks such as mudrocks and sandstone. Vast deposits in the

Figure 2
Phosphorite of the Bone Valley Formation in the IMC Four Corners Mine, Polk County, Florida.

Permian-aged Phosphoria Formation of Montana, Wyoming, and Idaho formed in this manner.

Upwelling accounts for the origin of most of Earth's bedded phosphorites, but other processes are responsible for some. For instance, *phosphatization* is a process in which carbonate grains such as animal skeletons and ooids are replaced by phosphate, and guano deposits consist of calcium phosphate from bird and bat excrement. Another type of phosphorous-rich deposit is essentially a placer deposit in which the skeletons of vertebrate animals are found in large numbers (vertebrate skeletons and teeth are made up mostly of hydroxyapatite $(Ca_5(PO_4)_3OH)$. A good example is the Miocene to Pliocene (15- to 3-million-year-old) Bone Valley Formation of Florida consisting of a complex of rocks deposited in continental and transitional environments (Figure 2). The phosphorous in this formation comes from skeletons of sharks, manatees, horses, rhinoceroses, and other vertebrate animals.

The United States is the world leader in production and consumption of phosphate rock, most of it (88%) coming from deposits in Florida and North Carolina, but some mined in Idaho and Utah. In 1998, 93% of all phosphorite mined in this country was used to make chemical fertilizers and animal-feed supplements. Phosphorous from phosphorite also has several other uses, such as in metallurgy, preserved foods, ceramics, and matches.

Chapter Summary

1. Scientific investigations of the oceans began during the late 1700s. Present-day research vessels are equipped to investigate the seafloor by sampling and drilling, echo sounding, and seismic profiling.

2. Continental margins separate the continents above sea level from the deep-ocean basin. They consist of a continental shelf, continental slope, and in some cases a continental rise.

3. Continental shelves slope gently seaward and vary from a few tens of meters to more than 1000 km wide.

4. The continental slope begins at an average depth of 135 m where the inclination of the seafloor increases rather abruptly from less than 1 degree to several degrees.

5. Submarine canyons are characteristic of the continental slope, but some of them extend well up onto the shelf and lie offshore from large streams. River erosion of the shelf during the Pleistocene Epoch may account for some submarine canyons, but many have no association with rivers on land and were probably eroded by turbidity currents.

6. Turbidity currents commonly move through submarine canyons and deposit an overlapping series of submarine fans that constitute a large part of the continental rise.

7. Active continental margins are characterized by a narrow shelf and a slope that descends directly into an oceanic trench with no rise present. These margins are also characterized by seismic activity and volcanism.

8. Passive continental margins lack volcanism and exhibit little seismic activity. The continental shelf along these margins is broad, and the slope merges with a continental rise. Abyssal plains are commonly present seaward of the rise.

9. Oceanic trenches are long, narrow features where oceanic crust is subducted. They are characterized by low heat flow, negative gravity anomalies, and the greatest oceanic depths.

10. Oceanic ridges consisting of mountainous topography are composed of volcanic rocks, and many ridges possess a central rift caused by tensional forces. Basaltic volcanism and shallow-focus earthquakes occur at ridges. Oceanic ridges nearly encircle the globe, but they are interrupted and offset by large fractures in the seafloor.

11. Other important features on the seafloor include seamounts that rise more than a kilometer high and guyots, which are submerged flat-topped seamounts. Many aseismic ridges are oriented more or less perpendicular to oceanic ridges and consist of a chain of seamounts and/or guyots.

12. Deep-sea sediments consist mostly of fine-grained particles derived from continents and oceanic islands and the microscopic shells of organisms. The primary types of deep-sea sediments are pelagic clay and ooze.

13. Reefs are wave-resistant structures composed of animal skeletons, particularly corals. Three types of reefs are recognized: fringing, barrier, and atoll.

14. Deep-sea drilling and the study of fragments of seafloor in mountain ranges on land reveal that the oceanic crust is composed in descending order of pillow lava, sheeted dikes, and gabbro.

15. The United States claims rights to all resources within 200 nautical miles (371 km) of its shorelines. Numerous resources including various metals are formed within this Exclusive Economic Zone.

Important Terms

abyssal plain
active continental margin
aseismic ridge
continental margin
continental rise
continental shelf
continental slope
Exclusive Economic Zone (EEZ)

guyot
oceanic ridge
oceanic trench
ooze
ophiolite
passive continental margin
pelagic clay
reef

seamount
seismic profiling
submarine canyon
submarine hydrothermal vent
submarine fan
turbidity current

Review Questions

1. Although submarine canyons are best developed on continental slopes, they are also found on (in):

   a. _____ aseismic ridges;
   b. _____ continental shelves;
   c. _____ oceanic ridges;
   d. _____ oceanic trenches;
   e. _____ abyssal plains.

2. Submarine fans constitute a large part of:

   a. _____ barrier reefs;
   b. _____ deep seafloor sediments;
   c. _____ the Mid-Atlantic Ridge;
   d. _____ fractures in the seafloor;
   e. _____ continental rises.

3. The greatest oceanic depths are found at:

   a. _____ oceanic trenches;
   b. _____ abyssal plains;
   c. _____ aseismic ridges;
   d. _____ guyots;
   e. _____ passive continental margins.

4. Which one of the following is characteristic of a passive continental margin?

   a. _____ volcanism;
   b. _____ wide continental shelf;
   c. _____ hydrothermal vents;
   d. _____ ophiolites;
   e. _____ oceanic trench.

5. The continental slope may have a slope of 25 degrees in some places, but it averages about:

   a. _____ 15 degrees;
   b. _____ less than 1 degree;
   c. _____ 4 degrees;
   d. _____ 10 degrees;
   e. _____ 20 degrees.

6. Turbidity current deposits typically show or possess:

   a. _____ sheeted dikes;
   b. _____ hydrothermal vent deposits;
   c. _____ many coral skeletons;
   d. _____ graded bedding;
   e. _____ a large component of calcareous ooze.

7. Which one of the following statements is correct?

   a. _____ Oceanic crust is composed mostly of basalt and gabbro.
   b. _____ Most passive continental margins are found around the margins of the Pacific Ocean basin.
   c. _____ The Exclusive Economic Zone (EEZ) extends seaward to the shelf–slope break.
   d. _____ Most intermediate- and deep-focus earthquakes take place at or near passive continental margins.
   e. _____ The greatest source of deep-sea sediments is meteorite dust.

8. Sediment that settles from suspension far from land is:

   a. _____ abyssal;
   b. _____ pelagic;
   c. _____ volcanic;
   d. _____ generally gravel sized;
   e. _____ sulfide.

9. Outgassing is the process that probably accounts for:

   a. _____ the fact that oceanic crust is denser than continental crust;
   b. _____ the salt content of ocean waters;
   c. _____ Earth's surface waters;
   d. _____ fractures in the seafloor;
   e. _____ the orientation of aseismic ridges perpendicular to oceanic ridges.

10. A circular reef enclosing a lagoon is a(n):

    a. _____ barrier reef;
    b. _____ atoll;
    c. _____ seamount;
    d. _____ ophiolite;
    e. _____ submarine fan.

11. A flat-topped seamount rising more than 1 km above the seafloor is known as a(n):

    a. _____ guyot;
    b. _____ seismic profile;
    c. _____ turbidity current deposit;
    d. _____ reef;
    e. _____ oceanic ridge.

12. Which of the following is a resource derived from seawater or seafloor deposits?

    a. _____ basalt;
    b. _____ pelagic clay;
    c. _____ siliceous ooze;
    d. _____ phosphorite;
    e. _____ gabbro.

13. How do mid-oceanic ridges form, and how do they differ from mountain ranges on land?

14. Explain what calcareous and siliceous oozes are and how they are deposited.

15. Illustrate and label an ideal sequence of rocks in an ophiolite. Also, for each rock type explain how and where it formed.

16. Why are abyssal plains common around the margins of the Atlantic Ocean basin but rare around the margins of the Pacific?

17. How do sulfide mineral deposits form on the seafloor?

18. Describe the sequence of events leading first to a fringing reef, then a barrier reef, and finally to an atoll.

19. What are the characteristics of an active continental margin? Why are margins of this type so common along the margins of the Pacific but not the Atlantic?

20. Where are fractures in the seafloor found? Do any of them extend onto land, and if so, where?

21. What is the Exclusive Economic Zone (EEZ)? What types of metal deposits are found within it?

22. Describe a submarine canyon. Also, explain how scientists think they form, and what supporting evidence they have for their interpretation.

23. What are the characteristics of a passive continental margin?

24. How are echo sounding and seismic profiling used to study the seafloor?

1. Hydrothermal vents on the seafloor are difficult to locate, because they lie far below the surface of the oceans. How would you propose looking for and locating such vents? Also, how can you account for the fact that metal deposits formed at hydrothermal vents can be found hundreds of kilometers from active vents?

2. The most distant part of an aseismic ridge is 1000 km from an oceanic ridge system, and its age is 30 million years. How fast, on the average, did the plate with this ridge move in centimeters per year?

3. During the Pleistocene Epoch, or what is called the Ice Age, sea level was as much as 130 m lower than at present. What effect did this lower sea level have on rivers? Is there any evidence from the continental shelves that might bear on this question?

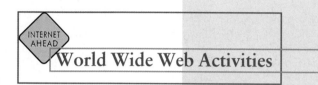

World Wide Web Activities

For these Web site addresses, along with current updates and exercises, log on to
http://www.brookscole.com/geo/

➤ WORLD DATA CENTER-A   MARINE GEOLOGY & GEOPHYSICS

This site mainly provides databases and database searches on the world's oceans. It also provides links to other related sites concerned with marine geology.

1. Check some of the database sites listed for information and maps about ocean sediments and rocks and sediment thicknesses.
2. Click on the *Ocean Drilling* site and then click on the *Ocean Drilling Program*. What is the history of the Ocean Drilling Program? What are the current legs of the ODP, and where is drilling taking place?
3. Go to *General News and Information* and click on *Other Related Sites* for links to other sites concerned with the world's oceans.

➤ LAMONT–DOHERTY EARTH OBSERVATORY OF COLUMBIA UNIVERSITY

This site contains extensive amounts of information and images of their many and varied ongoing research projects in marine geology and geophysics.

1. Click on the *U.S. Seascapes* site for some spectacular images of the U.S. continental margins. Compare these images with the information presented in this chapter about continental margins.
2. Click on the *Ridge Multibeam Synthesis*. Click on *Seafloor Movie* for movies of seafloor spreading.

➤ **AMNH—PAST DAY LOGS**

This site, maintained by the American Museum of Natural History, has an excellent animation of how black smokers work. The first screen one sees is *American Museum of Natural History Expeditions*. On this screen click *Black Smokers* and see how hydrothermal vent systems operate. Also, click on *World Ridge system* and then click *Life Forms* to see creatures living near hydrothermal vents. Also, click on *Underwater Tools* to see how underwater research is carried on.

➤ **USGS BAMG SEAFLOOR MAPPING SERVER**

This site is presented by the U.S. Geological Survey Coastal and Marine Geology Program and the Woods Hole Field Center at Woods Hole, Massachusetts. Click on and see what the GLORIA Mapping Program is. Also, click *GLORIA II Sidescan Sonar System*.
1. What does GLORIA mean, and what kind of technology does it use?
2. Read the section that tells how GLORIA creates seafloor images.
3. See the online mosaics of the U.S. Atlantic continental margin.

➤ **NOVA ONLINE | JOIN US/FEEDBACK**

When the main page is on your screen, go to search and enter *black smokers* and then investigate topics such as *NOVA Online | Into the Abyss | Billowing black smoker chimney in the Mothra field,* and *NOVA Online | Teacher's Guide | Volcanoes of the Deep,* and many others.

➤ **GAS (METHANE) HYDRATES—A NEW FRONTIER**

The U.S. Geological Survey Marine and Coastal Geology Program maintains this site that tells of methane hydrates in seafloor sediments. See the recent map by the USGS of the methane hydrates off the North and South Carolina coasts.
1. What is methane hydrate, and under what conditions is it stable?
2. How do methane hydrates contribute to submarine landslides?
3. What environmental problem might occur if methane is used from methane hydrates?

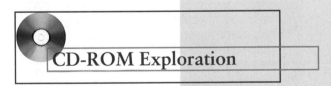

CD-ROM Exploration

➤ Exploring your *Earth Systems Today* CD-ROM will add to your understanding of the material in this chapter.

TOPIC: EARTH'S PROCESSES

MODULE: PLATE TECTONICS

Explore activities in this module to see if you can discover the following for yourself:

Explore links within this activity that allow you to visit mid-oceanic ridges (including the Gorda Ridge off the Washington–Oregon coast and the East Pacific Rise).

Explore links within this activity that take you to a triple-junction site study (i.e., junction of the Nazca, Antarctic, and South American plates) and links that show you several kinds of instrumentation being deployed to study underwater plate boundaries like the ridge and trench.

Chapter 12

The northwest-southeast trending, snow-covered, linear mountain range is the King Ata Tag Mountains located in extreme western-most China and part of Tajikistan. Highest peaks are just over 6130 m above sea level. The town or village of Muji is located in the larger river valley that is south of this mountain range.

OBJECTIVES

At the end of this chapter, you will have learned that

- Plate tectonics is the unifying theory of geology and has revolutionized the science.
- The hypothesis of continental drift was based on considerable geologic, paleontologic, and climatologic evidence.
- The hypothesis of seafloor spreading accounts for continental movement, and that thermal convection cells provide a mechanism for plate movement.

- The three types of plate boundaries are divergent, convergent, and transform, and along these boundaries new plates are formed, consumed, or slide past one another.
- Interaction along plate boundaries accounts for most of Earth's seismic and volcanic activity.
- The rate of movement and motion of plates can be calculated in several ways.

- Some type of convective heat system is involved in plate movement.
- Plate movement affects the distribution of natural resources.

Plate Tectonics: A Unifying Theory

Prologue

Two tragic events during 1985 serve to remind us of the dangers of living near a convergent plate margin. On September 19, a magnitude 8.1 earthquake killed more than 9000 people in Mexico City. Two months later and 3200 km to the south, a minor eruption of Colombia's Nevado del Ruiz volcano partially melted its summit glacial ice, causing a mudflow that engulfed Armero and several other villages and killed more than 23,000 people. These two tragedies resulted in more than 32,000 deaths, tens of thousands of injuries, and billions of dollars in property damage.

Both events occurred along the eastern portion of the Ring of Fire, a chain of intense seismic and volcanic activity that encircles the Pacific Ocean basin (see Figure 9.7). Some of the world's greatest disasters occur along this ring because of volcanism and earthquakes generated by plate convergence. The Mexico City earthquake resulted from subduction of the Cocos plate at the Middle America Trench. Sudden movement of the Cocos plate beneath Central America generated seismic waves that traveled in all directions. The violent shaking experienced in Mexico City, 350 km away, and elsewhere was caused by these seismic waves.

The strata underlying Mexico City consist of unconsolidated sediment deposited in a large ancient lake. Such sediment amplifies the shaking during earthquakes, with the unfortunate consequence that buildings constructed there are commonly more heavily damaged than those built on solid bedrock (see Perspective 9.2, Figure 3).

Less than two months after the Mexico City earthquake, Colombia experienced its greatest recorded natural disaster. Nevado del Ruiz is one of several active volcanoes resulting from the rise of magma generated where the Nazca plate is subducted beneath South America (see Figure 4.20). A minor eruption of Nevado del Ruiz partially melted the glacial ice on the mountain; the meltwater rushed down the valleys, mixed with the sediment, and turned it into a deadly viscous mudflow.

The city of Armero, Colombia, lies in the valley of the Lagunilla River, one of several river valleys inundated by mudflows. Of the city's 23,000 inhabitants, 20,000 died, and most of the city was destroyed. Another 3000 people were killed in nearby valleys. A geologic hazard assessment study completed one month before the eruption showed that Armero was in a high-hazard mudflow area!

These two examples vividly illustrate some of the dangers of living in proximity to a convergent plate boundary. Subduction of one plate beneath another repeatedly triggers large earthquakes, the effects of which are frequently felt far from their epicenters. Since 1900 earthquakes have killed more than 120,000 people in Central and South America alone. Even though volcanic eruptions in this region have not caused nearly as many casualties as earthquakes, they have nevertheless caused tremendous property damage and have the potential for triggering devastating events such as the 1985 Colombian mudflow.

Because the Ring of Fire is home to millions of people, can anything be done to decrease the devastation that inevitably results from the earthquake and volcanic activity in that region? Given our present state of knowledge, most of the disasters could not have been accurately predicted, but better planning and advance preparations by the nations bordering the Ring of Fire could have prevented much tragic loss of life. As long as people live near convergent plate margins, disasters will continue. By studying and understanding geologic activity along plate margins, however, geologists can help to minimize the destruction.

INTRODUCTION

The decade of the 1960s was a time of geologic as well as social and cultural revolution. The ramifications of the newly proposed plate tectonic theory radically changed the way in which geologists viewed our planet. Earth could now be treated as a system in which seemingly unrelated geologic phenomena were related and interconnected (Table 12.1). No longer could Earth be regarded as an unchanging planet on which continents and ocean basins remained fixed through time; instead, it was recognized as a dynamically changing planet.

Plate tectonics has been the dominant process affecting the evolution of Earth. The interactions between moving plates determines the locations of continents, ocean basins, and mountain systems, which in turn affects atmospheric and oceanic circulation patterns that ultimately determine global climates. Plate movements have also profoundly influenced the geographic distribution, evolution, and extinction of plants and animals.

Although plate tectonic processes are extremely slow by human standards, they nevertheless have profound effects on our lives. Geologists now realize that most earthquakes and volcanic eruptions occur at or near plate boundaries and are not merely random occurrences. Furthermore, the formation and distribution of many geologic resources, such as metal ores, are related to plate tectonic processes, and geologists are now incorporating plate tectonic theory into their prospecting efforts.

Plate tectonic theory has led to a greater understanding of how Earth has evolved and continues to do so. This powerful, unifying theory accounts for many apparently unrelated geologic events, allowing geologists to view Earth history in terms of interrelated events that are part of a global panorama of dynamic change through time. For example, the Paleozoic history of the Appalachian Mountains is no longer considered an isolated regional event but rather part of a global interaction between plates that culminated in the formation of a large supercontinent at the end of the Paleozoic Era.

Table 12.1

Plate Tectonics and Earth Systems	
SOLID EARTH	Plate tectonics is driven by convection in the mantle and in turn drives mountain-building processes and associated igneous and metamorphic activity.
ATMOSPHERE	Arrangement of continents affects solar heating and cooling and thus winds and weather systems. Rapid plate spreading and hot spot activity may release volcanic carbon dioxide and affect global climate.
HYDROSPHERE	Continental arrangement affects ocean currents. Rate of spreading affects volume of mid-ocean ridges and hence sea level. Placement of continents may contribute to onset of ice ages.
BIOSPHERE	Movement of continents creates corridors or barriers to migration and thus creates ecological niches. Habitats may be transported into more or less favorable climates.
EXTRATERRESTRIAL	Arrangement of continents affects free circulation of ocean tides and influences tidal slowing of Earth's rotation.

We will first review the various hypotheses that preceded plate tectonic theory, examining the evidence that led some people to accept the idea of continental movement and others to reject it. Because plate tectonic theory has evolved from numerous scientific inquiries and observations, only the more important ones will be covered.

WHAT WERE SOME EARLY IDEAS ABOUT CONTINENTAL DRIFT?

The idea that Earth's geography was different during the past is not new. During the fifteenth century, Leonardo da Vinci observed that "above the plains of Italy where flocks of birds are flying today fishes were once moving in large schools." In 1620 Francis Bacon commented on the similarity of the shorelines of western Africa and eastern South America but did not make the connection that the Old and New Worlds might once have been sutured together. Alexander von Humboldt made the same observation in 1801, although he attributed these similarities to erosion rather than the splitting apart of a larger continent.

One of the earliest specific references to continental drift is in the 1858 book *Creation and Its Mysteries Revealed* by Antonio Snider-Pellegrini. He suggested that all continents were linked together during the Pennsylvanian Period and later split apart. He based his conclusions on the similarities between plant fossils in the Pennsylvanian-aged coal beds of Europe and North America and attributed the separation of the continents to the biblical deluge.

During the late nineteenth century, the Austrian geologist Edward Suess noted the similarities between the Late Paleozoic plant fossils of India, Australia, Africa, Antarctica, and South America as well as evidence of glaciation in the rock sequences of these southern continents. In 1885 he proposed the name Gondwanaland (or **Gondwana** as we will use here) for a supercontinent composed of these southern landmasses. Gondwana is a province in east-central India where evidence exists for extensive glaciation as well as abundant fossils of the *Glossopteris* flora (Figure 12.1), an association of Late Paleozoic plants found only in India and the Southern Hemisphere continents. Suess thought the distribution of plant fossils and glacial deposits was a consequence of extensive land bridges that once con-

nected the continents and later sank beneath the ocean. The distribution of glacial deposits was also consistent with this interpretation.

In 1910 the American geologist Frank Taylor published a pamphlet presenting his own theory of continental drift. He explained the formation of mountain ranges as a result of the lateral movement of continents. He also envisioned the present-day continents as parts of larger polar continents that eventually broke apart and migrated toward the equator after Earth's rotation supposedly slowed due to gigantic tidal forces. According to Taylor, these tidal forces were generated when Earth captured the Moon about 100 million years ago.

Although we now know that Taylor's mechanism is incorrect, one of his most significant contributions was his suggestion that the Mid-Atlantic Ridge, discovered by the 1872–1876 British HMS *Challenger* expeditions, might mark the site along which an ancient continent broke apart to form the present-day Atlantic Ocean.

ALFRED WEGENER AND THE CONTINENTAL DRIFT HYPOTHESIS

Alfred Wegener, a German meteorologist (Figure 12.2), is generally credited with developing the hypothesis of **continental drift**. In his monumental book *The Origin of Continents and Oceans* (first published in 1915), Wegener proposed that all landmasses were originally united in a single supercontinent that he named **Pangaea**, from

(a)

(b)

Figure 12.1
Glossopteris leaves from (a) the Upper Permian Dunedoo Formation and (b) the Upper Permian Illawarra Coal Measures, Australia. Fossils of the *Glossopteris* flora are found on all five of the Gondwana continents. Their distribution on these presently widely separated continents led some scientists to conclude that these continents were at one time much closer together.

Figure 12.2
Alfred Wegener, a German meteorologist, proposed the continental drift hypothesis in 1912 based on a tremendous amount of geologic, paleontologic, and climatologic evidence. He is shown here waiting out the Arctic winter in an expedition hut.

the Greek meaning "all land." Wegener portrayed his grand concept of continental movement in a series of maps showing the breakup of Pangaea and the movement of the various continents to their present-day locations. Wegener had amassed a tremendous amount of geologic, paleontologic, and climatologic evidence in support of continental drift, but the initial reaction of scientists to his then-heretic ideas can best be described as mixed.

Opposition to Wegener's ideas became particularly widespread in North America after 1928 when the American Association of Petroleum Geologists held an international symposium to review the hypothesis of continental drift. After each side had presented its arguments, the opponents of continental drift were clearly in the majority, even though the evidence in support of continental drift, most of which came from the Southern Hemisphere, was impressive and difficult to refute. One problem with the hypothesis, however, was its lack of a mechanism to explain how continents, composed of granitic rocks, could seemingly move through the denser basaltic oceanic crust.

Nevertheless, the eminent South African geologist Alexander du Toit further developed Wegener's arguments and gathered more geologic and paleontologic evidence in support of continental drift. In 1937 du Toit published *Our Wandering Continents* in which he contrasted the glacial deposits of Gondwana with coal deposits of the same age found in the continents of the Northern Hemisphere. To resolve this apparent climatologic paradox, du Toit moved the Gondwana continents to the South Pole and brought the northern continents together such that the coal deposits were located at the equator. He named this northern landmass **Laurasia.** It consisted of present-day North America, Greenland, Europe, and Asia (except for India).

Despite what seemed to be overwhelming evidence, most geologists still refused to accept the idea that continents moved. Not until the 1960s, when oceanographic research provided convincing evidence that the continents had once been joined together and subsequently separated, did the hypothesis of continental drift finally become widely accepted.

WHAT IS THE EVIDENCE FOR CONTINENTAL DRIFT?

The evidence used by Wegener, du Toit, and others to support the hypothesis of continental drift includes the fit of the shorelines of continents, the appearance of the same rock sequences and mountain ranges of the same age on continents now widely separated, the matching of glacial deposits and paleoclimatic zones, and the similarities of many extinct plant and animal groups whose fossil remains are found today on widely separated continents.

Continental Fit

Wegener, like some before him, was impressed by the close resemblance between the coastlines of continents on opposite sides of the Atlantic Ocean, particularly between South America and Africa. He cited these similarities as partial evidence that the continents were at one time joined together as a supercontinent that subsequently split apart. As his critics pointed out, though, the configuration of coastlines results from erosional and depositional processes and therefore is continually being modified. So even if the continents had separated during the Mesozoic Era, as Wegener proposed, it is not likely that the coastlines would fit exactly.

A more realistic approach is to fit the continents together along the continental slope where erosion would be minimal. Recall from Chapter 11 that the true margin of a continent—that is, where continental crust changes to oceanic crust—is beneath the continental slope. In 1965 Sir Edward Bullard, an English geophysicist, and two associates showed that the best fit between the continents occurs at a depth of about 2000 m (Figure 12.3). Since then, other reconstructions using the latest ocean basin data have confirmed the close fit between continents when they are reassembled to form Pangaea.

Similarity of Rock Sequences and Mountain Ranges

If the continents were at one time joined together, then the rocks and mountain ranges of the same age in adjoining locations on the opposite continents should closely match. Such is the case for the Gondwana continents (Figure 12.4). Marine, nonmarine, and glacial rock sequences of Pennsylvanian to Jurassic age are almost identical for all five Gondwana continents, strongly indicating that they were joined together at one time.

The trends of several major mountain ranges also support the hypothesis of continental drift. These mountain ranges seemingly end at the coastline of one continent only to apparently continue on another continent across the ocean. The folded Appalachian Mountains of North America, for example, trend northeastward through the eastern United States and Canada and terminate abruptly at the Newfoundland coastline.

Figure 12.3
The best fit between continents occurs along the continental slope, where erosion would be minimal.

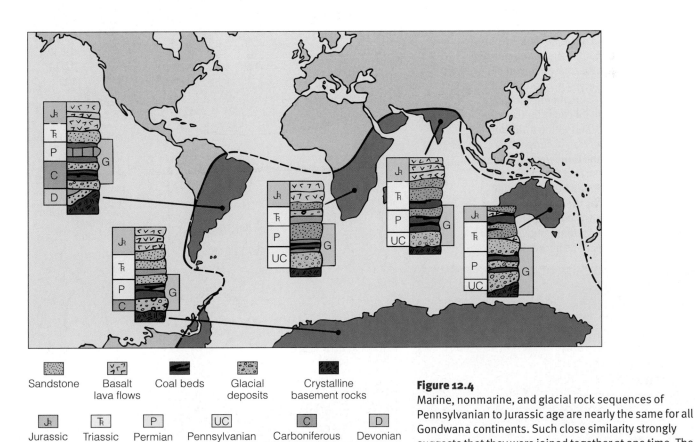

Sandstone   Basalt lava flows   Coal beds   Glacial deposits   Crystalline basement rocks

Jurassic   Triassic   Permian   Pennsylvanian   Carboniferous (Mississippian and Pennsylvanian)   Devonian

Figure 12.4
Marine, nonmarine, and glacial rock sequences of Pennsylvanian to Jurassic age are nearly the same for all Gondwana continents. Such close similarity strongly suggests that they were joined together at one time. The range indicated by G is that of the *Glossopteris* flora.

Figure 12.5

When continents are brought together, their mountain ranges form a single continuous range of the same age and style of deformation throughout. Such evidence indicates the continents were at one time joined together and were subsequently separated.

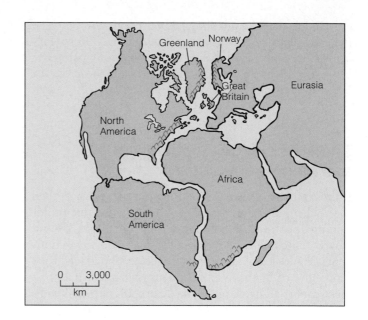

Mountain ranges of the same age and deformational style occur in eastern Greenland, Ireland, Great Britain, and Norway. Even though these mountain ranges are currently separated by the Atlantic Ocean, they form an essentially continuous mountain range when the continents are positioned next to each other (Figure 12.5).

Glacial Evidence

During the Late Paleozoic Era, massive glaciers covered large continental areas of the Southern Hemisphere. Evidence for this glaciation includes layers of till (sediments deposited by glaciers) and striations (scratch

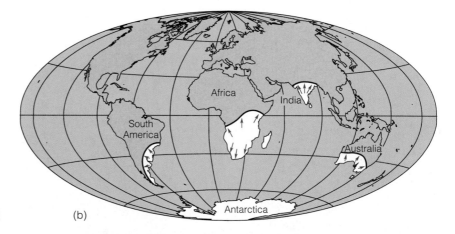

Figure 12.6

(a) Permian-aged glacial striations in bedrock exposed at Hallet's Cove, Australia, indicate the direction of glacial movement more than 200 million years ago. (b) If the continents did not move in the past, then Late Paleozoic glacial striations preserved in bedrock in Australia, India, and South America indicate that glacial movement for each continent was from the oceans onto land within a subtropical to tropical climate. Such an occurrence is highly unlikely. (c) If the Gondwana continents are brought together so that South Africa is located at the South Pole, then the glacial movement indicated by the striations makes sense. In this situation, the glacier, located in a polar climate, moved radially outward from a thick central area toward its periphery.

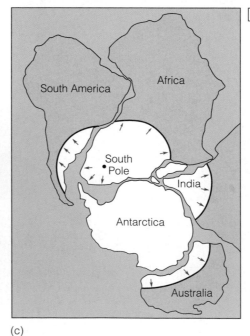

☐ Glaciated area
Arrows indicate the direction of glacial movement based on striations preserved in bedrock.

(a)

(c)

marks) in the bedrock beneath the till. Fossils and sedimentary rocks of the same age from the Northern Hemisphere, however, give no indication of glaciation. Fossil plants found in coals indicate that the Northern Hemisphere had a tropical climate during the time that the Southern Hemisphere was glaciated.

All the Gondwana continents except Antarctica are currently located near the equator in subtropical to tropical climates. Mapping of glacial striations in bedrock in Australia (Figure 12.6a), India, and South America indicates that the glaciers moved from the areas of the present-day oceans onto land (Figure 12.6b). This would be highly unlikely because large continental glaciers (such as occurred on the Gondwana continents during the Late Paleozoic Era) flow outward from their central area of accumulation toward the sea.

If the continents did not move during the past, one would have to explain how glaciers moved from the oceans onto land and how large-scale continental glaciers formed near the equator. But if the continents are reassembled as a single landmass with South Africa located at the south pole, the direction of movement of Late Paleozoic continental glaciers makes sense. Furthermore, this geographic arrangement places the northern continents nearer the tropics, which is consistent with the fossil and climatologic evidence from Laurasia (Figure 12.6c).

Fossil Evidence

Some of the most compelling evidence for continental drift comes from the fossil record (Figure 12.7). Fossils of the *Glossopteris* flora are found in equivalent Pennsylvanian- and Permian-aged coal deposits on all five Gondwana continents. The *Glossopteris* flora is characterized by the seed fern *Glossopteris* (Figure 12.1) as well as by many other distinctive and easily identifiable plants. Pollen and spores of plants can be dispersed over great distances by wind, but *Glossopteris*-type

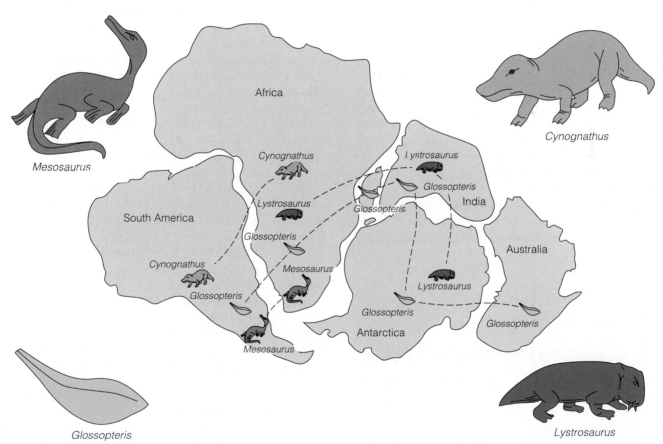

Figure 12.7
Some of the animals and plants whose fossils are found today on the widely separated continents of South America, Africa, India, Australia, and Antarctica. These continents were joined together during the Late Paleozoic to form Gondwana, the southern landmass of Pangaea. *Glossopteris* and similar plants are found in Pennsylvanian- and Permian-aged deposits on all five continents. *Mesosaurus* is a freshwater reptile whose fossils are found in Permian-aged rocks in Brazil and South Africa. *Cynognathus* and *Lystrosaurus* are land reptiles who lived during the Early Triassic Period. Fossils of *Cynognathus* are found in South America and Africa, and fossils of *Lystrosaurus* have been recovered from Africa, India, and Antarctica.

plants produced seeds too large to have been carried by winds. Even if the seeds had floated across the ocean, they probably would not have remained viable for any length of time in saltwater.

The present-day climates of South America, Africa, India, Australia, and Antarctica range from tropical to polar and are much too diverse to support the type of plants that compose the *Glossopteris* flora. Wegener therefore reasoned that these continents must once have been joined such that these widely separated localities were all in the same latitudinal climatic belt (Figure 12.7).

The fossil remains of animals also provide strong evidence for continental drift. One of the best examples is *Mesosaurus,* a freshwater reptile whose fossils are found in Permian-aged rocks in certain regions of Brazil and South Africa and nowhere else in the world (Figures 12.7 and 12.8). Because the physiology of freshwater and marine animals is completely different, it is hard to imagine how a freshwater reptile could have swum across the Atlantic Ocean and found a freshwater environment nearly identical to its former habitat. Moreover, if *Mesosaurus* could have swum across the ocean, its fossil remains should be widely dispersed. It is more logical to assume that *Mesosaurus* lived in lakes in what are now adjacent areas of South America and Africa but were then united into a single continent.

Lystrosaurus and *Cynognathus* are both land-dwelling reptiles that lived during the Triassic Period; their fossils are found only on the present-day continental fragments of Gondwana (Figure 12.7). Because they are both land animals, they certainly could not have swum across the oceans currently separating the Gondwana continents. Therefore, the continents must once have been connected.

The evidence favoring continental drift seemed overwhelming to Wegener and his supporters, yet the lack of a suitable mechanism to explain continental movement prevented its widespread acceptance. Not until new evidence from studies of Earth's magnetic field and oceanographic research showed that the ocean basins were geologically young features did renewed interest in continental drift occur.

Paleomagnetism and Polar Wandering

Some of the most convincing evidence for continental drift came from the study of paleomagnetism, a relatively new discipline during the 1950s. During that time, some geologists were researching past changes of Earth's magnetic field in order to better understand the present-day magnetic field. As so often happens in science, these studies led to other discoveries. In this case, they led to the discovery that the ocean basins are geologically young and that the continents have indeed moved during the past, just as Wegener and others had proposed.

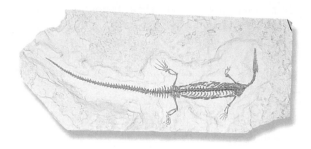

Figure 12.8
Mesosaurus, a Permian-aged freshwater reptile whose fossil remains are found in Brazil and South Africa, indicating these two continents were joined together at the end of the Paleozoic Era.

Recall from Chapter 10 that Earth's magnetic poles correspond closely to the location of the geographic poles (see Figure 10.22). When a magma cools, the iron-bearing minerals align themselves with Earth's magnetic field when they reach the Curie point, thus recording both the direction and the intensity of the magnetic field. This information can be used to determine the location of Earth's magnetic poles and the latitude of the rock when it formed.

As paleomagnetic research progressed in the 1950s, some unexpected results emerged. When geologists measured the magnetism of recent rocks, they found it was generally consistent with Earth's current magnetic field. The paleomagnetism of ancient rocks, though, showed different orientations. For example, paleomagnetic studies of Silurian lava flows in North America indicated that the north magnetic pole was located in the western Pacific Ocean at that time, whereas the paleomagnetic evidence from Permian lava flows pointed to yet another location in northern Asia. When plotted on a map, the paleomagnetic readings of numerous lava flows from all ages in North America trace the apparent movement of the magnetic pole through time (Figure 12.9). This paleomagnetic evidence from a single continent could be interpreted in three ways: The continent remained fixed and the north magnetic pole moved; the north magnetic pole stood still and the continent moved; or both the continent and the north magnetic pole moved.

On analysis, magnetic minerals from European Silurian and Permian lava flows pointed to different magnetic pole locations than those of the same age from North America (Figure 12.9). Furthermore, analysis of lava flows from all continents indicated each continent had its own series of magnetic poles. Does this mean there were different north magnetic poles for each continent? That would be highly unlikely and difficult to reconcile with the theory accounting for Earth's magnetic field.

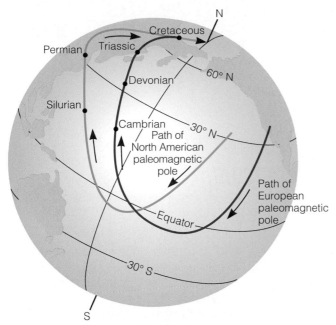

Figure 12.9
The apparent paths of polar wandering for North America and Europe. The apparent location of the north magnetic pole is shown for different periods on each continent's polar wandering path.

The best explanation for such data is that the magnetic poles have remained at their present locations near the geographic north and south poles and the continents have moved. When the continental margins are fitted together so that the paleomagnetic data point to only one magnetic pole, we find, just as Wegener did, that the rock sequences and glacial deposits match, and that the fossil evidence is consistent with the reconstructed paleogeography (see Perspective 12.1).

WHAT IS SEAFLOOR SPREADING?

A renewed interest in oceanographic research led to extensive mapping of the ocean basins during the 1960s. Such mapping revealed an oceanic ridge system more than 65,000 km long, constituting the most extensive mountain range in the world. Perhaps the best-known part of the ridge system is the Mid-Atlantic Ridge, which divides the Atlantic Ocean basin into two nearly equal parts (Figure 12.10).

In 1962, as a result of his discovery of guyots (see Figure 11.19) and other oceanographic research he conducted in the 1950s, Harry Hess of Princeton University proposed the theory of **seafloor spreading** to account for continental movement. Hess suggested that continents do not move across oceanic crust, but rather that the continents and oceanic crust move together. He suggested that the seafloor separates at oceanic ridges where upwelling magma forms new crust. As the magma cools, the newly formed oceanic crust moves laterally away from the ridge, thus explaining how volcanic islands that formed at or near ridge crests later become guyots.

As a mechanism to drive this system, Hess revived the idea (proposed in the 1930s and 1940s by Arthur Holmes and others) of **thermal convection cells** in the mantle; that is, hot magma rises from the mantle, intrudes along rift zone fractures defining oceanic ridges, and thus forms new crust. Cold crust is subducted back into the mantle at deep-sea trenches, where it is heated and recycled, thus completing a thermal convection cell (see Figure 1.14).

How could Hess's hypothesis be confirmed? If new crust is forming at oceanic ridges and Earth's

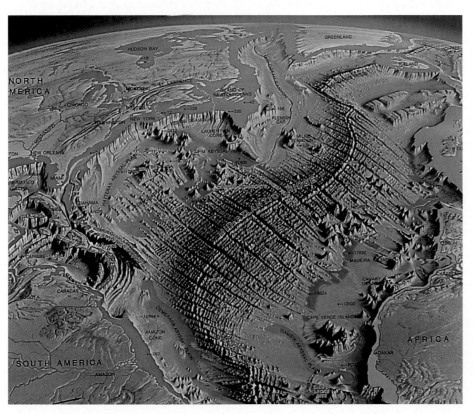

Figure 12.10
Artistic view of what the Atlantic Ocean basin would look like without water. The major feature is the Mid-Atlantic Ridge.

Perspective 12.1

Paleogeographic Reconstructions and Maps

The key to any reconstruction of world paleogeography is the correct positioning of the continents in terms of latitude and longitude as well as orientation of a paleocontinent relative to the paleo-north pole. The main criteria used for paleogeographic reconstructions are paleomagnetism, biogeography, tectonic patterns, and climatology.

Paleomagnetism provides the only source of quantitative data on the orientations of the continents. For the Paleozoic Era, the paleomagnetic data are often inconsistent and contradictory due to secondary magnetizations acquired through the effects of metamorphism or weathering.

The distribution of faunas and floras provides a useful check on the latitudes determined by paleomagnetism and can provide additional limits on longitudinal separation of continents. As is well known, the distribution of plants and animals is controlled by both climatic and geographic barriers. Such information can be used to position continents and ocean basins in a way that accounts for the biogeographic patterns indicated by fossil evidence.

Tectonic activity is indicated by deformed sediments associated with andesitic volcanics and ophiolites. Such features allow geologists to recognize ancient mountain ranges and zones of subduction. These mountain ranges may subsequently have been separated by plate movement, so the identification of large, continuous mountain ranges provides important information about continental positions in the geologic past.

Climate-sensitive sedimentary rocks are used to interpret past climatic conditions. Desert dunes are typically well sorted and cross-bedded on a large scale and associated with other deposits that indicate an arid environment. Coal forms in freshwater swamps where climatic conditions promote abundant plant growth. Evaporites result when evaporation exceeds precipitation, such as in desert regions or along hot, dry shorelines. Tillites result from glacial activity and indicate cold, wet environments.

Paleogeographic features can be determined by associations of sedimentary rocks and sedimentary structures. For example, large-scale cross-beds may indicate aeolian or windblown conditions such as in deserts. Delta complexes and deep-sea fans have characteristic internal features and three-dimensional forms that can be recognized in the geologic record, just as coal and associated deposits usually follow a particular sequence. These features can be used to interpret such geographic features as lakes, streams, swamps, and shallow and deep marine areas.

Former mountain ranges can be recognized by folded and faulted sedimentary rocks associated with metamorphic and igneous rocks. Furthermore, the presence of andesites and ophiolites can also be used as evidence of former mountain building.

By combining all relevant geologic, paleontologic, and climatologic information, geologists can construct paleogeographic maps. Such maps are simply interpretations of the geography of an area for a particular time in the geologic past. The majority of paleogeographic maps show the distribution of land and sea, probable climatic regimes, and such geographic features as mountain ranges, swamps, and glaciers (Figure 1).

---

magnetic field is periodically reversing itself, then these magnetic reversals should be preserved as magnetic anomalies in the rocks of the oceanic crust (Figure 12.11).

Around 1960 scientists from the Scripps Institution of Oceanography in California gathered magnetic data that indicated an unusual pattern of alternating positive and negative magnetic anomalies for the Pacific Ocean seafloor off the West Coast of North America. The pattern consisted of a series of roughly north–south parallel stripes, but they were broken and offset by essentially east–west fractures. Not until 1963 did F. Vine

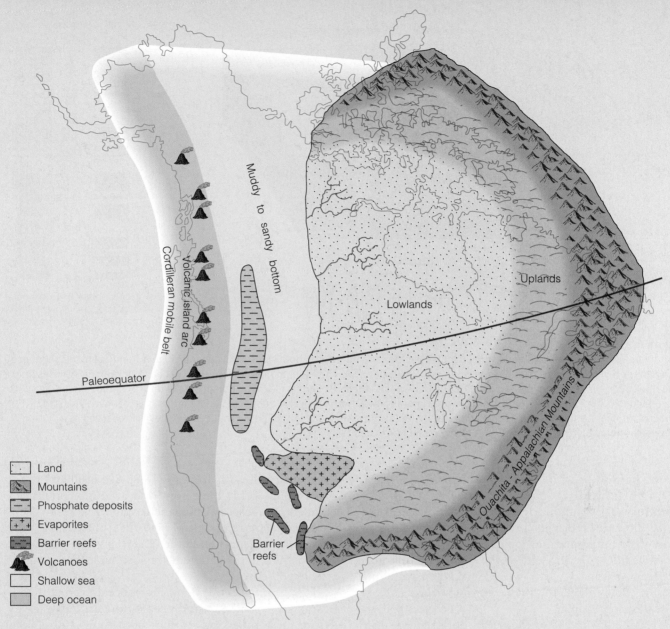

Figure 1
Paleogeography of North America during the Permian Period.

and D. Matthews of Cambridge University and L. W. Morley, a Canadian geologist, independently arrive at a model that explained this pattern of magnetic anomalies.

These three geologists proposed that when magma intruded along the crests of oceanic ridges, it recorded the magnetic polarity at the time it cooled. As the ocean floor moved away from these oceanic ridges, repeated intrusions would form a symmetric series of magnetic stripes, recording periods of normal and reverse polarity (Figure 12.11). Shortly thereafter, the Vine, Matthews, and Morley proposal was supported by evidence from magnetic readings across the Reykjanes

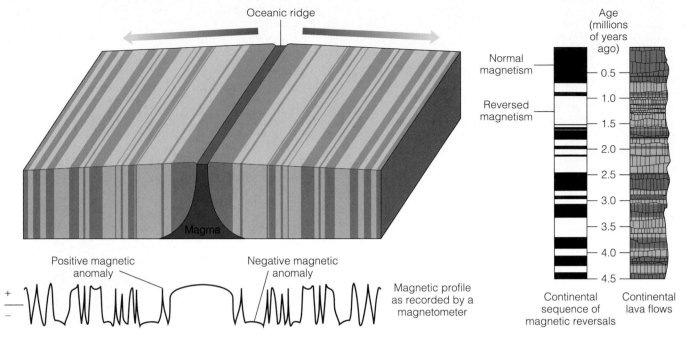

Figure 12.11

The sequence of magnetic anomalies preserved within the oceanic crust on both sides of an oceanic ridge is identical to the sequence of magnetic reversals already known from continental lava flows. Magnetic anomalies are formed when basaltic magma intrudes into oceanic ridges; when the magma cools below the Curie point, it records Earth's magnetic polarity at the time. Seafloor spreading splits the previously formed crust in half so that it moves laterally away from the oceanic ridge. Repeated intrusions record a symmetric series of magnetic anomalies that reflect periods of normal and reversed polarity. The magnetic anomalies are recorded by a magnetometer, which measures the strength of the magnetic field.

Ridge, part of the Mid-Atlantic Ridge south of Iceland. A group from the Lamont-Doherty Geological Observatory at Columbia University found that magnetic anomalies in this area did form stripes that were distributed parallel to and symmetric about the oceanic ridge. By the end of the 1960s, comparable magnetic anomaly patterns were found surrounding most oceanic ridges, and thus confirming Hess's theory of seafloor spreading.

Magnetic surveys of the ocean floor demonstrate that the youngest oceanic crust is adjacent to the spreading ridges and that the age of the crust increases with distance from the ridge axis, as would be expected according to the seafloor spreading hypothesis (Figure 12.12). Furthermore, the oldest oceanic crust is less than 180 million years old, whereas the oldest continental crust is 3.96 billion years old; this difference in age provides confirmation that the ocean basins are geologically young features whose openings and closings are partially responsible for continental movement.

Deep-Sea Drilling and the Confirmation of Seafloor Spreading

For many geologists, the paleomagnetic data amassed in support of continental drift and seafloor spreading were convincing. Results from the Deep-Sea Drilling Project (see Chapter 11) have confirmed the interpretations made from earlier paleomagnetic studies. Cores of deep-sea sediments and seismic profiles obtained by the *Glomar Challenger* and other research vessels have provided much of the data that support the seafloor spreading hypothesis.

According to this hypothesis, oceanic crust is continuously forming at mid-oceanic ridges, moves away from these ridges by seafloor spreading, and is consumed at subduction zones. If this is the case, oceanic crust should be youngest at the ridges and become progressively older with increasing distance away from them. Moreover, the age of the oceanic crust should be symmetrically distributed about the ridges. As we have just noted, paleomagnetic data confirm these statements. Furthermore, fossils from sediments overlying the oceanic crust and radiometric dating of rocks found on oceanic islands both substantiate this predicted age distribution.

Sediments in the open ocean accumulate, on average, at a rate of less than 0.3 cm per 1000 years. If the ocean basins were as old as the continents, we would expect deep-sea sediments to be several kilometers thick. However, data from numerous drill holes indicate that deep-sea sediments are at most only a few hundred meters thick and are thin or absent at oceanic

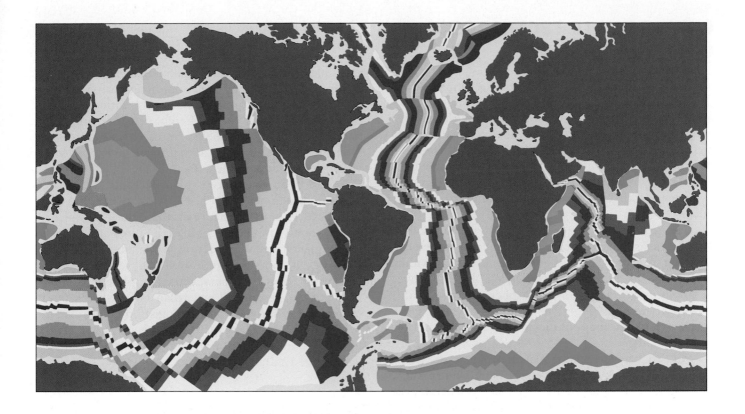

■	Pleistocene to Recent (0–1.6 MYA)	■	Paleocene (58–66 MYA)
□	Pliocene (1.6–5 MYA)	□	Late Cretaceous (66–88 MYA)
▨	Miocene (5–24 MYA)	▨	Middle Cretaceous (88–118 MYA)
▨	Oligocene (24–37 MYA)	▨	Early Cretaceous (118–144 MYA)
■	Eocene (37–58 MYA)	▨	Late Jurassic (144–161 MYA)

Figure 12.12
The age of the world's ocean basins established from magnetic anomalies demonstrates that the youngest oceanic crust is adjacent to the spreading ridges and that its age increases away from the ridge axis.

ridges. Their near-absence at the ridges should come as no surprise, because these are the areas where new crust is continuously produced by volcanism and sea-floor spreading. Accordingly, sediments have had little time to accumulate at or very close to spreading ridges where the oceanic crust is young, but their thickness increases with distance away from the ridges (Figure 12.13).

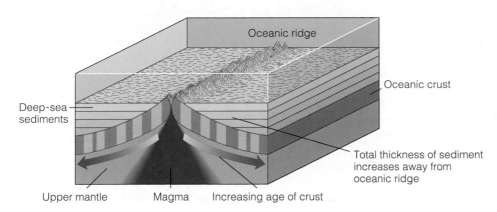

Figure 12.13
The total thickness of deep-sea sediments increases away from oceanic ridges. This is because oceanic crust becomes older away from oceanic ridges, and there has been more time for sediment to accumulate.

WHY IS PLATE TECTONICS A UNIFYING THEORY?

Plate tectonic theory is based on a simple model of Earth. The rigid lithosphere, consisting of both oceanic and continental crust, as well as the underlying upper mantle, consists of many variable-sized pieces called **plates** (Figure 12.14). The plates vary in thickness; those composed of upper mantle and continental crust are as much as 250 km thick, whereas those of upper mantle and oceanic crust are up to 100 km thick.

The lithosphere overlies the hotter and weaker semiplastic asthenosphere. It is thought that movement resulting from some type of heat transfer system within the asthenosphere causes the overlying plates to move. As plates move over the asthenosphere, they separate, mostly at oceanic ridges; in other areas such as at oceanic trenches, they collide and are subducted back into the mantle.

An easy way to visualize plate movement is to think of a conveyor belt moving luggage from an airplane's cargo hold to a baggage cart. The conveyor belt represents convection currents within the mantle, and the luggage represents Earth's lithospheric plates. The luggage is moved along by the conveyor belt until it is dumped into the baggage cart in the same way plates are moved by convection cells until they are subducted into Earth's interior. Although this analogy allows one to visualize how the mechanism of plate movement takes place, it must be remembered that there are limitations to this analogy. The major limitation is that unlike the luggage, plates consist of continental and oceanic crust, which have different densities, and only oceanic crust is subducted into Earth's interior. Nonetheless, this analogy does provide an easy way to visualize plate movement.

Most geologists accept plate tectonic theory, in part because the evidence for it is overwhelming and because it is a unifying theory that accounts for a variety of apparently unrelated geologic features and events. Consequently, geologists now view many geologic processes, such as mountain building, seismicity, and volcanism, from the perspective of plate tectonics. Furthermore, because all inner planets have had a similar origin and early history, geologists are interested in determining whether plate tectonics is unique to Earth or if it operates in the same way on other planets (see Perspective 12.2).

The Supercontinent Cycle

As a result of plate movement, all the continents came together to form the supercontinent Pangaea by the

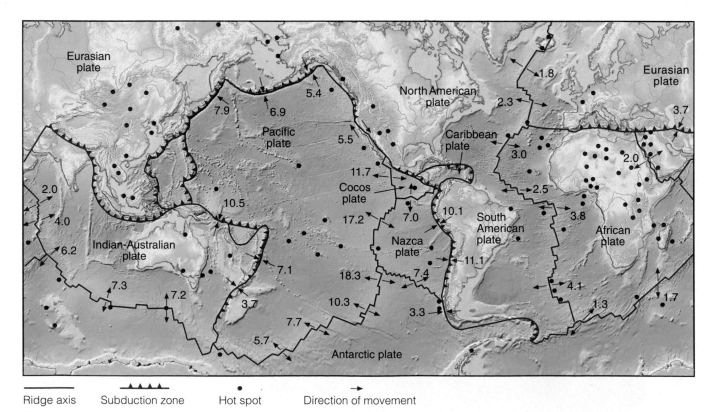

Ridge axis   Subduction zone   Hot spot   Direction of movement

Figure 12.14
A map of the world showing the plates, their boundaries, relative motion and rates of movement in centimeters per year, and hot spots.

end of the Paleozoic Era. Pangaea began fragmenting during the Triassic Period and continues to do so, thus accounting for the present distribution of continents and ocean basins. It has been proposed that supercontinents consisting of all or most of Earth's landmasses form, break up, and re-form in a cycle spanning about 500 million years.

The supercontinent cycle hypothesis is an expansion on the ideas of the Canadian geologist J. Tuzo Wilson. During the early 1970s, Wilson proposed a cycle (now known as the Wilson cycle) that includes continental fragmentation, the opening and closing of an ocean basin, and reassembly of the continent. According to the *supercontinent cycle hypothesis,* heat accumulates beneath a supercontinent because rocks of continents are poor conductors of heat. As a result of the heat accumulation, the supercontinent domes upward and fractures. Basaltic magma rising from below fills the fractures. As a basalt-filled fracture widens, it begins subsiding and forms a long, narrow ocean such as the present-day Red Sea. Continued rifting eventually forms an expansive ocean basin such as the Atlantic.

According to proponents of the supercontinent cycle, one of the most convincing arguments for their hypothesis is the "surprising regularity" of mountain building caused by compression during continental collisions. These mountain-building episodes occur about every 400 to 500 million years and are followed by an episode of rifting about 100 million years later. In other words, a supercontinent fragments and its individual plates disperse following a rifting episode, an interior ocean forms, and then the dispersed fragments reassemble to form another supercontinent.

The supercontinent cycle is yet another example of how interrelated the various systems and subsystems of Earth are and how they operate over vast periods of geologic time.

WHAT ARE THE THREE TYPES OF PLATE BOUNDARIES?

Because it appears that plate tectonics has operated since at least the Proterozoic, it is important that we understand how plates move and interact with each other and how ancient plate boundaries are recognized. After all, the movement of plates has had a profound effect on the geologic and biologic history of this planet.

Geologists recognize three major types of plate boundaries: *divergent, convergent,* and *transform* (Table 12.2). It is along these boundaries that new plates are formed, consumed, or slide laterally past one another. Interaction of plates at their boundaries accounts for most of Earth's seismic and volcanic activity and, as will be apparent in the next chapter, the origin of mountain systems.

Table 12.2

Types of Plate Boundaries			
TYPE	EXAMPLE	LANDFORMS	VOLCANISM
DIVERGENT			
Oceanic	Mid-Atlantic Ridge	Mid-oceanic ridge with axial rift valley	Basalt
Continental	East African Rift Valley	Rift valley	Basalt and rhyolite, no andesite
CONVERGENT			
Oceanic–oceanic	Aleutian Islands	Volcanic island arc, offshore oceanic trench	Andesite
Oceanic–continental	Andes	Offshore oceanic trench, volcanic mountain chain, mountain belt	Andesite
Continental–continental	Himalayas	Mountain belt	Minor
TRANSFORM	San Andreas fault	Fault valley	Minor

Perspective 12.2

Tectonics of the Terrestrial Planets

The four inner, or terrestrial, planets—Mercury, Venus, Earth, and Mars—all had a similar early history involving accretion, differentiation into a metallic core and silicate mantle and crust, and formation of an early atmosphere by outgassing. Their early history was also marked by widespread volcanism and meteorite impacts, both of which helped modify their surfaces. The volcanic and tectonic activity and resultant surface features (other than meteorite craters) of these planets are clearly related to the way they transport heat from their interiors to their surfaces.

Earth appears to be unique in that its surface is broken up into a series of plates. The creation and destruction of these plates at spreading ridges and subduction zones transfer the majority of Earth's internally produced heat. In addition, movement of the plates—together with life-forms, the formation of sedimentary rocks, and water—is responsible for the cycling of carbon dioxide between the atmosphere and lithosphere and thus the maintenance of a habitable climate on Earth.

Heat is transferred between the interior and surface of both Mercury and Mars mainly by lithospheric conduction. This method is sufficient for these planets because both are significantly smaller than Earth or Venus. Because Mercury and Mars have a single, globally continuous plate, they have exhibited fewer types of volcanic and tectonic activity than has Earth. The initial interior warming of Mercury and Mars produced tensional features such as normal faults (see Chapter 13) and widespread volcanism, while their subsequent cooling produced folds and faults resulting from compressional forces, as well as volcanic activity.

Mercury's surface is heavily cratered and shows little in the way of primary volcanic structures. It does, however, have a global system of lobate scarps (see Figure 20.9). These have been interpreted as evidence that Mercury shrank soon after its crust hardened, resulting in crustal cracking.

Mars has numerous features that indicate an extensive early period of volcanism. These include Olympus Mons, the solar system's largest volcano (see Figure 20.11), lava flows, and uplifted regions thought to have resulted from mantle convection. In addition to volcanic features, Mars also displays abundant evidence of tensional tectonics, including numerous faults and large fault-produced valley structures. Although Mars was tectonically active during the past, no evidence indicates that plate tectonics comparable to that on Earth has ever occurred there.

Venus underwent essentially the same early history as the other terrestrial planets, including a period of volcanism, but it is more Earth-like in its tectonics than either Mercury or Mars. Initial radar mapping in 1990 by the *Magellan* spacecraft revealed a surface of extensive lava flows, volcanic domes, folded mountain ranges, and an extensive and intricate network of faults, all of which attest to an internally active planet (Figure 1).

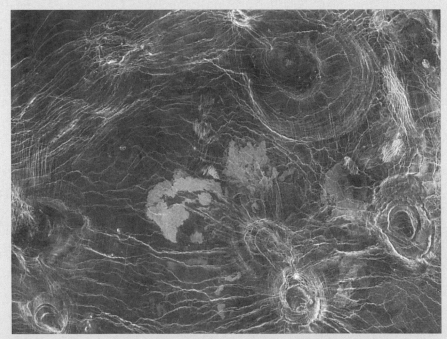

Figure 1
This radar image of Venus made by the *Magellan* spacecraft reveals circular and oval-shaped volcanic features. A complex network of cracks and fractures extends outward from the volcanic features. Geologists think these features were created by blobs of magma rising from the interior of Venus with dikes filling some of the cracks.

Divergent Boundaries

Divergent plate boundaries or *spreading ridges* occur where plates are separating and new oceanic lithosphere is forming. Divergent boundaries are places where the crust is extended, thinned, and fractured as magma, derived from the partial melting of the mantle, rises to the surface. The magma is almost entirely basaltic and intrudes into vertical fractures to form dikes and pillow lava flows (see Figure 4.8). As successive injections of magma cool and solidify, they form new oceanic crust and record the intensity and orientation of Earth's magnetic field (Figure 12.11). Divergent boundaries most commonly occur along the crests of oceanic ridges; for example, the Mid-Atlantic Ridge. Oceanic ridges are thus characterized by rugged topography with high relief resulting from displacement of rocks along large fractures, shallow-focus earthquakes, high heat flow, and basaltic flows or pillow lavas.

Divergent boundaries are also present under continents during the early stages of continental breakup (Figure 12.15). When magma wells up beneath a continent, the crust is initially elevated, stretched, and thinned, producing fractures and rift valleys (Figure 12.15a). During this stage, magma typically intrudes into the faults and fractures forming sills, dikes, and lava flows; the latter often cover the rift valley floor (Figure 12.15b). The East African rift valleys are an excellent example of this stage of continental breakup (Figure 12.16).

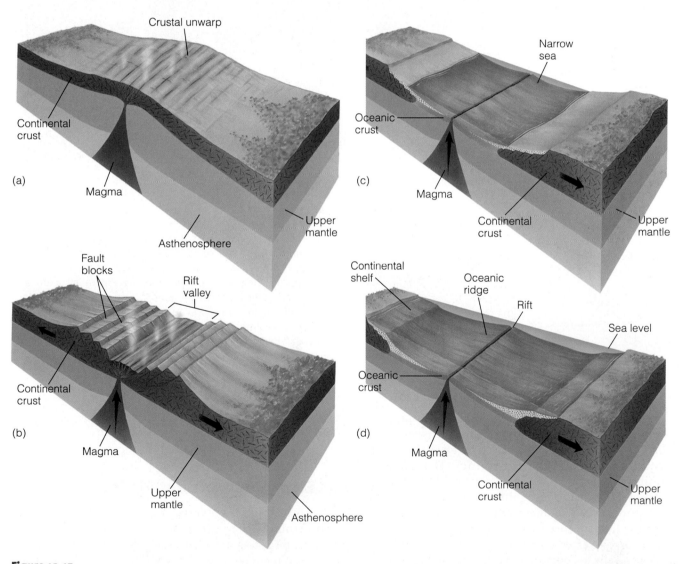

Figure 12.15
History of a divergent plate boundary. (a) Rising magma beneath a continent pushes the crust up, producing numerous cracks and fractures. (b) As the crust is stretched and thinned, rift valleys develop, and lava flows onto the valley floors. (c) Continued spreading further separates the continent until a narrow seaway develops. (d) As spreading continues, an oceanic ridge system forms, and an ocean basin develops and grows.

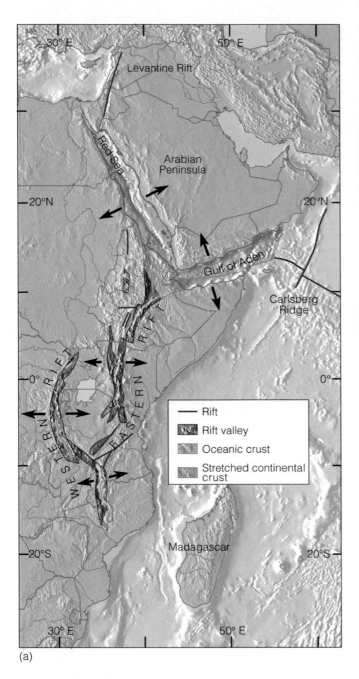

(a)

(b)

Figure 12.16

(a) The East African Rift Valley is being formed by the separation of eastern Africa from the rest of the continent along a divergent plate boundary. The Red Sea represents a more advanced stage of rifting, in which two continental blocks are separated by a narrow sea. (b) View looking down the Great Rift Valley of Africa. Little Magadi, seen in the background, is one of numerous soda lakes forming in the valley. Because of high evaporation rates and lack of any drainage outlets, these lakes are very saline. The Great Rift Valley is part of the system of rift valleys resulting from stretching of the crust as plates move away from each other in eastern Africa.

If spreading proceeds, some rift valleys continue to lengthen and deepen until the continental crust eventually breaks, and a narrow linear sea forms, separating two continental blocks (Figure 12.15c). The Red Sea separating the Arabian Peninsula from Africa (Figure 12.16a) and the Gulf of California, which separates Baja California from mainland Mexico, are good examples of this more advanced stage of rifting.

As a newly created narrow sea continues enlarging, it may eventually become an expansive ocean basin such as the Atlantic, which separates North and South America from Europe and Africa by thousands of kilometers (Figure 12.15d). The Mid-Atlantic Ridge is the boundary between these diverging plates; the American

plates are moving westward, and the Eurasian and African plates are moving eastward.

An Example of Ancient Rifting

What features in the geologic record can geologists use to recognize ancient rifting? Associated with regions of continental rifting are faults, dikes, sills, lava flows, and thick sedimentary sequences within rift valleys. The Triassic fault basins of the eastern United States are a good example of ancient continental rifting (Figure 12.17a). These fault basins mark the zone of rifting that occurred when North America split apart from

(a)

(b)

Figure 12.17

(a) Areas where Triassic fault-block basin deposits crop out in eastern North America. (b) Palisades of the Hudson River. This sill was one of many that were intruded into the fault-block basin sediments during the Late Triassic rifting that marked the separation of North America from Africa.

Africa. They contain thousands of meters of continental sediment and are riddled with dikes and sills.

Convergent Boundaries

Whereas new crust is formed at divergent plate boundaries, older crust must be destroyed and recycled in order for the entire surface area of Earth to remain constant. Otherwise, we would have an expanding Earth. Such plate destruction occurs at **convergent plate boundaries** where two plates collide.

At a convergent boundary, the leading edge of one plate descends beneath the margin of the other by **subduction.** A dipping plane of earthquake foci, referred to as a *Benioff zone,* defines subduction zones (see Figure 9.8). Most of these planes dip from oceanic trenches beneath adjacent island arcs or continents, marking the surface of slippage between the converging plates. As the subducting plate moves down into the asthenosphere, it is heated and eventually incorporated into the mantle. When both of the converging plates are continental, subduction does not occur because continental crust is not dense enough to be subducted into the mantle.

Convergent boundaries are characterized by deformation, volcanism, mountain building, metamorphism, seismicity, and important mineral deposits. Three types of convergent plate boundaries are recognized: oceanic–oceanic, oceanic–continental, and continental–continental.

Oceanic–Oceanic Boundaries When two oceanic plates converge, one is subducted beneath the other along an **oceanic–oceanic plate boundary** (Figure 12.18a). The subducting plate bends downward to form the outer wall of an oceanic trench. A *subduction complex,* composed of wedge-shaped slices of highly folded and faulted marine sediments and oceanic lithosphere scraped off the descending plate, forms along the inner wall. As the subducting plate descends into the asthenosphere, it is heated and partially melted, generating magma, commonly of andesitic composition. This magma is less dense than the surrounding mantle rocks and rises to the surface of the nonsubducted plate, forming a curved chain of volcanoes called a **volcanic island arc** (any plane intersecting a sphere makes an arc). This arc is nearly parallel to the oceanic trench and is separated from it by a distance of up to several hundred kilometers—the distance depending on the angle of dip of the subducting plate (Figure 12.18a).

In those areas where the rate of subduction is faster than the forward movement of the overriding plate, the lithosphere on the landward side of the volcanic island arc may be subjected to tensional stress and stretched and thinned, resulting in the formation of a *back-arc basin.* This back-arc basin may grow by spreading if magma breaks through the thin crust and forms new oceanic crust (Figure 12.18a). A good example of a back-arc basin associated with an oceanic–oceanic plate boundary is the Sea of Japan between the Asian continent and the islands of Japan.

Most present-day active volcanic island arcs are in the Pacific Ocean basin and include the Aleutian Islands, the Kermadec–Tonga arc, and the Japanese and Philippine Islands (Figure 12.18b). The Scotia and Antillean (Caribbean) island arcs are in the Atlantic Ocean basin.

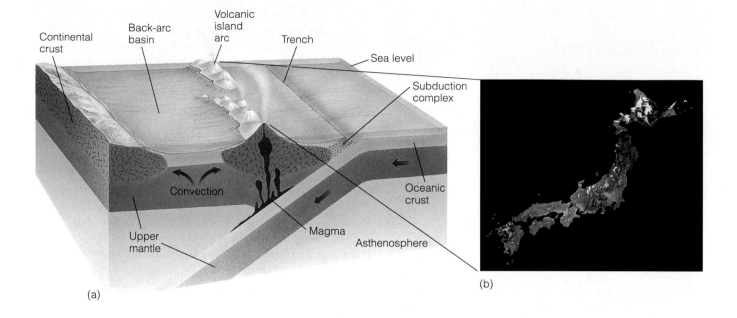

Figure 12.18
Oceanic–oceanic plate boundary. (a) An oceanic trench forms where one oceanic plate is subducted beneath another. On the nonsubducted plate, a volcanic island arc forms from the rising magma generated from the subducting plate. (b) Satellite image of Japan. The Japanese Islands are a volcanic island arc resulting from the subduction of one oceanic plate beneath another oceanic plate.

Oceanic–Continental Boundaries   When an oceanic and continental plate converge, the denser oceanic plate is subducted under the continental plate along an **oceanic–continental plate boundary** (Figure 12.19a). Just as at oceanic–oceanic plate boundaries, the descending oceanic plate forms the outer wall of an oceanic trench.

As the cold, wet, and slightly denser oceanic plate descends into the hot asthenosphere, melting occurs and magma is generated. This magma rises beneath the overriding continental plate and can extrude at the surface, producing a chain of andesitic volcanoes (also called a *volcanic arc*), or intrude into the continental margin as plutons, especially batholiths. An excellent example of an oceanic–continental plate boundary is the Pacific coast of South America where the oceanic Nazca plate is currently being subducted under South America (Figure 12.19b; see also Chapter 13).

Continental–Continental Boundaries   Two continents approaching each other will initially be separated by an ocean floor that is being subducted under one continent. The edge of that continent will display the features characteristic of oceanic–continental convergence. As the ocean floor continues to be subducted, the two continents will come closer together until they eventually collide. Because continental lithosphere, which consists of continental crust and the upper mantle, is less dense than oceanic lithosphere (oceanic crust and upper mantle), it cannot sink into the asthenosphere. Although one continent may partly slide under the other, it cannot be pulled or pushed down into a subduction zone (Figure 12.20a).

When two continents collide, they are welded together along a zone marking the former site of subduction. At this **continental–continental plate boundary,** an interior mountain belt is formed consisting of deformed sedimentary rocks, igneous intrusions, metamorphic rocks, and fragments of oceanic crust. In addition, the entire region is subjected to numerous earthquakes. The Himalayas in central Asia are the result of a continental–continental collision between India and Asia that began about 40 to 50 million years ago and is still continuing (Figure 12.20b; and see Chapter 13).

Recognizing Ancient Convergent Plate Boundaries

How can former subduction zones be recognized in the geologic record? One clue is provided by igneous rocks.

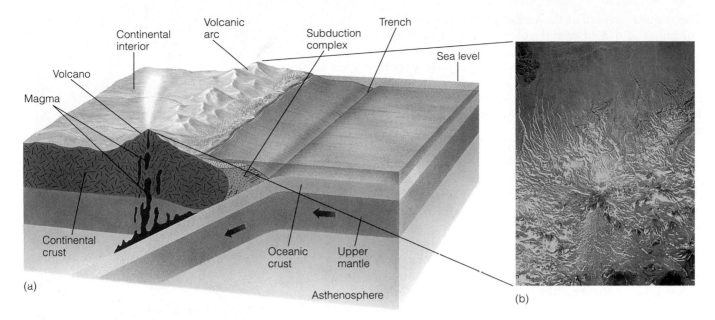

Figure 12.19

Oceanic–continental plate boundary. (a) When an oceanic plate is subducted beneath a continental plate, an andesitic volcanic mountain range is formed on the continental plate as a result of rising magma. (b) A portion of the Andes mountain range in the Potosi Region of Boliva as photographed by the crew of the space shuttle *Atlantis* in 1992. The Andes are one of the best examples of continuing mountain building at an oceanic-continental plate boundary.

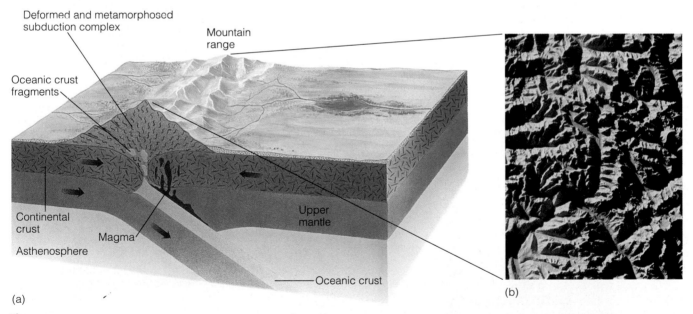

Figure 12.20

Continental–continental plate boundary. (a) When two continental plates converge, neither is subducted because of their great thickness and low and equal densities. As the two continental plates collide, a mountain range is formed in the interior of a new and larger continent. (b) Vertical view of the Himalayas, the youngest and highest mountain system in the world. The Himalayas began forming when India collided with Asia 40 to 50 million years ago.

The magma erupted at the surface, forming island arc volcanoes and continental volcanoes, is of andesitic composition. Another clue can be found in the zone of intensely deformed rocks between the deep-sea trench where subduction is taking place and the area of igneous activity. Here, sediments and submarine rocks are folded, faulted, and metamorphosed into a chaotic mixture of rocks termed a *mélange*.

During subduction, pieces of oceanic lithosphere are sometimes incorporated into the mélange and accreted onto the edge of the continent. Such slices of oceanic crust and upper mantle are called **ophiolites** (Figure 12.21). They consist of a layer of deep-sea sediments that include graywackes (poorly sorted sandstones containing abundant feldspars and rock fragments, usually in a clay-rich matrix), black shales, and cherts. These deep-sea sediments are underlain by pillow lavas, a sheeted dike complex, massive gabbro, and layered gabbro, all of which form the oceanic crust. Beneath the gabbro is peridotite, which probably represents the upper mantle. Ophiolites are key features in recognizing plate convergence along a subduction zone.

Elongate belts of folded and faulted marine sedimentary rocks, andesites, and ophiolites are found in the Appalachians, Alps, Himalayas, and Andes Mountains. The combination of such features is good evidence that these mountain ranges resulted from deformation along convergent plate boundaries.

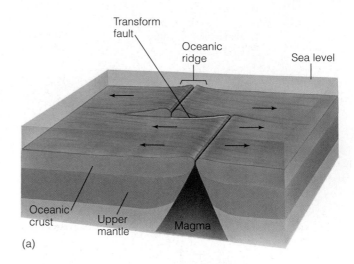

(a)

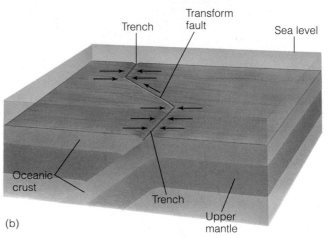

(b)

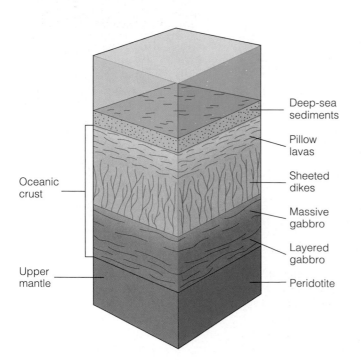

Figure 12.21
Ophiolites are sequences of rock on land consisting of deep-sea sediments, oceanic crust, and upper mantle.

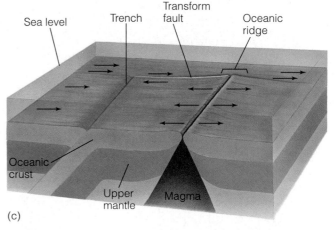

(c)

Figure 12.22
Horizontal movement between plates occurs along a transform fault. (a) The majority of transform faults connect two oceanic ridge segments. Note that relative motion between the plates only occurs between the two ridges. (b) A transform fault connecting two trenches. (c) A transform fault connecting a ridge and a trench.

Transform Boundaries

The third type of plate boundary is a **transform plate boundary.** These occur along fractures in the seafloor, known as *transform faults,* where plates slide laterally past one another roughly parallel to the direction of plate movement. Although lithosphere is neither created nor destroyed along a transform boundary, the movement between plates results in a zone of intensely shattered rock and numerous shallow-focus earthquakes.

Transform faults are particular types of faults that "transform" or change one type of motion between plates into another type of motion. The majority of transform faults connect two oceanic ridge segments, but they can also connect ridges to trenches and trenches to trenches (Figure 12.22). The majority of transform faults are in oceanic crust and are marked by distinct fracture zones, but they may also extend into continents.

One of the best-known transform faults is the San Andreas fault in California. It separates the Pacific plate from the North American plate and connects spreading ridges in the Gulf of California and the Juan de Fuca and Pacific plates off the coast of northern California (Figure 12.23). Many of the earthquakes affecting California are the result of movement along this fault.

Unfortunately, transform faults generally do not leave any characteristic or diagnostic features except for the obvious displacement of the rocks that they are associated with. This disreplacement is commonly large, on the order of tens to hundreds of kilometers. Such large displacements in ancient rocks can sometimes be related to transform fault systems.

HOW ARE PLATE MOVEMENT AND MOTION DETERMINED?

How fast and in what direction are Earth's various plates moving, and do they all move at the same rate? Rates of plate movement can be calculated in several ways. The least accurate method is to determine the age of the sediments immediately above any portion of the oceanic crust and divide that age by the distance from the spreading ridge. Such calculations give an average rate of movement.

A more accurate method of determining both the average rate of movement and relative motion is by dating the magnetic anomalies in the crust of the seafloor. The distance from an oceanic ridge axis to any magnetic anomaly indicates the width of new seafloor that formed during that time interval. Thus, for a given interval of time, the wider the strip of seafloor, the faster the plate has moved. In this way not only can the present average rate of movement and relative motion be determined

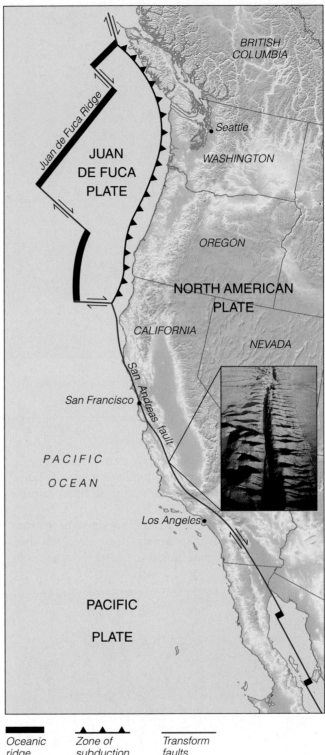

| Oceanic ridge | Zone of subduction | Transform faults |

Figure 12.23
Transform plate boundary. The San Andreas fault is a transform fault separating the Pacific plate from the North American plate. Movement along this fault has caused numerous earthquakes. The inset photograph shows a segment of the San Andreas fault as it cuts through the Carrizo Plain, California.

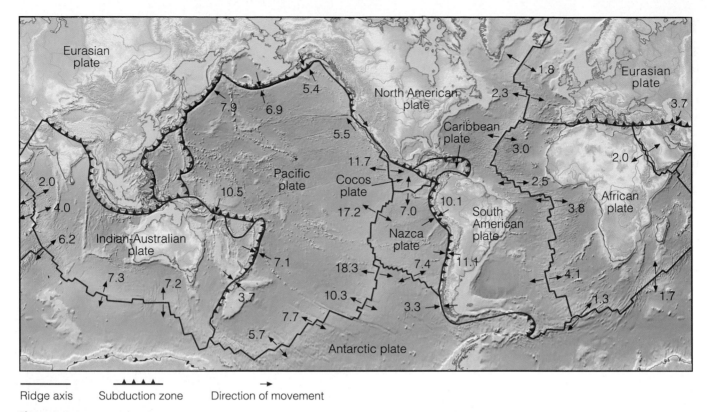

Ridge axis   Subduction zone   Direction of movement

Figure 12.24
This map shows the average rate of movement in centimeters per year and relative motion of Earth's plates.

(Figure 12.24), but the average rate of movement during the past can also be calculated by dividing the distance between anomalies by the amount of time elapsed between anomalies.

Not only can geologists calculate the average rate of plate movement from magnetic anomalies, but they can also determine plate positions at various times in the past, using magnetic anomalies. Because magnetic anomalies are parallel and symmetrical with respect to spreading ridges, all one has to do to determine the position of continents when particular anomalies formed is to move the anomalies back to the spreading ridge, which will also move the continents with them (Figure 12.25). Unfortunately, subduction destroys oceanic

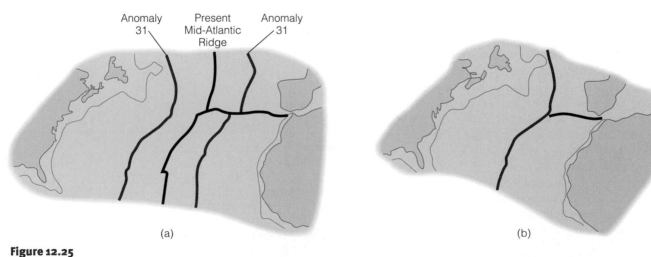

(a)   (b)

Figure 12.25
Reconstructing plate positions using magnetic anomalies. (a) The present North Atlantic, showing the present ridge and magnetic anomaly 31, which formed 67 million years ago. (b) The Atlantic 67 million years ago. Anomaly 31 marks the plate boundary 67 million years ago. By moving the anomalies back together, along with the plates they are on, we reconstruct the former positions of the continents.

crust and the magnetic record it carries. Thus we have an excellent record of plate movements since the breakup of Pangaea, but not as good an understanding of plate movement before that time.

From the information in Figure 12.24, it is obvious that the rate of movement varies among plates. The southeastern part of the Pacific plate and the Cocos plates are the two fastest moving plates, and the Arabian and southern African plates are the slowest.

The average rate of movement as well as the relative motion between any two plates can also be determined by satellite–laser ranging techniques. Laser beams from a station on one plate are bounced off a satellite (in geosynchronous orbit) and returned to a station on a different plate. As the plates move away from each other, the laser beam takes more time to go from the sending station to the stationary satellite and back to the receiving station. This difference in elapsed time is used to calculate the rate of movement and relative motion between plates. In addition, rates of movement and relative motion have also been calculated by measuring the difference between arrival times of radio signals from the same quasar to receiving stations on different plates. The rate of plate movement determined by these two techniques correlates closely with those determined from magnetic anomalies.

Hot Spots and Absolute Motion

Plate motions derived from magnetic anomalies, satellite–laser ranging techniques, and radio signals from quasars,

give only the relative motion of one plate with respect to another. To determine absolute motion, we must have a fixed reference from which the rate and direction of plate movement can be determined. **Hot spots,** which may provide reference points, are locations where stationary columns of magma, originating deep within the mantle (mantle plumes), slowly rise to the surface and form volcanoes or flood basalts (Figure 12.14).

One of the best examples of hot spot activity is that over which the Emperor Seamount–Hawaiian Island chain formed (Figure 12.26). Currently, the only active volcanoes in this island chain are on the island of Hawaii and Loihi Seamount. The rest of the islands and seamounts of the chain are also of volcanic origin and are progressively older west-northwest along the Hawaiian chain and north-northwest along the Emperor Seamount chain.

These islands and seamounts are progressively older as you move toward the north and northwest because the Pacific plate has moved over an apparently stationary mantle plume. Thus, a line of volcanoes was formed near the middle of the Pacific plate, marking the direction of the plate's movement. In the case of the Emperor Seamount–Hawaiian Island chain, the Pacific plate first moved north-northwesterly and then west-northwesterly over a single mantle plume.

Mantle plumes and hot spots are useful to geologists in helping explain some of the geologic activity occurring within plates as opposed to that occurring at or near plate boundaries. In addition, if mantle plumes are essentially fixed with respect to Earth's rotational axis—and some recent evidence suggests they might not

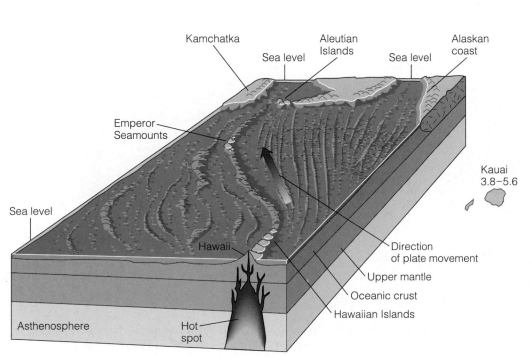

Figure 12.26
The Emperor Seamount–Hawaiian Island chain formed as a result of movement of the Pacific plate over a hot spot. The line of the volcanic islands traces the direction of plate movement. The numbers indicate the age of the islands in millions of years.

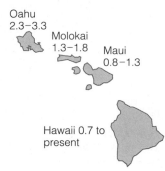

Kauai
3.8–5.6

Oahu
2.3–3.3

Molokai
1.3–1.8

Maui
0.8–1.3

Hawaii 0.7 to present

be—they may prove useful as reference points for determining paleolatitude.

WHAT IS THE DRIVING MECHANISM OF PLATE TECTONICS?

A major obstacle to the acceptance of continental drift was the lack of a driving mechanism to explain continental movement. When it was shown that continents and ocean floors moved together and not separately and that new crust formed at spreading ridges by rising magma, most geologists accepted some type of convective heat system as the basic process responsible for plate motion. The question still remains, however: What exactly drives the plates?

Two models involving thermal convection cells have been proposed to explain plate movement (Figure 12.27). In one model, thermal convection cells are restricted to the asthenosphere; in the second model, the entire mantle is involved. In both models, spreading ridges mark the ascending limbs of adjacent convection cells, and trenches are present where convection cells descend back into Earth's interior. The locations of spreading ridges and trenches are therefore determined by the convection cells themselves, and the lithosphere is considered to be the top of the thermal convection cell. Each plate thus corresponds to a single convection cell.

Although most geologists agree that Earth's internal heat plays an important role in plate movement, problems are inherent in both models. The major problem associated with the first model is the difficulty in explaining the source of heat for the convection cells and why they are restricted to the asthenosphere. In the second model, the source of heat comes from the outer core, but it is still not known how heat is transferred from the outer core to the mantle. Nor is it clear how convection can involve both the lower mantle and the asthenosphere.

Some geologists think that, besides thermal convection, plate movement also occurs, in part, because of a mechanism involving "slab-pull" or "ridge-push" (Figure 12.28). Both mechanisms are gravity driven but still depend on thermal differences within Earth. In slab-pull, the subducting cold slab of lithosphere is denser than the surrounding warmer asthenosphere and thus pulls the rest of the plate along with it as it descends into the asthenosphere. As the lithosphere moves downward, there is a corresponding upward flow back into the spreading ridge.

Operating in conjunction with slab-pull is the ridge-push mechanism. As a result of rising magma, the oceanic ridges are higher than the surrounding oceanic crust. It is thought that gravity pushes the oceanic lithosphere away from the higher spreading ridges and toward the trenches.

Currently, geologists are fairly certain that some type of convective system is involved in plate movement, and the extent to which other mechanisms such as slab-pull and ridge-push are involved is still unresolved. Consequently, a comprehensive theory of plate movement has not yet been developed, and much still remains to be learned about Earth's interior.

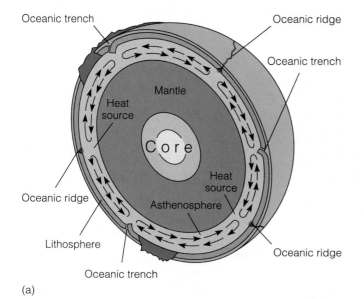

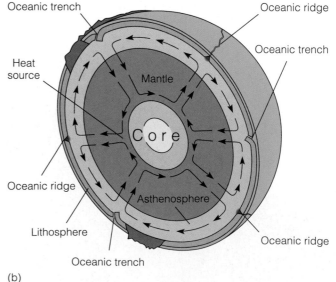

Figure 12.27
Two models involving thermal convection cells have been proposed to explain plate movement. (a) In one model, thermal convection cells are restricted to the asthenosphere. (b) In the other model, thermal convection cells involve the entire mantle.

Guest Essay

Daniel Sarewitz

Geology Meets Public Policy

Daniel Sarewitz is the Director of the Institute for Environmental Education at the Geological Society of America (GSA). From 1989 to 1993 he worked on Capitol Hill, first as GSA's Congressional Science Fellow and then as science consultant on the U.S. House of Representatives Committee on Science, Space, and Technology. He is the author of *Frontiers of Illusion: Science, Technology, and the Politics of Progress* (Temple University Press, 1996).

T he importance of geoscience for public policy grows every year. As the population of Earth tops 6 billion, geoscience can provide the information and insight needed to balance the needs of our burgeoning population with the realities of the finite planet on which we live. The success of this balancing act will depend upon our ability to make political choices that are consistent with our knowledge of the natural world. Geoscientists can play a crucial role in helping to make these decisions.

Global climate change, energy and water policy, nuclear and solid waste disposal, and protection from natural hazards are just a few of the issues that require a careful reconciling of science and policy to ensure the future well-being of society. This melding will not take place if scientists remain in the laboratory and deliver their results to the public in the form of technical papers that only experts can understand. Scientists and policymakers must both communicate clearly so that each understands the needs and limitations of the other. Politicians must understand that science cannot eliminate all uncertainty from controversial problems such as global warming, while scientists must realize that "good data" are not the only legitimate input from science into policy making.

At times, the quality of political debate over scientific issues is astonishingly ill-informed. But even if elected officials were all well-versed in science, legitimate tradeoffs would still have to be made between political and scientific considerations. For example, a member of Congress from a state whose economy depends on high-sulfur coal production might feel obliged to vote against regulations that prohibit using such coal to generate electricity, even though he or she understands that burning this coal contributes to acid rain. All the same, to make wise decisions, members of Congress must be able to weigh the relative importance of political pressures and scientific information. They cannot do so without the advice of staff who are scientifically literate.

While science may help politicians to determine the scope of a problem, science can rarely dictate the "best" solution to the problem. Global warming could be stemmed by limiting the combustion of oil and coal; damage from earthquakes and floods could be reduced by prohibiting development along faults and flood plains. To geoscientists, these are obvious statements; to politicians, they are impossible or impractical solutions.

Indeed, these hypothetical cases illustrate why the scientists' most effective role in the policy process is not simply to provide facts to policymakers: facts alone are rarely useful. Rather, geoscientists should work with policymakers to explore the implications of different policy alternatives—implications which are as "real" as the scientific facts themselves. Geoscientists are well-suited to this type of analysis because they are accustomed to solving complex problems through an interpretive approach. Whereas disciplines such as physics and molecular biology attempt to understand nature by reducing it to its component parts, geoscience is distinguished for its holistic or systems-oriented approach.

Traditionally, the academic system that educates geoscientists has failed to portray public policy as a legitimate and rewarding path of professional endeavor, but the need for geoscientists on Capitol Hill and in the other branches of government is increasing as the challenges of environmental protection, natural resource management, and natural hazard mitigation continue to unfold. The geoscience community should recognize that it can make a key contribution to the formulation of public policy and that careers in public policy represent a legitimate—and growing—area of professional opportunity. ∎

HOW DOES PLATE TECTONICS AFFECT THE DISTRIBUTION OF NATURAL RESOURCES?

B esides being responsible for the major features of Earth's crust, plate movements also affect the formation and distribution of natural resources. Consequently, geologists are using plate tectonic theory in their search for petroleum, natural gas, and mineral deposits and in explaining the occurrence of these natural resources.

It is becoming increasingly clear that if we are to keep up with the continuing demands of a global industrialized society, the application of plate tectonic theory to the origin and distribution of natural resources is essential.

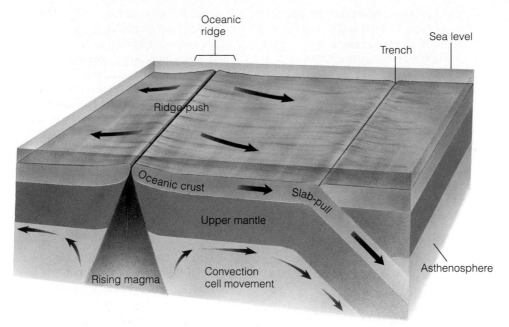

Figure 12.28
Plate movement is also thought to occur because of gravity-driven "slab-pull" or "ridge-push" mechanisms. In slab-pull, the edge of the subducting plate descends into the interior, and the rest of the plate is pulled downward. In ridge-push, rising magma pushes the oceanic ridges higher than the rest of the oceanic crust. Gravity thus pushes the oceanic lithosphere away from the ridges and toward the trenches.

Mineral Deposits

Many metallic mineral deposits such as copper, gold, lead, silver, tin, and zinc are related to igneous and associated hydrothermal activity, so it is not surprising that a close relationship exists between plate boundaries and the occurrence of these valuable deposits.

The magma generated by partial melting of a subducting plate rises toward the surface, and as it cools, it precipitates and concentrates various metallic ores. Many of the world's major metallic ore deposits are associated with convergent plate boundaries including those in the Andes of South America, the Coast Ranges and Rockies of North America, Japan, the Philippines, Russia, and a zone extending from the eastern Mediterranean region to Pakistan. In addition, the majority of the world's gold is associated with sulfide deposits located at ancient convergent plate boundaries in such areas as South Africa, Canada, California, Alaska, Venezuela, Brazil, southern India, Russia, and western Australia.

The copper deposits of western North and South America are an excellent example of the relationship between convergent plate boundaries and the distribution, concentration, and exploitation of valuable metallic ores (Figure 12.29a). The world's largest copper deposits are found along this belt. The majority of the copper deposits in the Andes and the southwestern United States were formed less than 60 million years ago when oceanic plates were subducted under the North and South American plates. The rising magma and associated hydrothermal fluids carried minute amounts of copper, which was originally widely disseminated but eventually became concentrated in the cracks and fractures of the surrounding andesites. These low-grade copper deposits contain from 0.2 to 2% copper and are extracted from large open-pit mines (Figure 12.29b).

Divergent plate boundaries also yield valuable resources. The island of Cyprus in the Mediterranean is rich in copper and has been supplying all or part of the world's needs for the last 3000 years. The concentration of copper on Cyprus formed as a result of precipitation adjacent to hydrothermal vents along a divergent plate boundary. This deposit was brought to the surface when the copper-rich seafloor collided with the European plate, warping the seafloor and forming Cyprus.

Studies indicate that minerals of such metals as copper, gold, iron, lead, silver, and zinc are currently forming as sulfides in the Red Sea. The Red Sea is opening as a result of plate divergence and represents the earliest stage in the growth of an ocean basin (see Figures 12.15c and 12.16a).

Petroleum

Although large concentrations of petroleum occur in many areas of the world, more than 50% of all proven reserves are in the Persian Gulf region. Why is there so much oil in the Persian Gulf region? The answer lies in the paleogeography and plate movements of this region during the Mesozoic and Cenozoic eras. During the Mesozoic Era, and particularly the Cretaceous Period when most of the petroleum formed, the Persian Gulf area was a broad, stable marine shelf extending eastward from Africa. This passive continental margin lay near the equator where countless microorganisms lived in the surface waters. The remains of these organisms accumulated with the bottom sediments and were buried, beginning the complex process of petroleum generation and the formation of source beds (see Chapter 6).

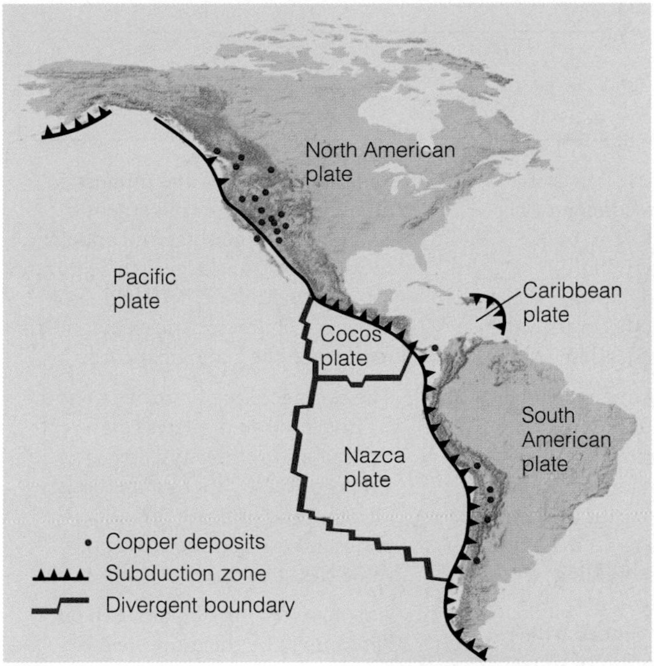

Figure 12.29
(a) Important copper deposits are located along the west coasts of North and South America. (b) Bingham Mine in Utah is a huge open-pit copper mine with reserves estimated at 1.7 billion tons. More than 400,000 tons of rock are removed each day.

(a)

(b)

As a consequence of rifting in the Red Sea and Gulf of Aden during the Cenozoic Era, the Arabian plate is moving northeast away from Africa and subducting beneath Iran. As the sediments of the passive continental margin were initially subducted during the early stages of collision between Arabia and Iran, the heating broke down the organic molecules and led to the formation of petroleum. The tilting of the Arabian block to the northeast allowed the newly formed petroleum to migrate upward into the interior of the Arabian plate. The continued subduction and collision with Iran folded the rocks, creating traps for petroleum to accumulate. Thus the vast area south of the collision zone produces oil.

WHAT WOULD YOU DO?

You are part of a mining exploration team that is exploring a promising and remote area of central Asia. You know that former convergent and divergent plate boundaries frequently are sites of ore deposits. What evidence would you look for to determine if the area you're exploring might be an ancient convergent or divergent plate boundary? Is there anything you can do before visiting the area that might help you determine what the geology of the area is?

Chapter Summary

1. The earliest maps showing the similarity between the east coast of South America and the west coast of Africa provided people with the first evidence that the continents may once have been united and subsequently separated from each other.

2. Alfred Wegener is generally credited with developing the hypothesis of continental drift. He provided abundant geologic and paleontologic evidence to show that the continents were once united into one supercontinent he named Pangaea. Unfortunately, Wegener could not explain how the continents moved, and most geologists ignored his ideas.

3. The hypothesis of continental drift was revived during the 1950s when paleomagnetic studies indicated the presence of multiple magnetic north poles instead of just one as there is today. This paradox was resolved by moving the continents into different positions, making the paleomagnetic data consistent with a single magnetic north pole.

4. Magnetic surveys of the oceanic crust reveal magnetic anomalies in the rocks indicating that Earth's magnetic field has reversed itself in the past. Because the anomalies are parallel and form symmetric belts adjacent to the oceanic ridges, new oceanic crust must have formed as the seafloor was spreading.

5. Seafloor spreading has been confirmed by dating the sediments overlying the oceanic crust and by radiometric dating of rocks on oceanic islands. Such dating reveals that the oceanic crust becomes older with distance from spreading ridges.

6. Plate tectonic theory became widely accepted by the 1970s because of the overwhelming evidence supporting it and because it provides geologists with a powerful theory for explaining such phenomena as volcanism, seismicity, mountain building, global climatic changes, past and present animal and plant distribution, and the distribution of some mineral resources.

7. The supercontinent cycle indicates that all or most of Earth's landmasses form, break up, and re-form in a cycle spanning about 500 million years.

8. Three types of plate boundaries are recognized: divergent boundaries, where plates move away from each other; convergent boundaries, where two plates collide; and transform boundaries, where two plates slide past each other.

9. Ancient plate boundaries can be recognized by their associated rock assemblages and geologic structures. For divergent boundaries, these may include rift valleys with thick sedimentary sequences and numerous dikes and sills. For convergent boundaries, ophiolites and andesitic rocks are two characteristic features. Transform faults generally do not leave any characteristic or diagnostic features in the geologic record.

10. The average rate of movement and relative motion of plates can be calculated in several ways. The results of these different methods all agree and indicate that the plates move at different average velocities.

11. Absolute motion of plates can be determined by the movement of plates over mantle plumes. A mantle plume is an apparently stationary column of magma that rises to the surface where it becomes a hot spot and forms a volcano.

12. Although a comprehensive theory of plate movement has yet to be developed, geologists think that some type of convective heat system is involved.

13. A close relationship exists between the formation of some mineral deposits and petroleum and plate boundaries. Furthermore, the formation and distribution of some natural resources are related to plate movements.

Important Terms

continental–continental plate boundary
continental drift
convergent plate boundary
divergent plate boundary
Glossopteris flora
Gondwana
hot spot

Laurasia
oceanic–continental plate boundary
oceanic–oceanic plate boundary
ophiolite
Pangaea
plate
plate tectonic theory

seafloor spreading
subduction
thermal convection cell
transform plate boundary
transform fault
volcanic island arc

Review Questions

1. The author of the 1915 book that included the first synthesis of information supporting continental drift was:

   a. _____ Alfred Wegener;
   b. _____ Alexander du Toit;
   c. _____ J. Tuzo Wilson;
   d. _____ Fred Vine;
   e. _____ Harry Hess.

2. The Northern Hemisphere landmass of Pangaea is called:

   a. _____ Laurentia;
   b. _____ Gondwana;
   c. _____ Laurasia;
   d. _____ America;
   e. _____ none of these.

3. Which of the following types of evidence for continental drift is most closely associated with the Southern Hemisphere?

   a. _____ glacial features;
   b. _____ similarity of rock sequences;
   c. _____ continental fit;
   d. _____ fossils;
   e. _____ all of the previous.

4. Studies of the ocean floor have shown that the following statement is true about the ocean floor:

   a. _____ Sediment is the same thickness everywhere.
   b. _____ Crust is the same age everywhere.
   c. _____ Crust progressively decreases in age toward continents.
   d. _____ Sediment progressively thins toward spreading ridges.
   e. _____ None of these.

5. The base of a tectonic plate is the same as:

   a. _____ base of crust;
   b. _____ base of upper mantle;
   c. _____ base of continental crust;
   d. _____ answers (a) and (b);
   e. _____ none of these.

6. Convergent plate boundaries are places where:

   a. _____ new oceanic lithosphere forms;
   b. _____ mountains are built;
   c. _____ rift valleys occur;
   d. _____ heat flow is high;
   e. _____ none of these.

7. Along what type of boundary does large lateral displacement of plates occur?

   a. _____ divergent;
   b. _____ convergent;
   c. _____ transform;
   d. _____ answers (a) and (b);
   e. _____ answers (a) and (c).

8. The Red Sea–Gulf of Aden area is an example of a(n) _____ plate boundary.

   a. _____ oceanic–oceanic;
   b. _____ oceanic–continental;
   c. _____ continental–continental;
   d. _____ divergent;
   e. _____ transform.

9. Trenches are associated with _____ plate boundaries.

   a. _____ convergent;
   b. _____ divergent;
   c. _____ transform;
   d. _____ answers (a) and (b);
   e. _____ answers (a) and (c).

10. The fault that connects spreading ridges in the Gulf of California and on the Juan de Fuca plate is an example of a _____ plate boundary.

    a. _____ continental–continental;
    b. _____ oceanic–continental;
    c. _____ divergent;
    d. _____ transform;
    e. _____ none of these.

11. A fixed reference point for studies of absolute plate motion is provided by:

    a. _____ hot spots;
    b. _____ satellite–laser ranging techniques;
    c. _____ dating magnetic reversals in seafloor crust;
    d. _____ ages of seafloor sediment;
    e. _____ none of these.

12. Many of the world's major metallic ore deposits are associated with what type of plate boundaries?

    a. _____ convergent;
    b. _____ divergent;
    c. _____ transform;
    d. _____ answers (a) and (b);
    e. _____ answers (a) and (c).

13. What is the supercontinent cycle? Who proposed this concept, and what kinds of geologic data were needed to support such a concept?

14. What is(are) the difference(s) between the concepts of polar wandering and continental drift?

15. In what way do magnetic anomalies show that the seafloor is really spreading? What was the significance of the discovery of seafloor magnetic anomalies in the history of development of plate tectonics?

16. What do geologists know now and want to find out in the future about the driving mechanism behind plate tectonics?

17. What is the relationship between "continental drift" and "seafloor spreading"?

18. Regarding deep-sea sediments and fossils, how did geologists use these to help confirm ideas about seafloor spreading?

19. Make a list or table of the various geologic features that are associated with each of the three types of modern plate boundaries.

20. Make a list or table of the various geologic features that are associated with convergent and divergent plate boundaries.

21. Alfred Wegener's early ideas were very good, but not well received by many scientists. What was lacking in his theory? How was this problem eventually solved?

22. Explain briefly the relationships among mantle plumes, hot spots, and seamounts. Explain briefly how the Emperor Seamounts and the Hawaiian Islands relate to the plate tectonics history of the Pacific. According to Figure 12.14, how many hot spots exist now?

23. What about the geographic distribution of fossils such as *Glossopteris, Lystrosaurus, Mesosaurus,* and *Cynognathus* provides a compelling argument in favor of plate tectonic motion through time?

24. Using Figure 12.14, locate the world's convergent plate boundaries. List their locations and then briefly discuss the implications of what you have found for the geologic future of the region that we call the Pacific Ocean.

25. How do we know that magnetic anomalies within seafloor crust are the same features as magnetic reversals in sequences of continental lava flows?

26. What is the supercontinent cycle hypothesis? How is Earth's internal heat, the ability of continental rock to conduct heat, and supercontinental breakup related in this hypothesis?

Points to Ponder

1. Estimate the age of seafloor crust, and age and thickness of the oldest sediment off the eastern coast of the United States (e.g., Virginia). In so doing, refer to Figure 12.12 for ages and the deep-sea sediment accumulation rate stated in this chapter. Considering what you learned in a previous chapter about absolute dating, how do we know the age of the seafloor crust in this area?

2. Considering the relative transform fault motion of the Pacific plate and North American plate (Figure 12.23), describe the expected changes in geography regarding southern California and adjacent Baja California, Mexico, as plate motion continues through geological time.

3. Using the age dates in Figure 12.26, explain how you would go about calculating the rate of Pacific plate motion over the past several million years.

4. What does plate tectonics on Earth tell us about what is happening inside our planet? How is internal heat of Earth related to plate tectonics? Must Earth have an asthenosphere for plate tectonics to work?

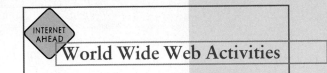
➤ ACTIVE TECTONICS

This site functions mainly to disseminate information and links to other sites related to plate tectonics. Check out some of the other links listed.

➤ PLATE TECTONICS

This site contains pages of tutorial material from the University of Nevada's Seismological Laboratory at Reno. It contains the same information as found in this chapter but also includes many images not seen in this chapter and is worth the visit.

➤ TECTONIC PLATE MOTION

Maintained by NASA, this site contains a wealth of information on how the various space geodetic technologies are used for calculating and tracking current plate movements.
1. What are the different ways plate movements can be determined and calculated?
2. Click on the various geographic sites listed to see what the average velocity and direction of movement of various plates are.

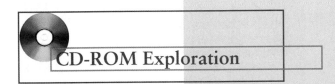

➤ Exploring your *Earth Systems Today* CD-ROM will add to your understanding of the material in this chapter.

TOPIC: EARTH'S PROCESSES

MODULE: PLATE TECTONICS

Explore activities in this module to see if you can discover the following for yourself:

Using the functions in this activity, examine the distribution of volcanoes and earthquakes in relation to plate boundaries. Also, look at the types of plate boundaries and their distribution.

Using this activity, play the Expedition game, which allows you to resolve placement of a plate boundary using an exploration budget. Draw your own plate boundaries based on the data that you collect.

Chapter 13

View of Mount Everest, Earth's highest mountain.

OBJECTIVES

At the end of this chapter, you will have learned that

- Rock deformation results from a variety of forces acting on them.
- Several criteria are used to differentiate among the various types of geologic structures such as folds and faults.
- Geologic structures are important in several human endeavors.

- Deformation is one process important in the origin of mountains.
- Several types of mountains are recognized, each of which differs depending on origin.
- Most of Earth's large mountain ranges on land formed, and in some cases continue to form, at the three types of convergent plate boundaries.

- Terranes have special significance in mountain building.
- Continents formed and evolved in specific ways and continue to do so.

Deformation, Mountain Building, and the Evolution of Continents

Prologue

The forces necessary to form lofty mountains are difficult to comprehend, yet when compared with Earth's size even the highest mountains are very small features. In fact, the greatest difference in elevation on Earth is about 20 km, which if depicted to scale on a 1-m diameter globe would be less than 2 mm. A common analogy used to illustrate just how small mountains are in relation to the entire planet is to compare Earth with a billiard ball or ball bearing. Both are exceptionally smooth, but if they were expanded to the size of Earth they would have deeper canyons and higher mountains. Thus, a globe depicting Earth features at true scale would be quite smooth.

Nevertheless, from a human perspective, mountains are indeed impressive features. For instance, Mount Whitney in California, the highest mountain peak in the continental United States, towers more than 4400 m (14,495 ft) above sea level. Alaska's Denali (Mount McKinley) at 6193 m (20,320 ft) (see Figure 3.1d) compares with peaks in the Andes of South America, though more than 49 Andean peaks exceed 6000 m high! At one time, it was thought that Earth's highest peak was in the Andes, but in 1852 British surveyors in India determined that Mount Everest in the Himalayas stood 8840 m (29,002 ft) high (see chapter opening photo).

Earth statistics such as the length of the longest river, the greatest oceanic depth, and the coldest temperature are fascinating to many people. For some of these statistics, the figures given are very accurate; for a mountain peak's elevation above sea level, however, the figure is a close approximation because measuring a mountain's height is not as easy as it might seem. First, one has to be clear on what is meant by *highest*. Mount Everest is the highest above sea level, the usual designation for *highest*; but because of Earth's equatorial bulge, the summit of Chimborazo in the Andes of Ecuador is most distant from Earth's center. It stands 6269 m (20,561 ft) above sea level, but its peak is about 2 km more distant from Earth's center than Mount Everest's. Highest might also be measured from a mountain's base, in which case Mauna Kea on the island of Hawaii would be the record holder. It rises more than 10,203 m (33,476 ft) above its base on the seafloor, but nearly 60% of it is beneath sea level.

In the case of Mount Everest, its height was originally determined by using an instrument known as a *theodolite* to measure vertical angles between the instrument and the peak from two locations of known elevation. More precise surveys done by Indian and Chinese groups during the 1950s and 1960s gave values within 1 foot of one another, and an elevation of 8848 m (29,028 ft) is now accepted. However, even this figure may have to be adjusted slightly because sea level is not the same everywhere; it simply conforms to the intensity of Earth's gravitational attraction, which differs from area to area. To account for variations in gravitational attraction, scientists refer to the *geoid,* an imaginary sea-level surface extending continuously through the continents. The problem is the geoid is poorly known beneath mountains.

The estimate of 8848 m (29,028 ft) made during the 1950s has recently been challenged. In November 1999, mountaineers and scientists announced measurements made with the aid of global positioning system (GPS) satellites indicate that the peak is 8850 m (29,035 ft) high. However, they measured to the top of the snow-covered peak, not to the rock below the snow, and their measurements remain uncertain because of the same difficulties of estimating the position of the geoid. Nevertheless, the National Geographic Society has accepted the new measurement and plans to use it on future maps and globes.

Occasionally one hears that K2 (Mount Godwin-Austen) on the India–Pakistan border is higher than Mount Everest. This challenge arose a decade ago when satellite data were briefly interpreted to indicate that K2 was higher than Everest. However, more accurate studies revealed that even though it rises an impressive 8613 m (28,250 ft), it is still in second place.

INTRODUCTION

In previous chapters we noted that rocks are not necessarily as solid as the expression "solid as a rock" implies. They decompose in response to chemical activities, and physical agents disaggregate them; heat, pressure, and chemical fluids bring about changes during metamorphism; and under high temperature and pressure deep within Earth they behave very differently than they do at the surface. Indeed, many rocks show the effects of deformation, meaning that dynamic forces caused fracturing, contortion of rock layers, or both (Figure 13.1). Some manifestations of the dynamic forces at work within Earth are seismic activity and the ongoing evolution of mountains in Asia, South America, and elsewhere. In short, Earth remains an active planet with a variety of processes being driven by internal heat, particularly the movement of lithospheric plates. In fact, most of Earth's present-day seismic activity, volcanism, rock deformation, and mountain building takes place at divergent, convergent, and transform plate boundaries.

A variety of processes account for the origin of mountains, some of which involved little or no deformation, but in most mountains the rocks have been complexly deformed by compression at convergent plate boundaries. The Alps in Europe, the Appalachians of North America, the Himalayas of Asia, and the Andes in South America owe their existence, and in some cases continuing evolution, to deformation at convergent plate boundaries. In short, rock deformation and mountain building are closely related phenomena and are thus both discussed in this chapter.

Much of this chapter is devoted to a review of the various types of geologic structures resulting from deformation, their descriptive terminology, and the forces responsible for them. But the study of rock deformation also has several applications. For one thing, geologic structures, such as folds and fractures, provide a record of the kinds and intensities of forces that operated during the past. When geologists interpret these structures and their origin, they can make inferences about Earth history that enable us to satisfy our curiosity about the past and to more effectively find and extract various natural resources. Understanding the nature of geologic structures helps geologists find and recover petroleum and natural gas (see Figure 6.29), and understanding deformation is essential in many mining endeavors. Local geologic structures must also be considered when selecting sites for dams, large bridges, and nuclear power plants, especially if the sites are in areas of active deformation as in Japan, the Philippines, northern Iraq and Iran, and large parts of western North America.

Fold

Fold

(a)

(b)

Figure 13.1
Many rocks show the effects of deformation. (a) Folded rock layers. (b) Rock deformation resulting in numerous closely spaced fractures.

ROCK DEFORMATION—WHAT DOES IT MEAN, AND HOW DOES IT OCCUR?

Deformation is a general term encompassing all changes in shape or volume (or both) of rocks. That is, rocks might be fractured or crumpled into folds as the result of **stress,** which is the result of force applied to a given area of rock. If the intensity of the stress is greater than the rock's internal strength, it will undergo **strain,** which is simply deformation caused by stress.

The terminology is a bit confusing at first, but keep in mind that *deformation* and *strain* are synonyms, and stress is the force that causes strain or deformation. The following example and reference to Figure 13.2 will perhaps clarify the meaning of stress and the distinction between stress and strain.

Remember that stress is the force applied to a given area of rock, which can be expressed in pounds per square inch (lbs/in.^{2}) or kilograms per square centimeter (kg/cm^2), or any other convenient expression of force per unit area. For instance, the stress, or force, exerted by a person walking on an ice-covered pond is a

Figure 13.2

Stress and strain exerted on an ice-covered pond. The vertical object (a) has a density of 1 g/cm³ and a volume of 5000 cm³. The area of the object on the ice is 100 cm², so the stress exerted on the ice is 50 g/cm². In (b), the object is on its side and thus has an area of 500 cm² in contact with the ice. Accordingly, the stress exerted on the ice is only 10 g/cm². The total stress at (a) and (b) is exactly the same, but in (b) it is spread over a larger area.

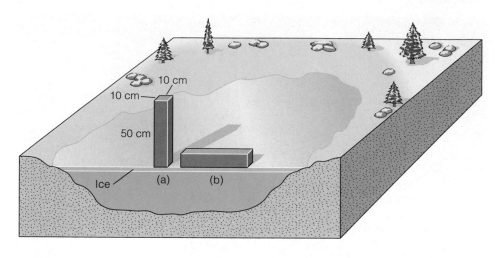

function of the person's weight and the area beneath his or her feet. The ice's internal strength resists the stress unless the stress is too great, in which case it might begin to crack as it is strained, or deformed. To avoid breaking through the ice and a very chilly swim, the person might lie down; this does not reduce the total force but does distribute it over a larger area, thus reducing the stress per unit area.

Although stress is force per unit area it comes in three varieties, *compression, tension,* and *shear,* depending on the direction of the applied forces. In **compression,** rocks are squeezed or compressed by external forces directed toward one another. Rock layers subjected to compression are commonly shortened in the direction of stress by either folding or faulting (Figure 13.3a). **Tension** results from forces acting in opposite directions along the same line, and tends to

lengthen rocks or pull them apart (Figure 13.3b). In **shear stress,** forces act parallel to one another but in opposite directions, resulting in deformation by displacement of adjacent layers along closely spaced planes (Figure 13.3c).

Three types of strain are recognized depending on the response of rocks to stress. **Elastic strain** takes place when deformed rocks return to their original shape when the stress is relaxed. A somewhat analogous situation is found when squeezing a rubber ball that regains its original shape when no longer squeezed. As one might expect, most rocks are not very elastic, but when loaded with glacial ice, Earth's crust responds by elastic strain and is depressed into the mantle. Recall that the crust has responded elastically by isostatic rebound in large parts of Scandinavia and Canada following the last Ice Age (see Figure 10.20).

Figure 13.3

Stress and possible types of resulting deformation. (a) Compression causes shortening of rock layers by folding or faulting. (b) Tension lengthens rock layers and causes faulting. (c) Shear stress causes deformation by displacement along closely spaced planes.

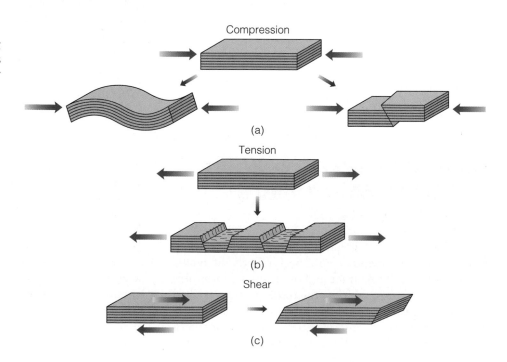

When stress is applied to rocks, they respond first by elastic strain, but when strained beyond their elastic limit, they respond by **plastic strain** as when they are folded, or they behave as brittle solids and **fracture** (Figure 13.4). In either case, the rocks cannot recover their original shape. So we can summarize by saying that elastic strain results in no permanent deformation, whereas rocks experiencing plastic strain or fracture are permanently deformed. One more point should be made regarding stress and strain. Rocks are considerably stronger in compression than they are in tension (see Perspective 13.1 on page 388).

Many rocks show the effects of plastic deformation that must have occurred deep within the crust where the temperature and pressure are high. In Chapters 7 and 10 we noted that the behavior of rocks under these conditions is very different from their behavior near the surface. At or near the surface, they behave as brittle solids, whereas under conditions of high temperature and high pressure, they more commonly deform plastically rather than fracture. The foci of most earthquakes are at depths of less than 30 km, indicating that deformation by fracturing becomes increasingly difficult with depth, and no fracturing is known at depths greater than 700 km.

The type of strain depends on the kind of stress applied, the amount of pressure, the temperature, the rock type, and length of time the rock is subjected to stress. A small stress applied over a long period of time, as on the marble slab shown in Figure 13.5, will cause plastic deformation. By contrast, a large stress applied rapidly to the same object, as when struck by a hammer, will probably result in fracture. Rock type is important because not all rocks respond to stress in the same way. Rocks are considered either *ductile* or *brittle* depending on the amount of plastic strain they exhibit. Brittle rocks show little or no plastic strain before fracture, whereas ductile rocks exhibit a great deal (Figure 13.4).

STRIKE AND DIP— DETERMINING THE ORIENTATION OF ROCK LAYERS

During the 1660s, Nicholas Steno, a Danish anatomist, proposed several principles essential for deciphering Earth history from the record preserved in rocks. One is the *principle of original horizontality,* which means that when sediments are deposited, they accumulate in horizontal or nearly horizontal layers. Thus, when we observe steeply inclined beds of sedimentary rocks we are justified in inferring that they were deposited horizontally, lithified, and then tilted into their present position (Figure 13.6). Some igneous rocks, especially ashfalls and lava flows, also form nearly horizontal layers. If rock layers have been deformed by folding or faulting or both, geologists use the concept of *strike* and

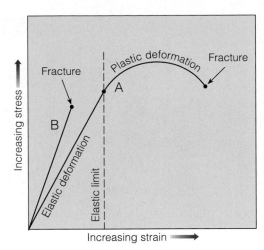

Figure 13.4
Strain, or deformation, in response to stress. Rocks initially respond to stress by elastic deformation and return to their original shape when the stress is released. If the elastic limit is exceeded as in curve A, rocks deform plastically, which is permanent deformation. The amount of plastic deformation rocks exhibit before fracturing depends on their ductility: if they are ductile, they show considerable plastic deformation (curve A), but if they are brittle, they show little or no plastic deformation before failing by fracture (curve B).

Figure 13.5
This marble slab in the Rock Creek Cemetery, Washington, D.C., bent under its own weight in about 80 years. *Source: Photo courtesy of John S. Shelton.*

dip to describe their orientation with respect to a horizontal plane.

By definition, **strike** is the direction of a line formed by the intersection of a horizontal plane with an inclined plane. For example, the surfaces of the rock layers in Figure 13.7 are inclined planes, whereas the water surface represents a horizontal plane. The direction of the line formed at the intersection of the water surface with the inclined surface of a rock layer is the strike. The strike line's orientation is determined by using a compass to measure its angle with respect to north.

(a)   (b)

Figure 13.6
(a) These layers of sedimentary rock in Valley of the Gods, Utah, are horizontal as when they were deposited and lithified. (b) We can infer that these sandstone beds in Colorado were deposited horizontally, lithified, and then deformed and tilted into their present position.

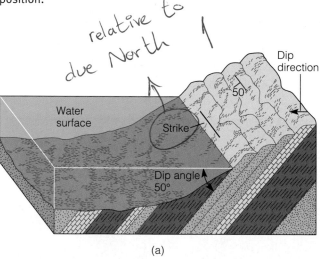 

relative to due North

(a)   (b)

Figure 13.7
Strike and dip. (a) The intersection of a horizontal plane (the water surface) and an inclined plane (the surface of the rock layer) forms a line known as the *strike*. The dip is the maximum angular deviation of the inclined layer from horizontal. Notice the strike and dip symbol with 50 adjacent to it indicating the angle of dip. (b) Natural example of strike and dip. The strike of the dipping rock surface is marked by its intersection with the water surface.

Dip is a measure of an inclined plane's inclination from horizontal, so it must be measured at right angles to the strike direction (Figure 13.7).

Geologic maps showing the age, aerial distribution, and geologic structures of rocks in an area employ a special symbol to indicate strike and dip. A long line oriented in the appropriate direction indicates strike and a short line perpendicular to the strike line shows the direction of dip (Figure 13.7). Adjacent to the strike and dip symbol is a number corresponding to the dip angle. The usefulness of strike and dip symbols will become apparent in the following sections on folds and faults.

What Would You Do?

The types of stresses as well as elastic versus plastic strain might seem rather esoteric, but perhaps they have some practical applications. What relevance do you think understanding these concepts has to some professions, other than geologists, and what professions might these be? We also contend with stress and strain in our daily lives. Can you think of examples? For example, what happens when a car smashes into a tree?

DEFORMATION AND GEOLOGIC STRUCTURES

Recall that *deformation* and its synonym *strain* refer to changes in the shape or volume of rocks. During deformation rocks might be crumpled into folds, or they might be fractured, or perhaps folded and fractured. Any of these features resulting from deformation is referred to as a *geologic structure*. Various geologic structures, or simply structures, are present almost everywhere that rock exposures can be observed, and many can be detected far below the surface by drilling and several geophysical techniques.

Folded Rock Layers

Geologic structures known as *folds* in which planar features such as bedding are bent are quite common. Most folding takes place in response to compression, as when you place your hands on a tablecloth and move them toward one another, thereby producing a series of up- and down-arched folds in the fabric. Rock layers in Earth's crust respond similarly to compression, but unlike the tablecloth, folding in rock layers is permanent. That is, plastic strain has taken place—so once folded, the rocks stay folded. Most folding probably takes place deep in the crust where rocks are more ductile than they are at or near the surface. The configuration of folds and the intensity of folding varies considerably, but only three basic types of folds are recognized—*monoclines, anticlines,* and *synclines.*

Monoclines, Anticlines, and Synclines A simple bend or flexure in otherwise horizontal or uniformly dipping rock layers is a **monocline** (Figure 13.8a). The large monocline in Figure 13.8b formed when the Bighorn Mountains in Wyoming were uplifted along a fault. The fault did not penetrate to the surface, so as uplift proceeded the near-surface rock layers were bent so that they now appear draped over the margin of the uplifted block.

Monoclines are not rare, but they are not nearly as common as anticlines and synclines. An **anticline** is an up-arched or convex upward fold with the oldest rock layers in its core, whereas a **syncline** is a down-arched or concave upward fold in which the youngest rock layers

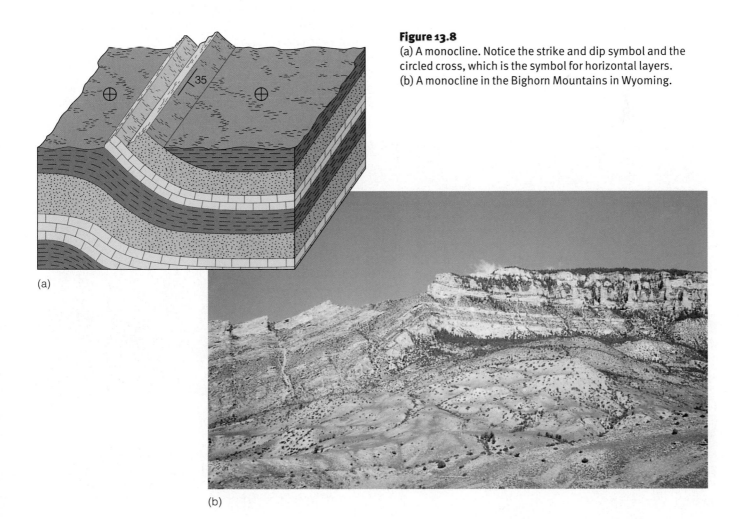

(a)

(b)

Figure 13.8
(a) A monocline. Notice the strike and dip symbol and the circled cross, which is the symbol for horizontal layers.
(b) A monocline in the Bighorn Mountains in Wyoming.

Perspective 13.1

Compression, Tension, and Stone as Building Material

Because it is readily available in many areas and it is quite durable, stone has been used as building material for thousands of years. What could be more permanent? After all, the tremendous weight of mountains is held up by stone, implying that it has great strength. And strong it is, but its strength varies depending on how it is used in construction. For example, experiments and observations clearly show that stone is much stronger in compression than it is in tension. Accordingly, structures of stone such as houses, fireplaces and chimneys, fences, and monuments including pyramids are invariably built so that stresses are directed downward from one structural element to another (Figure 1a). In other words, the stones in the lower part of the structure support the weight of those above. In this configuration stone is indeed strong; however, even if held together by mortar, stone structures do poorly during earthquakes (see Chapter 9).

Given the strength of stone, one might think it could be used in horizontal structural members such as beams spanning the space between pillars. The problem is that a stone beam is compressed on its upper side but subjected to tension on its underside (Figure 1b). The stress on the top of the beam is about the same as it is on the bottom, but the top part (under compression) contributes nothing to the beam's overall strength. If our hypothetical beam were composed of limestone, for example, its upper part (in compression) would be about 30 times stronger than its lower part (in tension). Even stone such as granite that invokes ideas of strength and permanence is about 26 times stronger in compression than in tension.

An example might help clarify any confusion about the strength of stone in compression versus tension. Consider a thin piece of rock laying on a flat surface

(a)

(b)

Figure 1
(a) In this stone chimney all stresses act vertically with lower stones supporting the weight of those above. (b) This Greek ruin shows how closely spaced columns were used to support massive stone beams.

as opposed to one supported only at its ends (Figure 2). If you were to simply press down on both with your hand, which do you think would break most easily? Certainly it would be the one supported at both ends, because as it bent a tension fracture would develop on its underside and travel upward to the top, causing failure.

The same considerations of compression and tension hold true for rock-like substances such as concrete. Accordingly, slabs of concrete as in buildings or bridges are reinforced with steel rods or steel mesh on their tension sides, that is, in the lower part of the slab. Placing rods or mesh on the compression side or near the central part of the beam, known as the *neutral plane,* where no force is exerted, do little to add strength to the beam. (As a matter of fact, one can drill holes at the neutral plane and not decrease a beam's strength.) So why not simply make the beam thicker and do away with the need for reinforcement? Unfortunately, this also makes the beam heavier, and it must bear its own additional weight in addition to any structural elements above it. For instance, if the width and height of a beam are doubled but its length remains the same, its weight increases by a factor of 4, and if length is doubled too, the weight increases eightfold.

Apparently builders have been aware of the limitations of using stone in large structures since at least the time of the ancient Greeks. Notice that in their large structures, numerous closely spaced, weight-bearing columns are still standing after thousands of years. They realized, probably as a result of trial and error, that long horizontal beams did not work well, and thus spans between columns were short. But some of the short horizontal beams now lie in ruins; many probably collapsed when subjected to added stress during earthquakes.

Stone as a building material also has limitations when erecting long structures such as aqueducts and bridges. One cannot simply make long horizontal stretches of stone unless the stone is supported by many vertical columns, and even this arrangement is not particularly sound. The Romans solved this problem by developing the

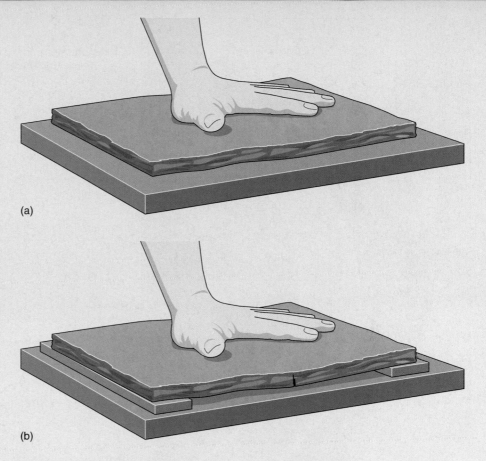

(a)

(b)

Figure 2
Rock strength in compression and tension. In (a) the slab of rock is compressed but does not fracture because it is supported along its entire length. The same amount of force applied in (b) causes fracture because the underside of the slab is unsupported and experiences tension. That is, it bends or deforms elastically, but because rock has little elasticity, it fractures more easily in tension.

Roman arch, in which all the structural elements are in compression. In this arrangement, stress is directed downward and outward toward the adjacent arch, the outermost ones of which were anchored to bedrock or some kind of supporting structure (Figure 3a). Unfortunately for the Romans, though, the arches could not be very large, because they had to be supported by some kind of framework until the last stone or keystone was in place.

Even the magnificent Gothic cathedrals of Europe depend on a modification of the Roman arch. To build cathedrals with large, open interiors, the high, steep *Gothic arch* was developed. It too resulted in all elements within the arch being in compression, and just as in the Roman arch, stress was also directed outward. But large

cathedrals could not be built to direct lateral stresses into bedrock, so the cathedral builders constructed *buttresses* to oppose the outward thrust from the arches (Figure 3b).

Because of their tremendous weight and huge compressive stresses, buildings of stone or brick are limited to only several stories. In contrast, structural steel is comparatively lightweight and equally strong in both compression and tension. With its development, builders could make truly large structures such as skyscrapers and long suspension bridges. Stone, however, remains popular in some aspects of construction such as decorative facing stones on buildings, and for various other uses such as for mantelpieces, tabletops, and paving stones for walkways.

(a)

Figure 3
(a) Roman arches were constructed so that each arch was in compression and the outermost arches were anchored to bedrock. Arches could not be very large, though, because they were not self-supporting until a keystone was in place. (b) This railroad viaduct in Emmersdorf, Austria, uses a series of Roman arches for support. (c) Part of Notre Dame Cathedral in Paris, France. Built between 1163 and 1330, its huge arches are supported by buttresses.

(b)

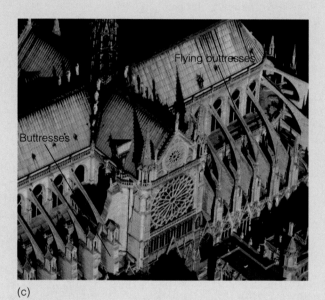

(c)

are in its core (Figure 13.9). These definitions are sufficient for most purposes, but they fail us for anticlines and synclines that in addition to their initial folding have also been tipped on their sides or turned upside down. For these, however, the age of the rocks in the core still serves to distinguish one from the other. We will have more to say about such complex folding later in this section.

Anticlines and synclines also possess an axial plane connecting the points of maximum curvature in each folded layer (Figure 13.10); the axial plane divides folds into halves, each half being a *limb*. Because folds most often are found as a series of anticlines alternating with synclines, an anticline and an adjacent syncline generally share a limb. It is important to remember that anticlines and synclines are simply folded rock layers and do not necessarily correspond to high and low areas at Earth's surface. In fact, folds might underlie rather flat areas (Figure 13.11).

Folds are commonly exposed to view in areas of deep erosion, but even where eroded, strike and dip and the relative ages of the folded rock layers can easily distinguish anticlines and synclines from one another. Notice in Figure 13.12 that in the surface view of the anticline, each limb dips outward or away from the center of the fold where the oldest exposed rocks are. In an eroded syncline, though, each limb dips inward to the fold's center, where the youngest exposed rocks are found.

Figure 13.9
Folded rocks in the Calico Mountains of southeastern California. Three folds are visible from left to right and consist of a syncline, an anticline, and another syncline. We can infer that compression was responsible for these folds.

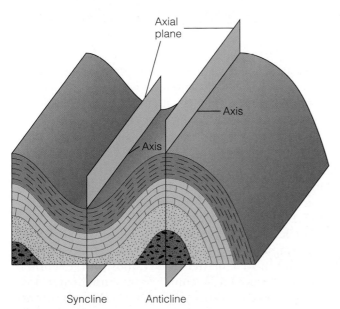

Figure 13.10
Syncline and anticline showing the axial plane, axis, and fold limbs.

Thus far, we have described upright folds in which the axial plane is vertical and each fold limb dips at the same angle (Figure 13.10). If the axial plane is inclined, however, the limbs dip at different angles, and the fold is characterized as *inclined* (Figure 13.13a). In an *overturned fold*, both limbs dip in the same direction. In other words, one fold limb has been rotated more than 90 degrees from its original position so that it is now upside down (Figure 13.13b). Folds in which the axial plane is horizontal are referred to as *recumbent* (Figure 13.13c). Overturned and recumbent folds are particularly common in mountain ranges that formed by compression at convergent plate boundaries (discussed later in this chapter).

We noted in our discussion of anticlines and synclines that the up- and down-arched criteria are not sufficient for folds tipped on their sides or upside down. In Figure 13.13c, for instance, one cannot determine which fold is an anticline or a syncline with the information available. Even strike and dip will not help in this case, but the relative ages of the folded rock layers will resolve the problem. Recall that an anticline has the oldest rocks in its core, so the fold nearest the surface is an anticline and the lower fold is a syncline.

(a)

(b)

Figure 13.11
Folds and their relationships to topography. (a) Cross section illustrating that anticlines and syncline do not necessarily correspond to high and low areas of the surface. Notice that folds even underlie the rather flat area. (b) Notice that a syncline is at the peak of this mountain in Kootenay National Park, British Columbia, Canada. Lower on the left flank of the mountain, an anticline and another syncline are also visible.

Plunging Folds   Folds may be further characterized as *nonplunging* or *plunging*. In the former, the fold *axis,* a line formed by the intersection of the axial plane with the folded beds, is horizontal (Figure 13.10). But it is much more common for the axis to be inclined so that it appears to plunge beneath the surrounding rocks; folds possessing an inclined axis are **plunging folds** (Figure 13.14a). To differentiate plunging anticlines from plunging synclines, geologists use exactly the same criteria used for nonplunging folds: That is, all rocks dip away from the fold axis in plunging anticlines; in plunging synclines, all rocks dip inward toward the axis. The

oldest exposed rocks are in the center of an eroded plunging anticline, whereas the youngest exposed rocks are in the center of an eroded plunging syncline (Figure 13.14b).

In Chapter 6 we noted that anticlines form one type of structural trap for petroleum and natural gas (see Figure 6.29). As a matter of fact, most of the world's petroleum production comes from anticlinal traps, although several other types are also important. Accordingly, geologists are particularly interested in correctly identifying the geologic structures in areas of potential petroleum and natural gas production.

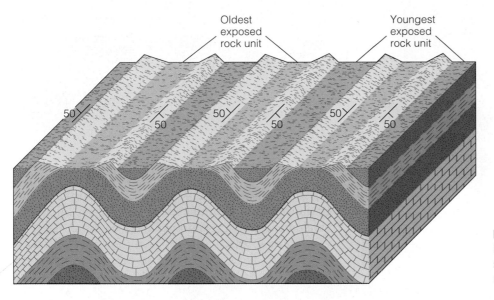

Figure 13.12
Identifying eroded anticlines and synclines by strike and dip and the relative ages of the folded rock layers.

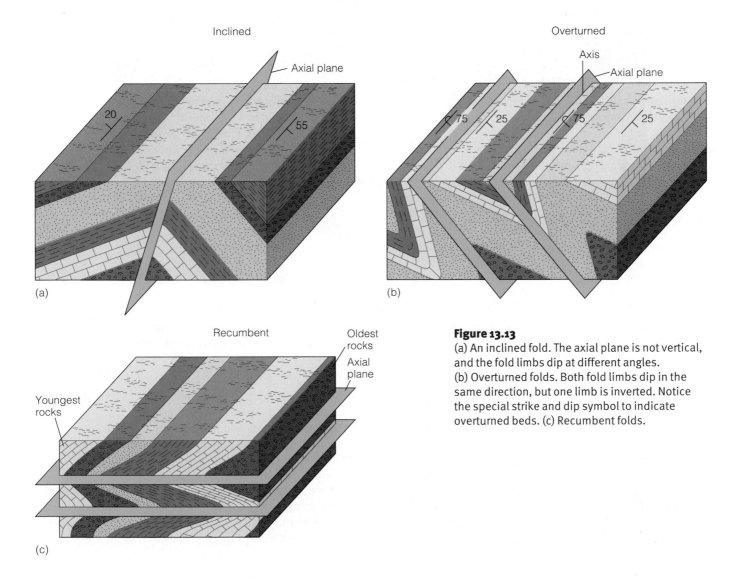

(a)

(b)

(c)

Figure 13.13
(a) An inclined fold. The axial plane is not vertical, and the fold limbs dip at different angles.
(b) Overturned folds. Both fold limbs dip in the same direction, but one limb is inverted. Notice the special strike and dip symbol to indicate overturned beds. (c) Recumbent folds.

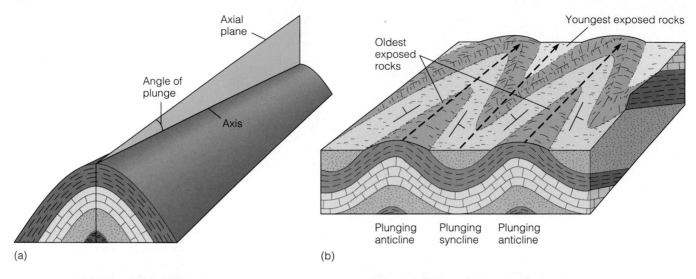

(c)

Figure 13.14

Plunging folds. (a) A schematic illustration of a plunging fold. (b) A block diagram showing surface and cross-sectional views of plunging folds. The long arrow at the center of each fold shows the direction of plunge. (c) Surface view of the eroded, plunging Sheep Mountain anticline in Wyoming. This anticline plunges toward the observer. *Source: Photo courtesy of John S. Shelton.*

Domes and Basins   **Domes** and **basins** are the circular to oval equivalents of anticlines and synclines, respectively (Figure 13.15). They tend to be rather equidimensional, whereas anticlines and synclines are elongate structures. Essentially the same criteria used to recognize anticlines and synclines are used to differentiate domes and basins. In an eroded dome, the oldest exposed rocks are in the center, whereas in a basin the opposite is true. All rocks in a dome dip away from a central point (as opposed to dipping away from a fold

axis, which is a line). By contrast, all rocks in a basin dip inward toward a central point (Figure 13.15).

The terms *dome* and *basin* are also used to designate high and low areas on Earth's surface, but as with anticlines and synclines, domes and basins resulting from deformation do not necessarily correspond to mountains or valleys. In several instances in the following discussions and chapters, we will mention basins in other contexts and have already used the word *dome,* as in "exfoliation dome," so here we will modify these two terms with *struc-*

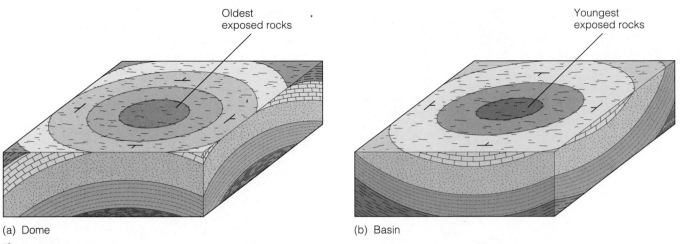

Oldest
exposed rocks

(a) Dome

Youngest
exposed rocks

(b) Basin

Figure 13.15
Block diagrams of (a) a dome and (b) a basin. Notice that in a dome the oldest exposed rocks are in the center and all rocks dip away from a central point, whereas in a basin the youngest exposed rocks are in the center and all rocks dip inward toward a central point.

tural. Accordingly, a structural dome or structural basin is an area of deformation corresponding to our previous definition, with no reference to surface elevations.

Some structural domes and basins are small structures that are easily recognized by their surface exposure patterns, but many are so large that they can be visualized only on geologic maps or aerial photographs. Many of these large-scale structures formed in the continental interior, not by compression but as a result of vertical uplift of parts of the crust with little additional folding and faulting.

The Black Hills of South Dakota are a large eroded structural dome with a core of ancient rocks surrounded by progressively younger rocks. The rocks dip outward rather uniformly from the Black Hills, which have been uplifted so that they now stand more than 2000 m above the adjacent plains. One of the best-known structural basins in the United States is the Michigan Basin. Most of the Michigan Basin is buried beneath younger rocks, so it is not directly observable at the surface. Nevertheless, strike and dip of exposed rocks near the basin margin and thousands of drill holes for oil and gas clearly show that the rock layers beneath the surface are deformed into a large structural basin.

Joints

Besides plastic deformation resulting in folding, rocks may also be permanently deformed by fracturing, which yields either joints or faults. **Joints** are fractures along which no movement has occurred or movement is perpendicular to the fracture surface (Figure 13.16). In other words, a joint may open up, but the rocks on opposite sides of the fracture show no movement parallel to the fracture. This lack of movement parallel to joint surfaces is what distinguishes joints from *faults,*

which do show movement parallel to the fracture surface. Coal miners originally used the term *joint* long ago for cracks in rock that appeared to be surfaces where adjacent blocks were "joined" together.

Joints are the commonest structures in rocks. Recall that rocks near the surface are brittle and therefore commonly fail by fracturing when subjected to stresses. Hence, nearly all near-surface rocks are jointed to some degree. Joints can form in response to compression, tension, and shearing. They vary from minute fractures to those of regional extent and are often arranged in parallel or nearly parallel sets. It is common for a region to have two or perhaps three prominent sets. Regional mapping reveals that joints and joint sets are usually related to other geologic structures such as faults and large folds. Weathering and erosion of jointed rocks in Utah have produced the spectacular scenery of Arches National Park (see Chapter 5 Prologue).

It may seem odd that joints, which indicate brittle behavior, are so common in folded rocks that have been deformed plastically. Of course, some joints form before folding when the rocks are near the surface where they are brittle. But even rocks that exhibit considerable plastic deformation, such as occurs during folding, can be fractured (Figure 13.4). The crest of an anticline provides a good example of how this might occur. Although anticlines are produced by compression, the rock layers are arched so that tension occurs perpendicular to fold crests, and joints form parallel to the long axis of the fold in the upper part of a folded layer (Figure 13.17).

We have already discussed two other types of joints in earlier chapters: columnar joints and sheet jointing. Columnar joints form in some lava flows and in some plutons. Recall from Chapters 3 and 4 that as cooling magma contracts, it develops tensional stresses that form polygonal fracture patterns (see Figure 4.7 and

(a)

(b)

(c)

Figure 13.17
Folding and the origin of joints parallel to the crest of an anticline.

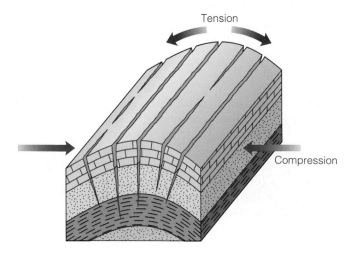

Tension

Compression

Perspective 4.1). Sheet jointing forms in response to pressure release (see Figure 5.7).

Faults

As we mentioned in the previous section, a **fault** is a fracture along which blocks on opposite sides of the fracture move parallel with the fracture surface, which is a **fault plane** (Figure 13.18a and b). One manifestation of faulting is a *fault scarp,* or a cliff formed as a result of vertical movement (Figure 13.18b). Fault scarps are usually quickly modified by erosion and obscured. When rocks on opposite sides of a fault plane move past one another, they may be scratched and polished by friction along the fault plane (Figure 13.18b) or crushed and shattered into angular blocks forming a *fault breccia* (Figure 13.18c).

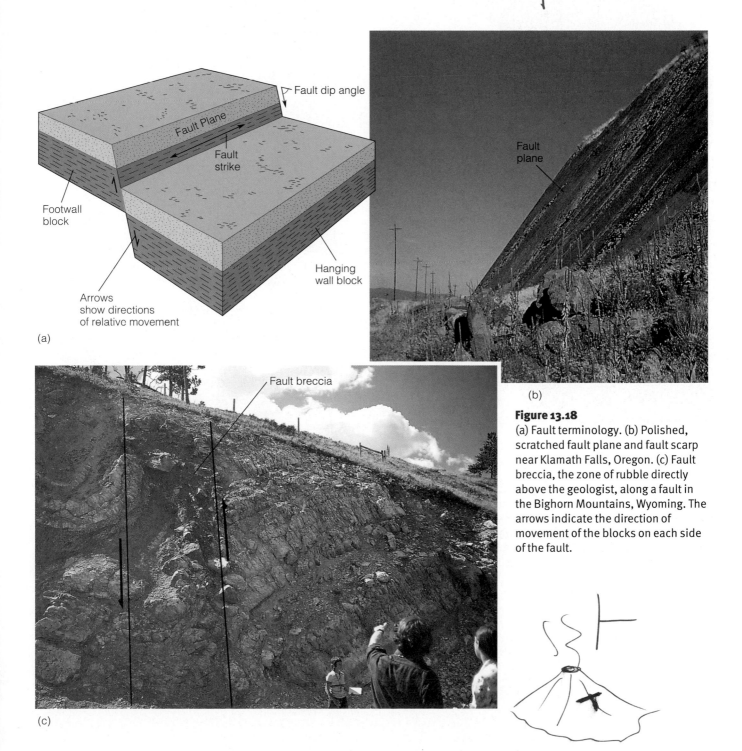

(a)

(b)

Figure 13.18
(a) Fault terminology. (b) Polished, scratched fault plane and fault scarp near Klamath Falls, Oregon. (c) Fault breccia, the zone of rubble directly above the geologist, along a fault in the Bighorn Mountains, Wyoming. The arrows indicate the direction of movement of the blocks on each side of the fault.

(c)

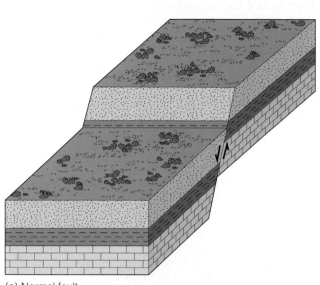

(a) Normal fault

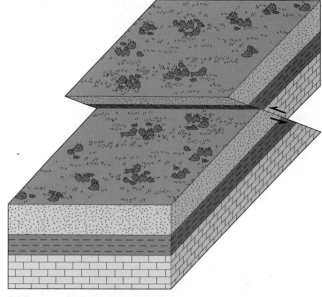

(b) Reverse fault

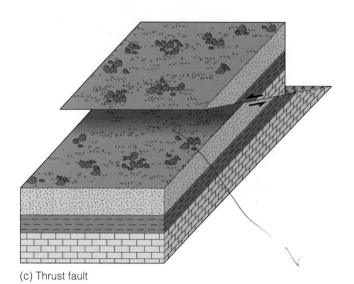

(c) Thrust fault

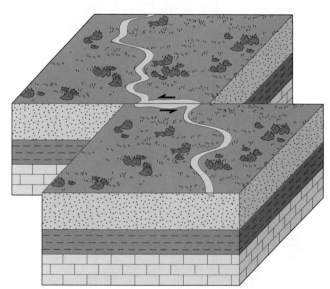

(d) Strike-slip fault

Figure 13.19
Types of faults. (a), (b), and (c) are dip-slip faults. (a) Normal fault—hanging wall block moves down relative to footwall block. (b) and (c) Reverse and thrust faults—hanging wall block moves up relative to footwall block. (d) Strike-slip fault—all movement is parallel to the strike of the fault. (e) Oblique-slip fault—combination of dip-slip and strike-slip movements.

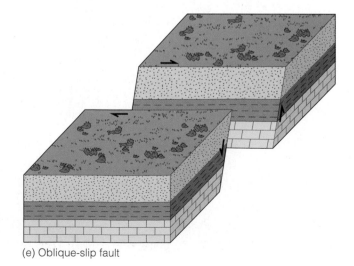

(e) Oblique-slip fault

Notice in Figure 13.18a that the blocks adjacent to the fault plane are designated *hanging wall block* and *footwall block*. The **hanging wall block** is the block of rock overlying the fault, whereas the **footwall block** lies beneath the fault plane. These two blocks can be recognized on any fault except one that is vertical.

To differentiate among the various types of faults, one must be able to identify the hanging wall and footwall blocks and understand the concept of relative movement. Geologists refer to *relative movement* because one usually cannot tell which block actually moved or if both blocks moved. In Figure 13.18a, for instance, it cannot be determined whether the hanging wall block moved up, the footwall block moved down, or both blocks moved. Nevertheless, the hanging wall block appears to have moved down relative to the footwall block.

Like rock layers, fault planes can also be characterized by their strike and dip (Figure 13.18). Two basic types of faults are recognized according to whether the blocks on opposite sides of the fault plane have moved parallel to the direction of dip (dip-slip faults) or along the direction of strike (strike-slip faults).

Dip-Slip Faults In **dip-slip faults,** all movement is parallel with the dip of the fault plane (Figure 13.19a–c). In other words, all movement is such that one block moves up or down relative to the block on the opposite side of the fault plane. Two types of dip-slip faults are recognized: normal and reverse. In Figure 13.19a, the hanging wall block appears to have moved down relative to the footwall block; dip-slip faults with this type of relative movement are **normal faults** (Figure 13.20a).

Normal faults are caused by tensional stresses, such as those that occur when the crust is stretched and thinned by rifting. The mountain ranges in the Basin and Range Province, a large area in the western United States and northern Mexico, are bounded on one or both sides by large normal faults. A normal fault is present along the east side of the Sierra Nevada in California,

Figure 13.20
(a) Two small normal faults cutting through layers of volcanic ash in Oregon. (b) Notice that the sandstone layers to the right of the hammer are cut by a reverse fault. Compare the sense of movement of the hanging wall blocks in these two images.

(a)

(b)

Figure 13.21

Right-lateral offset of a gully by the San Andreas fault in southern California. The gully is offset about 21 m. *Source: Photo courtesy of John S. Shelton.*

where uplift of the block west of the fault has elevated the mountains more than 3000 m above the lowlands to the east.

The second type of dip-slip fault is a **reverse fault** (Figure 13.19b), although a reverse fault involving a fault plane with a dip of less than 45 degrees is called a **thrust fault** (Figure 13.19c). Reverse and thrust faults are easily distinguished from normal faults because the hanging wall block moves up relative to the footwall block (Figure 13.20b). Reverse and thrust faults are caused by compression. Many examples of these faults are present in mountain ranges that formed by compression at convergent plate boundaries (discussed later in this chapter).

Strike-Slip Faults Shearing forces are responsible for **strike-slip faulting,** a type of faulting involving horizontal movement in which blocks on opposite sides of a fault plane slide sideways past one another (Figure 13.19d). In other words, all movement is in the direction of the fault plane's strike.

One of the best-known strike-slip faults is the San Andreas fault of California.* Movement on this fault caused the October 29, 1989, earthquake that damaged areas in Oakland, San Francisco, and several communities to the south and resulted in a 10-day delay of the World Series.

Strike-slip faults are further characterized as right-lateral or left-lateral, depending on the apparent direction of offset. In Figure 13.19d, for example, an observer looking at the block on the opposite side of the fault determines whether it appears to have moved to the right or to the left. In this example, movement appears to have been to the left, so the fault is characterized as a *left-lateral strike-slip fault*. Had this been a *right-lateral strike-slip fault,* the block across the fault from the observer would appear to have moved to the right. The San Andreas fault is a right-lateral strike-slip fault (Figure 13.21).

---

*Recall from Chapter 12 that the San Andreas fault is also called a *transform fault,* in plate tectonics terminology.

Oblique-Slip Faults It is possible for movement on a fault to show components of both dip-slip and strike-slip. Strike-slip movement may be accompanied by a dip-slip component giving rise to a combined movement that includes left-lateral and reverse, or right-lateral and normal. Faults showing both dip-slip and strike-slip movement are **oblique-slip faults.** The fault shown in Figure 13.19e has a combined movement that includes right-lateral and normal.

Various faults as well as folds and joints are commonly depicted on *geologic maps*. Geologists, engineers, city and regional planners, and people in various other professions use these maps for numerous purposes (see Perspective 13.2).

DEFORMATION AND THE ORIGIN OF MOUNTAINS

We noted in the Introduction that several processes are responsible for the origin of mountains, but that the truly large mountains of the continents are produced mostly by compression-induced deformation at convergent plate boundaries. Before discussing mountain building, however, it is necessary to define mountain and briefly discuss the types of mountains. *Mountain* is used to designate any area of land that stands significantly higher than the surrounding country. Some mountains are single, isolated peaks, but more commonly they are parts of linear associations of peaks and/or ridges known as *mountain ranges* that are related in age and origin. A *mountain system* includes several or many mountain ranges. Mountain systems such as the Rocky Mountains and Appalachians are complex linear zones of deformation and crustal thickening characterized by many of the geologic structures previously discussed. In several mountain systems, mountain-building processes remain active today.

Types of Mountains

Mountains develop in a variety of ways, some involving little or no deformation. For instance, differential weathering and erosion have yielded high areas with adjacent low areas in the southwestern United States, and a single volcanic mountain might develop over a hot spot. Or, more commonly, a series of volcanoes forms as a plate moves over a hot spot, as in the case of the Hawaiian Islands (see Figure 12.26). And keep in mind that the oceanic ridge system consists of mountainous topography that exceeds the size of any mountains on land. But these mountains form by volcanism at divergent plate boundaries and show features produced by tensional stresses.

Mountainous topography also forms where crust that has been intruded by batholiths is subsequently uplifted and eroded (Figure 13.22). The Sweetgrass Hills of northern Montana consist of resistant plutonic rocks exposed following uplift and erosion of the softer overlying sedimentary rocks. Elevated areas of plutonic rocks have also been exposed by erosion at Stone Mountain, Georgia (see Figure 5.8b), and in the Llano uplift in Texas.

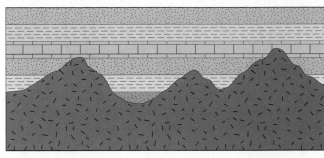
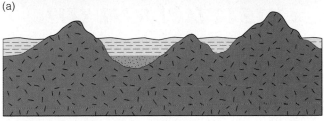

(a)

(b)

Figure 13.22
(a) Pluton intruded into sedimentary rocks. (b) Erosion of the softer overlying rocks reveals the pluton and forms small mountains.

Block-faulting is yet another way for mountains to form. Block-faulting involves movement on normal faults so that one or more blocks are elevated relative to adjacent areas (Figure 13.23). A classic example is the large-scale block-faulting currently occurring in the

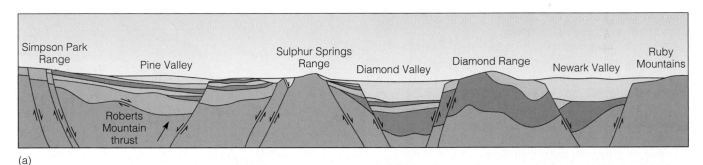

Simpson Park Range | Pine Valley | Sulphur Springs Range | Diamond Valley | Diamond Range | Newark Valley | Ruby Mountains

Roberts Mountain thrust

(a)

(b)

(c)

Figure 13.23
Block-faulting and the origin of horsts and grabens. (a) Cross section of part of the Basin and Range Province in Nevada. The mountain ranges (horsts) and valleys (grabens) are bounded by normal faults. (b) View of the Humboldt Range in Nevada, which was uplifted along normal faults. (c) View of the Sierra Nevada in California at the western margin of the Basin and Range Province. The Sierra Nevada is bounded on its east side by normal faults and rises more than 3000 m above the basins to the east.

Geologic Maps: Their Construction and Uses

Most people are familiar with roadmaps depicting communities, highways, waterways, and political boundaries such as county and state lines. Maps such as these always have a scale so we can determine distances between points of interest and a legend explaining any symbols or colors used on the map. Maps using lines of equal elevation, or what are called *topographic maps,* are also used by people in a variety of professions and by backpackers and other outdoor enthusiasts (see Appendix D). On geologic maps, lines, symbols, and colors are used to depict the distribution of various rock types, show age relationships among rocks, and delineate geologic structures such as folds and faults (Figure 1). Geologic maps are generally printed on a base map; that is, a map showing locations and perhaps elevations. In short, geologic maps provide in graphic form considerable information about the composition, structure, and distribution of geologic materials in a given region, and like other maps they have a scale and legend.

The area depicted on a geologic map depends on the purpose of the study that led to the map's construction in the first place. Geologic maps are available for entire continents or countries, but because they display such large areas, they cannot show small-scale details. These maps are useful for determining the regional distribution of rocks of a given age and for delineating very large geologic structures, but geo-

logic maps showing smaller areas as in Figure 1 are generally of more immediate use.

Geologic maps are constructed in a rather straightforward manner, using information gathered during field studies (Figure 2). Surface exposures of rocks are investigated and pertinent information recorded, such as strike and dip, composition, and relative-age

relationships. Notice in Figure 2 that surface exposures, or outcrops, of the various rocks are discontinuous, the most common situation encountered by geologists. Nevertheless, the exposures can be used to infer what lies beneath the covered areas between the outcrops. Thus, the geologic map shows the rocks of this area as if there were no soil cover (Figure 2c).

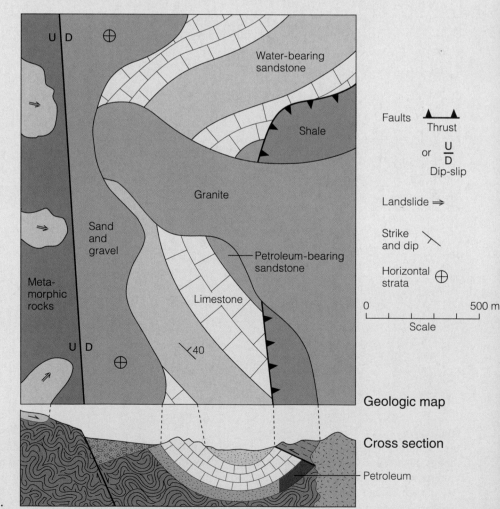

Figure 1
Geologic map and cross section showing the rocks and geologic structures of an area.

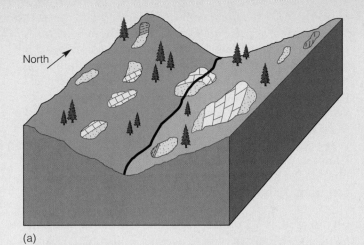

North

(a)

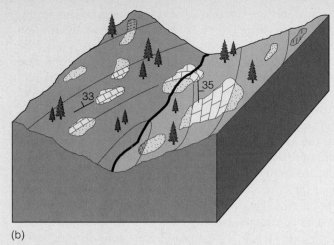

33   35

(b)

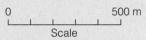

0   500 m

Scale

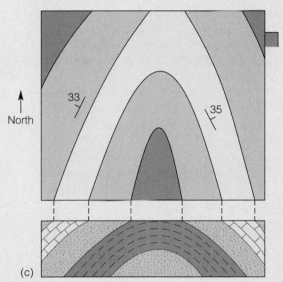

North

33   35

(c)

Figure 2
Construction of a geologic map and cross section from surface rock exposures (outcrops). (a) Valley with outcrops. In much of the area, the rocks are covered by soil. (b) Data from the outcrops are used to infer what is present in the covered areas between outcrops. The lines shown represent boundaries between different types of rock. (c) A geologic map (top) showing the areas as if the soil had been removed. Strike and dip would be recorded at many places, but only two are shown here. The orientation of the rocks as recorded by strike and dip is used to construct the cross section (bottom).

Geologists, of course, are the primary users of geologic maps. Once geologic structures and the types and relative ages of rocks in an area are known, geologists can interpret geologic processes and the geologic history. Furthermore, because geologic maps use strike-and-dip symbols and other symbols depicting geologic structures, cross sections of mapped areas can be constructed to illustrate three-dimensional relationships among rock units and geologic structures (Figure 1). This step is particularly important for many economic ventures because cross sections may indicate where an oil well should be drilled or where other resources might be present beneath the surface.

Besides geologists, people in many other disciplines use geologic maps. You, for example, may one day be a member of a safety planning board of a city with known active faults. Geologic maps might therefore be useful in devel-

oping zoning regulations and construction codes. Or perhaps you will become a land-use planner or a member of a county commission charged with selecting a site for a sanitary landfill or securing an adequate supply of groundwater for your community. Geologic maps would almost certainly be important in making these decisions.

Geologic maps are also used extensively in planning and constructing dams, choosing the best routes for highways through mountainous regions, and selecting sites for nuclear reactors. All such structures must be situated on stable foundations. For example, it would be unwise to build a nuclear reactor in an area of active faulting or to construct a dam across a valley with an active fault or with rocks too weak to support a large structure.

In areas of continuing volcanic activity, geologic maps provide essential information about the kinds of volcanic processes that might occur, such as lava flows, ashfalls, or mudflows. Recall from Chapter 4 that one way in which volcanic hazards are assessed is by mapping and

determining the geologic history of a region. Studies and maps prepared by geologists of the U.S. Geological Survey two years before the eruption of Mount St. Helens in Washington in 1980 allowed local officials to deal more effectively with the consequences of the eruption when it occurred.

Figure 1, a geologic map of a hypothetical area, depicts not only bedrock and various geologic structures but also recent landslide and stream deposits. Given the position of the fault, which is still active, and the landslide-prone area, zoning this area for housing developments would not seem prudent, although it might be perfectly satisfactory for agriculture or some other kind of land use. The map also shows the distribution of unconsolidated layers of sand, and these might be important sources of groundwater. The identification of a petroleum-bearing sandstone might be important to the economic development of this region. In short, this map provides a wealth of information that can be used for a variety of purposes.

Basin and Range Province of the western United States, a large area centered on Nevada but extending into several adjacent states and northern Mexico. Here, the crust is being stretched in an east–west direction and tensional stresses produce north–south oriented, range-bounding faults. Differential movement on these faults has yielded uplifted blocks called *horsts* and down-dropped blocks called *grabens* (Figure 13.23). Erosion of the horsts has yielded the mountainous topography now present, and the grabens have filled with sediments eroded from the horsts.

The processes just discussed can certainly yield mountains. However, the truly large mountain systems of the continents, such as the Alps of Europe and the Appalachians in North America, were produced by compression along convergent plate boundaries.

HOW DO MOUNTAINS FORM?

During an episode of mountain building, termed an *orogeny*, intense deformation takes place, generally accompanied by metamorphism and the emplacement of plutons, especially batholiths. The processes whereby mountains form is still not fully understood, but their origin is related to plate movements. In fact, the advent of plate tectonic theory has completely changed the way geologists view the origin of mountain systems.

Any theory accounting for orogenies must adequately explain the characteristics of mountain systems such as their geometry and location; they tend to be long and narrow and to be at or near plate margins. Mountain systems also show intense deformation, especially compression-induced overturned and recumbent folds and reverse and thrust faults. Furthermore, the deeper, interior parts or cores of mountain systems are characterized by granitic plutons and regional metamorphism. The presence of deformed shallow and deep marine sedimentary rocks that have been elevated far above sea level is another feature.

Deformation and associated activities at convergent plate boundaries are certainly important processes in mountain building. They account for a mountain range's geometry, as well as its complex geologic structures, plutons, and metamorphism. Yet the present-day topographic expression of mountains is also related to several surface processes such as mass wasting (gravity-driven processes, including landslides), glaciers, and running water. In other words, erosion also plays an important role in the evolution of mountains.

Plate Boundaries and Mountain Building

Most of Earth's geologically recent and present-day orogenic activity is concentrated in two major zones or belts: the *Alpine–Himalayan orogenic belt* and the *circum-Pacific orogenic belt* (Figure 13.24). Both belts consist of a number of smaller segments known as *orogens,* many of which are areas of active mountain building today. In fact, we can explain most of Earth's past and present-day orogenic activity in terms of the geologic activity at convergent plate boundaries.

Most orogenies occur in response to compressive stresses at convergent plate boundaries. Recall from Chapter 12 that three varieties of convergent plate boundaries are recognized: oceanic–oceanic, oceanic–continental, and continental–continental.

Orogenies at Oceanic–Oceanic Plate Boundaries

Orogenies occurring where oceanic lithosphere is subducted beneath oceanic lithosphere are characterized by the formation of a volcanic island arc and by deformation, igneous activity, and metamorphism. The subducted plate forms the outer wall of an oceanic trench, and the inner wall of the trench consists of a subduction complex, or *accretionary wedge,* composed of wedge-shaped slices of highly folded and faulted marine sedimentary rocks and oceanic lithosphere scraped from the descending plate (Figure 13.25). This subduction complex is elevated as a result of uplift along faults as subduction continues. In addition, plate convergence results in low-temperature, high-pressure metamorphism characteristic of the blueschist facies (see Figure 7.19).

Deformation also occurs in the island arc system where it is caused largely by the emplacement of plutons of intermediate and felsic composition, and many rocks show evidence of high-temperature, low-pressure metamorphism. As a result, the overriding oceanic plate is thickened as it is intruded by plutons and becomes more continental. The overall effect of island arc orogenesis is the origin of two more or less parallel orogenic belts consisting of a deformed volcanic island arc underlain by batholiths and a seaward belt of deformed trench rocks (Figure 13.25). The Aleutian Islands of Alaska are a good example of this type of deformation.

If a back-arc basin is present, as it is west of Japan, volcanic rocks and sediments derived from the island arc and the adjacent continent are also deformed as the plates continue to converge. They are intensely folded and displaced toward the adjacent continent along low-angle thrust faults. Eventually, the entire island arc complex with its core of metamorphic and plutonic rocks is fused to the edge of the continent, and the back-arc basin sediments are thrust onto the continent and form a thick stack of thrust sheets (Figure 13.25).

Orogenies at Oceanic–Continental Plate Boundaries

The Andes of western South America are perhaps the best example of continuing orogeny at an oceanic–continental plate boundary where oceanic lithosphere is subducted (Figure 13.26). Among the

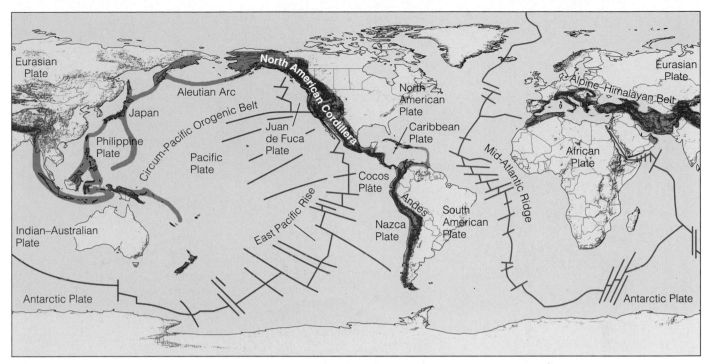

Figure 13.24
Most of Earth's geologically recent and present-day orogenic activity is concentrated in the circum-Pacific and Alpine–Himalayan orogenic belts.

ranges of the Andes are the highest mountain peaks in the Americas and many active volcanoes. Western South America is an extremely active segment of the circum-Pacific earthquake belt. Furthermore, one of Earth's great oceanic trench systems, the Peru–Chile Trench, lies just off the west coast (see Figure 11.14).

Prior to 200 million years ago (MYA), the western margin of South America was a passive continental margin, where sediments accumulated on the continental shelf, slope, and rise, much as they currently do along the East Coast of North America (Figure 13.26a). When Pangaea began fragmenting in response to rifting along what is now the Mid-Atlantic Ridge, the South American plate moved westward, and eastward-moving oceanic lithosphere began subducting beneath the continent (Figure 13.26b). What had been a passive continental margin was now an active one.

As subduction proceeded, rocks of the continental margin and trench were folded and faulted and are now part of an accretionary wedge along the west coast of South America (Figure 13.26c). Accretionary wedges here and elsewhere commonly contain fragments of oceanic crust and upper mantle called *ophiolites* (see Figure 11.25). A well-known accretionary wedge in North America is the Franciscan Complex, a 7000-m-thick, complex assemblage of various rock types exposed in the Coast Ranges of California.

Subduction also resulted in partial melting of the descending plate, producing an andesitic volcanic arc of composite volcanoes near the edge of the continent.

More viscous felsic magmas, mostly of granitic composition, were emplaced as large plutons beneath the volcanic arc (Figure 13.26). The coastal batholith of Peru, for example, consists of perhaps 800 individual plutons that were emplaced over several tens of millions of years.

As a result of the events just described, the Andes Mountains consist of a central core of granitic rocks capped by andesitic volcanoes. To the west of this central core along the coast are the deformed rocks of the accretionary wedge. And to the east of the central core are sedimentary rocks that have been intensely folded and thrust eastward onto the continent (Figure 13.26). Present-day subduction, volcanism, and seismicity along South America's west coast indicate that the Andes Mountains are still actively forming.

Orogenies at Continental–Continental Plate Boundaries The Himalayas of Asia began forming when India collided with Asia about 40 to 50 MYA. Prior to that time, India was far south of Asia and separated from it by an ocean basin (Figure 13.27a). As the Indian plate moved northward, a subduction zone formed along the southern margin of Asia where oceanic lithosphere was consumed (Figure 13.27a). Partial melting generated magma, which rose to form a volcanic arc, and large granite plutons were emplaced into what is now Tibet. At this stage, the activity along Asia's southern margin was similar to what is now occurring along the west coast of South America.

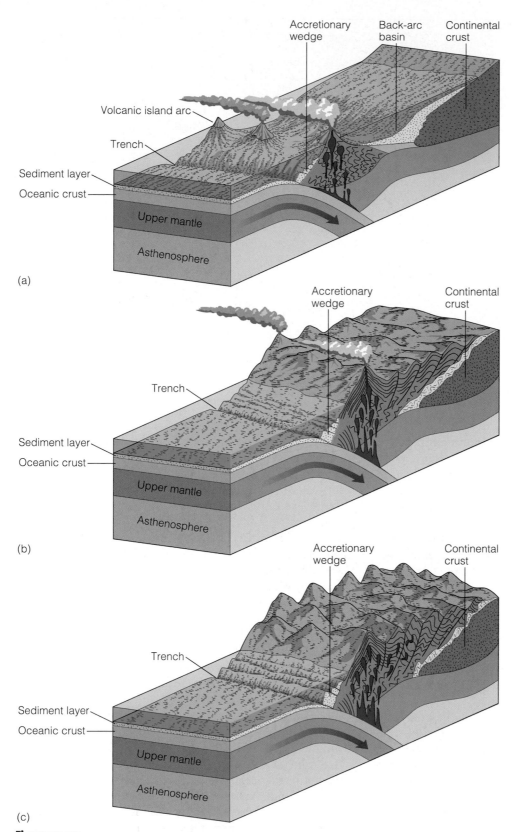

Figure 13.25
Orogeny and the origin of a volcanic island arc at an oceanic–oceanic plate boundary. (a) Subduction of an oceanic plate beneath an island arc. (b) Continued subduction, emplacement of plutons, and beginning of deformation by thrusting and folding of back-arc basin sediments. (c) Thrusting of back-arc basin sediments onto the adjacent continent and suturing of the island arc to the continent.

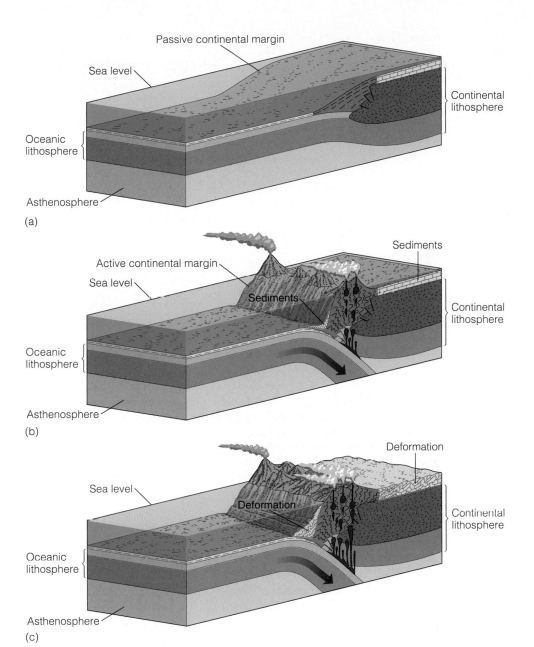

(a)

(b)

(c)

Figure 13.26
Generalized diagrams showing three stages in the development of the Andes of South America. (a) Prior to 200 million years ago, the west coast of South America was a passive continental margin. (b) An orogeny began when the west coast of South America became an active continental margin. (c) Continued deformation, volcanism, and plutonism.

The ocean separating India from Asia continued to close, and India eventually collided with Asia (Figure 13.27b). As a result, two continental plates became welded, or sutured, together. Thus, the Himalayas are now within a continent rather than along a continental margin (Figures 13.24 and 13.27b). The exact time of India's collision with Asia is uncertain, but between 40 and 50 MYA, India's rate of northward drift decreased abruptly—from 15 to 20 cm per year to about 5 cm per year. Because continental lithosphere is not dense enough to be subducted, this decrease in rate seems to mark the time of collision and India's resistance to subduction. Consequently, India's leading margin has moved beneath Asia, causing crustal thickening, thrust-ing, and uplift. Sedimentary rocks that had been deposited in the sea south of Asia were thrust northward, and two major thrust faults carried rocks of Asian origin onto the Indian plate (Figure 13.27c and d). Rocks deposited in the shallow seas along India's northern margin now form the higher parts of the Himalayas.

As the Himalayas were uplifted, they were also eroded but at a rate insufficient to match the uplift. Much of the debris shed from the rising mountains was transported to the south and deposited as a vast blanket of sediment on the Ganges Plain and as huge submarine fans in the Arabian Sea and the Bay of Bengal. Since its collision with Asia, India has been thrust about 2000 km beneath Asia and is still moving north at about 5 cm per year.

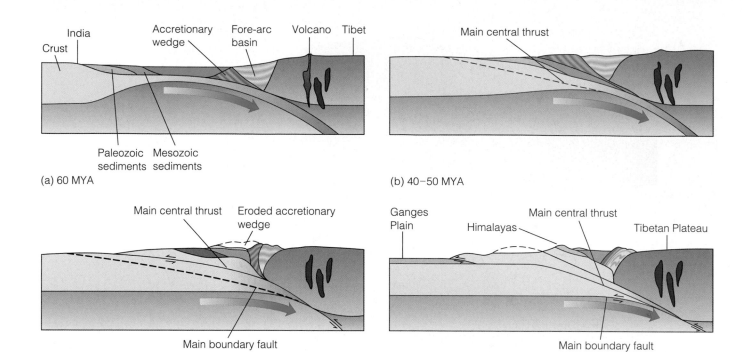

Figure 13.27
Simplified cross sections showing the collision of India with Asia and the origin of the Himalayas. (a) The northern margin of India before its collision with Asia. Subduction of oceanic lithosphere beneath southern Tibet as India approached Asia. (b) About 40 to 50 MYA, India collided with Asia, but because India was too light to be subducted, it was underthrust beneath Asia. (c) Continued convergence accompanied by thrusting of rocks of Asian origin onto the Indian subcontinent. (d) Since about 10 MYA, India has moved beneath Asia along the main boundary fault. Shallow marine sedimentary rocks that were deposited along India's northern margin now form the higher parts of the Himalayas. Sediment eroded from the Himalayas has been deposited on the Ganges Plain.

A number of other mountain systems also formed as a result of collisions between two continental plates. The Urals in Russia and the Appalachians of North America formed by such collisions. Also, west of the Himalayas, the Arabian plate is colliding with Asia along the Zagros Mountains of Iran.

TERRANES AND THE ORIGIN OF MOUNTAINS

In the preceding sections, we discussed orogenies along convergent plate boundaries resulting in *continental accretion*. Much of the material accreted to continents during these events is simply eroded older continental crust, but a significant amount of new material is added to continents as well—igneous rocks that formed by subduction and partial melting and the suturing of an island arc to a continent, for example. Although subduction is the predominant influence on tectonic history in many regions of orogenies, other processes are also involved in mountain building and continental accretion, especially the accretion of blocks of rock known as *terranes*.

During the late 1970s and 1980s, geologists discovered that portions of many mountain systems are composed of small accreted lithospheric blocks that are clearly of foreign origin. These **terranes*** differ completely in their fossil content, structural trends, and paleomagnetic properties from the rocks of the surrounding mountain system. In fact, they are so different from adjacent rocks that most geologists think they formed elsewhere and were carried great distances as parts of other plates until they collided with continents.

Geologic evidence indicates that much of the entire Pacific Coast from Alaska to Baja California consists of accreted terranes or igneous intrusions. They are composed of volcanic island arcs, oceanic ridges, seamounts, and small fragments of continents that were scraped off and accreted to the continent's margin as the oceanic plate with which they were carried was subducted under the continent. And numerous ridges, plateaus, and seamounts in today's oceans are potential terranes, especially in the Pacific Ocean (see Figure 11.14). According to one estimate, more than 100 different-sized

---

*Some geologists prefer the terms *suspect, exotic,* or *displaced terrane.* Notice also the spelling of *terrane* as opposed to the more familiar *terrain,* a geographic term indicating a particular area of land.

terranes have been added to the western margin of North America during the last 200 million years (Figure 13.28). Actually, the terranes illustrated in Figure 13.28 are composed of smaller terranes that cannot be shown at this scale, the Franciscan Complex of the West Coast being an excellent example (see Perspective 13.3).

The basic plate tectonic reconstruction of orogenies and continental accretion remains unchanged, but the details of these reconstructions are decidedly different in view of terrane accretion. For example, growth along active continental margins is faster than along passive continental margins because of the accretion of terranes. Furthermore, these accreted terranes are often new additions to a continent rather than reworked older continental material.

Most of the terranes identified so far are in mountains of the North American Pacific Coast region, but a number of such plates are suspected to be present in other mountain systems as well. They are more difficult to recognize in older mountain systems, such as the Appalachians, because of greater deformation and erosion. Nevertheless, about a dozen terranes have been identified in the Appalachians, although their boundaries are hard to discern.

HOW DID THE CONTINENTS FORM AND EVOLVE?

Rocks 3.8 billion years old that are thought to represent continental crust are known from several areas, including Minnesota, Greenland, and South Africa. Most geologists agree that even older continental crust probably existed, and, in fact, rocks dated at 3.96 billion years have been discovered in Canada.

According to one model for the origin of continents, the earliest crust was thin and unstable and was composed of ultramafic igneous rock. This early ultramafic crust was disrupted by upwelling basaltic magmas at ridges and was consumed at subduction zones. Ultramafic crust would therefore have been destroyed because its density was great enough to make recycling by subduction very likely. Apparently, only crust of a more granitic composition, which has a lower density, is resistant to destruction by subduction.

A second stage in crustal evolution began when partial melting of earlier formed basaltic crust resulted in the formation of andesitic island arcs, and partial melting of lower crustal andesites yielded granitic magmas that were emplaced in the crust that had formed earlier. As plutons were emplaced in these island arcs, they became more like continental crust. By 3.96 to 3.8 billion years ago (BYA), plate motions accompanied by sub-

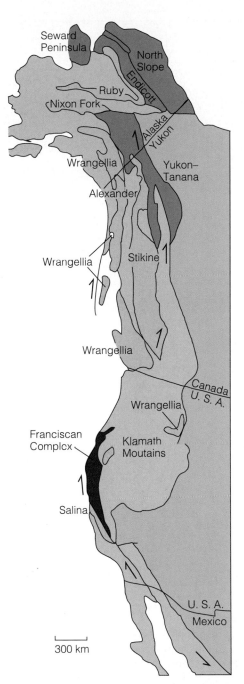

Figure 13.28
Some of the accreted lithospheric blocks called *terranes* that form the western margin of North America. The light brown blocks probably originated as parts of continents other than North America. The reddish brown blocks are possibly displaced parts of North America. The Franciscan Complex consists of a variety of seafloor rocks.

duction and collisions of island arcs had formed several granitic continental nuclei (Figure 13.29).

Shields and Cratons

Each continent is characterized by one or more areas of exposed ancient rocks called a **shield** (see Figure 7.2).

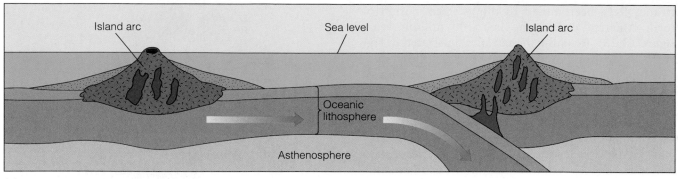

(a)

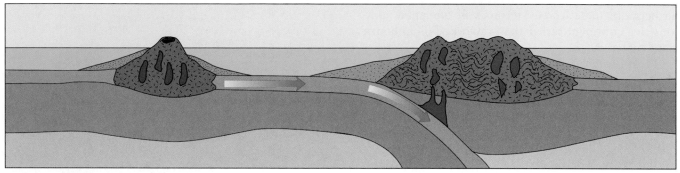

(b)

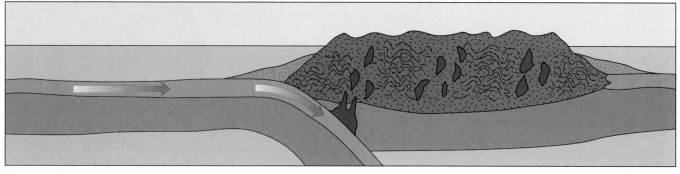

(c)

Figure 13.29

Three stages in the origin of granitic continental crust. Andesitic island arcs formed by the partial melting of basaltic oceanic crust are intruded by granitic magmas. As a result of plate movements, island arcs collide and form larger units or cratons. (a) Two island arcs on separate plates move toward one another. (b) The island arcs shown in (a) collide, forming a small craton, and another island arc approaches this craton. (c) The island arc shown in (b) collides with the craton.

Extending outward from these shields are broad platforms of ancient rocks buried beneath younger sediments and sedimentary rocks. The shields and buried platforms are collectively called **cratons,** so shields are simply the exposed parts of cratons. Cratons are considered to be the ancient, stable interior parts of continents.

All present-day continents have a similar history in that each has a craton and exposed shield area (some have more than one shield) around which new continental crust was added by a process known as *accretion.* That is, orogenies at the margins of cratons resulted in deformation of material eroded from and deposited adjacent to cratons as well as additions of new material by plutonism, volcanism, and collisions of island arcs, oceanic ridges, and seamounts with cratons. All these events are related to plate tectonic activity, especially deformation at convergent plate boundaries. In the following section, we concentrate on the geologic evolution of North America. We should point out, however, that although we use the term *North America,* it really had no meaning until rather recently, geologically speaking, because long ago it looked very different from the way it looks now.

The Geologic Evolution of North America

In North America, the *Canadian Shield* includes much of Canada; a large part of Greenland; parts of the Lake Superior region in Minnesota, Wisconsin, and Michigan; and parts of the Adirondack Mountains of New York (see Figure 7.2). In general, the Canadian Shield is a vast area of subdued topography, numerous lakes, and exposed ancient metamorphic, volcanic, plutonic, and sedimentary rocks.

Each continent evolved by accretion along the margins of ancient cratons. To this extent, all continents developed similarly, but the details of each continent's history differ. Here we will concentrate on the evolution of North America. Several cratons that would eventu-ally become part of the North American craton had formed by 2.5 BYA, but these were independent minicontinents that were later assembled into a larger craton. The Slave, Hearn, and Superior cratons illustrated in Figure 13.30, for instance, were separate landmasses until they collided to form a larger craton.

A major episode in the Precambrian evolution of North America took place between 2.0 and 1.8 BYA when several major orogens developed. These orogens are zones of complexly deformed rocks, many of which have been metamorphosed and intruded by plutons and thus represent areas of ancient mountain building. Smaller cratons were sutured along these orogenic belts so that by 1.8 BYA much of what is now Greenland, central Canada, and the north-central United States formed a large craton (Figure 13.30a).

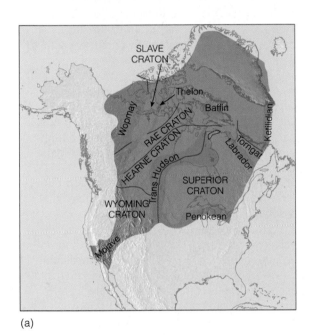

(a)

▮ 900 million – 1.2 billion

▯ 1.6 billion – 1.75 billion

▤ 1.75 billion – 1.8 billion

▦ 1.8 billion – 2.0 billion

▨ 2.5 billion – 3.0 billion

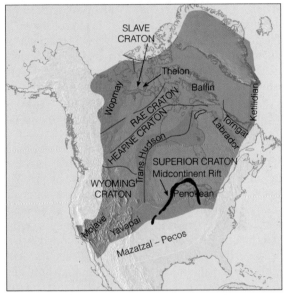

(b)

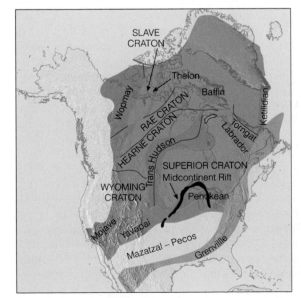

(c)

Figure 13.30
Three stages in the early evolution of the North American craton. (a) By about 1.8 BYA, North America consisted of the elements shown here. The various orogens formed when older cratons collided to form a larger craton. (b) and (c) Continental accretion along the southern and eastern margins of North America. By about 1 BYA, North America had the size and shape shown diagrammatically in (c).

Perspective 13.3

Terranes and the Geology of San Francisco, California

San Francisco is one of America's most beautiful cities; its residents and tourists not only can enjoy sweeping vistas of the city and its surroundings, but also take pleasure from its cultural diversity, museums, parks, and fine restaurants. In addition, the city has an interesting history of geologic events, not the least of which was the disastrous 1906 earthquake. Just offshore to the west of the city lies the famous San Andreas fault, which continually spawns small earthquakes and has the potential to cause large ones. In short, the San Francisco area, and indeed the entire West Coast of North America, is in a geologically active region and has been for many millions of years.

Reference to Figure 13.28 shows that a large part of coastal California is made up of the Franciscan Complex, which in turn consists of numerous terranes, five of which are present in San Francisco and nearby areas. Each of these terranes was accreted to the West Coast as the result of subduction of the Farralon plate beneath the North American plate during the Jurassic and Cretaceous Periods. This event also gave rise to the present-day San Andreas fault (discussed later in this chapter). The geologic map in Figure 1a shows the five terranes in this area, which are progressively older toward the northeast.

At this point it is instructive to point out that the Alcatraz terrane is the oldest and lies above successively younger terranes. Recall from Chapter 8 that in an undisturbed sequence of rocks, the youngest layer is at the top (principle of superposition). Yet the rocks here are in precisely the reverse order. However, the key word is "undisturbed." The rocks composing the five terranes were in effect scraped off against the continental margin as they were carried beneath one another in succession during subduction of the Farralon plate (Figure 2). Once subduction ceased, uplift and erosion exposed the rocks at the surface in their present form.

Notice also in Figure 1a that two of the terranes are characterized as *mélange,* a geologic term for blocks and fragments of rock in a clay matrix formed by intense shearing during subduction. A common rock in these mélanges is serpentinite, a greenish rock formed by hydrothermal alteration of ultramafic rocks of the mantle. Although both mélanges are exposed in San Francisco, similar ones are more easily observed at many locations along the Pacific Coast north of the city.

The other three terranes consist of a variety of rock types, and all are well exposed in San Francisco and nearby areas. Rocks of the San Bruno Mountain terrane consist mostly of thickly bedded graywacke, a type of sandstone containing considerable feldspar minerals, rock fragments, and clay (Figure 1c). It appears to have been deposited in an oceanic trench, much like the present-day Peru–Chile Trench along the west coast of South America.

Rocks of the Marin Headlands terrane are present in San Francisco, but more easily seen across the Bay in the Marin Headlands (Figure 1d). The rocks are varied, consisting of pillow lava, thinly bedded chert and mud, and sandstone, all of which have been complexly deformed. Many of the chert beds, for example, have been intensely folded (Figure 1d). The pillow lavas formed at a spreading ridge where they were incorporated into new oceanic crust, whereas the chert was deposited on the deep seafloor, and the sandstone was derived from the land and accumulated on top of the chert beds. All these rocks were deformed during subduction and accretion of the Marin Headlands terrane.

The oldest terrane, the Alcatraz terrane, can be seen in several locations, especially on Alcatraz Island (Figure 1e). The name Alcatraz is well known because from 1934 until 1963 the island was the site of a maximum-security federal prison where such infamous criminals as Al Capone were incarcerated. The rocks are mostly thick-bedded graywackes, much like those of the San Bruno Mountain terrane, but with one important difference. These graywackes contain little feldspar, whereas abundant feldspars are present in those of the San Bruno Mountain terrane. The reason for this difference is that the Alcatraz terrane being the oldest was emplaced before the feldspar-rich granitic rocks of the Sierra Nevada were exposed at the surface. In contrast, the much younger San Bruno Mountain terrane formed when the Sierra granitic rocks were eroded, yielding particles that were transported to the ocean. One of the distinctive aspects of San Francisco is its steep hills, most of which are underlain by resistant rocks of the Alcatraz terrane.

Legend:
- San Bruno Mountain terrane
- City College melange
- Marin Headlands terrane
- Hunters Point melange
- Alcatraz terrane

Sacramento
Oakland
CALIF.
NEV.

Marin Headlands

Alcatraz Island

B

0 — 4
km

North

San Francisco

A

San Andreas Fault

(a)

(b) A — B

(c)

(d)

(e)

Figure 1
(a) Geologic map showing the five terranes underlying San Francisco and nearby areas. (b) Cross section showing the vertical relationships among the terranes. Notice that the Alcatraz terrane, which is the oldest one, is the uppermost in this sequence. (c) San Bruno Mountain terrane. (d) Marin Headlands terrane. (e) Alcatraz terrane.

Following the origin of the five terranes discussed earlier, subduction ceased when the North American plate collided with the Pacific–Farralon ridge and the continent became bounded by a transform fault (see Figure 13.34 on page 417). Since that time, several sedimentary rock units have been deposited in continental and transitional environments, all of which have had a subsequent history of uplift, faulting, and erosion. During the Pleistocene Ice Age, sea level was much lower than at present, San Francisco Bay was above sea level, and the shoreline was nearly 30 km to the west. In fact, two large rivers joined just east of present-day San Francisco and cut a deep canyon that is now beneath the waters of the bay.

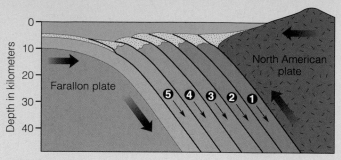

Figure 2
According to this model, the Alcatraz terrane (1) was subducted first, followed by the Hunters Point mélange (2), the Marin Headlands terrane (3), the City College mélange (4), and finally the San Bruno Mountain terrane (5). Uplift and erosion have exposed the rocks at the surface.

Following this initial stage of North America's evolution, continental accretion occurred along the southern and eastern margins of the craton (Figure 13.30b and c). By the end of this time, about 1 BYA, the size and shape of North America were approximately as shown in Figure 13.30c. No further episode of continental accretion occurred until the Paleozoic Era.

Orogeny and accretion during the last 545 million years occurred mostly along the eastern, southern, and western margins of the craton, giving rise to the present configuration of North America. During this time, the craton itself has been remarkably stable. It has periodically been invaded by the seas during marine transgressions followed by regressions, but it has been only mildly deformed into a number of large structural basins and domes.

In the east and south, the Appalachian and Ouachita Mountains formed during the Paleozoic Era in response to compression generated by the closure of an ocean basin during the amalgamation of Pangaea. What had been passive continental margins became active ones as the ocean basin began closing. By the end of the Paleozoic Era, a continent–continent collision occurred, causing deformation in what are now the Appalachian and Ouachita Mountains (Figure 13.31).

During the Late Triassic Period, the first stage in the breakup of Pangaea began: North America separated from Eurasia and North Africa. Along the east coast from Nova Scotia to North Carolina, block-faulting took place and formed numerous ranges with intervening valleys much like those of the present-day Basin and Range Province of the western United States (see Figure 13.23).

The valleys resulting from block-faulting filled with poorly sorted red-colored nonmarine detrital sediments, some of which are well known for dinosaur footprints (Figure 13.32). Rifting was accompanied by widespread volcanism, which resulted in extensive lava flows and numerous dikes and sills. Erosion of the block-fault mountains during the Jurassic and Cretaceous Periods produced a broad, low-lying erosion surface; renewed uplift and erosion during the Cenozoic Era account for the present-day topography of the Appalachian Mountains.

The North American Cordillera is a complex mountainous region in western North America, extending from Alaska into central Mexico (Figure 13.33). It has a long, complex geologic history involving accretion of island arcs along the continental margin, orogeny at an oceanic–continental boundary, vast outpourings of basaltic lavas, and block-faulting. Although the Cordillera has a long history of deformation, the most recent episode of large-scale deformation was the Laramide orogeny, which began 85 to 90 MYA. Like many other orogenies, it occurred along an oceanic–continental boundary. The main Laramide orogeny was centered in the Rocky Mountains of present-day Colorado and Wyoming, but deformation occurred far to the north in Canada and Alaska and as far south as Mexico City.

The Laramide orogeny ceased about 40 MYA, but since that time the Rocky Mountains have continued to evolve. The mountain ranges formed during the orogeny were eroded, and the valleys between ranges filled with sediments. Many of the ranges were nearly buried in their own erosional debris, and their present-day elevations

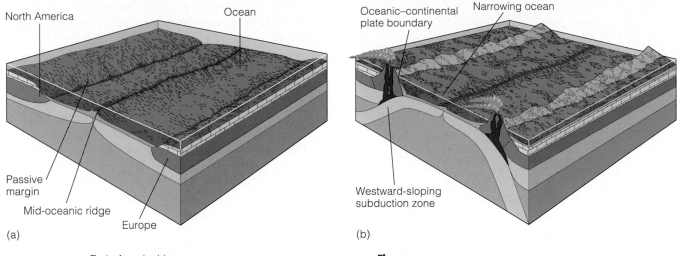

North America
Ocean
Passive margin
Mid-oceanic ridge
Europe

(a)

Oceanic–continental plate boundary
Narrowing ocean
Westward-sloping subduction zone

(b)

Early Appalachian mountains
North America
Europe
Suture zone

(c)

Figure 13.31
Evolution of the Appalachian Mountains. (a) During an Early Paleozoic stage, an ocean was opening up along a divergent boundary. Both the east coast of North America and the west coast of Europe were passive continental plate margins. (b) Beginning in the Ordovician Period, the North American and European passive margins became oceanic–continental plate boundaries, resulting in orogenic activity. (c) By the Late Paleozoic, North America and Europe collided.

(a)

(b)

Figure 13.32
(a) Areas where Triassic block-fault basin deposits are exposed in eastern North America. (b) Reptile tracks in Triassic basin rocks were uncovered during the excavation for a new state building in Hartford, Connecticut. Because the tracks were so spectacular, the building site was moved, and the excavation was designated as a state park.

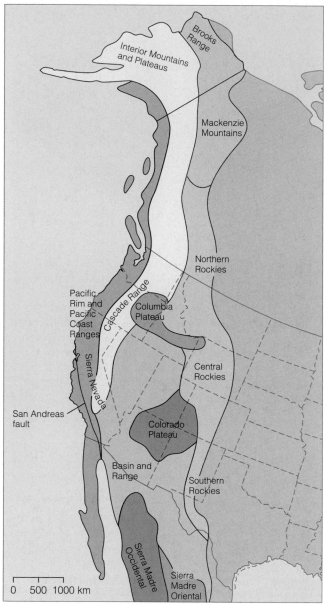

Figure 13.33
The North American Cordillera is a complex mountainous region extending from Alaska into central Mexico. It consists of a number of elements as shown here.

that forms the east face of the Sierra Nevada (Figure 13.23c). This escarpment resulted from movement on a normal fault that has elevated the Sierra Nevada 3000 m above the basins to the east.

In the Pacific Northwest, an area of about 200,000 km² is covered by the Cenozoic Columbia River basalts (see Figure 4.17). Issuing from long fissures, these flows overlapped to produce an aggregate thickness of about 1000 m.

The present-day elements of the Pacific Coast section of the Cordillera developed as a result of the westward drift of North America, the partial consumption of the oceanic Farallon plate, and the collision of North America with the Pacific–Farallon ridge. During the Early Cenozoic, the entire Pacific Coast was bounded by a subduction zone that stretched from Mexico to Alaska (Figure 13.34). Most of the Farallon plate was consumed at this subduction zone, and now only two small remnants exist—the Juan de Fuca and Cocos plates (Figure 13.34). As discussed earlier, the continuing subduction of these small plates accounts for seismicity and volcanism in the Cascade Range of the Pacific Northwest and Central America, respectively. Westward drift of the North American plate also resulted in its collision with the Pacific–Farallon ridge and the origin of the Queen Charlotte and San Andreas transform faults (Figure 13.34; see also Figure 13.21).

WHAT WOULD YOU DO?

As a member of a planning commission, you are charged with developing zoning regulations for an area with known active faults, steep hills, and deep soils. A number of contractors as well as developers and citizens of your community are demanding action because they want to begin several badly needed housing developments. How might geologic maps and an appreciation of geologic structures influence you in this endeavor? Do you think, considering the probable economic gains for your community, that the regulations you draft should be rather lenient or very strict? If the latter, how would you explain why you favored regulations that would involve additional cost for houses?

are the result of renewed uplift, which is continuing in such areas as the Teton Range of Wyoming.

In other parts of the Cordillera, the Colorado Plateau was uplifted far above sea level, but the rocks were little deformed. In the Basin and Range Province, block-faulting began during the Middle Cenozoic and continues to the present (see Figure 13.23). At its western edge, the province is bounded by a large escarpment

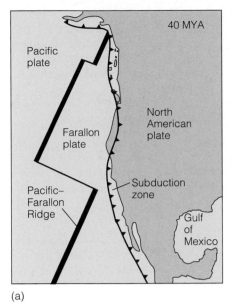
(a)

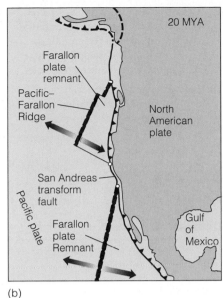
(b)

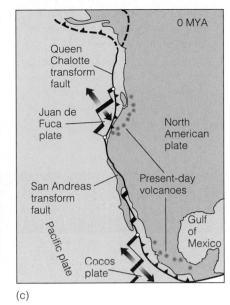
(c)

Figure 13.34
Three stages in the evolution of the San Andreas fault. (a) The westward moving North American plate and its relationships to the Pacific-Farallon Ridge 40 million years ago. (b) By about 20 million years ago, the Queen Charlotte fault was present along the west coast of Canada, and the San Andreas fault was forming along California's coast. (c) Now only two small remnants of the Farallon Plate remain, and most of North America's west coast is bounded by transform faults.

Chapter Summary

1. Contorted and fractured rocks have been deformed or strained by applied stresses.

2. Stress is characterized as compression, tension, or shear. Elastic strain is not permanent, meaning that when the stress is removed, the rocks return to their original shape or volume. Plastic strain and fracture are both permanent types of deformation.

3. The orientation of deformed rock layers is described by strike and dip. Strike and dip can also be used for other planar features such as fault planes.

4. Anticlines and synclines are up- and down-arched folds, respectively. They can be identified by the strike and dip of the folded rocks and by the relative age of the rocks in the cores of the folds.

5. Structural domes and basins are the circular to oval equivalents of anticlines and synclines but are commonly much larger structures.

6. Two types of structures resulting from fracturing are recognized: Joints are fractures along which the only movement, if any, is perpendicular to the fracture surface, and faults are fractures along which the blocks on opposite sides of the fracture move parallel to the fracture surface.

7. Joints, which are the commonest geologic structures, form in response to compression, tension, and shear.

8. On dip-slip faults, all movement is up or down the dip of the fault plane. Two varieties of dip-slip faults are recognized: Normal faults (hanging wall down) form in response to tension, whereas reverse faults (hanging wall up) are caused by compression.

9. Strike-slip faults are those on which all movement is in the direction of strike of the fault plane. They are characterized as right-lateral or left-lateral, depending on the apparent direction of offset of one block relative to the other.

10. Some faults show components of both dip-slip and strike-slip movement; they are called oblique-slip faults.

11. Mountains can form in a variety of ways, some of which involve little or no folding or faulting. Mountain systems consisting of several mountain ranges result from deformation related to plate movements.

12. Most orogenies occur where plates converge and one plate is subducted beneath another or where two continental plates collide.

13. A volcanic island arc, deformation, igneous activity, and metamorphism characterize orogenies occurring at oceanic–oceanic plate boundaries. Subduction of oceanic lithosphere at an oceanic–continental plate boundary also results in orogeny.

14. Some mountain systems, such as the Himalayas, are within continents far from a present-day plate boundary. Such mountains formed when two continental plates collided and became sutured.

15. Geologists now realize that mountain building also occurs when terranes collide with continents.

16. A craton is the stable core of a continent. Broad areas in which the cratons of continents are exposed are called *shields;* each continent has at least one shield area.

17. Cratons formed as a result of accretion, a process involving the addition of eroded continental material, igneous rocks, and island arcs to the margin of a craton during orogenies.

18. In North America, the craton evolved during Precambrian time by collisions of smaller cratons along belts of deformation known as *orogens* and by accretion along its southern and eastern margins. Since then, orogenies have resulted in continental accretion along the craton's margins.

Important Terms

anticline
basin
compression
craton
deformation
dip
dip-slip fault
dome
elastic strain
fault
fault plane

footwall block
fracture
hanging wall block
joint
monocline
normal fault
oblique-slip fault
orogeny
plastic strain
plunging fold
reverse fault

shear stress
shield
strain
stress
strike
strike-slip fault
syncline
tension
terrane
thrust fault

Review Questions

1. Rocks that show little or no plastic strain are said to be:

   a. _____ brittle;
   b. _____ elastic;
   c. _____ ductile;
   d. _____ folded;
   e. _____ compressed.

2. A fault along which the hanging wall block appears to have moved up relative to the footwall block is a _____ fault.

   a. _____ strike-slip;
   b. _____ transform;
   c. _____ normal;
   d. _____ transverse;
   e. _____ reverse.

3. Rocks that have been sheared show:

   a. _____ dip-slip movement;
   b. _____ no deformation;
   c. _____ elastic strain;

   d. _____ movement along closely spaced planes;
   e. _____ none of these.

4. Strike is defined as the:

   a. _____ number of degrees a plane is inclined from horizontal;
   b. _____ intensity of folding and faulting in a region;
   c. _____ line formed by the intersection of a horizontal plane with an inclined plane;
   d. _____ amount of stress stored in rocks;
   e. _____ process whereby new materials are accreted to continents.

5. An oval to circular fold with all rocks dipping inward toward a central point is a(n):

   a. _____ structural basin;
   b. _____ recumbent syncline;

   c. _____ plunging fold;
   d. _____ fault scarp;
   e. _____ compression joint.

6. Most folding of rock layers results from:

   a. _____ shear stress;
   b. _____ compression;
   c. _____ fracturing;
   d. _____ convection;
   e. _____ tension.

7. The fault illustrated in Figure 13.19e shows components of both _____ and _____ faulting.

   a. _____ thrust/reverse;
   b. _____ reverse dip-slip/left-lateral strike-slip;
   c. _____ low-angle thrusting/normal;
   d. _____ hanging wall block uplift/footwall block subsidence;

e. _____ right-lateral strike-slip/normal.

8. Anticlines and synclines with inclined axes are said to be _____ folds.

   a. _____ overturned;
   b. _____ recumbent;
   c. _____ plunging;
   d. _____ jointed;
   e. _____ reversed.

9. The geologic term used for an episode of mountain building is:

   a. _____ orogeny;
   b. _____ terrane;
   c. _____ craton;
   d. _____ shield;
   e. _____ accretion.

10. The Andes of South America and the Himalayas in Asia are still forming as the result of plates colliding at _____ and _____ plate boundaries.

    a. _____ divergent/transform;
    b. _____ oceanic–oceanic/continental–continental;
    c. _____ transform/passive continental margin;
    d. _____ oceanic–continental/continental–continental;
    e. _____ strike-slip/compression-tension.

11. The ancient stable core of a continent is known as its:

    a. _____ platform;
    b. _____ craton;
    c. _____ active interior;
    d. _____ thrust belt;
    e. _____ basin.

12. Which one of the following is correct?

    a. _____ Most mountain building is caused by compression at divergent plate boundaries;
    b. _____ The San Andreas fault is a large normal fault along the east side of the Sierra Nevada;
    c. _____ In a recumbent anticline the youngest rock layers are in the core of the fold;
    d. _____ A monocline is a type of fault along which all movement is in the direction of dip;
    e. _____ Movement on joints, if any, is perpendicular to the joint surface.

13. Explain what is meant by the term *terrane* and how terranes are incorporated into continents. Where would one go to see examples of terranes?

14. How do compression, tension, and shear differ from one another? What kinds of deformation does each cause?

15. Discuss the features of mountains formed at an oceanic–oceanic plate boundary, and give an example of where such activity is currently taking place.

16. Explain how rock type, time, pressure, and temperature influence the type of deformation in rocks.

17. Discuss two ways in which mountains can form with little or no folding and faulting.

18. Draw simple cross sections showing a normal and a reverse fault. Also, label the hanging wall and footwall blocks.

19. What is meant by the elastic limit of rocks, and what happens when rocks are strained beyond their elastic limit?

20. How can you determine whether a recumbent fold is an anticline or a syncline? Illustrate.

21. How is it possible for the same kind of rock to behave both elastically and plastically?

22. What are the similarities and differences between a syncline and a structural basin?

23. Give an example of how you would explain stress and strain to someone unfamiliar with the concepts.

24. Give a brief account of how the San Andreas fault originated.

Points to Ponder

1. During the Paleozoic Era, eastern North America experienced considerable deformation; during the Mesozoic and Cenozoic Eras, deformation has occurred mostly in the western part of the continent. What accounts for this shifting pattern of deformation?

2. Over 5 million years, rocks are displaced 6000 m along a normal fault. What was the average yearly movement on this fault? Is this average likely to represent the actual rate of displacement on this fault? Explain.

3. What kinds of evidence would indicate that mountain building occurred in the Canadian Shield where no mountains are now present?

➤ STRUCTURAL GEOLOGY

There are not a lot of resources on the WWW devoted only to structural geology. This site, maintained by Steven Henry Schimmrich, contains a listing of data sets, bibliographies, organizations, computer software, and online courses in structural geology. Check out the various research projects listed under the Research Information section.

➤ SALT TECTONICS

This site gives an overview of the research being conducted on salt tectonics by Giovanni Guglielmo, at the Applied Geodynamics Laboratory of the Bureau of Economic Geology at the University of Texas at Austin. Much of it will probably not be of interest to general geology students, but for those planning to major in geology, it shows what is currently being done in the field of salt tectonics.

Click on any of the sites listed under Examples of My Current Research. Some of these will require software to run, but it can be downloaded from the links provided.

➤ STRUCTURAL GEOLOGY PHOTO GALLERY GS 326, CORNELL UNIVERSITY

This site, maintained by R. W. Allmendinger, contains a sampling of slides of various geologic structures from field trips and class lectures. Click on any of the designations such as *normal and thrust faults,* then click the thumbnail on the left of the images to see a full-screen photo.

➤ KECK GEOLOGY CONSORTIUM STRUCTURAL GEOLOGY SLIDE SET

This slide set was compiled by H. Robert Burger of Smith College with the support of the W. M. Keck Foundation of Los Angeles, California. The database was created at the Department of Earth and Ocean Sciences, University of British Columbia, Vancouver, BC, Canada. The site has a searchable database. Simply enter a keyword such as *anticline, syncline,* or *normal fault* in the "submit query" space. When the image appears, click to enlarge.

➤ USGS GEOLOGY IN THE PARKS

The U.S. Geological Survey and National Park Service maintain this site with information on various aspects of the national parks and monuments. Under the menu *The Basics,* click on *Geologic Maps* to see examples of geologic maps and explanations of the colors, symbols, and map key.

1. What is a contact, and what two types of contacts are recognized?
2. Why are some contacts represented by solid lines, whereas others are shown as dashed or dotted lines?
3. Why might rocks of the same type—sandstone, for example—be shown on geologic maps by different colors?

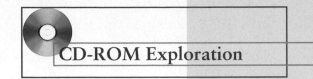

CD-ROM Exploration

➤ Exploring your *Earth Systems Today* CD-ROM will add to your understanding of the material in this chapter.

TOPIC: EARTH'S PROCESSES

MODULE: GEOLOGIC TIME

Explore activities in this module to see if you can discover the following for yourself:

Using the feature that allows you to see changes in the geography of Earth through time, examine changes in continental position through time. Note how continents situated at the margins of plates can experience intensive deformation and mountain building during plate tectonic motion. Describe several of these deformation and mountain-building events and their ages. In particular, observe what happens during development of the Appalachian mountains of the eastern United States.

Chapter 14

Residents of Caracas, Venezuela, clean up the debris from massive flooding and mudslides that devastated large areas of Venezuela during December, 1999.

OBJECTIVES

At the end of this chapter, you will have learned that

- It is important to understand the different types of mass wasting, because mass wasting affects us all and is economically significant in terms of the destruction it causes.
- Factors such as slope angle, weathering and climate, water content, vegetation, and overloading are interrelated, and all contribute to mass wasting.
- Mass movements can be triggered by such factors as overloading, soil saturation, and ground shaking.
- There are two major types of mass wasting: rapid mass movements and slow mass movements.
- The different types of rapid mass movements are rockfalls, slumps, rock slides, mudflows, debris flows, and quick clays, and each type has recognizable characteristics.
- The different types of slow mass movements are earthflows, solifluction, and creep, and each type has recognizable characteristics.
- One can minimize the effects of mass wasting by conducting geologic investigations of an area and stabilizing slopes to prevent and ameliorate movement.

Mass Wasting

Prologue

Triggered by relentless torrential rains that began on December 15, 1999, Venezuela wa devastated by floods and mudslides that have been described as the country's deadliest natural disaster of the century (see chapter opening photo). And although a reliable death toll is impossible to calculate, it is estimated that between 10,000 and 30,000 people were killed, 100,000 to 150,000 were left homeless, 35,000 to 40,000 homes were destroyed or buried by mudslides, and between $10 billion to $20 billion in damage was done before the rains and slides abated. Even weeks later, there were still at least 60,000 homes in areas of high risk from new mudslides. It is easy to throw around numbers of dead and homeless; however, the human side of the disaster is vividly brought home by the statement of a mother who described standing helplessly by and watching her four small children buried alive in the family car as it was carried away by a raging mudslide.

Sparing no one, mudslides engulfed and buried not only homes, buildings, and roads, but also entire communities, leaving some areas covered with as much as 7 m of mud. In addition, flooding and the accompanying mudslides swept away large parts of many of Venezuela's northern coastal communities, leaving huge areas uninhabitable.

It was not unexpected there would be mudslides in the areas where they occurred, but the magnitude of the disaster surprised many. This disaster was caused by a combination of nature, politics, and economics. Prior to the flooding and mudslides, Venezuela needed approximately 100,000 new homes annually to keep up with demand, yet only about half that number had been build since the early 1980s. This forced many low- and moderate-income families to find housing in geologically unsuitable areas that were prone to flooding and landslides. Thus, when the rains came, most of these structures and those in the path of the floods and mudslides were destroyed. The government, under newly elected president Hugo Chévez, has promised to build new homes and an economic infrastructure for many of the displaced families in sparsely inhabited inland areas. Whether he and the government will make good on this promise remains to be seen.

What makes this terrible tragedy important is how it illustrates the close link between geology and individuals, governments, and society in general, a theme we stressed in Chapter 1. The underlying causes of the mudslides are not unique to Venezuela, but can be found anywhere in the world. By being able to recognize and understand these causes and what the ultimate result may be, we can then find ways to reduce these hazards and minimize the damage, both in terms of human suffering and misery, as well as in property damage. The important lessons to be learned from this tragedy are how geology impacts all our lives and how interconnected the various systems and subsystems of Earth are.

INTRODUCTION

The topography of land areas is the result of the interaction among Earth's internal processes, type of rocks exposed at the surface, effects of weathering, and the erosional agents of water, ice, and wind. The specific type of landscape developed depends, in part, on which agent of erosion is dominant. Landslides (mass movements), which can be very destructive, are part of the normal adjustment of slopes to changing surface conditions.

Geologists use the term *landslide* in a general sense to cover a wide variety of mass movements that may cause loss of life, property damage, or a general disruption of human activities. In 218 B.C., avalanches in the European Alps buried 18,000 people; an earthquake-generated landslide in Hsian, China, killed an estimated 1 million people in 1556; and 7000 people died when mudflows and avalanches destroyed Huaraz, Peru, in 1941. What makes these mass movements so terrifying, and yet so fascinating, is that they almost always occur with little or no warning and are over in a very short time, leaving behind a legacy of death and destruction (Table 14.1).

Table 14.1

Selected Landslides, Their Cause, and the Number of People Killed

DATE	LOCATION	TYPE	DEATHS
218 B.C.	Alps (European)	Avalanche—destroyed Hannibal's army	18,000
1512	Alps (Biasco)	Landslide—temporary lake burst	>600
1556	China (Hsian)	Landslides—earthquake triggered	1,000,000
1689	Austria (Montaton Valley)	Avalanche	>300
1806	Switzerland (Goldau)	Rock slide	457
1881	Switzerland (Elm)	Rockfall	115
1892	France (Haute-Savoie)	Icefall, mudflow	150
1903	Canada (Frank, Alberta)	Rock slide	70
1920	China (Kansu)	Landslides—earthquake triggered	~200,000
1936	Norway (Loen)	Rockfall into fiord	73
1941	Peru (Huaraz)	Avalanche and mudflow	7,000
1959	USA (Madison Canyon, Montana)	Landslide—earthquake triggered	26
1962	Peru (Mt. Huascarán)	Ice avalanche and mudflow	~4,000
1963	Italy (Vaiont Dam)	Landslide—subsequent flood	~2,000
1966	Brazil (Rio de Janeiro)	Landslides	279
1966	United Kingdom (Aberfan, South Wales)	Debris flow—collapse of mining-waste tip	144
1970	Peru (Mt. Huascarán)	Rockfall and debris avalanche—earthquake triggered	25,000
1971	Canada (St. Jean-Vianney, Quebec)	Quick clays	31
1972	USA (West Virginia)	Landslide and mudflow—collapse of mining-waste tip	400
1974	Peru (Mayunmarca)	Rock slide and debris flow	430
1978	Japan (Myoko Kogen Machi)	Mudflow	12
1979	Indonesia (Sumatra)	Landslide	80
1980	USA (Washington)	Avalanche and mudflow	63
1981	Indonesia (West Irian)	Landslide—earthquake triggered	261
1981	Indonesia (Java)	Mudflow	252
1983	Iran (Northern area)	Landslide and avalanche	90
1987	El Salvador (San Salvador)	Landslide	1,000
1988	Chile (Tupungatito area)	Mudflow	41
1989	Tadzhikistan	Mudflow—earthquake triggered	274
1989	Indonesia (West Irian)	Landslide—earthquake triggered	120
1991	Guatemala (Santa Maria)	Landslide	33
1994	Colombia (Paez River Valley)	Avalanche—earthquake triggered	>300
1995	Brazil (Northeastern area)	Mudflow	15
1996	Brazil (Recife)	Mudflow	49
1997	Peru (Ccocha and Pumaranra)	Mudflow	33
1999	Venezuela	Mudflow	>10,000

Source: Data from J. Whittow, Disasters: The Anatomy of Environmental Hazards (Athens: University of Georgia Press, 1979); Geotimes; and Earth.

Why is it important to study mass wasting? Because mass wasting affects all of us, no matter where we live (see the Prologue). In the United States alone, mass wasting occurs in all 50 states and is economically significant in terms of destruction in more than 25 states (Figure 14.1). Furthermore, between 25 and 50 people, on average, are killed each year by landslides in the United States, and the annual cost in damages from them exceeds $1.5 billion. Almost all major landslides have natural causes, yet many of the smaller ones are the result of human activity and could have been prevented or their damage minimized.

Mass wasting (also called *mass movement*) is defined as the downslope movement of material under the direct influence of gravity. Most types of mass wasting are aided by weathering and usually involve surficial material. The material moves at rates ranging from almost imperceptible, as in the case of creep, to extremely fast, as in a rockfall or slide. Although water can play an important role, the relentless pull of gravity is the major force behind mass wasting.

Mass wasting is an important geologic process that can occur at any time and almost any place. Although most people associate mass wasting with steep and unstable slopes, it can also occur on near-level land, given the right geologic conditions. Furthermore, the rapid types of mass wasting, such as avalanches and mudflows, typically get the most publicity, but the slow, imperceptible types, such as creep, usually do the greatest amount of property damage.

WHAT FACTORS INFLUENCE MASS WASTING?

When the gravitational force acting on a slope exceeds its resisting force, slope failure (mass wasting) occurs. The resisting forces helping to maintain slope stability include the slope material's strength and

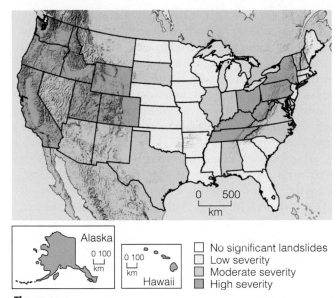

Figure 14.1
Severity of landslides in the United States. Note that areas of greatest severity occur in the coastal mountain ranges of the West Coast, the Rocky Mountains, and Appalachian Mountains.

cohesion, the amount of internal friction between grains, and any external support of the slope (Figure 14.2). These factors collectively define a slope's **shear strength.**

Opposing a slope's shear strength is the force of gravity. Gravity operates vertically but has a component acting parallel to the slope, thereby causing instability (Figure 14.2). The greater a slope's angle, the greater the component of force acting parallel to the slope, and the greater the chance for mass wasting. The steepest angle that a slope can maintain without collapsing is its *angle of repose*. At this angle, the shear strength of the slope's material exactly counterbalances the force of gravity. For unconsolidated material, the angle of repose normally ranges from 25 to 40 degrees. Slopes steeper than 40 degrees usually consist of unweathered rock.

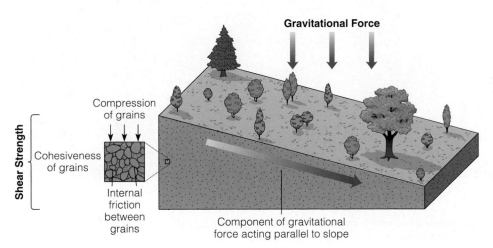

Figure 14.2
A slope's shear strength depends on the slope material's strength and cohesiveness, the amount of internal friction between grains, and any external support of the slope. These factors promote slope stability. The force of gravity operates vertically but has a component acting parallel to the slope. When this force, which promotes instability, exceeds a slope's shear strength, slope failure occurs.

All slopes are in a state of *dynamic equilibrium,* which means that they are constantly adjusting to new conditions. Although we tend to view mass wasting as a disruptive and usually destructive event, it is one of the ways that a slope adjusts to new conditions. Whenever a building or road is constructed on a hillside, the equilibrium of that slope is affected. The slope must then adjust, perhaps by mass wasting, to this new set of conditions.

Many factors can cause mass wasting: a change in slope angle, weakening of material by weathering, increased water content, changes in the vegetation cover, and overloading. Although most of these are interrelated, we will examine them separately for ease of discussion, but will also show how they individually and collectively affect a slope's equilibrium.

Slope Angle

Slope angle is probably the major cause of mass wasting. Generally speaking, the steeper the slope, the less stable it is. Therefore steep slopes are more likely to experience mass wasting than gentle ones.

A number of processes can oversteepen a slope. One of the most common is undercutting by stream or wave action (Figure 14.3). This removes the slope's base, increases the slope angle, and thereby increases the

gravitational force acting parallel to the slope. Wave action, especially during storms, often results in mass movements along the shores of oceans or large lakes (Figure 14.4).

Excavations for roads and hillside building sites are another major cause of slope failure (Figure 14.5). Grading the slope too steeply, or cutting into its side, increases the stress in the rock or soil until it is no longer strong enough to remain at the steeper angle and mass movement ensues. Such action is analogous to undercutting by streams or waves and has the same result, thus explaining why so many mountain roads are plagued by frequent mass movements.

Weathering and Climate

Mass wasting is more likely to occur in loose or poorly consolidated slope material than in bedrock. As soon as rock is exposed at Earth's surface, weathering begins to disintegrate and decompose it, reducing its shear strength and increasing its susceptibility to mass wasting. The deeper the weathering zone extends, the greater the likelihood of some type of mass movement.

Recall that some rocks are more susceptible to weathering than others and that climate plays an important role in the rate and type of weathering. In the tropics, where temperatures are high and considerable

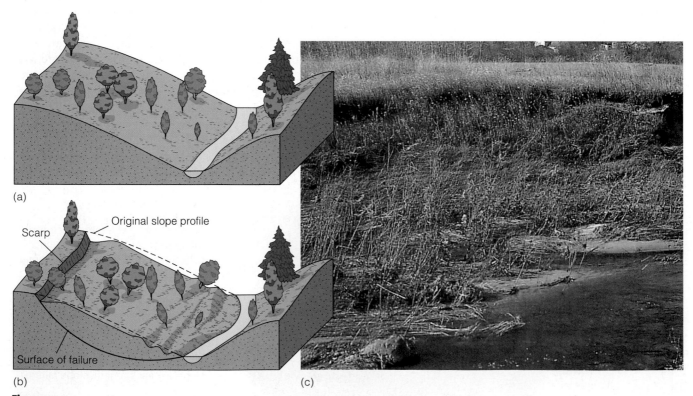

Figure 14.3
Undercutting by stream erosion (a) removes a slope's base, which increases the slope angle and (b) can lead to slope failure. (c) Undercutting by stream erosion caused slumping along this stream near Weidman, Michigan.

Figure 14.4
This sea cliff north of Bodega Bay, California, was undercut by waves during the winter of 1997–1998. As a result, part of the land slid into the ocean, damaging several houses.

Figure 14.5
(a) Highway excavations disturb the equilibrium of a slope by (b) removing a portion of its support as well as oversteepening it at the point of excavation. (c) Such action can result in frequent landslides. (d) Cutting into the hillside to construct this portion of the Pan-American Highway in Mexico resulted in a rockfall that completely blocked the road.

(a)

(b)

(c)

(d)

rainfall occurs, the effects of weathering extend to depths of several tens of meters, and mass movements most commonly occur in the deep weathering zone. In arid and semiarid regions, the weathering zone is usually considerably shallower. Nevertheless, intense, localized cloudbursts can drop large quantities of water on an area in a short time. With little vegetation to absorb this water, runoff is rapid and frequently results in mudflows.

Water Content

The amount of water in rock or soil influences slope stability. Large quantities of water from melting snow or heavy storms greatly increase the likelihood of slope failure. An excellent example of this is the landslides in Washington, Oregon, and California, which resulted from the widespread rainstorms during December 1996 and January 1997. The additional weight that water adds to a slope can be enough to cause mass movement. Furthermore, water percolating through a slope's material helps decrease friction between grains, contributing to a loss of cohesion. For example, slopes composed of dry clay are usually quite stable, but when wetted they quickly lose cohesiveness and internal friction and become an unstable slurry. This occurs because clay, which can hold large quantities of water, consists of platy particles that easily slide over each other when wet. For this reason, clay beds are frequently the slippery layer along which overlying rock units slide downslope (see Perspective 14.1).

Vegetation

Vegetation affects slope stability in several ways. By absorbing the water from a rainstorm, vegetation decreases water saturation of a slope's material that would otherwise lead to a loss of shear strength. Vegetation root systems also help stabilize a slope by binding soil particles together and holding the soil to bedrock.

The removal of vegetation by either natural or human activity is a major cause of many mass movements. Summer brush and forest fires in southern California frequently leave the hillsides bare of vegetation. Fall rainstorms saturate the ground, causing mudslides that do tremendous damage and cost millions of dollars to clean up (Figure 14.6). The soils of many hillsides in New Zealand are sliding because deep-rooted native bushes have been replaced by shallow-rooted grasses used for sheep grazing. When heavy rains saturate the soil, the shallow-rooted grasses cannot hold the slope in place, and parts of it slide downhill.

Overloading

Overloading is almost always the result of human activity and typically results from dumping, filling, or piling up of material. Under natural conditions, a material's load is carried by its grain-to-grain contacts, with the friction between the grains maintaining a slope. The additional weight created by overloading increases the water pressure within the material, which in turn decreases its shear strength, thereby weakening the slope material. If enough material is added, the slope will eventually fail, sometimes with tragic consequences.

Geology and Slope Stability

The relationship between topography and the geology of an area is important in determining slope stability (Figure 14.7). If the rocks underlying a slope dip in the

Figure 14.6
A California Highway Patrol officer stands on top of a 2-m-high wall of mud that rolled over a patrol car near the Golden State Freeway on October 23, 1987. Flooding and mudslides also trapped other vehicles and closed the freeway.

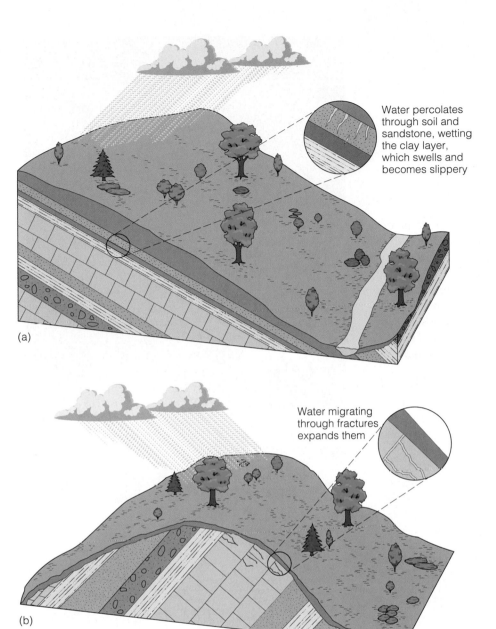

Water percolates through soil and sandstone, wetting the clay layer, which swells and becomes slippery

(a)

Water migrating through fractures expands them

(b)

Figure 14.7
(a) Rocks dipping in the same direction as a hill's slope are particularly susceptible to mass wasting. Undercutting of the base of the slope by a stream removes support and steepens the slope at the base. Water percolating through the soil and into the underlying rock increases its weight and, if clay layers are present, wets the clay, making them slippery. (b) Fractures dipping in the same direction as a slope are enlarged by chemical weathering, which can remove enough material to cause mass wasting.

same direction as the slope, mass wasting is more likely to occur than if the rocks are horizontal or dip in the opposite direction. When the rocks dip in the same direction as the slope, water can percolate along the various bedding planes and decrease the cohesiveness and friction between adjacent rock layers (Figure 14.7a). This is particularly true when clay layers are present, because clay becomes slippery when wet.

Even if the rocks are horizontal or dip in a direction opposite to that of the slope, joints may dip in the same direction as the slope. Water migrating through them weathers the rock and expands these openings until the weight of the overlying rock causes it to fall (Figure 14.7b).

Triggering Mechanisms

The factors discussed thus far all contribute to slope instability, most—though not all—rapid mass movements are triggered by a force that temporarily disturbs slope equilibrium. The most common triggering mechanisms are strong vibrations from earthquakes and excessive amounts of water from a winter snow melt or a heavy rainstorm (Figure 14.8).

Volcanic eruptions, explosions, and even loud claps of thunder may also be enough to trigger a landslide if the slope is sufficiently unstable. Many *avalanches*, which are rapid movements of snow and ice down steep mountain slopes, are triggered by the sound of a loud gunshot or, in rare cases, even a person's shout.

Figure 14.8
Heavy winter rains caused this 200,000-yd³ landslide in March 1995 at La Conchita, California, 75 miles northeast of Los Angeles. Although no casualties occurred, nine homes were destroyed or badly damaged.

WHAT ARE THE DIFFERENT TYPES OF MASS WASTING?

Geologists recognize a variety of mass movements (Table 14.2). Some are of one distinct type, while others are a combination of different types. It is not uncommon for one type of mass movement to change into another along its course. A landslide, for example, may start out as a slump at its head and, with the addition of water, become an earthflow at its base. Even though many slope failures are combinations of different materials and movements, classifying them according to their dominant behavior is still convenient.

Mass movements are generally classified on the basis of three major criteria (Table 14.2): (1) rate of movement (rapid or slow); (2) type of movement (primarily falling, sliding, or flowing); and (3) type of material involved (rock, soil, or debris).

Rapid mass movements involve a visible movement of material. Such movements usually occur quite suddenly, and the material moves very quickly downslope.

Table 14.2

Classification of Mass Movements and Their Characteristics			
TYPE OF MOVEMENT	SUBDIVISION	CHARACTERISTICS	RATE OF MOVEMENT
Falls	Rockfall	Rocks of any size fall through the air from steep cliffs, canyons, and road cuts	Extremely rapid
Slides	Slump	Movement occurs along a curved surface of rupture; most commonly involves unconsolidated or weakly consolidated material	Extremely slow to moderate
	Rock slide	Movement occurs along a generally planar surface	Rapid to very rapid
Flows	Mudflow	Consists of at least 50% silt- and clay-sized particles and up to 30% water	Very rapid
	Debris flow	Contains larger-sized particles and less water than mudflows	Rapid to very rapid
	Earthflow	Thick, viscous, tongue-shaped mass of wet regolith	Slow to moderate
	Quick clays	Composed of fine silt and clay particles saturated with water; when disturbed by a sudden shock, lose their cohesiveness and flow like a liquid	Rapid to very rapid
	Solifluction	Water-saturated surface sediment	Slow
	Creep	Downslope movement of soil and rock	Extremely slow
Complex movements		Combination of different movement types	Slow to extremely rapid

Perspective 14.1

The Tragedy at Aberfan, Wales

The debris brought out of underground coal mines in southern Wales typically consists of a wet mixture of various sedimentary rock fragments. This material is usually dumped along the nearest valley slope where it builds up into large waste piles called *tips*. A tip is fairly stable as long as the material composing it is relatively dry and its sides are not too steep.

Between 1918 and 1966, seven large tips composed of mine debris had been built at various elevations on the valley slopes above the small coal-mining village of Aberfan. Shortly after 9:00 A.M. on October 21, 1966, the 250-m-high, rain-soaked Tip No. 7 collapsed, and a black sludge flowed down the valley with the roar of a loud train (Figure 1). Before it came to a halt 800 m from its starting place, the flow had destroyed two farm cottages, crossed a canal, and buried Pantglas Junior School, suffocating virtually all the children of Aberfan. A total of 144 people died in the flow, among them 116 children who had gathered for morning assembly in the school.

After the disaster, everyone asked, "Why did this tragedy occur, and could it have been prevented?" The subsequent investigation revealed that no stability studies had ever been made on

the tips and that repeated warnings about potential failure of the tips, as well as previous slides, had all been ignored.

In 1939, 8 km to the south, a tip constructed under conditions almost identical to those of Tip No. 7 collapsed. Luckily, no one was injured, but unfortunately, the failure was soon forgotten and the Aberfan tips continued to grow. In 1944 Tip No. 4 failed, and again no one was injured.

In 1958 Tip No. 7 was sited solely on the basis of available space, with no regard to the area's geology. Although springs and seeps were known to be a potential source of instability, the tip was still located over a spring. Despite previous tip failures and warnings of slope failure by tip workers and others, mine debris was being piled onto Tip No. 7

until the day of the disaster.

What exactly caused Tip No. 7 and the others to fail? The official investigation revealed that the foundation of the tips had become saturated with water from the springs over which they were built. In the case of the collapsed tips, pore pressure from the water exceeded the friction between grains, and the entire mass liquefied like a "quicksand." Behaving as a liquid, the mass quickly moved downhill spreading out laterally. As it flowed, water escaped from the mass, and the sedimentary particles regained their cohesion.

Following the inquiry, it was recommended that a National Tip Safety Committee be established to assess the dangers of existing tips and advise on the construction of new tip sites.

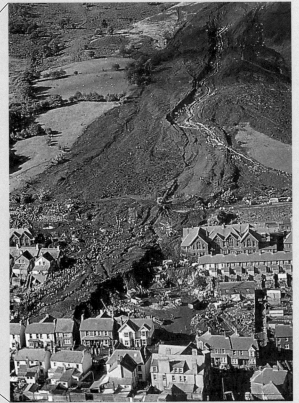

Figure 1
Location map and aerial view of the Aberfan tip disaster in which 144 people died.

Rapid mass movements are potentially dangerous and frequently result in loss of life and property damage. Most rapid mass movements occur on relatively steep slopes and can involve rock, soil, or debris.

Slow mass movements advance at an imperceptible rate and are usually only detectable by the effects of their movement, such as tilted trees and power poles or cracked foundations. Although rapid mass movements are more dramatic, slow mass movements are responsible for the downslope transport of a much greater volume of weathered material.

Falls

Rockfalls are a common type of extremely rapid mass movement in which rocks of any size fall through the air (Figure 14.9). Rockfalls occur along steep canyons, cliffs, and road cuts and build up accumulations of loose rocks and rock fragments, called *talus,* at their base (see Figure 5.6).

Rockfalls result from failure along joints or bedding planes in the bedrock and are commonly triggered by natural or human undercutting of slopes, or by earthquakes. Many rockfalls in cold climates are the result of frost wedging. Chemical weathering caused by water percolating through the fissures in carbonate rocks (limestone, dolostone, and marble) is also responsible for many rockfalls.

Rockfalls range in size from small rocks falling from a cliff to massive falls involving millions of cubic meters of debris that destroy buildings, bury towns, and block highways (Figure 14.10). Rockfalls are a particularly common hazard in mountainous areas where roads have been built by blasting and grading through steep hillsides of bedrock. Anyone who has ever driven through the Appalachians, the Rocky Mountains, or the Sierra Nevada is familiar with the "Watch for Falling Rocks" signs posted to warn drivers of the danger. Slopes particularly prone to rockfalls are sometimes covered with wire mesh in an effort to prevent dislodged rocks from falling to the road below (Figure 14.11). Another tactic is to put up wire mesh fences along the base of the slope to catch or slow down bouncing or rolling rocks.

Slides

A **slide** involves movement of material along one or more surfaces of failure. The type of material may be soil, rock, or a combination of the two, and it may break apart during movement or remain intact. A slide's rate of movement can vary from extremely slow to very rapid (Table 14.2).

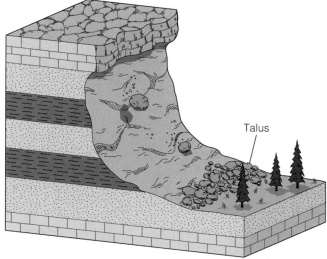

Figure 14.9
Rockfalls result from failure along cracks, fractures, or bedding planes in the bedrock and are common features in areas of steep cliffs.

Figure 14.10
Rockfall in Jefferson County, Colorado. All eastbound traffic and part of the westbound lane of Interstate 70 were blocked by the rockfall. Heavy rainfall and failure along joints and foliation planes in Precambrian gneiss caused this rockfall.

Figure 14.11
Wire mesh has been used to cover this steep slope in Hawaii. This is a common practice in mountainous areas to prevent rocks from falling on the road.

Two types of slides are generally recognized: (1) slumps or rotational slides, in which movement occurs along a curved surface, and (2) rock or block slides, which move along a more or less planar surface.

A **slump** involves the downward movement of material along a curved surface of rupture and is characterized by the backward rotation of the slump block (Figure 14.12). Slumps usually occur in unconsolidated or weakly consolidated material and range in size from small individual sets, such as occur along stream banks, to massive, multiple sets that affect large areas and cause considerable damage.

Slumps can be caused by a variety of factors, but the most common is erosion along the base of a slope, which removes support for the overlying material. This local steepening may be caused naturally by stream erosion along its banks (Figure 14.12) or by wave action at the base of a coastal cliff. Slope oversteepening can also be caused by human activity, such as the construction of highways and housing developments. Slumps are particularly prevalent along highway cuts and fills where they are generally the most frequent type of slope failure observed.

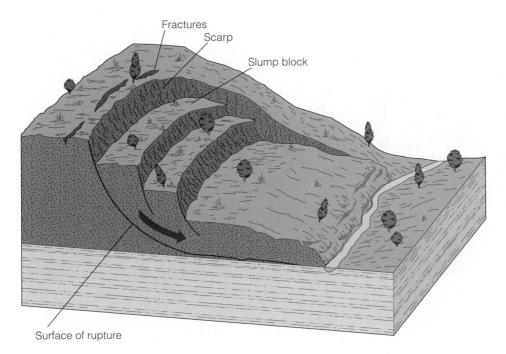

Figure 14.12
In a slump, material moves downward along the curved surface of a rupture, causing the slump block to rotate backward. Most slumps involve unconsolidated or weakly consolidated material and are typically caused by erosion along the slope's base.

Although many slumps are merely a nuisance, large-scale slumps involving populated areas and highways can cause extensive damage. Such is the case in coastal southern California where slumping and sliding have been a constant problem. Many areas along the coast are underlain by poorly to weakly consolidated silts, sands, and gravels interbedded with clay layers, some of which are weathered ashfalls. In addition, southern California is tectonically active so that many of these deposits are cut by faults and joints, which allow the infrequent rains to percolate downward rapidly, wetting and lubricating the clay layers.

Southern California lies in a semiarid climate and is dry most of the year. When it does rain, typically between November and March, large amounts of rain can fall in a short time. Thus, the ground quickly becomes saturated, leading to landslides along steep canyon walls as well as along coastal cliffs (Fig-ure 14.13). Most of the slope failures along the southern California coast are the result of slumping. These slumps have destroyed many expensive homes and forced many roads to be closed and relocated.

A **rock** or *block* **slide** occurs when rocks move downslope along a more or less planar surface. Most rock slides take place because the local slopes and rock layers dip in the same direction (Figure 14.14), although they can also occur along fractures parallel to a slope. Rock slides are common occurrences along the southern California coast. At Point Fermin, seaward-dipping rocks with interbedded slippery clay layers are undercut by waves, causing numerous slides (Figure 14.15a).

Farther south, in the town of Laguna Beach, startled residents watched as a rock slide destroyed or damaged 50 homes on October 2, 1978 (Figure 14.15b). Just as at Point Fermin, the rocks at Laguna Beach dip about 25 degrees in the same direction as the slope of the

Figure 14.13
Undercutting of steep sea cliffs by wave action resulted in massive slumping in the Pacific Palisades area of southern California on March 31 and April 3, 1958. Highway 1 was completely blocked. Note the heavy earth-moving equipment for scale. *Source: Photo courtesy of John S. Shelton.*

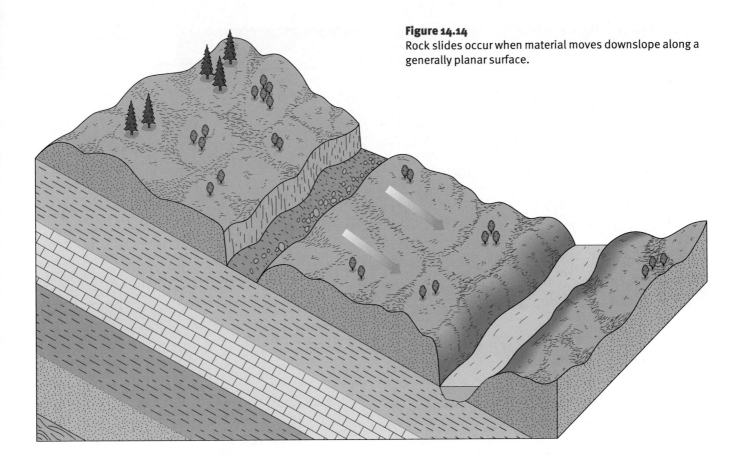

canyon walls and contain clay beds that "lubricate" the overlying rock layers, causing the rocks and the houses built on them to slide. In addition, percolating water from the previous winter's heavy rains wet a subsurface clayey siltstone, thus reducing its shear strength and helping activate the slide. Although the 1978 slide covered only about five acres, it was part of a larger ancient slide complex.

Not all rock slides are the result of rocks dipping in the same direction as a hill's slope. The rock slide at Frank, Alberta, Canada, on April 29, 1903, illustrates how nature and human activity can combine to create a situation with tragic results (Figure 14.16).

It would appear at first glance that the coal-mining town of Frank, lying at the base of Turtle Mountain, was in no danger from a landslide (Figure 14.16). After all, many of the rocks dipped away from the mining valley, unlike the situations at Point Fermin and Laguna Beach. The joints in the massive limestone composing Turtle Mountain, however, dip steeply toward the valley and are essentially parallel with the slope of the mountain itself. Furthermore, Turtle Mountain is supported by weak limestones, shales, and coal layers that underwent slow plastic deformation from the weight of the overlying massive limestone. Coal mining along the base of the valley also contributed to the stress on the rocks by removing some of the underlying support. All these factors, as well as frost action and chemical weath-

ering that widened the joints, finally resulted in a massive rock slide. Almost 40 million m³ of rock slid down Turtle Mountain along joint planes, killing 70 people and partially burying the town of Frank.

Flows

Mass movements in which material flows as a viscous fluid or displays plastic movement are termed *flows*. Their rate of movement ranges from extremely slow to extremely rapid (Table 14.2). In many cases, mass movements begin as falls, slumps, or slides and change into flows farther downslope.

Of the major mass movement types, **mudflows** are the most fluid and move most rapidly (at speeds of up to 80 km/hr). They consist of at least 50% silt- and clay-sized material combined with a significant amount of water (up to 30%). Mudflows are common in arid and semiarid environments where they are triggered by heavy rainstorms that quickly saturate the regolith, turning it into a raging flow of mud that engulfs everything in its path. Mudflows can also occur in mountain regions (Figure 14.17) and in areas covered by volcanic ash where they can be particularly destructive (see Chapter 4). Because mudflows are so fluid, they generally follow preexisting channels until the slope decreases or the channel widens, at which point they fan out.

(a)

Pacific Palisades

Santa
Monica

Los Angeles

Palos
Verdes
Peninsula

Long Beach

Pacific
Ocean

Point
Fermin

Laguna
Beach

Figure 14.15
(a) A combination of interbedded clay beds that become slippery when wet, rocks dipping in the same direction as the slope of the sea cliffs, and undercutting of the sea cliffs by wave action has caused numerous rock slides and slumps at Point Fermin, California.
(b) Farther south at Laguna Beach, the same combination of factors apparently activated a rock slide that destroyed numerous homes and cars on October 2, 1978.

(b)

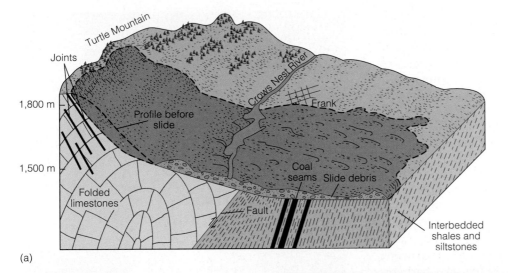

(a)

Figure 14.16
(a) The tragic Turtle Mountain rock slide that killed 70 people and partially buried the town of Frank, Alberta, Canada, on April 29, 1903, was caused by a combination of factors. These included joints that dipped in the same direction as the slope of Turtle Mountain, a fault partway down the mountain, weak shale and siltstone beds underlying the base of the mountain, and mined-out coal seams. (b) Results of the 1903 rock slide at Frank.

(b)

As urban areas in arid and semiarid climates continue to expand, mudflows and the damage they create are becoming problems. Mudflows are common, for example, in the steep hillsides around Los Angeles where they have damaged or destroyed many homes.

Debris flows are composed of larger-sized particles than those in mudflows and do not contain as much water. Consequently, they are usually more viscous than mudflows, typically do not move as rapidly, and rarely are confined to preexisting channels. Debris flows can be just as damaging, though, because they can transport large objects (Figure 14.18).

Earthflows move more slowly than either mudflows or debris flows. An earthflow slumps from the upper part of a hillside, leaving a scarp, and flows slowly downslope as a thick, viscous, tongue-shaped mass of wet regolith (Figure 14.19). Like mudflows and debris flows, earthflows can be of any size and are frequently destructive. They occur most commonly in humid climates on grassy soil-covered slopes, following heavy rains.

Some clays spontaneously liquefy and flow like water when they are disturbed. Such **quick clays** have caused serious damage and loss of lives in Sweden, Norway, eastern Canada (Figure 14.20), and Alaska (Table 14.1). Quick clays are composed of fine silt and clay particles made by the grinding action of glaciers. Geologists think these fine sediments were originally deposited in a marine environment where their pore space was filled with saltwater. The ions in saltwater helped establish strong bonds between the clay particles, thus stabilizing and strengthening the clay. When the clays were subsequently uplifted above sea level, the saltwater was flushed out by fresh groundwater, reducing the effectiveness of the ionic bonds between the clay particles and thereby reducing the overall strength and cohesiveness of the clay. Consequently, when the clay is disturbed by a sudden shock or shaking, it essentially turns to a liquid and flows.

An example of the damage that can be done by quick clays occurred in the Turnagain Heights area of Anchorage, Alaska, in 1964 (Figure 14.21). Underlying most of the Anchorage area is the Bootlegger Cove Clay, a massive clay unit of poor permeability. Because the Bootlegger Cove Clay forms a barrier preventing groundwater from flowing through the adjacent glacial deposits to the sea, considerable hydraulic pressure builds up behind the clay. Some of this water has flushed out the salt-water in the clay and has saturated the lenses of sand and silt associated with the clay beds. When the magnitude 8.6 Good Friday earthquake struck on March 27, 1964, the shaking turned parts of the Bootlegger Cove Clay into a quick clay and precipitated a series of massive slides in the coastal bluffs that destroyed most of the homes in the Turnagain Heights subdivision (Figure 14.21).

Solifluction is the slow downslope movement of water-saturated surface sediment. Solifluction can occur in any climate where the ground becomes saturated with water, but is most common in areas of permafrost.

Figure 14.17
A mudflow near Estes Park, Colorado.

Figure 14.18
A debris flow and damaged house in lower Ophir Creek, western Nevada. Note the many large boulders that are part of the debris flow.

Figure 14.19
(a) Earthflows form tongue-shaped masses of wet regolith that move slowly downslope. They occur most commonly in humid climates on grassy soil-covered slopes. (b) An earthflow near Baraga, Michigan.

(a)

Scarp

(b)

Figure 14.20
Quick-clay slide at Nicolet, Quebec, Canada. The house on the slide (to the right of the bridge) traveled several hundred feet with relatively little damage.

Figure 14.21
(a) Groundshaking by the 1964 Alaska earthquake turned parts of the Bootlegger Cove Clay into a quick clay, causing numerous slides. (b) Low-altitude photograph of the Turnagain Heights subdivision of Anchorage shows some of the numerous landslide fissures that developed as well as the extensive damage to buildings in the area. The remains of the Four Seasons apartment building can be seen in the background.

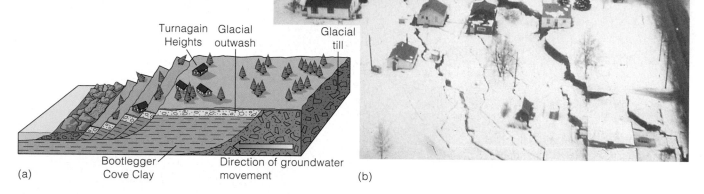

(a)

(b)

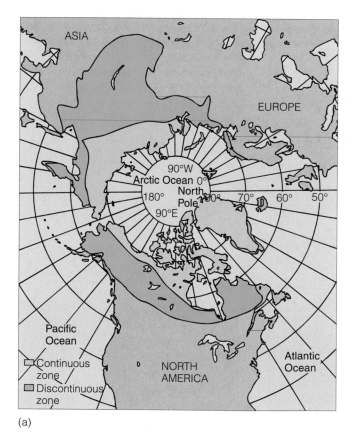

(a)

Figure 14.22
(a) Distribution of permafrost areas in the Northern Hemisphere. (b) Solifluction flows near Suslositna Creek, Alaska, show the typical lobate topography that is characteristic of solifluction conditions.

(b)

Permafrost, ground that remains permanently frozen, covers nearly 20% of the world's land surface (Figure 14.22a). During the warmer season when the upper portion of the permafrost thaws, water and surface sediment form a soggy mass that flows by solifluction and produces a characteristic lobate topography (Figure 14.22b).

As might be expected, many problems are associated with construction in a permafrost environment. A good example is what happens when an uninsulated building is constructed directly on permafrost. In this instance, heat escapes through the floor, thaws the ground below, and turns it into a soggy, unstable mush. Because the ground is no longer solid, the building settles unevenly into the ground, and numerous structural problems result (Figure 14.23).

Construction of the Alaska pipeline from the oil fields in Prudhoe Bay to the ice-free port of Valdez raised numerous concerns about the effect it might have on the permafrost and the potential for solifluction. Some thought that oil flowing through the pipeline would be warm enough to melt the permafrost, causing the pipeline to sink farther into the ground and possibly rupture. After numerous studies were conducted, scientists concluded that the pipeline, completed in 1977, could safely be buried for more than half of its 1280-km length; where melting of the permafrost might cause structural problems to the pipe, it was insulated and installed above ground.

Creep, the slowest type of flow, is the most widespread and significant mass-wasting process in terms of the total amount of material moved downslope and the monetary damage it does annually. Creep involves extremely slow downhill movement of soil or rock. Although it can occur anywhere and in any climate, it is most effective and significant as a geologic agent in humid regions. In fact, it is the most common form of mass wasting in the southeastern United States and the southern Appalachian Mountains.

Because the rate of movement is essentially imperceptible, we are frequently unaware of creep's existence until we notice its effects: tilted trees and power poles, broken streets and sidewalks, or cracked retaining walls or foundations (Figure 14.24). Creep usually involves the whole hillside and probably occurs, to some extent, on any weathered or soil-covered, sloping surface.

Creep is not only difficult to recognize but also to control. Although engineers can sometimes slow or stabilize creep, many times the only course of action is to simply avoid the area if at all possible or, if the zone of creep is relatively thin, design structures that can be anchored into the solid bedrock.

WHAT WOULD YOU DO?

You are a member of a planning board for your seaside community. A developer wants to rezone some coastal property to build twenty condominiums. This would be a boon to the local economy because it would provide jobs and increase the tax base. However, because the area is somewhat hilly and fronts the ocean, you are concerned about how safe the buildings would be in this area. What types of studies would need to be done before any rezoning could take place? Is it possible to build safe structures along a hilly coastline? What specifically would you ask the environmental consulting firm the planning board has hired to look for in terms of actual or potential geologic hazards if the condominiums are built?

Figure 14.24
(a) Some evidence of creep: (A) curved tree trunks;
(B) displaced monuments; (C) tilted power poles;
(D) displaced and tilted fences; (E) roadways moved
out of alignment; (F) hummocky surface. (b) Creep has
bent these sandstone and shale beds of the Haymond
Formation near Marathon, Texas. (c) Trees bent by creep,
Wyoming. (d) Stone wall tilted due to creep, Champion,
Michigan.

Complex Movements

Recall that many mass movements are combinations of
different movement types. When one type is dominant,
the movement can be classified as one of those described
thus far. If several types are more or less equally involved, it is called a **complex movement.**

The most common type of complex movement is
the slide-flow, in which there is sliding at the head and
then some type of flowage farther along its course. Most
slide-flow landslides involve well-defined slumping
at the head, followed by a debris flow or earth-flow
(Figure 14.25). Any combination of different mass-
movement types can, however, be classified as a complex movement.

A *debris avalanche* is a complex movement that
often occurs in very steep mountain ranges. Debris avalanches typically start out as rockfalls when large quan-

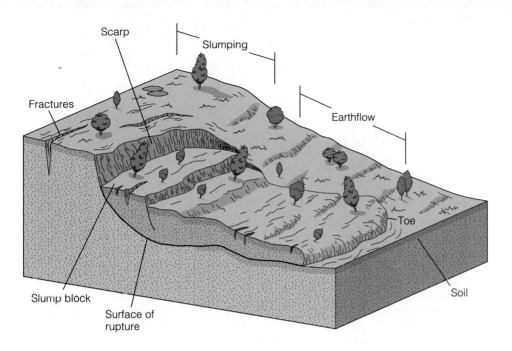

Figure 14.25
A complex movement in which slumping occurs at the head, followed by an earthflow.

tities of rock, ice, and snow are dislodged from a mountainside, frequently as a result of an earthquake. The material then slides or flows down the mountainside, picking up additional surface material and increasing in speed. The 1970 Peru earthquake set in motion the debris avalanche that destroyed the towns of Yungay and Ranrahirca, Peru, and killed more than 25,000 people (Figure 14.26).

HOW CAN WE RECOGNIZE AND MINIMIZE THE EFFECTS OF MASS MOVEMENTS?

The most important factor in eliminating or minimizing the damaging effects of mass wasting is a thorough geologic investigation of the region in question. In this way, former landslides and areas susceptible to mass movements can be identified and perhaps avoided (see Perspective 14.2). By assessing the risks of possible mass wasting before construction begins, steps can be taken to eliminate or minimize the effects of such events.

Identifying areas with a high potential for slope failure is important in any hazard-assessment study; these studies include identifying former landslides as well as sites of potential mass movement. Scarps, open fissures, displaced or tilted objects, a hummocky surface, and sudden changes in vegetation are some of the features indicating former landslides or an area susceptible to slope failure. The effects of weathering, erosion, and vegetation may, however, obscure the evidence of previous mass wasting.

Figure 14.26
An earthquake 65 km away triggered a landslide on Nevado Huascarán, Peru, that destroyed the towns of Yungay and Ranrahirca and killed more than 25,000 people.

Perspective 14.2

The Vaiont Dam Disaster

On October 9, 1963, more than 240 million m³ of rock and soil slid into the Vaiont Reservoir, in Italy, triggering a destructive flood that killed nearly 3000 people (Figure 1). To fully appreciate the immensity of this catastrophe, consider the following: Within a period of 15 to 30 seconds, the slide filled the reservoir with a mass of debris 2 km long and as high as 175 m above the reservoir level. The impact of the debris created a wave of water that overflowed the dam by 100 m and was still more than 70 m high 1.6 km downstream. The slide also set off a blast of wind that shook houses, broke windows, and even lifted the roof off one house in the town of Casso, which is 260 m above the reservoir on the opposite side of the valley; it also set off shock waves recorded by seismographs throughout Europe. Considering the forces generated by the slide, it is a tribute to the designer and construction engineer that the dam itself survived the disaster (Figure 2)!

The dam was built in a glacial valley underlain by thick layers of folded and faulted limestones and clay layers that were further weakened by jointing (Figure 3). Signs of previous slides in the area were obvious, and the few boreholes in the valley slopes revealed clay layers and small-scale slide planes. Despite the geologic evidence of previous mass wasting and objections to the site by some of the early investigators, construction of the 265-m-high Vaiont Dam began.

A combination of both adverse geologic features and conditions resulting from the dam's construction contributed to the massive landslide. Among the geologic causes were the rocks themselves, which were weak to begin with and dipped in the same direction as the valley walls of the reservoir. Fractured limestones make up the bulk of the rocks and are interbedded with numerous clay beds that are particularly prone to slippage. Active solution of the limestones by slightly acid groundwater further weakened them by developing and expanding an extensive network of cracks, joints, fissures, and other openings.

During the two weeks before the slide occurred, heavy rains saturated the ground, adding extra weight and reducing the shear strength of the rocks. Besides water from the rains, water from the reservoir infiltrated the rocks of the lower valley walls, further reducing their strength.

Soon after the dam was completed, a relatively small slide of 1 million m³ of material occurred on the south side of the reservoir. Following this slide, it was decided to limit the amount of water in the reservoir and to install monitoring devices throughout the potential slide area. Between 1960 and 1963, the eventual slide area moved an average of about 1 cm per week. On September 18, 1963, numerous monitoring stations reported movement had increased to about 1 cm per day. It was assumed that these were individual blocks moving, but it was actually the entire slide area!

Heavy rains fell between September 28 and October 9, increasing the amount of subsurface water. By October 8, the creep rate had increased to

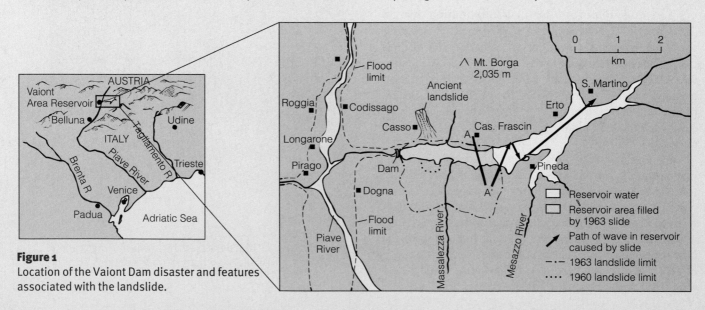

Figure 1
Location of the Vaiont Dam disaster and features associated with the landslide.

almost 39 cm per day. Engineers finally realized that the entire slide area was moving, and quickly began lowering the reservoir level. On October 9, the rate of movement in the slide area had increased still further, in some locations up to 80 cm per day, and there were reports that the reservoir level was actually rising. This was to be expected if the south bank was moving into the reservoir and displacing water. Finally, at 10:41 P.M. that night, during yet another rainstorm, the south bank of the Vaiont valley slid into the reservoir.

The lesson to be learned from this disaster is that before construction on any dam begins, a complete and systematic appraisal of the area must be conducted. Such a study should examine the geology of the area, identify past mass movements, assess their potential for recurrence, and evaluate the effects that the project will have on the rocks, including how it will alter their shear strength over time. Without these precautions, similar disasters will occur and lives will needlessly be lost.

Figure 2
Aerial view of the Vaiont Dam.

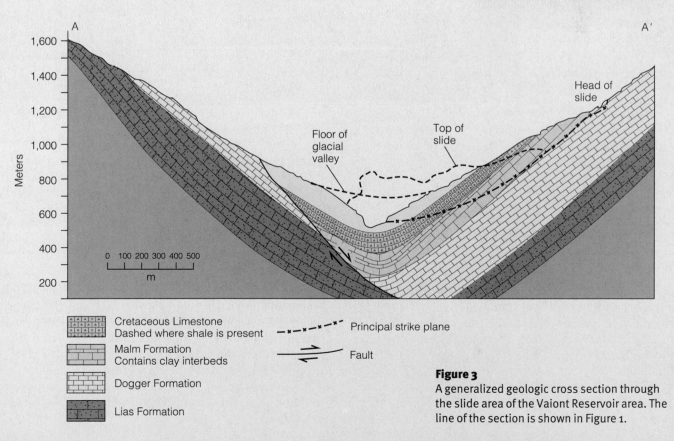

Cretaceous Limestone
Dashed where shale is present

Malm Formation
Contains clay interbeds

Dogger Formation

Lias Formation

Principal strike plane

Fault

Figure 3
A generalized geologic cross section through the slide area of the Vaiont Reservoir area. The line of the section is shown in Figure 1.

William Goodman

Engineering Geology: Using Geologic Principles for Practical Solutions

William Goodman is the principal geologist of a geotechnical consulting firm working primarily in California. He graduated with a B.S. degree in geology from California State University–Fullerton in 1983. He is currently a Registered Geologist in California and Arizona and a Certified Engineering Geologist.

C urrently, I am a consulting engineering geologist, applying basic geologic principles to solve problems related to construction. The varied nature of the projects I work on is very stimulating, but the time frame in which private work is completed is more accelerated than academic studies would allow. High-quality technical work is a top priority and is integrated with good consulting service and sound business practice.

A residential project site that was underlain by layered sedimentary rock provides a good example of how basic geologic principles are used to solve a practical problem. During the preliminary investigation, a typical strati-graphic section was determined based on data collected during background research, aerial photograph review, field mapping, and subsurface exploration (drilling and trenching). During analysis, the stratigraphic section was utilized to predict where certain "problem" layers would occur, and remedial recommendations were presented prior to mass grading. During grading, the field geologist confirmed the preliminary interpretations and refined them where changes occurred. On this particular project, a potentially active fault transected the site. At one location, this fault intersected a landslide. Exposures during grading revealed that the landslide actually cut off the fault, indicating the fault was older than the landslide. Within the landslide graben, an old organic soil horizon (paleosol) was carbon dated to be 12,500 years old. Since this soil layer was buried in the graben of the landslide after the failure occurred, the fault must be older than 12,500 years and not active according to the definition of a Recent fault.

On one large planned community project, I encountered a variety of geologic problems that affected different aspects of the development. The problems included landslides, deeply incised alluvial-filled channels, sandstone layers, expansive claystones, and diatomaceous siltstones. The topographic relief at the site was over 400 feet. Storm-drain-outlet structures were built that had to function beneath 20-year flood levels. They have been tested numerous times by nature since completion and performed as designed. Residential structures were built over compacted fills as thick as 200 vertical feet. A 30-inch-diameter major water trunk line was relocated around the development. The electrical transmission lines crossing the site were removed from the towers and put in an underground vault. The cemented rock encountered was disposed of by burying it in deep fills or crushing it to gravel size and using it to bridge removal bottoms in saturated alluvium. Sand, the other by-product of crushing, was used as backfill for utility trenches. Hard rock in cut areas was undercut and replaced with compacted fill to mitigate problems during foundation, utility, and pool excavation. Select grading techniques were utilized to minimize the waiting period for settlement of deep fills prior to building and exclusion of highly expansive clays at finish grade in fill areas.

Other short-term projects on which I've worked provide examples of the types of challenges engineering geologists encounter: investigation of earthquake-damaged homes in the Northridge area (seismic shaking, liquefaction); investigation of a landslide in San Clemente that destroyed five homes and closed the Atchison, Topeka, and Santa Fe Railroad for two weeks and the Pacific Coast Highway for more than one year; investigation of a distressed property in the Newport Harbor area where a seawall was in jeopardy of failure; investigation for a major transportation corridor planned across hillside terrain in southern Orange County; environmental impact study and report for proposed improvements for a 23-mile segment of a major highway in Orange County; numerous groundwater seepage investigations in occupied residential areas; and numerous forensic geotechnical investigations involving distressed residential structures. ■

Soil and bedrock samples are also studied, both in the field and laboratory, to assess such characteristics as composition, susceptibility to weathering, cohesiveness, and ability to transmit fluids. These studies help geologists and engineers predict slope stability under a variety of conditions.

The information derived from a hazard-assessment study can be used to produce *slope-stability maps* of the area (Figure 14.27). These maps allow planners and developers to make decisions about where to site roads, utility lines, and housing or industrial developments based on the relative stability or instability of a particular location. In addition, the maps indicate the extent of an area's landslide problem and the type of mass movement that may occur. This information is important for grading slopes or building structures, to prevent or minimize slope-failure damage.

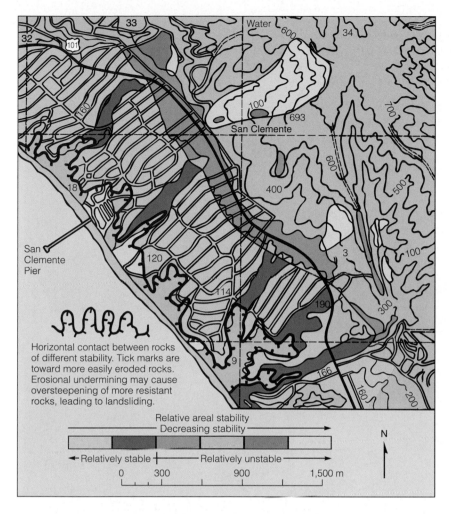

Horizontal contact between rocks of different stability. Tick marks are toward more easily eroded rocks. Erosional undermining may cause oversteepening of more resistant rocks, leading to landsliding.

Relative areal stability
Decreasing stability

← Relatively stable ┼ Relatively unstable →

0   300   900   1,500 m

N

Figure 14.27
Relative slope stability map of part of San Clemente, California, showing areas delineated according to relative stability.

WHAT WOULD YOU DO?

You've just found your dream parcel of land in the hills of northern Baja, California, where you plan to retire someday. Because you want to make sure the area is safe to build a house, you decide to do your own geologic investigation of the area to make sure there aren't any obvious geologic hazards. What specific things would you look for that might indicate mass wasting may have taken place in the past? Even if there is no obvious evidence of rapid mass wasting, what features would you look for that might indicate a problem with slow types of mass wasting such as creep?

Although most large mass movements usually cannot be prevented, geologists and engineers can employ various methods to minimize the danger and damage resulting from them. Because water plays such an important role in many landslides, one of the most effective and inexpensive ways to reduce the potential for slope failure or to increase existing slope stability is through surface and subsurface drainage of a hillside. Drainage serves two purposes. It reduces the weight of the material likely to slide and increases the shear strength of the slope material by lowering pore pressure.

Surface waters can be drained and diverted by ditches, gutters, or culverts designed to direct water away from slopes. Drainpipes perforated along one surface and driven into a hillside can help remove subsurface water (Figure 14.28). Finally, planting vegetation on hillsides helps stabilize slopes by holding the soil together and reducing the amount of water in the soil.

Another way to help stabilize a hillside is to reduce its slope. Recall that overloading or oversteepening by grading are common causes of slope failure. By reducing the angle of a hillside, the potential for slope failure is decreased. Two methods are usually employed to reduce a slope's angle. In the *cut-and-fill* method, material is removed from the upper part of the slope and used as fill at the base, thus providing a flat surface for construction and reducing the slope (Figure 14.29). The second method, which is called *benching*, involves cutting a

Figure 14.28
(a) Driving drainpipes that are perforated on one side into a hillside, with the perforated side up, can remove some subsurface water and help stabilize the hillside. (b) A drainpipe driven into the hillside at Point Fermin, California, helps remove subsurface water and stabilize the slope.

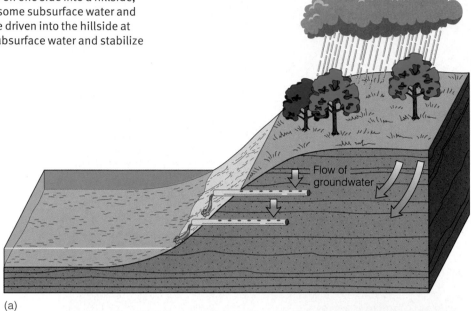

(a)

(b)

series of benches or steps into a hillside (Figure 14.30). This process reduces the overall average slope, and the benches serve as collecting sites for small landslides or rockfalls that might occur. Benching is most commonly used on steep hillsides in conjunction with a system of surface drains to divert runoff.

In some situations, retaining walls can be constructed to provide support for the base of the slope (Figure 14.31). These are usually anchored well into bedrock, backfilled with crushed rock, and provided with drain holes to prevent the buildup of water pressure in the hillside.

Rock bolts, similar to those employed in tunneling and mining, can sometimes be used to fasten potentially unstable rock masses into the underlying stable bedrock (Figure 14.32). This technique has been used

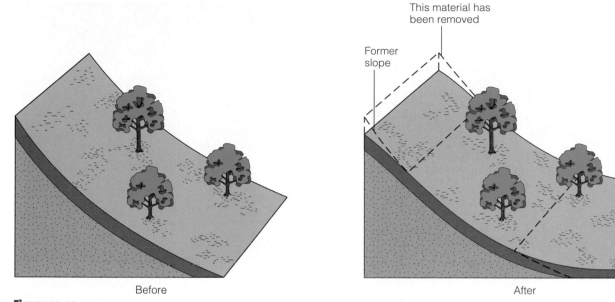

Figure 14.29
One common method used to help stabilize a hillside and reduce its slope is the cut-and-fill method. Here, material from the steeper upper part of the hillside is removed, thereby reducing the slope angle, and is used to fill in the base. This provides some additional support at the base of the slope.

Figure 14.30
(a) Another common method used in stabilizing a hillside and reducing its slope is benching. This process involves making several cuts along a hillside to reduce the overall slope. Furthermore, individual slope failures are now limited in size, and the material collects on the benches. (b) Benching is used in nearly all road cuts.

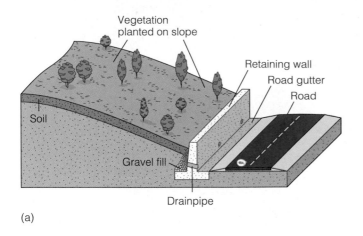

(a)

(b)

Figure 14.31
(a) Retaining walls anchored into bedrock, backfilled with gravel, and provided with drainpipes can support a slope's base and reduce landslides. (b) Steel retaining wall built to stabilize the slope and keep falling and sliding rocks off the highway.

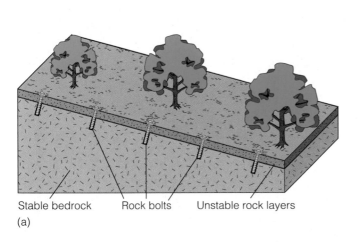

(a)

(b)

Figure 14.32
(a) Rock bolts secured in bedrock can help stabilize a slope and reduce landslides. (b) Rock bolts and wire mesh are used to help secure rock on a steep hillside in Brisbane, Australia.

successfully on the hillsides of Rio de Janeiro, Brazil, and to help secure the slopes at the Glen Canyon Dam on the Colorado River.

Recognition, prevention, and control of landslide-prone areas is expensive, but not nearly as expensive as the damage can be when such warning signs are ignored or not recognized. The collapse of Tip No. 7 at Aberfan, Wales (see Perspective 14.1), and the Vaiont Dam disaster (see Perspective 14.2) are two tragic examples in which the warning signs of impending disaster were ignored.

Chapter Summary

1. Mass wasting is the downslope movement of material under the influence of gravity. It occurs when the gravitational force acting parallel to a slope exceeds the slope's strength.

2. Mass wasting frequently results in the loss of life, as well as causing millions of dollars in damage annually.

3. Mass wasting can be caused by many factors including slope angle, weathering of slope material, water content, overloading, and removal of vegetation. Usually several of these factors in combination contribute to slope failure.

4. Mass movements are generally classified on the basis of their rate of movement (rapid versus slow), type of movement (falling, sliding, or flowing), and type of material (rock, soil, or debris).

5. Rockfalls are a common mass movement in which rocks free-fall.

6. Two types of slides are recognized. Slumps are rotational slides involving movement along a curved surface; they are most common in poorly consolidated or unconsolidated material. Rock slides occur when movement takes place along a more or less planar surface; they usually involve solid pieces of rock.

7. Several types of flows are recognized on the basis of their rate of movement (rapid or slow), type of material (rock, sediment, or soil), and amount of water.

8. Mudflows consist of mostly clay- and silt-sized particles and contain more than 30% water. They are most common in semiarid and arid environments and generally follow preexisting channels.

9. Debris flows are composed of larger particles and contain less water than mudflows. They are more viscous and do not flow as rapidly as mudflows.

10. Earthflows move more slowly than either debris flows or mudflows; they move downslope as thick, viscous, tongue-shaped masses of wet regolith.

11. Quick clays are clays that spontaneously liquefy and flow like water when they are disturbed.

12. Solifluction is the slow downslope movement of water-saturated surface material and is most common in areas of permafrost.

13. Creep, the slowest type of flow, is the imperceptible downslope movement of soil or rock. Creep is the most widespread of all types of mass wasting.

14. Complex movements are combinations of different types of mass movements in which one type is not dominant. Most complex movements involve sliding and flowing.

15. The most important factor in reducing or eliminating the damaging effects of mass wasting is a thorough geologic investigation to outline areas susceptible to mass movements.

16. Slopes can be stabilized by retaining walls, draining excess water, regrading slopes, and planting vegetation.

Important Terms

complex movement
creep
debris flow
earthflow
mass wasting
mudflow

permafrost
quick clay
rapid mass movement
rockfall
rock slide
shear strength

slide
slow mass movement
slump
solifluction

1. Which of the following is a factor influencing mass wasting?

   a. _____ weathering and climate;
   b. _____ water content;
   c. _____ slope angle;
   d. _____ overloading;
   e. _____ all of these.

2. The resisting forces (i.e., shear strength) helping to maintain slope stability include:

   a. _____ material strength;
   b. _____ material cohesion;
   c. _____ amount of internal friction between grains;
   d. _____ external support;
   e. _____ all of these.

3. Mass wasting is less likely to occur on slopes that have the following geologic characteristics:

   a. _____ horizontal beds or layers;
   b. _____ beds or layers that dip in the opposite direction to slope;
   c. _____ beds or layers that dip in the same direction as slope;
   d. _____ answers (a) and (b);
   e. _____ none of these.

4. The greatest amount of property damage is probably caused by which type of mass wasting?

   a. _____ slow, imperceptible types;
   b. _____ extremely fast types;
   c. _____ types that proceed at intermediate speeds;
   d. _____ answers (b) and (c);
   e. _____ none of these.

5. An example of a type of mass wasting that involves one or more surfaces of failure is:

   a. _____ rockfall;
   b. _____ slump;
   c. _____ mudflow;
   d. _____ creep;
   e. _____ none of these.

6. Which types of mass wasting produce a deposit called *talus?*

   a. _____ slumps;
   b. _____ rock slides;
   c. _____ rockfalls;
   d. _____ answers (a) and (b);
   e. _____ none of these.

7. Another geological term for slump is:

   a. _____ rock slide;
   b. _____ rotational slide;
   c. _____ rockfall;
   d. _____ creep;
   e. _____ none of these.

8. Which is the most rapid mass-wasting flow?

   a. _____ mudflow;
   b. _____ earthflow;
   c. _____ solifluction;
   d. _____ slump;
   e. _____ none of these.

9. Which of the following can help minimize the occurrence of mass wasting?

   a. _____ rock bolting;
   b. _____ benching;
   c. _____ cut and fill;
   d. _____ answers (b) and (c);
   e. _____ all of these.

10. The slow, downslope movement of water-saturated surface sediment is a(n):

    a. _____ slump;
    b. _____ earthflow;
    c. _____ debris flow;
    d. _____ rock slide;
    e. _____ none of these.

11. An area that was the site of mass wasting in the past may be identified by:

    a. _____ ground with a hummocky surface;
    b. _____ scarps and fissures;
    c. _____ abrupt changes in type of vegetation;
    d. _____ none of these;
    e. _____ all of these.

12. The most common type of complex movement in mass wasting starts with _____ at the head.

    a. _____ debris avalanche;
    b. _____ earthflow;
    c. _____ debris flow;
    d. _____ sliding;
    e. _____ none of these.

13. How does vegetation affect slope stability?

14. What are the various triggering mechanisms for mass wasting?

15. Why is benching used to help minimize mass wasting and its effects on road cuts? Why are some road cuts particularly susceptible to mass wasting?

16. What is the relationship between overloading, water pressure, and shear strength in mass wasting?

17. Briefly distinguish rock fall, slide, and flow as types of mass wasting. What are their respective characteristics and how are they different from one another?

18. How are mudflows and debris flows different with respect to particle size, water content, and speed of movement? Which is more fluid (i.e., least viscous), and why?

19. What is the relationship between rockfalls and talus? Describe how and where talus forms.

20. How does the introduction of fresh water and subsequent violent shaking act to initiate quick clays?

21. What is the relationship between solifluction and permafrost?

22. Describe an example that illustrates complex slide-flow movement.

23. List at least seven distinctly different types of evidence for creep. Select one and describe in detail how this type of evidence is formed or developed.

24. What is(are) the specific effect(s) of inserting perforated drainpipes into a hillside as a means of minimizing the effect of mass wasting? What specific factors influencing mass wasting are mitigated as a result of this procedure?

25. How can information on the geology of an area be used to help make predictions about its long-term slope stability? Cite one or two examples.

Points to Ponder

1. What potential value would a slope stability map be to a person seeking to purchase new home property? Using the slope stability map illustrated in Figure 14.27, locate the line on this map that shows a horizontal contact between rocks of different stability. What is the potential for mass wasting along this line and why?

2. What is the relationship between geological planes of weakness (e.g., bedding planes) and water in causing mass wasting? Using Figure 14.14 as an example, explain how the geological planes of weakness in this slope plus water from rainfall influenced development of the depicted slide.

3. Vegetation affects slope stability in various ways. How would removing preexisting vegetation tend to affect most slopes in humid regions with respect to mass wasting?

4. If an area has a documented history of mass wasting that has endangered or taken human life, how should people and governments keep such events from happening again? Are most major large mass-wasting events preventable or predictable?

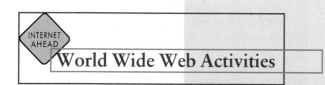

World Wide Web Activities

For these Web site addresses, along with current updates and exercises, log on to
http://www.brookscole.com/geo/

➤ U.S. GEOLOGICAL SURVEY EARTH SCIENCE IN THE PUBLIC SERVICE: GEOLOGIC HAZARDS

The home page of this site contains links to earthquakes, landslides, other geologic hazards, and geomagnetism.

1. Click on the *Landslides* site. At this site, click on *The National Landslide Information Center (NLIC)* site, which contains links to landslide information, landslide publications, and interesting landslide images and topics. Click on the *Landslide information* site. What is the latest landslide, and what type of damage did it do?

2. Click on the *Interesting Landslide Topics* site. Under the Landslide Information heading, click on the *What to Do and Look For* site. What features might be noticed prior to major landsliding?

3. Click on the *Activities of the NLIC* site. What does the NLIC do?

➤ JCP GEOLOGISTS, INC.

This site, maintained by a company that provides disclosure reports for the real estate profession in California, contains interesting and useful geologic information and links to other geologic sites. Click on the *Geologic Information* site. Under the Geologic Hazard Information heading, click on the *Landslides* site. Read the information presented. Can landslides be predicted? What are some of the ways landslides might be prevented? What are some of the ways that the buying and selling of property are affected by landslides?

Chapter 15

Laurel Falls in the Great Smokey Mountains National Park, Tennessee North Carolina.

OBJECTIVES

At the end of this chapter, you will have learned that

- Running water, one part of the hydrologic cycle, does considerable geologic work.
- Water is continuously cycled from the oceans to land and back to the oceans.
- Running water transports large quantities of sediment and deposits

sediment in or adjacent to braided and meandering streams.

- Alluvial fans (on land) and deltas (in a standing body of water) are deposited when a stream's capacity to transport sediment decreases.
- Flooding is a natural part of stream activity, and not all floods can be controlled.

- Streams continually adjust to changes.
- The concept of a graded stream is an ideal, and the graded condition of a stream is temporary.
- Most valleys form and change in response to stream activity coupled with other geologic processes such as mass wasting.

Running Water

Prologue

A t 4:07 P.M. on May 31, 1889, the residents of Johnstown, Pennsylvania, heard "a roar like thunder," and within 10 minutes a flood destroyed the town and left at least 2200 people dead. A huge wall of water, 18 m high in places, roared down the narrow valley at more than 60 km/hr, sweeping up debris, houses, and hundreds of people. A vivid account notes that "Thousands of people desperately tried to escape the wave. Those caught by the wave found themselves swept up in a torrent of oily, muddy water, surrounded by tons of grinding debris, which crushed some, provided rafts for others. Many became helplessly entangled in miles of barbed wire from the destroyed wire works."* Eighty people who survived the initial flood by floating on debris died when the debris caught fire after it lodged against a bridge.

In the aftermath of the flood, an estimated 2200 people had died, one in three bodies recovered was never identified, and many of the flood victims were never found. Ninety-nine of Johnstown's families had perished in the flood, and 98 children were left without parents. Survivors salvaged what they could from the wreckage and made temporary shelters from whatever was available. The town, of course, was in complete ruins and would not fully recover for five years (Figure 15.1). Within days of the flood, assistance began pouring in from concerned citizens elsewhere, volunteers, and the Red Cross.

*National Park Service—U.S. Department of Interior, Johnstown Information Service Online.

(a)

(b)

Figure 15.1
Johnstown, Pennsylvania, before (a) and after (b) the May 31, 1889, flood.

Several factors account for the destruction and tragic loss of life in this flood. The town was established in 1784 and began to prosper during the 1830s and 1850s, and by the 1880s about 30,000 people were living in Johnstown and nearby communities. Unfortunately, Johnstown was built on a floodplain at the junction of the Little Conemaugh River and Stony Creek, and as the town expanded, the rivers were constricted, causing increased problems with flooding. Another factor contributing to the disastrous flood was the 20 to 25 cm of rain that fell in 24 hours. The storm began in Kansas and Nebraska and moved east, causing havoc in its path. In fact, the U.S. Signal Corps issued a storm warning on May 29 for the Mid-Atlantic States.

The large amount of rainfall alone was serious enough, but the most important factor in Johnstown's destruction was the failure of the South Fork Dam about 22 km upstream on the Little Conemaugh River. Measuring about 22 m high and 270 m long, it was not particularly large as dams go, but it impounded a 4.8-km-long lake more than 130 m higher than Johnstown. The lake was built during the 1840s, but purchased and repaired by the South Fork Fishing and Hunting Club. Unfortunately, it was poorly maintained thereafter, and its poor condition contributed to its failure when it was filled to overflowing by excess rain.

The final ingredient in this catastrophe is the fact that a warning of the dam's imminent collapse was ignored by residents of one nearby community, apparently because rumors of its collapse had circulated for years. And the warnings never reached Johnstown because the telegraph lines were down. Although no one doubts that the poor condition of the South Fork Dam was at least partly responsible for the disaster, none of the survivors successfully sued the South Fork Hunting and Fishing Club.

As tragic as the Johnstown flood was, it was certainly not the last time a dam failed resulting in fatalities. Just one week after it was filled in March 1928, the St. Francis Dam in California collapsed (Figure 15.2). A 55-m-high wall of water rushed forth carrying pieces of the dam weighing more than 9000 metric tons nearly a kilometer. At least 450 people died because the dam had been built on rock incapable of supporting such a large structure. The death toll was actually much higher, because the flooded canyon was occupied by hundreds of illegal immigrants, many of whom could not be accounted for after the flood.

On February 26, 1972, at least 125 people perished and 4000 were left homeless at Buffalo Creek, West Virginia, when a dam constructed of waste from coal mining collapsed. During particularly heavy rains the dam failed, releasing huge volumes of water. More than 600 survivors sued the company responsible for the dam and won a settlement of $13.5 million.

The Teton Dam in Fremont County, Idaho, failed shortly after it was completed in 1976, resulting in several deaths and considerable property damage. Leakage was noticed after the dam's completion but was not considered serious because dams of this type (earth-fill dams) usually leak. As the leak became larger, though, the alarm was raised and most people were evacuated from the path of the subsequent flood. The magnitude of this flood is illustrated by the fact that a tractor weighing more than 16 metric tons was carried 11 km downstream!

(a)

(b)

Figure 15.2
(a) The St. Francis Dam in southern California in March 1928. (b) On March 13, 1928, only the central part of the dam remained. The catastrophic flooding following the dam's collapse killed at least 450 people.

INTRODUCTION

W hen considering interactions between the solid Earth and the major systems acting on its surface, certainly the hydrosphere has an important impact, although the biosphere–solid Earth and atmosphere–solid Earth interactions are also significant (see Table 1.1). Of course, the hydrosphere consists of several elements such as water vapor in the atmosphere, groundwater (see Chapter 16), water frozen in glaciers (see Chapter 17), water in the oceans (see Chapters 11 and 19), and that small but important amount of water on the land surface. Here our principal interest is in water confined to channels on land, or simply running water.

The incredible power of running water is vividly illustrated by the examples in the Prologue. And although floods caused either by human carelessness or purely natural causes continue to be a threat, there are also many benefits from running water. Water in streams is one source of fresh (nonsaline) water for industry, agriculture, and domestic use, and about 8% of all electricity used in North America is generated by falling water at hydroelectric plants (Figure 15.3a) (see Perspective 15.1). Large waterways throughout the world continue to be major avenues of commerce (Figure 15.3b), and when Europeans first explored the interior of North America they did so by following large waterways such as the St. Lawrence, Mississippi, and Missouri Rivers. And even though rivers flood periodically, some floods are actually beneficial. For instance, the agricultural lands along the Nile River in Egypt depend on annual flood-derived deposits to maintain their fertility.

Among the terrestrial planets, Earth is unique in having abundant liquid water. Both Mercury and Earth's moon are too small to retain any water, and Venus, because of its high temperature, is too hot to have any surface water. At present, Mars has only some frozen water and trace amounts of water vapor in its atmosphere, but studies of *Mariner* and *Viking* spacecraft images reveal areas with winding valleys that apparently formed by running water during the planet's early history (Figure 15.4). In contrast, 71% of Earth's surface is covered by oceans and seas (see Figure 11.4), its atmosphere contains a small but important quantity of water vapor, and water is present in streams, lakes, swamps, glaciers, and beneath the land surface as groundwater.

(a)

(b)

Figure 15.3

(a) Falling water at Hoover Dam on the Colorado River in Nevada is used to generate electricity. At 221 m it was the highest dam in the United States when completed in 1936. It is now second highest. In addition to electrical power generation, the reservoir also functions in flood control, irrigation, and recreation. (b) Inland waterways are important avenues of commerce. These freight barges on the Tennessee River are traveling north into Kentucky Lake, where they will bypass Kentucky Dam via locks.

Figure 15.4
Mars has no liquid water at present, but some evidence indicates it did during its early history. This view of the Martian surface shows outflow channels extending from areas known as *chaotic terrain* (CT), consisting of what appears to be loosely piled rubble. Arrows show inferred directions of flow.

systems. Natural changes, too, affect stream dynamics. When more rain falls in a stream's drainage area due to a long-term climatic change, more water flows in the stream's channel, and greater energy is available for erosion and sediment transport. In short, streams adjust to any change occurring within their overall systems.

According to one estimate, Earth has 1.36 billion km³ of water, most of which (97.2%) is in the oceans. Some 2.15% is frozen in glaciers on land, especially in Antarctica and Greenland, with the remaining 0.65% constituting all water in streams, lakes, ponds, swamps, groundwater, and the atmosphere (Figure 15.5). Thus, only a tiny proportion of the total water on Earth is in streams, but running water is nevertheless the most important geologic process modifying Earth's land surface in most areas.

Despite the importance of running water in erosion, sediment transport, and deposition, its role is limited in some areas. Antarctica and Greenland, for instance, are nearly completely ice covered, so erosion by glaciers is currently important, whereas running water has only a minimal effect in some ice-free coastal regions. And parts of some deserts are little affected by running water. Sections of the Atacama Desert of Chile, for example, have had no rain during historic time, but even in most deserts the effects of running water are conspicuous, although channels are dry most of the time.

Much of the following discussion of running water is necessarily descriptive, but one should always be aware that streams are dynamic systems that must continually respond to change. For example, paving in urban areas increases surface runoff to streams, while other human actions such as building dams and impounding reservoirs also alter the dynamics of stream

THE HYDROLOGIC CYCLE

Although the quantity of water in streams is small at any one time, during the course of a year very large volumes of water move through stream channels. In fact, water is continually recycled from the oceans,

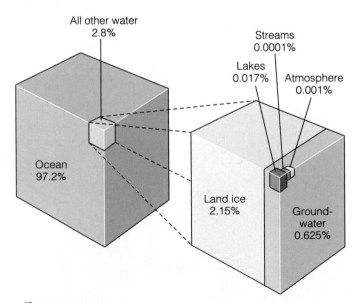

Figure 15.5
The relative amounts of water on Earth. Of the estimated 1.36 billion km³ of water, 97.2% is in the oceans; most of the rest, 2.15%, is in glaciers on land.

through the atmosphere, to the continents, and back to the oceans. This continuous recycling of water is the **hydrologic cycle** (Figure 15.6), which will also be relevant to our discussions of groundwater in Chapter 16 and glaciers in Chapter 17.

The hydrologic cycle, which is powered by solar radiation, is possible because water changes from liquid to gas (water vapor) easily under Earth surface conditions. Huge quantities of water evaporate from the oceans as the surface waters are heated by solar energy. The amount of ocean water evaporated yearly corresponds to a layer about 1 m thick from all the oceans. In fact, approximately 85% of all water entering the atmosphere is derived from the oceans; the remaining 15% comes from evaporation of water on land.

When water evaporates, the vapor rises into the atmosphere where the complex processes of condensation and cloud formation take place. About 80% of all precipitation falls directly into the oceans, in which case the hydrologic cycle is limited to a three-step process of evaporation, condensation, and precipitation (Figure 15.6).

About 20% of all precipitation falls on the continents as rain or snow. In this case the hydrologic cycle involves evaporation, condensation, movement of water vapor from the oceans to land, precipitation, and then surface runoff and infiltration into the groundwater reservoir. In short, this is a more complex sequence of events than for rain falling directly into the oceans. Some of the precipitation on land evaporates as it falls

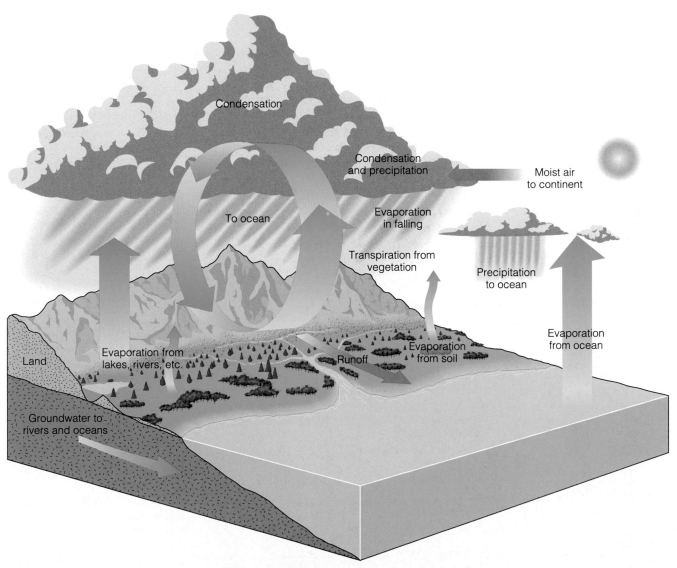

Figure 15.6
During the hydrologic cycle, water evaporates from the oceans and rises as water vapor to form clouds that release their precipitation over oceans or over land. Much of the precipitation falling on land returns to the oceans by surface runoff, thus completing the cycle.

Perspective 15.1

Dams, Reservoirs, and Hydroelectric Power

F lip a switch and our homes, offices, and factories are illuminated; turn a dial, and we might heat our homes or cook our food; and some of our public transportation relies on an unseen but important energy source—electricity. In fact, about 40% of the total energy in industrialized nations is converted to electricity. Most of it is generated at plants burning fossil fuels (oil, natural gas, and especially coal), but nuclear power plants are important in some countries. Geothermal energy (see Chapter 16), wind, and tidal power (see Chapter 19) account for only a small percentage of electricity.

Hydroelectric power, that is, electricity generated by moving water, accounts for only about 8% of all electricity produced in North America, but its importance varies considerably from area to area. For instance, less than 2% of the electricity generated in Ohio, Texas, and Florida comes from hydroelectric plants, whereas Washington, Oregon, and Idaho derive more than 80% of their electricity from this source.

A spinning turbine connected to a generator containing an electromagnet inside a coil of wire produces electricity at all power plants. However, the energy to turn the turbines differs. In fossil-fuel-burning plants and nuclear plants,

steam is the energy source, whereas moving water is used in hydroelectric plants. To provide the necessary water, a dam is built impounding a reservoir where the water is higher than the power-generating plant. Water moves through a large pipe called a *penstock* and encounters the blades of a turbine (Figure 1). In short, the potential energy of the water in the reservoir is converted to electrical energy at the power plant. Regardless of the energy source, once electricity is generated it is transmitted to areas of use by power lines.

Falling water was first used to generate electricity in 1882 at Appleton, Wisconsin, and since then hydroelectric plants have become common features on the world's waterways. Currently the largest one in terms of generating capacity is on the Parana River on the Paraguay–Brazil border, and an even larger one is under construction in the People's Republic of China. Several similar facilities are also present in North America. A large region in the Pacific Northwest depends on electricity generated at the Grand Coulee Dam in Washington, and hydroelectric

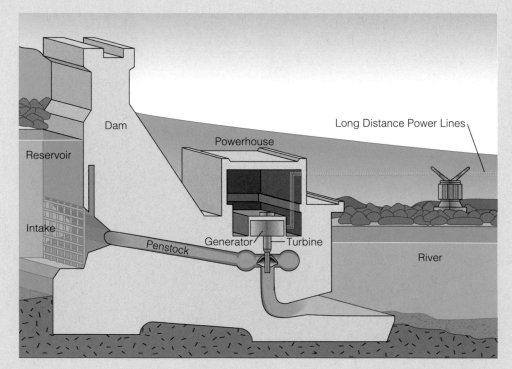

Figure 1
Electricity is generated at a hydroelectric dam when water rushes through the penstock, where it spins a turbine connected by a shaft to an electromagnet within a generator. The spinning electromagnet inside a coil of wire generates electricity.

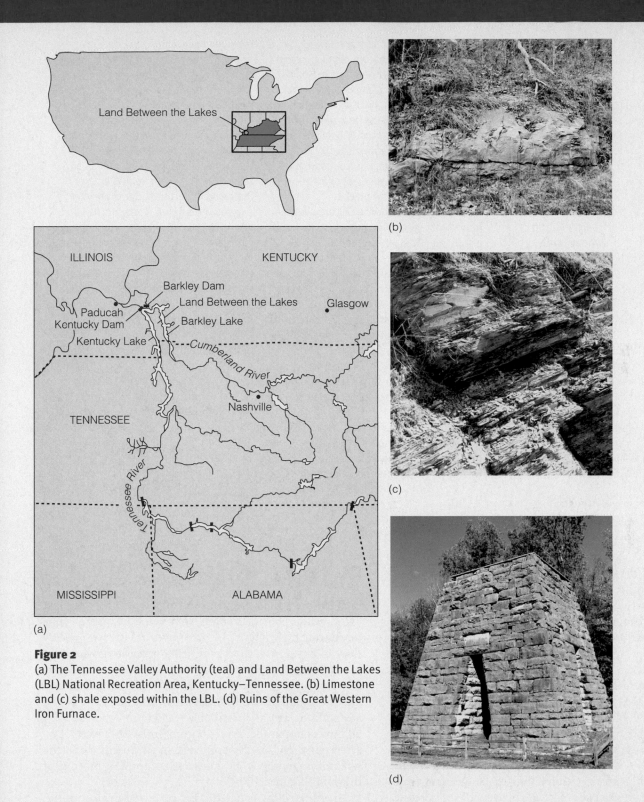

Figure 2
(a) The Tennessee Valley Authority (teal) and Land Between the Lakes (LBL) National Recreation Area, Kentucky–Tennessee. (b) Limestone and (c) shale exposed within the LBL. (d) Ruins of the Great Western Iron Furnace.

plants on the Montreal River in Canada supply power to a vast area. Here, however, we concentrate on the Tennessee Valley Authority, an agency responsible for supplying electricity to all of Tennessee and parts of six adjacent states.

Established in 1933, the Tennessee Valley Authority (TVA) was part of President Franklin D. Roosevelt's New Deal designed to help restore the economy during the Great Depression. He directed Congress to create "a corporation clothed with the power of government but possessed of the flexibility and initiative of a private enterprise."* Athough the TVA has a variety of responsibilities, such as flood control

*A Short History of the TVA, at http://www.tva.gov/heritage/hert_history.htm

and river navigation, generating electricity is one of its primary functions. Currently it produces enough energy for three cities the size of New York City at 11 coal-burning plants, 29 hydroelectric dams, and 4 nuclear power plants. Fossil-fuel-burning plants account for 68% of the total electricity generated, whereas 19% comes from hydroelectric dams.

The TVA also administers the nearly 700 km² Land Between the Lakes (LBL) National Recreation Area, an inland peninsula formed between Kentucky Lake and Lake Barkley where dams were constructed across the Tennessee and Cumberland Rivers (Figure 2). The two rivers, rarely more than about 10 km apart and only 1.6 km apart at one point, parallel one another for a considerable distance. Even though both rivers have dams across them, navigation is still possible because a canal connects the two lakes and shipping can bypass Kentucky Dam via locks and proceed down the Tennessee River to the Ohio River.

LBL is interesting for several reasons. Its scenery is certainly appealing, as are its opportunities for several outdoor activities such as boating, fishing, camping, and walking its nature trails. Soil and dense vegetation cover most of the rocks underlying the area, but exposures are present at several lakeshore localities and in some road excavations. Limestone and shale (Figure 2b and c) are the most common rocks, but a conspicuous conglomerate composed of chert pebbles is visible in a few places.

During the 1800s, the LBL area was an important center of iron production, having all the necessary elements such as iron ore, limestone, and timber for charcoal. Furthermore, the nearby Tennessee and Cumberland Rivers provided access to distant markets. At one time, 14 iron-producing furnaces were operating in the area (Figure 2d), but the Civil War interrupted production. And although iron production resumed after the war, by the 1880s the area's resources were largely depleted and the last furnace ceased operating in 1927.

One might be curious about why the TVA and other agencies and governments do not simply increase their hydroelectric output. After all, hydroelectric power generation has several appealing aspects, not the least of which is that it is a renewable resource.

Some countries, such as New Zealand, have enough hydroelectric generating capacity to meet all their needs, although they use geothermal energy too. However, not all areas have this potential; suitable sites for dams and reservoirs might not be present, for instance. In addition, dams are very expensive to build, reservoirs fill with sediment, and during droughts enough water might not be present to keep reservoirs sufficiently full. And, of course, people must be relocated from areas where reservoirs are impounded and from the discharge areas downstream from dams. Even though some floods still take place despite the presence of dams, many floods are indeed prevented, resulting in habitat degradation downstream. So although hydroelectric dams remain important, we cannot realistically look forward to very much additional use of this energy source.

and reenters the hydrologic cycle as vapor. In addition, water evaporated from lakes, ponds, swamps, and streams also reenters the cycle as vapor, as does moisture evaporated from plants by *transpiration* (Figure 15.6).

Each year about 36,000 km³ of the precipitation falling on land returns to the oceans by **runoff,** the surface flow of streams. Of course, the amount of precipitation and runoff varies widely across the continents; some areas are extremely arid, whereas others receive hundreds of centimeters of precipitation yearly. In any case, the water returning to the oceans by runoff enters Earth's ultimate reservoir where it begins the hydrologic cycle again.

Some of the precipitation falling on land is temporarily stored in lakes, snow fields, and glaciers or seeps below the surface where it is temporarily stored as groundwater. This water is effectively removed from the system for up to thousands of years, but eventually, glaciers melt, lakes feed streams, and groundwater flows into streams or directly into the oceans (Figure 15.6). Our concern here is with the comparatively small quantity returning to the oceans as runoff, for the energy of running water is responsible for a great many surface features.

RUNNING WATER

Unlike solids or even semisolids such as wet clay, water has no strength, so it will flow on any slope no matter how slight. In other words, it responds by flow to any stress, the stress being generated by a component of gravity operating parallel to a slope. The flow of water, or any other fluid, can be characterized as *laminar* or *turbulent*. In laminar flow, lines of flow called streamlines are all parallel with one another so that all flow is in parallel layers with no mixing between layers (Figure 15.7a). By contrast, in turbulent flow, the streamlines intertwine, causing a complex mixing of the fluid (Figure 15.7b).

Laminar flow is most easily observed in viscous fluids such as cold motor oil or syrup. One can also see laminar flow in parking lots where a thin film of water containing oil moves slowly over the surface. Turbulent flow, on the other hand, occurs in all streams. The primary control on the type of flow is velocity; roughness of the surface over which flow occurs also plays a role. Flow is laminar when water moves very slowly, as when groundwater moves through the tiny pores in sediments

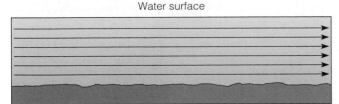

Water surface

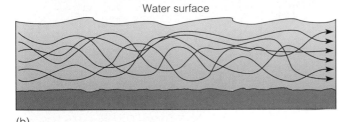

Water surface

Figure 15.7
(a) In laminar flow, streamlines are parallel to one another, and little or no mixing takes place between adjacent layers in the fluid. (b) In turbulent flow, streamlines are complexly intertwined, indicating mixing between layers. Most flow in streams is turbulent.

and soil. In streams, however, flow is usually fast enough and the channel walls and bed rough enough so that flow is fully turbulent. Laminar flow is so slow, and generally so shallow, that it causes little or no erosion. Turbulent flow is much more energetic and is capable of considerable erosion and sediment transport, and thus of much more interest in this chapter.

Sheet Flow Versus Channel Flow

The amount of runoff in any area during a rainstorm depends on **infiltration capacity,** the maximum rate at which soil or other surface materials can absorb water. Infiltration capacity depends on several factors, including the intensity and duration of rainfall. Loosely packed, dry soils absorb water faster than tightly packed, wet soils. Hard, dry surfaces, such as those that develop during droughts, also have low-infiltration capacities. Therefore, when they do receive rain, there is still considerable runoff. And of course, soil or sediment absorbs water faster than rock.

If rain is absorbed as fast as it falls, no surface runoff takes place. Should the infiltration capacity be exceeded or should surface materials become saturated, excess water collects on the surface and, if a slope exists, moves downhill. Even on steep slopes, flow is initially slow and hence causes little or no erosion. As the water moves downslope, though, it accelerates and may move by *sheet flow,* a more or less continuous film of water flowing over the surface. Sheet flow is not confined to depressions, and it accounts for *sheet erosion,* a particular problem on some agricultural lands (see Chapter 5).

In *channel flow,* surface runoff is confined to long, troughlike depressions. Channels vary from tiny rills containing a trickling stream of water to the Amazon River of South America, which is 6450 km long and at one place 2.4 km wide and 90 m deep. Channelized flow is described by various terms, including rill, brook, creek, stream, and river, most of which are distinguished by size and volume. The term **stream** carries no connotation of size and is used here to refer to all runoff confined to channels regardless of size.

Streams receive water from several sources, including sheet flow and rain falling directly into their channels. Far more important, though, is the water supplied by soil moisture and groundwater, both of which flow downslope and discharge into streams (Figure 15.6). In humid areas where groundwater is plentiful, streams may maintain a fairly stable flow year-round, even during dry seasons, because they are continually supplied by groundwater. In contrast, the amount of water in streams of arid and semiarid regions fluctuates much more widely because these streams depend more on infrequent rainstorms and surface runoff for their water supply.

Stream Gradient

Streams flow downhill from a source area to a lower elevation where most of them empty into another stream, a lake, or the sea. Exceptions are streams in arid regions that diminish in a downstream direction by evaporation and infiltration until they disappear. Streams in areas with numerous caverns may flow into a surface opening and disappear below ground, although they commonly reappear elsewhere (see Chapter 16). The slope that a stream flows over is its **gradient.** For example, if the source (headwaters) of a stream is 1000 m above sea level and the stream flows 500 km to the sea, it drops 1000 m vertically over a horizontal distance of 500 km (Figure 15.8). Its gradient is calculated by dividing the vertical drop by the horizontal distance; in this example, it is 1000 m/500 km = 2 m/km.

Gradients vary considerably, even along the course of a single stream (Figure 15.8). Generally, streams are steeper in their upper reaches where their gradients may be tens of meters per kilometer, but in their lower reaches gradients as little as a few centimeters per kilometer are common. Some streams in mountainous regions have particularly steep gradients of several hundred meters per kilometer.

Velocity and Discharge

Stream velocity and discharge are closely related aspects of streams. **Velocity** is simply a measure of the downstream distance water travels per unit of time. It is usually expressed in feet per second (ft/sec) or meters per

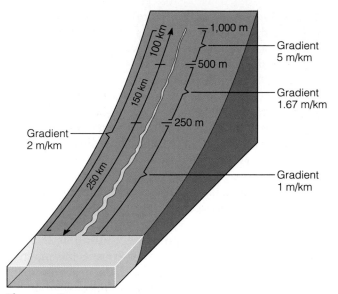

Figure 15.8
The average gradient of this stream is 2 m/km, but gradient can be calculated for any segment of a stream, as shown in this example. Notice that the gradient is steepest in the headwaters area and decreases in a downstream direction.

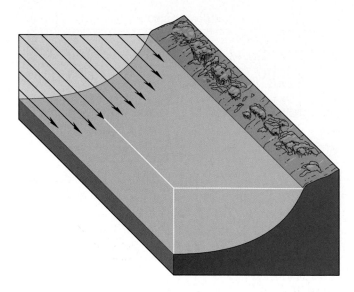

Figure 15.9
Flow velocity varies as a result of friction with a stream's banks and bed. The maximum flow velocity is near the center and top in a straight channel. The lengths of the arrows in this illustration are proportional to velocity.

second (m/sec) and varies considerably among streams and even within the same stream. Variations in flow velocity occur not only with distance along a stream channel but also across a channel's width. For example, a stream's bed and banks cause frictional resistance to flow, whereas the water some distance away is unaffected by friction and has a higher velocity (Figure 15.9). However, even though flow is slower near a stream's bed and banks, it is more turbulent.

Other controls on velocity include channel shape and roughness. Broad, shallow channels and narrow, deep channels have proportionally more water in contact with their perimeters than do channels with semicircular cross sections (Figure 15.10). Consequently, the

water in semicircular channels flows more rapidly because it encounters less frictional resistance. In many streams, the maximum flow velocity is near the surface at the center of the channel; it is slightly below the surface because of frictional resistance from the air above.

Channel roughness is a measure of the frictional resistance within channels. Frictional resistance to flow is greater in a channel containing large boulders than in those with banks and beds composed of sand or clay. In channels with abundant vegetation, flow is slower than in barren channels of comparable size.

The most obvious control on velocity is gradient, and one might think that the steeper the gradient, the

Figure 15.10
All of these channels have the same cross-sectional area, but each has a different shape. The semicircular channel has the least perimeter in contact with water and causes the least frictional resistance to flow. If other variables, such as channel roughness, are the same in all these channels, flow velocity will be greatest in the semicircular one.

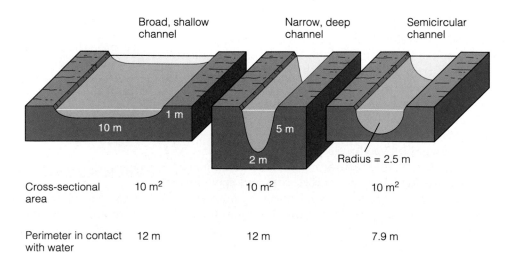

	Broad, shallow channel	Narrow, deep channel	Semicircular channel
Cross-sectional area	10 m²	10 m²	10 m²
Perimeter in contact with water	12 m	12 m	7.9 m

greater the flow velocity. In fact, the average velocity generally increases in a downstream direction, even though the gradient decreases in the same direction. Three factors contribute to this: First, velocity increases continuously, even as gradient decreases, in response to the acceleration of gravity unless other factors retard flow. Second, in their upstream reaches, streams commonly have boulder-strewn, broad, shallow channels, so flow resistance is high and average velocity is correspondingly slower. Downstream, however, channels generally become more semicircular, and the bed and banks are usually composed of finer-grained materials, thus reducing the effects of friction. Third, the number of tributary streams joining a larger stream increases in a downstream direction. Thus, the total volume of water (discharge) increases, and increasing discharge results in increased velocity.

To determine **discharge,** the total volume of water in a stream moving past a particular point in a given period of time, one must know the dimensions of the water-filled part of a channel—that is, its cross-sectional area (A)—and its flow velocity (V). Discharge (Q) can then be calculated by the formula $Q = VA$, and it is generally expressed in cubic feet per second (ft^3/sec) or cubic meters per second (m^3/sec). The Mississippi River has an average discharge of 18,000 m^3/sec (Table 15.1), but, as for all streams, the discharge varies, being greatest during floods and lowest during long, dry spells.

RUNNING WATER AND EROSION

Streams possess two kinds of energy, potential and kinetic. *Potential energy* is the energy of position, such as that possessed by water behind a dam or at a high elevation. In stream flow, potential energy is converted to *kinetic energy,* the energy of motion. Most of this kinetic energy is dissipated as heat within streams by fluid turbulence, but a small amount, perhaps 5%, is available to erode and transport sediment. Erosion involves the physical removal of dissolved substances, minerals, and loose particles of soil and rock from a source area. Thus, the sediment transported in a stream consists of both dissolved materials and solid particles.

Because the *dissolved load* of a stream is invisible, it is commonly overlooked, but it is an important part of the total sediment load. Some of it is acquired from the streambed and banks where soluble rocks such as limestone and dolostone are exposed, but much of it is carried into streams by sheet flow and by groundwater.

The solid sediment carried in streams ranges from clay-sized particles to large boulders. Much of this sediment is supplied to streams by mass wasting, but some is derived directly from the streambed and banks (Figure 15.11). The power of running water, called **hydraulic action,** is sufficient to set particles in motion.

Table 15.1

Some Large Rivers of the World				
RIVER	CONTINENT	LENGTH (km)	DRAINAGE AREA (km²)	AVERAGE DISCHARGE AT MOUTH (m³/sec)
Amazon	South America	6,450	6,150,000	200,000
Zaire (Congo)	Africa	4,675	3,720,000	44,000
Orinoco	South America	2,580	970,000	36,000
Ganges and	Asia	2,500	1,480,000	31,000
Brahmaputra		2,900		
Rio Negro	South America	2,255	1,000,000	30,000
Yangtze	Asia	5,530	1,810,000	29,000
Mississippi	North America	3,770	3,240,000	18,000
Zambezi	Africa	2,740	1,420,000	16,000
Mekong	Asia	4,190	786,000	15,000
St. Lawrence	North America	1,290	1,030,000	9,900
Mackenzie	North America	4,245	1,810,000	9,700
Ohio	North America	2,110	526,000	8,000
Columbia	North America	2,000	850,000	7,500
Yukon	North America	3,190	850,000	6,400
Nile	Africa	6,705	3,030,000	1,000
Murray	Australia	2,580	1,060,000	700

Figure 15.11
(a) This stream acquires some of its sediment load by undercutting its bank.
(b) The Snake River in Idaho gets some of its sediment from these talus cones that accumulated as a result of mass wasting.

(a)

(b)

Everyone has seen the results of hydraulic action, although perhaps not in streams. For example, if the flow from a garden hose is directed onto loose soil, a hole is soon gouged out by hydraulic action.

Another process of erosion in streams is **abrasion,** in which exposed rock is worn and scraped by the impact of solid particles. If running water contains no sediment, no abrasion of rock surfaces will result, but if it is transporting sand and gravel, the impact of these particles abrades exposed rock surfaces. *Potholes* in the beds of streams are one obvious manifestation of abrasion (Figure 15.12). These circular to oval holes form where eddying currents containing sand and gravel swirl around and erode depressions into rock.

HOW DOES RUNNING WATER TRANSPORT AND DEPOSIT SEDIMENT?

Streams transport a **dissolved load** consisting of materials taken into solution during chemical weathering, and they also transport sedimentary particles. The smallest of these particles, mostly silt and clay, are kept suspended by fluid turbulence and transported as a **suspended load** (Figure 15.13). The Mississippi River transports nearly 200 million metric tons of suspended load past Vicksburg, Mississippi, each year, and the Yellow River of China carries almost four times as much suspended load per year. Particles transported in suspension are deposited only where turbulence is minimal, as in lakes and lagoons.

Streams also transport a **bed load** consisting of larger particles such as sand and gravel (Figure 15.13). Because fluid turbulence is insufficient to keep sand and gravel suspended, they move along the streambed. However, part of the bed load can be suspended at least temporarily as when an eddying current swirls across a streambed and lifts sand grains into the water. These particles move forward at approximately the flow velocity, but at the same time they settle toward the streambed where they come to rest, to be moved again later by the same process. This process of intermittent bouncing and skipping along the streambed is called *saltation* (Figure 15.13).

(a)

Figure 15.12
(a) These circular depressions in the bed of the Chippewa River in Ontario, Canada, are potholes. They measure about 1 m across, but two potholes at the top center of the image have merged to form a larger, composite pothole.
(b) Close-up view of a 2-m-diameter pothole in Lucerne, Switzerland. Notice the large boulders in the pothole.

(b)

Particles too large to be suspended even temporarily are transported by rolling or sliding (Figure 15.13). Obviously, greater flow velocity is required to move particles of these sizes. The maximum-sized particles that a stream can carry define its *competence,* a factor related to flow velocity. Figure 15.14 shows the velocities required to erode, transport, and deposit particles of various sizes. As expected, high velocity is necessary to erode and transport gravel-sized particles, whereas sand is eroded and transported at lower velocities. Notice, though, that high velocity is needed to erode clay because clay deposits are very cohesive: the tiny clay particles adhere to one another and can be disrupted only by energetic flow conditions. Once eroded, however, very little energy is needed to keep the clay particles in motion.

Capacity is a measure of the total load a stream can carry. It varies as a function of discharge; with greater discharge, more sediment can be carried. Capacity and competence may seem quite similar, but they are actually related to different aspects of stream transport. For instance, a small, swiftly flowing stream may have the competence to move gravel-sized particles but not to transport a large volume of sediment, so it has a low capacity. A large, slow-flowing stream, on the other hand, has a low competence, but may have a very large suspended load and hence a large capacity.

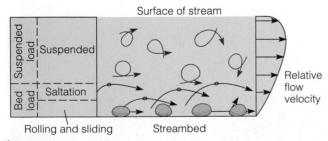

Figure 15.13
Methods of sediment transport by running water. The arrows in the velocity profile at the right are proportional to flow velocity, indicating that the water flows fastest near the surface and slowest along the streambed.

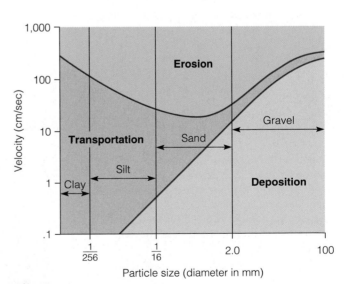

Figure 15.14
Sediment erosion, transport, and deposition by running water are related to particle size and flow velocity.

STREAMS AND THEIR DEPOSITS

Streams can transport sediment a considerable distance from the source area. For instance, some of the sediment deposited in the Gulf of Mexico by the Mississippi River came from such distant sources as Pennsylvania, Minnesota, and southern Alberta, Canada. Along the way, sediments might be deposited in a variety of environments, such as stream channels, the floodplains adjacent to channels, and the points where streams flow into lakes or seas or flow from mountain valleys onto adjacent lowlands (see Figure 6.5). As a matter of fact, much of the detrital sediment on continental shelves is transported to the seas by streams, and huge quantities of land-derived sediment eventually come to rest on the continental slopes and rises (see Chapter 11).

Streams do most of their erosion, sediment transport, and deposition when they flood. Consequently, stream deposits, collectively called **alluvium,** do not represent the continuous day-to-day activity of streams but rather those periodic, large-scale events of sedimentation during floods.

The Deposits of Braided Streams

Braided streams possess an intricate network of dividing and rejoining channels separated from one another by sand or gravel bars (Figure 15.15). Seen from above, the channels resemble the complex strands of a braid. Braided stream channels develop when a stream is supplied with more sediment than it can transport, which over time is deposited as sand and gravel bars within its channel. During high-water stages, the bars are submerged, but when the water is low, they are exposed and divide a single channel into multiple channels. Braided streams have broad, shallow channels, and are characterized as bed-load transport streams because they transport and deposit mostly sand and gravel (Figure 15.15).

(a)

(b)

Figure 15.15
Braided streams and their deposits. (a) A braided stream near Santa Fe, New Mexico. The deposits of this stream are composed mostly of sand. (b) A braided stream with gravel bars near Chester, California.

Braided streams are common in arid and semiarid regions where little vegetation exists and erosion rates are high. Streams with easily eroded banks are also likely to become braided. In fact, a stream that is braided where its banks are easily eroded may have a single sinuous or meandering channel when it flows into an area of more resistant materials. Streams fed by melting glaciers are also commonly braided because melting glacial ice yields so much sediment (see Chapter 17).

Meandering Streams and Their Deposits

Meandering streams have a single, sinuous channel with broadly looping curves known as *meanders* (Figure 15.16). These channels are semicircular in cross section along straight reaches, but at meanders they are markedly asymmetric, being deepest near the outer bank, which commonly descends vertically into the channel (Figure 15.17). The outer bank is called the *cut bank* because greater velocity and turbulence on that side

of the channel erode it. Despite the known dynamics of erosion along the outer banks of meanders, it is remarkable that houses and other structures are built overlooking cut banks. The life expectancy of these structures is short because during a single flood, several meters of erosion might take place along a cut bank. In contrast, flow velocity is at a minimum near the inner bank, which slopes gently into the channel (Figure 15.17).

As a result of the unequal distribution of flow velocity across meanders, the cut bank is eroded, and deposition takes place along the opposite side of the channel. The net effect is that a meander migrates laterally, and the channel maintains a more or less constant width because erosion on the cut bank is offset by an equal amount of deposition on the opposite side of the channel. The deposit formed in this manner is a **point bar**; it consists of cross-bedded sand or, in some cases, gravel (Figure 15.17).

It is not uncommon for meandering streams to become so sinuous that the thin neck of land separating adjacent meanders is eventually cut off during a flood. The valley floors of meandering streams are commonly

Figure 15.16
Aerial view of a meandering stream. The broad, flat area adjacent to the stream channel is the floodplain. Notice the crescent-shaped lakes—these are cutoff meanders, which are known as *oxbow lakes. Source: Photo courtesy of John S. Shelton.*

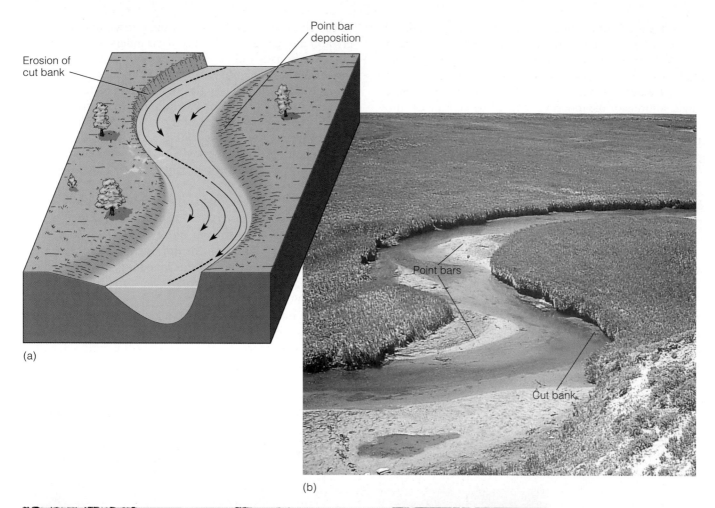

Erosion of cut bank

Point bar deposition

(a)

Point bars

Cut bank

(b)

Point bar

(c)

Figure 15.17
(a) In a meandering channel, flow velocity is greatest near the outer bank. The dashed line follows the path of maximum velocity, and the solid arrows are proportional to velocity. Because of varying velocity across the channel, the outer bank or cut bank is eroded but a point bar is deposited on the opposite side of the channel. (b) Two small point bars of sand in a meandering stream. Notice how they are inclined into the deeper part of the channel. Also note the cut bank. (c) This point bar is composed of gravel.

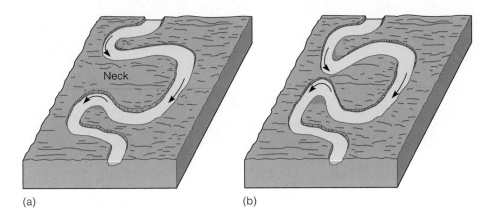

(a)   (b)

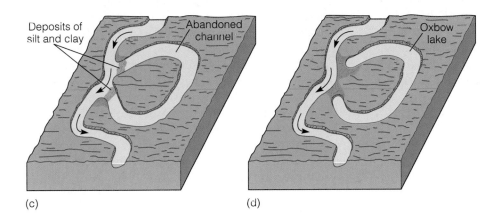

(c)   (d)

Figure 15.18
Four stages in the origin of an oxbow lake. In (a) and (b), the meander neck becomes narrower. (c) The meander neck is cut off, and part of the channel is abandoned. (d) When it is completely isolated from the main channel, the abandoned meander is an oxbow lake.

marked by crescent-shaped **oxbow lakes,** which are actually cutoff meanders (Figures 15.16 and 15.18). These oxbow lakes may persist as lakes for some time but are eventually filled with organic matter and fine-grained sediment carried by floods. Even after they are filled, they remain visible on floodplains.

One immediate effect of meander cutoff is an increase in flow velocity; following the cutoff, the stream abandons part of its old course and flows a shorter distance, thereby increasing its gradient. Numerous cutoffs would, of course, significantly shorten a meandering stream, but streams usually establish new meanders elsewhere when old ones are cut off.

Deposits on Floodplains

Most streams periodically receive more water than their channel can carry, so they spread across low-lying, relatively flat **floodplains** adjacent to their channels (Figure 15.16). Even small streams commonly have a floodplain, but this feature is usually proportional to the size of the stream; thus, small streams have narrow floodplains, whereas the lower Mississippi and other large streams have floodplains many kilometers wide.

Streams restricted to deep, narrow valleys usually have little or no floodplain.

Some floodplains are composed mostly of sand and gravel that were deposited as point bars. When a meandering stream erodes its cut bank and deposits on the opposite bank, it migrates laterally across its floodplain. As lateral migration occurs, a succession of point bars develops by *lateral accretion* (Figure 15.19). That is, the deposits build laterally as a result of repeated episodes of sedimentation on the inner banks of meanders.

Many floodplains are dominated by *vertical accretion* of fine-grained sediments. When a stream overflows its banks and floods, the velocity of the water spilling onto the floodplain diminishes rapidly because of greater frictional resistance to flow as the water spreads out as a broad, shallow sheet. In response to the diminished velocity, ridges of sandy alluvium known as **natural levees** are deposited along the margins of the stream channel (Figure 15.20). Natural levees are built up by repeated deposition of sediment during numerous floods. These natural levees separate most of the floodplain from the stream channel, so floodplains are commonly poorly drained and swampy. In fact, tributary streams may parallel the main stream for many kilometers until

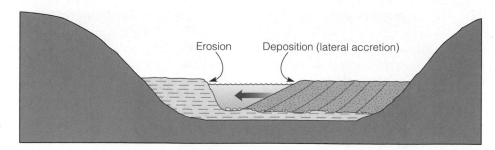

Figure 15.19
Floodplain deposits forming by lateral accretion of point bars.

they find a way through the natural levee system (Figure 15.20).

Floodwaters spilling from a main channel carry large quantities of silt- and clay-sized sediment beyond the natural levees and onto the floodplain. During the waning stages of a flood, floodwaters may flow very slowly or not at all, and the suspended silt and clay eventually settle as layers of mud that build upward by deposition during successive floods, a process known as vertical accretion (Figure 15.20).

Deltas

The fundamental process of delta formation is rather simple: When a stream flows into another body of water, its flow velocity decreases rapidly and it deposits its sediment. As a result, a **delta** forms, causing the local shoreline to build out, or *prograde* (Figure 15.21), unless the stream-deposited sediment is swept along the shoreline or into

deeper water by marine currents. Deltas in lakes are common, but marine deltas are much larger, far more complex, and more important as areas of natural resources.

The simplest prograding deltas exhibit a characteristic vertical sequence in which *bottomset beds* are successively overlain by *foreset beds* and *topset beds* (Figure 15.21a). This sequence develops when a stream enters another body of water and the finest sediments are carried some distance beyond the stream's mouth, where they settle from suspension and form bottomset beds. Nearer the stream's mouth, foreset beds are formed as sand and silt are deposited in gently inclined layers. The topset beds consist of coarse-grained sediments deposited in a network of *distributary channels* traversing the top of the delta.

Many small deltas in lakes have the three-part division described above, but large marine deltas are usually much more complex. Depending on the relative importance of stream, wave, and tidal processes, three major

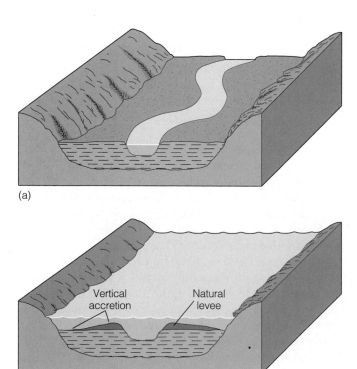

(a)

(b)

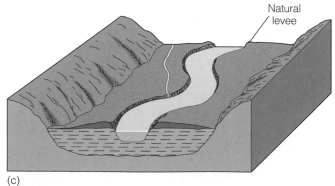

(c)

Figure 15.20
Three stages in the formation of vertical accretion deposits on a floodplain. (a) Stream at low-water stage. (b) Flooding stream and deposition of natural levees. The levees form after many such episodes of flooding. (c) After flooding. Notice the tributary stream, which parallels the main stream until it finds a way through the natural levees.

types of marine deltas are recognized (Figure 15.22). *Stream-dominated deltas,* such as the Mississippi River delta, consist of long fingerlike sand bodies, each deposited in a distributary channel that progrades far seaward. These deltas are commonly called *bird's-foot deltas* because the projections resemble the toes of a bird. In contrast, the Nile delta of Egypt is *wave-dominated,* although it also possesses distributary channels; the seaward margin of the delta consists of a series of barrier islands formed by reworking of sediments by waves, and the entire margin of the delta progrades seaward. *Tide-dominated deltas,* such as the Ganges–Brahmaputra of Bangladesh, are continually modified into tidal sand bodies that parallel the direction of tidal flow.

Coal can form in a variety of sedimentary environments, such as the freshwater marshes between distribu-tary channels of deltas (Figure 15.21a). These marshes are dominated by nonwoody plants whose remains accumulate to form peat, the first stage in the origin of coal (see Chapter 6). If peat is buried, the volatile components of the plants are driven off, leaving mostly carbon that eventually forms coal.

Delta progradation is one way that potential reservoirs for oil and gas form. Because the sediments of distributary sand bodies are porous and in proximity to organic-rich marine sediments, they commonly contain oil and gas that can be extracted in economic quantities. Much of the oil and gas production of the Gulf Coast of Texas comes from buried delta deposits. Some of the older deposits of the Niger River delta of Africa and the Mississippi River delta are also known to contain vast reserves of oil and gas.

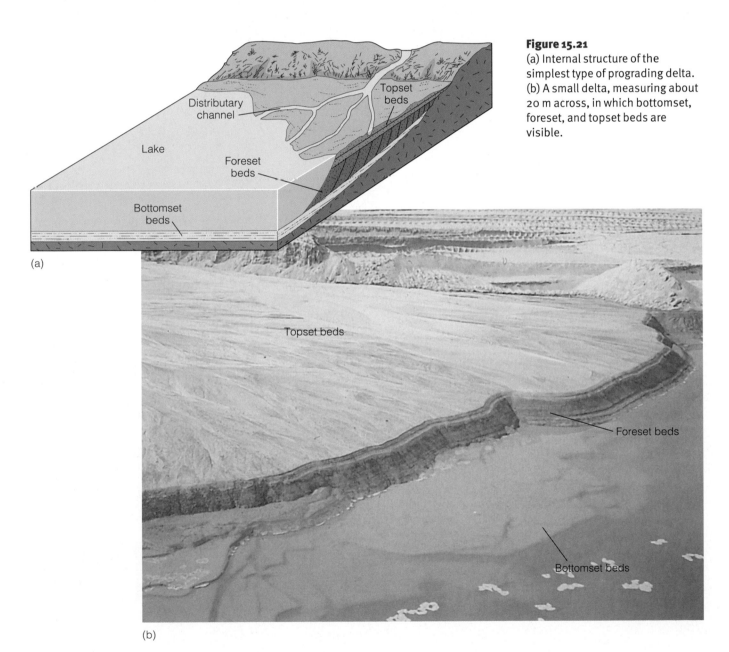

Figure 15.21
(a) Internal structure of the simplest type of prograding delta. (b) A small delta, measuring about 20 m across, in which bottomset, foreset, and topset beds are visible.

Alluvial Fans

Lobate deposits of alluvium known as **alluvial fans** form best on lowlands adjacent to highlands in arid and semi-arid regions where little or no vegetation exists to stabilize surface materials (Figure 15.23). When periodic rainstorms occur, surface materials are quickly saturated and runoff begins. During a particularly heavy rain, all surface flow in a drainage area is funneled into a mountain canyon leading to an adjacent lowland. The stream is confined in the mountain canyon, but as it discharges from the canyon onto the lowland area, it quickly spreads out, its velocity diminishes, and deposition of a fan-shaped body of alluvium ensues.

The alluvial fans that develop by the process just described are mostly accumulations of sand and gravel, a large proportion of which is deposited by streams. In some cases, the water flowing through a mountain canyon picks up so much sediment that it becomes a viscous mudflow (see Chapter 14). Consequently, mudflow deposits make up a large part of some alluvial fans.

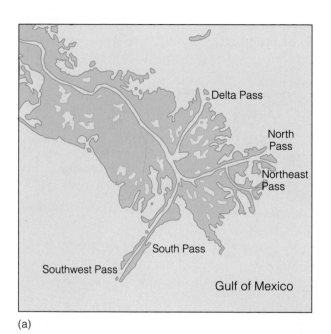

(a)

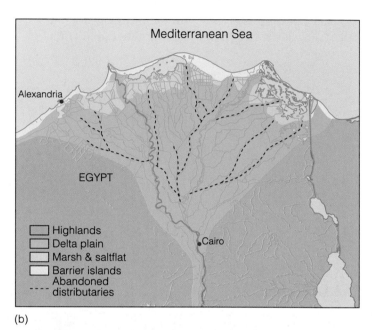

(b)

Figure 15.22

(a) The Mississippi River delta of the U.S. Gulf Coast is stream-dominated, and (b) the Nile delta of Egypt is wave-dominated. (c) The Ganges–Brahmaputra delta of Bangladesh is tide-dominated.

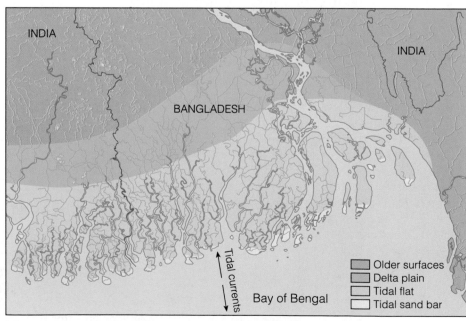

(c)

Figure 15.23
(a) Alluvial fans form where streams discharge from mountain canyons onto adjacent lowlands. (b) Alluvial fans adjacent to the Panamint Range on the margin of Death Valley, California.
Source: Photo courtesy of John S. Shelton.

Alluvial fan

Channels are dry most of the time

(a)

(b)

CAN FLOODS BE PREDICTED AND CONTROLLED?

When a stream receives more water than its channel can handle, it floods, occupying part or all of its floodplain. Indeed, floods are so common that, unless they cause considerable property damage or fatalities, they rarely rate more than a passing note in the news (Figure 15.24). The most extensive recent flooding in the United States was the Flood of '93 in the Midwest (see Perspective 15.2), but since then several other areas have experienced serious flooding. For instance, a series of storms from late December 1996 to early January 1997 caused widespread flooding in several western states and resulted in at least 28 deaths and billions of dollars in property damage. And in April 1997, the spring thaw of a record snowfall caused the Red River

to flood, resulting in the evacuation of all 50,000 residents of Grand Forks, North Dakota, and an additional 28,000 people from areas adjacent to the river in Manitoba, Canada. Floods and mudslides triggered by heavy rain killed tens of thousands during December 1999 in Venezuela (see Chapter 14).

Annual property damage from flooding in the United States exceeds $100 million. Despite the completion of more and more flood-control projects, the amount of property damage is not decreasing. In fact, the combination of fertile soils, level surfaces for construction, and proximity to water for agriculture, industry, and domestic uses makes floodplains popular sites for settlement. These human activities generally increase the potential for flooding. Urbanization greatly increases surface runoff because surface materials are compacted or covered by concrete and asphalt, reducing their infiltration capacity. Storm drains in urban areas

Guest Essay

Bonnie Robinson

Remediation and Management of Hazardous Wastes*

Bonnie Robinson earned an A.B. degree in geology from Oberlin College in 1974. She worked as a petroleum geologist for 13 years prior to joining the U.S. Environmental Protection Agency in Washington, D.C. In 1993 she was elected vice president of the American Association of Petroleum Geologists Division of Environmental Geosciences.

I n my current position with the U.S. Environmental Protection Agency's (EPA) Office of Solid Waste (OSW), I am involved in the corrective action program that provides requirements for the remediation of hazardous releases from hazardous-waste-management facilities regulated under the Resource, Conservation and Recovery Act (RCRA). A key goal of this program is to ensure that RCRA facilities do not become Superfund cleanup sites. My duties include providing technical assistance to EPA headquarters and regional offices, state and tribal agencies, and the public regarding the geologic and hydrogeologic analysis of environmental activities undertaken by the EPA. Many primary areas of interest involve the contamination, remediation, and monitoring of groundwater.

I have also been involved in the development of EPA guidance documents concerning petroleum- and mining-waste management. EPA's Office of Solid Waste conducted studies of the characteristics of the wastes, waste-handling methods and their impact on human health and the environment, and legal and administrative controls on the generation, handling, and disposal of the wastes.

The complexity of the oil and gas industry, the wide range of environmental settings affected, and the variety of state regulatory programs present interesting challenges. For example, oil and gas production is scattered throughout more than 30 states, where over 26,000 companies are involved in the exploration and production of oil and gas. Each year thousands of new wells are drilled, and thousands of well sites are abandoned. The major wastes generated at these locations consist of water extracted with the oil and gas, drilling fluids, and a variety of lesser wastes. These wastes often contain varying amounts of potentially hazardous constituents.

One of the key issues facing the EPA is how to determine the most efficient alternatives for improving petroleum- and mining-waste management without significant adverse impacts on production. Continued domestic production of these resources is vital to the nation's interest, but it must be balanced with adequate environmental protection.

Knowledge of science and technology, or science literacy, is essential for intelligent decision making regarding critical national issues. Opportunities exist for full participation by minorities and women, who are severely underrepresented in science and technology. It is vital that we encourage, develop, and utilize this pool of talent. ∎

*Opinions expressed in this paper are solely those of the author and do not necessarily represent those of the U.S. Environmental Protection Agency.

Figure 15.24
Flooding of New Richmond, Ohio, by the Ohio River on March 4, 1997. Heavy thunderstorms on March 1 and 2 caused the Ohio River and several of its tributaries to flood resulting in 18 Ohio counties being declared federal and state disaster areas. About 20,000 people were evacuated, 5 died, and property damage was estimated at $180 million. *Source: © 1997 Jamie Sabau.*

quickly carry water to nearby streams, many of which flood much more commonly than they did in the past.

To monitor stream behavior, the U.S. Geological Survey maintains more than 11,000 stream gauging stations, and various state agencies also monitor streams. Data collected at gauging stations can be used to construct a *hydrograph* showing how a stream's discharge varies over time (Figure 15.25). Hydrographs are useful in planning irrigation and water-supply projects, and they give planners a better idea of what to expect during floods.

Stream gauge data are also used to construct *flood-frequency curves* (Figure 15.26). To construct such a curve, the peak discharges for a stream are first arranged in order of volume; the flood with the greatest discharge has a magnitude rank of 1, the second largest is 2, and so on (Table 15.2). The *recurrence interval*—that is, the time period during which a flood of a given magnitude or larger can be expected over an average of many years—is determined by the equation shown in Table 15.2. For this stream, floods with magnitude ranks of 1 and 23 have recurrence intervals of 77.00 and 3.35 years, respectively. Once the recurrence interval has been calculated, it is plotted against discharge, and a line is drawn through the data points (Figure 15.26).

According to Figure 15.26, the 10-year flood for the Rio Grande near Lobatos, Colorado, has a discharge of 245 m³/sec. This means that, on average, we can expect one flood of this size or greater to occur within a 10-year interval. One cannot, however, predict that such a flood will take place in any particular year, only that it has a probability of 1 in 10 (1/10) of occurring in any year. Furthermore, 10-year floods are not necessarily separated by 10 years. That is, two such floods could occur in the same year or in successive years, but over a period of centuries their average occurrence would be once every 10 years.

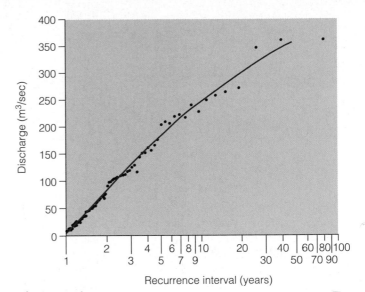

Figure 15.26
Flood-frequency curve for the Rio Grande near Lobatos, Colorado. The curve was constructed from the data in Table 15.2.

Unfortunately, stream gauge data in North America have been available for only a few decades, and rarely for more than a century. Accordingly, we have a good idea of stream behavior over short periods, the 2- and 5-year floods, for example, but our knowledge of long-term behavior is limited by the short record-keeping

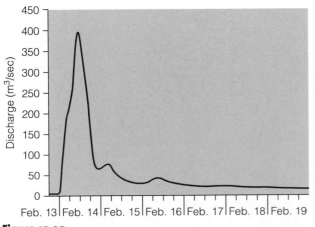

Figure 15.25
Hydrograph for Sycamore Creek near Ashland City, Tennessee, for the February 1989 flood.

Table 15.2

Selected Data and Recurrence Intervals for the Rio Grande near Lobatos, Colorado

Year	Discharge (m³/sec)	Rank	Recurrence Interval
1900	133	23	3.35
1901	103	35	2.20
1902	16	69	1.12
1903	362	2	38.50
1904	22	66	1.17
1905	371	1	77.00
1906	234	10	7.70
1907	249	7	11.00
1908	61	45	1.71
1909	211	13	5.92
.	The greatest yearly discharge is given a magnitude rank (m) ranging from 1 to N ($N = 76$ in this example), and the recurrence interval (R) is calculated by the equation $R = (N + 1)/m$.		
1974	22	64	1.20
1975	68	43	1.79

Source: U.S. Geological Survey Open-File Report 79–681.

The Flood of '93

According to one news report, the flooding between late June and August 1993 in the midwestern United States caused an estimated $15 to $20 billion in property damage. Fifty people died as a result of the flooding, 70,000 were left homeless, and more than 200 counties in several states including every county in Iowa were declared disaster areas. In addition to Iowa, flooding also occurred in South Dakota, North Dakota, Minnesota, Wisconsin, Nebraska, Missouri, Illinois, and Kansas.

One of the areas flooded earliest was Davenport, Iowa, where the Mississippi River crested on July 7 at 1.8 m above flood stage. By July 13, it crested almost 10 m above flood stage at Quincy, Illinois. The Des Moines River crested at 8 m above flood stage on July 11 and flooded Des Moines, Iowa, and surrounding areas. By the end of the first week in July, more than 40,000 km² of farmland had been flooded, with more rain expected; by August more than 92,000 km² had been flooded.

Despite heroic efforts of thousands of volunteers and the National Guard to stabilize levees by sandbagging them, levees failed everywhere or were simply overtopped by rising floodwaters. Grafton, Illinois, at the juncture of the Mississippi and Illinois Rivers was 80% underwater. Grafton, which lies on the floodplain and has no protective levee system, has been flooded six times in the last 20 years. In the aftermath of the flood, the mayor and many of Grafton's residents decided to move much of the town to higher ground at an estimated cost of $25 million.

Much of St. Charles County, Missouri, near the confluence of the Mississippi and Missouri Rivers, was so extensively flooded that 8000 people were evacuated from this county alone. In the small town of Portage des Sioux in St. Charles County, only eight homes were not flooded or were only slightly damaged by flooding. Even the 5.5-m-high pedestal of a statue near the Mississippi was swamped by floodwaters (Figure 1).

By mid-July flooding was extensive, and still more rain was expected. On July 16, an additional 17.8 cm of rain fell in North Dakota and Minnesota, and 15.2 cm more fell in South Dakota, further swelling already flooding rivers. Obviously, the direct cause of flooding was too much water for the Mississippi and Missouri Rivers and their tributaries to handle. But the reason so much water was present was the unusual behavior of the jet stream.

The *jet stream* is a narrow band of strong winds in the atmosphere. It is usually over the Midwest during the spring and then shifts north into Canada during the summer. In 1993 it remained over the Midwest, and as hot, moisture-laden air moved up from the south, thunderstorms developed over the Midwest (Figure 2). In short, these storms are usually distributed over a much larger region and spread to the Northeast, but in this case they simply dumped their precipitation over a much smaller region, resulting in as much as one and one-half to two times the normal amount.

One important lesson learned from the Flood of '93 is that despite our best efforts at flood control, some floods will occur anyway. Th 29 dams on the Mississippi and 36 reservoirs on upstream tributaries, as well as about 5800 km of levees, had little effect on holding floodwaters in check. Reservoirs have a limited capacity and when full are of no further use in controlling floods. In fact, some reservoirs are kept at high levels for recreation, which is inconsistent with their role in flood control.

No doubt debate will now focus on the utility of levees in flood control. Levees are effective in protecting many areas during floods, yet in some cases they actually exacerbate the problem by restricting the flow that would have formerly spread over a floodplain. They are certainly expensive to build and maintain, and their overall effectiveness has been and will continue to be questioned.

The U.S. Army Corps of Engineers has spent about $25 billion in the twentieth century to build 500 dams and more than 16,000 km of levees. No one doubts that some of these projects have been successful, at least within their design limits. But critics charge that such flood-control projects actually make the problem of flooding worse, particularly because development in flood-prone areas commonly follows the completion of flood-control projects, and nothing can be done to prevent some floods.

(a)

Figure 1
(a) Portage des Sioux, St. Charles County, Missouri, on July 16, 1993. The channel of the Mississippi River is at the far right. (b) Floodwaters in Portage des Sioux covered the 5.5-m-high pedestal of this statue on the bank of the Mississippi River.

(b)

Figure 15.2
The dominant weather pattern for June and July 1993. The jet stream remained over the Midwest during the summer rather than shifting north over Canada as it usually does. Thunderstorms developed in the convergence zone where warm, moist air and cool, dry air met.

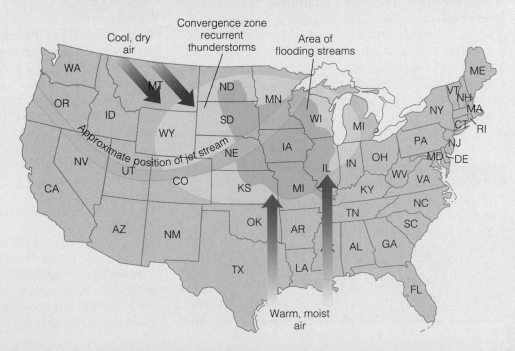

period. Accordingly, predictions of 50- or 100-year floods from Figure 15.26 are unreliable. In fact, the largest magnitude flood shown in Figure 15.26 may have been a unique event for this stream that will never be repeated. On the other hand, it may actually turn out to be a magnitude 2 or 3 flood when data for a longer time are available.

Although flood-frequency curves have limited applicability, they are nevertheless helpful in making flood-control decisions. Careful mapping of floodplains can identify areas at risk for floods of a given magnitude. For a particular stream, planners must decide what magnitude of flood to protect against because the cost goes up faster than the increasing sizes of floods would indicate.

Federal, state, and local agencies and land-use planners use flood-frequency analyses to develop recommendations and regulations concerning construction on and use of floodplains. Geologists and engineers are interested in these analyses for planning appropriate flood-control projects. They must decide, for example, where dams and basins should be constructed to contain the excess water of floods.

Flood control has been practiced for thousands of years. Common practices are to construct dams that impound reservoirs and to build levees along stream banks. Levees raise the banks of a stream, thereby restricting flow during floods. Unfortunately, deposition within the channel results in raising the streambed, making the levees useless unless they too are raised. Levees along the banks of the Huang He in China caused the streambed to rise more than 20 m above its surrounding floodplain in 4000 years. When the Huang He breached its levees in 1887, more than 1 million people were killed.

Dams and levees alone are insufficient to control large floods, so in many areas floodways are also used. These usually consist of a channel constructed to divert part of the excess water in a stream around populated areas or areas of economic importance. Reforestation of cleared land also helps reduce the potential for flooding because vegetated soil helps prevent runoff by absorbing more water.

WHAT WOULD YOU DO?

Flooding causes more damage and fatalities than any other natural disaster, so it seems prudent to construct flood-control projects. One way to protect structures and residents of floodplains is to build artificial levees along the margins of a stream. Explain, in terms of cost, benefits, and limitations, whether or not it would be advisable to build levees.

When flood-control projects are well planned and constructed, they are functional. What many people fail to realize is that these projects are designed to contain floods of a given size; should larger floods occur, streams spill onto floodplains anyway. Furthermore, dams occasionally collapse (see the Prologue), and reservoirs eventually fill with sediment unless dredged. In short, flood-control projects are not only initially expensive but also require constant, costly maintenance. Such costs must be weighed against the cost of damage if no control projects were undertaken.

DRAINAGE BASINS AND DRAINAGE PATTERNS

Thousands of streams, which are parts of larger drainage systems, flow either directly or indirectly into the oceans. Any stream system has a main channel and numerous smaller *tributary streams* that supply water to it. The Mississippi and all of its tributaries, or any other drainage system for that matter, carry surface runoff from an area known as the **drainage basin** (Table 15.1). Individual drainage basins are separated from adjacent ones by topographically higher areas called **divides** (Figure 15.27). Some divides are rather modest rises, such as that separating the Great Lakes's drainage basin from that of the Mississippi River, whereas others, such as the Continental Divide along the crest of the Rocky Mountains, are more impressive.

Various **drainage patterns** are recognized based on the regional arrangement of channels in a drainage system. The most common is *dendritic drainage,* consisting of a network of channels resembling tree branching (Figure 15.28a). Dendritic drainage develops on gently sloping surface materials that respond more or less homogeneously to erosion. Areas of flat-lying sedimentary rocks and some terrains of igneous or metamorphic rocks usually display a dendritic drainage pattern.

In marked contrast to dendritic drainage in which tributaries join larger streams at various angles, *rectangular drainage* is characterized by channels with right-angle bends and tributaries that join larger streams at right angles (Figure 15.28b). The positions of the channels are strongly controlled by geologic structures, particularly regional joint systems that intersect at right angles. Rectangular drainage develops because streams more easily erode and establish channels along the traces of joints.

In some parts of the eastern United States, such as Virginia and Pennsylvania, erosion of folded sedimentary rocks develops a landscape of alternating parallel ridges and valleys. The ridges consist of more resistant rocks such as sandstone, whereas the valleys overlie less resistant rocks such as shale. Main streams follow the

(a)

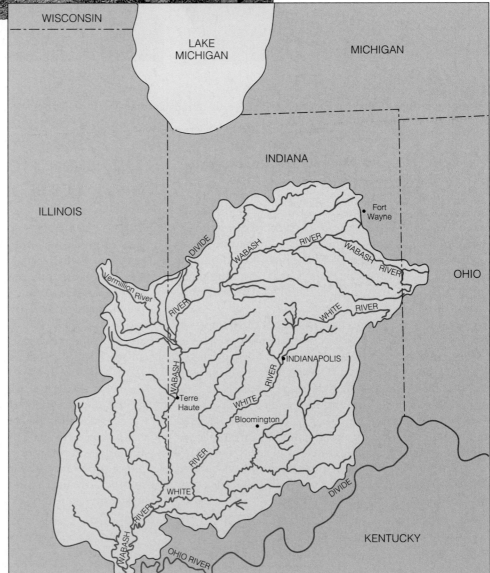

(b)

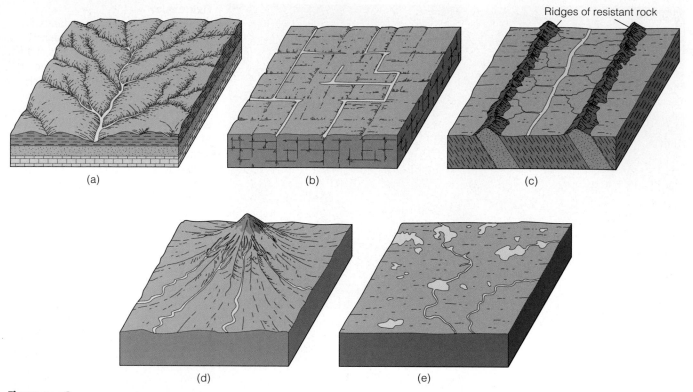

Figure 15.28
Examples of drainage patterns: (a) dendritic drainage, (b) rectangular drainage, (c) trellis drainage, (d) radial drainage, and (e) deranged drainage.

trends of the valleys, and short tributaries flowing from the adjacent ridges join the main stream at nearly right angles, hence the name *trellis drainage* (Figure 15.28c).

In *radial drainage,* streams flow out in all directions from a central high area (Figure 15.28d). Radial drainage develops on large, isolated volcanic mountains and where the crust has been arched up by the intrusion of plutons such as laccoliths. Structural domes that also form topographically high areas, such as the Black Hills in South Dakota, also have radial drainage.

In some areas, streams flow in and out of swamps and lakes with irregular flow directions. Drainage patterns characterized by such irregularity are called *deranged* (Figure 15.28e). The presence of deranged drainage indicates that the drainage system developed recently and has not yet formed an organized system. In parts of Minnesota, Wisconsin, and Michigan that were glaciated until about 10,000 years ago, the previously established drainage systems were obliterated by glaciers. Following the final retreat of the glaciers, drainage systems became established but have not yet become fully organized, thus leaving many swampy areas undrained.

THE SIGNIFICANCE OF BASE LEVEL

Streams require a slope in order to flow, so they can erode downward only to the level of the body of water into which they discharge. A stream flowing into the sea, for example, cannot erode its valley lower than sea level—if it could, it would have to flow uphill to reach the sea. The lower limit to which streams can erode is called **base level** (Figure 15.29). Theoretically, a stream could erode its entire valley to very near sea level, so sea level is commonly referred to as *ultimate base level.** Streams never reach ultimate base level along their entire length, though, because they must have some gradient in order to maintain flow.

Besides ultimate base level, streams have *local* or *temporary base levels.* Where a stream flows across particularly resistant rock, a waterfall may develop, forming a local or temporary base level (Figure 15.29). A lake or another stream can serve as a local base level for the upstream segment of a stream (Figure 15.30).

---

*Streams flowing into depressions below sea level, such as Death Valley in California, have a base level corresponding to the lowest point of the depression and are not limited by sea level.

Figure 15.29
(a) Sea level is ultimate base level, and a resistant rock layer forms a local base level. (b) Cumberland Falls on the Big South Fork River in Cumberland Falls State Resort Park, Kentucky. The rock layer the falls plunge over is a local base level. At 38 m wide and 18 m high, Cumberland Falls is the second highest waterfall east of the Rocky Mountains.

Local base level

Rock resistant to erosion

Ultimate base level

(a)

(b)

Local base level

Ultimate base level

Figure 15.30
Ultimate and local base level.

When sea level rises or falls with respect to the land or when the land that a stream flows over is uplifted or subsides, changes in base level occur. For example, during the Pleistocene Epoch when extensive glaciers were present on the Northern Hemisphere continents, sea level was about 130 m lower than at present. Accordingly, streams deepened their valleys by adjusting to a new, lower base level. In addition, many streams extended their valleys onto the exposed continental shelves. Rising sea level at the end of the Pleistocene caused base level to rise, and the stream valleys on the continental shelf were flooded (see Chapter 11).

Streams also adjust to human intervention but not always in anticipated or desirable ways. Geologists and engineers are well aware that building a dam and impounding a reservoir creates a local base level (Figure 15.31a). Where a stream enters a reservoir, its flow velocity diminishes rapidly and deposition occurs, so unless dredged, reservoirs are eventually filled with sediment. Another consequence of building a dam is that the water discharged at the dam is largely sediment-free, but it still possesses energy to transport sediment. Commonly, such streams simply acquire a new sediment load by vigorously eroding downstream from the dam.

Draining a lake along a stream's course may seem like a small change that is well worth the time and expense to expose dry land for agriculture or commercial development. But unless one anticipates the stream's probable response, dire consequences can result. Remember that a lake is a temporary base level, so draining it lowers the base level for that part of the stream above the lake, and the stream will very likely respond by rapid downcutting (Figure 15.31b).

WHAT IS A GRADED STREAM?

A stream's *longitudinal profile* shows the elevations of a channel along its length as viewed in cross section (Figure 15.32). The longitudinal profiles of many streams show a number of irregularities such as lakes and waterfalls, which are local base levels (Figure 15.32a). Over time these irregularities tend to be eliminated; where the gradient is steep, erosion decreases it, and where the gradient is too low to maintain sufficient flow velocity for sediment transport, deposition occurs, steepening the gradient. In short, streams tend to develop a smooth, concave longitudinal profile of equilibrium, meaning that all parts of the system are dynamically adjusting to one another (Figure 15.32b).

Streams possessing an equilibrium profile are said to be **graded streams;** that is, a delicate balance exists between gradient, discharge, flow velocity, channel

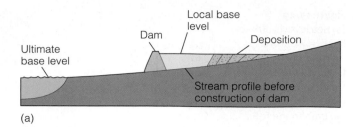

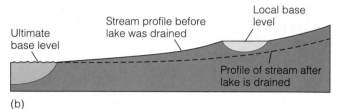

Figure 15.31
(a) Constructing a dam and impounding a reservoir creates a local base level. A stream deposits much of its sediment load where it flows into a reservoir. (b) A stream adjusts to a lower base level when a lake is drained.

characteristics, and sediment load so that neither significant erosion nor deposition takes place within the channel. Such a delicate balance is rarely attained, so the concept of a graded stream is an ideal. Nevertheless, many streams do indeed approximate the graded condition, although not along their entire courses and usually only temporarily.

Even though the concept of a graded stream is an ideal, we can generally anticipate the responses of a graded stream to changes altering its equilibrium. A change in base level would cause a stream to adjust as previously discussed. Increased rainfall in a stream's drainage basin would result in greater discharge and flow velocity. In short, the stream would now possess greater energy—energy that must be dissipated within the stream system by, for example, a change in channel shape. A change from a semicircular to a broad, shallow channel would dissipate more energy by friction. In contrast, the stream may respond by eroding a deeper valley and effectively reduce its gradient until it is once again graded.

Vegetation inhibits erosion by having a stabilizing effect on soil and other loose surface materials. So a decrease in vegetation in a drainage basin might lead to higher erosion rates, causing more sediment to be washed into a stream than it can effectively carry. Accordingly, the stream may respond by deposition within its channel, which increases its gradient until it is sufficiently steep to transport the greater sediment load.

Figure 15.32

(a) An ungraded stream has irregularities in its longitudinal profile. (b) Erosion and deposition along the course of a stream tend to eliminate irregularities and yield the smooth, concave profile typical of a graded stream. (c) The actual measured profiles of three streams. All profiles are concave but differ in the degree of concavity because each stream has a different slope. The vertical scale is exaggerated 275 times in this illustration.

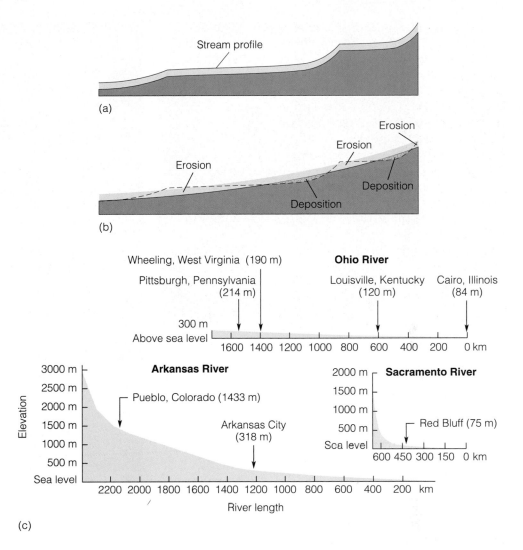

HOW DO VALLEYS FORM AND EVOLVE?

Low areas, known as **valleys,** bounded by higher land are common landforms, and with few exceptions they form and evolve mostly as a result of stream processes. The shapes and sizes of valleys vary considerably: Some are small, steep-sided *gullies,* others are broad and have gently sloping walls, and some *canyons* such as the Grand Canyon are steep-walled and deep (Figure 15.33). Particularly narrow and deep valleys are known as *gorges.* Several processes—in particular downcutting, lateral erosion, sheet wash, headward erosion, and mass wasting—contribute to the origin and evolution of valleys.

When a stream possesses more energy than it needs to transport its sediment load, some of its excess energy is expended by eroding a deeper valley. If downward erosion were the only process operating, valleys would be narrow, steep-sided gorges, but in most cases they are widened as their walls are undercut by streams. Such undermining, termed *lateral erosion,* creates unstable conditions so that part of a bank or valley wall moves downslope by mass-wasting processes. Furthermore, sheet wash and erosion of rill and gully tributaries carry materials from the valley walls to the main stream.

In addition to becoming deeper and wider, stream valleys are commonly lengthened in an upstream direction by *headward erosion* as drainage divides are eroded by entering runoff water (Figure 15.34a). In some cases, headward erosion eventually breaches the drainage divide and diverts part of the drainage of another stream by a process called *stream piracy* (Figure 15.34b). Once stream piracy has occurred, both drainage systems must adjust; one now has more water, greater discharge, and greater potential to erode and transport sediment, whereas the other is diminished in all of these aspects.

(a)

(b)

Figure 15.33
Valleys of various sizes and shapes form mostly as a result of erosion by running water and mass-wasting processes. (a) Many valleys have gently sloping walls much like this one. (b) Canyon de Chelly in Arizona is broad and has steep walls.

Stream Terraces

Adjacent to many streams are **stream terraces,** which are erosional remnants of floodplains formed when the streams were flowing at a higher level. They consist of a fairly flat upper surface and a steep slope descending to the level of the lower, present-day floodplain (Figure 15.35). In some cases, a stream has several step-like surfaces above its present-day floodplain, indicating that stream terraces formed several times.

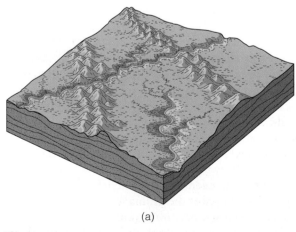

(a)

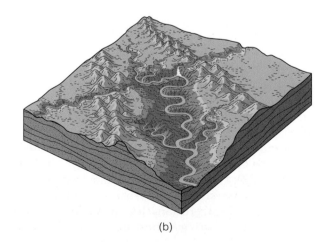

(b)

Figure 15.34
Two stages in the evolution of a valley. (a) The stream widens its valley by lateral erosion and mass wasting while simultaneously extending its valley by headward erosion. (b) As the larger stream continues to erode headward, stream piracy takes place when it captures some of the drainage of the smaller stream. Notice also that the larger valley is wider in (b) than it was in (a).

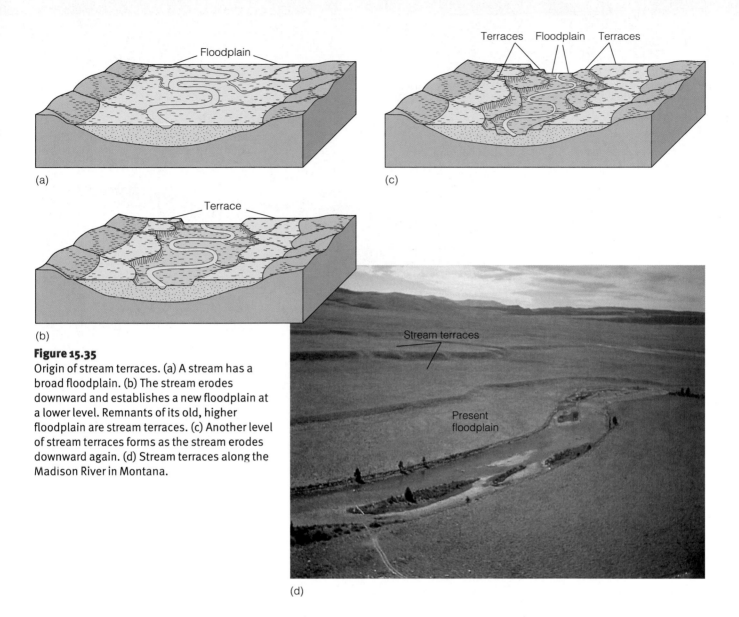

Figure 15.35
Origin of stream terraces. (a) A stream has a broad floodplain. (b) The stream erodes downward and establishes a new floodplain at a lower level. Remnants of its old, higher floodplain are stream terraces. (c) Another level of stream terraces forms as the stream erodes downward again. (d) Stream terraces along the Madison River in Montana.

Although all stream terraces result from erosion, they are preceded by an episode of floodplain formation and deposition of sediment. Subsequent erosion causes the stream to cut downward until it is once again graded (Figure 15.35), and then it begins eroding laterally and establishes a new floodplain at a lower level. Several such episodes account for the multiple terrace levels seen adjacent to some streams. Stream terraces are commonly cut into previously deposited sediment, but some are cut into bedrock where the terrace surface is generally covered by a thin veneer of sediment.

Renewed erosion and the formation of stream terraces are usually attributed to a change in base level. Either uplift of the land that a stream flows over or lowering of sea level yields a steeper gradient and increased flow velocity, thus initiating an episode of downcutting. When the stream reaches a level at which it is once again graded, downcutting ceases.

Incised Meanders

Some streams are restricted to deep, meandering canyons cut into bedrock, where they form features called **incised meanders.** The San Juan River in Utah, for example, occupies a meandering canyon more than 390 m deep (Figure 15.36). Such streams, being restricted by solid rock walls, are generally ineffective in eroding laterally; thus, they lack a floodplain and occupy the entire width of the canyon floor. Some incised meandering streams do erode laterally, thereby cutting off meanders and producing natural bridges (see Perspective 15.3).

Understanding how a stream can cut down into rock is not difficult, but understanding how a stream can form a meandering pattern in bedrock is another matter. Because lateral erosion is inhibited once downcutting begins, one must infer that the meandering course was established when the stream flowed across

Perspective 15.3

Natural Bridges

T he term *natural bridge* has been used to describe a variety of features including spans of rock resulting from wave erosion, the partial collapse of cavern roofs, and weathering and erosion along closely spaced, parallel joints as in Arches National Park in Utah (see Figure 13.16a). Here we are concerned only with natural bridges that span a valley eroded by running water.

The best place to observe this type of natural bridge is in Natural Bridges National Monument in southwestern Utah. Three natural bridges are present within the monument, and all originated in the same way. Of these three, Sipapu Bridge is the largest (Figure 1); it stands 67 m above White Canyon and has a span of 81.5 m.

The process by which these natural bridges formed is well understood, and, as a matter of fact, it is still going on. In the first stage, a meandering

Figure 1
Sipapu Bridge in Natural Bridges National Monument, Utah.

stream was incised into bedrock (Figure 2). In Natural Bridges National Monument, this rock unit is the Cutler Formation, which consists of sandstone

Figure 15.36
The Goose Necks of the San Juan River in Utah are incised meanders. This meandering canyon is more than 390 m deep. *Source: Photo courtesy of John S. Shelton.*

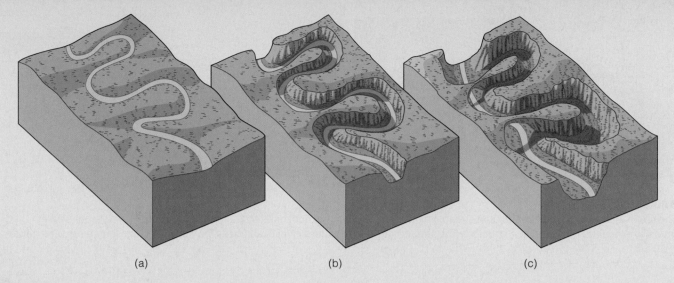

Figure 2
Origin of a natural bridge. (a) A meandering stream flows across a gently sloping surface. (b) Incised meanders develop as the stream erodes down into solid rock. (c) A thin wall of rock between meanders is eventually breached, forming a natural bridge.

formed from windblown sand deposited during the Permian Period. When local meandering streams became incised, lateral erosion created a thin wall of rock between adjacent meanders that was eventually breached (Figure 2). As the breach was subsequently enlarged, the stream abandoned its old meander, and the stream flow was diverted. As we dis-cussed previously, oxbow lakes are formed by a similar process (Figure 15.18). The only significant dif-ference is that the streams forming nat-ural bridges are incised.

Natural bridges are temporary fea-tures. Once formed, they are destroyed by other processes. Rocks fall from the undersides of bridges, their surfaces are weathered and eroded, and eventu-ally they collapse. The monument con-tains several examples of collapsed bridges, but new ones are in the process of forming.

an area covered by alluvium. For instance, suppose a stream near base level has established a meandering pat-tern. If the land the stream flows over is uplifted, erosion is initiated, and the meanders become incised into the underlying bedrock.

Superposed Streams

Streams flow downhill in response to gravity, so their courses are determined by preexisting topography. Yet a number of streams seem, at first glance, to have defied this fundamental control. For example, the Delaware, Potomac, and Susquehanna Rivers in the eastern United States have valleys that cut directly through ridges lying in their paths. All these streams are **superposed**. To understand how superposition occurs, it is necessary to know the geologic histories of these streams.

During the Mesozoic Era, the Appalachian Moun-tain region was eroded to a sediment-covered plain crossed by numerous streams generally flowing eastward. During the Cenozoic Era, regional uplift commenced, and the streams eroded downward and were superposed directly upon resistant rock. Instead of changing course, they cut narrow, steep-walled canyons called *water gaps* (Figure 15.37).

VALLEY DEVELOPMENT— A SUMMARY

According to one concept, stream erosion of an area uplifted above sea level yields a distinctive series of landscapes. When erosion begins, streams erode downward; their valleys are deep, narrow, and

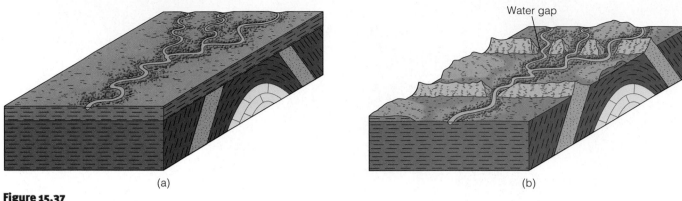

(a)

(b)

Figure 15.37
The origin of a superposed stream. (a) A stream begins cutting down into horizontal strata. (b) The horizontal layer is removed by erosion, exposing the underlying structure. The stream cuts narrow valleys in resistant rocks that form the ridges, forming water gaps.

V-shaped in cross section; and a number of irregularities are present in their profiles (Figure 15.38a). As streams cease eroding downward, they start eroding laterally, establishing a meandering pattern and a broad flood-plain (Figure 15.38b). Finally, with continued erosion, a vast, rather featureless plain develops (Figure 15.38c).

Many streams do indeed show an association of features typical of these stages. For instance, the Colorado River flows through the Grand Canyon and closely matches the features in the initial stage shown in Figure 15.38a. Streams in many areas approximate the second stage of development, and certainly the lower Mississippi closely resembles the last stage. Nevertheless, the idea of a sequential development of stream-eroded landscapes has been largely abandoned because there is no reason to think that streams necessarily follow this idealized cycle. Indeed, a stream on a gently sloping surface near sea level could develop features of the last stage very early in its history. In addition, as long as the rate of uplift exceeds the rate of downcutting, a stream will continue to erode downward and be confined to a narrow canyon.

Although it is doubtful that streams go through the three-stage sequence just outlined, they do indeed evolve in response to changing conditions in the variables that influence them. Of course, human intervention can change a stream's capacity for erosion and sediment transport, and increase the frequency of flooding. However, natural processes such as uplift or subsidence, climatic change, cutting a deeper outlet and draining a lake, or encountering a particularly resistant layer of rock can also influence stream activity. In short, they are dynamic systems continually adjusting to changes.

Figure 15.38
Idealized stages in the development of a stream and its associated landforms. According to this idea, an uplifted area begins eroding as in (a), and with time a landscape evolves as illustrated in (b) and finally in (c).

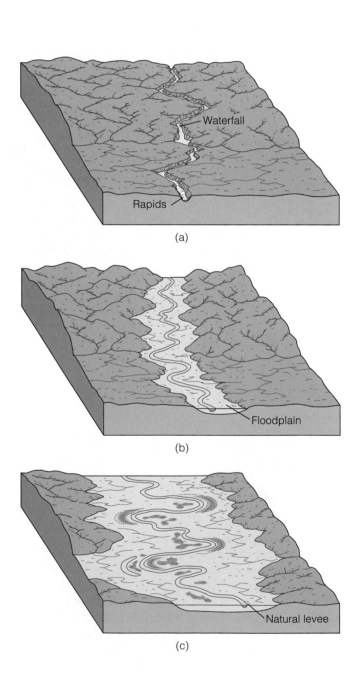

Chapter Summary

1. Water is continuously evaporated from the oceans, rises as water vapor, condenses, and falls as precipitation. About 20% of all precipitation falls on land and eventually returns to the oceans, mostly by surface runoff.

2. Running water moves by either laminar or turbulent flow. In the former, streamlines parallel one another, whereas in the latter they are complexly intertwined. Most flow in streams is turbulent.

3. Runoff can be characterized as either sheet flow or channel flow. Running water in a channel of any size is called a *stream*.

4. Gradient generally varies from steep to gentle along the course of a stream, being steep in upper reaches and gentle in lower reaches.

5. Flow velocity and discharge are related. A change in one of these parameters causes the other to change as well.

6. Streams and their tributaries carry runoff from drainage basins, which are separated from one another by divides.

7. Streams erode by hydraulic action, abrasion, and dissolution of soluble rocks.

8. The coarser part of a stream's sediment load is transported as bed load and the finer part as suspended load. Streams also transport a dissolved load of materials in solution.

9. Competence is a measure of the maximum-sized particles a stream can carry and is related to velocity. Capacity is a function of discharge and is a measure of the total load transported by a stream.

10. Braided streams are characterized by a complex of dividing and rejoining channels. Braiding occurs when sediment transported by a stream is deposited within channels as sand and gravel bars.

11. Meandering streams have a single, sinuous channel with broad looping curves. Meanders migrate laterally as the cut bank is eroded and point bars form on the inner bank. Oxbow lakes are cutoff meanders in which fine-grained sediments and organic matter accumulate.

12. Floodplains, rather flat areas paralleling stream channels, may be composed mostly of point bar deposits formed by lateral accretion or mud accumulated by vertical accretion during numerous floods.

13. Alluvial deposits at a stream's mouth are known as *deltas*. Many small deltas in lakes conform to the three-part division of bottomset, foreset, and topset beds, but large marine deltas are more complex. Marine deltas are characterized as stream-, wave-, or tide-dominated.

14. Lobate alluvial deposits on land consisting mostly of sand and gravel are alluvial fans. They form best in arid and semiarid regions where erosion rates are high.

15. Sea level is ultimate base level, the lowest level to which streams can erode. However, streams commonly have local base levels such as lakes, other streams, or the points where they flow across particularly resistant rocks.

16. Streams tend to eliminate irregularities in their channels so that they develop a smooth, concave profile of equilibrium. Such streams are graded. In a graded stream, a balance exists between gradient, discharge, flow velocity, channel characteristics, and sediment load so that little or no erosion or deposition occurs within its channel.

17. Stream valleys develop by a combination of processes including downcutting, lateral erosion, mass wasting, sheet wash, and headward erosion.

18. Renewed downcutting by a stream possessing a floodplain commonly results in the formation of stream terraces, which are remnants of an older floodplain at a higher level.

19. Incised meanders are generally attributed to renewed downcutting by a meandering stream so that it now occupies a deep, meandering valley.

20. Streams flowing through valleys cut into ridges directly in their paths are superposed, meaning that they once flowed on a higher surface and eroded downward into resistant rocks.

Important Terms

abrasion
alluvial fan
alluvium
base level
bed load
braided stream
delta
discharge
dissolved load
divide

drainage basin
drainage pattern
floodplain
graded stream
gradient
hydraulic action
hydrologic cycle
incised meander
infiltration capacity
meandering stream

natural levee
oxbow lake
point bar
runoff
stream
stream terrace
superposed stream
suspended load
valley
velocity

1. One way streams erode is by hydraulic action, which can be described as the:

   a. _____ capacity of water to dissolve minerals;

   b. _____ direct impact of water;

   c. _____ amount of bed load in a stream;

   d. _____ depth of a stream valley;

   e. _____ velocity of flow.

2. In which one of the following areas would a radial drainage pattern likely develop?

   a. _____ stream terrace;

   b. _____ delta;

   c. _____ drainage divide;

   d. _____ point bar;

   e. _____ composite volcano.

3. The phenomenon of vertical accretion takes place when:

   a. _____ a stream flows over very resistant rocks;

   b. _____ point bars are deposited along a stream's cut bank;

   c. _____ mud settles from water on a floodplain;

   d. _____ a valley is deepened by lateral erosion;

   e. _____ bed load is deposited on a point bar.

4. A meandering stream is one having:

   a. _____ a single, sinuous channel;

   b. _____ long, straight reaches and waterfalls;

   c. _____ numerous sand and gravel bars;

   d. _____ a dendritic drainage pattern;

   e. _____ alluvial fans along its course.

5. The capacity for Earth materials to absorb water is known as:

   a. _____ competence;

   b. _____ infiltration capacity;

   c. _____ meandering pattern;

   d. _____ dissolved load;

   e. _____ base level.

6. The bed load of a stream consists of:

   a. _____ dissolved materials and alluvium;

   b. _____ natural levees;

   c. _____ sediment deposited during the waning stages of floods;

   d. _____ particles carried by rolling, sliding, and saltation;

   e. _____ clay and silt.

7. The quantity of water moving past a given place in a given amount of time is a stream's:

   a. _____ gradient;

   b. _____ velocity;

   c. _____ capacity;

   d. _____ runoff;

   e. _____ discharge.

8. The deposits of braided streams consist mostly of:

   a. _____ suspended load;

   b. _____ terraces;

   c. _____ sand and gravel;

   d. _____ point bars;

   e. _____ dissolved load.

9. An erosional remnant of a floodplain now higher than the current level of a stream is a:

   a. _____ delta;

   b. _____ alluvial fan;

   c. _____ lake;

   d. _____ terrace;

   e. _____ base level.

10. Streams carrying sand and gravel effectively erode by:

    a. _____ abrasion;

    b. _____ solution;

    c. _____ deposition;

    d. _____ stream piracy;

    e. _____ lateral accretion.

11. An oxbow lake is a:

    a. _____ deposit formed where a stream leaves a canyon;

    b. _____ mound of sediment on the margins of a channel;

    c. _____ crescent-shaped lake on a floodplain;

    d. _____ type of valley eroded when meanders are incised;

    e. _____ type of drainage developed on gently dipping rock layers.

12. A particularly narrow, deep valley is known as a(an):

    a. _____ canyon;

    b. _____ gully;

    c. _____ alluvial fan;

    d. _____ gorge;

    e. _____ graded stream.

13. Explain how a stream can lengthen its channel at both its upstream and downstream ends.

14. Describe the hydrologic cycle, and explain what makes it operate.

15. How is it possible for a meandering stream to erode laterally and yet maintain a more or less constant channel width?

16. What factors influence infiltration capacity, and why is infiltration capacity important when considering runoff?

17. What are laminar and turbulent flow and what factors determine which will take place? Also, explain why one is more important than the other in erosion and transport.

18. How do alluvial fans and deltas compare and differ?

19. Describe how a point bar and a natural levee develop.

20. Is the following statement correct: "The steeper the gradient, the greater the flow velocity"? Explain.

21. Explain the concept of a graded stream, and describe conditions that might upset the graded condition.

22. Describe or illustrate the sequence of events leading to the origin of an oxbow lake.

23. Describe stream terraces and explain how multiple levels of terraces might form adjacent to a stream.

24. What processes are responsible for deepening and widening valleys?

Points to Ponder

1. A stream 2000 m above sea level at its source flows 1500 km to the sea. What is the stream's gradient? Do you think the gradient you calculated will be correct for all segments of this stream? Explain.

2. What long-term changes may occur in the hydrologic cycle? How might human activities bring about such changes?

3. According to one estimate, 10.76 km³ of sediment is eroded from the continents each year, much of it by running water. Given that the volume of the continents above sea level is 92,832,194 km³, they should be eroded to sea level in only 8,627,527 years. Although the calculation is correct, there is something seriously wrong with this line of reasoning. What is it?

4. Calculate the daily discharge for a stream 148 m wide, 2.6 m deep, with a flow velocity of 0.3 m/sec.

5. Why is Earth the only planet with abundant liquid water?

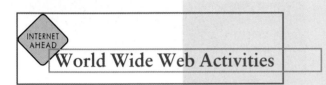

World Wide Web Activities

INTERNET AHEAD

For these Web site addresses, along with current updates and exercises, log on to
http://www.brookscole.com/geo/

➤ **U.S. GEOLOGICAL SURVEY WATER RESOURCES OF THE UNITED STATES**

This site contains a wealth of information about streams, flooding, flood forecasting, flood warnings, and groundwater.

1. Click on the *Real-Time Hydrologic Data* site under the USGS Water Resources Sites of Regional and State Offices. At that site, click on your state on the map. What is the current status of the major rivers in your state?

2. Under the Publications heading, click on *On-line Reports*. What are some of the current publications available?

3. Under the Data heading, click on the *Water-Use Data* site. Click on the *Periodic Water Fact* icon for the latest information about a new and interesting fact about water resources.

4. Under the Data heading, click on the *Water-Use Data* site. Click on the *Color maps of water use by State for 1990* icon. Click on any of the *National Water-Use Maps* sites for color maps of water usage in the United States in 1990. What was the domestic, industrial, and agricultural surface water usage for your state?

➤ **LINKS TO INFORMATION SOURCES FOR HYDROGEOLOGY, HYDROLOGY, AND ENVIRONMENTAL SCIENCES**

This site contains links to information sources for hydrogeology, hydrology, and environmental sciences. The links are listed alphabetically. Check out several sites for information about rivers, flooding, surface water usage, both in the United States and elsewhere in the world.

➤ **WW2010 (WEATHER WORLD)**

This site has descriptions of all aspects of the hydrologic cycle, including Earth's water budget, evaporation, runoff, and groundwater. It has a good summary and animation of the hydrologic cycle. This site is maintained by the Department of Atmospheric Sciences, University of Illinois.

➤ **FLOODPLAIN MANAGEMENT**

This site, maintained by the Floodplain Management Association, contains information on all aspects of flooding and floodplain management. Scroll down and click on *Flood Basins,* then go to and read *Basic Facts of Flooding*. Click on *Learning Center* and see Overview of Floods, Cause of Floods, Frequency of Floods, and How Much Risk Is Acceptable.

Chapter 16

A variety of cave deposits are present in Blanchard Springs Caverns, one of many caves in the Ozark Plateau.

OBJECTIVES

At the end of this chapter, you will have learned that

- Groundwater is one reservoir of the hydrologic cycle and represents approximately 22% of the world's supply of freshwater.
- Porosity and permeability are largely responsible for the amount, availability, and movement of groundwater.
- The water table separates the zone of aeration from the underlying zone of saturation and is a subdued replica of the overlying land surface.
- Groundwater moves downward due to the force of gravity.

- An artesian system is one in which groundwater is confined and builds up high hydrostatic pressure; there are several important artesian systems in the United States as well as many other parts of the world.
- Groundwater is an important agent of erosion and is responsible for sinkholes and karst topography in some areas.
- Groundwater is also an important agent of deposition and is responsible for a variety of cave features.

- Modifications of the groundwater system may result in lowering of the water table, saltwater incursion, subsidence, and contamination.
- Hot springs and geysers result when groundwater is heated, typically in regions of recent volcanic activity.
- Geothermal energy is a desirable and relatively nonpolluting alternate form of energy.

Groundwater

Prologue

For more than two weeks in February 1925, Floyd Collins, an unknown farmer and cave explorer, became a household word (Figure 16.1). News about attempts to rescue him from a narrow underground fissure near Mammoth Cave, Kentucky, captured the attention of the nation. Unfortunately, the story ended in tragedy when 17 days after he became trapped, rescue workers recovered his body.

The saga of Floyd Collins is rooted in what is known as the Great Cave War of Kentucky. The western region of Kentucky is riddled with caves formed by groundwater weathering and erosion. Many of them were developed as tourist attractions to help supplement meager farm earnings. The largest and best known is Mammoth Cave (see Perspective 16.1 on page 502). So spectacular is Mammoth, with its numerous caverns, underground rivers, and dramatic cave deposits, that it soon became the standard by which all other caves were measured.

As Mammoth Cave drew more and more tourists, rival cave owners became increasingly bold in attempting to lure visitors to their caves and curio shops. Signs pointing the way to Mammoth Cave frequently disappeared, while "official" cave information booths redirected unsuspecting tourists away from Mammoth Cave. It was in this environment that Floyd Collins grew up.

Seven years before his tragic death, Collins had discovered Crystal Cave on the family farm and opened it up for visitors. But like most of the caves in the area, Crystal Cave attracted few tourists—they visited Mammoth Cave instead. During one of his visits to Crystal Cave, Collins was trapped for about 20 hours before being rescued by his brothers. And on another occasion he was temporarily trapped when he jumped into a pit he could not climb back out of until he piled up enough stones to reach some handholds.

Perhaps it was the thought of discovering a cave rivaling Mammoth or even connecting to it that drove Collins to his fateful exploration of Sand Cave on January 30, 1925. As he inched his way back up through a narrow fissure he had crawled down, he dislodged from the ceiling a small oblong piece of limestone that immediately pinned his left ankle. Try as he might, he was trapped in total darkness 17 m below ground. As he lay half on his left side, Collins's left arm was partially wedged under him, while his right arm was held fast by an overhanging ledge. During his struggles to free himself, he dislodged enough silt and small rocks to bury his legs, further immobilizing him and adding to his anguish.

The next day, several neighbors reached Collins and were able to talk to him, feed him, encourage him, and try to make him more comfortable, but they could not get him out. Word of his plight quickly spread, and the area soon swarmed with reporters. Eventually, volunteers were able to excavate an area around Collins's upper body, but could not free his pinned legs. While an anxious country waited, rescue attempts led by Floyd's brother Homer continued.

Three days after he had become trapped, a harness was put around Collins's chest, and rescuers tried to pull him free. After numerous attempts to yank him out, workers had to abandon that plan because Collins was unable to bear the pain.

Meanwhile, at the surface a carnival-like atmosphere developed as a horde of up to 20,000 spectators converged on the scene. Moonshiners (people who illegally distill and sell liquor) set up stands and did a booming business. Unfortunately, drunks interfered with the rescue attempts and started fights. Finally, the National Guard had to be called out to maintain order. This atmosphere also led to rumors and exaggerations in some newspapers, when they reported that the 12.2-kg stone trapping Collins's leg weighed more than 1800 kg and that he was trapped 100 m underground rather than 17 m.

Two days after the attempt to pull Collins out of the fissure failed, part of the passageway used by the rescuers collapsed, sealing Collins's fate. The only hope now was to dig a vertical relief shaft from which a lateral tunnel could be dug to reach Collins. On February 16, rescuers finally reached the chamber where Collins's lifeless body lay entombed. After his body was brought out, he was buried near Crystal Cave, where his grave is appropriately marked by a beautiful stalagmite and pink granite headstone.

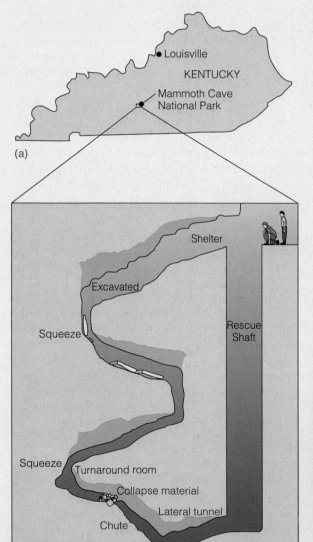

(a)

Figure 16.1
(a) Location of the cave in which Floyd Collins was trapped.
(b) Cross section showing the fissure where Collins was trapped, the rescue shaft that was sunk, and the lateral tunnel that finally reached him. (c) Collins looking out of a fissure near the cave where he ultimately died.

(b)

(c)

INTRODUCTION

Even the most casual observations make us aware of some aspects of Earth's hydrosphere. Everyone has seen precipitation in one form or another, and all are aware of running water in channels and many have seen water in the oceans and perhaps even frozen in glaciers. Few, however, have much experience with what is called **groundwater,** water filling open spaces in rocks, sediment, and soil beneath the surface. As a natural resource it is just as important as surface water, and its recovery is essential to many human endeavors.

Groundwater rights have always been significant in North America, especially in semiarid regions where surface water is insufficient to maintain agricultural productivity. Many legal battles have resulted from claims and counterclaims about groundwater rights. Groundwater also played a pivotal role in the development of the U.S. railway system during the nineteenth century, when locomotives needed a reliable source of water, much of it coming from groundwater tapped by wells.

Today, the study of groundwater and its movement has become increasingly important as the demand for freshwater by agricultural, industrial, and domestic users has reached an all-time high. More than 65% of the groundwater used in the United States each year goes for irrigation, with industrial use second, followed by domestic needs. These demands have severely depleted the groundwater supply in many areas and have led to such problems as ground subsidence and salt-water contamination. In other areas, pollution from landfills, toxic waste, and agriculture has rendered the groundwater supply unsafe.

As the world's population and industrial development expand, the demand for water, particularly groundwater, will increase. Not only is it important to locate new groundwater sources, but, once found, these sources must be protected from pollution and managed properly to ensure that users do not withdraw more water than can be replenished.

GROUNDWATER AND THE HYDROLOGIC CYCLE

Groundwater is one reservoir in the hydrologic cycle, representing approximately 22% (8.4 million km^3) of the world's supply of freshwater (see Figure 15.6). Just as for all other water in the hydrologic cycle, the ultimate source of groundwater is the oceans, but its more immediate source is the precipitation that infiltrates the ground and seeps down through the voids in soil, sediment, and rocks. It might also come from water infiltrating from streams, lakes, swamps, artificial recharge ponds, and water treatment systems.

Regardless of its source, groundwater moving through the tiny openings between soil and sediment particles and the spaces in rocks filters out many impurities such as disease-causing microorganisms and many pollutants. Not all soils and rocks are good filters, though, and sometimes so much undesirable material may be present that it contaminates the groundwater. Groundwater movement and its recovery at wells depends on two critical aspects of the materials it moves through—*porosity* and *permeability*.

WHAT PROPERTIES OF EARTH MATERIALS ALLOW THEM TO ABSORB WATER?

Porosity and permeability are important physical properties of Earth materials and are largely responsible for the amount, availability, and movement of groundwater. Water soaks into the ground because soil, sediment, or rock has open spaces or pores. **Porosity** is simply the percentage of a material's total volume that is pore space. Whereas porosity most often consists of the spaces between particles in soil, sediments, and sedimentary rocks, other types of porosity include cracks, fractures, faults, and vesicles in volcanic rocks (Figure 16.2).

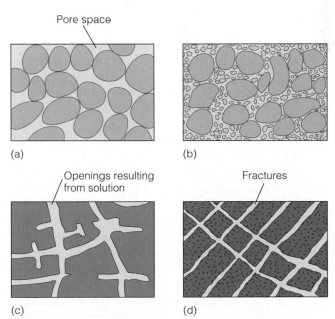

Figure 16.2
A rock's porosity is dependent on the size, shape, and arrangement of the material composing the rock. (a) A well-sorted sedimentary rock has high porosity, whereas (b) a poorly sorted one has lower porosity. (c) In soluble rocks such as limestone, porosity can be increased by solution, whereas (d) crystalline metamorphic and igneous rocks can be rendered porous by fracturing.

Porosity varies among different rock types and depends on the size, shape, and arrangement of the material composing the rock (Table 16.1). Most igneous and metamorphic rocks, as well as many limestones and dolostones, have very low porosity because they are composed of tightly interlocking crystals. Their porosity can be increased, however, if they have been fractured or weathered by groundwater. This is particularly true for massive limestone and dolostone whose fractures can be enlarged by acidic groundwater.

By contrast, detrital sedimentary rocks composed of well-sorted and well-rounded grains can have high porosity because any two grains touch only at a single point, leaving relatively large open spaces between the grains (Figure 16.2a). Poorly sorted sedimentary rocks, on the other hand, typically have low porosity because smaller grains fill in the spaces between the larger grains, further reducing porosity (Figure 16.2b). In addition, the amount of cement between grains can also decrease porosity.

Porosity determines the amount of groundwater Earth materials can hold, but it does not guarantee that the water can be easily extracted. So in addition to being porous, Earth materials must also have the capacity to transmit fluids, a property known as **permeability**. Thus, both porosity and permeability play important roles in groundwater movement and recovery. Permeability is dependent not only on porosity but also on the size of the pores or fractures and their interconnections. For example, deposits of silt or clay are typically more porous than sand or gravel, but they have low permeability because the pores between the particles are very small, and the molecular attraction between the particles and water is great, thereby preventing movement of the water. In contrast, pore spaces between grains in sandstone and conglomerate are much larger, and the molecular attraction on the water is therefore low. Chemical and biochemical sedimentary rocks, such as limestone and dolostone, and many igneous and metamorphic rocks that are highly fractured can also be very permeable provided that the fractures are interconnected.

The contrasting porosity and permeability of familiar substances is well demonstrated by sand versus clay. Simply pour some water on sand and it rapidly sinks in, whereas water poured on clay simply remains on the surface. Furthermore, wet sand dries quickly, but once clay absorbs water, it may take days to dry out, because of its low permeability. Neither sand nor clay makes a good substance in which to grow crops or gardens, but a mixture of the two plus some organic matter in the form of humus makes an excellent soil for farming and gardening (see Chapter 5).

A permeable layer transporting groundwater is an **aquifer,** from the Latin *aqua* meaning "water." The most effective aquifers are deposits of well-sorted and

Table 16.1

Porosity Values for Different Materials

Material	Percentage Porosity
Unconsolidated sediment	
Soil	55
Gravel	20–40
Sand	25–50
Silt	35–50
Clay	50–70
Rocks	
Sandstone	5–30
Shale	0–10
Solution activity in limestone, dolostone	10–30
Fractured basalt	5–40
Fractured granite	10

Source: U.S. Geological Survey, Water Supply Paper 2220 (1983) and others.

well-rounded sand and gravel. Limestones in which fractures and the surfaces between layers have been enlarged by solution are also good aquifers. Shales and many igneous and metamorphic rocks make poor aquifers because they are typically impermeable, unless fractured, and rocks and any other materials that prevent the movement of groundwater are **aquicludes.**

WHAT IS THE WATER TABLE?

Some of the precipitation on land evaporates, and some enters streams and returns to the oceans by surface runoff, and the remainder seeps into the ground. As this water moves down from the surface, a small amount adheres to the material it moves through and halts its downward progress. With the exception of this *suspended water,* however, the rest seeps further downward and collects until it fills all the available pore spaces. Thus, two zones are defined by whether their pore spaces contain mostly air, the **zone of aeration,** and the underlying **zone of saturation.** The surface separating these two zones is the **water table** (Figure 16.3). The base of the zone of saturation varies from place to place but usually extends to a depth where an impermeable layer is encountered or to a depth where confining pressure closes all open space. Extending irregularly upward a few centimeters to several meters from the zone of saturation is the *capillary fringe.* Water moves upward in this region because of surface tension, much as water moves upward through a paper towel.

In general, the configuration of the water table is a subdued replica of the overlying land surface; that is, it

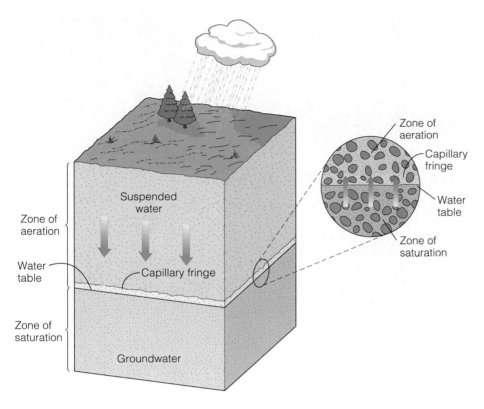

Figure 16.3
The zone of aeration contains both air and water within its open spaces, whereas all open spaces in the zone of saturation are filled with groundwater. The water table is the surface separating the zones of aeration and saturation. Within the capillary fringe, water rises by surface tension from the zone of saturation into the zone of aeration.

rises beneath hills and has its lowest elevations beneath valleys. Several factors contribute to the surface configuration of a region's water table. These include regional differences in the amount of rainfall, permeability, and the rate of groundwater movement. During periods of high rainfall, groundwater tends to rise beneath hills because it cannot flow fast enough into the adjacent valleys to maintain a level surface. During droughts, the water table falls and tends to flatten out because it is not being replenished. In arid and semiarid regions, however, the water table is usually quite flat regardless of the overlying land surface and is below the level of river valleys.

HOW DOES GROUNDWATER MOVE?

Gravity provides the energy for the downward movement of groundwater. Water entering the ground moves through the zone of aeration to the zone of saturation (Figure 16.4). When water reaches the water table, it continues to move through the zone of saturation from areas where the water table is high toward areas where it is lower, such as streams, lakes, or swamps. Only some of the water follows the direct route along the slope of the water table. Most of it takes

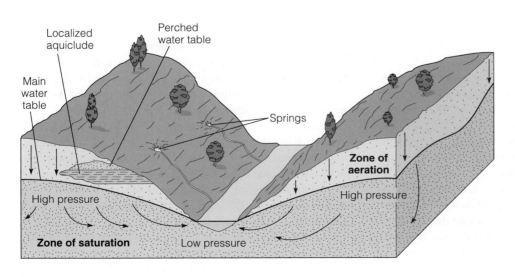

Figure 16.4
Groundwater moves down through the zone of aeration to the zone of saturation. There, some of it moves along the slope of the water table, and the rest moves through the zone of saturation from areas of high pressure toward areas of low pressure. Some water might collect over a local aquiclude, such as a shale layer, thus forming a perched water table.

Figure 16.5
Storm Water Basin No. 129, Garden City Park, Long Island, New York, is one of many recharge basins operated by the Nassau County Department of Public Works.

longer curving paths down and then enters a stream, lake, or swamp from below, because it moves from areas of high pressure toward areas of lower pressure within the saturated zone.

Groundwater velocity varies greatly and depends on many factors. Velocities range from 250 m per day in some extremely permeable material to less than a few centimeters per year in nearly impermeable material. In most ordinary aquifers, the average velocity of groundwater is a few centimeters per day.

WHAT ARE SPRINGS, WATER WELLS, AND ARTESIAN SYSTEMS?

We can think of the water in the zone of saturation much like a reservoir whose surface rises or falls depending on additions versus natural and artificial withdrawals. *Recharge*—that is, additions to the zone of saturation—comes from rainfall or melting snow, or water might be added artificially at wastewater-treatment plants or recharge ponds constructed for just this purpose (Figure 16.5). But if groundwater is discharged naturally or withdrawn at wells without sufficient recharge, the water table drops, just as a savings account diminishes if withdrawals exceed deposits. Withdrawals from the groundwater system take place where groundwater flows laterally into streams, lakes, or swamps, where it discharges at the surface as *springs,* and where it is withdrawn from the system at water wells.

Springs

Places where groundwater flows or seeps out of the ground as **springs** have always fascinated people. The water flows out of the ground for no apparent reason and from no readily identifiable source. So it is not surprising that springs have long been regarded with superstition and revered for their supposed medicinal value and healing powers. Nevertheless, there is nothing mystical or mysterious about springs.

Although springs can occur under a wide variety of geologic conditions, they all form in basically the same way (Figure 16.6). When percolating water reaches the water table or an impermeable layer, it flows laterally, and if this flow intersects the surface, the water discharges as a spring (Figure 16.7). The Mammoth Cave area in Kentucky is underlain by fractured limestones whose fractures have been enlarged into caves by solution activity (see Perspective 16.1). In this geologic environment, springs occur where the fractures and caves intersect the ground surface, allowing groundwater to exit onto the surface. Most springs are along valley walls where streams have cut valleys below the regional water table.

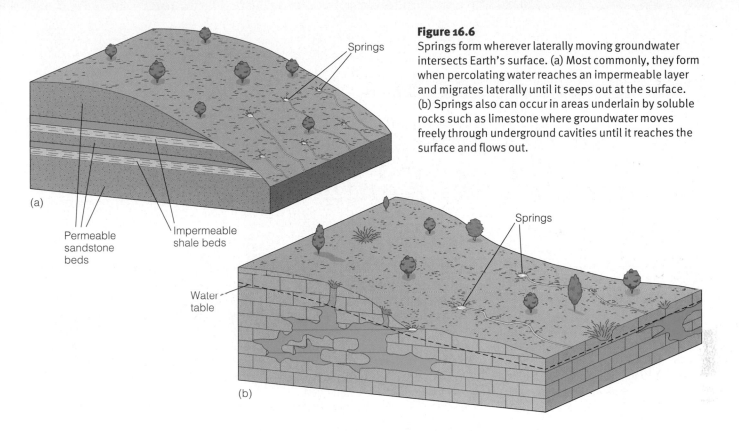

Figure 16.6
Springs form wherever laterally moving groundwater intersects Earth's surface. (a) Most commonly, they form when percolating water reaches an impermeable layer and migrates laterally until it seeps out at the surface. (b) Springs also can occur in areas underlain by soluble rocks such as limestone where groundwater moves freely through underground cavities until it reaches the surface and flows out.

(a)

Permeable sandstone beds

Impermeable shale beds

Springs

Springs

Water table

(b)

Springs can also develop wherever a perched water table intersects the surface (Figure 16.4). A *perched water table* may occur wherever a local aquiclude is present within a larger aquifer, such as a lens of shale within sandstone. As water migrates through the zone of aeration, it is stopped by the local aquiclude, and a localized zone of saturation "perched" above the main water table forms. Water moving laterally along the perched water table may intersect the surface to produce a spring.

Water Wells

Water wells are simply openings made by drilling or digging down into the zone of saturation. Once the zone of saturation has been penetrated, water percolates into the well, filling it to the level of the water table. A few wells are free flowing (see the next section), but at most the water must be brought to the surface. In some parts of the world, water is raised to the surface with nothing more than a bucket on a rope or a hand-operated pump.

(a)

(b)

Figure 16.7
Springs. (a) This spring issues from rocks at the base of the mountain in the background. (b) A good view of surface runoff and the discharge of groundwater at Burney Falls State Park, California. The water on either side of the falls that discharges from the cliff face flows through a porous zone between lava flows.

Perspective 16.1

Mammoth Cave National Park, Kentucky

Within the limestone region of western Kentucky lies the largest cave system in the world. In 1941 approximately 51,000 acres were set aside and designated as Mammoth Cave National Park. In 1981 it became a World Heritage Site.

From ground level, the topography of the area is unimposing with numerous sinkholes, lakes, valleys, and disappearing streams. Beneath the surface are more than 540 km of interconnecting passageways whose spectacular geologic features have been enjoyed by numerous cave explorers and tourists.

Based on carbon 14 dates from some of the many artifacts found in the cave (such as woven cord and wooden bowls), Mammoth Cave had been explored and used by Native Americans for about 4000 years prior to its rediscovery in 1799 by a bear hunter named Robert Houchins. During the War of 1812, approximately 180 metric tons of saltpeter (a potassium nitrate mineral), used in the manufacture of gunpowder, were mined from Mammoth Cave. At the end of the war, the saltpeter market collapsed, and Mammoth Cave was developed as a tourist attraction, easily

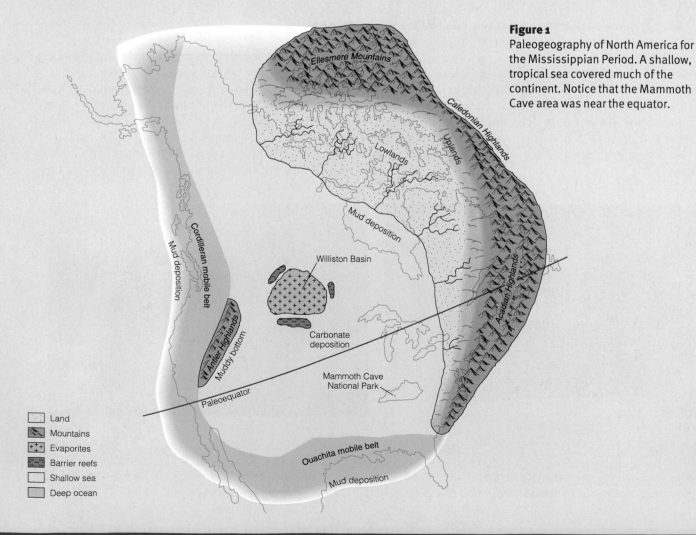

Figure 1
Paleogeography of North America for the Mississippian Period. A shallow, tropical sea covered much of the continent. Notice that the Mammoth Cave area was near the equator.

Land
Mountains
Evaporites
Barrier reefs
Shallow sea
Deep ocean

overshadowing the other caves in the area. During the next 150 years, the discovery of new passageways and caverns helped establish Mammoth Cave as the world's premier cave and the standard against which all others were measured.

Although Mammoth Cave formed in much the same way as other caves (see Figure 16.17), its history actually began during the Mississippian Period (320 to 360 million years ago). During the Mississippian, much of North America was submerged beneath a shallow, tropical sea in which countless skeletons of marine organisms accumulated, forming what we now call the St. Genevieve Limestone (Figure 1). During the following Pennsylvanian Period (286 to 320 million years ago), the area was largely above sea level and the site

of sand deposition in streams flowing from the north.

The actual formation of caves began rather recently, geologically speaking, perhaps 3 million years ago. In any event, chemical weathering (solution) and erosion of the St. Genevieve Limestone by groundwater yielded a complex network of openings, passageways, and huge chambers that constitute present-day Mammoth Cave. Flowing through the various caverns is the Echo River, a system of streams that eventually joins the Green River at the surface.

The colorful cave deposits are the primary reason millions of tourists have visited Mammoth Cave. Here can be seen numerous cave deposits, as well as spectacular travertine flowstone deposits (Figure 2). Most of the cave

deposits are composed of calcium carbonate ($CaCO_3$) precipitated from solution in the form of icicle-like structures (stalactites) hanging from the ceiling, pillarlike structures built up from the floor (stalagmites), and smooth curtains of the same material sometimes referred to as *flowstone*. In addition, in some of the drier parts of the cave system beautiful gypsum ($CaSO_4 \cdot 2H_2O$) flowers are found (Figure 2). Other attractions include the Giant's Coffin, a 15-m collapsed block of limestone, and giant rooms such as Mammoth Dome, which is about 58 m high. The cave is also home to more than 200 species of insects and other animals, including about 45 blind species; some of these can be seen on the Echo River Tour, which conveys visitors 5 km along the underground stream.

Figure 2
(a) Frozen Niagara is a spectacular example of a travertine flowstone deposit. (b) These beautiful gypsum flowers formed in a drier part of Mammoth Cave.

(a)

(b)

(a)

(b)

Figure 16.8
(a) During the past, windmills like this one were used extensively to pump water to the surface. This one still functions and supplies the water for the small cemetery in the foreground. (b) Electric pumps such as this one in a farmer's field have largely replaced windmills.

In many parts of the United States and Canada, one can see windmills from times past that employed wind power to pump water. Most of these are no longer in use, having been replaced by more efficient electric pumps (Figure 16.8).

When a well is pumped, the water table in the area around the well is lowered because water is removed from the aquifer faster than it can be replenished. A **cone of depression** thus forms around the well, varying in size according to the rate and amount of water being withdrawn (Figure 16.9). If water is pumped out of a well faster than it can be replaced, the cone of depression grows until the well goes dry. This lowering of the water table normally does not pose a problem for the average domestic well, provided that the well is drilled sufficiently deep into the zone of saturation. The tremendous amounts of water used by industry and irrigation, however, may create a large cone of depression that lowers the water table sufficiently to cause shallow wells in the immediate area to go dry (Figure 16.9). This situation is not uncommon and frequently results in lawsuits by the owners of the shallow dry wells. Furthermore, lowering of the regional water table is becoming a serious problem in many areas, particularly in the southwestern United States where rapid growth has placed tremendous demands on the groundwater system. Unrestricted withdrawal of groundwater cannot continue indefinitely, and the rising costs and decreasing supply of groundwater should soon limit the growth of this region of the United States.

People in rural areas and those without access to a municipal water system are well aware of the problems of locating an adequate groundwater supply. The distribution and type of rocks present, their porosity and permeability, fracture patterns, and so on are all factors that determine whether a water well will be successful (Figure 16.10).

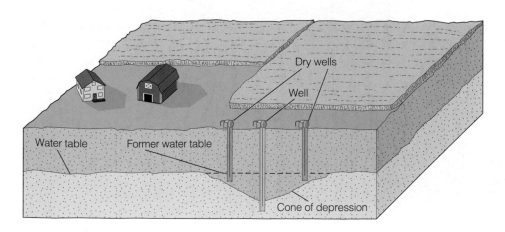

Figure 16.9
A cone of depression forms whenever water is withdrawn from a well. If water is withdrawn faster than it can be replenished, the cone of depression will grow in depth and circumference, lowering the water table in the area and causing nearby shallow wells to go dry.

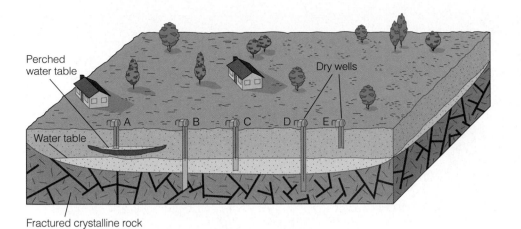

Perched water table

Water table

Dry wells

A B C D E

Fractured crystalline rock

Figure 16.10
Many factors determine whether a water well will be successful. Wells A and E were drilled to the same depth. Well A was successful because it tapped a perched water table, whereas well E did not. To be successful, it will have to be drilled below the water table like well C. Well B tapped a fracture below the water table and was successful, whereas well D missed the fractures and was dry.

Artesian Springs and Wells

The word *artesian* comes from the French town and province of Artois (called Artesium during Roman times) near Calais, where the first European artesian well was drilled in 1126 and is still flowing today. The term **artesian system** can be applied to any system in which groundwater is confined and builds up high hydrostatic (fluid) pressure. Water in such a system is able to rise above the level of the aquifer if a well is drilled through the confining layer, thereby reducing the pressure and forcing the water upward (Figure 16.11). For an artesian system to develop, three geologic conditions must be present (Figure 16.12): (1) The aquifer must be confined above and below by aquicludes to prevent water from escaping; (2) the rock sequence is usually tilted and exposed at the surface, enabling the aquifer to be recharged; and (3) there is sufficient precipitation in the recharge area to keep the aquifer filled.

The elevation of the water table in the recharge area and the distance of the well from the recharge area determine the height to which artesian water rises in a well. The surface defined by the water table in the recharge area, called the *artesian-pressure surface,* is indicated by the sloping dashed line in Figure 16.12. If there were no friction in the aquifer, well water from an artesian aquifer would rise exactly to the elevation of the artesian-pressure surface. Friction, however, slightly reduces the pressure of the aquifer water and consequently the level to which artesian water rises. This is why the pressure surface slopes.

An artesian well will flow freely at the ground surface only if the wellhead is at an elevation below the artesian-pressure surface. In this situation, the water flows out of the well because it rises toward the artesian-pressure surface, which is at a higher elevation than the wellhead (Figure 16.11). In a nonflowing artesian well, the wellhead is above the artesian-pressure surface, and thus the water will rise in the well only as high as the artesian-pressure surface.

In addition to artesian wells, many artesian springs also exist. Such springs form if a fault or fracture intersects the confined aquifer, allowing water to rise above the aquifer. Oases in deserts are commonly artesian springs.

Because the geologic conditions necessary for artesian water can occur in a variety of ways, artesian systems are common in

Figure 16.11
Artesian well at Deep Well Ranch, South Fork of the Madison River, Gallatin County, Montana.

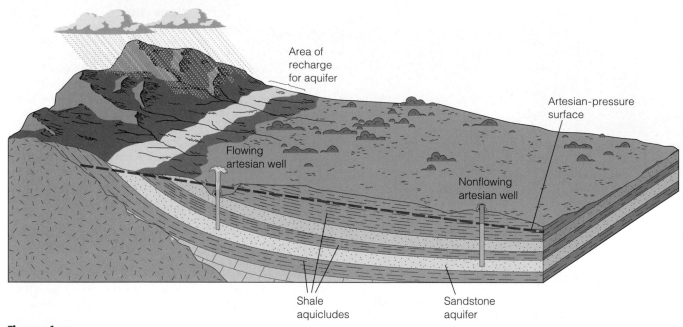

Figure 16.12
An artesian system must have an aquifer confined above and below by aquicludes, the aquifer must be exposed at the surface, and there must be sufficient precipitation in the recharge area to keep the aquifer filled. The elevation of the water table in the recharge area, which is indicated by a sloping dashed line (the artesian-pressure surface), defines the highest level to which well water can rise. If the elevation of a wellhead is below the elevation of the artesian-pressure surface, the well will be free-flowing because the water will rise toward the artesian-pressure surface, which is at a higher elevation than the wellhead. If the elevation of a wellhead is at or above that of the artesian-pressure surface, the well will be nonflowing.

many areas of the world underlain by sedimentary rocks. One of the best-known artesian systems in the United States underlies South Dakota and extends southward to central Texas. The majority of the artesian water from this system is used for irrigation. The aquifer of this artesian system, the Dakota Sandstone, is recharged where it is exposed along the margins of the Black Hills of South Dakota. The hydrostatic pressure in this system was originally great enough to produce free-flowing wells and to operate waterwheels. The extensive use of water for irrigation over the years has reduced the pressure in many of the wells so that they are no longer free-flowing and the water must be pumped.

Another example of an important artesian system is the Floridan aquifer system. Here Tertiary-aged carbonate rocks are riddled with fractures, caves, and other openings that have been enlarged and interconnected by solution activity. These carbonate rocks are exposed at the surface in the northwestern and central parts of the state where they are recharged, and they dip toward both the Atlantic and Gulf coasts where they are covered by younger sediments. The carbonates are interbedded with shales forming a series of confined aquifers and aquicludes. This artesian system is tapped in the southern part of the state where it is an important

source of freshwater and one that is being rapidly depleted.

One final comment on artesian wells/springs and the water derived from them. It is not unusual for advertisers to tout the quality of artesian water as somehow superior to other groundwater. Some artesian water might in fact be of excellent quality, but its quality is not dependent on the fact that water rises above the surface of an aquifer. Rather, its quality is a function of dissolved minerals and any introduced substances, so it really is no different from any other groundwater. The myth of its superiority probably arises from the fact that people have always been fascinated by water that flows freely from the ground.

WHAT WOULD YOU DO?

Your well is going dry periodically and never produces the amount you would like. Is your well too shallow, drilled into Earth materials with low permeability, or are nearby pumps used for irrigation causing your problems? What would you do to resolve this problem?

Guest Essay

Chris Weiland

Restoring Earth

Chris Weiland is a consultant in Las Vegas, Nevada.

Soils and near-surface rocks are one of the main pathways that pollutants may follow after a spill or leak occurs. Pollutants can migrate to groundwater supplies, infiltrate surface waters, or even affect crops. Because of this, the need for geologists and hydrogeologists is critical when cleaning up environmental contamination is required.

Geologists perform important work in three major areas of environmental remediation. They are instrumental in determining where contaminants are and how they are moving through the soils and bedrock. Geologists then use computer models to help assess where (and how fast) pollutants may migrate and, with risk assessors, whether they will pose a risk to people or ecosystems. Finally, geologists work with engineers to help design and implement remedial actions.

The discovery of environmental contamination starts an oftentimes lengthy process of investigation, planning, and remediation. Geologists and hydrogeologists team with chemists, biologists, and environmental scientists to investigate the type of the contaminants and their path through the soil. The team may use geophysics, air photo interpretation, remote sensing, soil-vapor monitoring, drilling and sampling, and well installation to find where contaminants have gone and how they are traveling. During the course of my work, I have drilled soil borings and installed wells in factories, backyards, pastures and farm fields, the desert, the Everglades, and temperate forests. Many of these sites were selected on the basis of my interpretations of surface geology, geomorphology, and human impacts from air photos or from site-reconnaissance surveys. This is, for me, the most interesting part of the job because it involves field work and travel. I also get to deal with a wide variety of people, from scientists and engineers to farmers and ranchers to worried homeowners.

Next comes the process of estimating the risk the contaminants pose to people, animals, and ecosystems. Is there immediate risk of exposure, or will exposures occur later because the pollutants reach water supplies? Are the pollutants likely to create immediate health hazards from exposures to small doses, or will health impacts require long exposure? Answers to these questions determine whether responses to the release must be immediate or if there is time to plan and construct a permanent remedial solution. To get these answers, we work with computer models of groundwater flow and contaminant transport to estimate the most likely routes and rates of movement. With toxicologists, chemists, and ecologists, we look at every possible way that contaminants could affect people or surface ecosystems.

Restoration of a contaminated area involves designing a system that will either treat the contaminants in place or remove the contaminated soil or groundwater for surface treatment. Geologists play an important role in determining how best to treat the contaminants. If in-place treatment is the best way to reduce contamination, geologists and hydrogeologists will be responsible for designing the well or gallery systems needed to reach the polluted area. If treatment on the surface is best, geologists and hydrogeologists will assist engineers in designing the means to extract soils or groundwater for treatment. The aim of restoration is to reduce or eliminate toxicity without creating greater hazards. From a health-risk basis and because of the regulations governing hazardous wastes, it is often easiest and best to fix or destroy the contaminants in the soil. There are many old and new technologies available to do this, but one of the most exciting is the use of existing soil microbes to metabolize organic compounds. Called *bioremediation,* it has been known for a long time that this process works in shallow soils, but recent discoveries of microbes in deep bedrock holds out the possibility that we may be able to enhance these processes to treat deep contaminants as well. I have been involved in two restoration projects that used bioremediation to destroy solvents or fuels in place. Bioremediation often occurs naturally but at a slow pace. We designed groundwater removal and air-injection systems to oxygenate large volumes of contaminated soil, which thus speeds the destruction of contaminants. The sites we worked on were cleaned up in a matter of one to two years from the time we installed the aeration systems. While this seems long, it is much faster than strictly natural destruction and even many engineered restorations, and it disturbs the surface much less than on-surface treatment systems.

Earth can go a long way toward cleaning up itself if the environmental insult is stopped and if it is not too grievous. It is very gratifying for me to participate in helping this process proceed. ∎

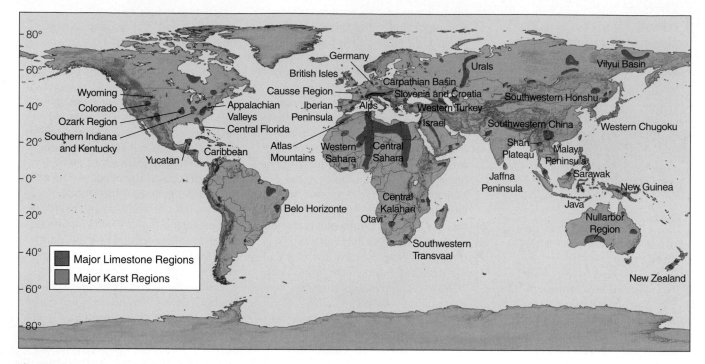

Figure 16.13
The distribution of the major limestone and karst areas of the world.

HOW DOES GROUNDWATER ERODE AND DEPOSIT MATERIAL?

When rainwater begins seeping into the ground, it immediately starts to react with the minerals it contacts, weathering them chemically. In an area underlain by soluble rock, groundwater is the principal agent of erosion and is responsible for the formation of many important features of the landscape.

Limestone, a common sedimentary rock composed primarily of the mineral calcite ($CaCO_3$), underlies large areas of Earth's surface (Figure 16.13). Although limestone is practically insoluble in pure water, it readily dissolves if a small amount of acid is present. Carbonic acid (H_2CO_3) is a weak acid that forms when carbon dioxide combines with water ($H_2O + CO_2 \rightarrow H_2CO_3$) (see Chapter 5). Because the atmosphere contains a small amount of carbon dioxide (0.03%) and carbon dioxide is also produced in soil by the decay of organic matter, most groundwater is slightly acidic. When groundwater percolates through the various openings in limestone, the slightly acidic water readily reacts with the calcite to dissolve the rock by forming soluble calcium bicarbonate, which is carried away in solution (see Chapter 5).

Sinkholes and Karst Topography

In regions underlain by soluble rock, the ground surface may be pitted with numerous depressions that vary in size and shape. These depressions, called **sinkholes** or merely *sinks,* mark areas with underlying soluble rock (Figure 16.14). Most sinkholes form in one of two ways. The first is when soluble rock below the soil is dissolved by seeping water, and openings in the rock are enlarged and filled in by the overlying soil. As the groundwater continues to dissolve the rock, the soil is eventually removed, leaving shallow depressions with gently sloping sides. When adjacent sinkholes merge, they form a network of larger, irregular, closed depressions called *solution valleys.*

Sinkholes also form when a cave's roof collapses, usually producing a steep-sided crater. Sinkholes formed in this way are a serious hazard, particularly in populated areas. In regions prone to sinkhole formation, the depth and extent of underlying cave systems must be mapped before any development to ensure that the underlying rocks are thick enough to support planned structures.

Karst topography, or simply *karst,* develops largely by groundwater erosion in many areas underlain by soluble rocks (Figure 16.15). The name *karst* is derived from the plateau region of the border area of Slovenia,

(a)

Figure 16.14
(a) This sinkhole formed on May 8 and 9, 1981, in Winter Park, Florida. It formed in previously dissolved limestone following a drop in the water table. The 100-m-wide, 35-m-deep sinkhole destroyed a house, numerous cars, and a municipal swimming pool. (b) A small sinkhole in Montana now occupied by a lake. The water enters the lake from a hot spring so it remains warm year round. In fact, tropical fish have been introduced into the lake and thus live in an otherwise inhospitable climate.

(b)

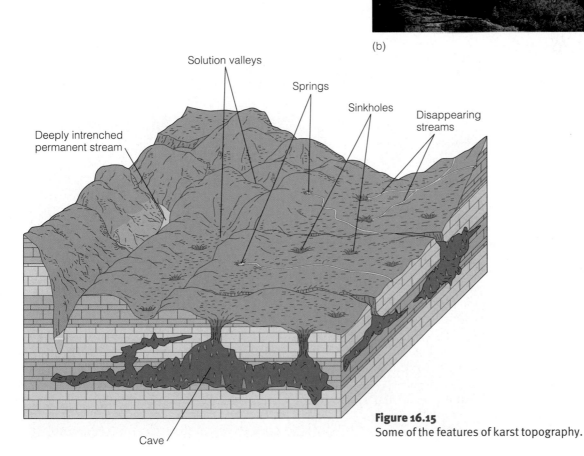

Figure 16.15
Some of the features of karst topography.

Croatia, and northeastern Italy where this type of topography is well developed. In the United States, regions of karst topography include large areas of southwestern Illinois, southern Indiana, Kentucky, Tennessee, northern Missouri, Alabama, and central and northern Florida (Figure 16.13).

Karst topography is characterized by numerous caves, springs, sinkholes, solution valleys, and disappearing streams (Figure 16.15). *Disappearing streams* are so named because they typically flow only a short distance at the surface and then disappear into a sinkhole. The water continues flowing underground through various fractures or caves until it surfaces again at a spring or other stream. Another feature commonly encountered in karst regions is *terra rossa,* a residual red, clayey soil resulting from the solution of limestone.

Karst topography varies from the spectacular high-relief landscapes of China to the subdued and pock-marked landforms in Kentucky (Figure 16.16). Common to all karst topography, though, is the presence of thick-bedded, readily soluble rock at the surface or just below the soil, and enough water for solution activity to occur. Karst topography is therefore typically restricted to humid and temperate climates.

Caves and Cave Deposits

Caves are perhaps the most spectacular examples of the combined effects of weathering and erosion by groundwater. As groundwater percolates through carbonate rocks, it dissolves and enlarges fractures and openings to form a complex interconnecting system of crevices, caves, caverns, and underground streams. A **cave** is usually defined as a naturally formed subsurface opening that is generally connected to the surface and is large enough for a person to enter. A *cavern* is a very large cave or a system of interconnected caves.

More than 17,000 caves are known in the United States. Most of them are small, but some are quite large and spectacular. Some of the more famous ones are Mammoth Cave, Kentucky (see Perspective 16.1); Carlsbad Caverns, New Mexico; Lewis and Clark Caverns, Montana; Wind Cave and Jewel Cave, South Dakota; Lehman Cave, Nevada; and Meramec Caverns, Missouri, which Jesse James and his outlaw band often used as a

(a)

Figure 16.16
(a) The Stone Forest, 125 km southeast of Kunming, People's Republic of China, is a high-relief karst landscape formed by the dissolution of carbonate rocks.
(b) Solution valleys, sinkholes, and sinkhole lakes dominate the subdued karst topography east of Bowling Green, Kentucky. *Source: Photo courtesy of John S. Shelton.*

(b)

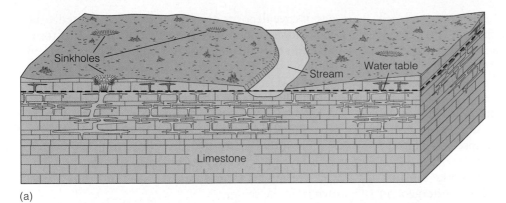

(a)

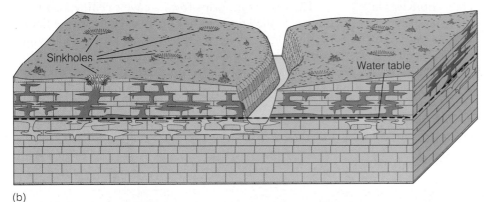

(b)

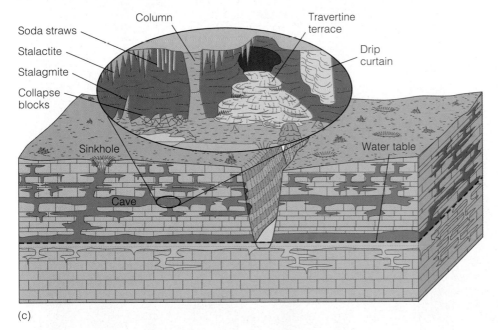

(c)

Figure 16.17
The formation of caves. (a) As groundwater percolates through the zone of aeration and flows through the zone of saturation, it dissolves the carbonate rocks and gradually forms a system of passageways. (b) Groundwater moves along the surface of the water table, forming a system of horizontal passageways through which dissolved rock is carried to the surface streams, thus enlarging the passageways. (c) As the surface streams erode deeper valleys, the water table drops, and the abandoned channelways form an interconnecting system of caves and caverns.

hideout. The United States has many famous caves, but Canada also has many caves, including 536-m-deep Arctomys Cave in Mount Robson Provincial Park, British Columbia, the deepest known cave in North America.

Caves and caverns form as a result of the dissolution of carbonate rocks by weakly acidic groundwater (Figure 16.17). Groundwater percolating through the zone of aeration slowly dissolves the carbonate rock and enlarges its fractures and bedding planes. On reaching the water table, the groundwater migrates toward the region's surface streams. As the groundwater moves through the zone of saturation, it continues to dissolve the rock and gradually forms a system of horizontal passageways through which the dissolved rock is carried to the streams. As the surface streams erode deeper valleys, the water table drops in response to the lower elevation of the streams. The water that flowed through the system of horizontal passageways now percolates to the lower water table where a new system of passageways begins to form. The abandoned channelways now form an interconnecting system of caves and caverns. Caves eventually become unstable and collapse, littering the floor with fallen debris.

Recall our discussion of disappearing streams, and that many of these streams flow underground for some distance before resurfacing. The Echo River in Mammoth Cave is a well-known example (see Perspective 16.1). Underground waterfalls are also known in a few caves. For instance, Ruby Falls plunges 44 m over a precipice in a cave beneath Lookout Mountain at Chattanooga, Tennessee.

When most people think of caves, they think of the seemingly endless variety of colorful and bizarre-shaped deposits found in them. Although a great many different types of cave deposits exist, most form in essentially the same manner and are collectively known as **dripstone.** As water seeps through a cave, some of the dissolved carbon dioxide in the water escapes, and a small amount of calcite is precipitated. In this manner, the various dripstone deposits are formed.

Stalactites are icicle-shaped structures hanging from cave ceilings that form as a result of precipitation from dripping water (Figure 16.18). With each drop of water, a thin layer of calcite is deposited over the previous layer, forming a cone-shaped projection that grows down from the ceiling.

The water that drips from a cave's ceiling also precipitates a small amount of calcite when it hits the floor. As additional calcite is deposited, an upward-growing projection called a *stalagmite* forms (Figure 16.18). If a stalactite and stalagmite meet, they form a *column.* Groundwater seeping from a crack in a cave's ceiling may form a vertical sheet of rock called a *drip curtain,* while water flowing across a cave's floor may produce *travertine terraces* (Figure 16.17).

Figure 16.18
Stalactites are the icicle-shaped structures hanging from the cave's ceiling, whereas the upward-pointing structures on the floor are stalagmites. Several columns are present where the stalactites and stalagmites have met in this chamber of Luray Caves, Virginia.

HOW DO HUMANS IMPACT THE GROUNDWATER SYSTEM, AND WHAT ARE THE EFFECTS?

Groundwater is a valuable natural resource that is rapidly being exploited with little regard to the effects of overuse and misuse. Currently, about 20% of all water used in the United States is groundwater. This percentage has remained rather constant for several decades but unless this resource is used more wisely, sufficient amounts of clean groundwater will not be available in the future. Modifications of the groundwater system may have many consequences, including (1) lowering the water table, which causes wells to dry up; (2) loss of hydrostatic pressure, which causes once free-flowing wells to require pumping; (3) saltwater incursion; (4) subsidence; and (5) contamination.

Lowering the Water Table

Withdrawing groundwater at a significantly greater rate than it is replaced by either natural or artificial recharge can have serious effects. For example, the High Plains aquifer is one of the most important aquifers in the United States. It underlies more than 450,000 km², including most of Nebraska, large parts of Colorado and Kansas, portions of South Dakota, Wyoming, and New Mexico, as well as the panhandle regions of Oklahoma and Texas, and it accounts for approximately 30% of the groundwater used for irrigation in the United States (Figure 16.19). Irrigation from the High Plains aquifer is largely responsible for the region's high agricultural productivity, which includes a significant percentage of the nation's corn, cotton, and wheat, and half of U.S. beef cattle. Large areas of land (about 18 million acres) are currently irrigated with water pumped from the High Plains aquifer. Irrigation is popular because yields from irrigated lands can be triple what they would be without irrigation.

Although the High Plains aquifer has contributed to the high productivity of the region, it cannot continue providing the quantities of water that it has in the past. In some parts of the High Plains, from 2 to 100 times more water is being pumped annually than is being recharged. Consequently, water is being removed from the aquifer faster than it is being replenished, causing the water table to drop significantly in many areas (Figure 16.19).

What will happen to this region's economy if long-term withdrawal of water from the High Plains aquifer greatly exceeds its recharge rate so that it can no longer supply the quantities of water necessary for irrigation? Solutions range from going back to farming without irrigation to diverting water from other regions such as the Great Lakes. Farming without irrigation would

result in greatly decreased yields and higher costs and prices for agricultural products, while the diversion of water from elsewhere would cost billions of dollars and the price of agricultural products would still rise.

Another excellent example of what we might call deficit spending with regard to groundwater took place in California during the drought of 1987 to 1992. During that time, the state's aquifers were overdrawn at the rate of 10 million acre/feet per year (an acre/foot is the amount of water that covers one acre one foot deep). In short, during each year of the drought California was withdrawing more than 12 km³ of groundwater than was being replaced. Unfortunately, excessive depletion of the groundwater reservoir has other consequences, such as subsidence involving sinking or settling of the ground surface (discussed in a later section).

Water supply problems certainly exist in many areas, but on the positive side, water use in the United States actually declined during the five years following 1980 and has remained nearly constant since then, even though the population has increased (Figure 16.20). This

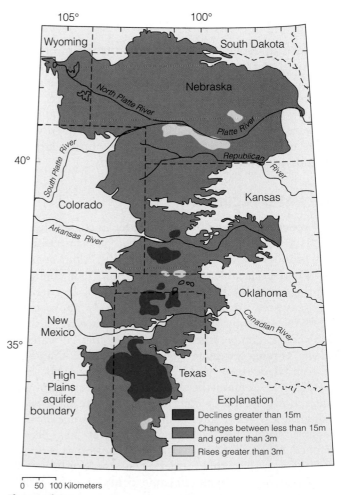

Figure 16.19
Geographic extent of the High Plains aquifer and changes in water level from predevelopment through 1993.

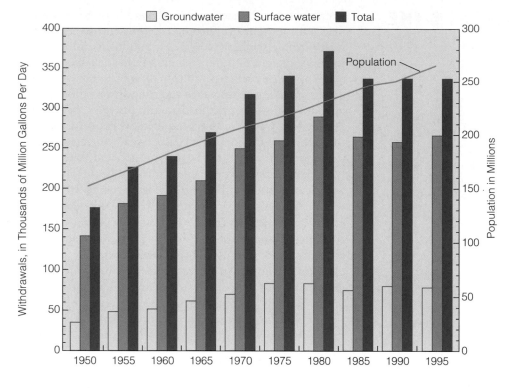

Figure 16.20
Trends in water use in the United States from 1950 to 1995. Notice the steady increase until 1980, then the decrease to the present level of use.

downturn in demand resulted largely from improved techniques in irrigation, more efficient industrial water use, and a general public awareness of water problems coupled with conservation practices. Nevertheless, the rates of withdrawal of groundwater from some aquifers still exceed their rate of recharge, and population growth in the arid to semiarid Southwest are putting larger demands on an already limited water supply.

Saltwater Incursion

The excessive pumping of groundwater in coastal areas can result in **saltwater incursion** such as occurred on Long Island, New York, during the 1960s. Along coastlines where permeable rocks or sediments are in contact with the ocean, the fresh groundwater, being less dense than seawater, forms a lens-shaped body above the underlying saltwater (Figure 16.21a). The weight of the freshwater exerts pressure on the underlying saltwater. As long as rates of recharge equal rates of withdrawal, the contact between the fresh groundwater and the seawater will remain the same. If excessive pumping occurs, however, a deep cone of depression forms in the fresh groundwater (Figure 16.21b). Because some of the pressure from the overlying freshwater has been removed, saltwater forms a *cone of ascension* as it rises to fill the pore space that formerly contained freshwater. When this occurs, wells become contaminated with saltwater and remain contaminated until recharge by freshwater restores the former level of the fresh-ground-water water table.

Saltwater incursion is a major problem in many rapidly growing coastal communities. As the population in these areas grows, greater demand for groundwater creates an even greater imbalance between recharge and withdrawal.

Not only is saltwater incursion a major concern for some coastal communities, it is also becoming a problem in the Salinas Valley, California, which produces fruits and vegetables valued at about $1.7 billion annually. Here, in an area encompassing about 160,000 acres of rich farmland 160 km south of San Francisco, saltwater incursion caused by overpumping of several shallow-water aquifers is threatening the groundwater supply used for irrigation. Because of droughts in recent years and increased domestic needs caused by a burgeoning population, overpumping has resulted in increased seepage of saltwater into the groundwater system such that large portions of some of the aquifers are now too salty even for irrigation. At some locations in the Salinas Valley, seawater has migrated more than 11 km inland during the past 13 years. If this incursion is left unchecked, the farmlands of the valley could become too salty to support most agriculture. Uncontaminated deep aquifers could be tapped to replace the contaminated shallow groundwater, but doing so is too expensive for most farmers and small towns.

To counteract the effects of saltwater incursion, recharge wells are often drilled to pump water back into the groundwater system (Figure 16.21c). Recharge ponds that allow large quantities of fresh surface water to infiltrate the groundwater supply may also be con-

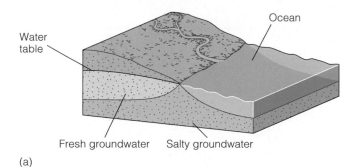

(a)

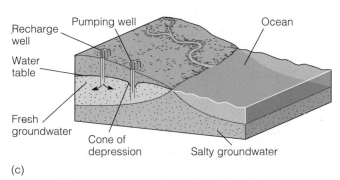

(b)

(c)

Figure 16.21
Saltwater incursion. (a) Because freshwater is not as dense as saltwater, it forms a lens-shaped body above the underlying saltwater. (b) If excessive pumping occurs, a cone of depression develops in the fresh groundwater, and a cone of ascension forms in the underlying salty groundwater that may result in saltwater contamination of the well. (c) Pumping water back into the groundwater system through recharge wells can help lower the interface between the fresh groundwater and the salty groundwater and reduce saltwater incursion.

structed (Figure 16.5). Both of these methods are successfully used on Long Island, New York, which has had a saltwater incursion problem for several decades.

Subsidence

As excessive amounts of groundwater are withdrawn from poorly consolidated sediments and sedimentary

rocks, the water pressure between grains is reduced, and the weight of the overlying materials causes the grains to pack closer together, resulting in *subsidence* of the ground. As more and more groundwater is pumped to meet the increasing needs of agriculture, industry, and population growth, subsidence is becoming more prevalent.

The San Joaquin Valley of California is a major agricultural region that relies largely on groundwater for irrigation. Between 1925 and 1977, groundwater withdrawals in parts of the valley caused subsidence of nearly 9 m (Figure 16.22). Other areas in the United States that have experienced subsidence include New Orleans, Louisiana, and Houston, Texas, both of which have subsided more than 2 m, and Las Vegas, Nevada, where 8.5 m of subsidence has taken place (Table 16.2).

Figure 16.22
The dates on this power pole dramatically illustrate the amount of subsidence in the San Joaquin Valley, California. Because of groundwater withdrawals and subsequent sediment compaction, the ground subsided nearly 9 meters between 1925 and 1977. For a time, surface water use reduced subsidence, but during the drought of 1987 to 1992 it started again as more groundwater was withdrawn.

Table 16.2

Subsidence of Cities and Regions Due Primarily to Groundwater Removal		
Location	Maximum Subsidence (m)	Area Affected (km2)
Mexico City, Mexico	8.0	25
Long Beach and Los Angeles, California	9.0	50
Taipei Basin, Taiwan	1.0	100
Shanghai, China	2.6	121
Venice, Italy	0.2	150
New Orleans, Louisiana	2.0	175
London, England	0.3	295
Las Vegas, Nevada	8.5	500
Santa Clara Valley, California	4.0	600
Bangkok, Thailand	1.0	800
Osaka and Tokyo, Japan	4.0	3,000
San Joaquin Valley, California	9.0	9,000
Houston, Texas	2.7	12,100

Source: Data from R. Dolan and H. G. Goodell, "Sinking Cities," American Scientist 74 (1986): 38–47; and J. Whittow, Disasters: The Anatomy of Environmental Hazards (Athens: University of Georgia Press, 1979).

Elsewhere in the world, the tilt of the Leaning Tower of Pisa is partly due to groundwater withdrawal (Figure 16.23). The tower started tilting soon after construction began in 1173 because of differential compaction of the foundation. During the 1960s, the city of Pisa withdrew ever-larger amounts of groundwater, causing the ground to subside further; as a result, the tilt of the tower increased until it was considered in danger of falling over. Strict control of groundwater withdrawal and stabilization of the foundation have reduced the amount of tilting to about 1 mm per year, thus ensuring that the tower should stand for several more centuries.

A spectacular example of continuing subsidence is taking place in Mexico City, which was built on a former lakebed. As groundwater is removed for the increasing needs of the city's 15.6 million people, the water table has been lowered up to 10 meters. As a result, the fine-grained lake deposits are compacting, and Mexico City is slowly and unevenly subsiding. Its opera house has settled more than 3 m, and half of the first floor is now below ground level. Other parts of the city have subsided as much as 7.5 m, creating similar problems for other structures (Figure 16.24). The fact that 72% of the city's water comes from the aquifer beneath the metropolitan area ensures that problems of subsidence will continue.

The extraction of oil can also cause subsidence. Long Beach, California, has subsided 9 m as a result of

Figure 16.23
The Leaning Tower of Pisa, Italy. The tilting is partly the result of subsidence due to the removal of groundwater.

34 years of oil production. More than $100 million of damage was done to the pumping, transportation, and harbor facilities in this area because of subsidence and encroachment of the sea (Figure 16.25). Once water was pumped back into the oil reservoir, thus stabilizing it, subsidence virtually stopped.

Groundwater Contamination

A major problem facing our society is the safe disposal of the numerous pollutant by-products of an industrialized economy. We are becoming increasingly aware that streams, lakes, and oceans are not unlimited reservoirs for waste, and that we must find new safe ways to dispose of pollutants.

The most common sources of groundwater contamination are sewage, landfills, toxic-waste-disposal sites (see Perspective 16.2), and agriculture. Once pollutants get into the groundwater system, they spread wherever groundwater travels, which can make their containment difficult. Furthermore, because groundwater moves so slowly, it takes a long time to cleanse a groundwater reservoir once it has become contaminated.

In many areas, septic tanks are the most common way of disposing of sewage. A septic tank slowly releases sewage into the ground where it is decomposed by oxidation and microorganisms and filtered by the sediment as it percolates through the zone of aeration. In most situations, by the time the water from the sewage reaches the zone of saturation, it has been cleansed of any impurities and is safe to use (Figure 16.26a). If the water table is close to the surface or if the rocks are very permeable, water entering the zone of saturation may still be contaminated and unfit to use.

Figure 16.24
Excessive withdrawal of groundwater from beneath Mexico City has resulted in subsidence and uneven settling of buildings. The right side of this church (Our Lady of Guadalupe) has settled slightly more than a meter.

Figure 16.25
The withdrawal of petroleum from the oil field in Long Beach, California, resulted in up to 9 m of ground subsidence because of sediment compaction. It was not until water was pumped back into the reservoir to replace the petroleum that ground subsidence essentially ceased. (2 to 29 feet = 0.6 to 8.8 meters)

Landfills are also potential sources of groundwater contamination (Figure 16.26b). Not only does liquid waste seep into the ground, but rainwater also carries dissolved chemicals and other pollutants down into the groundwater reservoir. Unless the landfill is carefully designed and lined with an impermeable layer such as clay, many toxic compounds such as paints, solvents, cleansers, pesticides, and battery acid will find their way into the groundwater system.

Toxic-waste sites where dangerous chemicals are either buried or pumped underground are an increasing source of groundwater contamination. The United States alone must dispose of several thousand metric tons of hazardous chemical waste per year. Unfortunately, much of this waste has been and still is being improperly dumped and is contaminating the surface water, soil, and groundwater.

Examples of indiscriminate dumping of dangerous and toxic chemicals can be found in every state. Per-haps the most famous is Love Canal, near Niagara Falls, New York. During the 1940s, the Hooker Chemical Company dumped approximately 19,000 tons of chemical waste into Love Canal. In 1953 it covered one of the dump sites with dirt and sold it for $1 to the Niagara Falls Board of Education, which built an elementary school and playground on the site. Heavy rains and snow during the winter of 1976–1977 raised the water table and turned the area into a muddy swamp in the spring of 1977. Mixed with the mud were thousands of different toxic, noxious chemicals that formed puddles in the playground, oozed into people's basements, and covered gardens and lawns. Trees, lawns, and gardens began to die, and many of the residents of the area suffered from serious illnesses. The cost of cleaning up the Love Canal site and relocating its residents will eventually exceed $100 million, and the site and neighborhood are now vacant.

Figure 16.26
(a) A septic system slowly releases sewage into the zone of aeration. Oxidation, bacterial degradation, and filtering usually remove impurities before they reach the water table. However, if the rocks are very permeable or the water table is too close to the septic system, contamination of the groundwater can result.
(b) Unless there is an impermeable barrier between a landfill and the water table, pollutants can be carried into the zone of saturation and contaminate the groundwater supply.

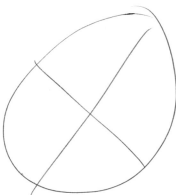

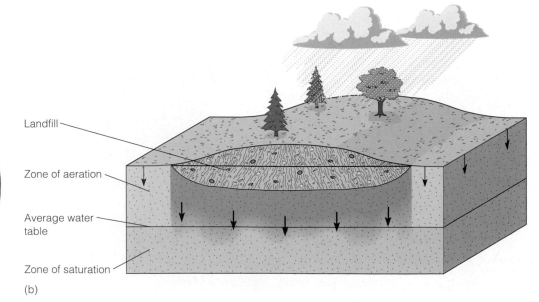

(a)

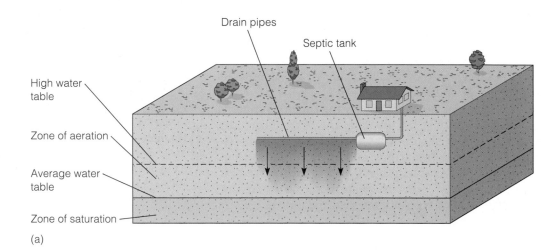

(b)

Groundwater Quality

Finding groundwater is rather easy, because it is present beneath the land surface nearly everywhere, although the depth to the water table varies considerably. But just finding water is not enough. Sufficient amounts in porous and permeable materials must be located if it is to be withdrawn for agricultural, industrial, or domestic use. The availability of groundwater was important in the westward expansion in both Canada and the United States, and now more than one-third of all water for irrigation comes from the groundwater system. More than 90% of the water used for domestic purposes in rural America and the water for a number of large cities comes from groundwater, and, as one would expect, quality is more important here than it is for most other purposes. Discounting contamination by humans from landfills, septic systems, toxic-waste sites, and industrial effluents, groundwater quality is mostly a function of (1) the kinds of materials making up an aquifer, (2) the residence time of water in an aquifer, and (3) the solubility of rocks and minerals. These factors account for the amount of dissolved materials in groundwater, such as calcium, iron, fluoride, and several others. Most pose no health problems, but some have undesirable effects such as deposition of minerals in water pipes and water heaters, offensive taste or smell, or they may stain clothing and fixtures, or inhibit the effectiveness of detergents.

Everyone has heard of *hard water* and knows that *soft water* is more desirable for most purposes. But just what is hard water, and how does it become hard? Recall from Chapter 5 that some rocks, especially limestone, composed of calcite ($CaCO_3$), and dolostone, composed of dolomite [$CaMg(CO_3)_2$], are readily soluble in water containing a small amount of carbonic acid. In fact, the amount of dissolved calcium (Ca^{+2}) and magnesium (Mg^{+2}) is what determines whether water is hard or soft. Water with less than 60 milligrams per liter (mg/L) is soft, whereas 61–120 mg/L indicate moderately hard water, values from 121 to 180 mg/L characterize hard water, and any water with more than 180 mg/L is very hard. One negative aspect of hard water is the precipitation of scale (Ca and Mg salts) in water pipes, water heaters, dishwashers, and even on glasses and dinnerware. The gray, hard scaly deposits in the bottoms of teakettles result from the same process. Furthermore, soaps and detergents do not lather well in hard water, and dirt and soap combine to form an unpleasant scum.

Hard water is a problem in many areas, especially those underlain by limestone and dolostone. To remedy this problem many households have a water softener, a device in which sodium (Na^+) replaces calcium and magnesium using an ion exchanger or a mineral sieve. In either case, the amount of calcium and magnesium is reduced and the water is more desirable for most domestic purposes. However, people on low-sodium diets, such as those with hypertension (high blood pressure), are cautioned not to drink softened water because it contains more sodium.

Iron in water is common in some areas, especially places underlain by rocks with abundant iron-bearing minerals. If present, it imparts a disagreeable taste and might leave red stains in toilets, sinks, and other plumbing fixtures, and light-colored clothing washed in iron-rich water takes on a rusty color. If the water is especially rich in iron, a thick, red slime forms in toilet tanks and must be cleaned out periodically. Even though iron poses no health threat, it is so disagreeable that most people remove it. Water softeners will remove some of it, but in many cases some kind of iron filter must be placed on the incoming water line before it reaches the softener. In any case, the problem is usually easily solved, but of course it adds another expense to household maintenance.

Not all dissolved materials in groundwater are undesirable, at least in small quantities. Fluoride (F^-), for instance, if present in amounts of 1.0 to 1.5 parts per million (ppm), combines with the calcium phosphate composing teeth and makes them more resistant to decay. Too much fluoride—more than 4.0 ppm—though, gives children's teeth a dark, blotchy appearance. Fluoride in natural waters is rare, so not many communities benefit from its presence, and artificially adding fluorine to water has met with considerable opposition because of possible health risks.

HYDROTHERMAL ACTIVITY— WHAT IS IT, AND WHERE DOES IT OCCUR?

Hydrothermal is a term referring to hot water. Some geologists restrict the meaning to encompass only water heated by magma, but here we use it to mean any hot subsurface water and surface activity resulting from its discharge. One manifestation of hydrothermal activity in areas of active or recently active volcanism is the discharge of gases, much of it as steam, at vents known as *fumeroles* (see Figure 4.3). Of more immediate concern here, however, is the groundwater that rises to the surface as *hot springs* or *geysers*. It may be heated by its proximity to magma or by Earth's geothermal gradient, because it circulates deeply.

Hot Springs

A **hot spring** (also called a *thermal spring* or *warm spring*) is any spring in which the water temperature is greater than 37°C, the temperature of the human body

Perspective 16.2

Radioactive Waste Disposal

One problem of the nuclear age is finding safe storage sites for radioactive waste from nuclear power plants, the manufacture of nuclear weapons, and the radioactive by-products of nuclear medicine. Radioactive waste can be grouped into two categories: low-level and high-level waste. Low-level wastes are low enough in radioactivity that, when properly handled, they do not pose a significant environmental threat. Most low-level wastes can be safely buried in controlled dump sites where the geology and groundwater system are well known and careful monitoring is provided.

High-level radioactive waste, such as the spent uranium-fuel assemblies used in nuclear reactors and the material used in nuclear weapons, is extremely dangerous because of high amounts of radioactivity; it therefore presents a major environmental problem. Currently, some 28,000 metric tons of spent uranium fuel are being stored in 70 "temporary" sites in 35 states, while awaiting shipment to a permanent site in Nevada. Furthermore, the Department of Energy (DOE) estimates that by the year 2000 the nation will have produced almost 50,000 metric tons of highly radioactive waste that must be disposed of safely.

In 1986 Congress chose Yucca Mountain, Nevada, as the only candidate to house the nation's ever-increasing amounts of civilian high-level radioactive waste (Figure 1). Congress also authorized the DOE to study the suitability of the site. Such a facility must be able to isolate high-level waste from the environment for at least 10,000 years, which is the minimum time the waste will remain dangerous. The Yucca Mountain site will have a capacity of 70,000 metric tons of waste and will not be completely filled until around the year 2030, at which time its entrance shafts will be sealed and backfilled (Figure 2).

The canisters holding the waste are designed to remain leakproof for at least 300 years, so there is some possibility that leakage could occur over the next 10,000 years. The DOE thinks, however, that the geology of the area will prevent radioactive isotopes from entering the groundwater system. Under an Environmental Protection Agency (EPA) regulation, a radioactive dump site must be located so that the travel time for groundwater from the site to the outside environment is at least 1000 years.

The radioactive waste at the Yucca Mountain repository will be buried in a volcanic tuff at a depth of about 300 m. The water table in the area will be an additional 200 to 420 m below the dump site. Thus, the canisters will be

Figure 1
The location and aerial view of Nevada's Yucca Mountain.

stored in the zone of aeration, which was one of the reasons Yucca Mountain was selected. Only about 15 cm of rain fall in this area per year, and only a small amount of this percolates into the ground. Most of the water that does seep into the ground evaporates before it migrates very far, so the rock at the depth the canisters are buried will be very dry, helping prolong their lives.

Geologists think that the radioactive waste at Yucca Mountain is most likely to contaminate the environment if it is in liquid form; if liquid, it could seep into the zone of saturation and enter the groundwater supply. But because of the low moisture in the zone of aeration, there is little water to carry the waste downward, and it will take more than 1000 years to reach the zone of saturation. In fact, the DOE estimates that the waste will take longer than 10,000 years to move from the repository to the water table.

Some geologists are concerned that the climate will change during the next 10,000 years. If the region should become more humid, more water will percolate through the zone of aeration. This will increase the corrosion rate of the canisters and could cause the water table to rise, thereby decreasing the travel time between the repository and the zone of saturation. This area of the

country was much more humid during the Ice Age, 1.6 million to 10,000 years ago (see Chapter 17).

Another concern is the seismic activity in the area. It is in fact riddled with faults. At least 27 earthquakes with magnitudes greater than 3 occurred in the area between 1852 and 1991. The most recent, which occurred on June 29, 1992, had a magnitude of 5.6 and an epicenter only 32 km from Yucca Mountain. Nevertheless, the DOE is convinced that earthquakes pose little danger to the underground repository itself because the disruptive effects of an earthquake are usually confined to the surface.

Finally, some suggest that the DOE has not thoroughly evaluated the economic potential of the area. Exploration around the Yucca Mountain site has indicated geologic evidence for various metals and possibly oil and gas in the area. Should human intrusion occur during the 10,000 to 100,000 years that the site is supposed to be isolated, dangerous radiation could be released into the environment. Others think the economic potential is being sufficiently evaluated and that the area in question has a low economic potential.

Although it appears that Yucca Mountain meets all the requirements for a safe high-level radioactive waste

dump, the site is still controversial. In fact, the government has spent $3 billion since 1986 studying Yucca Mountain, and dug a huge U-shaped tunnel through the site in early 1997 to ensure the subsurface geology is suitable for a high-level radioactive waste-disposal site. Nevertheless, there are those critics of the project who believe the site is not safe and should be abandoned.

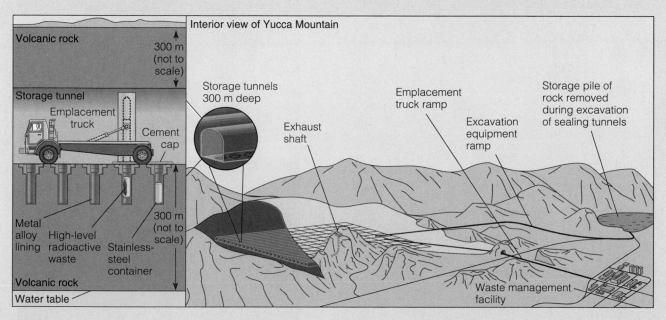

Figure 2
Schematic diagram of the proposed Yucca Mountain high-level radioactive waste dump facility.

(Figure 16.27a). Some hot springs are much hotter, with temperatures up to the boiling point in many instances (Figure 16.27b). Another type of hot spring, called a *mud pot,* results when chemically altered rocks yield clays that bubble as hot water and steam rise through them (Figure 16.27c). Of the approximately 1100 known hot springs in the United States, more than 1000 of them are in the far West, with the others in the Black Hills of South Dakota, Georgia, the Appalachian region, and the Ouachita region of Arkansas. At Hot Springs National Park in Arkansas, 47 hot springs with temperatures of 62°C are found on the slopes of Hot Springs Mountain. Hot springs are also common in other parts of the world. One of the most famous is at Bath, England, where shortly after the Roman conquest of Britain in A.D. 43, numerous bathhouses and a temple were built around the hot springs (Figure 16.28).

The heat for most hot springs comes from magma or cooling igneous rocks. The geologically recent igneous activity in the western United States accounts for the large number of hot springs in that region. The water in some hot springs, however, circulates deep into Earth, where it is warmed by the normal increase in temperature, the geothermal gradient. For example, the

Figure 16.27
Hot springs. (a) Hot spring in the West Thumb Geyser Basin, Yellowstone National Park, Wyoming. (b) The water in this hot spring at Bumpass Hell in Lassen Volcanic National Park, California, is boiling. (c) Mud pot at the Sulfur Works, also in Lassen Volcanic National Park. (d) The Park Service warns of the dangers in hydrothermal areas, but some people ignore the warnings and are injured or killed.

(a)

(b)

(c)

(d)

Figure 16.28
One of the many bathhouses in Bath, England, that were built around hot springs shortly after the Roman conquest in A.D. 43.

spring water of Warm Springs, Georgia, is heated in this manner. This hot spring was a health and bathing resort long before the Civil War; later, with the establishment of the Georgia Warm Springs Foundation, it was used to help treat polio victims.

Geysers

Hot springs that intermittently eject hot water and steam with tremendous force are known as **geysers**. The word comes from the Icelandic *geysir,* which means "to gush" or "to rush forth." One of the most famous geysers in the world is Old Faithful in Yellowstone National Park in Wyoming (Figure 16.29a). With a thunderous roar, it erupts a column of hot water and steam every 30 to 90 minutes. Other well-known geyser areas are found in Iceland and New Zealand.

Geysers are the surface expression of an extensive underground system of interconnected fractures within hot igneous rocks (Figure 16.30). Groundwater percolating down into the network of fractures is heated as it comes into contact with the hot rocks. Because the

(b)

(a)

Figure 16.29
Geysers in Yellowstone National Park, Wyoming. (a) Old Faithful Geyser erupts every 30 to 90 minutes, spewing water from 32 to 56 m high. (b) A small geyser erupting in Norris Geyser Basin.

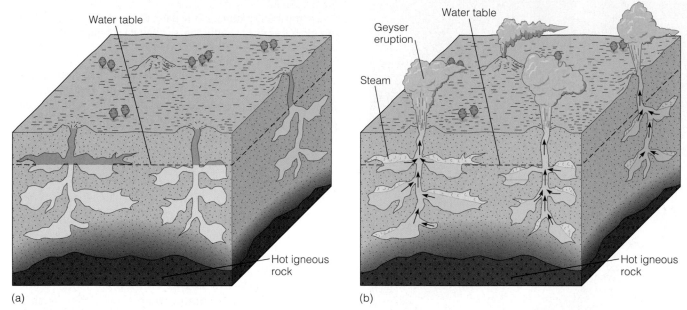

Figure 16.30

The eruption of a geyser. (a) Groundwater percolates down into a network of interconnected openings and is heated by the hot igneous rocks. The water near the bottom of the fracture system is under greater pressure than that near the top and consequently must be heated to a higher temperature before it will boil. (b) Any rise in temperature of the water above its boiling point or a drop in pressure will cause the water to change to steam, which quickly pushes the water above it up and out of the ground, producing a geyser eruption.

water near the bottom of the fracture system is under greater pressure than that near the top, it must be heated to a higher temperature before it will boil. Thus, when the deeper water is heated to near the boiling point, a slight rise in temperature or a drop in pressure, such as from escaping gas, will cause it to change instantly to steam. The expanding steam quickly pushes the water above it out of the ground and into the air, thereby producing a geyser eruption. After the eruption, relatively cool groundwater starts to seep back into the fracture system where it is heated to near its boiling temperature and the eruption cycle begins again. Such a process explains how geysers can erupt with some regularity.

Hot-spring and geyser water typically contains large quantities of dissolved minerals because most minerals dissolve more rapidly in warm water than in cold water. Due to this high-mineral content, the waters of many hot springs are believed by some to have medicinal properties. Numerous spas and bathhouses have been built at hot springs throughout the world to take advantage of these supposed healing properties.

When the highly mineralized water of hot springs or geysers cools at the surface, some of the material in solution is precipitated, forming various types of deposits. The amount and type of precipitated mineral depend on the solubility and composition of the material that the groundwater flows through. If the groundwater contains dissolved calcium carbonate ($CaCO_3$), then *traver-*

tine or *calcareous tufa* (both of which are varieties of limestone) are precipitated. Spectacular examples of hot-spring travertine deposits are found at Pamukhale in Turkey and at Mammoth Hot Springs in Yellowstone National Park (Figure 16.31a). Groundwater containing dissolved silica will, upon reaching the surface, precipitate a soft, white, hydrated mineral called *siliceous sinter* or *geyserite,* which can accumulate around a geyser's opening (Figure 16.31b).

Geothermal Energy

Geothermal energy is defined as any energy produced from Earth's internal heat. In fact, the term *geothermal* comes from *geo,* meaning Earth, and *thermal* meaning heat. Several forms of internal heat are known, such as hot dry rocks and magma, but so far only hot water and steam are used (see Perspective 16.3). Approximately 1 to 2% of the world's current energy needs could be met by geothermal energy. In those areas where it is plentiful, geothermal energy can supply most, if not all, of the energy needs, sometimes at a fraction of the cost of other types of energy.

As oil reserves decline, geothermal energy is becoming an attractive alternative, particularly in parts of the western United States, such as the Salton Sea area of southern California, where geothermal exploration and development have begun.

(a)

Figure 16.31
Hot-spring deposits in Yellowstone National Park, Wyoming.
(a) Minerva Terrace formed when calcium-carbonate-rich hot-
spring water cooled, precipitating travertine. (b) Liberty Cap is a
geyserite mound formed by numerous geyser eruptions of
silicon-dioxide-rich hot-spring water.

(b)

WHAT WOULD YOU DO?

Americans generate tremendous amounts of waste,
and some, such as battery acid, paint, cleaning
agents, insecticides, and pesticides, can easily contaminate the groundwater system. Your community
is planning to construct a city dump to contain waste
products, but simply wants to dig a hole, dump waste
in, and then bury it. What do you think of this plan?
Are you skeptical of this plan's merits, and if so, what
would you suggest to remedy any potential problems?

Perspective 16.3

Energy from Earth's Internal Heat

In Perspective 15.1, we noted that most electricity is generated at coal-burning power plants, whereas other energy sources account for much smaller amounts. One of these, geothermal energy—that is, energy derived from Earth's heat—is a vast resource that at present contributes only slightly to our energy needs. Nevertheless, people have used geothermal energy in one form or another throughout history. In ancient times, people camped on the warm ground in geothermal areas and some even cooked food in hot springs. Geothermal energy was used to heat buildings in the Roman city of Pompeii before it was destroyed in A.D. 79. Even monkeys in Japan visit hot springs in winter to soak in the warm, soothing waters, and many hot springs around the world have been developed for their presumed therapeutic value (see Figure 16.28).

Earth's internal heat is used as an energy source in two ways. The first, *direct use,* involves simply piping hot water or steam from geothermal areas to houses, office buildings, schools, and greenhouses for space heating. Indeed, many communities around the world use geothermal energy in this manner. In Iceland, geothermal energy has been successfully used since 1928. Steam and hot water from wells drilled in geothermal areas are pumped into buildings for heating and hot water in Reykjavik, Iceland's capital. Fruits and vegetables are grown year-round in hothouses heated from geothermal wells. Direct heating in this manner is significantly cheaper than fuel oil or electrical heating and much cleaner.

The city of Rotorua, New Zealand, is world famous for its volcanoes, hot springs, geysers, and geothermal fields. Since the first well was sunk in the 1930s, more than 800 wells have been drilled to tap the hot water and steam. Geothermal energy in Rotorua is used in a variety of ways, including home, commercial, and greenhouse heating. Geothermal heating is also used extensively in some U.S. communities such as in greenhouses near Susanville, California; for buildings in Boise, Idaho; and at Klamath Falls, Oregon, where geothermal energy is used for, among other things, melting snow and ice on sidewalks. However, direct use can be employed only locally; that is, very near the geothermal area, because the energy cannot be shipped great distances as can conventional energy sources such as petroleum, natural gas, or coal.

If *hydrothermal reservoirs,* consisting of hot water and steam in porous rocks, are tapped and used to generate electricity, the electricity can and is transported via power lines to locations far from the generating facilities. Several countries already have geothermal power plants, and others are contemplating building them. However, the world's largest geothermal development is The Geysers in Sonoma and Lake Counties, California, where Pacific Gas and Electric has been producing electricity since 1960 (Figure 1). In addition to two other areas in California, geothermal plants are also operating in Hawaii, Nevada, and Utah, and similar plants are planned for Alaska, Oregon, and New Mexico. And an operating plant in Cerro Prieto, Mexico, sends half the electricity generated to the San Diego, California, area.

The technology using hydrothermal reservoirs is rather straightforward. At The Geysers geothermal field in California, for instance, wells are drilled into the numerous near-vertical fractures underlying the region. As pressure on the rising hot groundwater decreases, the water changes to steam that is piped directly to electricity-generating turbines and generators (Figure 2). In this aspect of power generation, a geothermal plant is like any other. The difference is the energy source.

Geologists recognize five types of geothermal energy, but only direct use and tapping hydrothermal reservoirs are currently used. Other types of geo-

Figure 1
Steam rising from geothermal power plants at The Geysers in Sonoma County, California.

thermal energy have promise but so far technologies do not exist to exploit them. For instance, water could be pumped under pressure into hot, dry rock to widen fractures, thus creating a hot-water reservoir. Near-surface *magma* might somehow be tapped and energy extracted, and geopressurized brine, hot water with dissolved methane, has the potential for generating electricity. Cooperative efforts by government and industry are now attempting to develop technologies capable of using these other sources of geothermal energy.

Geothermal energy is cost effective and much cleaner than coal-burning power plants. Emissions of carbon dioxide, a greenhouse gas, are minimal, and nearly all sulfur gases are removed from their fumes. However, geothermal waters are acidic and very corrosive, so turbines must either be built of expensive corrosion-resistant alloys or frequently replaced. Furthermore, geothermal energy is not inexhaustible. The steam and hot water removed for geothermal power cannot be easily replaced, and eventually pressure in the wells drops until the geo-

thermal field must be abandoned. Indeed, five of the first plants built in The Geysers area of California have been closed, and generating capacity at others has declined. Careful management and injection of water to replace that pumped out should considerably prolong the life of the field.

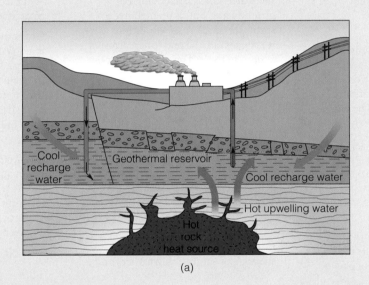

(a)

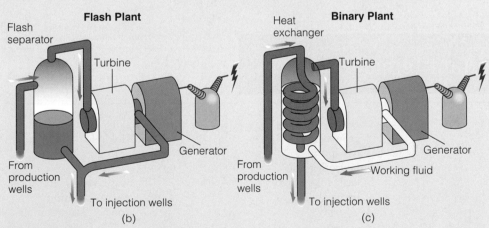

(b)   (c)

Figure 2
(a) Diagram showing how geothermal energy is used to generate electricity at a power plant. Whether the power facility is a flash plant or binary plant depends on the temperature of the hydrothermal fluid. (b) In a flash plant, used for water more than 200°C, water entering a low-pressure tank flashes to steam and the steam drives a turbine. (c) For water less than 200°C, hot water vaporizes another fluid that drives a turbine. In both (b) and (c) the spinning turbine connected to a generator containing an electromagnet inside a coil of wire produces electricity.

Chapter Summary

1. Groundwater consists of all subsurface water trapped in the pores and other open spaces in rocks, sediment, and soil.

2. About 22% of the world's supply of freshwater is groundwater, which constitutes one reservoir in the hydrologic cycle.

3. For groundwater to move through materials, they must be porous and permeable. Any material that transmits groundwater is an aquifer, whereas materials that prevent groundwater movement are aquicludes.

4. The zone of saturation (in which pores are filled with water) is separated from the zone of aeration (in which pores are filled with air and water) by the water table. The water table is a subdued replica of the surface in most places.

5. Groundwater moves slowly through the pore spaces in the zone of aeration and moves through the zone of saturation to outlets such as streams, lakes, and swamps.

6. Springs are found wherever the water table intersects the surface. Some springs are the result of a perched water table, that is, a localized aquiclude within an aquifer and above the regional water table.

7. Water wells are made by digging or drilling into the zone of saturation. When water is pumped out of a well, a cone of depression forms. If water is pumped out faster than it can be recharged, the cone of depression deepens and enlarges and may locally drop to the base of the well, resulting in a dry well.

8. In an artesian system, confined groundwater builds up high hydrostatic pressure. Three conditions must generally be met for an artesian system to form: The aquifer must be confined above and below by aquicludes; the aquifer is usually tilted and exposed at the surface so it can be recharged; and precipitation must be sufficient to keep the aquifer filled.

9. Karst topography results from groundwater weathering and erosion and is characterized by sinkholes, caves, solution valleys, and disappearing streams.

10. Caves form when groundwater in the zone of saturation weathers and erodes soluble rock such as limestone. Cave deposits, called dripstone, result from the precipitation of calcite.

11. Modifications of the groundwater system can cause serious problems. Excessive withdrawal of groundwater might result in dry wells, loss of hydrostatic pressure, saltwater incursion, and ground subsidence.

12. Groundwater contamination is becoming a serious problem and can result from sewage, landfills, and toxic waste.

13. Groundwater might be heated by magma or be heated by the geothermal gradient as it circulates deeply. In either case, the water commonly rises to the surface, thus accounting for hydrothermal activity in the form of hot springs, geysers, and several other features.

14. Geothermal energy comes from the steam and hot water trapped within the crust. It is a relatively nonpolluting form of energy that is used as a source of heat and to generate electricity.

Important Terms

aquiclude
aquifer
artesian system
cave
cone of depression
dripstone
geothermal energy

geyser
groundwater
hot spring
hydrothermal
karst topography
permeability
porosity

saltwater incursion
sinkhole
spring
water table
water well
zone of aeration
zone of saturation

Review Questions

1. Which one of the following is a cave deposit?

   a. _____ chamber;
   b. _____ stalagmite;
   c. _____ sinkhole;
   d. _____ artesian spring;
   e. _____ aquiclude.

2. Which one of the following statements is correct?

   a. _____ The water table dips beneath hills and rises beneath valleys;
   b. _____ Groundwater erosion yields geysers and hot springs;
   c. _____ The water from an artesian well is better than water from other wells;
   d. _____ A perched water table is found above the main water table;
   e. _____ In the United States, karst topography is best developed in the Pacific Northwest.

3. A layer of Earth materials that inhibits the movement of groundwater is a(n):

   a. _____ aquiclude;
   b. _____ geyser;
   c. _____ dripstone;
   d. _____ cone of depression;
   e. _____ capillary fringe.

4. One of the following conditions must exist for an artesian system to form. Which one is it?

   a. _____ Groundwater must circulate near magma;
   b. _____ The water table must be at or very near the surface;
   c. _____ The rocks below the surface must be especially resistant to solution;
   d. _____ Water must rise very high in the capillary fringe;
   e. _____ An aquifer must be confined above and below by aquicludes.

5. A cone of depression forms when:

   a. _____ water is withdrawn from a well faster than it can be replaced;
   b. _____ a stream flows into a sinkhole;
   c. _____ a spring forms where a perched water table intersects the surface;
   d. _____ water in the zone of aeration is replaced by water from the zone of saturation;
   e. _____ the ceiling of a cave collapses, forming a steep-sided crater.

6. A hot spring that periodically erupts is known as a:

   a. _____ geothermal gradient;
   b. _____ zone of saturation;
   c. _____ geyser;
   d. _____ mud pot;
   e. _____ travertine terrace.

7. In which one of the following states is geothermal energy used to generate electricity?

   a. _____ Texas;
   b. _____ New York;
   c. _____ Idaho;
   d. _____ California;
   e. _____ Arkansas.

8. What is the correct order, from highest to lowest, of groundwater usage in the United States?

   a. _____ agricultural, industrial, domestic;
   b. _____ industrial, domestic, agricultural;
   c. _____ domestic, agricultural, industrial;
   d. _____ agricultural, domestic, industrial;
   e. _____ industrial, agricultural, domestic.

9. Two features typical of areas of karst topography are:

   a. _____ sinkholes and disappearing streams;
   b. _____ dripstone and a cone of ascension;
   c. _____ geysers and hot springs;
   d. _____ saltwater incursion and pollution;
   e. _____ hydrothermal activity and springs.

10. The porosity of Earth materials is defined as:

    a. _____ their ability to transmit fluids;
    b. _____ the depth of the zone of saturation;
    c. _____ the percentage of void spaces;
    d. _____ their solubility in the presence of weak acids;
    e. _____ the temperature of groundwater.

11. When water is pumped from wells in some coastal areas, a problem arises known as:

    a. _____ permeability decrease;
    b. _____ saltwater incursion;
    c. _____ geothermal depression;
    d. _____ artesian recharge;
    e. _____ dripstone deposition.

12. Hydrothermal is a term referring to:

    a. _____ calcareous tufa;
    b. _____ sinkhole formation;
    c. _____ hot water;
    d. _____ artesian wells;
    e. _____ groundwater contamination.

13. Explain how groundwater weathers and erodes? Also, what features are yielded by this activity?

14. Discuss two ways in which the groundwater system can become contaminated.

15. Explain how some Earth materials can be porous and permeable, whereas others are porous but not very permeable. Give examples of each.

16. What is a saltwater incursion, and how does it take place?

17. How do sinkholes and caves form? What depositional features might you see in a cave?

18. Why does groundwater move so much more slowly than surface water?

19. Discuss the role of groundwater in the hydrologic cycle.

20. Explain fully why pumping from one well might make nearby wells go dry.

21. Describe the configuration of the water table beneath a humid area and an arid region. Why do they differ?

22. Diagram the conditions necessary for an artesian system to form. Do all artesian wells flow freely at the surface?

23. Describe three features you might see in an active hydrothermal area. Where in the United States would you go to see such activity?

24. Explain what geothermal energy is, and briefly discuss its potential to decrease our dependence on traditional energy sources.

Points to Ponder

1. Describe some of the ways of quantitatively measuring the rate of groundwater movement.

2. One concern geologists have about using Yucca Mountain as a repository for nuclear waste is that the climate may change during the next 10,000 years and become more humid, thus allowing more water to percolate through the zone of aeration. What would the average rate of groundwater movement have to be during the next 10,000 years to reach the canisters containing radioactive waste buried at a depth of 300 m?

3. Why isn't geothermal power a virtually unlimited source of energy?

4. Why should we be concerned about how fast the groundwater supply is being depleted in some areas?

World Wide Web Activities

For these Web site addresses, along with current updates and exercises, log on to
http://www.brookscole.com/geo/

➤ THE VIRTUAL CAVE

This Web site contains images of various cave features from around the world. In addition, it contains links to and information on other cave sites elsewhere in the world.
1. Click on any of the headings for cave features. View the images and read the text about each image.
2. Click on the site link to learn about caves in your area or near where you may be traveling. What are some of the caves near you?

➤ U.S. GEOLOGICAL SURVEY WATER RESOURCES OF THE UNITED STATES

This Web site contains a wealth of information about streams, flooding, flood forecasting, flood warnings, and groundwater.
1. Under the Publications heading, click on *On-line Reports*. What are some of the current publications available?
2. Under the Data heading, click on the *Water-Use Data* site. Click on the *Periodic Water Fact* icon for the latest information about new and interesting facts about water resources.
3. Under the Data heading, click on the *Water-Use Data* site. Click on the *3-D graphic of total water use in the U.S.* icon. Where is water withdrawal greatest in the United States? Does this surprise you?
4. Under the Data heading, click on the *Water-Use Data* site. Click on the *Color maps of water use by state for 1990* icon. Click on any of the *National Water-Use Maps* sites for color maps of U.S. groundwater usage in 1990. What was the domestic, industrial, and agricultural groundwater withdrawal for your state?
5. Click on the *California–Nevada Ground Water Atlas* site for information on groundwater in this region, including quality and problems with subsidence. This is the first segment of the Ground Water Atlas of the United States, and it has some great maps.

YUCCA MOUNTAIN PROJECT STUDIES

This Web site is devoted to the Yucca Mountain Project, which is the only current site being considered for the storage of the nation's high-level radioactive waste. It includes information, maps, and images.

1. Click on the *What's News (News and hot topics)* site. Check out what has happened at Yucca Mountain in the previous year.
2. Click on the *Frequently Asked Questions* site. Click on the *Why Study Yucca Mountain?* site. After reading this information, why is it important we study Yucca Mountain, and why should the average citizen be concerned about this site?

LINKS TO INFORMATION SOURCES FOR HYDROGEOLOGY, HYDROLOGY, AND ENVIRONMENTAL SCIENCES

This site contains links, which are listed alphabetically, to information sources for hydrogeology, hydrology, and environmental sciences. Check out several sites for information about groundwater and groundwater usage, both in the United States and elsewhere in the world.

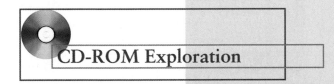

CD-ROM Exploration

Exploring your *Earth Systems Today* CD-ROM will add to your understanding of the material in this chapter.

TOPIC: SURFICIAL PROCESSES AND HYDROSPHERE

MODULE: GROUNDWATER

Explore activities in this module to see if you can discover the following for yourself:

Using this activity, examine the flow of water from a recharge area to an artesian well, a pumping well, and a stream.

Using this activity, examine the relationship between atmospheric precipitation and stream flow and also groundwater flow through the zone of aeration, capillary fringe, and zone of saturation.

Using this activity, study the development of a cone of depression. How do pumping rate and permeability affect development of a cone of depression?

Chapter 17

Pense La Glacier in the Zanskar Valley, India.

OBJECTIVES

At the end of this chapter, you will have learned that

- Glaciers constitute one reservoir in the hydrologic cycle.
- As snow accumulates, it is converted to glacial ice, and that once formed a glacier moves by plastic flow and basal slip.
- The concept of a glacial budget is important in considering the dynamics of any glacier.

- Several factors determine the rate at which a glacier moves.
- Glaciers effectively erode, transport, and deposit sediment.
- Erosion and deposition by glaciers yield a variety of distinctive landforms.
- Glaciers directly or indirectly affected many areas during the Ice Age, and are responsible for some of the distinctive landscapes present today.

- Loading of Earth's crust by glaciers caused it to be depressed into the mantle.
- Several ideas seek to account for the onset of ice ages and the advance and retreat of glaciers during an ice age.

Glaciers and Glaciation

Prologue

Since the end of the Great Ice Age about 10,000 years ago, Earth has experienced several climatic fluctuations. During the Holocene Maximum about 6000 years ago, temperatures were slightly warmer on average than at present, and some arid regions such as the Sahara Desert of North Africa supported lush vegetation, swamps, and lakes. Then followed a time of cooler temperatures, but from about A.D. 1000 to 1300, Europe experienced what is known as the Medieval Warm Period during which wine grapes grew more than 480 km north of their present limits. However, a cooling trend beginning about 1300 led to the Little Ice Age, a time of expansion of glaciers in mountain valleys, overall colder winters, cooler, wetter summers, and the persistence of sea ice at high latitudes for long periods.

During the preceding Medieval Warm Period, the climates of Europe and North America were rather mild, and the North Atlantic Ocean was warmer and more storm free than it is now. Taking advantage of the mild conditions, the Norse, commonly called Vikings, discovered and settled Iceland, and by A.D. 1200 about 80,000 people were living there. They also sailed to Greenland and North America, where they established two colonies on the former and one on the latter. Unfortunately for these settlers, the climate changed, the North Atlantic became stormier, and sea ice moved farther south and persisted longer each year. The poor sea conditions coupled with political problems in Norway brought a halt to all shipping across the North Atlantic, and the colonies in Greenland and North America eventually disappeared.

The Little Ice Age was also a time of expansion of small glaciers in mountain valleys. In Europe and Iceland, glaciers moved far down their valleys, reaching their greatest historic extent by the early 1800s. A small ice cap formed in Iceland where none had existed previously, and glaciers in Alaska, Canada, and the mountains of the western United States expanded to their greatest limits during historic time. During the early 1600s, people in the European Alps watched as the Mer de Glace (Sea of Ice Glacier) advanced and overwhelmed farms and villages. Elsewhere glaciers advanced across roadways and pastures, destroyed some villages in Scandinavia, and threatened other villages.

Advancing glaciers caused some problems, but their overall impact on humans was minimal. Far more important from the human perspective was that during much of the Little Ice Age, summers in the Northern Hemisphere were cooler and wetter. Particularly hard hit were the Scandinavian countries and the Norse colonies in Greenland and Iceland, but at times much of Northern Europe was affected (Figure 17.1). The harsher summer weather resulted in shorter growing seasons, reduced crop yields, food shortages, and a number of famines in Iceland, Scotland, and mainland Europe. Iceland's population was down by half to about 40,000 by 1700. Between 1610 and 1870, sea ice was seen near Iceland for as much as three months a year, and each year the sea ice persisted for longer

(a)

Figure 17.1
(a) During the Little Ice Age, many glaciers in Europe, such as this one in Switzerland, extended much farther down their valleys than they do now. This painting, *The Unterer Grindelwald,* was painted in 1826 by Samuel Birmann (1793–1847). (b) The canals of Holland are frozen in this mid-seventeenth-century painting, *The Village of Nieukoop in Winter* by Jan-Abrahamsz Beerstraten. These canals rarely freeze today.

periods, poor growing seasons and food shortages followed.

Although cool, wet summers caused most of the problems, winter temperatures were colder too, and sea ice was a problem in the North Atlantic. Occasionally, Eskimos paddled their kayaks as far south as Scotland, and the canals of Holland froze over during some winters. In 1607 the first "Frost Fair" was held in London on the River Thames, which had begun to freeze over nearly every winter. Even New York Harbor in the newly founded United States froze over. And 1816 is known as the "year without a summer," when unusually cold temperatures persisted into June and July in New England and northern Europe; both areas experienced frost and even some snow during these months. Eruptions of Tambora in Indonesia in 1815 and Mayon Volcano in the Philippines in 1814 contributed to the cool spring and summer of 1816.

Exactly when the Little Ice Age ended is debatable. Some authorities put the end at 1880, whereas others

(b)

think it ended as early as 1850. In either case, during the mid- to late 1800s, sea ice was retreating northward, glaciers began retreating back up their valleys, and summer weather became more moderate.

INTRODUCTION

The Little Ice Age was just one of several climatic fluctuations that took place during the last few thousands of years, but one that certainly had a dramatic impact on people living in the affected areas. Even regions far removed from the areas of glacier expansion and cooler, wetter summers experienced changes, although these changes were more beneficial such as experiencing more rainfall in areas now arid. Unfortunately, our period of recordkeeping is too short to resolve the question of whether the last Ice Age and the Little Ice Age are truly events of the past or simply parts of long-term climatic events and likely to occur again. In any event, even though glaciers are much more restricted now, they are still with us and capable of considerable geologic work.

Most people have some idea of what a glacier is and have heard of the Ice Age, a time when glaciers were much more widespread than they are now. By definition, a **glacier** is a mass of ice consisting of compacted and recrystallized ice on land that flows under its own weight. Accordingly, sea ice in the north polar region or the ice shelves adjacent to Antarctica are not glaciers, nor are drifting icebergs, even though they may have broken off from glaciers that flow into the sea. Snowfields in high mountains might persist for years, but these are not glaciers either because they are not actively flowing.

Presently, glaciers of one kind or another cover nearly 15 million km², or about one-tenth, of Earth's land surface (Table 17.1). As a matter of fact, enough glacial ice is present to cover the United States and Canada to a depth of 1.4 to 1.7 km! Small glaciers are especially abundant in the mountains of the western United States, especially Alaska, western Canada, the Andes in South America, the Alps of Europe, the Himalayas of Asia, and in other high mountains. Even Mount Kilamanjaro in Africa is high enough to have glaciers, although it is close to the equator. In fact, Australia is the only continent lacking glaciers. These small glaciers in mountains are impressive and picturesque, but the truly vast glaciers are in Antarctica and Greenland. More than 96% of all glacial ice on this planet is in Greenland (12%) and Antarctica (84.5%), both of which are nearly covered by glaciers (Table 17.1).

At first glance glaciers appear static. Even a brief visit to a glacier may not dispel this idea, because although they move, they do so slowly. Nevertheless, just like other geologic agents such as running water, glaciers are dynamic systems that continually adjust to changes. For instance, the ability of streams to erode and transport sediment varies depending on velocity and discharge. Likewise, the amount of ice in a glacier or the amount of water present determines rate of movement and the capacity to erode and transport sediment. So glaciers also respond to changes, they just do so more slowly than streams.

Among the various processes modifying Earth's surface processes, glaciers are particularly effective at erosion, transport, and deposition. They deeply scour the land they move over, producing a number of easily recognized erosional landforms, and eventually deposit huge quantities of sediment. Many examples of landscapes that continue to be modified by glaciers can be found, but most landscapes formed by glaciation developed during the Pleistocene Epoch or what is commonly called the Ice Age (1.6 million to 10,000 years ago). Vast glaciers were much more extensive during the Ice Age than they are now, especially on the Northern Hemisphere continents.

GLACIERS—PART OF THE HYDROLOGIC CYCLE

Recall from Chapter 1 that Earth is a dynamic planet possessing several major systems that interact in complex ways. One of its systems is the hydrosphere, consisting of all surface water in the oceans, lakes, streams, the atmosphere, and frozen in glaciers (see Table 1.1). It is the comparatively small amount of

Table 17.1

Present-Day Ice-Covered Areas	
Antarctica	12,653,000 km²
Greenland	1,802,600
Northeast Canada	153,200
Central Asian ranges	124,500
Spitsbergen group	58,000
Other Arctic islands	54,000
Alaska	51,500
South American ranges	25,000
West Canadian ranges	24,900
Iceland	11,800
Scandinavia	5,000
Alps	3,600
Caucasus	2,000
New Zealand	1,000
USA (other than Alaska)	650
Others	about 800
	14,971,550
Total volume of present ice: 28 to 35 million km³	

Source: C. Embleton and C. A. King, Glacial Geomorphology (New York: Halsted Press, 1975).

water in glaciers, about 2.15% of the total but more than 75% of Earth's freshwater, that interests us here.

Glaciers constitute one reservoir in the hydrologic cycle where water is stored for long periods, but even this water eventually returns to its original source, the oceans (see Figure 15.6). Many glaciers at high latitudes, as in Alaska and northern Canada, flow directly into the seas where they melt, or icebergs break off (a process known as *calving*) and drift out to sea where they eventually melt. At lower elevations or areas remote from the sea, glaciers flow to lower elevations where melting takes place and the water yielded either enters the groundwater system (another reservoir in the hydrologic cycle) or it returns to the seas by surface runoff. In some parts of the western United States and Canada, glaciers are important freshwater reservoirs that release water to streams during the dry season.

Melting is the most important process in returning glacial ice to the hydrologic cycle, but glaciers also lose water by *sublimation,* a process in which ice changes to water vapor without an intermediate liquid phase. Sublimation is easy to understand if one thinks of ice cubes stored in a container in a freezer where no melting can take place. Because of sublimation, the older ice cubes at the bottom of the container are much smaller than the more recently formed ones. In any event, the water vapor thus derived from glaciers enters the atmosphere where it may condense and fall again as rain or snow, but in the long run that water too returns to the oceans.

HOW DOES GLACIAL ICE ORIGINATE?

Ice is a mineral in every sense of the word; it is a crystalline solid possessing characteristic physical and chemical properties. Accordingly, geologists consider glacial ice a rock, although it is a type of rock that is easily deformed. It forms in a fairly straightforward manner (Figure 17.2). In any area where more winter snow falls than can melt during the spring and summer seasons, a net accumulation occurs. Freshly fallen snow consists of about 80% air and 20% solids, but it compacts as it accumulates, partly thaws, and refreezes; in the process, the original snow layer is converted to a granular type of ice known as **firn.** Deeply buried firn is further compacted and is finally converted to **glacial ice,** consisting of about 90% solids (Figure 17.2).

When accumulated snow and ice reach a critical thickness of about 40 m, the pressure on the ice at depth is sufficient to cause deformation and flow, even though it remains solid. Once the critical thickness is reached and flow begins, the moving mass of ice becomes a gla-

cier. In polar regions where little snow melts in summer, glaciers can exist at or very near sea level, but at lower latitudes they are found only at higher elevations.

Good examples of elevation as a control on glaciers are found in the western United States. Many small glaciers are present in California—in fact, more than in any other state except Alaska and Washington—but all are found only at elevations exceeding 3900 m. Another contributing factor to the presence of glaciers in the high mountains of the Pacific Coast states is the fact that they receive such large quantities of snow. Indeed, Mount Baker in Washington had almost 29 m of snow during the winter of 1998–1999, and 10 m or more of snow is common on the higher peaks in Washington, Oregon, and California. Glaciers are also present in the coastal mountains of Canada, which receive considerable snow and are farther north. Some of the higher mountains in the Rocky Mountain states and Canadian provinces also support glaciers.

At this time it is useful to recall some terms from Chapter 13 on deformation. Remember that stress is force per unit area and strain is defined as a change in shape or volume or both of solids. When snow reaches the critical thickness of about 40 m, the stress on the ice at depth is great enough to induce **plastic flow,** the primary way that glaciers move. However, glaciers also move by **basal slip,** involving the mass of ice sliding over the underlying surface (Figure 17.3). Basal slip is facilitated by the presence of water, which reduces frictional resistance between the underlying surface and the glacier. The total movement of a glacier, then, results from a combination of plastic flow and basal slip, although the former occurs continuously whereas the latter varies

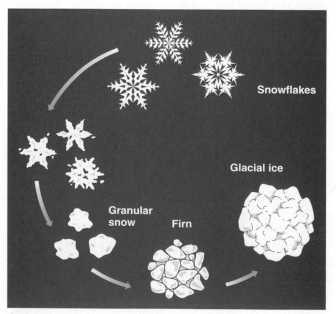

Figure 17.2
The conversion of freshly fallen snow to firn and glacial ice.

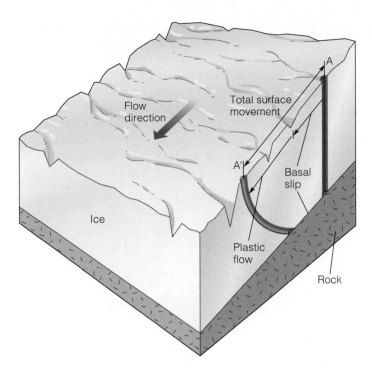

Figure 17.3
Part of a glacier showing movement by a combination of plastic flow and basal slip. Plastic flow involves internal deformation within the ice, whereas basal slip is sliding over the underlying surface. If a glacier is solidly frozen to its bed, it moves only by plastic flow. Notice that the top of the glacier moves farther in a given time than the bottom does.

depending on the season. Indeed, if the glacier is solidly frozen to the underlying surface, it moves only by plastic flow.

WHAT KINDS OF GLACIERS EXIST?

Geologists generally recognize two basic types of glaciers: valley and continental. A **valley glacier,** as its name implies, is confined to a mountain valley or perhaps to an interconnected system of mountain valleys (Figure 17.4). Large valley glaciers commonly have several smaller tributary glaciers, much as large streams have tributaries. Valley glaciers flow from higher to lower elevations and are invariably small in comparison to continental glaciers, even though some are more than 100 km long, several kilometers wide, and several hundred meters thick. The Bering Glacier in Alaska, for instance, is about 200 km long, whereas the Salmon Glacier in the Canadian Rocky Mountains is nearly 500 m thick.

When a valley glacier flows from a mountain valley onto a wider plain and spreads out, or where two or more valley glaciers coalesce at the base of a mountain range, they form much more extensive ice covers, known as *piedmont glaciers*. With an area of about 8000 km², Alaska's Malaspina Glacier is the world's largest piedmont glacier.

Continental glaciers, also called *ice sheets,* cover vast areas (at least 50,000 km²) and are unconfined by

Figure 17.4
A large valley glacier in Alaska. Notice the tributaries to the large glacier.

topography (Figures 17.5 and 17.6). In contrast to valley glaciers, which flow downhill within the confines of a valley, continental glaciers flow out in all directions from a central area of accumulation. Valley glaciers flow in the direction of an existing slope, whereas the direction a continental glacier flows is determined by variations in ice thickness.

Currently, continental glaciers exist in only two areas, Greenland and Antarctica. In both areas, the ice is more than 3000 m thick in the central areas, becomes thinner toward the margins, and covers all but the highest mountains (Figures 17.5 and 17.6). The areal extent of the continental glacier in Greenland is more than 1,800,000 km², and the East and West Antarctic Glaciers merge to form a continuous ice sheet covering 12,653,000 km² (Table 17.1). During the Pleistocene Epoch, continental glaciers covered large parts of the Northern Hemisphere continents. Many of the erosional and depositional landforms in much of Canada and the northern tier of the United States formed as a result of continental glaciation during the Pleistocene.

Although valley and continental glaciers are easily differentiated by their size and location, an intermediate variety called an *ice cap* is also recognized. Ice caps are

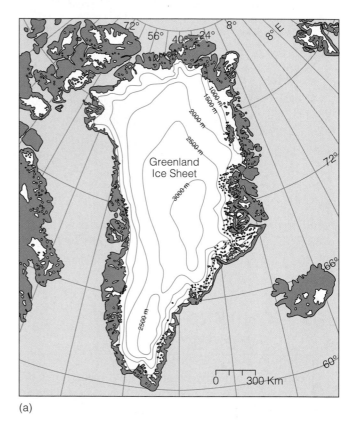

(a)

(a)

(b)

Figure 17.5

(a) The West Antarctic and much larger East Antarctic ice sheet merge to form a nearly continuous ice cover averaging 2160 m thick and reaching a maximum thickness of 4000 m. (b) View of the margin of the ice sheet in Antarctica.

(b)

Figure 17.6

(a) The ice sheet in Greenland is much smaller than the one in Antarctica (compare with Figure 17.5). Nevertheless, it covers more than 1,800,000 km² and has a maximum thickness of about 3350 m. (b) Aerial view of the eastern margin of the Greenland ice sheet.

similar to but smaller than continental glaciers, covering less than 50,000 km². The 6000 km² Penny Ice Cap on Baffin Island, Canada, is a good example. Some ice caps form when growing valley glaciers overtop the divides and passes between adjacent valleys and coalesce to form a continuous ice cap. They also form on fairly flat terrain including some of the islands of the Canadian Arctic and Iceland.

ACCUMULATION AND WASTAGE—THE GLACIAL BUDGET

Just as a savings account grows and shrinks as funds are deposited and withdrawn, glaciers expand and contract in response to accumulation and wastage. Their behavior can be described in terms of a **glacial budget,** which is essentially a balance sheet of accumulation and wastage. The upper part of a valley glacier is a **zone of accumulation** where additions exceed losses, and the glacier's surface is perennially covered by snow. In contrast, the lower part of the same glacier is a **zone of wastage,** where losses from melting, sublimation, and calving of icebergs exceed the rate of accumulation (Figure 17.7).

At the end of winter, a glacier's surface is usually completely covered with the accumulated seasonal snowfall. During spring and summer, the snow begins to melt, first at lower elevations and then progressively higher up the glacier. The elevation to which snow recedes during a wastage season is called the *firn limit* (Figure 17.7). One can easily identify the zones of accumulation and wastage by noting the position of the firn limit.

Observations of a single glacier reveal that the position of the firn limit usually changes from year to year. If it does not change or shows only minor fluctuations, the glacier is said to have a balanced budget; that is, additions in the zone of accumulation are exactly balanced by losses in the zone of wastage, and the distal end, or *terminus,* of the glacier remains stationary (Figure 17.8a). When the firn limit moves down the glacier, the glacier has a positive budget; its additions exceed its losses, and its terminus advances (Figure 17.8b). If the budget is negative, the glacier recedes, and its terminus retreats up the glacial valley (Figure 17.8c). But even though a glacier's terminus may be receding, the glacial ice continues to move toward the terminus by plastic flow and basal slip. If a negative budget persists long enough, though, a glacier recedes and thins until it no longer flows and becomes a *stagnant glacier.*

Although we used a valley glacier as our example, the same budget considerations control the flow of continental glaciers as well. The entire Antarctic ice sheet, for example, is in the zone of accumulation, but it flows into the ocean where wastage occurs.

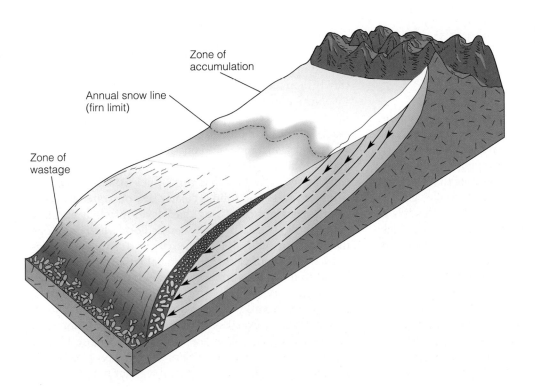

Zone of accumulation

Annual snow line (firn limit)

Zone of wastage

Figure 17.7
The glacial budget is the annual balance between additions in the zone of accumulation and losses in the zone of wastage. Ice and rock debris, progressively buried by newly formed ice in the zone of accumulation, eventually reach the surface in the zone of wastage as the overlying ice melts.

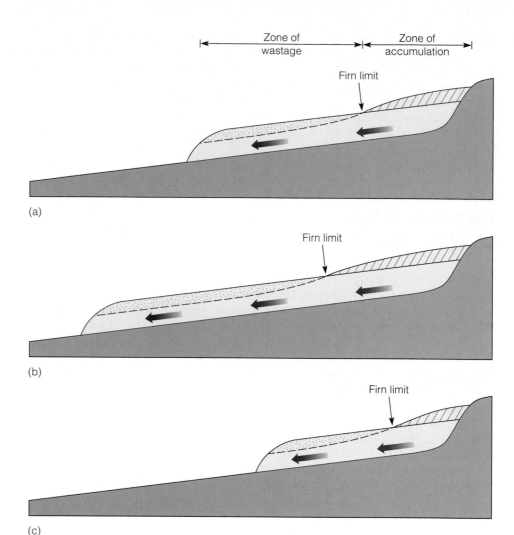

Figure 17.8
Response of a hypothetical glacier to changes in its budget. (a) If losses in the zone of wastage, shown by stippling, equal additions in the zone of accumulation, shown by cross-hatching, the terminus of the glacier remains stationary.
(b) Gains exceed losses, and the glacier's terminus advances.
(c) Losses exceed gains, and the glacier's terminus retreats, although the glacier continues to flow.

HOW FAST DO GLACIERS MOVE?

In general, valley glaciers move more rapidly than continental glaciers, but the rates for both vary, ranging from centimeters to tens of meters per day. Valley glaciers moving down steep slopes flow more rapidly than glaciers of comparable size on gentle slopes, assuming that all other variables are the same. The main glacier in a valley glacier system contains a greater volume of ice and thus has a greater discharge and flow velocity than its tributaries (Figure 17.4). Temperature exerts a seasonal control on valley glaciers because, although plastic flow remains rather constant year-round, basal slip is more important during warmer months when meltwater is more abundant.

Flow rates also vary within the ice itself. For example, flow velocity generally increases in the zone of accumulation until the firn limit is reached; from that point, the velocity becomes progressively slower toward the glacier's terminus. Valley glaciers are similar to streams, in that the valley walls and floor cause frictional resistance to flow, so the ice in contact with the walls and floor moves more slowly than the ice some distance away (Figure 17.9).

Notice in Figure 17.9 that flow velocity increases upward until the top few tens of meters of ice are reached, but little or no additional increase occurs after that point. This upper ice constitutes the rigid part of the glacier that is moving as a result of basal slip and plastic flow below. The fact that this upper 40 m or so of ice behaves as a brittle solid is clearly demonstrated by large fractures called *crevasses* that develop when a valley glacier flows over a step in its valley floor where the slope increases or where it flows around a corner (Figure 17.10). In either case, the glacial ice is stretched (subjected to tension), and large crevasses develop, but they extend downward only to the zone of plastic flow. In some cases, a valley glacier descends over such a steep precipice that crevasses break up the ice into a jumble of blocks and spires, and an *ice fall* develops (Figure 17.10).

Flow rates of valley glaciers are also complicated by *glacial surges*, which are bulges of ice that move through

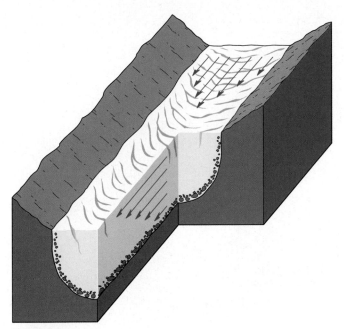

Figure 17.9
Flow velocity in a valley glacier varies both horizontally and vertically. Velocity is greatest at the top center of the glacier because friction with the walls and floor of the trough causes flow to be slower adjacent to these boundaries. The lengths of the arrows in the figure are proportional to velocity.

a glacier at a velocity several times faster than the normal flow. Although surges are best documented in valley glaciers, they take place in ice caps and perhaps continental glaciers as well (see Perspective 17.1).

Continental glaciers ordinarily flow at a rate of centimeters to meters per day. Nevertheless, even a rather modest rate of a meter or so per day has a great cumulative effect after several decades. One reason continental glaciers move comparatively slowly is that they exist at higher latitudes and are frozen to the underlying surface most of the time, which limits the amount of basal slip. Some basal slip does occur even beneath the Antarctic ice sheet, but most of its movement is by plastic flow. Nevertheless, some parts of continental glaciers manage to achieve extremely high flow rates. Near the margins of the Greenland ice sheet, the ice is forced between mountains in what are called *outlet glaciers*. In some of these outlets, flow velocities exceeding 100 m per day have been recorded.

In parts of the continental glacier covering West Antarctica, ice streams have been identified in which flow rates are considerably greater than in adjacent glacial ice. Drilling has revealed a 5-m-thick layer of water-saturated sediment beneath these ice streams, which apparently acts to facilitate movement of the ice

Figure 17.10
Crevasses in a glacier in Alaska.

Perspective 17.1

Surging Glaciers

Normally, the snout or terminus of a glacier possessing a positive budget advances a few meters to a few tens of meters per year. Under some circumstances, though, a glacier suddenly advances as much as several tens of meters per day for a few months or years and then returns to normal. In 1986 the terminus of Hubbard Glacier in Alaska began advancing at about 10 m per day (Figure 1), and more recently, in 1993, Alaska's Bering Glacier advanced more than 1.5 km in just three weeks.

The 1993 surge of the Bering Glacier, the largest surging glacier in the world, continued for 17 months, by which time the glacier's terminus had been displaced 9 km. For the next seven months the terminus retreated slightly, and then the glacier began to surge again, advancing 750 m in less than two weeks.

Figure 1
During a 1986 surge, the terminus of Hubbard Glacier in Alaska advanced across Russell Fiord, the shallow embayment at the right back. Environmentalists saved some of the marine mammals trapped in the former bay but hundreds of seals and porpoises died.

above. Some geologists think that geothermal heat from active volcanism melts the underside of the ice, thus accounting for the layer of water-saturated sediment.

WHAT WOULD YOU DO?

Many scientists are convinced that global warming is a real phenomenon. That is, the concentration of greenhouse gases in the atmosphere has brought about worldwide temperature increases, a trend they think will continue. One indication of global warming is shrinking glaciers; that is, their termini are retreating. Let's say you make visits to glaciers in Canada and Alaska over several years, and you notice that the ice in these glaciers continues to advance despite global warming and retreating termini. How can you explain these observations? What will happen if the termini of these glaciers continue to retreat?

HOW GLACIERS ERODE AND TRANSPORT SEDIMENT

Glaciers are currently limited in areal extent, but during the Pleistocene Epoch, they covered much larger areas and were more important than their present distribution would indicate. Glaciers as moving solids can erode and transport huge quantities of materials, especially unconsolidated sediment and soil. Important erosional processes associated with glaciers include bulldozing, plucking, and abrasion. *Bulldozing,* although not a formal geologic term, is fairly self-explanatory: A glacier simply shoves or pushes unconsolidated materials in its path. This effective process was aptly described in 1744 during the Little Ice Age by an observer in Norway:

> When at times [the glacier] pushes forward a great sound is heard, like that of an organ and it pushes in front of it unmeasurable masses of soil, grit and rocks

As during the previous surge, many deep cracks appeared on the glacier's surface when the new surge began. Bering Glacier is also interesting because during 1994 it produced two *glacial outburst floods,* which take place when water is suddenly released from cavities within or beneath the glacier.

Surging glaciers, constituting a tiny fraction of all glaciers, have been known for more than 100 years, but they are still not fully understood. None are present in the United States outside Alaska, and the only ones in Canada are in the Yukon Territory and the Queen Elizabeth Islands of the Arctic. Others are present in the Andes of South America, the Himalayas in Asia, Iceland, and some of Russia's Arctic islands. Many of these surging glaciers are difficult to study because they are in remote areas, although some of them surge at somewhat regular intervals. In one case, geologists accurately predicted the 1982–1983 surge of the Variegated Glacier of Alaska, based on its past history of surges.

Predicting glacier surges is of more than academic interest. Although few people live in areas directly threatened by an advancing ice front, surges can advance across highways and endanger facilities such as the Alaska pipeline. Geologists are currently monitoring the Black Rapids Glacier in Alaska for just this reason. Furthermore, during surges huge quantities of meltwater are released, which threaten downstream areas with flooding. And in some cases, a surging glacier advances across a stream valley, thus forming an ice-dammed lake. Unfortunately, ice dams are unstable and commonly collapse, thereby rapidly releasing water that causes catastrophic flooding.

Besides their direct or indirect threats to humans and their structures, surging glaciers have caused numerous animal deaths and disruptions of habitats. During the 1986 surge of the Hubbard Glacier in Alaska, glacial ice advanced across a shallow marine embayment, thus isolating it from the open sea (Figure 1). Hundreds of seals and porpoises were trapped in the former bay and died, although environmentalists did manage to save some of them.

The onset of a surge is marked by a noticeable thickening in the upper part of a glacier. As this thickened bulge of ice begins moving at several times the normal velocity toward the glacier's terminus, thousands of crevasses appear in the surface. Commonly the bulge reaches the terminus and then advances rapidly, as much as 20 km in two years in one case. Rarely, the bulge stops before reaching the terminus.

Even though some surges take place following a period of unusually heavy snowfall in the zone of accumulation, plastic flow alone cannot account for the accelerated flow rates in most surging glaciers. Accordingly, basal slip must be more important during these events.

Two theories, neither being mutually exclusive, have been proposed to account for accelerated basal slip. One theory holds that thickening in the zone of accumulation is accompanied by thinning in the zone of wastage, thereby increasing the glacier's slope. Eventually, the stress at the bottom of the upper part of a glacier causes channels below the glacier to close up, and this forces water out under the entire glacier, causing accelerated sliding. According to the other theory, pressure beneath a glacier resting on soft sediment forces fluids through the sediment layer, which is easily deformed, thus allowing the overlying glacier to slide more effectively.

---

bigger than any house could be, which it then crushes small like sand.*

Plucking, also called *quarrying,* results when glacial ice freezes in the cracks and crevices of a bedrock projection and eventually pulls it loose. One manifestation of plucking is a landform called a *roche moutonnée,* a French term for "rock sheep." As shown in Figure 17.11, a glacier smooths the "upstream" side of an obstacle, such as a small hill, and plucks pieces of rock from the "downstream" side by repeatedly freezing and pulling away from the obstacle.

Bedrock over which sediment-laden glacial ice moves is effectively eroded by **abrasion** and commonly develops a **glacial polish,** a smooth surface that glistens in reflected light (Figure 17.12a). Abrasion also yields **glacial striations,** consisting of rather straight scratches rarely more than a few millimeters deep on rock surfaces (Figure 17.12b). Abrasion also thoroughly pulverizes rocks, yielding an aggregate of clay- and silt-sized particles having the consistency of flour—hence the name *rock flour.* Rock flour is so common in streams discharging from glaciers that the water generally has a milky appearance (Figure 17.12c).

Continental glaciers can derive sediment from mountains projecting through them, and windblown dust settles on their surfaces. Otherwise, most of their sediment is obtained from the surface they move over and is transported in the lower part of the ice sheet. In contrast, valley glaciers carry sediment in all parts of the ice, but it is concentrated at the base and along the margins. Some of the marginal sediment is derived by abrasion and plucking, but much of it is supplied by mass-wasting processes, as when soil, sediment, or rock falls or slides onto the glacier's surface.

---

*Quoted in C. Officer and J. Page, *Tales of the Earth* (New York: Oxford University Press, 1993), p. 99.

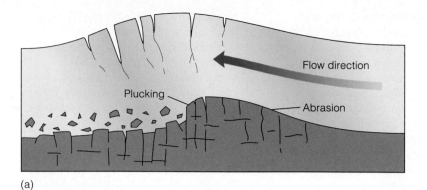

(a)

Figure 17.11
(a) Origin of a roche moutonnée. As the ice moves over a hill, it smooths the "upstream" side by abrasion and shapes the "downstream" side by plucking. (b) A roche moutonnée.

(b)

Erosion by Valley Glaciers

Some of the world's most inspiring scenery results from erosion by valley glaciers. Many mountain ranges are scenic to begin with, but when modified by valley glaciers they take on a unique appearance of angular ridges and peaks in the midst of broad valleys (see Perspective 17.2). Several of the national parks and monuments in the western United States and Canada owe their scenic appeal to erosion by valley glaciers. The erosional landforms resulting from valley glaciation are easily recognized and enable us to appreciate the tremendous erosive power of moving ice.

U-Shaped Glacial Troughs   A **U-shaped glacial trough** is one of the most distinctive features of valley glaciation (Figure 17.13c). Mountain valleys eroded by running water are typically V-shaped in cross section; that is, they have valley walls that descend steeply to a narrow valley bottom (Figure 17.13a). In contrast, valleys scoured by glaciers are deepened, widened, and straightened so that they have very steep or vertical walls but broad, rather flat valley floors; thus, they exhibit a U-shaped profile (Figure 17.14).

Many glacial troughs contain triangular-shaped *truncated spurs,* which are cutoff or truncated ridges that extend into the preglacial valley (Figure 17.13c). Another common feature is a series of steps or rock basins in the valley floor where the glacier eroded rocks of varying resistance; many of the basins now contain small lakes.

During the Pleistocene, when glaciers were extensive, sea level was about 130 m lower than at present, so glaciers flowing into the sea eroded their valleys to much greater depths than they do now. When the glaciers melted at the end of the Pleistocene, sea level rose, and the ocean filled the lower ends of the glacial troughs so that now they are long, steep-walled embayments called **fiords** (Figure 17.15).

(a)

(b)

Figure 17.12
When sediment-laden ice moves over rocks, it abrades them and imparts a sheen known as glacial polish (a) as on this gneiss in Michigan. Glacial polish is also visible in (b) and so are straight scratches called glacial striations. The rock is basalt at Devil's Postpile National Monument, California. (c) The water in this stream in Switzerland is discolored by rock flour, small particles yielded by glacial abrasion.

(c)

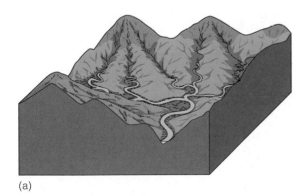

(a)

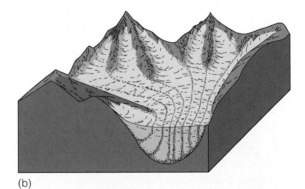

(b)

Figure 17.13
Erosional landforms produced by valley glaciers. (a) A mountain area before glaciation. (b) The same area during the maximum extent of valley glaciers. (c) After glaciation.

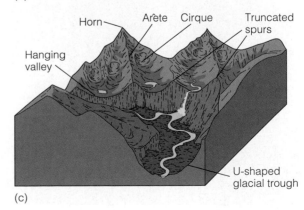

(c)

Horn · Arête · Cirque · Truncated spurs · Hanging valley · U-shaped glacial trough

(a)

(b)

Figure 17.14
U-shaped glacial troughs. (a) This glacial trough is in the Bighorn Mountains on the Wyoming–Montana border. (b) A similar feature in southern Germany. The lake in this glacial trough is impounded behind a glacial deposit known as an end moraine.

Figure 17.15
Geirangerfjorden, a fiord in Norway.

Figure 17.16
Bridalveil Falls in Yosemite National Park, California, plunges about 190 m from a hanging valley. The valley in the foreground is a huge U-shaped glacial trough.

Fiords are restricted to high latitudes where glaciers can be maintained even at low elevations, such as Alaska, western Canada, Scandinavia, Greenland, southern New Zealand, and southern Chile. Lower sea level during the Pleistocene was not entirely responsible for the formation of all fiords. Unlike running water, glaciers can erode a considerable distance below sea level. In fact, a glacier 500 m thick can stay in contact with the seafloor and effectively erode it to a depth of about 450 m before the buoyant effects of water cause the glacial ice to float! The depth of some fiords is impressive; some in Norway and southern Chile are about 1300 m deep.

Hanging Valleys Waterfalls can form in several ways, but some of the world's highest and most spectacular are found in recently glaciated areas. Bridalveil Falls in Yosemite National Park, California, plunge from a **hanging valley,** which is a tributary valley whose floor is at a higher level than that of the main valley (Figure 17.16). Where the two valleys meet, the mouth of the hanging valley is perched far above the main valley's floor (Figure 17.13c). Accordingly, streams flowing through hanging valleys plunge over vertical or steep precipices.

Although not all hanging valleys form by glacial erosion, many do. As Figure 17.13 shows, the large glacier in the main valley vigorously erodes, whereas the smaller glaciers in tributary valleys are less capable of large-scale erosion. When the glaciers disappear, the smaller tributary valleys remain as hanging valleys.

Cirques, Arêtes, and Horns Perhaps the most spectacular erosional landforms in areas of valley glaciation are at the upper ends of glacial troughs and along the divides separating adjacent glacial troughs. Valley glaciers form and move out from steep-walled, bowl-shaped depressions called **cirques** at the upper end of their troughs (Figure 17.13c). Cirques are typically steep-walled on three sides, but one side is open and leads into the glacial trough. Some cirques slope continuously into the glacial trough, but many have a lip or threshold at their lower end (Figure 17.17).

The details of cirque origin are not fully understood, but they apparently form by erosion of a preexisting depression on a mountain side. As snow and ice accumulate in the depression, frost wedging and plucking enlarge it until it takes on the typical cirque shape. In cirques with a lip or threshold, the glacial ice apparently not only moves out but rotates as well, scouring out a depression rimmed by rock. These depressions commonly contain a small lake known as a *tarn* (Figure 17.17).

Cirques become wider and are cut deeper into mountainsides by headward erosion as a result of abrasion, plucking, and several mass-wasting processes. For example, part of a steep cirque wall may collapse, while frost wedging continues to pry loose rocks that tumble downslope, so a combination of processes can erode a small mountainside depression into a large cirque; the largest one known is the Walcott Cirque in Antarctica, which is 16 km wide and 3 km deep.

The fact that cirques expand laterally and by headward erosion accounts for the origin of two other distinctive erosional features, arêtes and horns. **Arêtes—** narrow, serrated ridges—can form in two ways. In many cases, cirques form on opposite sides of a ridge, and headward erosion reduces the ridge until only a thin partition of rock remains (Figure 17.13c). The same effect occurs when erosion in two parallel glacial troughs reduces the intervening ridge to a thin spine of rock (Figure 17.18).

(a)

Figure 17.17
(a) This bowl-shaped depression on Mount Wheeler in Great Basin National Park, Nevada, is a cirque. Notice that it has steep walls on three sides but opens out into a glacial trough in the foreground. (b) Many cirques contain small lakes known as tarns, such as these on Mount Whitney in California. *Source: Photo courtesy of John S. Shelton.*

(b)

The most majestic of all mountain peaks are **horns,** steep-walled, pyramidal peaks formed by headward erosion of cirques. For a horn to form, a mountain peak must have at least three cirques on its flanks, all of which erode headward (Figure 17.13c). Excellent examples of horns include Mount Assiniboine in the Canadian Rockies, the Grand Teton in Wyoming, and the most famous of all, the Matterhorn in Switzerland (Figure 17.19).

Continental Glaciers and Erosional Landforms

Areas eroded by continental glaciers tend to be smooth and rounded because these glaciers bevel and abrade high areas that projected into the ice. Rather than yielding the sharp, angular landforms typical of valley glaciation, they produce a landscape of rather flat topography interrupted by rounded hills.

In a large part of Canada, particularly the vast Canadian Shield region, continental glaciation has stripped off the soil and unconsolidated surface sediment, revealing extensive exposures of striated and polished bedrock. These areas are thus characterized by deranged drainage (see Figure 15.28e), numerous lakes and swamps, low relief, extensive bedrock exposures, and little or no soil. They are generally referred to as *ice-scoured plains* (Figure 17.20). Similar though smaller bedrock exposures are also widespread in the northern United States from Maine through Minnesota.

Figure 17.18
This knifelike ridge between glacial troughs in California is an arête.

Figure 17.19
The Matterhorn in Switzerland is a well-known horn.

Perspective 17.2

Waterton Lakes National Park, Alberta, and Glacier National Park, Montana

W aterton Lakes National Park in Alberta, Canada, and Glacier National Park in Montana lie adjacent to one another and in 1932 were designated an international peace park, the first of its kind. Both parks have spectacular scenery; interesting wildlife such as mountain goats, bighorn sheep, and grizzly bears; and an impressive geologic history. The present-day landscapes resulted from deformation and uplift from Cretaceous to Eocene times, followed by deep erosion by streams and glaciers. Park visitors can see the results of the phenomenal forces at work during deformation by visiting sites where a large fault is visible, and

glacial landforms such as U-shaped glacial troughs, arêtes, cirques, and horns are some of the finest in North America (Figure 1).

Most of the rocks exposed in Glacier National Park belong to the Late Proterozoic-aged Belt Supergroup,* whereas those in Waterton Lakes National Park are assigned to the Purcell Supergroup. The names differ north and south of the border, but the rocks are the same. These Belt-Purcell rocks are nearly 4000 m thick and were

---

Supergroup is a geologic term for two or more groups that in turn are composed of two or more formations.

deposited between 1.45 billion and 850 million years ago. The rocks themselves are interesting, and some are attractive, especially red and green rocks consisting mostly of mud and thick limestone formations. In addition, many of the rocks contain a variety of sedimentary structures such as mud cracks, ripple marks, and cross-bedding that help geologists interpret how they were deposited in the first place. Dark-colored mudstones and sandstones were deposited during the Cretaceous Period when a marine transgression took place that covered a large part of North America, including the area of the present-day parks.

(a)

(b)

Figure 1
The sharp angular peaks and ridges and rounded valleys are typical of areas eroded by valley glaciers such as (a) Glacier National Park, Montana, and (b) Waterton Lakes National Park, Alberta.

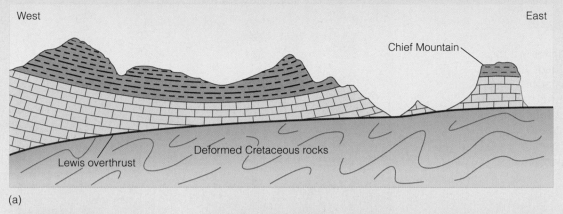

West   East

Chief Mountain

Deformed Cretaceous rocks

Lewis overthrust

(a)

Lewis overthrust

(b)

Chief Mountain

(c)

Figure 2

(a) Diagrammatic view of the Lewis overthrust in Glacier National Park, Montana. Rocks of Late Proterozoic age now rest on deformed Cretaceous sedimentary rocks. (b) View from Marias Pass reveals the fault as a light-colored line on the mountainside. (c) Erosion has isolated Chief Mountain from the rest of the slab of overthrust rock.

The most impressive geologic structure in the parks is the Lewis overthrust,* a large fault along which Belt-Purcell rocks have moved at least 75 km eastward so that they now rest on much younger Cretaceous-aged rocks (Figure 2). If you take the trail from Marias Pass in Glacier National Park to get a closer look at the fault, you can see intense deformation of the rocks lying below the fault. In any case, this large slab of ancient rocks has been deeply dissected by streams and glaciers, giving rise to the parks' present landscapes.

During the Pleistocene Epoch, glaciers formed and grew, overtopping the divides between valleys and thus form-

ing an ice cap that nearly buried the entire area. In fact, several episodes of Pleistocene glaciation took place, but the evidence for the most recent one is most obvious. These glaciers flowed outward in all directions, and in the east they merged with the continental glacier covering most of Canada and the northern states. Much of the parks' landscapes developed during these glacial episodes as valleys were gouged deeper and widened, and cirques, arêtes, and horns developed (Figure 1).

Today only a few dozen small glaciers remain active in the parks. But just like earlier glaciers, they continue to erode, transport, and deposit sediment, only at a considerably reduced rate. In fact, many of the 150 or so glaciers present in Glacier National Park in 1850 are now gone or remain simply as

patches of stagnant ice. And even among the others it is difficult to determine exactly how many are active because they are so small and move so slowly, only a few meters per year. They did, however, expand markedly during the Little Ice Age, but have since retreated (Figure 3). For example, Grinnell Glacier covered only 0.88 km² (217 acres) in 1993 as opposed to 2.33 km² (576 acres) in 1850 (Figure 3), and during the same time Sperry Glacier was down to 0.87 km² from 3.76 km².

It seems that these small glaciers, which are very sensitive to climatic changes, are shrinking as a result of the 1°C increase in average summer temperatures in this region since 1900. According to one U.S. Geological Survey report, expected increased warming would eliminate the glaciers

---

*An overthrust fault is simply a very low angle thrust fault along which movement is usually measured in kilometers.

by 2030, and certainly by 2100 even if no additional warming takes place.

None of the active glaciers in either park can be reached by road, but several are visible from a distance. Nevertheless, Pleistocene glaciers and the remaining active ones were responsible for much of the striking scenery. Now weathering, mass wasting, and streams are modifying the glacial landscape.

Grinnell Glacier 1850–1981

1850
1937
1968
1981

Figure 3
Grinnell Glacier in Glacier Nationàl Park, Montana. In 1850, at the end of the Little Ice Age, the glacier extended much further and covered about 576 acres. By 1981, its terminus had retreated to the position shown, and in 1993 it covered only about 217 acres.

Figure 17.20
An ice-scoured plain in the Northwest Territories of Canada.

GLACIAL DEPOSITS

Both valley and continental glaciers are effective at erosion and transport, but eventually they deposit their sediment load as **glacial drift**, a general term for all deposits resulting from glacial activity. A vast sheet of Pleistocene glacial drift is present in the northern tier of the United States and adjacent parts of Canada. Smaller but similar deposits are also found where valley glaciers existed or remain active. The appearance of these deposits may not be inspiring as are some landforms resulting from glacial erosion, but they are important as reservoirs of groundwater and in many areas they are exploited for their sand and gravel. As a matter of fact, sand and gravel constitute a large part of the mineral extraction economies of several states and provinces.

All glacial drift has been carried far from its source area, but one conspicuous element of drift is boulders scattered around that are obviously not derived from the area in which they now rest. These **glacial erratics**, as they are called, have simply been eroded and transported from some distant region and then deposited (Figure 17.21). A good example is the popular decorative stone (called pudding stone) in Michigan consisting of quartzite containing conspicuous pieces of red jasper that was eroded from surface exposures in Ontario, Canada.

As noted, *glacial drift* is a general term and geologists generally define two types of drift: till and stratified drift. **Till** consists of sediments deposited directly by glacial ice. These deposits are not sorted by particle size or density, and they exhibit no layering or stratification. The till of both valley and continental glaciers is similar, but that of continental glaciers is much more extensive and generally has been transported much farther.

As opposed to till, **stratified drift** is layered—that is, stratified—and it invariably exhibits some degree of sorting by particle size. As a matter of fact, most of the deposits recognized as stratified drift are actually layers of sand and gravel or mixtures thereof that accumulated in braided stream channels. In Chapter 15 we mentioned that streams issuing from melting glaciers are commonly braided because they receive more sediment than they can effectively transport.

Landforms Composed of Till

Landforms composed of till include several types of moraines and elongated hills known as *drumlins*.

End Moraines   The terminus of either a valley or a continental glacier may become stabilized in one position for some period of time, perhaps a few years or even decades. Stabilization of the ice front does not mean that the glacier has ceased flowing, only that it has a balanced budget (Figure 17.8). When an ice front is stationary, flow within the glacier continues, and any sediment transported within or upon the ice is dumped as a pile of rubble at the glacier's terminus (Figures 17.22 and 17.23). These deposits are **end moraines**, which continue to grow as long as the ice front remains stationary. End moraines of valley glaciers are commonly crescent-shaped ridges of till spanning the valley occupied by the glacier. Those of continental glaciers similarly parallel the ice front, but are much more extensive.

Following a period of stabilization, a glacier may advance or retreat, depending on changes in its budget. If it advances, the ice front overrides and modifies its former moraine. Should it have a negative budget, though, the ice front retreats toward the zone of accumulation. As the ice front recedes, till is deposited as it is liberated from the melting ice and forms a layer of **ground moraine** (Figure 17.22b). Ground moraine has an irregular, rolling topography, whereas end moraine consists of long ridgelike accumulations of sediment.

(a)

(b)

Figure 17.21
Glacial erratics at (a) Hammond, New York, and (b) on Beaver Island in Lake Michigan. Both boulders have been transported far from their sources and are unlike the rocks at their present locations.

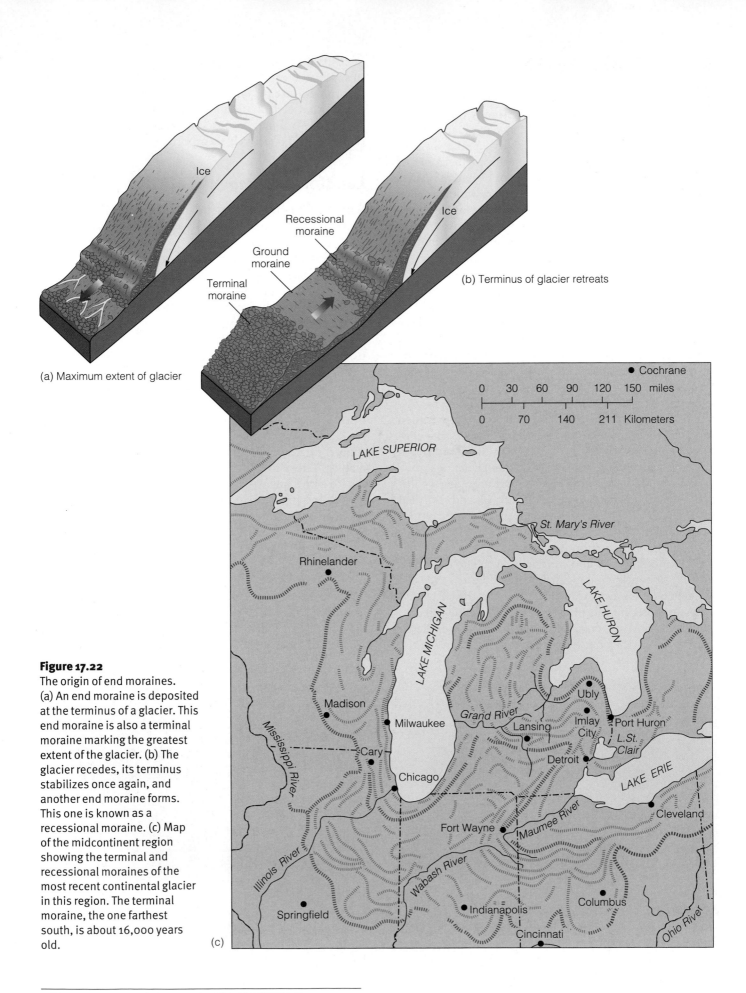

(a) Maximum extent of glacier

(b) Terminus of glacier retreats

Figure 17.22
The origin of end moraines.
(a) An end moraine is deposited at the terminus of a glacier. This end moraine is also a terminal moraine marking the greatest extent of the glacier. (b) The glacier recedes, its terminus stabilizes once again, and another end moraine forms. This one is known as a recessional moraine. (c) Map of the midcontinent region showing the terminal and recessional moraines of the most recent continental glacier in this region. The terminal moraine, the one farthest south, is about 16,000 years old.

End moraine

(a)

(b)

Figure 17.23
(a) An end moraine deposited by a valley glacier. This particular end moraine is also a terminal moraine because it is the one most distant from the glacier's source. (b) Close-up of an end moraine. Notice that the deposit is not sorted by particle size, and it shows no layering or stratification.

After a glacier has retreated for some time, its terminus may once again stabilize, and it deposits another end moraine. Because the ice front has receded, such moraines are called **recessional moraines** (Figure 17.22b). During the Pleistocene Epoch, continental glaciers in the midcontinent region extended as far south as southern Ohio, Indiana, and Illinois. Their outermost end moraines, marking the greatest extent of the glaciers, go by the special name **terminal moraine** (valley glaciers also deposit terminal moraines). As the glaciers retreated from the positions where their terminal moraines were deposited, they temporarily ceased retreating numerous times and deposited dozens of recessional moraines (Figure 17.22c).

Lateral and Medial Moraines Valley glaciers transport considerable sediment along their margins. Much of this sediment is abraded and plucked from the valley walls, but a significant amount falls or slides onto the glacier's surface by mass-wasting processes. In any case, this sediment is deposited as long ridges of till called **lat-**eral moraines along the margin of the glacier (Figure 17.24).

Where two lateral moraines merge, as when a tributary glacier flows into a larger glacier, a **medial moraine** forms (Figure 17.24). A large glacier will often have several dark stripes of sediment on its surface, each of which is a medial moraine. Although medial moraines are identified by their position on a valley glacier, they are, in fact, formed from the coalescence of two lateral moraines. One can generally determine how many tributaries a valley glacier has by the number of its medial moraines (Figure 17.24).

Drumlins In many areas where continental glaciers deposited till, the till has been reshaped into elongated hills known as **drumlins.** Some drumlins measure as much as 50 m high and 1 km long, but most are much smaller. From the side, a drumlin looks like an inverted spoon with the steep end on the side from which the glacial ice advanced, and the gently sloping end pointing in

(a)

(b)

Figure 17.24
(a) Lateral and medial moraines on a glacier in Alaska. Notice that where the two large tributary glaciers converge two lateral moraines merge to form a medial moraine. (b) The two parallel ridges extending from this mountain valley are lateral moraines.

the direction of ice movement (Figure 17.25). Thus, drumlins can be used to determine the direction of ice movement. Drumlins are rarely found as single, isolated hills; instead, they occur in *drumlin fields* that contain hundreds or thousands of drumlins. Drumlin fields are found in several states and Ontario, Canada, but perhaps the finest example is near Palmyra, New York.

No one fully understands how drumlins originate. According to one hypothesis, they form when till beneath a glacier is reshaped into streamlined hills as the ice moves over it by plastic flow. Another hypothesis holds that huge floods of glacial meltwater modify till into drumlins.

Landforms Composed of Stratified Drift

As already noted, stratified drift is a type of glacial deposit exhibiting sorting and layering, both indications that it was deposited by running water. Stratified drift is deposited by both valley and continental glaciers, but as one would expect, it is more extensive in areas of continental glaciation.

Outwash Plains and Valley Trains Glaciers discharge meltwater laden with sediment most of the time, except perhaps during the coldest months. This meltwater forms a series of braided streams that radiate out from the front of continental glaciers over a wide region. So much sediment is supplied to these streams that much of it is deposited within the channels as sand and gravel bars. The vast blanket of sediments so formed is an **outwash plain** (Figures 17.26 and 17.27).

Occasionally, huge quantities of meltwater are discharged from glaciers when, for instance, a volcano erupts beneath them. A recent example is an eruption beneath Iceland's largest ice cap, which melted the base of the glacier. Meltwater accumulated in a caldera, also beneath the ice cap, and on November 5, 1996, suddenly burst forth in Iceland's largest flood in 60 years. Huge blocks of ice estimated to weigh 900 metric tons were ripped from the glacier and deposited nearly 5 km away. Damage to roads, utility lines, and bridges amounted to about $15 million, but no lives were lost.

Valley glaciers also discharge large amounts of meltwater and, like continental glaciers, have braided

Figure 17.25
These streamlined hills are drumlins.

(a)

(b)

Figure 17.26
(a) Outwash plain deposits in Michigan. (b) Braided streams discharging from a valley glacier, such as these in the Yukon Territory, Canada, deposit a valley train made of stratified drift. *Source: Photo courtesy of John S. Shelton.*

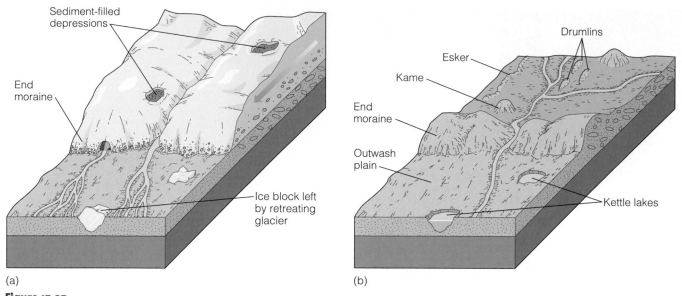

(a)

(b)

Figure 17.27
Two stages in the origin of kettles, kames, eskers, and outwash plains: (a) during glaciation and (b) after glaciation.

streams extending from them. However, these streams are confined to the lower parts of glacial troughs, and their long, narrow deposits of stratified drift are known as **valley trains** (Figure 17.26b).

Outwash plains, valley trains, and some end moraines commonly contain numerous circular to oval depressions, many of which contain small lakes. These depressions are *kettles*; they form when a retreating ice sheet or valley glacier leaves a block of ice that is subsequently partly or wholly buried (Figure 17.27). When the ice block eventually melts, it leaves a depression; if the depression extends below the water table, it be-

Figure 17.28
(a) This small hill in Wisconsin is a kame. (b) This sinuous ridge near Dahlen, North Dakota is an esker.

(a)

(b)

comes the site of a small lake. Some outwash plains have so many kettles that they are called *pitted outwash plains*.

Kames and Eskers   **Kames** are conical hills as much as 50 m high composed of stratified drift (Figures 17.27 and 17.28a). Many kames form when a stream deposits sediment in a depression on a glacier's surface; as the ice melts, the deposit is lowered to the surface. They also form in cavities within or beneath stagnant ice.

Long sinuous ridges of stratified drift, many of which meander and have tributaries, are **eskers** (Figures 17.27 and 17.28b). Most eskers have sharp crests and sides that slope at about 30 degrees. Some are quite high, as much as 100 m, and can be traced for more than 500 km. Eskers are usually found in areas once covered by continental glaciers, but they are also associated with large valley glaciers. The sorting and stratification of the sediments within eskers clearly indicate deposition by running water. The properties of ancient eskers and observations of present-day glaciers indicate that they form in tunnels beneath stagnant ice (Figure 17.27). Excellent examples of eskers can be seen at Kettle Moraine State Park in Wisconsin and in several other states, but the most extensive eskers in the world are in northern Canada.

Deposits in Glacial Lakes

Numerous lakes exist in areas of glaciation. Some formed as a result of glaciers scouring out depressions; others occur where a stream's drainage was blocked; and others are the result of water accumulating behind moraines or in kettles. Regardless of how they formed, glacial lakes, like all lakes, are areas of deposition. Sediment may be carried into them and deposited as small deltas, but of special interest are the fine-grained deposits. Mud deposits in glacial lakes are commonly finely laminated (having layers less than 1 cm thick) and consist of alternating light and dark layers. Each light–dark couplet is a *varve* (Figure 17.29), which represents an annual episode of deposition; the light layer formed during the spring and summer and consists of silt and clay; the dark layer formed during the winter when the smallest particles of clay and organic matter settled from suspension as the lake froze over. The number of varves indicates how many years a glacial lake has existed.

Another distinctive feature of glacial lakes containing varved deposits is the presence of *dropstones* (Figure 17.29). These are pieces of gravel, some of boulder size, in otherwise very fine-grained deposits. The presence of varves indicates that currents and turbulence in such lakes was minimal, otherwise clay and organic matter would not have settled from suspension. How then can we account for dropstones in a low-energy environment? Most of them were probably carried into

Figure 17.29
Glacial varves with a dropstone. Each varve, a dark–light couplet, is an annual deposit.

the lakes by icebergs that eventually melted and released sediment contained in the ice.

THE ICE AGE—WHAT WAS IT, AND WHEN DID IT TAKE PLACE?

In hindsight, it is hard to believe that so many competent naturalists of the last century were skeptical that widespread glaciers existed on the northern continents during the not too distant past. Many naturalists invoked the biblical flood to account for the large boulders throughout Europe that occur far from their sources. Others believed that the boulders were rafted to their present positions by icebergs floating in floodwaters. It was not until 1837 that the Swiss naturalist Louis Agassiz argued convincingly that the displaced boulders, many coarse-grained sedimentary deposits, polished and striated bedrock, and many of the valleys of Europe resulted from huge ice masses moving over the land.

We know today that the Pleistocene Ice Age began about 1.6 million years ago (MYA) and consisted of several intervals of glacial expansion separated by warmer interglacial periods. At least four major episodes of Pleistocene glaciation have been recognized in North America, and six or seven major glacial advances and retreats are recognized in Europe. It now appears that at least 20 warm–cold cycles can be detected in deep-sea

Dennis C. Trabant

Understanding Ice: A Career in Glaciology

Dennis C. Trabant received a B.S. in geology from Kansas State University and an M.S. in geology/glaciology from the University of Alaska. He is a glaciologist for the U.S. Geological Survey and is currently on the staff of the Alaska Volcano Observatory.

The U.S. Geological Survey (USGS) studies glaciers to understand glacier-related hydrologic processes for improving predictions of water resources, glacier-related hazards, and the consequences of global change. When I joined the USGS in 1973, my first duties involved maintaining the long-term climate and glacier mass-balance observations on two glaciers in Alaska. One of my later projects was to develop a coordinated ice-motion and mass-balance monitoring methodology for research on surging and calving glaciers. For reasons that we do not fully understand, surging glaciers periodically increase their flow speeds by 10 times or more and then return to more normal flow rates for periods of 20 to 50 years. Columbia Glacier is a large calving glacier along the south coast of Alaska. I was a member of the team that predicted the retreat of the glacier and the associated increased iceberg hazard to oil tankers leaving the southern terminal of the Alaska pipeline. These predictions were partly based on analysis of the team's ice-motion and mass-balance measurements. During the Columbia Glacier study, I introduced the use of beacon radio tracking in glacier research for determining glacier motion and internal temperature. Beacon radios are better known for their use in animal tracking studies.

I have been involved in volcano–glacier hazards analysis since a posteruptive snow- and ice-hazard assessment of Mount St. Helens shortly after its reawakening in 1980. Following the eruption of Mount St. Helens, heightened interest in volcano–hydrologic hazards led to glacier ice-volume assessment work in the Cascade Mountains of Washington and Oregon.

The consequences of global change are revealed in the long-term data series from two stations in Alaska. Gulkana Glacier in interior Alaska is approaching a newly reestablished mass-balance equilibrium, while Wolverine Glacier on the coast is growing under the influence of a small climate warming. The association of glacier growth and climate warming is contrary to the generally accepted idea that glacier volume decreases with warming. The measurements at the coastal glacier apply for only a small temperature rise and are caused by the increased storminess associated with warmer waters at the surface of the Gulf of Alaska.

I joined the staff of the USGS's Alaska Volcano Observatory during the eruptions of Redoubt volcano in 1989 and 1990. These eruptions mobilized 113 to 121 × 10^6 m³ of perennial snow and ice, producing lahars and floods that swept the full 35-km length of the river valley to the sea. The Redoubt work led to understanding that mechanical entrainment of ice and snow in volcanic flows dominates the first phase of the conversion of the snow and ice to the liquid state and flood generation.

Recent work has established a relation between synthetic aperture radar (SAR) image patterns and glacier-bed topography. Although SAR wavelengths do not penetrate to glacier beds, glacier-bed topography is reflected on the ice surface by crevasse size and orientation, which influence SAR image patterns. The indirect relation has proved to be useful in identifying subglacial bedrock structures.

I am currently developing hazard assessments for glacier-clad volcanoes and have recently applied a GIS volume-modeling approach for evaluating the volume of perennial snow and ice volume and its distribution with altitude and aspect on volcanoes. I have recently helped document a surge of the Bering Glacier, Alaska—the world's largest temperate-ice surge-type glacier—using time-lapse cameras and satellite telemarketing lake-level and water-quality parameters from proglacial lakes. My interest there is the relation between surge motion and subglacial hydrology.

For those interested in pursuing glaciology as a possible career, glaciology has no fixed curriculum requirements. Most glaciologists have geology or geophysics backgrounds, but others come from subatomic physics, pure mathematics, electrical engineering, and oceanography. My own career has given me an opportunity to travel throughout Alaska, the western United States, and abroad and to spend time in environments that few people are privileged to visit. ∎

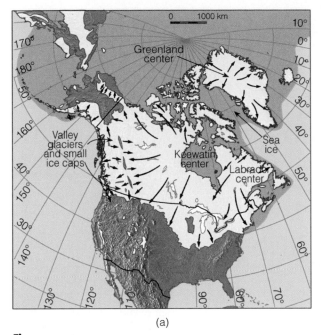

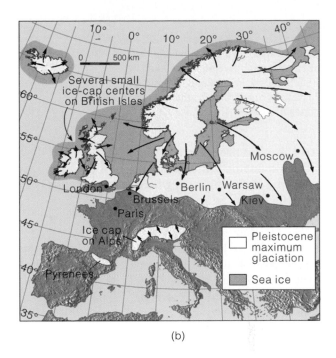

Figure 17.30
(a) Centers of ice accumulation and maximum extent of Pleistocene glaciation in North America. (b) Centers of ice accumulation and directions of ice movement in Europe during the maximum extent of Pleistocene glaciation.

cores. In view of these data, the traditional four-part subdivision of the Pleistocene of North America must be modified. Based on the best available evidence, it appears that the Pleistocene ended about 10,000 years ago (YA). But geologists do not know if the present interglacial period will persist indefinitely, or whether we will enter another glacial interval.

The onset of glacial conditions really began about 40 MYA when surface ocean waters at high southern latitudes suddenly cooled. By about 38 MYA, glaciers had formed in Antarctica, but a continuous ice sheet did not develop there until 15 MYA. Following a brief warming trend during the Late Tertiary Period, ice sheets began forming in the Northern Hemisphere, and by 1.6 MYA the Pleistocene Ice Age was under way. At their greatest extent, Pleistocene glaciers covered about three times as much of Earth's surface as they do now and were up to 3 km thick (Figure 17.30). Large areas of North America were covered by glacial ice as were Greenland, Scandinavia, Great Britain, Ireland, and a large part of northern Russia. Mountainous areas also experienced an expansion of valley glaciers and the development of ice caps.

Pleistocene Climates

As one would expect, the climatic effects responsible for Pleistocene glaciation were worldwide. Contrary to popular belief, though, the world was not as frigid as it is commonly portrayed in cartoons and movies. During times of glacier growth, those areas in the immediate vicinity of the glaciers experienced short summers and long, wet winters.

Areas outside the glaciated regions experienced varied climates. During times of glacial growth, lower ocean temperatures reduced evaporation so that most of the world was drier than it is today, but some areas that are arid now were much wetter. For example, as the cold belts at high latitudes expanded, the temperate, subtropical, and tropical zones were compressed toward the equator, and the rain that now falls on the Mediterranean shifted so that it fell on the Sahara of North Africa, enabling lush forests to grow in what is now desert. California and the arid southwestern United States were also wetter because a high-pressure zone over the northern ice sheet deflected Pacific winter storms south.

Following the Pleistocene, mild temperatures prevailed between 8000 and 6000 YA. After this warm period, conditions gradually became cooler and moister, favoring the growth of valley glaciers on the Northern Hemisphere continents. Careful studies of the deposits at the margins of present-day glaciers reveal that during the last 6000 years (a time called the *Neoglaciation*), glaciers expanded several times. The last expansion, which took place between 1500 and the mid-to-late 1800s, was the Little Ice Age (see the Prologue).

Pluvial and Proglacial Lakes

During the Pleistocene, many of the basins in the western United States contained large lakes that formed as a

Figure 17.31

Pleistocene lakes in the western United States. Lake Missoula was a proglacial lake but all the others were pluvial lakes. The Great Salt Lake in Utah and Pyramid Lake in Nevada are shrunken remnants of once much more extensive lakes.

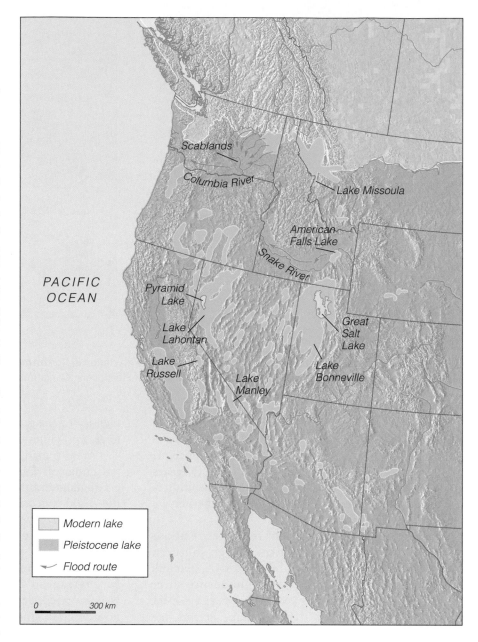

result of greater precipitation and overall cooler temperatures (especially during the summer), which lowered the evaporation rate (Figure 17.31). The largest of these *pluvial lakes,* as they are called, was Lake Bonneville, which attained a maximum size of 50,000 km² and a depth of at least 335 m (Figure 17.31). The vast salt deposits of the Bonneville Salt Flats west of Salt Lake City in Utah formed as parts of this ancient lake dried up: Great Salt Lake is simply the remnant of this once much larger lake.

Another large pluvial lake existed in Death Valley, California (see Perspective 18.1), which is now the hottest, driest place in North America. During the Pleistocene, Death Valley received enough rainfall to maintain a lake 145 km long and 178 m deep. When the lake evaporated, the dissolved salts were precipitated on the valley floor; some of these evaporite deposits, especially borax, are important mineral resources (see Chapter 18).

In contrast to pluvial lakes, which form far from glaciers, *proglacial lakes* are formed by meltwater accumulating along the margins of glaciers. In fact, one shoreline of proglacial lakes is the ice front itself (see Perspective 17.3). Lake Agassiz, named in honor of the naturalist Louis Agassiz, was a large proglacial lake covering about 250,000 km², mostly in Manitoba, Saskatchewan, and Ontario, Canada, but extending into North Dakota and Minnesota. It persisted until the glacial ice along its northern margin melted, at which time the lake was able to drain northward into Hudson Bay.

Before the Pleistocene, no large lakes existed in the Great Lakes region, which was then an area of generally flat lowlands with broad stream valleys draining to the north. As the glaciers advanced southward, they eroded the stream valleys more deeply, forming what were to become the basins of the Great Lakes—four of the five Great Lakes basins were eroded below sea level.

At their greatest extent, the glaciers covered the entire Great Lakes region and extended far to the south. As the ice sheet retreated north during the late Pleistocene, the ice front periodically stabilized, and numerous recessional moraines were deposited (see Figure 17.22c). By about 14,000 YA, parts of the Lake Michigan and Lake Erie basins were ice free, and glacial meltwater began forming proglacial lakes (Figure 17.32). As the retreat of the ice sheet continued—although periodically interrupted by minor readvances of the ice front—the Great Lakes basins were uncovered, and the lakes expanded until they eventually reached their present size and configuration. Currently, the Great Lakes contain nearly 23,000 km³ of water, about 18% of the water in all freshwater lakes.

13,000 years ago

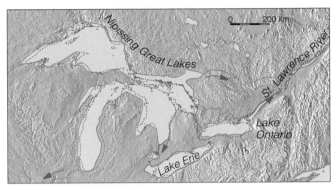

11,500 years ago

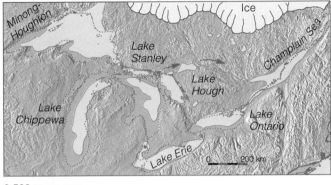

9,500 years ago

6,000 years ago

Figure 17.32

Four stages in the evolution of the Great Lakes. As the glacial ice retreated northward, the lake basins began filling with meltwater. The dotted lines indicate the present-day shorelines of the lakes.

Glaciation and Changes in Sea Level

More than 70 million km³ of snow and ice covered the continents during the maximum glacial coverage of the Pleistocene. The storage of ocean waters in glaciers lowered sea level 130 m and exposed extensive areas of the present-day continental shelves, which were quickly covered with vegetation. Indeed, a land bridge existed across the Bering Strait from Alaska to Siberia. Native Americans crossed the Bering land bridge, and various animals migrated between the continents; the American bison, for example, migrated from Asia. The British Isles were connected to Europe during the glacial intervals because the shallow floor of the North Sea was above sea level. When the glaciers disappeared, these areas were again flooded, drowning the plants and forcing the animals to migrate farther inland.

Lowering of sea level during the Pleistocene also affected the base level of most major streams. When sea level dropped, streams eroded deeper valleys as they sought to adjust to a new lower base level (see Chapter 15). Stream channels in coastal areas were extended and deep-

ened along the emergent continental shelves. When sea level rose at the end of the Pleistocene, the lower ends of river valleys along the East Coast of North America were flooded and are now important harbors (see Chapter 19).

A tremendous quantity of water is still stored on land in present-day glaciers (see Figure 15.5). If these glaciers should completely melt, sea level would rise about 70 m, flooding all coastal areas of the world where many of the large population centers are located.

WHAT WOULD YOU DO?

Suppose that for some reason rainwater becomes slightly more acidic than it is now. Not acid enough to be a health threat, but certainly acid enough to have an effect on chemical weathering. Given that vast exposures of limestone are present on continents, how might this increased acidity have an ultimate effect on glaciers?

Perspective 17.3

Glacial Lake Missoula and the Channeled Scablands

The term *scabland* is used in the Pacific Northwest to describe areas where the surface deposits have been scoured, thus exposing the underlying rock. An area of scablands exists in part of eastern Washington where numerous deep and generally dry channels are present. Some of these channels, cut into basalt lava flows, are more than 70 m deep, and their floors are covered by gigantic "ripple marks" as much as 10 m high and 70 to 100 m apart. In addition, a number of high hills in the area are arranged so that they appear to have been islands in a large, braided stream.

In 1923 J Harlan Bretz proposed that the channeled scablands of eastern Washington formed during a single, gigantic flood of glacial meltwater lasting only a few days. Bretz's unorthodox explanation was rejected by most geologists who preferred a more traditional interpretation based on normal stream erosion over a long period of time.

The problem with Bretz's hypothesis was that he could not identify an adequate source for his floodwater. He knew that glaciers had advanced as far south as Spokane, Washington, but he could not explain how so much ice melted so rapidly. The answer to Bretz's dilemma came from western Montana where an enormous ice-dammed lake (Lake Missoula) had formed. Lake Missoula filled when an advancing glacier plugged the Clark Fork Valley at Ice Cork, Idaho, causing the water to fill the valleys of western Montana (Figure 1). At its highest level, Lake Missoula covered about 7800 km² and contained an estimated 2090 km³ of water (about 42% of the volume of present-day Lake Michigan). The shorelines of Lake Missoula are still clearly visible on the mountainsides around Missoula, Montana (Figure 2a).

When the ice dam impounding Lake Missoula collapsed, the water rushed out at tremendous velocity and drained south and southwest across Idaho and into Washington. The maximum rate of flow was probably 11 million m³/sec, about 55 times greater than the average discharge of the Amazon River. When these raging floodwaters reached eastern Washington, they stripped away the soil and most of the surface sediment, carving out huge valleys in bedrock. The currents were so powerful and turbulent they plucked out and moved pieces of basalt measuring 10 m across. Within the channels, sand and gravel were shaped into huge ridges, the so-called giant ripple marks (Figure 2b).

Bretz originally thought that one massive flood formed the channeled scablands, but geologists now know that Lake Missoula formed, flooded, and reformed at least four times and perhaps as many as seven times. The largest lake formed 18,000 to 20,000 years ago, and its draining produced the last great flood.

How long did the flood last, and did humans witness it? According to one estimate, approximately one month passed from the time the ice dam first broke and water rushed out onto the scablands to the time the scabland streams returned to normal flow. No one knows for sure if anyone witnessed the flood. The oldest known evidence of humans in the region is from a site in southeastern Washington dated at 10,130 years ago, long after the last flood from Lake Missoula. However, it is now generally accepted that Native Americans were present in North America at least 15,000 years ago, but whether anyone was in this area 18,000 years ago is unknown.

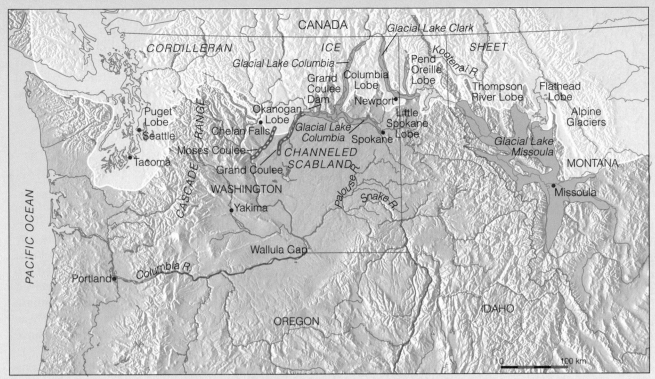

Figure 1
Location of glacial Lake Missoula and the channeled scablands of eastern Washington.

(a)   (b)

Figure 2
(a) The horizontal lines on Sentinel Mountain at Missoula, Montana, are wave-cut shorelines of glacial Lake Missoula. (b) These gravel ridges are the so-called giant ripple marks that formed when glacial Lake Missoula drained across this area near Camas Hot Springs, Montana.

GLACIERS AND ISOSTASY

In Chapter 10 we discussed the concept of isostasy and noted that the crust responds to an increased load by subsiding, and it rises when a load on it is reduced. No one questions that isostatic rebound has taken place in areas formerly covered by continental glaciers. In fact, a number of features in these areas clearly shows the effects of isostatic adjustment of the crust.

When the Pleistocene ice sheets formed and increased in size, the weight of the ice caused the crust to respond by slowly subsiding deeper into the mantle. In some places, the surface was depressed as much as 300 m below preglacial elevations. As the ice sheets disappeared, the downwarped areas gradually rebounded to their former positions. As noted in Chapter 10, parts of Scandinavia are still rebounding at a rate of about 1 m per century. And more than 100 m of isostatic rebound has taken place in northeastern Canada during the last 6000 years.

We noted previously that the Great Lakes evolved as the glaciers retreated to the north. As one would expect, isostatic rebound began as the ice front retreated north. Rebound began first in the southern part of the region because that area was free of ice first. Furthermore, the greatest loading by glaciers, and hence the greatest crustal depression, occurred farther north in Canada in the zones of accumulation. For these reasons, rebound has not been evenly distributed over the entire glaciated area: it increases in magnitude from south to north. As a result of this uneven isostatic rebound, coastal features in the Great Lakes region, such as old shorelines, are now elevated higher above their former levels in the north and thus slope to the south.

WHAT CAUSES ICE AGES?

So far we have examined the effects of glaciation but have not addressed the central questions of what causes large-scale glaciation and why there have been so few episodes of widespread glaciation. For more than a century, scientists have been attempting to develop a comprehensive theory explaining all aspects of ice ages, but they have not yet been completely successful. One reason for their lack of success is that the climatic changes responsible for glaciation, the cyclic occurrence of glacial–interglacial episodes, and short-term events such as the Little Ice Age operate on vastly different time scales.

Only a few periods of glaciation are recognized in the geologic record, each separated from the others by long intervals of mild climate. Such long-term climatic changes probably result from slow geographic changes related to plate tectonic activity. Moving plates can carry continents to high latitudes where glaciers can exist, provided that they receive enough precipitation as snow. Plate collisions, the subsequent uplift of vast areas far above sea level, and the changing atmospheric and oceanic circulation patterns caused by the changing shapes and positions of plates also contribute to long-term climatic change.

The Milankovitch Theory

Changes in Earth's orbit as a cause of intermediate-term climatic events was first proposed during the mid-1800s, but it was made popular during the 1920s by the Serbian astronomer Milutin Milankovitch. He proposed that minor irregularities in Earth's rotation and orbit are sufficient to alter the amount of solar radiation received at any given latitude and hence bring about climate changes. Now called the **Milankovitch theory,** it was initially ignored but has received renewed interest since the 1970s and is widely accepted.

Milankovitch attributed the onset of the Pleistocene Ice Age to variations in three aspects of Earth's orbit. The first is *orbital eccentricity,* which is the degree to which Earth's orbit departs from a circle (Figure 17.33a). When the orbit is nearly circular, both the Northern and Southern Hemispheres have similar contrasts between the seasons. However, if the orbit is more elliptic, hot summers and cold winters will occur in one hemisphere, whereas warm summers and cool winters will take place in the other hemisphere. Calculations indicate a roughly 100,000-year cycle between times of maximum eccentricity, which corresponds closely to the 20 warm–cold climatic cycles that took place during the Pleistocene.

Milankovitch also pointed out that the angle between Earth's axis and a line perpendicular to the plane of the ecliptic shifts about 1.5 degrees from its current value of 23.5 degrees during a 41,000-year cycle (Figure 17.33b). Although changes in *axial tilt* have little effect on equatorial latitudes, they strongly affect the amount of solar radiation received at high latitudes and the duration of the dark period at and near Earth's poles. Coupled with the third aspect of Earth's orbit, precession of the equinoxes, high latitudes might receive as much as 15% less solar radiation, certainly enough to affect glacial growth and melting.

The last aspect of Earth's orbit that Milankovitch cited is *precession of the equinoxes,* which refers to a change in the time of the equinoxes. At present, the equinoxes take place on about March 21 and September 21 when the sun is directly over the equator. But as Earth rotates on its axis, it also wobbles as its axial tilt varies 1.5 degrees from its current value, thus changing the time of the equinoxes. Taken alone, the time of the

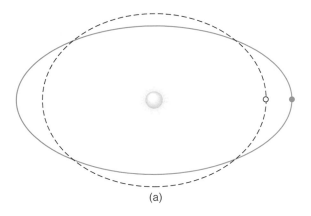

(a)

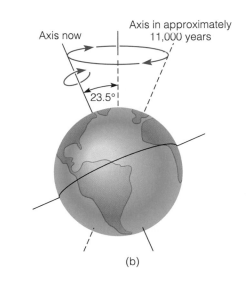

Axis now

Axis in approximately 11,000 years

23.5°

(b)

Conditions now

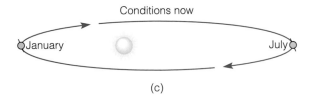

January

July

(c)

Conditions in about 11,000 years

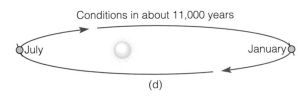

July

January

(d)

Figure 17.33
(a) Earth's orbit varies from nearly a circle (dashed line) to an ellipse (solid line) and back again in about 100,000 years. (b) Earth moves around its orbit while spinning about its axis, which is tilted to the plane of the ecliptic at 23.5 degrees and points toward the North Star. Earth's axis of rotation slowly moves and traces out the path of a cone in space. (c) At present, Earth is closest to the Sun in January when the Northern Hemisphere experiences winter. (d) In about 11,000 years, as a result of precession, Earth will be closer to the Sun in July, when summer occurs in the Northern Hemisphere.

equinoxes has little climatic effect, but changes in Earth's axial tilt also change the time of *aphelion* and *perihelion*, which are, respectively, when Earth is farthest from and closest to the Sun during its orbit (Figure 17.33c and d). Earth is now at perihelion, closest to the Sun, during Northern Hemisphere winters, but in about 11,000 years perihelion will be in July. Accordingly, Earth will be at aphelion, farthest from the Sun, in January and have colder winters.

Continuous variations in Earth's orbit and axial tilt cause the amount of solar heat received at any latitude to vary slightly through time. The total heat received by the planet changes little, but according to Milankovitch, and now many scientists agree, these changes cause complex climatic variations and provided the triggering mechanism for the glacial–interglacial episodes of the Pleistocene.

Short-Term Climatic Events

Climatic events having durations of several centuries, such as the Little Ice Age, are too short to be accounted for by plate tectonics or Milankovitch cycles. Several hypotheses have been proposed, including variations in solar energy and volcanism.

Variations in solar energy could result from changes within the Sun itself or from anything that would reduce the amount of energy Earth receives from the Sun. The latter could result from the solar system passing through clouds of interstellar dust and gas or from substances in the atmosphere reflecting solar radiation back into space. Records kept over the past 75 years indicate that during this time the amount of solar radiation has varied only slightly. Although variations in solar energy may influence short-term climatic events, such a correlation has not been demonstrated.

During large volcanic eruptions, tremendous amounts of ash and gases are spewed into the atmosphere where they reflect incoming solar radiation and thus reduce atmospheric temperatures. Small droplets of sulfur gases remain in the atmosphere for years and can have a significant effect on climate. Several large-scale volcanic events have been recorded, such as the 1815 eruption of Tambora, and are known to have had climatic effects. However, no relationship between periods of volcanic activity and periods of glaciation has yet been established.

Chapter Summary

1. Glaciers are masses of ice on land that move by plastic flow and basal slip. Glaciers currently cover about 10% of the land surface and contain 2.15% of all water on Earth.

2. Valley glaciers are confined to mountain valleys and flow from higher to lower elevations, whereas continental glaciers cover vast areas and flow out in all directions from a zone of accumulation.

3. A glacier forms when winter snowfall in an area exceeds summer melt and therefore accumulates year after year. Snow is compacted and converted to glacial ice, and when the ice is about 40 m thick, pressure causes it to flow.

4. The behavior of a glacier depends on its budget, which is the relationship between accumulation and wastage. If a glacier possesses a balanced budget, its terminus remains stationary; a positive or negative budget results in advance or retreat of the terminus, respectively.

5. Glaciers move at varying rates depending on slope, discharge, and season. Valley glaciers tend to flow more rapidly than continental glaciers.

6. Glaciers are powerful agents of erosion and transport because they are solids in motion. They are particularly effective at eroding soil and unconsolidated sediment, and they can transport any size sediment supplied to them.

7. Continental glaciers transport most of their sediment in the lower part of the ice, whereas valley glaciers carry sediment in all parts of the ice.

8. Erosion of mountains by valley glaciers produces several sharp, angular landforms including cirques, arêtes, and horns. U-shaped glacial troughs, fiords, and hanging valleys are also products of valley glaciation.

9. Continental glaciers abrade and bevel high areas, forming a smooth, rounded landscape, known as an ice-scoured plain.

10. Depositional landforms include moraines, which are ridgelike accumulations of till. Several types of moraines are recognized, including terminal, recessional, lateral, and medial moraines.

11. Drumlins are composed of till that was apparently reshaped into streamlined hills by continental glaciers or floods of glacial meltwater.

12. Stratified drift consists of sediments deposited by meltwater streams issuing from glaciers; it is found in outwash plains and valley trains. Ridges called eskers and conical hills called kames are also composed of stratified drift.

13. During the Pleistocene Epoch, glaciers covered about 30% of the land surface, especially in the Northern Hemisphere and Antarctica. Several intervals of widespread glaciation, separated by interglacial periods, occurred in North America.

14. Areas far beyond the ice were affected by Pleistocene glaciation; climate belts were compressed toward the equator, large pluvial lakes existed in what are now arid regions, and sea level was as much as 130 m lower than at present.

15. Loading of Earth's crust by Pleistocene glaciers caused isostatic subsidence. When the glaciers melted, isostatic rebound began and continues in some areas.

16. Major glacial intervals separated by tens or hundreds of millions of years probably occur as a result of the changing positions of plates, which in turn cause changes in oceanic and atmospheric circulation patterns.

17. Currently, the Milankovitch theory is widely accepted as the explanation for glacial–interglacial intervals.

18. The reasons for short-term climatic changes, such as the Little Ice Age, are not understood. Two proposed causes for these events are changes in the amount of solar energy received by Earth and volcanism.

Important Terms

abrasion
arête
basal slip
cirque
continental glacier
drumlin
end moraine
esker
fiord
firn
glacial budget
glacial drift

glacial erratic
glacial ice
glacial polish
glacial striation
glacier
ground moraine
hanging valley
horn
kame
lateral moraine
medial moraine
Milankovitch theory

outwash plain
plastic flow
recessional moraine
stratified drift
terminal moraine
till
U-shaped glacial trough
valley glacier
valley train
zone of accumulation
zone of wastage

1. A pyramid-shaped peak known as a *horn* forms:

   a. _____ by headward erosion of a group of cirques on a mountain peak;

   b. _____ when two or more valley glaciers merge to form a much larger glacier;

   c. _____ as till is modified by floods of glacial meltwater;

   d. _____ when a continental glacier freezes to its underlying surface and ceases moving;

   e. _____ when the terminus of a glacier recedes and deposits a recessional moraine.

2. The number of medial moraines on a valley glacier generally indicates the number of its:

   a. _____ valley trains;

   b. _____ terminal moraines;

   c. _____ eskers;

   d. _____ drumlins;

   e. _____ tributary glaciers.

3. The two areas presently with continental glaciers are _____ and _____ :

   a. _____ Glacier National Park/ Waterton Lakes National Park;

   b. _____ Antarctica/Greenland;

   c. _____ Norway/Sweden;

   d. _____ Canada/Siberia;

   e. _____ North Pole/Mount Baker.

4. A glacially transported boulder now resting in an area far from its source is a glacial _____ :

   a. _____ kame;

   b. _____ fiord;

   c. _____ erratic;

   d. _____ cirque;

   e. _____ esker.

5. A cirque can be described as a(an):

   a. _____ deposit consisting of outwash and till;

   b. _____ erosional remnant of a mountain;

   c. _____ deep scratch caused by abrasion;

   d. _____ bowl-shaped depression at the upper end of a glacial trough;

   e. _____ type of lake in which dark- and light-colored laminae are deposited.

6. Glaciers move mostly by:

   a. _____ isostatic rebound;

   b. _____ plastic flow;

   c. _____ lateral compression;

   d. _____ basal slip;

   e. _____ surging.

7. Crevasses in glaciers extend down to:

   a. _____ the bottom of the glacier;

   b. _____ variable depths depending on ice thickness;

   c. _____ about 300 m;

   d. _____ the area of accumulation;

   e. _____ the zone of plastic flow.

8. If a glacier has a balanced budget:

   a. _____ it stops moving;

   b. _____ its terminus remains stationary;

   c. _____ its rate of wastage exceeds its rate of accumulation;

   d. _____ the glacier's length decreases;

   e. _____ crevasses no longer form.

9. A knifelike ridge separating adjacent cirques or glacial troughs is a(an):

   a. _____ horn;

   b. _____ pitted outwash plain;

   c. _____ arête;

   d. _____ medial moraine;

   e. _____ firn.

10. When freshly fallen snow compacts and partly melts and refreezes, it forms granular ice known as:

    a. _____ till;

    b. _____ kame;

    c. _____ firn;

    d. _____ outwash;

    e. _____ drift.

11. The subdued landscape resulting from erosion by a continental glacier is a(an):

    a. _____ drumlin field;

    b. _____ valley train;

    c. _____ glacial striation;

    d. _____ ice-scoured plain;

    e. _____ zone of glacial abrasion.

12. During the Ice Age or _____ Epoch, glaciers covered about _____% of the land surface:

    a. _____ Cretaceous/75%;

    b. _____ Proterozoic/10%;

    c. _____ Mesozoic/15%;

    d. _____ Pleistocene/30%;

    e. _____ Paleozoic/50%.

13. How do the erosional landforms of continental glaciers differ from those of valley glaciers? Give examples of landforms formed by each, and where you might see such examples.

14. How does glacial ice originate, and why is it considered a rock?

15. Draw a diagram showing how a glacier can have four medial moraines on its surface.

16. Explain in terms of the glacial budget how a once active glacier becomes stagnant.

17. What are basal slip and plastic flow, what causes each, and how do they vary seasonally?

18. What is the firn limit on a glacier, and how does its position indicate whether a glacier has a negative, positive, or balanced budget?

19. Explain or diagram how a terminal moraine and a recessional moraine originate.

20. How do an arête and a hanging valley form?

21. Why are so many fiords present in Alaska and the Scandinavian countries?

22. What kinds of evidence would you look for to demonstrate that an ice-free area was once covered by a continental glacier?

23. What are eskers composed of, and how do they originate?

24. How does the Milankovitch theory explain the cyclic nature of glaciation during the Pleistocene Epoch?

Points to Ponder

1. In a roadside rock exposure, you observe a deposit of alternating light and dark laminated mud containing a few large boulders. Explain the sequence of events responsible for deposition.

2. A glacier has a cross-sectional area of 400,000 m² and a flow velocity of 2 m/day. How long would it take for a discharge of 1 km³ to occur?

3. We can be sure that the ancient shorelines of the Great Lakes were horizontal when they were formed, yet now they are not only elevated above their former level but also tilt toward the south. How do you account for these observations?

4. How might human activities affect the amount of glacial ice on Earth?

5. In North America, valley glaciers are common in Alaska and western Canada, and small ones are present in the mountains of the western United States; none, however, occur east of the Rocky Mountains. How can you explain this distribution of glaciers?

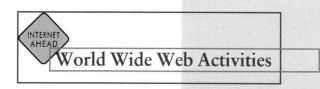

For these Web site addresses, along with current updates and exercises, log on to
http://www.brookscole.com/geo/

➤ GLACIER RESEARCH GROUP

This site is maintained by the Climate Change Research Center, University of New Hampshire, Institute for the Study of Earth, Oceans, and Space. It contains an overview of what it is and what it does, as well as a listing of its current projects and links to other glacially related sites. Click on the *GRG Photo Gallery* site under The Group heading. What are some of the projects the GRG is involved in? How do these relate to glaciers and glaciation?

➤ U.S. GEOLOGICAL SURVEY ICE AND CLIMATE PROJECT

This site contains links to the various Ice and Climate Projects being undertaken by the USGS. Click on the *Benchmark Glaciers* site under the Ice and Climate Projects links heading. What are the three benchmark glaciers being studied and where are they located? What measurements are being taken on the three glaciers? How do these glaciers compare to each other?

➤ ILLINOIS STATE MUSEUM

This site, maintained by the Illinois State Museum, contains information about programs, exhibits, collections, and a calendar of events at the museum.
1. Click on the *Exhibits* icon, then click on the *Ice Ages*. Here you can learn about the ice ages. What are ice ages? When did they occur? What causes ice ages?
2. Click on the *Exhibits* icon, then click on *The Midwestern U.S. 16,000 Years Ago*. This site shows what the Midwest looked like 16,000 years ago. Click on the *Wander Through the Exhibit* and find out more about the environment, plants, and animals of the midwestern United States during the Ice Age. Here you can click on the *Lists of the topics* covered at this site.

➤ ICE AGES AND GLACIATION

This site is maintained by the Department of Geological and Environmental Sciences at Hartwick College in New York. Take a virtual field trip through the Ice Age, and see maps and images of glacial features.

> ## WISCONSIN'S GLACIAL LANDSCAPE

The Ice Age Park and Trail Foundation maintains this site, which has information about the Ice Age in Wisconsin, along with maps and illustrations. When did the most recent continental glacier enter Wisconsin, how long was it present, and how much of the state did it cover? What is the Driftless Area of Wisconsin?

> ## GLACIAL LAKES

The U.S. Geological Survey maintains this site, which is devoted largely to glacial Lake Missoula. Click *Glacial Lake Missoula,* then click *Description: Glacial Lake Missoula and the Missoula Floods.* How many floods might have taken place? What caused the catastrophic floods of glacial Lake Missoula?

> ## GRAND TETON NATIONAL PARK

The Grand Teton National Park, Wyoming, site is maintained by the National Park Service. At the park's home page click *geology,* and then scroll down and click *Journey Through the Past: A Geology Tour.* Next, go to the section on *glaciation.* What evidence indicates that glaciers were once much more widespread in the park than at present? What depositional and erosional glacial landforms can be observed in the park? When was the onset of widespread glaciation in this area, and when did the last extensive glaciers begin to melt?

> ## SHAPES OF THE LAND

The Geological Survey of Canada maintains this site, which has descriptions of glacial features as well as good graphics and some images. Go to the menu under *How glaciers shape the land* . . . and click on *during and after the ice hits, morains,* and *eskers.*

> ## GEOMORPHOLOGY FROM SPACE

Geomorphology from Space "is an out of print 1986 NASA publication edited by Nicholas M. Short, Sr., and Robert W. Blair, Jr., designed for use by the remote sensing science and educational communities to study landforms and landscapes." This site has a menu listing a variety of topics from tectonics to oceans. Click *glacial* and see discussions, graphics, and images of valley and continental glaciers as well as ice caps and ice shelves.

> ## GLACIER

This site is presented by Rice University with photographs courtesy of David A. Mucciarone of Rice University. It has a menu with the following listings: "Introduction," "Expedition," "Weather," "Oceans," "Ice," and "Global Connections" (the last one is still under construction). Open any of the headings in the menu to see photographs of glaciers and learn about Antarctica and glaciers elsewhere. Under "Ice" see How Do Glaciers Form? What Causes Ice Ages? and Where Are Glaciers Today? What are temperate, subpolar, and polar glaciers?

Chapter 18

The Saharan community of El Gedida in western Egypt is slowly being overwhelmed by advancing sand.

OBJECTIVES

At the end of this chapter, you will have learned that

- Wind transports sediment and modifies the landscape through the processes of abrasion and deflation.
- Dunes and loess are the result of deposition of material by wind.
- Dunes form when wind flows over and around an obstruction.
- There are four major dune types: barchan, longitudinal, transverse, and parabolic.

- Loess is formed from windblown silt and clay and is derived from three main sources: deserts, Pleistocene glacial outwash deposits, and floodplains of rivers in semiarid regions.
- The global pattern of air-pressure belts and winds are responsible for Earth's atmospheric circulation patterns.
- Deserts are dry and receive less than 25 cm of rain per year, have high evaporation rates, typically

have poorly developed soils, and are mostly or completely devoid of vegetation.
- The majority of deserts are found in the dry climates of the low and middle latitudes.
- Deserts have many distinctive landforms, produced by both wind and running water.

The Work of Wind and Deserts

Prologue

During the last few decades, deserts have been advancing across millions of acres of productive land, destroying rangelands, croplands, and even villages. Such expansion, estimated at 70,000 km² per year, has exacted a terrible toll in human suffering. Because of the relentless advance of deserts, hundreds of thousands of people have died of starvation or been forced to migrate as "environmental refugees" from their homelands to camps where the majority are severely malnourished. This expansion of deserts into formerly productive lands is called **desertification.**

Most regions undergoing desertification lie along the margins of existing deserts. These margins have a delicately balanced ecosystem that serves as a buffer between the desert on one side and a more humid environment on the other. Their potential to adjust to increasing environmental pressures from natural causes or human activity is limited. Ordinarily, desert regions expand and contract gradually in response to natural processes such as climatic change, but much recent desertification has been greatly accelerated by human activities. In many areas, the natural vegetation has been cleared as crop cultivation has expanded into increasingly drier fringes to support the growing population. Because these areas are especially prone to droughts, crop failures are common occurrences, leaving the land bare and susceptible to increased wind and water erosion.

Because grasses constitute the dominant natural vegetation in most fringe areas, raising livestock is a common economic activity. Usually, these areas achieve a natural balance between vegetation and livestock as nomadic herders graze their animals on the available grasses. In many fringe areas, livestock numbers have been greatly increasing in recent years and now far exceed the land's capacity to support them. As a result, the vegetation cover that protects the soil has diminished, causing the soil to crumble. This leads to further drying of the soil and accelerated soil erosion by wind and water.

Drilling water wells also contributes to desertification because human and livestock activity around a well site strips away the vegetation. With its vegetation gone, the topsoil blows away, and the resultant bare areas merge with the surrounding desert. In addition, the water used for irrigation from these wells sometimes contributes to desertification by increasing the salt content of the soil. As the water evaporates, a small amount of salt is deposited in the soil and is not flushed out as it would be in an area that receives more rain. Over time, the salt concentration becomes so high that plants can no longer grow. Desertification resulting from soil salinization is a major problem in North Africa, the Middle East, southwest Asia, and the western United States.

Collecting firewood for heating and cooking is another major cause of desertification, particularly in many less-developed countries where wood is the major fuel source. In the Sahel of Africa (a belt 300 to 1100 km wide that lies south of the Sahara), the expanding population has completely removed all trees and shrubs in the areas surrounding many towns and cities. Journeys of several days on foot to collect firewood are common there. Furthermore, the use of dried animal dung to supplement firewood has exacerbated desertification because important nutrients in the dung are not returned to the soil.

The Sahel averages between 10 and 60 cm of rainfall per year, 90% of which evaporates when it falls. Because drought is common in the Sahel, the region can support only a limited population of livestock and humans. Traditionally, herders and livestock existed in a natural balance with the vegetation, following the rains north during the rainy season and returning south to greener rangeland during the dry seasons. Some areas were alternately planted and left fallow to help regenerate the soil. During fallow periods, livestock fed off the stubble of the previous year's planting, and their dung helped fertilize the soil.

With the emergence of new nations and increased foreign aid to the Sahel during the 1950s and 1960s, nomads and their herds were restricted, and large areas of grazing land were converted to cash crops such as peanuts and cotton that have a short growing season. Expanding human and animal populations and more intensive agriculture increased the demands on the land. These factors, combined with the worst drought of the twentieth century (1968–1973), brought untold misery to the people of the Sahel. Without rains, the crops failed and the livestock denuded the land of what little vegetation remained. As a result, nearly 250,000 people and 3.5 million cattle died of starvation, and the adjacent Sahara expanded southward as much as 150 km.

The tragedy of the Sahel and prolonged droughts in other desert fringe areas serve to remind us of the delicate equilibrium of ecosystems in such regions. Once the fragile soil cover has been removed by erosion, it takes centuries for new soil to form (see Chapter 5).

INTRODUCTION

Most people associate the work of wind with deserts. Wind is an effective geologic agent in desert regions, but it also plays an important role wherever loose sediment can be eroded, transported, and deposited, such as along shorelines (see Chapter 19) or on the plains (see Perspective 5.2). Therefore, we will examine the work of wind in general, as both an erosional and depositional geologic agent, and then discuss the distribution, characteristics, and landforms of deserts in particular.

Why should you study deserts, and why is it important to understand the processes under which they form, particularly since they tend to be unhospitable and generally uninhabited? One reason is that deserts cover large regions of Earth's surface. More than 40% of Australia is desert, and the Sahara occupies a vast part of northern Africa. Although sparsely populated, many peoples and cultures call desert regions home.

Furthermore, with the current debate about global warming, it is important to understand how desert processes operate and how global climate changes affect the various Earth systems and subsystems. By understanding how desertification operates, we can take steps to eliminate or reduce the destruction done, particularly in terms of human suffering. Understanding the underlying causes of climate change by examining ancient desert regions may provide insight into the possible duration and severity of present and future climatic changes. We have already seen the importance of this in the debate over storage of nuclear waste in Nevada's Yucca Mountain (see Perspective 16.2). One of the concerns is what will happen to the water table if the climate should become more humid.

More than 6,000 years ago, the Sahara was a fertile savanna supporting a diverse fauna and flora, including humans. Then the climate changed, and the area became a desert. How did this happen, and will this region change back again in the future? These are some of the questions geoscientists hope to answer by studying deserts.

And finally, many of the agents and processes that have shaped deserts do not appear to be limited to our planet. Many of the features found on Mars are apparently the result of the same wind-driven processes operating here on Earth (see Chapter 20).

WHAT WOULD YOU DO?

You have been asked to testify before a congressional committee charged with determining whether the National Science Foundation should continue to fund research devoted to the study of climate changes during the Cenozoic Era. Your specialty is the formation of deserts and desert landforms. What arguments would you make to convince the committee to continue funding research concerned with ancient climates?

HOW DOES WIND TRANSPORT SEDIMENT?

Wind is a turbulent fluid and therefore transports sediment in much the same way as running water. Although wind typically flows at a greater velocity than water, it has a lower density and, thus, can carry only clay- and silt-sized particles as *suspended load*. Sand and larger particles are moved along the ground as *bed load*.

Bed Load

Sediments too large or heavy to be carried in suspension by water or wind are moved as bed load either by *saltation* or by rolling and sliding. As we discussed in Chapter 15, saltation is the process by which a portion of the bed load moves by intermittent bouncing along a streambed. Saltation also occurs on land. Wind starts sand grains rolling and lifts and carries some grains short distances before they fall back to the surface. As the descending sand grains hit the surface, they strike other grains, causing them to bounce along by saltation (Figure 18.1). Wind-tunnel experiments have shown that once sand grains begin moving, they will continue to move, even if the wind drops below the speed necessary to start them moving! This happens because once saltation begins, it sets off a chain reaction of collisions between sand grains that keeps the grains in constant motion.

Saltating sand usually moves near the surface, and even when winds are strong, grains are rarely lifted higher than about a meter. If the winds are very strong, these wind-whipped grains can cause extensive abrasion. A car's paint can be removed by sandblasting in a short time, and its windshield will become completely frosted and translucent from pitting.

Suspended Load

Silt- and clay-sized particles constitute most of a wind's suspended load. Even though these particles are much smaller and lighter than sand-sized particles, wind usually starts the latter moving first. The reason for this phenomenon is that a very thin layer of motionless air lies next to the ground where the small silt and clay particles remain undisturbed. The larger sand grains, however, stick up into the turbulent air zone, where they can be moved. Unless the stationary air layer is disrupted, the silt and clay particles remain on the ground, providing a smooth surface. This phenomenon can be observed on a dirt road on a windy day. Unless a vehicle travels over the road, little dust is raised even though it is windy. When a vehicle moves over the road, it breaks the calm boundary layer of air and disturbs the smooth layer of dust, which is picked up by the wind and forms a dust cloud in the vehicle's wake.

In a similar manner, when a sediment layer is disturbed, silt- and clay-sized particles are easily picked up and carried in suspension by the wind, creating clouds

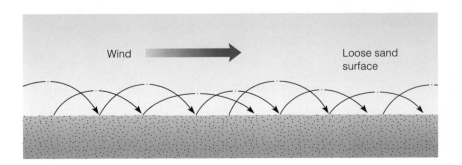

Figure 18.1
Most sand is moved near the ground surface by saltation. Sand grains are picked up by the wind and carried a short distance before falling back to the ground where they usually hit other grains, causing them to bounce and move in the direction of the wind.

Figure 18.2
A dust storm in Death Valley, California.

of dust or even dust storms (Figure 18.2). Once these fine particles are lifted into the atmosphere, they may be carried thousands of kilometers from their source. For example, large quantities of fine dust from the southwestern United States were blown east and fell on New England during the Dust Bowl of the 1930s.

HOW DOES WIND ERODE?

Although wind action produces many distinctive erosional features and is an extremely efficient sorting agent, running water is still responsible for most erosional landforms in arid regions, even though stream channels are typically dry. Wind erodes material in two ways: abrasion and deflation.

Abrasion

Abrasion involves the impact of saltating sand grains on an object and is analogous to sandblasting. The effects of abrasion are usu-

Figure 18.3
Wind abrasion has formed these structures by eroding the exposed limestone in Desierto Libico, Egypt.

ally minor because sand, the most common agent of abrasion, is rarely carried more than 1 m above the surface. Rather than creating major erosional features, wind abrasion typically modifies existing features by etching, pitting, smoothing, or polishing. Nonetheless, wind abrasion can produce many strange-looking and bizarre-shaped features (Figure 18.3).

Ventifacts are a common product of wind abrasion; these are stones whose surfaces have been polished, pitted, grooved, or faceted by the wind (Figures 18.4 and 18.7c). If the wind blows from different directions or if the stone is moved, the ventifact will have multiple facets. Ventifacts are most common in deserts, yet they

Figure 18.4

(a) A ventifact forms when wind-borne particles (1) abrade the surface of a rock (2) forming a flat surface. If the rock is moved, (3) additional flat surfaces are formed. (b) Large ventifacts lying on desert pavement in Death Valley National Monument, California.

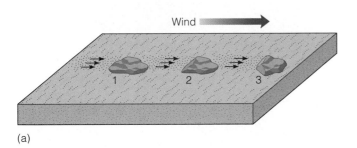

Wind ▸

(a)

(b)

can also form wherever stones are exposed to saltating sand grains, as on beaches in humid regions and some outwash plains in New England.

WHAT WOULD YOU DO?

As an expert in desert processes, you have been assigned the job of teaching the first astronaut crew that will explore Mars all about deserts and their landforms. Many Martian features display evidence of having formed as a result of wind processes, and many landforms are the same as those found in deserts on Earth. Describe how you would teach the astronauts to recognize wind-formed features and where you would take the astronauts in the field to show them the type of landforms they may find on Mars.

Yardangs are larger features than ventifacts and also result from wind erosion (Figure 18.5). They are elongated and streamlined ridges that look like an overturned ship's hull. They are typically found grouped in clusters aligned parallel to the prevailing winds. They probably form by differential erosion in which depressions, parallel to the direction of wind, are carved out of a rock body, leaving sharp, elongated ridges. These ridges may then be further modified by wind abrasion into their characteristic shape. Although yardangs are fairly common desert features, interest in them was renewed when images radioed back from Mars showed that they are also widespread features on the Martian surface.

Deflation

Another important mechanism of wind erosion is **deflation,** which is the removal of loose surface sediment by the wind. Among the characteristic features of deflation in many arid and semiarid regions are *deflation hollows,* or *blowouts.* These shallow depressions of variable dimensions result from differential erosion of surface materials (Figure 18.6). Ranging in size from several kilometers in diameter and tens of meters deep to small depressions only a few meters

Figure 18.5
Profile view of a streamlined yardang in the Roman playa deposits of the Kharga Depression, Egypt.

Figure 18.6
A deflation hollow in Death Valley, California.

wide and less than a meter deep, deflation hollows are common in the southern Great Plains region of the United States.

In many dry regions, the removal of sand-sized and smaller particles by wind leaves a surface of pebbles, cobbles, and boulders. As the wind removes the fine-grained material from the surface, the effects of gravity and occasional floodwaters rearrange the remaining coarse particles into a mosaic of close-fitting rocks called **desert pavement** (Figures 18.4b and 18.7). Once a desert pavement forms, it protects the underlying material from further deflation.

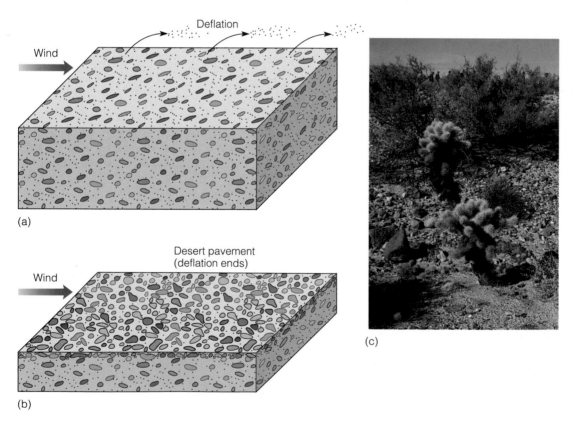

Figure 18.7
Deflation and the origin of desert pavement. (a) Fine-grained material is removed by wind, (b) leaving a concentration of larger particles that form desert pavement. (c) Desert pavement in the Mojave Desert, California. Several ventifacts can be seen in the lower left of the photo.

WHAT ARE THE DIFFERENT TYPES OF WIND DEPOSITS?

Although wind is of minor importance as an erosional agent, it is responsible for impressive deposits, which are primarily of two types. The first, dunes, occur in several distinctive types, all of which consist of sand-sized particles that are usually deposited near their source. The second is loess, which consists of layers of windblown silt and clay deposited over large areas downwind and commonly far from their source.

The Formation and Migration of Dunes

The most characteristic features associated with sand-covered regions are **dunes,** which are mounds or ridges of wind-deposited sand (Figure 18.8). Dunes form when wind flows over and around an obstruction, resulting in deposition of sand grains, which accumulate and build up deposits of sand. As they grow, these sand deposits become self-generating in that they form ever-larger wind barriers that further reduce the wind's velocity, resulting in further sand deposition, and growth of the dune.

Most dunes have an asymmetric profile, with a gentle windward slope and a steeper downwind, or leeward, slope that is inclined in the direction of the prevailing wind (Figure 18.9a). Sand grains move up the gentle windward slope by saltation and accumulate on the leeward side forming an angle between 30 and 34 degrees from the horizontal, which is the angle of repose of dry sand. When this angle is exceeded by accumulating sand, the slope collapses, and the sand slides down the leeward slope, coming to rest at its base. As sand moves from a dune's windward side and periodically slides down its leeward slope, the dune slowly migrates in the direction of the prevailing wind (Figure 18.9b). When preserved in the geologic record, dunes help geologists determine the prevailing direction of ancient winds (Figure 18.10).

Figure 18.8
Large sand dunes in Death Valley, California. Well-developed ripple marks can be seen on the surface of the dunes. The prevailing wind direction is from left to right.

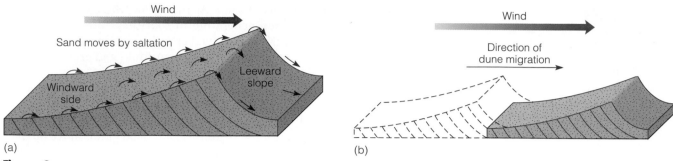

Figure 18.9
(a) Profile view of a sand dune. (b) Dunes migrate when sand moves up the windward side and slides down the leeward slope. Such movement of the sand grains produces a series of cross-beds that slope in the direction of wind movement.

Dune Types

Four major dune types are generally recognized (barchan, longitudinal, transverse, and parabolic), although intermediate forms between the major types also exist. The size, shape, and arrangement of dunes result from the interaction of such factors as sand supply, the direction and velocity of the prevailing wind, and the amount of vegetation. While dunes are usually found in deserts, they can also develop wherever there is an abundance of sand, such as along the upper parts of many beaches.

Barchan dunes are crescent-shaped dunes whose tips point downwind (Figure 18.11). They form in areas having a generally flat, dry surface with little vegetation, a limited supply of sand, and a nearly constant wind direction. Most barchans are small, with the largest reaching about 30 m high. Barchans are the most mobile of the major dune types, moving at rates that can exceed 10 m per year.

Longitudinal dunes (also called *seif dunes*) are long, parallel ridges of sand aligned generally parallel to the direction of the prevailing winds; they form where the sand supply is somewhat limited (Figure 18.12). Longitudinal dunes result when winds converge from slightly different directions to produce the prevailing wind. They range in size from about 3 m to more

than 100 m high, and some stretch for more than 100 km. These dunes are especially well developed in central Australia, where they cover nearly one-fourth of the continent. They also cover extensive areas in Saudi Arabia, Egypt, and Iran.

Transverse dunes form long ridges perpendicular to the prevailing wind direction in areas where abundant sand is available and little or no vegetation exists (Figure 18.13). When viewed from the air, transverse dunes have a wavelike appearance and are therefore sometimes called *sand seas*. The crests of transverse dunes can be as high as 200 m, and the dunes may be as much as 3 km wide. Some transverse dunes develop a clearly distinguishable barchan form and may separate into individual barchan dunes along the edges of the dune field where there is less sand. Such intermediate-form dunes are known as *barchanoid dunes* (Figure 18.14).

Figure 18.10
Cross-bedding in this sandstone in Zion National Park, Utah, helps geologists determine the prevailing direction of the wind that formed these ancient sand dunes.

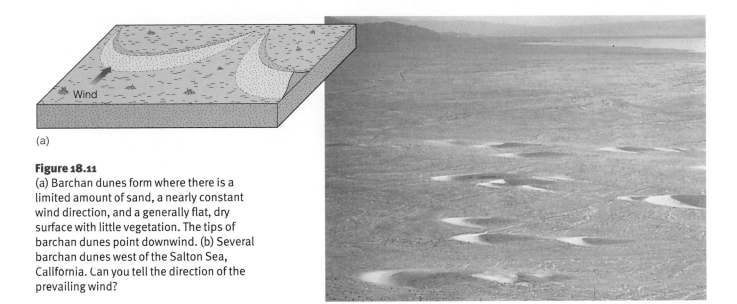

(a)

(b)

Figure 18.11
(a) Barchan dunes form where there is a limited amount of sand, a nearly constant wind direction, and a generally flat, dry surface with little vegetation. The tips of barchan dunes point downwind. (b) Several barchan dunes west of the Salton Sea, California. Can you tell the direction of the prevailing wind?

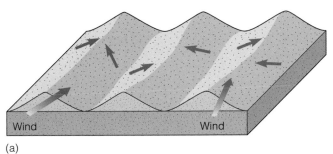

(a)

(b)

Figure 18.12
(a) Longitudinal dunes form long, parallel ridges of sand aligned roughly parallel to the prevailing wind direction. They typically form where sand supplies are limited.
(b) Longitudinal dunes, 15 m high, in the Gibson Desert, west central Australia, are shown in this image. The bright blue areas between the dunes are shallow pools of rainwater, while the darkest patches are areas where the Aborigines have set fires to encourage the growth of spring grasses.

(a)

(b)

Figure 18.13
(a) Transverse dunes form long ridges of sand that are perpendicular to the prevailing wind direction in areas of little or no vegetation and abundant sand. (b) Transverse dunes, Great Sand Dunes National Monument, Colorado. The prevailing wind direction is from lower left to upper right.

Figure 18.14
Barchanoid dunes at White Sands National Monument, New Mexico.

Parabolic dunes are most common in coastal areas with abundant sand, strong onshore winds, and a partial cover of vegetation (Figure 18.15). Although parabolic dunes have a crescent shape like barchan dunes, their tips point upwind. Parabolic dunes form when the vegetation cover is broken and deflation produces a deflation hollow or blowout. As the wind transports the sand out of the depression, it builds up on the convex downwind dune crest. The central part of the dune is excavated by the wind, while vegetation holds the ends and sides fairly well in place.

Another type of dune commonly found in the deserts of North Africa and Saudi Arabia is the *star dune,* so named because of its resemblance to a multi-pointed representation of a star (Figure 18.16). Star dunes are among the tallest in the world, rising, in some cases, more than 100 m above the surrounding desert

(a)

Figure 18.15
(a) Parabolic dunes typically form in coastal areas where there is a partial cover of vegetation, a strong onshore wind, and abundant sand.
(b) Parabolic dune developed along the Lake Michigan shoreline west of St. Ignace, Michigan.

(b)

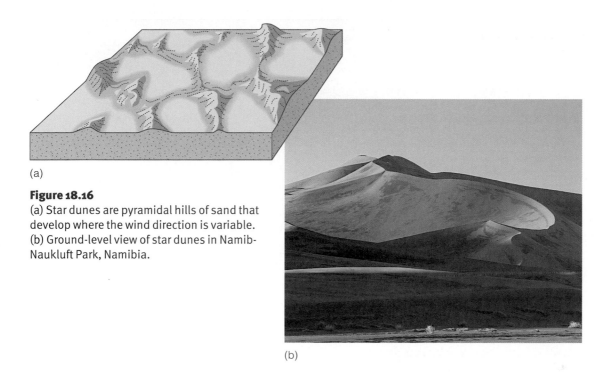

(a)

Figure 18.16
(a) Star dunes are pyramidal hills of sand that develop where the wind direction is variable.
(b) Ground-level view of star dunes in Namib-Naukluft Park, Namibia.

(b)

plain. They consist of pyramidal hills of sand, from which radiate several ridges of sand, and develop where the wind direction is variable. Star dunes can remain stationary for centuries at a time and have served as desert landmarks for many nomadic peoples.

Loess

Windblown silt and clay deposits composed of angular quartz grains, feldspar, micas, and calcite are known as **loess.** The distribution of loess shows that it is derived from three main sources: deserts, Pleistocene glacial outwash deposits, and the floodplains of rivers in semiarid regions. It must be stabilized by moisture and vegetation in order to accumulate. Consequently, loess is not found in deserts, even though they provide much of its material. Because of its unconsolidated nature, loess is easily eroded, and as a result, eroded loess areas are characterized by steep cliffs and rapid lateral and headward stream erosion (Figure 18.17).

At present, loess deposits cover approximately 10% of Earth's land surface and 30% of the United States (Figure 18.18). The most extensive and thickest loess deposits occur in northeast China where accumulations greater than 30 m are common. The extensive deserts in central Asia are the source for this loess. Other important loess deposits are on the North European Plain from Belgium eastward to Ukraine, in Central Asia, and in the Pampas of Argentina. In

Figure 18.17
These steep banks along the Yukon River, Yukon Territory, Canada, are formed of loess.

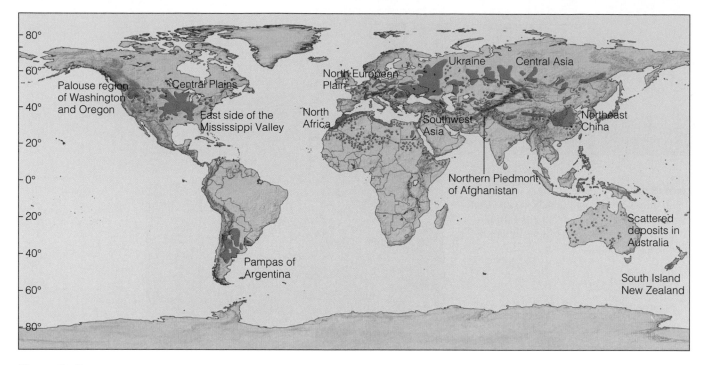

Figure 18.18
The distribution of Earth's major loess-covered areas.

the United States, they occur in the Great Plains, the Midwest, the Mississippi River Valley, and eastern Washington.

Loess-derived soils are some of the world's most fertile. It is therefore not surprising that the world's major grain-producing regions correspond to the distribution of large loess deposits such as the North European Plain, Ukraine, and the Great Plains of North America.

WHAT IS THE DISTRIBUTION OF AIR-PRESSURE BELTS AND GLOBAL WIND PATTERNS?

To understand the work of wind and the distribution of deserts, we need to consider the global pattern of air-pressure belts and winds, which are responsible for Earth's atmospheric circulation patterns. Air pressure is the density of air exerted on its surroundings (that is, its weight). When air is heated, it expands and rises, reducing its mass for a given volume and causing a decrease in air pressure. Conversely, when air is cooled, it contracts and air pressure increases. Therefore, those areas of Earth's surface that receive the most solar radiation, such as the equatorial regions, have low air pressure, while the colder areas, such as the polar regions, have high air pressure.

Air flows from high-pressure zones to low-pressure zones. If Earth did not rotate, winds would move in a straight line from one zone to another. Because Earth rotates, however, winds are deflected to the right of their direction of motion (clockwise) in the Northern Hemisphere and to the left of their direction of motion (counterclockwise) in the Southern Hemisphere. Such a deflection of air between latitudinal zones resulting from Earth's rotation is known as the **Coriolis effect.** Therefore, the combination of latitudinal pressure differences and the Coriolis effect produces a worldwide pattern of wind belts oriented east–west (Figure 18.19).

Earth's equatorial zone receives the most solar energy, which heats the surface air, causing it to rise. As the air rises, it cools and releases moisture that falls as rain in the equatorial region (Figure 18.19). The rising air is now much drier as it moves northward and southward toward each pole. By the time it reaches 20 to 30 degrees north and south latitudes, the air has become cooler and denser and begins to descend. Compression of the atmosphere warms the descending air mass and produces a warm, dry, high-pressure area, providing the perfect conditions for the formation of the low-latitude deserts of the Northern and Southern Hemispheres (Figure 18.20).

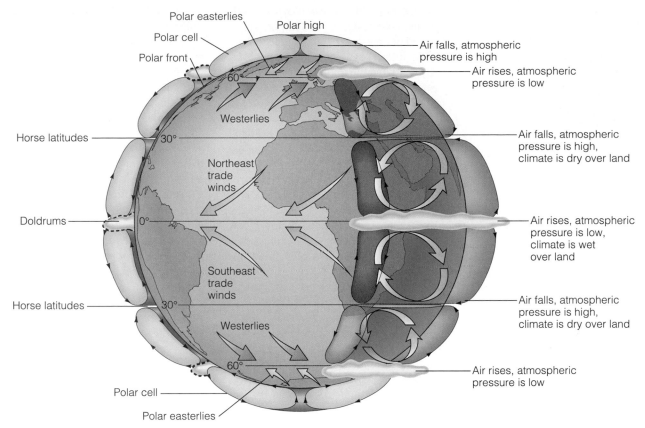

Figure 18.19
The general circulation of Earth's atmosphere.

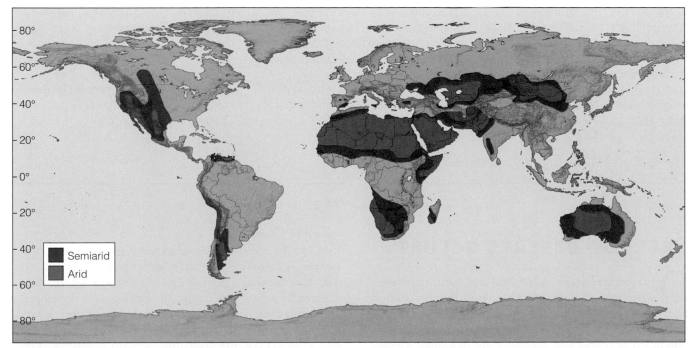

Figure 18.20
The distribution of Earth's arid and semiarid regions.

Controlling Dust Storms Through Revegetation

Nicholas Lancaster, the world's foremost expert on desert sand dunes, is a research professor with the Quaternary Sciences Center at the Desert Research Institute, part of the University and Community College System of Nevada. Scientists at the Desert Research Institute study air quality, water resources, surface processes, climate history and dynamics, and human adaptations in arid lands.

Transport of sand and dust by the wind is an important geologic process in many desert regions of the world. These areas are affected by dust storms that reduce visibility, close airports, and cause traffic accidents. Fine particles in the air can be hazardous to human health, causing respiratory diseases and possibly spreading fungal infections. Blowing sand can block roads and fill irrigation ditches, while migrating sand dunes can bury buildings, pipelines, and scarce agricultural land.

Dealing with these geologic hazards of living in desert regions requires a good understanding of the physical principles of sand and dust transport by wind and the formation and movement of sand dunes. I have traveled and conducted research in many desert regions, studying dune formation and movement as well as the ways in which sand and dust are moved by wind. My research, and that of many others, has shown that sand dunes form and move in very similar ways in widely separated desert regions. We can therefore take the results of studies conducted in one area and apply them to other locations to assist people in controlling blowing sand and dust.

One such area is Owens Lake in eastern California. Once the site of a shallow lake, the area is now a saline lake bed, or *playa,* from which dust storms arise on frequent occasions, affecting air quality throughout much of southern California. In fact, Owens Lake is regarded as the largest single source of windblown dust in the Americas. The drying up of Owens Lake is the result of diversion of water since the 1920s from the Owens River and its tributaries to the Los Angeles aqueduct, which now supplies a major part of the water used by the city of Los Angeles. Many of the dust storms at Owens Lake occur because windblown sand grains—lifted from the sandy sediments of the former delta of the Owens River and from old beach deposits— abrade the salt-encrusted playa sediments. The fine particles released by this natural sand blasting are raised into the air by the turbulence of the wind and fill the Owens Valley with dust.

The dust problem is so severe that California state agencies were instructed to find a way to control dust emissions from Owens Lake by the end of the century. One way being considered to reduce dust storms is revegetation of parts of the dry lake bed with natural salt-tolerant grasses. Even sparse desert vegetation can have an important effect in protecting the sediments from wind (and water) erosion. Vegetation helps protect the surface by slowing the wind speed close to the ground and reducing or preventing entirely wind transport of surface sediments. The question is: How much grass cover is needed to stop sand transport and dust emissions?

In the past two years, I have been conducting field studies of sand transport in areas of different natural- and planted-vegetation cover on sand sheets at Owens Lake. The goal is to find the ground cover of salt-tolerant grass that reduces sediment transport to a minimal level and to develop a model that can be used to predict the amount of grass required to stabilize sand surfaces. Our field studies show that vegetation increases the threshold wind velocity required for sand transport so that there is an exponential decrease in the rate of sand transport with increasing vegetation cover. The result is that sand transport is effectively eliminated when the cover of salt grass is greater than 15%. We can now use these findings to predict how much vegetation is required to stabilize sandy areas at Owens Lake, as well as sand dune areas worldwide. ∎

WHERE DO DESERTS OCCUR?

Dry climates occur in the low and middle latitudes where the potential loss of water by evaporation exceeds the yearly precipitation (Figure 18.20). Dry climates cover 30% of Earth's land surface and are subdivided into semiarid and arid regions. *Semiarid regions* receive more precipitation than arid regions, yet are moderately dry. Their soils are usually well developed and fertile and support a natural grass cover. *Arid regions,* generally described as **deserts,** are dry; on average, they receive less than 25 cm of rain per year, have high evaporation rates, typically have poorly developed soils, and are mostly or completely devoid of vegetation.

The majority of the world's deserts are found in the dry climates of the low and middle latitudes (Figure 18.20). In North America, most of the southwestern United States and northern Mexico are characterized by

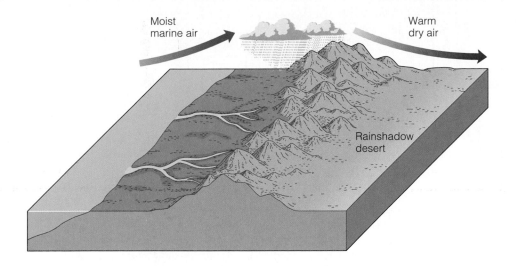

Figure 18.21
Many deserts in the middle and high latitudes are rainshadow deserts, so named because they form on the leeward side of mountain ranges. When moist marine air moving inland meets a mountain range, it is forced upward where it cools and forms clouds that produce rain. This rain falls on the windward side of the mountains. The air descending on the leeward side is much warmer and drier, producing a rainshadow desert.

this hot, dry climate, whereas in South America this climate is primarily restricted to the Atacama Desert of coastal Chile and Peru. The Sahara in northern Africa, the Arabian Desert in the Middle East, and the majority of Pakistan and western India form the largest essentially unbroken desert environment in the Northern Hemisphere. More than 40% of Australia is desert, and most of the rest of it is semiarid.

The remaining dry climates of the world are found in the middle and high latitudes, mostly within continental interiors in the Northern Hemisphere (Figure 18.20). Many of these areas are dry because of their remoteness from moist maritime air and the presence of mountain ranges that produce a **rainshadow desert** (Figure 18.21). When moist marine air moves inland and meets a mountain range, it is forced upward. As it rises, it cools, forming clouds and producing precipitation that falls on the windward side of the mountains. The air that descends on the leeward side of the mountain range is much warmer and drier, producing a rainshadow desert.

Three widely separated areas are included within the mid-latitude dry-climate zone (Figure 18.20). The largest of these is the central part of Eurasia extending from just north of the Black Sea eastward to north-central China. The Gobi Desert in China is the largest desert in this region. The Great Basin area of North America is the second largest mid-latitude dry-climate zone and results from the rainshadow produced by the Sierra Nevada (see Perspective 18.1). This region adjoins the southwestern deserts of the United States that formed as a result of the low-latitude subtropical high-pressure zone. The smallest of the mid-latitude dry-climate areas is the Patagonian region of southern and western Argentina. Its dryness results from the rainshadow effect of the Andes. The remainder of the world's deserts are found in the cold but dry high latitudes, such as Antarctica.

WHAT ARE THE CHARACTERISTICS OF DESERTS?

To people who live in humid regions, deserts may seem stark and inhospitable (see Perspective 18.2). Instead of a landscape of rolling hills and gentle slopes with an almost continuous cover of vegetation, deserts are dry, have little vegetation, and consist of nearly continuous rock exposures, desert pavement, or sand dunes. And yet despite the great contrast between deserts and more humid areas, the same geologic processes are at work, only operating under different climatic conditions.

Temperature, Precipitation, and Vegetation

The heat and dryness of deserts are well known. Many of the deserts of the low latitudes have average summer temperatures that range between 32° and 38°C. It is not uncommon for some low-elevation inland deserts to record daytime highs of 46° to 50°C for weeks at a time. The highest temperature ever recorded was 58°C in El Azizia, Libya, on September 13, 1922.

During the winter months when the angle of the Sun is lower and there are fewer daylight hours, daytime temperatures average between 10° and 18°C. Winter nighttime lows can be quite cold, with frost and freezing temperatures common in the more poleward deserts. Winter daily temperature fluctuations in low-latitude deserts are among the greatest in the world, ranging between 18° and 35°C. Temperatures have been known to fluctuate from below 0°C to more than 38°C in a single day!

Perspective 18.1

Death Valley National Park

Death Valley National Monument was established in 1933 and encompasses 7700 km² of southeastern California and part of western Nevada (Figure 1). It was changed to a national park in 1994. The hottest, driest, and lowest of the U.S. National Monuments and Parks, it receives less than 5 cm of rain per year and features normal daytime summer temperatures above 42°C. The highest temperature ever recorded was 57°C in the shade! The topographic relief in Death Valley is impressive. Telescope Peak near the southwestern border is 3368 m high, while the lowest point in the Western Hemisphere—86 m below sea level—is less than 32 km to the east at Badwater.

Within Death Valley and its bordering mountains are excellent examples of a wide variety of desert landforms and economically valuable evaporite deposits. In addition, numerous folds, faults, landslides, and considerable evidence of volcanic activity can be seen.

The geologic history of Death Valley is complex and still being worked out, but rocks from every geologic era can be found in the valley or the surrounding mountains. Although the geologic history of the region reaches back to the Precambrian, Death Valley itself formed less than 4 million years ago.

Death Valley formed during the Pliocene Epoch, when Earth's crust was stretched and rifted, forming various horsts and grabens. Death Valley continues to subside along normal faults and is sinking most rapidly along its western side. This movement has been so great that more than 3000 m of sediments are buried beneath the present valley floor.

During the Pleistocene, when the climate of this region was more humid than it is today, numerous lakes spread over the valley (see Chapter 17). Lake Manly, the largest of these lakes (145 km long and 178 m deep), dried up about 10,000 years ago when the climate became arid.

Volcanic activity has also been occurring during the last several thousand years. The most famous volcanic feature in Death Valley is Ubehebe Crater, an explosion crater that formed approximately 2000 years ago (Figure 2).

Besides the usual desert features, Death Valley also includes some unusual ones such as the Devil's Golf Course, a bed of solid rock salt displaying polygonal ridges and pinnacles that are almost impossible to traverse (Figure 3). The Harmony Borax Works was home to the famous 20-mule teams that hauled out countless wagons of borax (Figure 4). Borax, used for ceramic glazes, fertilizers, glass, solder, and pharmaceuticals, was leached from volcanic ash by hot groundwater and then accumulated in layers of lake sediment.

Besides the numerous geologic features that have made Death Valley famous, it is also home to more than 600 species of plants as well as numerous animal species.

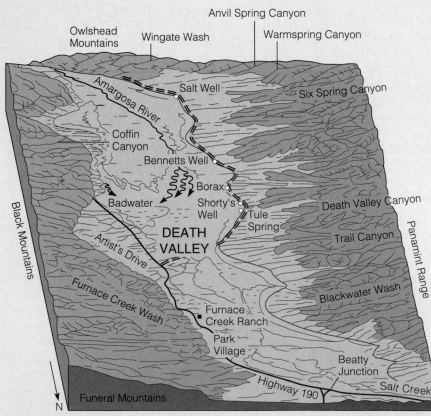

Figure 1
Death Valley National Park, California, encompasses 7700 km² of southeastern California and part of western Nevada.

Figure 2
Ubehebe Crater, an explosion crater, last erupted approximately 2000 years ago.

Figure 3
Devil's Golf Course consists of a layer of solid rock salt that has formed a network of polygonal ridges and pinnacles, making it very difficult to traverse.

Figure 4
Twenty-mule teams carried borax out of Death Valley.

Perspective 18.2

Petroglyphs

R ock art includes rock paintings (where paints made from natural pigments are applied to a rock surface) and petroglyphs (from the Greek *petro* meaning "rock" and *glyph* meaning "carving or engraving"), which are the abraded, pecked, incised, or scratch marks made by humans on boulders, cliffs, and cave walls. Rock art is found throughout the world and is a valuable archaeological resource that can be used to interpret and increase our knowledge of prehistoric peoples. It provides us with graphic evidence of cultural, social, and religious relationships and practices, as well as changing patterns of trade and communication between different ancient peoples. The oldest known rock art was made by hunters in western Europe and dates back to the Pleistocene Epoch.

In the arid Southwest and Great Basin area of North America where rock art is plentiful, rock paintings and petroglyphs extend back to about 2000 B.C. Here, rock art can be divided into two categories. *Representational art* deals with life-forms such as humans, birds, snakes, and humanlike supernatural beings. Rarely exact replicas, they are more or less stylized versions of the beings depicted. *Abstract art,* in contrast, bears no resemblance to any real-life images.

Petroglyphs are the most common form of rock art in North America. They are especially abundant in the Southwest and Great Basin area where they occur by the thousands, having been made by Native Americans from many cultures during the past several thousand years. Petroglyphs can be viewed in many of the National Parks and Monuments such as Petrified Forest National Park, Arizona; Dinosaur National Monument, Utah (Figure 1); Canyonlands National Park, Utah; and

Figure 1
Various petroglyphs exposed at an outcrop along Cub Creek Road, Dinosaur National Monument, Utah.

Petroglyph National Monument, New Mexico—to name a few.

Petroglyphs are made using a variety of techniques. The most common is pecking, in which the surface of the rock is struck by a harder stone, or for more precise control using a stone chisel hit by a hammer stone. This results in a series of dots, which can be made into lines by continued pecking. Differences in pecking techniques characterize different styles of petroglyphs. Another method used is incising or scratching. Here lines up to several centimeters deep are carved into the rock using a sharp tool. Sometimes a combination of techniques is used, including painting the figures after they have been carved in the rock.

In arid regions, many rock surfaces display a *patination,* or thin brown or black coating, known as rock varnish, composed of iron and manganese oxides. When this coating is broken by pecking, incising, or scratching, the underlying lighter colored natural-rock surface provides an excellent contrast for the petroglyphs (Figure 2).

One of the best places to view petroglyphs in North America is at Petroglyph National Monument, west of Albuquerque, New Mexico. Established by Congress in 1990, this national monument protects more than 15,000 petroglyphs. Although some may be as old as 2000 to 3000 years, the vast majority were created between A.D. 1300 and 1650 and reflect the complex culture and religion of the Pueblo peoples.

It is important to remember that petroglyphs are a fragile and nonrenewable cultural resource that cannot be replaced if they are damaged or destroyed. A commitment to their preservation is essential so that future generations can study them as well as enjoy their beauty and mystery.

Figure 2
Humanlike petroglyph exposed at an outcrop along Cub Creek Road, Dinosaur National Monument, Utah. Note the contrast between the fresh exposure of the rock where the upper part of the petroglyph's head has been removed, the weathered brown surface of the rest of the petroglyph, and the black rock varnish coating the rock surface.

Figure 18.22
Desert vegetation is typically sparse, widely spaced, and characterized by slow growth rates. The vegetation shown here in Organ Pipe National Monument, Arizona, includes saguaro and cholla cacti, paloverde trees, and jojoba bushes and is characteristic of the vegetation found in the Sonoran Desert of North America.

The dryness of the low-latitude deserts results primarily from the year-round dominance of the subtropical high-pressure belt, while the dryness of the mid-latitude deserts is due to their isolation from moist marine winds and the rainshadow effect created by mountain ranges. The dryness of both is further accentuated by their high temperatures.

Although deserts are defined as regions receiving, on average, less than 25 cm of rain per year, the amount of rain that falls each year is very unpredictable and unreliable. It is not uncommon for an area to receive more than an entire year's average rainfall in one cloudburst and then to receive very little rain for several years. Thus, yearly rainfall averages can be quite misleading.

Deserts display a wide variety of vegetation (Figure 18.22). Although the driest deserts, or those with large areas of shifting sand, are almost devoid of vegetation, most deserts support at least a sparse plant cover. Compared to humid areas, desert vegetation may appear monotonous. A closer examination, however, reveals an amazing diversity of plants that have evolved the ability to live in the near absence of water.

Desert plants are widely spaced, typically small, and grow slowly. Their stems and leaves are usually hard and waxy to minimize water loss by evaporation and protect the plant from sand erosion. Most plants have a widespread, shallow root system to absorb the dew that forms each morning in all but the driest deserts and to help anchor the plant in what little soil there may be. In extreme cases, many plants lie dormant during particularly dry years and spring to life after the first rain shower, with a beautiful profusion of flowers.

Weathering and Soils

Mechanical weathering is dominant in desert regions. Daily temperature fluctuations and frost wedging are the primary forms of mechanical weathering (see Chapter 5). The breakdown of rocks by roots and from salt crystal growth are of minor importance. Some chemical weathering does occur, but its rate is greatly reduced by aridity and the scarcity of organic acids produced by the sparse vegetation. Most chemical weathering takes place during the winter months when more precipitation occurs, particularly in the mid-latitude deserts.

An interesting feature seen in many deserts is a thin, red, brown, or black shiny coating on the surface of many rocks. This coating, called *rock varnish*, is composed of iron and manganese oxides (Figure 18.23). Because many of the varnished rocks contain little or no iron and manganese oxides, the varnish is thought to result either from windblown iron and manganese dust that settles on the ground or from the precipitated waste of microorganisms.

Desert soils, if developed, are usually thin and patchy because the limited rainfall and the resultant

Figure 18.23
The shiny black coating on this rock exposed at Castle Valley, Utah, is rock varnish. It is composed of iron and manganese oxides.

scarcity of vegetation reduce the efficiency of chemical weathering and hence soil formation. Furthermore, the sparseness of the vegetative cover enhances wind and water erosion of what little soil actually forms.

Mass Wasting, Streams, and Groundwater

When traveling through a desert, most people are impressed by such wind-formed features as moving sand, sand dunes, and sand and dust storms. They may also notice the dry washes and dry streambeds. Because of the lack of running water, most people would conclude that wind is the most important erosional agent in deserts. They would be wrong. Running water, even though it occurs infrequently, causes most of the erosion in deserts. The dry conditions and sparse vegetation characteristic of deserts enhance water erosion. If you look closely, you will see the evidence of erosion and transportation by running water nearly everywhere except in areas covered by sand dunes.

Most of a desert's average annual rainfall of 25 cm or less comes in brief, heavy, localized cloudbursts. During these times, considerable erosion takes place because the ground cannot absorb all the rainwater. With so little vegetation to hinder its flow, runoff is rapid, especially on moderately to steeply sloping surfaces, resulting in flash floods and sheetflows. Dry stream channels quickly fill with raging torrents of muddy water and mudflows, which carve out steepsided gullies and overflow their banks. During these times, a tremendous amount of sediment is rapidly transported and deposited far downstream.

Although water is the major erosive agent in deserts today, it was even more important during the Pleistocene Epoch when these regions were more humid (see Chapter 17). During that time, many of the major topographic features of deserts were forming. Today that topography is being modified by wind and infrequently flowing streams.

Most desert streams are poorly integrated and flow only intermittently. Many of them never reach the sea because the water table is usually far deeper than the channels of most streams, so they cannot draw upon groundwater to replace water lost to evaporation and absorption into the ground. This type of drainage in which a stream's load is deposited within the desert is called *internal drainage* and is common in most arid regions.

Although the majority of deserts have internal drainage, some deserts have permanent through-flowing streams such as the Nile and Niger Rivers in Africa, the Rio Grande and Colorado River in the southwestern United States, and the Indus River in Asia. These streams can flow through desert regions because their headwaters are well outside the desert and water is plentiful enough to offset losses resulting from evaporation and infiltration. Demands for greater amounts of water for agriculture and domestic use from the Colorado River, however, are leading to increased salt concentrations in its lower reaches and causing political problems between the United States and Mexico.

The water table in most desert regions is below the stream channels and is only recharged for a short time after a rainfall. In deserts with through-flowing streams, the water table slopes away from the streams. The through-flowing streams help recharge the groundwater supply and can support vegetation along their banks. Trees, which have high moisture requirements, are rare in deserts, but may occasionally occur along the banks of both ephemeral and permanent streams, where their roots can reach the higher water table.

Wind

Although running water does most of the erosional work in deserts, wind can also be an effective geologic agent capable of producing a variety of distinctive erosional and depositional features (Figure 18.3). It is effective in transporting and depositing unconsolidated sand, silt, and dust-sized particles. Contrary to popular belief, most deserts are not sand-covered wastelands but rather consist of vast areas of rock exposures and desert pavement. Sand-covered regions, or sandy deserts, constitute less than 25% of the world's deserts. The sand in these areas has accumulated primarily by the action of wind.

WHAT TYPES OF LANDFORMS ARE FOUND IN DESERTS?

Because of differences in temperature, precipitation, and wind, as well as the underlying rocks and recent tectonic events, landforms in arid regions vary considerably. Although wind is an important geologic agent in deserts, many distinctive landforms are produced and modified by running water.

After an infrequent and particularly intense rainstorm, excess water not absorbed by the ground may accumulate in low areas and form *playa lakes* (Figure 18.24a). These lakes are temporary, lasting from a few hours to several months. Most of them are shallow and have rapidly shifting boundaries as water flows in or leaves by evaporation and seepage into the ground. The water is often very saline.

When a playa lake evaporates, the dry lakebed is called a **playa,** or *salt pan,* and is characterized by

(a)

Figure 18.24
(a) Playa lake formed after a rainstorm filled Croneis Dry Lake, Mojave Desert, California.
(b) Racetrack Playa, Death Valley, California. The Inyo Mountains can be seen in the background.

(b)

mud cracks and precipitated salt crystals (Figure 18.24b). Salts in some playas are thick enough to be mined commercially. For example, borates have been mined in Death Valley, California, for more than 100 years (see Perspective 18.1).

Other common features of deserts, particularly in the Basin and Range region of the United States, are alluvial fans and bajadas. **Alluvial fans** form when sediment-laden streams flowing out from the generally straight, steep mountain fronts deposit their load on the relatively flat desert floor. Once beyond the mountain front where no valley walls contain streams, the sediment spreads out laterally, forming a gently sloping and poorly sorted fan-shaped sedimentary deposit (Figure 18.25). Alluvial fans are similar in origin and shape to deltas (see Chapter 15) but are formed entirely on land. Alluvial fans may coalesce to form a *bajada,* a broad alluvial apron that typically has an undulating surface resulting from the overlap of adjacent fans (Figure 18.26).

Large alluvial fans and bajadas are fre-

Figure 18.25
Aerial view of an alluvial fan, Death Valley, California.

Figure 18.26
Coalescing alluvial fans forming a bajada at the base of the Black Mountains, Death Valley, California.

quently important sources of groundwater for domestic and agricultural use. Their outer portions are typically composed of fine-grained sediments suitable for cultivation, and their gentle slopes allow good drainage of water. Many alluvial fans and bajadas are also the sites of large towns and cities, such as San Bernardino, California, Salt Lake City, Utah, and Teheran, Iran.

Most mountains in desert regions, including those of the Basin and Range region, rise abruptly from gently sloping surfaces called pediments. **Pediments** are erosional bedrock surfaces of low relief that slope gently away from mountain bases (Figure 18.27). Most pediments are covered by a thin layer of debris, alluvial fans, or bajadas.

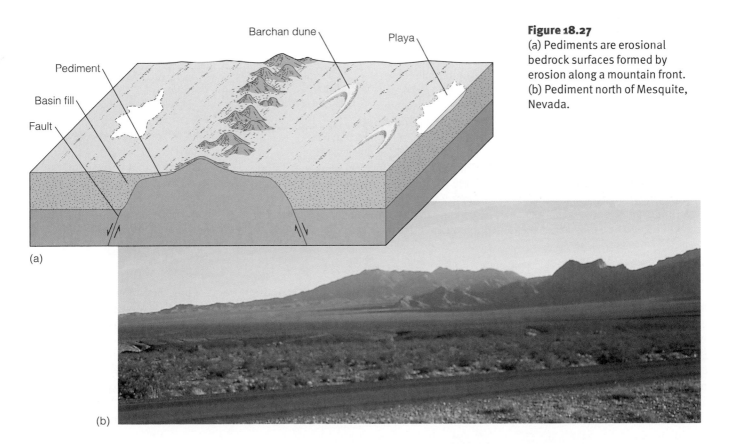

Figure 18.27
(a) Pediments are erosional bedrock surfaces formed by erosion along a mountain front. (b) Pediment north of Mesquite, Nevada.

Perspective 18.3

The Geologic History of Uluru and Kata Tijuta

Rising majestically above the surrounding flat desert of central Australia is Uluru (Ayers Rock) and its neighbor to the northwest, Kata Tijuta (The Olgas) (Figure 1). Uluru and Kata Tijuta are the Aboriginal names for what most people know as Ayers Rock and The Olgas. Uluru was named Ayers Rock in 1873 by the explorer W. C. Gosse, after Sir Henry Ayers, the then Premier of South Australia. Kata Tijuta was named Mount Olga in 1872 by Ernest Giles, in honor of the then King and Queen of Spain.

Archaeological evidence indicates that there has been human activity in this region of Australia for at least the past 22,000 years, and probably even earlier. Both Uluru and Kata Tijuta are an important part of Aboriginal culture, and some sites are so sacred that only Aboriginal people can visit them. When the Australian government returned title to Uluru National Park to the Aboriginal people in 1985, part of the agreement was to replace the European names of features and localities with Aboriginal (Anangu) ones.

The geologic history of Uluru and Kata Tijuta begins between about 900 million and 600 million years ago (Late Proterozoic), when much of central Australia was beneath sea level and part of the Amadeus Basin. This period of marine deposition was followed by the Petermann Ranges Orogeny that took place during the Cambrian, resulting in a large mountain range. As this mountain range eroded, huge alluvial fans formed at its base. The sediments comprising these alluvial fans were eventually lithified into sedimentary rocks that would later become Uluru and Kata Tijuta. The sedimentary rocks of Uluru and Kata Tijuta are coarse-grained and poorly sorted arkoses (see Figure 6.10b on page 165) (Uluru Arkose) and arkosic conglomerates (Mount Currie Conglomerate), indicative of alluvial fan deposition.

At some time about 500 million years ago, the region was again covered by warm, shallow seas, and the alluvial fan deposits were buried by sandstones, mudstones, and lime-

Figure 1
Location of Uluru and Kata Tijuta in central Australia.

The origin of pediments has been the subject of much controversy. Most geologists agree that they are erosional features developed on bedrock in association with the erosion and retreat of a mountain front (Figure 18.27a). The disagreement concerns how the erosion has occurred. Although not all geologists would agree, it appears that pediments are produced by the combined activities of lateral erosion by streams, sheet flooding, and various weathering processes along the retreating mountain front. Thus, pediments grow at the expense of the mountains, and they will continue to expand as the mountains are eroded away or partially buried.

Rising conspicuously above the flat plains of many deserts are isolated steep-sided erosional remnants called **inselbergs,** a German word meaning "island mountain" (see Perspective 18.3). Inselbergs have survived for a longer period of time than other

stones. Another orogeny, the Alice Springs orogeny, uplifted the area about 400 million years ago, resulting in much folding and faulting. Although the beds around Uluru were tilted almost vertically, the deformation in the Kata Tijuta region was not as severe. The area that is now Uluru National Park was raised above sea level during the Alice Springs orogeny and has remained above sea level ever since. Erosion of the area began to shape the uplifted rocks, giving form to what would eventually become Uluru and Kata Tijuta (Figure 2).

About 70 million years ago, a wide valley was eroded between Uluru and Kata Tijuta. The climate during this time was far different from the climate today. Instead of the arid conditions found presently, a rainy and more temperate climate prevailed, with year-round flowing streams and marshes a common feature of the landscape. Within the marshes, plant material accumulated, forming peat layers that later became converted to thin layers of low-grade coal.

By about 500,000 years ago, the climate became more arid, and has remained arid to the present day. Many of the sand dunes seen in Uluru Park today began forming about 30,000 years ago.

The spectacular and varied rock shapes of Uluru and Kata Tijuta are the result of millions of years of weathering and erosion by water, and to a minor extent, wind acting on the various fractures and joints caused by uplifting of the area. Differences in the composition of the arkoses and conglomerates that make up Uluru and Kata Tijuta have also played a role in the sculpting of these colorful and magnificent structures.

(a)

(b)

Figure 2
(a) Uluru. (b) Kata Tijuta.

mountains because of their greater resistance to weathering.

Other easily recognized erosional remnants common to arid and semiarid regions are mesas and buttes (Figure 18.28). A **mesa** is a broad, flat-topped erosional remnant bounded on all sides by steep slopes. Continued weathering and stream erosion will form isolated pillarlike structures known as **buttes.** Buttes and mesas consist of relatively easily weathered sedi-

mentary rocks capped by nearly horizontal, resistant rocks such as sandstone, limestone, or basalt. They form when the resistant rock layer is breached, allowing rapid erosion of the less resistant underlying sediment. One of the best-known areas of mesas and buttes in the United States is Monument Valley on the Arizona–Utah border.

(a)

(b)

Figure 18.28
(a) Spearhead Mesa, Monument Valley Navajo Tribal Park, Arizona/Utah. Other mesas and several buttes can be seen in the background, as well as a dune field in the foreground. (b) Left Mitten Butte and Right Mitten Butte in Monument Valley on the border of Arizona and Utah.

Chapter Summary

1. Wind can transport sediment in suspension or as bed load, which involves saltation and surface creep.

2. Wind erodes material either by abrasion or deflation. Abrasion is a near-surface effect caused by the impact of saltating sand grains. Ventifacts are common wind-abraded features.

3. Deflation is the removal of loose surface material by the wind. Deflation hollows resulting from differential erosion of surface material are common features of many deserts, as is desert pavement, which effectively protects the underlying surface from additional deflation.

4. The two major wind deposits are dunes and loess. Dunes are mounds or ridges of wind-deposited sand, whereas loess is wind-deposited silt and clay.

5. The four major dune types are barchan, longitudinal, transverse, and parabolic. The amount of sand available, the prevailing wind direction, the wind velocity, and the amount of vegetation present determine which type will form.

6. Loess is derived from deserts, Pleistocene glacial outwash deposits, and river floodplains in semiarid regions. Loess covers approximately

10% of Earth's land surface and weathers to a rich and productive soil.

7. Deserts are very dry (averaging less than 25 cm of rain per year), have poorly developed soils, and are mostly or completely devoid of vegetation.

8. The winds of the major east–west-oriented air-pressure belts resulting from rising and cooling air are deflected by the Coriolis effect. These belts help control the world's climate.

9. Dry climates are located in the low and middle latitudes where the potential loss of water by evaporation exceeds the yearly precipitation.

Dry climates cover 30% of Earth's surface and are subdivided into semiarid and arid regions.

10. The majority of the world's deserts are in the low-latitude dry climate zone between 20 and 30 degrees north and south latitudes. Their dry climate results from a high-pressure belt of descending dry air. The remaining deserts are in the middle latitudes, where their distribution is related to the rainshadow effect, and in the dry polar regions.

11. Deserts are characterized by lack of precipitation and high evaporation rates. Furthermore, rainfall is unpredictable and, when it does occur, tends to be intense and of short duration. As a consequence of such aridity, desert vegetation and animals are scarce.

12. Mechanical weathering is the dominant form of weathering in deserts. The sparse precipitation and slow rates of chemical weathering result in poorly developed soils.

13. Running water is the dominant agent of erosion in deserts and was even more important during the Pleistocene Epoch when wetter climates resulted in humid conditions.

14. Wind is an erosional agent in deserts and is very effective in transporting and depositing unconsolidated fine-grained sediments.

15. Important desert landforms include playas, which are dry lakebeds; when temporarily filled with water, they form playa lakes. Alluvial fans are fan-shaped sedimentary deposits that may coalesce to form bajadas.

16. Pediments are erosional bedrock surfaces of low relief gently sloping away from mountain bases. The origin of pediments is controversial, although most geologists think that they form by the combined activities of lateral erosion by streams, sheet flooding, and various weathering processes.

17. Inselbergs are isolated steep-sided erosional remnants that rise above the surrounding desert plains. Buttes and mesas are, respectively, pinnacle-like and flat-topped erosional remnants with steep sides.

Important Terms

abrasion
alluvial fan
barchan dune
butte
Coriolis effect
deflation
desert

desertification
desert pavement
dune
inselberg
loess
longitudinal dune
mesa

parabolic dune
pediment
playa
rainshadow desert
transverse dune
ventifact

Review Questions

1. A worldwide pattern of east–west-oriented wind belts is produced by:

   a. _____ Coriolis effect;
   b. _____ latitudinal air-pressure differences;
   c. _____ doldrums;
   d. _____ answers (a) and (b);
   e. _____ all of these.

2. Warm, dry high-pressure areas that provide perfect conditions for formation of low-latitude deserts occur at:

   a. _____ 20 to 30 degrees north latitude;
   b. _____ 20 to 30 degrees south latitude;
   c. _____ within 20 degrees north and south of the equator;
   d. _____ answers (a) and (b);
   e. _____ all of these.

3. The primary process for movement of sand as bed load by wind is:

   a. _____ rolling and sliding;
   b. _____ saltation;
   c. _____ suspension;
   d. _____ answers (a) and (b);
   e. _____ none of these.

4. Which size of particle tends to lie below the thin, motionless layer of air next to the ground, and thus is not commonly moved by wind?

   a. _____ sand;
   b. _____ silt;
   c. _____ clay;
   d. _____ answers (a) and (b);
   e. _____ answers (b) and (c).

5. These form when air-borne particles abrade the surface of a rock:

   a. _____ dunes;
   b. _____ playas;
   c. _____ ventifacts;
   d. _____ loess soils;
   e. _____ none of these.

6. The angle of repose of dry sand controls:

   a. _____ the slope of the leeward side of a sand dune;
   b. _____ the height of a sand dune;
   c. _____ the slope of the windward side of a sand dune;
   d. _____ answers (a) and (b);
   e. _____ answers (a) and (c).

7. The sand dune type that forms long, parallel ridges and occurs in areas of

limited sand supply owing to convergent wind directions is called:

  a. _____ barchan;
  b. _____ transverse;
  c. _____ longitudinal;
  d. _____ parabolic;
  e. _____ none of these.

8. What is the name of the type of non-vegetated, desert sand dune that occurs as a small crescent shape?

  a. _____ barchan;
  b. _____ transverse;
  c. _____ longitudinal;
  d. _____ parabolic;
  e. _____ none of these.

9. From where is loess potentially derived?

  a. _____ deserts;
  b. _____ floodplains of rivers in semiarid regions;
  c. _____ Pleistocene glacial outwash deposits;
  d. _____ answers (a) and (b);
  e. _____ all of these.

10. The air that descends on a rain-shadow desert is:

  a. _____ warm and moist;
  b. _____ cold and dry;
  c. _____ warm and dry;
  d. _____ cold and moist;
  e. _____ none of these.

11. Alluvial fans may coalesce to form:

  a. _____ playas;
  b. _____ bajadas;
  c. _____ inselbergs;
  d. _____ answers (a) and (b);
  e. _____ none of these.

12. What is the major erosive agent at work in desert areas today?

  a. _____ wind's abrasion;
  b. _____ wind's deflation;
  c. _____ water flow;
  d. _____ answers (a) and (b);
  e. _____ none of these.

13. What are the reasons for dryness in low-latitude deserts today? What is the primary cause of this dryness?

14. What roles do wind direction and sand supply have in determining what type of sand dune may be formed in a desert setting? In your answer, review all dune types.

15. What is the relationship between particle size and the way wind transports particles? How does the density difference between water and air affect what particle sizes are transported typically by each?

16. With particular reference to Figure 18.19, explain how Earth's rotation and heating plus cooling of air work together to produce Earth's east–west-oriented pattern of wind belts.

17. Explain how desert pavements develop and their importance in preserving the desert environment.

18. Describe the process of dune formation and migration? What roles are played by angle of repose, saltation, and gravity in this process? How does a dune "migrate"? How is this process related to origin of cross-bedding in sandstones?

19. What is the relationship between modern suspended load of wind and ancient loess deposits? Compare grain sizes of modern suspended load versus loess. Where does loess occur today? Where are suspended load sediments coming from today?

20. Describe the location and/or name the world's main low-latitude deserts. Why are low-latitude deserts so common?

21. What is the difference between mesas and buttes? How were these "erosional remnants" formed? What is the role of horizontal rocks capping these features? How are these features different from inselbergs?

22. What are the differences between the two types of wind erosion, abrasion and deflation? Name some specific features produced by each of these two types of erosion. Is wind an effective agent for erosion, compared to water? Why?

23. What are the differences between arid and semiarid regions?

24. What is internal drainage in deserts, and what is its relationship to desert groundwater? Explain what typically happens to water precipitated during a desert region cloudburst?

25. What do geologists think that pediments represent?

26. In a typical desert, how are temperature ranges, amount and frequency of precipitation, and type of vegetation related to one another?

27. What is rock varnish, and how is it thought to form?

Points to Ponder

1. Loess covers 10% of Earth's continental surface and 30% of the United States. What do you think was the source of all the silt and clay comprising these wind-blown deposits? The age of this loess is Pleistocene, so what geologic event during the Pleistocene could be related to loess formation?

2. Considering what we know about where and why deserts form today, how can this knowledge be applied to the study of sedimentary rocks, our record of the geologic past? Specifically, could cross-bedding in sandstones be used to suggest dune type, thus yielding information about past conditions such as sand supply and wind direction? Also, could the location of wind-deposited, cross-bedded sandstones help us locate former low-latitude, continental interior, and rainshadow areas? How would such studies be done and why?

3. Why are so many water-erosion features evident in desert settings, even though very little rainwater falls there and rainfall is so infrequent?

4. Regarding Figures 18.25 and 18.26, what is the relationship between the mountain stream at the head of each of the alluvial fans in these pictures and the formation of the alluvial fan itself? As noted in the text, some large cities are built on such alluvial fans. Describe a possible geologic hazard of living on the surface of an alluvial fan that might be associated with sudden, heavy precipitation in the mountains above the fan.

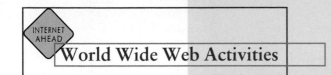

World Wide Web Activities

➤ THE NATIONAL PARK SERVICE—DEATH VALLEY NATIONAL PARK

This site contains general information about Death Valley National Park. What is the general climate of the park? What activities are available? Where is Death Valley National Park located?

➤ DEATH VALLEY NATIONAL PARK

This site, operated by DesertUSA: The Ultimate Desert Resource, is a compendium of information about Death Valley National Park. Click on *Description of the Park* under the Exploring Death Valley National Park heading. What is the physical environment of the area? What types of plants and animals live in this area? What is the geology of the area?

➤ DESERTUSA

This site, devoted to information about deserts, contains a tremendous amount of information about the fauna, flora, geology, physical environment, and other items of interest about deserts. Under the What's New heading, click on the *The Desert* site. Read about what deserts are. Click on one of the deserts shown on the map of the United States. Learn about the different deserts in the United States.

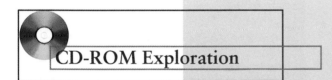

CD-ROM Exploration

➤ Exploring your *Earth Systems Today* CD-ROM will add to your understanding of the material in this chapter.

TOPIC: ATMOSPHERIC PROCESSES

MODULE: WEATHER AND CLIMATE

Explore activities in this module to see if you can discover the following for yourself:

What are Earth's primary climatic zones? Are deserts confined to one climatic zone? Use features in this module to help you understand the likely effects of global warming on desertification.

What are the cyclic weather effects of El Niño and La Niña? How does the rotation of Earth produce a Coriolis effect, and how does that in turn affect local weather? Select several places on Earth and answer these questions about each.

Chapter 19

A rocky shoreline along the Pacific Coast of California.

OBJECTIVES

At the end of this chapter, you will have learned that

- Wind-generated waves and their associated nearshore currents effectively modify shorelines by erosion and deposition.
- The gravitational attraction by the Moon and Sun coupled with Earth's rotation is responsible for the once- or twice-daily rise and fall of sea level known as *tides*.

- Seacoasts and lakeshores are both modified by waves and nearshore currents, but seacoasts also experience tides, which are insignificant in even the largest lakes.
- The concept of a nearshore sediment budget considers equilibrium, losses, and gains in the amount of sediment in a coastal area.

- Several distinctive erosional and deposition landforms are found along shorelines.
- Shoreline management is complicated by rising sea level.
- Several types of coasts are recognized based on criteria such as fluctuations in sea level, and erosion versus deposition.

Shorelines and Shoreline Processes

Prologue

Wind-generated waves, especially those formed during storms, are responsible for most geologic work on shorelines, but tsunami and landslide surges can have disastrous effects. As we explained in Chapter 9, tsunami are generated by fault displacement of the seafloor, submarine slides and slumps, and explosive volcanic eruptions.

In the open sea, tsunami may pass unnoticed, but when they enter shallow coastal waters, their wave height increases to as much as 65 m! The first indication of an approaching tsunami is a rapid withdrawal of the sea from coastal regions, followed a few minutes later by destructive waves. In many cases, tsunami come in as a rapidly rising tide, and their backwash, which undermines structures and carries loose objects out to sea, causes most of the damage and fatalities.

The largest waves occur in restricted bodies of water, such as bays or lakes, when water is suddenly displaced by large landslides or rockfalls. The largest of these so-called landslide surges took place on July 9, 1958, in Lituya Bay, Alaska, when an estimated 30.5 million m^3 of rock plunged into the bay from a height of more than 900 m. The sudden displacement of water caused a surge on the opposite side of the bay that rose 536 m above sea level! The wave then moved out of the harbor into the open ocean where it quickly dissipated.

Of considerably more concern to shoreline dwellers is coastal flooding caused by storm-generated waves. Flooding during hurricanes is caused by large waves being driven onshore and by intense rainfall, more than 60 cm in 24 hours in some cases. In addition, as a hurricane moves over the ocean, low atmospheric pressure beneath the eye of the storm causes the ocean surface to bulge upward as much as 0.5 m. When the eye reaches the shoreline, the bulge coupled with wind-driven waves piles up in a *storm surge* that can rise several meters above normal high tide and inundate areas several kilometers inland.

Several coastal areas in the United States have been devastated by storm surges. In 1900 Galveston, Texas, a town of 38,000 on a long, narrow barrier island a short distance from the mainland, was nearly destroyed, and between 6000 and 8000 people died. When the hurricane struck, storm waves surged inland, eventually covering the entire island. Buildings and other structures near the shoreline were battered to pieces, and "great beams and railway ties were lifted by the [waves] and driven like battering rams into dwellings and business houses"* farther inland. Finally, after the first four shoreline blocks were

---

*L. W. Bates, Jr., "Galveston—A City Built upon Sand," *Scientific American* 95 (1906): 64.

destroyed, the debris piled up high enough to form a protective barrier for the rest of the city. To protect the city from future flooding, a colossal two-part project was begun in 1902. First, a huge seawall was constructed to protect the city from waves (Figure 19.1a). Next, parts of the city were raised to the level of the top of the seawall. Buildings were elevated and supported on jacks while sand fill was pumped beneath them (Figure 19.1b).

More recently, in 1989, Charleston, South Carolina, and nearby areas were flooded by a storm surge generated by Hurricane Hugo (Figure 19.2). This storm surge, which ranged from 2.5 to 6 m high, devastated an extensive area of shoreline, caused 21 deaths, and resulted in more than $7 billion in property damage. In fact, coastal flooding is responsible for most of the damage and about 90% of all fatalities attributed to hurricanes.

One of the greatest natural disasters of the twentieth century occurred in 1970 when a storm surge estimated at 8 to 10 m high flooded the low-lying coastal areas of Bangladesh, drowning 300,000 people. Since 1970 the coastal areas of Bangladesh have been flooded several more times, the most recent and most tragic being on April 30, 1991, when an estimated 139,000 people were drowned and 10 million were left homeless.

The problem of coastal flooding during hurricanes is exacerbated by rising sea level, a topic that will be discussed in more detail in a later section. To gain some perspective on the magnitude of the problem, consider this: Of the nearly $2 billion paid out by the federal government's National Flood Insurance Program since 1974, most has gone to owners of beachfront homes.

(a)

Figure 19.1
(a) Construction of this seawall to protect Galveston, Texas, from storm waves began in 1902. Notice that the wall is curved to deflect waves upward. (b) Some of the nearly 3000 buildings in Galveston that were raised and supported on stilts until sand fill could be pumped beneath them.

(b)

(b)

(a)

Figure 19.2
On September 22, 1989, Hurricane Hugo generated a storm surge that caused considerable damage along the South Carolina coast. (a) Homes at Folly Beach near Charleston before the hurricane. The barrier along the beach was for protection from wave erosion. (b) Hurricane Hugo nearly leveled the barrier and destroyed several homes. In fact, boulders from the barrier were deposited in the living rooms of some of the homes still standing after the storm.

INTRODUCTION

A **shoreline** can be defined simply as the area of land in contact with the sea or a lake. We can expand this definition somewhat by noting that it consists of the area between low tide and the highest level on land affected by storm waves. Accordingly, a shoreline is a long, rather narrow zone where marine processes, particularly tides, waves, and nearshore currents, continually modify existing shoreline features. In short, shorelines, like streams and glaciers, are dynamic areas where energy is expended, erosion takes place, and sediment is transported and deposited. Our main interest here is with ocean shorelines, but waves and nearshore currents are also quite effective in large lakes, where shorelines exhibit many of the features seen along seashores. The most notable differences are that waves and nearshore currents are much more energetic along seashores, and even the largest lakes lack appreciable tides.

The geologic processes operating along shorelines provide another example of interactions among Earth systems, in this case between the hydrosphere and solid Earth (see Table 1.1). In this chapter we concentrate on that narrow zone between land and sea or a lake, but we must be aware that a much larger element of the hydrosphere is involved. After all, waves are generated in the oceans, and the gravitational attraction of the Moon and Sun affect all oceanic waters, not just those along shorelines. It is, however, along shorelines that much of the energy of waves and tides is expended.

The continents possess more than 400,000 km of shorelines. Some are rocky and steep, such as those in Maine, the maritime provinces of Canada, and much of North America's West Coast, whereas shorelines with broad sandy beaches are found in much of the eastern United States and along the Gulf Coast. Whatever their type, all shorelines, including those on lakes, are dynamic areas of change where shoreline processes constantly act on shoreline materials. On shorelines where energy levels are particularly high, erosion predominates, and the shoreline might retreat landward. Indeed, coastal erosion and landward migration of shorelines are problems in many areas. Yet some shorelines receive copious quantities of sediment and actually build outward. That is, they prograde seaward in much the same way that deltas do (see Figure 15.21), although deltas tend to be lobe shaped, whereas shorelines are linear.

Living near a shoreline has both positive and negative aspects. Many of the world's major centers of commerce are at or near seashores, or on lakeshores or streams where ships can navigate far inland as along the Mississippi and St. Lawrence Rivers. And certainly the scenery and recreational opportunities of shorelines have great appeal for many people. Communities such as Myrtle Beach, South Carolina,

and Fort Lauderdale, Florida, are heavily dependent on tourists visiting and enjoying their shorelines.

However, in many parts of the world, including most of the United States and much of Canada, sea level is rising, and buildings that were once far inland are in peril and many have been destroyed. Slumps and slides are common along steep shorelines eroded by waves (Figure 19.3), and narrow offshore barrier islands move landward because of erosion on their seaward sides and deposition on their landward sides. And hurricanes expend much of their energy on shorelines, resulting in extensive coastal flooding, numerous fatalities, and considerable property damage (see the Prologue).

Scientists from several disciplines have contributed to our understanding of shorelines and shoreline processes. Geologists, oceanographers, marine biologists, and coastal engineers are interested in all aspects of this dynamic area. Elected officials and city planners of coastal communities must become familiar with shoreline processes so they can develop policies and propose legislation that serve the public as well as protect the fragile shoreline environment. In short, the study of shorelines is not only interesting but also has many practical applications.

SHORELINES AND SHORELINE PROCESSES

In contrast to other geologic agents such as running water, wind, and glaciers that operate over vast areas, shoreline processes are restricted to a narrow zone at any particular time. But shorelines might

Figure 19.3
Erosion of sea cliffs during February 1998 storms resulted in the evacuation of the families from these houses near Bodega Bay, California. Attempts to stabilize the cliffs have failed.

migrate landward or seaward depending on changing sea level or uplift or subsidence of coastal regions. During a rise in sea level, for instance, the shoreline migrates landward, and as it does so, wave, tide, and nearshore current activity shifts landward as well; during times when sea level falls, just the opposite takes place. Recall from Chapter 6 that during marine transgressions and regressions that beach and nearshore sediments can be deposited over vast regions.

In the marine realm, several biological, chemical, and physical processes are operating continuously. Organisms change the local chemistry of seawater and contribute their skeletons to nearshore sediments, temperature and salinity changes, and internal waves occur in the oceans. However, the processes most important for modifying shorelines are purely physical ones, especially waves, tides, and nearshore currents. We cannot totally discount some of these other processes, though; offshore reefs, for instance, may protect a shoreline area from most of the energy of waves.

Tides

The surface of the oceans rises and falls twice daily in response to the gravitational attraction of the Moon and Sun. These regular fluctuations in the ocean's surface, or **tides,** result in most shoreline areas having two daily high tides and two low tides as sea level rises and falls anywhere from a few centimeters to more than 15 m (Figure 19.4). During rising or *flood tide,* more and more of the nearshore area is flooded until high tide is reached. *Ebb tide* occurs when currents flow seaward during a decrease in the height of the tide.

Both the Moon and the Sun have sufficient gravitational attraction to exert tide-generating forces strong enough to deform the solid body of Earth, but they have a much greater influence on the oceans. The Sun is 27 million times more massive than the Moon, but it is 390 times as far away and its tide-generating force is only 46% as strong as that of the Moon. Accordingly, the tides are dominated by the Moon, but the Sun does play an important role as well.

If we consider only the Moon acting on a spherical, water-covered Earth, the tide-generating forces produce two bulges on the ocean surface (Figure 19.5a). One bulge extends toward the Moon because it is on the side of Earth where the Moon's gravitational attraction is greatest. The other bulge is on the opposite side of Earth, where the Moon's gravitational attraction is least. These two bulges always point toward and away

(a)

(b)

Figure 19.4
(a) Low tide and (b) high tide.

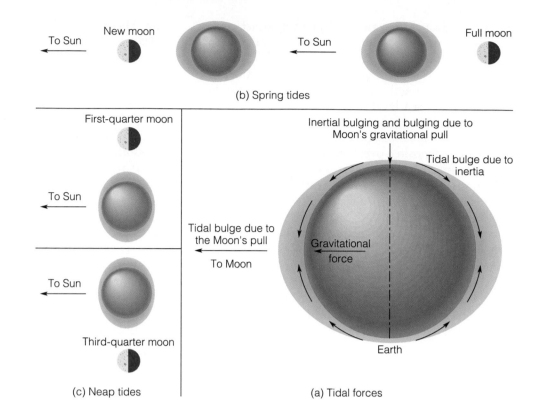

Figure 19.5
(a) Tides are caused by the gravitational pull of the Moon and, to a lesser degree, the Sun. The Earth–Moon–Sun alignments at the times of the (b) spring and (c) neap tides are shown.

New moon
To Sun
(b) Spring tides
To Sun
Full moon

First-quarter moon
To Sun

Inertial bulging and bulging due to Moon's gravitational pull
Tidal bulge due to inertia
Tidal bulge due to the Moon's pull
To Moon
Gravitational force
Earth
(a) Tidal forces

To Sun
Third-quarter moon
(c) Neap tides

from the Moon (Figure 19.5a), so as Earth rotates and the Moon's position changes, an observer at a particular shoreline location experiences the rhythmic rise and fall of tides twice daily. But the heights of two successive high tides may vary depending on the Moon's inclination with respect to the equator.

The Moon revolves around Earth every 28 days, so its position with respect to any latitude changes slightly each day. That is, as the Moon moves in its orbit and Earth rotates on its axis, it takes the Moon 50 minutes longer each day to return to the same position it was in the previous day. Thus, an observer would experience a high tide at 1:00 P.M. on one day, for example, and at 1:50 P.M. on the following day.

Tides are also complicated by the combined effects of the Moon and the Sun. Even though the Sun's tide-generating force is weaker than the Moon's, when the Moon and Sun are aligned every two weeks, their forces are added together and generate *spring tides,* which are about 20% higher than average tides (Figure 19.5b). When the Moon and Sun are at right angles to one another, also at two-week intervals, the Sun's tide-generating force cancels some of that of the Moon, and *neap tides* about 20% lower than average occur (Figure 19.5c).

Tidal ranges are also affected by shoreline configuration. Broad, gently sloping continental shelves as in the Gulf of Mexico have low tidal ranges, whereas steep, irregular shorelines experience a much greater rise and fall of tides. Tidal ranges are greatest in some narrow, funnel-shaped bays and inlets. For example, in the Bay of Fundy in Nova Scotia a tidal range of 16.5 m has been recorded, and ranges greater than 10 m occur in several other areas.

Tides have an important impact on shorelines because the area of wave attack constantly shifts onshore and offshore as the tides rise and fall. Tidal currents themselves have little modifying effect on shorelines, except in narrow passages where tidal current velocity is great enough to erode and transport sediment. Indeed, if it were not for strong tidal currents, some passageways would be blocked by sediments deposited by nearshore currents. However, tides are one potential source of energy (see Perspective 19.1).

Waves

Waves, or oscillations of a water surface, occur on all bodies of water but are most significant along seashores where they are responsible for most of the erosion, transport, and deposition of sediment. Many of the erosional and depositional features of the world's shorelines form and are modified by the energy of incoming waves.

Figure 19.6 shows a typical series of waves in deep water and the terminology applied to them. The highest part of a wave is its **crest,** and the low point between crests is the **trough. Wavelength** is the distance between successive wave crests (or troughs), and **wave height** is

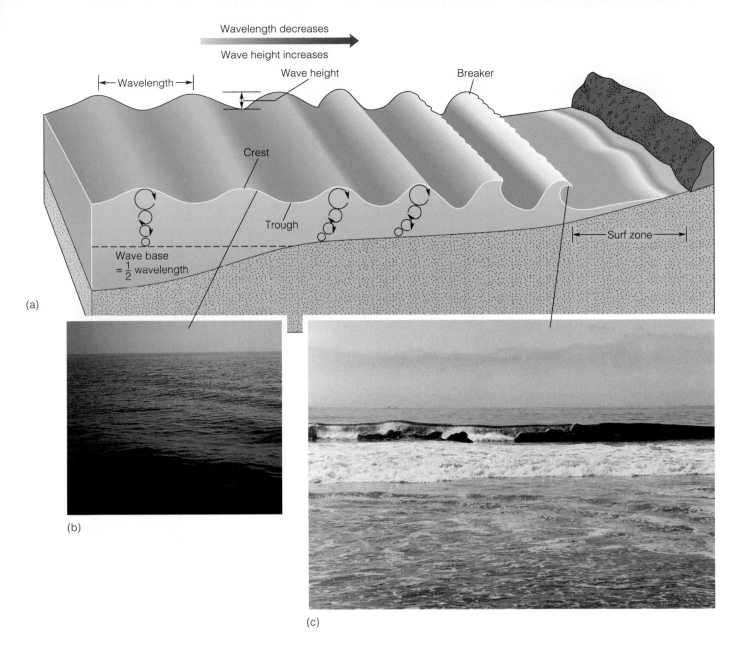

Figure 19.6
(a) Waves and the terminology applied to them. As swells (b) in deep water move toward shore, the orbital motion of water within them is disrupted when they enter water shallower than wave base. Wavelength decreases while wave height increases, causing the waves to oversteepen and plunge forward as breakers (c).

the vertical distance from trough to crest. The speed at which a wave advances, generally called *celerity* (C), can be calculated if one knows the wavelength (L) and the **wave period** (T), which is the time required for two successive wave crests (or troughs) to pass a given point:

$$C = L/T$$

The speed of wave advance (C) is actually a measure of the velocity of the waveform rather than a measure of the speed of the molecules of water. In fact, the water in waves moves forward and back as a wave passes but has little or no net forward movement. As waves move across a water surface, the water "particles" rotate in circular orbits and transfer energy in the direction of wave advance (Figure 19.6).

Perspective 19.1

Energy from the Oceans

The fact that several systems interact in complex ways has been stressed many times in previous chapters. Here too, we emphasize these interactions as we discuss the potential to extract energy from the oceans. Solar energy reaching Earth, for instance, is responsible for differential heating of the atmosphere and thus air circulation as wind. Some of the energy of wind is transferred to the oceans where it causes waves and is partly responsible for oceanic currents, although Earth's rotation also plays a role in currents. Gravitational attraction between Earth and the Sun and Moon generate tides, and coupled with Earth's rotation result in most coastal areas experiencing a twice-daily rise and fall of sea level. In short, the oceans possess a tremendous reservoir of largely untapped energy.

If we can effectively harness the energy possessed by the oceans, an almost limitless, largely nonpolluting energy supply would be ensured. Unfortunately, ocean energy is diffuse, meaning that the amount of energy for a given volume of water is small and thus difficult to concentrate and use. Several ways of using ocean energy are being considered or under development, and one is currently in use, although it accounts for only a tiny proportion of all energy production. Of the several sources of ocean energy—temperature differences with depth, waves, currents, and tides—only the latter shows much promise for the near future.

Ocean water at depth might be as much as 25°C colder than surface water, which is the basis for ocean thermal energy conversion (OTEC). OTEC exploits this temperature difference to run turbines and generate electricity.

The amount of energy available from this source is enormous, but a number of practical problems must be solved before it can be used. For one thing, any potential site must be close to land and also have a sufficiently rapid change in depth to result in the required temperature difference. Furthermore, enormous quantities of warm and cold seawater would have to circulate through an electrical-generating plant, thus requiring large surface areas to be devoted to this purpose. The concept of OTEC is more than a century old, but despite several decades of research no commercial OTEC plants are operating or even under construction, although small experimental ones have been tested in Hawaii and Japan.

Wind-generated ocean currents, such as the Gulf Stream that flows along the East Coast of North America, also possess energy that might be tapped to generate electricity. Unlike streams that can be dammed to impound a reservoir, any electrical-generating facility exploiting oceanic currents would have to concentrate their diffuse energy and contend with any unpredictable changes in direction. In addition, whereas hydroelectric generating plants on land depend on water moving rapidly from a higher elevation to the turbines, the energy of ocean currents comes from their flow velocity, that is at most a few kilometers per hour.

The most obvious form of energy in the oceans is waves. Harnessing wave energy and converting it to electricity is not a new idea, and it is used on an extremely limited scale. Unfortunately, the energy possessed by a wave is distributed along its crest and is difficult to concentrate. Furthermore, any facility would have to be designed to with-

stand the effects of storms and saltwater corrosion. No large-scale wave-energy plants are operating, but the Japanese have developed devices to power lighthouses and buoys.

Perhaps tidal power is the most promising form of ocean energy. In fact, it has been used for centuries in some coastal areas to run mills, but its use at present for electrical generation is limited. Most coastal areas experience a twice-daily rise and fall of tides, but only a few areas are suitable for exploiting this energy source. One limitation is that the tidal range must be at least 5 m, and there must also be a coastal region where water can be stored following high tide.

Suitable sites for using tidal power are limited not only by tidal range, but also by location. Many areas along the U.S. Gulf Coast would certainly benefit from a tidal power plant, but the tidal range of generally less than 1 m precludes the possibility of development. However, an area with an appropriate tidal range in some remote area such as southern Chile or the Arctic islands of Canada offer little potential because of their great distance from population centers. Accordingly, in North America only a few areas show much potential for developing tidal energy: Cook Inlet, Alaska, the Bay of Fundy on the U.S.–Canadian border, and some areas along the New England coast, for instance.

The idea behind tidal power is simple, although putting it into practice is not easy. First, a dam with sluice gates to regulate water flow must be built across the entrance to a bay or estuary. When the water level has risen sufficiently high during flood tide, the sluice gates are closed. Water held on the landward side of the dam is then

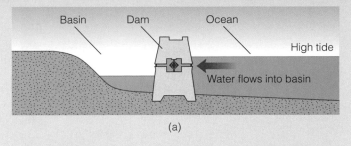

Basin   Dam   Ocean

High tide

Water flows into basin

(a)

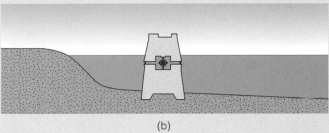

(b)

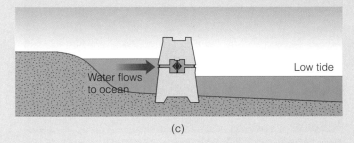

Low tide

Water flows
to ocean

(c)

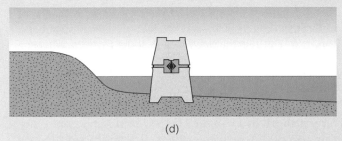

(d)

Figure 1
Rising and falling tides produce electricity by spinning turbines connected to generators, just as at hydroelectric plants. (a) Water flows from ocean to basin during high tide. (b) Basin full. (c) Water flows from basin to ocean during low tide. (d) Basin water is depleted, and cycle begins once more.

released and electricity is generated just as it is at a hydroelectric dam (see Figure 1, Perspective 15.1). Actually, a tidal power plant can operate during both flood and ebb tides (Figure 1).

The first tidal power-generating facility was constructed at the La Rance River estuary in France (Figure 2). This 240-megawatt (MW) plant began operations in 1966 and has been quite successful. In North America, a much smaller 20-MW tidal power plant has been operating in the Bay of Fundy, Nova Scotia, where the tidal range, the greatest in the world, exceeds 16 m. The Bay of Fundy is part of the more extensive Gulf of Maine where the United States and Canada have been considering a much larger tidal power-generating project for several decades.

Although tidal power shows some promise, it will not solve our energy needs even if developed to its fullest potential. Most analysts think that only 100 to 150 sites worldwide have sufficiently high tidal ranges and the appropriate coastal configuration to exploit this energy resource. This, coupled with the facts that construction costs are high and tidal energy systems can have disastrous effects on the ecology (biosphere) of estuaries, makes it unlikely that tidal energy will ever contribute more than a small percentage of all energy production.

Figure 2
The La Rance River estuary tidal power-generating plant in western France. This 850-m-long dam was built where the maximum tidal range is 13.4 m. Electricity is generated during both flood and ebb tides.

The diameters of the orbits followed by water particles in waves diminish rapidly with depth, and at a depth of about one-half wavelength ($L/2$), called **wave base,** they are essentially zero (Figure 19.6). At depths exceeding wave base, the water and seafloor, or lake floor, are unaffected by surface waves (Figure 19.7). The significance of wave base is explored more fully in later sections.

Wave Generation   Waves can be generated by several processes including displacement of water by landslides, displacement of the seafloor by faulting, and volcanic explosions. Most of the geologic work done on shorelines, though, is accomplished by wind-generated waves. When wind blows over water, some of its energy is transferred to the water, causing the water surface to oscillate. The mechanism that transfers energy from wind to water is related to the frictional drag resulting from one fluid (air) moving over another (water).

In an area of wave generation, as beneath a storm center at sea, sharp-crested, irregular waves called *seas* develop. Seas are an aggregate of waves of various heights and lengths, and one wave cannot be clearly distinguished from another. As seas move out from the area of wave generation, though, they are sorted into broad *swells* that have rounded, long crests and are all about the same size (Figure 19.6b).

As one would expect, the harder and longer the wind blows, the larger are the waves generated. Wind velocity and duration, however, are not the only factors controlling the size of waves. High-velocity wind blowing over a small pond will never generate large waves regardless of how long it blows. In fact, waves are present on ponds and most lakes only while the wind is blowing; once the wind stops, the water quickly smooths out. In contrast, the surface of the ocean is always in motion, and waves in the open seas with heights of 34 m have been recorded during storms.

The reason for the disparity between the wave sizes on ponds and lakes and on the oceans is the **fetch,** which is the distance the wind blows over a continuous water surface. Fetch is limited by the available water surface, so on ponds and lakes it corresponds to their length or width, depending on wind direction. A wind blowing the length of Lake Superior, for instance, can generate large waves, but even larger ones develop in the oceans. To produce waves of greater length and height, more energy must be transferred from wind to water; hence, large waves form beneath large storms at sea.

In the areas where storm waves are generated, waves with different lengths, heights, and periods might merge, making them smaller or larger. Occasionally, two wave crests merge to form *rogue waves* that might be three or four times higher than the average. Indeed, these rogue waves can rise unexpectedly out of an otherwise comparatively calm sea and threaten even the

Figure 19.7
Notice that the nearshore water in this lake is turbid, whereas water offshore is clear. The waves have a wavelength of about 2 m, so we can infer that wave base is about 1 m deep. Where the waves encounter wave base, they stir up the bottom sediments, thus accounting for the nearshore turbid water.

largest ships. In 1942 the HMS *Queen Mary,* while carrying 15,000 American soldiers, was hit broadside by a rogue wave and nearly capsized.

Shallow-Water Waves and Breakers   Waves moving out from the area of generation form swells and lose only a small amount of energy as they travel across the ocean. In deep-water swells, the water surface oscillates and water particles move in orbital paths, with little net displacement of water taking place in the direction of wave advance. When these waves enter progressively shallower water, the water is displaced in the direction of wave advance (Figure 19.6).

As deep-water waves enter shallow water, they are transformed from broad, undulating swells into sharp-crested waves. This transformation begins at a water depth of wave base; that is, it begins where wave base intersects the seafloor. At this point, the waves "feel" the bottom, and the orbital motions of water particles within waves are disrupted (Figure 19.6). As they move further shoreward, the speed of wave advance and wavelength decrease, and wave height increases. In effect, as waves enter shallower water, they become oversteepened; the wave crest advances faster than the wave form, until eventually the crest plunges forward as a **breaker** (Figure 19.6c). Breakers are commonly several times higher than deep-water waves, and when they plunge forward their kinetic energy is expended on the shoreline.

The waves just described are the classic plunging breakers that pound the shorelines of areas with steep offshore slopes, such as on the north shore of Oahu in the Hawaiian Islands (Figure 19.8a). In contrast, shorelines with gentle offshore slopes are characterized by spilling breakers, where the waves build up slowly and the wave's crest spills down the front of the wave (Figure 19.8b). Whether the breakers spill or plunge, the water rushes onto the shore, then returns seaward to become part of another breaking wave.

The size of breakers releasing their energy on shorelines varies enormously. Waves tend to be larger and more energetic in winter because storms are more common during that season. In addition, waves of various sizes merging in the area where they form not only create rogue waves but also account for variations in the size of waves breaking on the shore. For instance, when waves having different lengths merge, smaller waves result, whereas merging of waves with the same length forms larger waves. As a result, several small waves might break on a beach followed by a series of larger ones. Surfers commonly take advantage of this phenomenon by swimming out to sea during a relative calm where they wait for a large set of waves to ride toward shore.

Nearshore Currents and Sediment Transport

It is convenient to identify the *nearshore zone* as the area extending seaward from the upper limits of the shoreline to just beyond the area where waves break. It includes a *breaker zone* and a *surf zone,* which is where breaking waves rush forward onto the shore followed by seaward movement of the water as backwash (Figure 19.6). The width of the nearshore zone varies depending on the wavelength of approaching waves because long waves break at a greater depth, and thus farther offshore, than do short waves. Two types of currents are important in the nearshore zone, longshore currents and rip currents.

Wave Refraction and Longshore Currents Deep-water waves are characterized by long, continuous crests, but rarely are their crests parallel with the shoreline (Figure 19.9). One part of a wave enters shallow water where it encounters wave base and begins breaking before other parts of the same wave. As a wave begins breaking, its velocity diminishes, but the part of

(a)

(b)

Figure 19.8
(a) A plunging breaker on the north shore of Oahu, Hawaii.
(b) A spilling breaker.

Figure 19.9
Wave refraction. These waves are refracted and more nearly parallel the shoreline as they enter progressively shallower water. The refracted waves generate a longshore current that flows in the direction of wave approach, from right to left in this example. *Source: Photo courtesy of John S. Shelton.*

the wave still in deep water races ahead until it too encounters wave base. The net effect of this oblique approach is that waves bend so that they more nearly parallel the shoreline (Figure 19.9). This phenomenon is known as **wave refraction.**

Even though waves are refracted, they still usually strike the shoreline at some angle, causing the water between the breaker zone and the beach to flow parallel to the shoreline. These **longshore currents,** as they are called, are long and narrow and flow in the same general direction as the approaching waves (Figure 19.9). These currents are particularly important because they transport and deposit large quantities of sediment in the nearshore zone.

Rip Currents   Waves carry water into the nearshore zone, so there must be a mechanism for mass transfer of water back out to sea. One way in which water moves seaward from the nearshore zone is in **rip currents,** which are narrow surface currents that flow out to sea through the breaker zone (Figure 19.10). Surfers commonly take advantage of rip currents for an easy ride out beyond the breaker zone, but these currents pose a danger to inexperienced swimmers. Rip currents might flow at several kilometers per hour, so if a swimmer is caught in one, it is useless to try to swim directly back to shore. Instead, because rip currents are narrow and usually nearly perpendicular to the shore, one can swim parallel to the shoreline for a short distance and then turn shoreward with no difficulty.

Rip currents can be characterized as circulating cells fed by longshore currents. When waves approach a shoreline, the amount of water builds up until the excess moves out to sea through the breaker zone. Rip currents are fed by nearshore currents that increase in velocity from midway between each rip current (Figure 19.10).

WHAT WOULD YOU DO?

While swimming at your favorite beach, you suddenly notice that you are far down the beach from the point where you entered the water. Furthermore, the large waves you were swimming in have diminished considerably. You decide to swim back to shore and then walk back to your original starting point, but no matter how hard you swim, you are carried farther and farther offshore! What is happening, and what can you do to remedy the situation?

The configuration of the seafloor plays an important role in determining the location of rip currents. They commonly develop where wave heights are lower than in adjacent areas. Differences in wave height are commonly controlled by variations in water depth. For instance, if waves move over a depression, the height of the wave over the depression tends to be less than in adjacent areas.

DEPOSITION ALONG SHORELINES

Depositional features of shorelines include beaches, spits, baymouth bars, and barrier islands. The characteristics of beaches are determined by wave energy and shoreline materials, and they are continually modified by waves and longshore currents. Spits and baymouth bars both result from deposition by longshore currents, but the origin of barrier islands is controversial. Rip currents play only a minor role in the configuration of shorelines, but they do transport fine-grained sediment seaward through the breaker zone.

Beaches

Beaches are the most familiar of all coastal landforms, attracting millions of visitors each year and providing the economic base for many communities. They consist of a long, narrow strip of unconsolidated sediment, commonly sand, and are constantly changing. Depending on shoreline configuration and wave intensity, beaches may be discontinuous, existing only as *pocket beaches* in protected areas such as embayments, or they may be continuous for long distances (Figure 19.11).

By definition, a **beach** is a deposit of unconsolidated sediment extending landward from low tide to a change in topography such as a line of sand dunes, a sea cliff, or the point where permanent vegetation begins (Figure 19.12). Typically, a beach has several component parts, including a *backshore* that is usually dry, being covered by water only during storms or exceptionally high tides. The backshore consists of one or more **berms,** platforms composed of sediment deposited by waves. Berms are nearly horizontal or slope gently in a landward direction. The sloping area below the berm that is exposed to wave swash is the *beach face.* The beach face is part of the *foreshore,* an area covered by water during high tide but exposed during low tide (Figure 19.12).

Some sediment on beaches is derived from weathering and wave erosion of the shoreline, but most of it is transported to the coast by streams and redistributed along the shoreline by longshore currents. **Longshore drift** is the phenomenon involving sand transport along a shoreline by longshore currents (Figure 19.13). As previously noted, waves usually strike beaches at some angle, causing the sand grains to move up the beach face

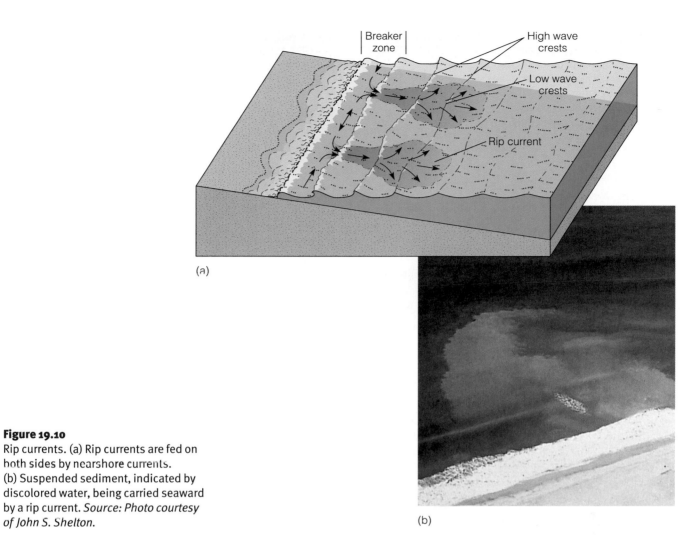

(a)

(b)

Figure 19.10
Rip currents. (a) Rip currents are fed on both sides by nearshore currents. (b) Suspended sediment, indicated by discolored water, being carried seaward by a rip current. *Source: Photo courtesy of John S. Shelton.*

(a)

(b)

Figure 19.11
(a) Small pocket beaches along the California coast near San Francisco. (b) The Grand Strand of South Carolina, shown here at Myrtle Beach, is 100 km of nearly continuous beach.

Figure 19.12
(a) Cross section of a typical beach showing its component parts. (b) The backshore area of a pocket beach along the Pacific Coast. Notice that the berm ends at the rocks on the right and the beach face slopes steeply seaward.

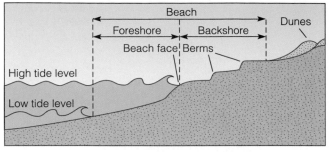

(a)

(b)

at a similar angle, but as the sand grains are carried seaward in the backwash they move perpendicular to the long axis of the beach. So individual sand grains move in a zigzag pattern in the direction of longshore currents. This movement is not restricted to the beach but extends seaward to the outer edge of the breaker zone (Figure 19.13a).

In an attempt to widen a beach or prevent erosion, shoreline residents often build structures known as *groins* that project seaward at right angles from the shoreline (Figure 19.13b). They interrupt the flow of longshore currents, causing sand to be deposited on their upcurrent side, widening the beach at that location. However, erosion inevitably occurs on the downcurrent side of a groin (Figure 19.13b).

Many beaches are sandy, but in areas of particularly vigorous wave activity, they might be gravel covered. Most beach sand is composed of quartz, but a number of other minerals and rock fragments may be present as well. One of the most common accessory minerals in beach sands is magnetite; because of its high specific gravity, magnetite is commonly separated from the other minerals and is visible as thin, black layers.

WHAT WOULD YOU DO?

In 1945, the ore of mercury (cinnabar, HgS) was found in beach sands in Marin County, California. Suppose you made this discovery and wanted to exploit it. How would you identify the source of the cinnabar? And once the source was found, what other considerations, geologic and otherwise, would be important in whether or not to open a mine?

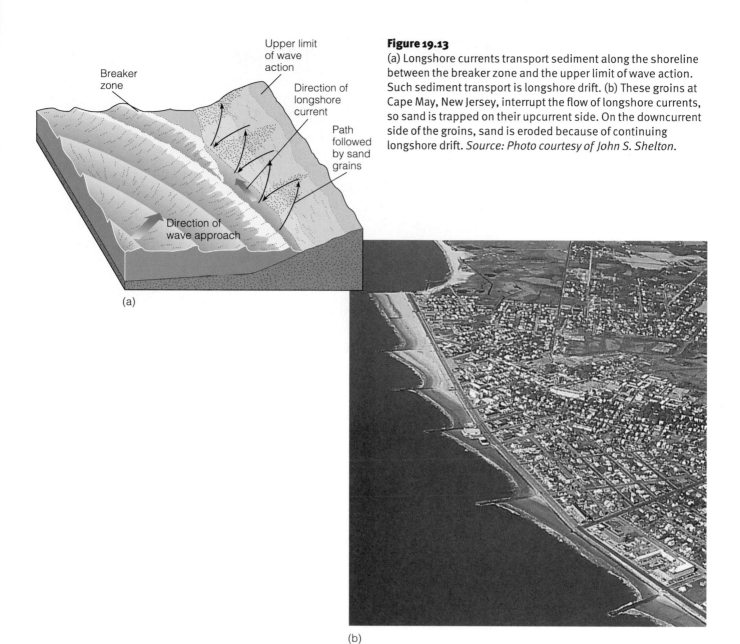

Breaker zone

Upper limit of wave action

Direction of longshore current

Path followed by sand grains

Direction of wave approach

(a)

(b)

Figure 19.13
(a) Longshore currents transport sediment along the shoreline between the breaker zone and the upper limit of wave action. Such sediment transport is longshore drift. (b) These groins at Cape May, New Jersey, interrupt the flow of longshore currents, so sand is trapped on their upcurrent side. On the downcurrent side of the groins, sand is eroded because of continuing longshore drift. *Source: Photo courtesy of John S. Shelton.*

Although quartz is the most common mineral in most beach sands, there are some notable exceptions. The black sand beaches of Hawaii are composed of sand-sized basalt rock fragments (Figure 19.14), and some Florida beaches are composed of the fragmented calcium carbonate shells of marine organisms. In short, beaches are composed of whatever material is available; quartz is most abundant simply because it is abundant in most areas and is the most durable and stable of the common rock-forming minerals (see Chapter 5).

Seasonal Changes in Beaches

A beach is an area where wave energy is dissipated, so the loose grains composing it are constantly affected by wave motion. However, the overall configuration of a beach remains unchanged as long as equilibrium conditions persist. The beach profile consisting of a berm or berms and a beach face shown in Figure 19.12 can be thought of as a profile of equilibrium; that is, all parts of the beach are adjusted to the prevailing conditions of wave intensity and nearshore currents.

Tides and longshore currents affect the configuration of beaches to some degree, but by far the most important agent modifying their equilibrium profile is storm waves. In many areas, beach profiles change with the seasons; so we recognize *summer beaches* and *winter beaches,* which have adjusted to the conditions prevailing at these times (Figure 19.15). Summer beaches are generally sand covered and possess a wide berm, a steep beach face, and a smooth offshore profile. During

Figure 19.14
Black sand beach on Maui in the Hawaiian Islands.

Figure 19.15
(a) Seasonal changes in beach profiles. (b) and (c) San Gregorio State Beach, California. These photographs were taken from nearly the same place but (c) was taken two years after (b). Much of the change from (b) to (c) can be accounted for by beach erosion during 1997–1998 winter storms.

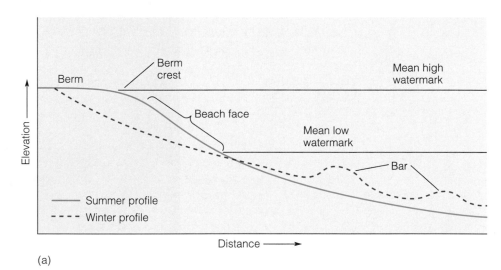

(a)

(b)

(c)

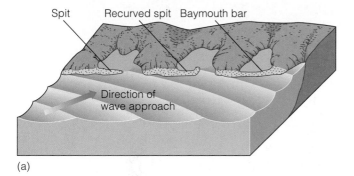

Spit Recurved spit Baymouth bar

Direction of
wave approach

(a)

Figure 19.16
(a) Spits form where longshore currents deposit sand in deeper water, as at the entrance to a bay. A baymouth bar is simply a spit that has grown until it extends across the mouth of a bay. (b) A spit at the mouth of the Russian River near Jenner, California. (c) A baymouth bar.

(b)

(c)

winter, however, when waves more vigorously attack the beach, a winter beach develops, which has little or no berm, a gentle slope, and offshore sand bars paralleling the shoreline (Figure 19.15). Seasonal changes can be so profound that sand moving onshore and offshore yields sand-covered beaches during summer and gravel-covered beaches during winter.

Seasonal changes in beach profiles are related to changing wave intensity. During the winter, energetic storm waves erode the sand from the beach and transport it offshore where it is stored in sand bars (Figure 19.15). The same sand that was eroded from the beach during the winter returns the next summer when it is driven onshore by the more gentle swells that occur during that season. The volume of sand in the system remains more or less constant; it simply moves offshore or onshore depending on the energy of waves.

The terms *winter* and *summer beach,* although widely used, are somewhat misleading. A "winter beach" profile can develop at any time of the year if a large storm occurs, and a "summer beach" profile

can develop during a prolonged calm period in the winter.

Spits and Baymouth Bars

Other than the beach itself, some of the most common depositional landforms on shorelines are spits and baymouth bars, both of which are variations of the same feature. A **spit** is simply a continuation of a beach forming a point, or "free end," that projects into a body of water, commonly a bay, and a **baymouth bar** is a spit that has grown until it completely closes off a bay from the open sea (Figure 19.16).

Both spits and baymouth bars form and grow as a result of deposition by longshore currents. Where currents are weak, as in the deeper water at the opening to a bay, longshore current velocity diminishes, and sediment is deposited, forming a sand bar. The free ends of many spits are curved by wave refraction or waves approaching from a different direction. Such spits are called *hooks* or *recurved spits* (Figure 19.16) (see Perspective 19.2).

Perspective 19.2

Cape Cod, Massachusetts

C ape Cod, some of which is designated as a National Seashore, resembles a large human arm extending into the Atlantic Ocean from the coast of Massachusetts (Figure 1). It projects 65 km east of the mainland to the "elbow" and then extends another 65 km north where it resembles a half-curled hand. Cape Cod, Nantucket Island, Martha's Vineyard, and the adjacent Elizabeth Islands all owe their existence to deposition by late Pleistocene glaciers and the subsequent modification of these glacial deposits by waves. The granite bedrock or foundation on which the Cape and nearby islands were built is at depths of 90 to 150 m.

Between about 23,000 and 16,000 years ago, during the greatest southward advance of the continental glaciers in this area, the ice front stabilized in the area of present-day Martha's Vineyard and Nantucket Island (Figure 2a). Recall that a stabilized terminus means a glacier has a balanced budget, but flow continues within the glacier, transporting and depositing sediment as an end moraine. The end moraine that makes up these islands is a terminal moraine because it is the most southerly of all the moraines deposited by this glacier.

As the climate became warmer, the Cape Cod Bay Lobe of this vast glacier began retreating north. About 14,000 years ago, it once again became stabilized, but this time in the area of present-day Cape Cod and the Elizabeth Islands where it deposited a large recessional moraine (Figure 2b). The ice front once again retreated north, and meltwater trapped between the ice front and the recessional moraine formed Glacial Lake Cape Cod, covering about 1000 km² (Figure 2c).

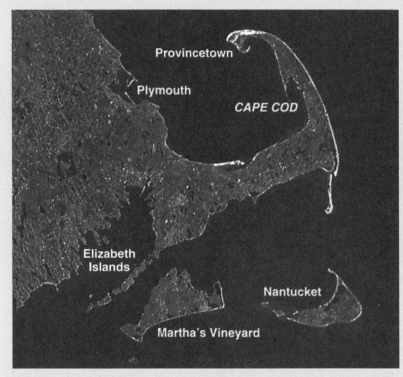

Figure 1
Satellite view of Cape Cod. Notice the long beaches, spits, and baymouth bars on the Cape, Martha's Vineyard, and Nantucket Island. The dark circular to oval areas on Cape Cod are kettle lakes, or ponds as they are called locally.

Deposits of mud, silt, and sand accumulated in this lake, and these, too, make up part of present-day Cape Cod.

When the continental glaciers withdrew entirely from this region, Cape Cod looked different from the way it does now. On its east side were several headlands and embayments, but by 6000 years ago sea level had risen enough so that waves began smoothing the shoreline by redistributing the sediment. The headlands were eroded, whereas the embayments were filled in, and the shoreline has eroded from 1 to 4 km westward from its former location. Even today many steep wave-cut cliffs

are present on the Cape, and those facing east are currently eroding at nearly 1 m per year. Most of this erosion takes place during storms, so cliff retreat is episodic rather than continuous.

Some of the sediment eroded from Cape Cod is transported offshore where it settles beyond the reach of waves, but much of it is transported by longshore currents and redistributed along the Cape. Hundreds of thousands of cubic meters of sand are transported along the shores of the Cape and deposited as spits and baymouth bars, many of which are still forming. Extending south from the "elbow" is a long spit, and another continues to form at the north where sand is transported around the end of the Cape and forms a hook (Figure 1). Wind also transports some of the sand inland, where it accumulates as a series of coastal dunes.

The fact that the northerly extending part of Cape Cod is migrating to the west is well documented. One observation that reveals this to be the case are the organic-rich marsh deposits that formed on the Cape's landward (west) side. They are now exposed on the beaches in many places on the seaward (east) side of the Cape (Figure 3).

Other distinctive features of Cape Cod are its numerous circular to oval ponds. These ponds occupy kettles that formed when large blocks of glacial ice were partly or completely buried by outwash deposits (see Figure 17.27). When these ice masses finally melted, they left depressions measuring up to 0.8 km in diameter that filled with water when the water table rose as a result of rising sea level.

Native Americans lived on Cape Cod for thousands of years before Europeans arrived. Unfortunately, by 1764 these earliest inhabitants had nearly ceased to exist, mostly because of diseases. The first European settlers

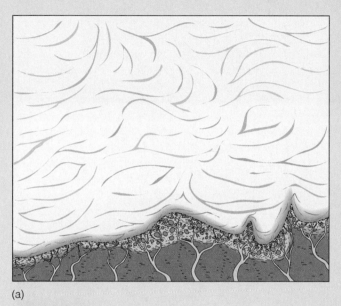
(a)

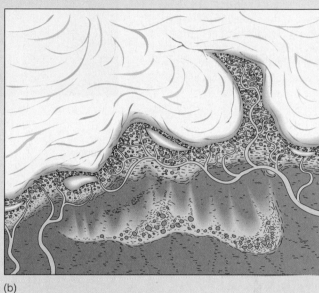
(b)

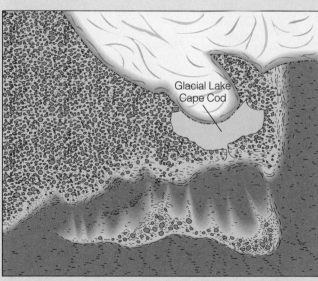
(c)

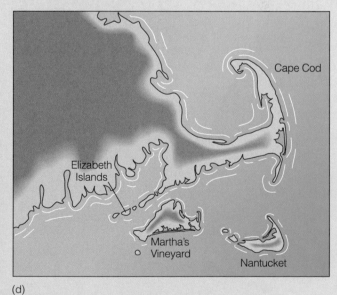
(d)

Figure 2
(a) Position of the Cape Cod Lobe of glacial ice 23,000 to 16,000 years ago when it deposited its terminal moraine that would become the foundations for Martha's Vineyard and Nantucket Island. (b) Position of the Cape Cod Lobe when the recessional moraine was deposited that now forms much of Cape Cod. (c) About 14,000 years ago Glacial Lake Cape Cod was present between the ice front and the recessional moraine. (d) About 5000 years ago, rising sea level covered the lowlands between the moraines and beaches and spits formed.

in this region were the Pilgrims. Despite the persistent myth that they first landed at Plymouth Rock, their first landfall was actually on Cape Cod near what is now Provincetown.

Figure 3
View looking south along the northerly extending part of Cape Cod. The dark substance at the shoreline to the left (east) is organic-rich sediment that was deposited on the west side of the Cape but is now exposed here because of westward migration of this part of the Cape.

Figure 19.17
(a) Wave refraction around an island causes longshore currents to converge and deposit a sand bar known as a tombolo that connects the island with the mainland.
(b) A small tombolo.
(c) Close-up view of the tombolo in (b).

Figure 19.18
Soon after this breakwater was constructed offshore at Santa Monica, California, a bulge appeared in the beach. Wave refraction around the breakwater resulted in the formation of a feature similar to a tombolo. *Source: Photo courtesy of John S. Shelton.*

A rarer type of spit, a **tombolo,** extends out into the sea and connects an island to the mainland. Tombolos develop on the shoreward sides of islands as shown in Figure 19.17. Wave refraction around an island causes converging currents that turn seaward and deposit a sand bar connecting the shore with the island. A similar feature may form when an artificial breakwater is constructed offshore (Figure 19.18).

Although spits, baymouth bars, and tombolos are most commonly found on irregular seacoasts, many examples of the same features are present in large lakes (Figure 19.19). Whether along seacoasts or lakeshores, these sand deposits present a continuing problem where bays must be kept open for pleasure boating or commercial shipping. The entrances to these bays must either be regularly dredged or protected. The most common way to protect entrances to bays is to build *jetties,* which are

(a)

(b)

Figure 19.19
Shoreline features of large lakes are similar to those of sea coasts. (a) An arch eroded by waves on Mackinac Island in Lake Huron. The arch is now above the lake level because the island has been uplifted as a result of isostatic rebound. (b) A small tombolo in Lake Superior at Marquette, Michigan.

structures extending seaward (or lakeward) that protect the bay from deposition by longshore currents (Figure 19.20).

Barrier Islands

Long, narrow islands composed of sand and separated from the mainland by a lagoon are **barrier islands** (Figures 19.20 and 19.21). On their seaward margins, barrier islands are smoothed by waves, but their lagoon sides are irregular. During large storms, waves overtop these islands and deposit lobes of sand in the lagoon. Once deposited, these lobes are modified only slightly because they are protected from further wave action. Windblown sand dunes are common and are generally the highest part of barrier islands.

The origin of barrier islands has been long debated and is still not completely resolved. It is known that they form on gently sloping continental shelves with abundant sand in areas where both tidal fluctuations and wave-energy levels are low. Although barrier islands are found in many areas, most of them are along the East Coast of the United States from New York to Florida and along the U.S. Gulf Coast. Barrier islands are not nearly as common in Canada, but some are found along the coasts of New Brunswick and Prince Edward Island. According to one model, barrier islands formed as spits that became detached from the land, whereas another model proposes that they formed as beach ridges on coasts that subsequently subsided (Figure 19.22).

Because sea level is currently rising, most barrier islands are migrating landward. Such migration is a natural consequence of evolution of these islands, but it is a problem for the island residents and communities. Barrier islands generally migrate rather slowly, but the rates for many are rapid enough to cause shoreline problems (discussed in a later section).

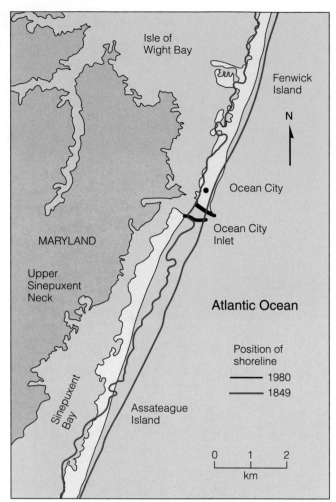

Figure 19.20
The seaward-projecting, heavy black lines represent jetties that were constructed in the 1930s to protect the Ocean City Inlet at Ocean City, Maryland. The jetties have protected the inlet, but they also disrupted the net southerly longshore drift. As a consequence, Assateague Island has been starved of sediment and migrated about 500 m landward and is now offset from Fenwick Island to the north. Both islands are barrier islands.

The Nearshore Sediment Budget

We can think of the gains and losses of sediment in the nearshore zone in terms of a **nearshore sediment budget** (Figure 19.23). If a nearshore system has a balanced budget, sediment is supplied as fast as it is removed, and the volume of sediment remains more or less constant, although sand may shift offshore and onshore with the changing seasons. A positive budget means gains exceed losses, whereas a negative budget results when losses exceed gains. If a negative budget prevails long enough, the nearshore system is depleted and beaches may disappear (Figure 19.23).

Erosion of sea cliffs provides some sediment to beaches, but in most areas probably no more than 5 to 10% of the total sediment supply is derived from this source. There are exceptions, though; almost all the sediment on the beaches of Maine is derived from the erosion of shoreline rocks. Most sediment on typical beaches is transported to the shoreline by streams and then redistributed along the shoreline by longshore drift. Thus, longshore drift also plays a role in the nearshore sediment budget because it continually moves sediment into and away from beach systems (Figure 19.23).

The primary ways that a nearshore system loses sediment include longshore drift, offshore transport, wind, and deposition in submarine canyons. Offshore transport mostly involves fine-grained sediment that is carried seaward, where it eventually settles in deeper water.

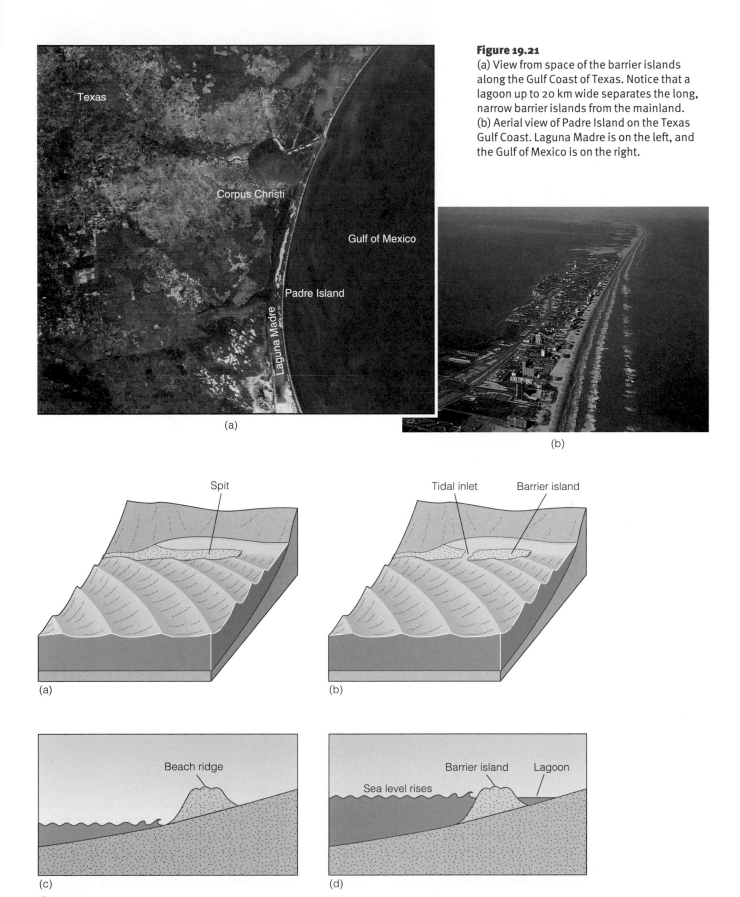

Figure 19.21
(a) View from space of the barrier islands along the Gulf Coast of Texas. Notice that a lagoon up to 20 km wide separates the long, narrow barrier islands from the mainland. (b) Aerial view of Padre Island on the Texas Gulf Coast. Laguna Madre is on the left, and the Gulf of Mexico is on the right.

Figure 19.22
Two models for the origin of barrier islands. (a) Longshore currents extend a spit along a coast. (b) During a storm, the spit is breached, forming a tidal inlet and a barrier island. (c) A beach ridge forms on land. (d) Sea level rises, partly submerging the beach ridge.

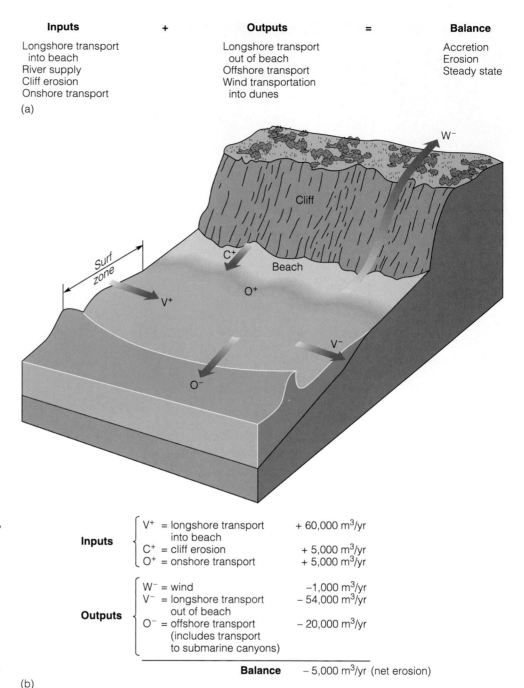

Inputs	+	Outputs	=	Balance
Longshore transport into beach		Longshore transport out of beach		Accretion
River supply		Offshore transport		Erosion
Cliff erosion		Wind transportation into dunes		Steady state
Onshore transport				

(a)

Figure 19.23
The nearshore sediment budget.
(a) The long-term sediment budget can be assessed by considering inputs versus outputs. If inputs and outputs are equal, a beach is in a steady state, or equilibrium. If outputs exceed inputs, however, the beach has a negative budget and erosion occurs. Accretion takes place when the beach has a positive budget with inputs exceeding outputs. (b) A hypothetical example of a negative nearshore sediment budget. In this example the beach is losing 5000 m³ a year to erosion.

Inputs	V^+ = longshore transport into beach	+ 60,000 m³/yr	
	C^+ = cliff erosion	+ 5,000 m³/yr	
	O^+ = onshore transport	+ 5,000 m³/yr	
Outputs	W^- = wind	−1,000 m³/yr	
	V^- = longshore transport out of beach	− 54,000 m³/yr	
	O^- = offshore transport (includes transport to submarine canyons)	− 20,000 m³/yr	
Balance		− 5,000 m³/yr (net erosion)	

(b)

Wind is an important process because it removes sand from beaches and blows it inland where it commonly piles up as sand dunes.

If the heads of submarine canyons are near shore, huge quantities of sand are funneled into them and deposited in deeper water. La Jolla and Scripps submarine canyons off the coast of southern California funnel off an estimated 2 million m³ of sand each year. In most areas, however, submarine canyons are too far offshore to interrupt the flow of sand in the nearshore zone.

It should be apparent from the preceding discussion that if a nearshore system is in equilibrium, its incoming supply of sediment exactly offsets its losses. Such a delicate balance tends to continue unless the system is somehow disrupted. One common change that affects this balance is the construction of dams across the streams supplying sand. Once dams have been built, all sediment from the upper reaches of the drainage systems is trapped in reservoirs and thus cannot reach the shoreline.

HOW ARE SHORELINES ERODED?

Along seacoasts where erosion rather than deposition predominates, beaches are lacking or poorly developed, and sea cliffs are commonly present (Figure 19.24). Sea cliffs are frequently pounded by waves, especially during storms, and the cliff face retreats landward as a result of corrosion, hydraulic action, and abrasion. *Corrosion* is an erosional process involving the wearing away of rock by chemical processes, especially the solvent action of seawater. The force of the water itself, called *hydraulic action,* is a particularly effective erosional process. Waves exert tremendous pressure on shorelines by direct impact but are most effective on sea cliffs composed of sediment or highly fractured rocks. *Abrasion* is an erosional process involving the grinding action of rocks and sand carried by waves. Remember from Chapter 15 that running water also erodes by hydraulic action and abrasion.

Wave-Cut Platforms and Associated Landforms

The rate at which sea cliffs erode and retreat landward depends on wave intensity and the resistance of the

(a)

(b)

(c)

Figure 19.24
(a) Wave erosion causes a sea cliff to migrate landward, leaving a gently sloping surface known as a wave-cut platform. A wave-built platform originates by deposition at the seaward margin of the wave-cut platform. (b) Sea cliffs and a wave-cut platform. (c) This gently sloping surface along the Pacific Coast of California is a marine terrace. Notice the old sea stacks rising above the terrace and present-day sea stacks formed by erosion of the sea cliff. *Source: Photo (b) courtesy of John S. Shelton.*

coastal rocks or sediments. Most sea cliff retreat takes place during storms and, as one would expect, occurs most rapidly in sea cliffs composed of sediment. For example, a sea cliff composed of glacial drift on Cape Cod, Massachusetts, retreats as much as 30 m per century, and some parts of the White Cliffs of Dover in Great Britain are retreating at a rate of more than 100 m per century. By comparison, sea cliffs consisting of dense igneous or metamorphic rocks retreat much more slowly.

Sea cliffs erode mostly as a result of hydraulic action and abrasion at their bases. As a sea cliff is undercut by erosion, the upper part is left unsupported and susceptible to mass-wasting processes. Thus, sea cliffs retreat little by little, and as they do, they leave a beveled surface

known as a **wave-cut platform** that slopes gently seaward (Figure 19.24). Broad wave-cut platforms exist in many areas, but invariably the water over them is shallow because the abrasive planing action of waves is effective to a depth of only about 10 m. Sediment eroded from sea cliffs is transported seaward until it reaches deeper water at the edge of the wave-cut platform. There it is deposited and forms a seaward extension of the wave-cut platform called a *wave-built platform* (Figure 19.24). Wave-cut platforms now above sea level are known as **marine terraces** (Figure 19.24c).

Sea cliffs do not retreat uniformly because some of the materials of which they are composed are more resistant to erosion than others. Those parts of a shoreline consisting of resistant materials might form

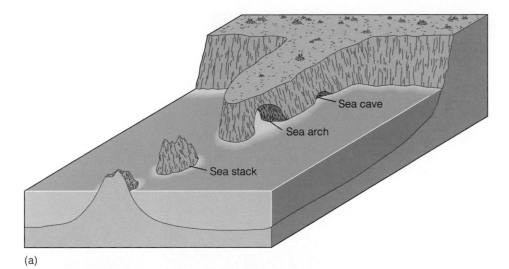

(b)

(c)

Figure 19.25
(a) Erosion of a headland and the origin of sea caves, sea arches, and sea stacks.
(b) This sea stack in Australia has an arch developed in it.
(c) Sea stacks on California's central coast.

seaward-projecting headlands, whereas less resistant materials erode more rapidly, yielding embayments where pocket beaches may be present. Wave refraction around headlands causes them to erode on both sides so that *sea caves* may form; if caves on the opposite sides of a headland join, they form a *sea arch* (Figure 19.25). Continued erosion generally causes the span of an arch to collapse, yielding isolated *sea stacks* on wave-cut platforms (Figure 19.25). In the long run, shoreline processes tend to straighten an initially irregular shoreline. They do so because wave refraction causes more wave energy to be expended on headlands and less on embayments. Headlands become eroded, and some of the sediment yielded by erosion is deposited in the embayments. The net effect of these processes is to straighten the shoreline (Figure 19.26).

Wave-cut platforms and their associated features are most common along seashores, but they are also present along the shores of large lakes. A number of these features are present in the Great Lakes, many of

which are now above lake level because of isostatic rebound (see Chapter 17).

HOW ARE COASTAL AREAS MANAGED AS SEA LEVEL RISES?

During the last century, sea level rose about 12 cm worldwide, and all indications are that it will continue to rise. The absolute rate of sea-level rise in a particular shoreline region depends on two factors. The first is the volume of water in the ocean basins, which is increasing as a result of the melting of glacial ice and the thermal expansion of near-surface seawater. Many scientists think that sea level will continue to rise because of global warming resulting from concentrations of greenhouse gases in the atmosphere.

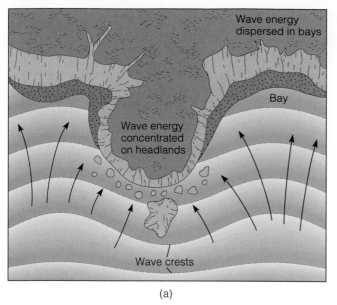

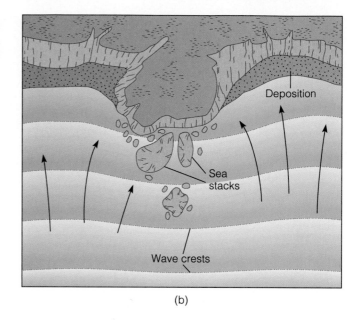

Figure 19.26
(a) Wave refraction acts to straighten shorelines by concentrating wave energy on headlands. (b) The same shoreline after extensive erosion of the headlands and deposition in the bays.

The second factor controlling sea level is the rate of uplift or subsidence of a coastal area. In some areas, uplift is occurring so fast that sea level is actually falling with respect to the land. In other areas, sea level is rising while the coastal region is simultaneously subsiding, resulting in a net change in sea level of as much as 30 cm per century. Perhaps such a "slow" rate of sea-level change seems insignificant; after all it amounts to only a few millimeters per year. But in gently sloping coastal areas, as in the eastern United States from New Jersey southward, even a slight rise in sea level would eventually have widespread effects.

Many of the nearly 300 barrier islands along the East and Gulf Coasts of the United States are migrating landward as sea level rises. During storms, large waves erode sand from the seaward sides of barrier islands and deposit it on their landward sides, resulting in a gradual landward shift of the entire island complex (Figure 19.27). During the last 120 years, Hatteras Island, North Carolina, has migrated nearly 500 m landward (see Perspective 19.3). At its current rate of landward migration, Assateague Island, Maryland, will eventually merge with the mainland (Figure 19.20).

Landward migration of barrier islands would pose few problems if it were not for the numerous communities, resorts, and vacation homes on them. Moreover, barrier islands are not the only threatened areas. For example, Louisiana's coastal wetlands, an important wildlife habitat and seafood-producing area, are currently being lost at a rate of about 90 km² per year. Much of this loss results from sediment compaction, but sea-level rise exacerbates the problem.

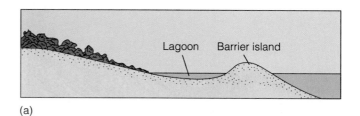

(a)

(b)

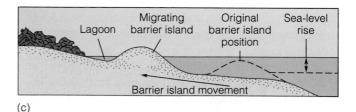

(c)

Figure 19.27
Rising sea level and the landward migration of barrier islands. (a) Barrier island before landward migration. (b) Sea level rises as the barrier island migrates landward when storm waves wash sand from its seaward side and deposits it in the lagoon. (c) Over time, the entire complex migrates landward.

(a)

(b)

Figure 19.28
The beach at Miami Beach, Florida, (a) before and (b) after the U.S. Army Corps of Engineers' beach nourishment project.

Rising sea level also directly threatens many beaches that communities depend on for revenue. The beach at Miami Beach, Florida, for instance, was disappearing at an alarming rate until the Army Corps of Engineers began replacing the eroded beach sand (Figure 19.28). The problem is even more serious in other countries. A rise in sea level of only 2 m would inundate large areas of the East and Gulf Coasts, but would cover 20% of the entire country of Bangladesh. Other problems associated with sea-level rise include increased coastal flooding during storms and saltwater incursions that may threaten groundwater supplies (see Chapter 16).

Because nothing can be done to prevent sea level from rising, engineers, scientists, planners, and political leaders must examine what can be done to prevent or minimize the effects of shoreline erosion. At present, only a few viable options exist. One is to put strict controls on coastal development. North Carolina, for example, permits large structures to be sited no closer to the shoreline than 60 times the annual erosion rate. Although a growing awareness of shoreline processes has resulted in similar legislation elsewhere, some states have virtually no restrictions on coastal development.

Regulating coastal development is commendable, but it has no impact on existing structures and coastal communities. A general retreat from the shoreline may be possible but expensive for individual dwellings and small communities, but it is impractical for large population centers. Such communities as Atlantic City, New Jersey, Miami Beach, Florida, and Galveston, Texas, have adopted one of two strategies to combat coastal erosion.

One option is to build protective barriers such as seawalls. Seawalls, like the one at Galveston (see the Prologue), can be effective, but they are tremendously expensive to construct and maintain. During a five-year period, more than $50 million was spent to replenish the beach sand and build a protective seawall at Ocean City, Maryland. Unfortunately, seawalls retard erosion only in the area directly behind them; Galveston Island west of the seawall has been eroded back about 45 m.

Armoring shorelines with seawalls and similar structures has had an unanticipated effect on Oahu in the Hawaiian Islands. Erosion of coastal rocks is the primary supply of sand to many Hawaiian beaches, so beaches maintain their width even as sea level rises. However, seawalls trap sediment on their landward sides, thus diminishing the amount of sand on the beaches. In four areas studied on Oahu, 24% of the beaches have either been lost or are much narrower than they were several decades ago. Seawalls protect beachfront homes, hotels, and apartment buildings effectively, but they are expensive to build and during large storms are commonly damaged, requiring repairs. Furthermore, they commonly lead to loss of sand on beaches. North Carolina, South Carolina, Rhode Island, Oregon, and Maine no longer allow construction of seawalls.

Perspective 19.3

The Outer Banks of North Carolina

Most of the world's barrier islands are found along the U.S. Atlantic Coast from New York southward and along the Gulf Coast of Texas. Among these are the Outer Banks, which are simply several barrier islands extending more than 240 km from the Virginia–North Carolina state line on the north to Shackleford Banks in the south (Figure 1). Their dimensions can be appreciated when viewed from space; they are much longer than wide. However, unlike most other barrier islands, they lie much further offshore from the mainland. In most barrier island systems the lagoon between the island and mainland is only a few kilometers wide and rarely more than about 20 km. In the case of the Outer Banks, though, a 48-km-wide lagoon is present (Figure 1).

The scenic appeal and recreational opportunities of the Outer Banks make them a favorite with tourists and local

Figure 1
This chain of barrier islands comprises the Outer Banks of North Carolina. Cape Hatteras, near the center of the photo, juts the farthest out into the Atlantic.

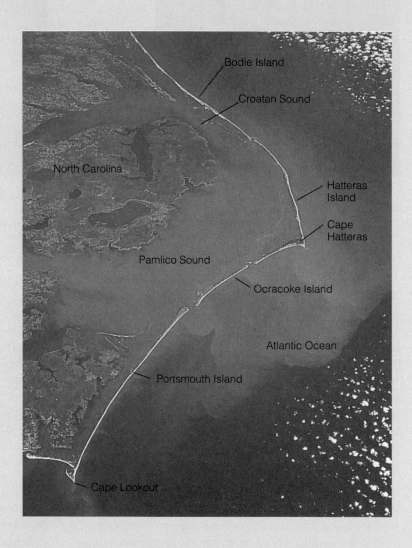

Another option, adopted by both Atlantic City and Miami Beach, is to pump sand onto the beaches to replace that lost to erosion (Figure 19.28). This, too, is expensive as the sand must be replenished periodically, because erosion is a continuing process. In many areas, groins are constructed to preserve beaches, but unless additional sand is artificially supplied to the beaches, longshore currents invariably erode sand from the downcurrent sides of the groins.

The futility of artificially maintaining beaches is aptly demonstrated by the efforts to protect homes on the northeast end of the Isle of Palms, a barrier island off the coast of

residents. Furthermore, the wildlife, grasslands, sand dunes, and salt marshes are of interest to many people, and the Wright Brothers National Memorial is present on Bodie Island. Those interested in the history of aviation will recall that Orville and Wilbur Wright made their first flight near Kitty Hawk on December 17, 1903. And since 1526, more than 600 ships have been lost on the dangerous shoals adjacent to the islands, particularly at Cape Lookout and Cape Hatteras. To reduce ship losses, a lighthouse funded by the federal government was erected at Cape Hatteras in 1802.

Other than the fact that the Outer Banks lie further offshore, they differ little from other barrier islands (see Figures 19.20 and 19.21). They are gently curved and smoothed by waves on their seaward (easterly) sides where broad, sandy beaches are present; sand dunes make up their higher parts; and a lagoon with lobes of sand washed over the islands during storms and salt marshes lie on their landward (westerly) sides.

A few barrier islands along the U.S. Gulf Coast are actually building or prograding out to sea. That is, they are migrating seaward. However, most barrier islands, including those making up the Outer Banks, migrate landward as they are eroded on their seaward sides and storm waves carry sand into the lagoons. In short, the entire barrier island complex migrates landward (see Figure 19.27). Most barrier island migration takes place during large storms, especially during hurricanes. These islands are especially susceptible to the effects of hurricanes, because they are the most exposed parts of coasts.

That the Outer Banks are actively migrating landward is well demonstrated by the fact that the National Park Service had the Cape Hatteras Lighthouse moved nearly 500 m inland and 760 m to the southwest (Figure 2). When the lighthouse was built in 1870, it was 457 m from the shoreline, but by the time it was moved in July 1999 it stood no more than 36.5 m inland!

Given that the annual erosion rate was 3 m per year, and that all previous attempts to stabilize the shoreline had failed, the Park Service found it prudent to act when it did. Shoreline erosion will continue, of course, but the Cape Hatteras Lighthouse, the world's tallest brick lighthouse, should be safe for several centuries.

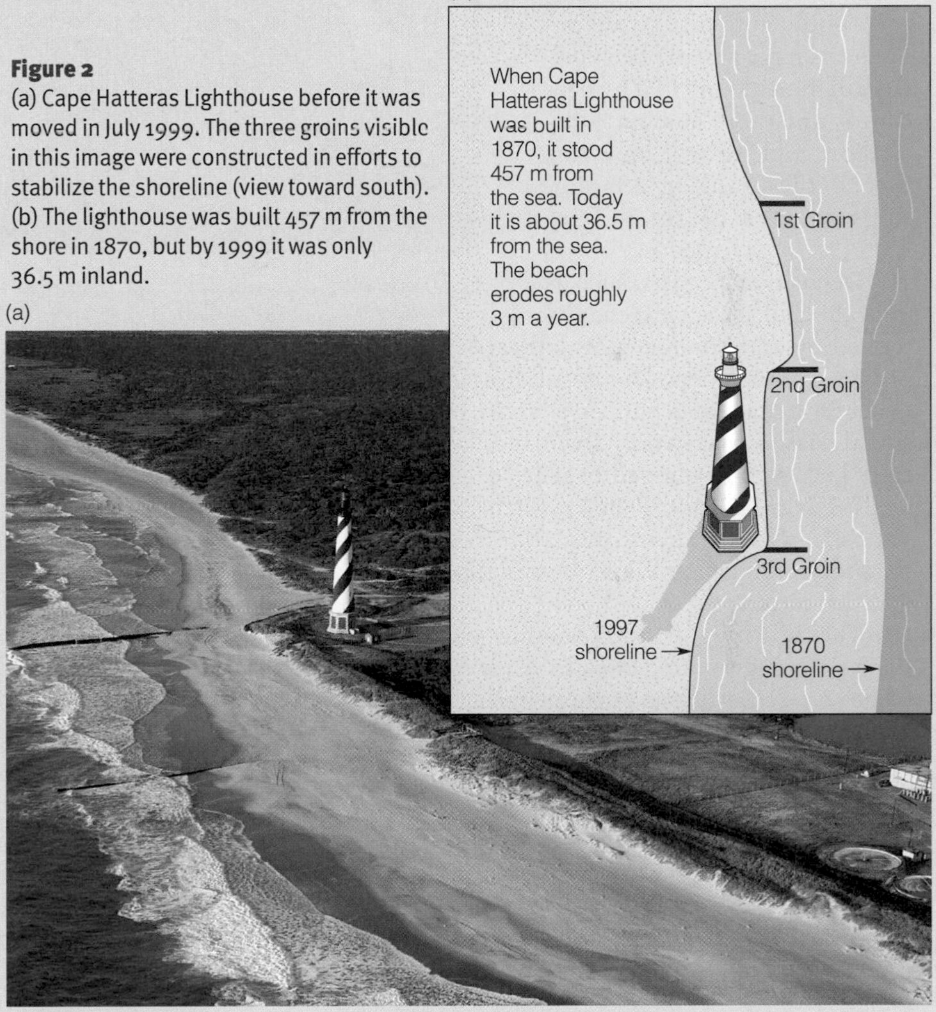

Figure 2
(a) Cape Hatteras Lighthouse before it was moved in July 1999. The three groins visible in this image were constructed in efforts to stabilize the shoreline (view toward south). (b) The lighthouse was built 457 m from the shore in 1870, but by 1999 it was only 36.5 m inland.

When Cape Hatteras Lighthouse was built in 1870, it stood 457 m from the sea. Today it is about 36.5 m from the sea. The beach erodes roughly 3 m a year.

1st Groin

2nd Groin

3rd Groin

1997 shoreline →

1870 shoreline →

South Carolina. Following each spring tide, heavy equipment excavates sand from a deposit known as an *ebb tide delta* and constructs a sand berm to protect the houses from the next spring tide (Figure 19.29). Two weeks later, the process must be repeated in a never-ending cycle of erosion and artificial replenishment of the beach.

TYPES OF COASTS

Coasts are difficult to classify because of variations in the factors controlling their development and differences in the composition and configuration of coasts. Rather than attempt to categorize all coasts, we

Figure 19.29
Heavy equipment builds a berm, an embankment of sand, on the seaward side of beach homes on the Isle of Palms, South Carolina, to protect them from erosion. During the next spring tide, the berms disappear and must be rebuilt.

will simply note that two types of coasts have already been discussed, those dominated by deposition and those dominated by erosion, and we will look further at the changing relationships between coasts and sea level.

Depositional coasts, such as the U.S. Gulf Coast, are characterized by an abundance of detrital sediment and the presence of such depositional landforms as wide, sandy beaches, deltas, and barrier islands. In contrast, erosional coasts are generally steep and irregular and typically lack well-developed beaches except in protected areas (Figure 19.11a). They are further characterized by sea cliffs, wave-cut platforms, and sea stacks. Many of the beaches along the West Coast of North America fall into this category.

The following section examines coasts in terms of their changing relationships to sea level. But note that while some coasts, such as those in southern California, are described as emergent (uplifted), these same coasts may be erosional as well. In other words, coasts commonly possess features allowing them to be classified in several ways.

Submergent and Emergent Coasts

If sea level rises with respect to the land or the land subsides, coastal regions are flooded and said to be **submergent** or *drowned* (Figure 19.30). Much of the East Coast of North America from Maine southward through South Carolina was flooded during the rise in sea level following the Pleistocene Epoch, so it is extremely irregular. Recall that during the expansion of glaciers during the Pleistocene, sea level was as much as 130 m lower than at present and that streams eroded their valleys more deeply and extended across continental shelves. When sea level rose, the lower ends of these valleys were drowned, forming *estuaries* such as Delaware and Chesapeake Bays (Figure 19.30). Estuaries are the seaward ends of river valleys where seawater and freshwater mix.

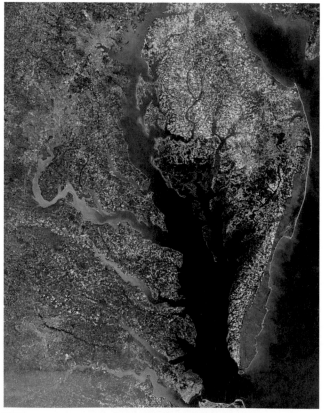

Figure 19.30
Submergent coasts tend to be extremely irregular with estuaries such as Chesapeake Bay. It formed when the East Coast of the United States was flooded as sea level rose following the Pleistocene Epoch.

Marine terrace

Figure 19.31
An emergent coast in California. Emergent coasts tend to be steep and straighter than submergent coasts. Notice the several sea stacks and the sea arch. Also, a marine terrace is visible in the distance.

Submerged coasts are also present at higher latitudes where Pleistocene glaciers flowed into the sea. When sea level rose, the lower ends of the glacial troughs were drowned, forming fiords (see Figure 17.15).

Emergent coasts are found where the land has risen with respect to sea level (Figure 19.31). Emergence can take place when water is withdrawn from the oceans, as occurred during the Pleistocene expansion of glaciers. Presently, coasts are emerging as a result of isostasy or tectonism. In northeastern Canada and the Scandinavian countries, the coasts are irregular because isostatic rebound is elevating formerly glaciated terrain from beneath the sea.

Coasts rising in response to tectonism, in contrast, tend to be rather straight because the seafloor topography being exposed as uplift proceeds is smooth. The west coasts of North and South America are rising as a result of plate tectonics. Distinctive features of these coasts include *marine terraces* (Figure 19.24c), which are old wave-cut platforms now elevated above sea level. Uplift in such areas appears to be episodic rather than continuous, as indicated by the multiple levels of terraces in some areas. In southern California, for example, several terrace levels are present, each of which probably represents a period of tectonic stability followed by uplift (Figure 19.32). The highest of these terraces is now about 425 m above sea level.

Figure 19.32
Marine terraces along the coast of California. Each terrace represents a time when that area was at sea level. The highest terrace is now about 425 m above sea level. *Source: Photo courtesy of John S. Shelton.*

Chapter Summary

1. Shorelines are continually modified by the energy of waves and longshore currents and, to a lesser degree, by tidal currents.

2. The gravitational attraction of the Moon and Sun causes the ocean surface to rise and fall as tides twice daily in most shoreline areas. Tidal currents usually have little effect on shorelines.

3. Waves are oscillations on water surfaces that transmit energy in the direction of wave movement. Surface waves affect the water and seafloor only to wave base, which is equal to one-half the wavelength.

4. Little or no net forward motion of water occurs in waves in the open sea. When waves enter shallow water, they are transformed into waves in which water does move in the direction of wave advance.

5. Wind-generated waves, especially storm waves, are responsible for most geologic work on shorelines, but waves can also be generated by faulting, volcanic explosions, and rockfalls.

6. Breakers form where waves enter shallow water and the orbital motion of water particles is disrupted. Waves plunge or spill onto the shoreline, expending their kinetic energy.

7. Waves approaching a shoreline at an angle generate a longshore current, which is capable of considerable erosion, transport, and deposition.

8. Rip currents are narrow surface currents that carry water from the nearshore zone seaward through the breaker zone.

9. Beaches, the most common shoreline depositional features, are continually modified by nearshore processes, and their profiles generally exhibit seasonal changes.

10. Spits, baymouth bars, and tombolos all form and grow as a result of longshore current transport and deposition.

11. Barrier islands are nearshore sediment deposits of uncertain origin. They parallel the mainland but are separated from it by a lagoon.

12. The volume of sediment in a nearshore system remains rather constant unless the system is somehow disrupted, as when dams are built across the streams supplying sand to the system.

13. Many shorelines are characterized by erosion rather than deposition. Such shorelines have sea cliffs and wave-cut platforms. Other features commonly present include sea caves, sea arches, and sea stacks.

14. Depositional coasts are characterized by long sandy beaches, deltas, and barrier islands.

15. Submergent and emergent coasts are defined on the basis of their relationships to changes in sea level.

Important Terms

barrier island
baymouth bar
beach
berm
breaker
crest
emergent coast
fetch
longshore current

longshore drift
marine terrace
nearshore sediment budget
rip current
shoreline
spit
submergent coast
tide
tombolo

trough
wave
wave base
wave-cut platform
wave height
wavelength
wave period
wave refraction

1. A barrier island:

    a. _____ forms when a rocky, steep shoreline is eroded;

    b. _____ is a long, narrow, sandy island separated from the mainland by a lagoon;

    c. _____ has sea stacks and arches rising above it;

    d. _____ originates when shoreline erosion yields a surface sloping gently seaward;

    e. _____ is likely to be found along an emergent coast.

2. Longshore currents are generated when:

    a. _____ isostatic rebound occurs more rapidly than sea-level rise;

    b. _____ the orbital motion of water in waves stirs up seafloor sediment;

    c. _____ the amount of sand reaching a beach diminishes;

    d. _____ waves approach a shoreline at an angle;

    e. _____ sea level rises with respect to a continent.

3. A sand deposit extending into the mouth of a bay is a:

    a. _____ headland;

    b. _____ sea stack;

    c. _____ spit;

    d. _____ wave-cut platform;

    e. _____ berm.

4. The continuous distance that wind blows over a water surface is known as the:

    a. _____ fetch;

    b. _____ baymouth bar;

    c. _____ beach;

    d. _____ wave period;

    e. _____ tombolo.

5. Which one of the following is a depositional landform?

    a. _____ sea cave;

    b. _____ baymouth bar;

    c. _____ marine terrace;

    d. _____ wave base;

    e. _____ headland.

6. Wave refraction is the phenomenon of:

    a. _____ excess water in the nearshore zone flowing out to sea through the breaker zone;

    b. _____ sediment deposition at the seaward margin of a wave-cut platform;

    c. _____ waves oversteepening and plunging forward on beaches;

    d. _____ offshore sediment transport;

    e. _____ waves bending so they more nearly parallel a shoreline.

7. The time it takes for the crests (or troughs) of two successive waves to pass a given point is known as the:

    a. _____ wave period;

    b. _____ wave celerity;

    c. _____ wave form;

    d. _____ wave height;

    e. _____ wave fetch.

8. Although there are some exceptions, most beaches receive most of their sediment from:

    a. _____ coastal submergence;

    b. _____ erosion of shoreline rocks;

    c. _____ streams;

    d. _____ erosion of reefs;

    e. _____ breakers.

9. The part of the backshore zone consisting of a horizontal or gently landward-sloping surface is a:

    a. _____ berm;

    b. _____ beach face;

    c. _____ crest;

    d. _____ tide;

    e. _____ spit.

10. Erosional renmants of a shoreline rising above a wave-cut platform are:

    a. _____ embayments;

    b. _____ headlands;

    c. _____ tombolos;

    d. _____ sea stacks;

    e. _____ beach cusps.

11. Waves approaching a shoreline at an angle generate or cause:

    a. _____ flood tides;

    b. _____ longshore currents;

    c. _____ headland retreat;

    d. _____ deposition of sea stacks;

    e. _____ coastal emergence.

12. The solvent activity of saltwater is _____, whereas the direct impact of water on a shoreline is _____:

    a. _____ corrosion/hydraulic action;

    b. _____ wave oscillation/fetch;

    c. _____ terracing/wave reflection;

    d. _____ rip current/longshore drift;

    e. _____ salt wedging/shoreline progradation.

13. What are the characteristics of emergent and submergent coasts?

14. Why does an observer at a shoreline location experience two high and two low tides daily?

15. Explain why quartz is the most common mineral in beach sand.

16. What is wave base, and how does it affect waves as they enter shallow water?

17. Explain how a longshore current is generated and how it transports and deposits sand.

18. Name and describe three coastal landforms resulting from shoreline deposition.

19. While driving along North America's West Coast, you notice a rather flat surface dipping gently seaward with several isolated masses of rock rising above it. What is this landform, and how did it originate?

20. What are the positive and negative aspects of building groins along a shoreline?

21. How and why do summer and winter beaches differ?

22. What are the similarities and differences between spits and tombolos?

23. What are rip currents, and how would you recognize one?

24. How do barrier islands migrate landward? What evidence indicates that some U.S. barrier islands are migrating? Give some specific examples.

Points to Ponder

1. An initially straight shoreline is composed of granite flanked on both sides by glacial drift. Diagram and explain this shoreline's probable reponse to erosion.

2. How might burning fossil fuels have some long-term effect on shoreline erosion?

3. Why are long, broad, sandy beaches more common in eastern North America than western North America?

4. A hypothetical nearshore area has a balanced sand budget, but a dam is constructed on the stream supplying most of the land-derived sand and a wall is built to protect sea cliffs from erosion. What will likely happen to the beach in this area?

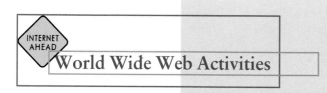

World Wide Web Activities

For these Web site addresses, along with current updates and exercises, log on to
http://www.brookscole.com/geo/

➤ COASTAL PROGRAM DIVISION COASTAL ZONE MANAGEMENT PROGRAM

This government site is dedicated to the coastal zone management program of the National Ocean Service. What is the mission of the Coastal Zone Management Program?

➤ U.S. GEOLOGICAL SURVEY: THE NATIONAL MARINE AND COASTAL GEOLOGY PROGRAM

As stated at the top of this Web site, The National Marine and Coastal Geology Program is "a plan for geologic research on environmental, hazards, and resources issues affecting the Nation's coastal realms and marine federal lands."

1. Click on the *U.S. Ocean margins; issues for the next decade* site under the Introduction of the Table of Contents. What are some of the issues facing the United States in the next decade concerning its coastal margins? Why should we as citizens be concerned about these issues?

2. Click on the *Mapping the U.S. Exclusive Zone* site under the Profile of the USGS Marine and Coastal Geology Program of the Table of Contents. What is the EEZ? What are the plans for it in the next decade?

3. Click on the *Coastal and Nearshore Erosion* site under the Description of the USGS Marine and Coastal Geology Program of the Table of Contents. What is the scope of the problem concerning coastal and nearshore erosion? What are some of the projects being undertaken in this area?

➤ THE NATURAL HISTORY OF NOVA SCOTIA

The Nova Scotia Museum of Natural History maintains this site, which gives an overview of the geology, land and seascapes, plants, animals, and ecology of Nova Scotia. Click "Topics," and scroll down and click "T7.3 Coastal Landforms" for images and discussions of various coastal features.

U.S. GEOLOGICAL SURVEY WOODS HOLE CENTER, COASTAL AND MARINE GEOLOGY

Scroll down to "Earth History," and click "Chesapeake Bay Boldie." What is a boldie, and how was one important in the geologic evolution of Chesapeake Bay?

PARK GEOLOGY TOUR—SHORELINE GEOLOGY

This is one of many sites maintained by the National Park Service. It has a menu listing many National Parks, National Monuments, and National Seashores and Lakeshores. Read about and see images of Acadia National Park, Maine; Cape Hatteras National Seashore, North Carolina; and Point Reyes National Seashore, California. How do the shoreline features in these areas compare, and what processes were important in their geologic evolution?

U.S. GEOLOGICAL SURVEY WESTERN REGION COASTAL AND MARINE GEOLOGY

Click "Table of Contents," which lists numerous topics of interest such as hazards, natural resources, seafloor mapping, and environmental quality. Click "Coastal Erosion" under the heading "Natural Hazards and Public Safety." Then click "Hurricane George's Coastal Erosion." What areas were impacted by the hurricane? Was there any coastal flooding? Next click "North Carolina Coastal Erosion Studies," and see what kinds of problems exist and what solutions have been proposed.

U.S. GEOLOGICAL SURVEY INDEX OF MARINE AND COASTAL GEOLOGY PROGRAM FACT SHEETS

Scroll down, and click "Limited Sand Resources for Eroding Beaches."
1. What concerns are expressed in the discussion?
2. What is beach nourishment, and how is it accomplished?
Return to the menu and click "The Chesapeake Bay: Geologic Product of Rising Sea Level."
1. Why is rising sea level a concern here?
2. What evidence indicates that sea level is rising, and how fast is it rising at this location?
Return to the menu and click "Coastal Erosion of Southern Lake Michigan." How is shoreline erosion in lakes similar to that occurring along sea coasts?

CD-ROM Exploration

Exploring your *Earth Systems Today* CD-ROM will add to your understanding of the material in this chapter.

TOPIC: SURFICIAL PROCESSES AND HYDROSPHERE

MODULE: WAVES, TIDES, AND CURRENTS

Explore activities in this module to see if you can discover the following for yourself:

Using activities in this module, explore the changes that occur in wind-generated waves as they move from deep to shallow water. Specifically, note how these waves behave in shallow water at the shoreline, and relate these to shoreline processes.

Using activities in this module, note how tidal fluctuations develop along shorelines. What effect might tides have on shoreline development and processes?

Chapter 20

With a diameter of 3126 km, Jupiter's moon Europa (center) is the sixth largest moon in the solar system. Evidence from the Galileo mission indicates that Europa has an icy outer covering and perhaps liquid water below. Also shown in this image, but not to scale, are Jupiter (left center) and three of its other moons including Io (upper left) and Ganymede and Callisto (right).

OBJECTIVES

At the end of this chapter, you will have learned that

- The Big Bang theory for the origin of the universe is currently widely accepted.
- The general characteristics of the solar system must be accounted for by any theory proposed to explain how it formed and evolved.

- The three basic types of meteorites provide evidence for the composition and age of the solar system.
- Several criteria such as density and composition are used to define terrestrial versus Jovian planets.
- Earth formed as a gaseous sphere that became a planet with a core, mantle, and crust.

- Scientists theorize that the Moon originated from material ejected from Earth following an asteroid impact.
- Once it formed, Earth became a dynamic planet with several major systems that even now continue to interact.

A History of the Universe, Solar System, and Planets

Prologue

Does life exist elsewhere in the universe, or is it unique to Earth? This question has intrigued people for centuries. The discovery of extraterrestrial life would certainly create a major scientific and theological upheaval as the fact that we are not alone sinks into the collective human consciousness.

The search for life elsewhere in the solar system and beyond goes back at least as far as Galileo. His discovery that other planets exist fueled people's imagination about whether life may also exist in these places. In 1894 the wealthy amateur astronomer Percival Lowell built his own observatory in Flagstaff, Arizona, to further the study of Mars, a planet especially interesting to him. Inspired by earlier reports of dark lines on the Martian surface and his own observations, Lowell published a book in 1895 speculating that an advanced civilization had built canals (the dark crisscrossing lines some people observed) to bring water from the melting polar ice caps to the rest of the planet in response to a global drought.

Lowell's writings were instrumental in firing public interest in Mars and Martians and the possibility of extraterrestrial life-forms. In 1898 H. G. Wells published his novel *The War of the Worlds* about an invasion of Earth by Martians. This theme was further amplified in a 1938 radio broadcast by Orson Welles in which Martians invaded New Jersey, and the 1953 movie titled *War of the Worlds*.

Based on what we know presently about Mars from telescope observations and the various space probes sent to study the planet, scientists are convinced that Mars does not harbor any higher forms of life, that is, complex organisms comparable to Earth's plants and animals. Yet the possibility remains that microorganisms may have lived on the Martian surface early in its history when water was present or are still living beneath its surface. Nevertheless, all attempts so far have failed to detect any organism.

Do any other planets in the solar system or elsewhere in the universe have the conditions necessary for life to evolve? If they do, they would have to be located at a distance from their sun where water is a liquid rather than a solid or gas. Water is an ideal medium in which carbon-based organic compounds can dissolve and react with each other in the various ways necessary for life to exist.

Most scientists think that if extraterrestrial life exists, it will be carbon based, because carbon is found in so many different and complex compounds, as well as being the fourth most abundant element in the universe. Carbon compounds are found in meteorites and comets, as well as in huge molecular clouds floating throughout the cosmos. In fact, some scientists think that life evolved from organic compounds that fell to Earth from space early in its history. Whether life originated from organic compounds from space or near hot-water volcanic vents in the ocean as other scientists think, we know organisms were present on Earth at least 3.5 billion years ago.

During the 1990s, several new planets were discovered circling stars within the Milky Way Galaxy. What makes these discoveries so exciting is that, if these stars in the Milky Way Galaxy have planets circling them, the odds are greatly increased that other stars in the 100-billion-star Milky Way Galaxy have not only large, inhospitable planets circling them but also smaller, more life-friendly planets. Based on what we know about how life might have originated, we can say the probability is high that other planets in our galaxy also contain life.

At an August 1996 news conference, NASA announced the first tangible evidence that life may have existed on Mars. A 1.9-kg meteorite, named ALH84001, contains within it what appear to be chemical and fossil remains of Martian life (Figure 20.1). This potato-sized piece of rock was blasted from the surface of Mars 3.6 billion years ago and circled the Sun for millions of years, until finally crashing into an ice field in Antarctica's Allan Hills some 13,000 years ago. It remained there until it was discovered in 1984.

After studying the Martian meteorite for two years, the NASA-led team of scientists found organic molecules called *polycyclic aromatic hydrocarbons,* which on Earth are the by-products of metabolic activity. In addition, traces of iron sulfide and magnetite were also found. These compounds are also associated with, but not necessarily the result of, bacterial activity. Besides the chemical evidence, microscopic structures superficially resembling bacteria were also recovered from inside the meteorite. Although these segmented, tubelike objects look like bacteria, they are 100 times smaller than fossil bacteria found on Earth.

As startling as these discoveries are, many scientists are very skeptical about accepting this evidence as proof that life exists or has existed elsewhere in the solar system.

The human fascination with Mars is understandable, given that the so-called Red Planet has always had an air of mystery about it. And perhaps it does harbor living microorganisms or evidence of ancient organisms now long extinct. In any event, the search goes on, but now another body in the solar system is receiving some attention. Europa, one of Jupiter's moons, with a diameter of 3126 km is the sixth largest moon in the solar system (see chapter opening photo). Because of Jupiter's tremendous gravitational attraction and interactions with other Jovian moons, Europa experiences what is called *tidal flexing,* resulting in stretching and compression of the entire moon and especially its icy outer layer. It possesses few craters, indicating that it is a geologically active body.

Data from *Galileo* missions indicate that Europa has a vast ocean beneath its outer icy cover, an ocean perhaps 200 km deep. If that is correct, Europa has more liquid water than contained in all of Earth's oceans. In any event, scientists know that Europa is too small to retain any heat from its time of origin, and radioactive decay is insufficient to explain its geologic activity. Tidal heating is more probable, thus accounting for the liquid water. Because liquid water is likely present, scientists speculate that Europa's oceans may harbor some kinds of organisms, perhaps in environments similar to those adjacent to black smokers on Earth's seafloor (see Figure 11.17). This question will remain unanswered until a future mission reaches this small celestial body.

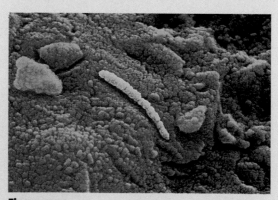

Figure 20.1
This microscopic tube-shaped structure within meteorite ALH84001, which was discovered in Antarctica, is thought to have come from Mars. Some scientists think it might represent fossilized remains of an organism that lived on Mars more than 3.6 billion years ago. However, many scientists are skeptical and think the object had an inorganic origin.

INTRODUCTION

When considered in the context of the universe, Earth is a medium-sized planet circling a rather average star among billions in the Milky Way Galaxy, which is but one of billions of galaxies. A number of other planets have recently been discovered around other star systems and perhaps life exists elsewhere in the universe, but of all the known planets and moons in the solar system only Earth is known to have life. Viewed from the blackness of nearby space, Earth is a brilliant, bluish, shimmering planet, wrapped in a veil of swirling white clouds. Vast oceans, seven continents, and numerous islands cover its surface, and organisms of one kind or another are present nearly everywhere. This unique planet, revolving around the Sun every 365.25 days, is a complex, evolving body. In short, it is a dynamic planet that has changed and continues to change in response to interactions among its major systems (see Figure 1.4).

Earth has not always looked the way it does today. Based on various kinds of evidence, many scientists think that it began as a homogeneous mass of rotating dust and gases that contracted, heated, and differentiated during its early history to form a medium-sized planet with a metallic core, a mantle composed of iron- and magnesium-rich rocks, and a thin crust (see Figure 1.13). Overlying this crust is an atmosphere now composed of 78% nitrogen and 21% oxygen.

As the third planet from the Sun, Earth seems to have formed at just the right distance from the Sun (150 millon km) so that it is neither too hot nor too cold to support life as we know it. Furthermore, its size is just right to hold an atmosphere. If Earth were smaller, its gravity would be so weak that it could retain little, if any, atmosphere, and any surface water would vaporize.

ORIGIN OF THE UNIVERSE—DID IT BEGIN WITH A BIG BANG?

Most scientists think that the universe originated about 15 billion years ago in what is popularly called the **Big Bang.** In a region infinitely smaller than an atom, both time and space were set at zero. Therefore, there is no "before the Big Bang," only what occurred after it. The reason for this is that space and time are unalterably linked to form a space–time continuum demonstrated by Einstein's theory of relativity. Without space, there can be no time.

How do we know the Big Bang took place approximately 15 billion years ago? Why couldn't the universe have always existed as we know it today? Two fundamental phenomena indicate that the Big Bang occurred. The first is that the universe is expanding; the second is that it is permeated by background radiation.

That the universe is expanding was first recognized by Edwin Hubble in 1929. By measuring the optical spectra of distant galaxies, Hubble noted that the velocity at which a galaxy moves away from Earth increases proportionally to its distance from Earth. He observed that the spectral lines (wavelengths of light) of the galaxies are shifted toward the red end of the spectrum; that is, the lines are shifted toward longer wavelengths. Such a redshift would be produced by galaxies receding from each other at tremendous speeds (Figure 20.2). This is an example of the **Doppler effect,** which is the change in the frequency of a sound, light, or other wave caused by

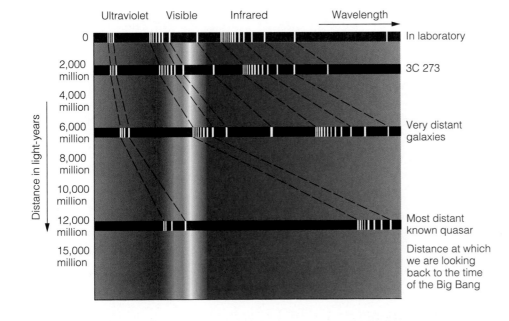

Figure 20.2
Spectra showing the redshift of the galaxies. Just as the Doppler effect makes a rapidly receding train whistle drop in pitch, the expansion of the universe stretches light waves from receding objects toward the longer wavelengths of the spectrum. This light is said to be *redshifted,* and the amount of the redshift is proportional to the object's distance from us. In this diagram the spectral lines of hydrogen are stretched in proportion to their original wavelengths. All lines in the spectrum of any object, however, exhibit the same ratio of wavelength increase to the original wavelength.

Figure 20.3

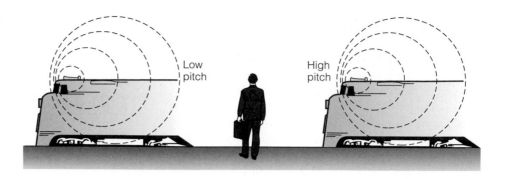

The Doppler effect applied to sound waves. The sound waves of an approaching whistle are slightly compressed so that the individual hears a shorter-wavelength, higher-pitched sound. As the whistle passes and recedes from the individual, the sound waves are slightly spread out, and a longer-wavelength, lower-pitched sound is heard.

movement of the wave's source relative to the observer (Figure 20.3).

One way to envision how velocity increases with increasing distance is by reference to the popular analogy of a rising loaf of raisin bread in which the raisins are uniformly distributed throughout the loaf (Figure 20.4). Suppose that before the dough begins to rise, the raisins are 1 cm apart. After 1 hour, the dough has risen to the point where the raisins are 2 cm apart. Any raisin is now 2 cm from its nearest neighbor and 4 cm away from the next one, 6 cm from the next, and so on. From the perspective of any raisin, its nearest neighbor has moved away from it at a speed of 1 cm per hour (it originally was 1 cm away and is now 2 cm away), the next raisin over has moved away at a speed of 2 cm/hr (it was 2 cm away, and is now 4 cm away), while the next one has moved away at 3 cm/h (it was originally 3 cm away and is now 6 cm away from the reference raisin), and so on. If the dough continues to rise until the distance between any two raisins is 4 cm, we would find that all the distances between raisin pairs are quadruple

what they were to start with. The speeds needed to accomplish this are directly proportional to the distances between any two raisins: The farther away a given raisin is to begin with, the farther it must move to maintain the regular spacing during the expansion, and hence the greater its velocity must be. In the same way that raisins move apart in a rising loaf of bread, galaxies are receding from each other at a rate proportional to the distance between them, which is exactly what astronomers see when they observe the universe.

A corollary to Hubble's expanding universe is that astronomers can use this expansion rate to calculate how long ago the galaxies were all together in a single point. Such calculations yield an estimated age for the universe of approximately 15 billion years.

The second important observation providing evidence of the Big Bang was made in 1965 by Arno Penzias and Robert Wilson of Bell Telephone Laboratories when they discovered that there is a pervasive background radiation of 2.7 degrees K above absolute zero (absolute zero equals $-273°C$) everywhere in the

Figure 20.4

The motion of raisins in a rising loaf of raisin bread illustrates the relationship that exists between distance and speed and is analogous to an expanding universe. In this diagram, adjacent raisins are located 2 cm apart before the loaf rises. After 1 hour, any raisin is now 4 cm away from its nearest neighbor and 8 cm away from the next raisin over, and so on. Therefore, from the perspective of any raisin, its nearest neighbor has moved away from it at a speed of 2 cm per hour, and the next raisin over has moved away from it at a speed of 4 cm per hour. In the same way that raisins move apart in a rising loaf of bread, galaxies are receding from each other at a rate proportional to the distance between them.

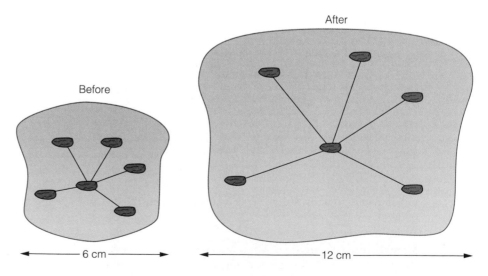

Table 20.1

The Four Basic Forces of the Universe

Four forces appear to be responsible for all interactions of matter:

1. *Gravity* is the attraction of one body toward another.

2. The *electromagnetic force* combines electricity and magnetism into the same force and binds atoms into molecules. It also transmits radiation across the various spectra at wavelengths ranging from gamma rays (shortest) to radio waves (longest) through massless particles called *photons*.

3. The *strong nuclear force* binds protons and neutrons together in the nucleus of an atom.

4. The *weak nuclear force* is responsible for the breakdown of an atom's nucleus, producing radioactive decay.

universe. This background radiation is thought to be the fading afterglow of the Big Bang.

At the time of the Big Bang, matter as we know it did not exist, and the universe consisted of pure energy. Within the first second after the Big Bang, the four basic forces—gravity, electromagnetic force, strong nuclear force, and weak nuclear force (Table 20.1)—had all separated, and the universe experienced enormous expansion. Matter and antimatter collided and annihilated each other. Fortunately, a slight excess of leftover matter would become the universe. By the time the universe was 3 minutes old, temperatures were cool enough for protons and neutrons to fuse together to form the nuclei of hydrogen and helium atoms. Approximately 300,000 years later, electrons joined with the previously formed nuclei to make complete atoms of hydrogen and helium. At the same time, photons (the energetic particles of light) separated from matter, and light burst forth for the first time.

After about 200 million years, as the universe continued expanding and cooling, stars and galaxies began forming, and the chemical makeup of the universe changed. Early in its history, the universe was 100% hydrogen and helium, whereas today it is 98% hydrogen and helium by weight.

Over the course of their history, stars undergo many nuclear reactions whereby lighter elements are converted into heavier elements by nuclear fusion in which atomic nuclei combine to form more massive nuclei. Such reactions, which convert hydrogen to helium, occur in the cores of all stars. The subsequent conversion of helium to heavier elements, such as carbon, depends on the mass of the star. When a star dies, often explosively, the heavier elements that were formed in its core are returned to interstellar space and are available for inclusion in new stars. When new stars form, they will have a small amount of these heavier elements, which may be converted to still heavier elements. In this way, the chemical composition of the galaxies, which are made up of billions of stars, is gradually enhanced in heavier elements.

THE SOLAR SYSTEM—ITS ORIGIN AND EARLY DEVELOPMENT

Having briefly examined the origin and history of the universe, we can now examine how the solar system, which is part of the Milky Way Galaxy, formed. The solar system consists of the Sun, 9 planets, 61 known moons, a tremendous number of asteroids—most of which orbit the Sun in a zone between Mars and Jupiter—and millions of comets and meteors, as well as interplanetary dust and gases (Table 20.2).

General Characteristics of the Solar System

Recall from earlier discussions that a *theory* is an explanation for some phenomenon or related phenomena. As such it must be testable in some fashion; that is, it must account for observations and experimental results. Furthermore, some experiments and/or observations could at least conceivably prove it incorrect, otherwise it would not be testable. Many scientific theories are well supported by evidence, but any theory is always open to question and might be modified or incorporated into a better theory as more evidence accumulates or more definitive experiments are performed. Thus, any theory formulated to explain the origin and evolution of the solar system must take into account its several general characteristics (Table 20.3).

One characteristic of the solar system is that all planets revolve around the Sun in the same direction, in nearly circular orbits, and in approximately the same plane (called the *plane of the ecliptic*), except for Pluto whose orbit is both highly elliptical and tilted 17 degrees to the orbital plane of the rest of the planets (Figure 20.5).

All planets except Venus and Uranus, and nearly all planetary moons, rotate counterclockwise when viewed

Table 20.2

Characteristics of the Sun, Planets, and Moon								
OBJECT	MEAN DISTANCE TO SUN (km × 10⁶)	ORBITAL PERIOD (DAYS)	ROTATIONAL PERIOD (DAYS)	TILT OF AXIS	EQUATORIAL DIAMETER (km)	MASS (kg)	MEAN DENSITY (g/cm³)	NUMBER OF SATELLITES
Sun	—	—	25.5	—	1,391,400	1.99×10^{30}	1.41	—
Terrestrial planets								
Mercury	57.9	88.0	58.7	28°	4,880	3.33×10^{23}	5.43	0
Venus	108.2	224.7	243	3°	12,104	4.87×10^{24}	5.24	0
Earth	149.6	365.3	1	24°	12,760	5.97×10^{24}	5.52	1
Mars	227.9	687.0	1.03	24°	6,787	6.42×10^{23}	3.96	2
Jovian planets								
Jupiter	778.3	4,333	0.41	3°	142,796	1.90×10^{27}	1.33	16
Saturn	1,428.3	10,759	0.43	27°	120,660	5.69×10^{26}	0.69	18
Uranus	2,872.7	30,685	0.72	98°	51,200	8.69×10^{25}	1.27	17
Neptune	4,498.1	60,188	0.67	30°	49,500	1.03×10^{26}	1.76	8
Pluto	5,914.3	90,700	6.39	122°	2,300	1.20×10^{22}	2.03	1
Moon	0.38 (from Earth)	27.3	27.32	7°	3,476	7.35×10^{22}	3.34	—

Table 20.3

General Characteristics of the Solar System

1. *Planetary orbits and rotation*
   - Planetary and satellite orbits lie in a common plane.
   - Nearly all the planetary and satellite orbital and spin motions are in the same direction.
   - The rotation axes of nearly all the planets and satellites are roughly perpendicular to the plane of the ecliptic.
2. *Chemical and physical properties of the planets*
   - The terrestrial planets are small, have a high density (4.0 to 5.5 g/cm³), and are composed of rock and metallic elements.
   - The Jovian planets are large, have a low density (0.7 to 1.7 g/cm³), and are composed of gases and frozen compounds.
3. *The slow rotation of the Sun*
4. *Interplanetary material*
   - The existence and location of the asteroid belt.
   - The distribution of interplanetary dust.

from a point in space above Earth's North Pole. Furthermore, the axes of rotation of the planets, except for those of Uranus and Pluto, are nearly perpendicular to the plane of the ecliptic (Figure 20.5b).

The nine planets can be divided into two groups based on their chemical and physical properties. The four inner planets—Mercury, Venus, Earth, and Mars—are all small and have high mean densities (Table 20.2), indicating that they are composed of rock and metallic elements. They are known as the **terrestrial planets** because they are similar to *terra,* which is Latin for "earth."

The next four planets—Jupiter, Saturn, Uranus, and Neptune—are known as **Jovian planets** because they all resemble Jupiter. The Jovian planets are large and have low mean densities, indicating they are composed of light-weight gases such as hydrogen and helium, as well as frozen compounds such as ammonia and methane. The outermost planet, Pluto, is small and has a low mean density of slightly more than 2.0 g/cm³.

The slow rotation of the Sun is another feature that must be accounted for in any comprehensive theory of the origin of the solar system. If the solar system formed from the collapse of a rotating cloud of gas and dust as

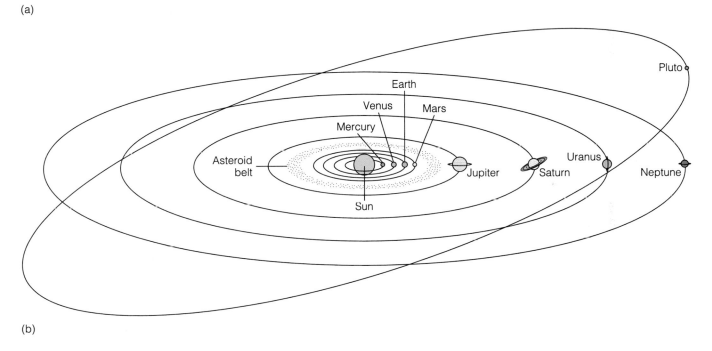

Figure 20.5
Diagrammatic representation of the solar system showing (a) the relative sizes of the planets and (b) their orbits around the Sun.

(a)

(b)

is currently accepted, the Sun, which was at the center of that cloud, should have a very rapid rate of rotation, instead of its leisurely 25-day rotation.

Finally, any theory of the origin of the solar system must accommodate the nature and distribution of the various interplanetary objects such as the asteroid belt, comets, and interplanetary gases and dust.

Current Theory of the Origin and Early History of the Solar System

Various scientific theories of the origin of the solar system have been proposed, modified, and discarded since the French scientist and philosopher René Descartes first suggested, in 1644, that the solar system formed from a gigantic whirlpool within a universal fluid. Most theories have involved an origin from a primordial rotating cloud of gas and dust. Through the forces of gravity and rotation, this cloud then shrank and collapsed into a rotating disk. Detached rings within the disk condensed into planets, and the Sun condensed in the center of the disk.

The problem with most of these theories is that they failed to explain the slow rotation of the Sun, which according to the laws of physics should be rotating rapidly. This problem was finally solved with the discovery of the solar wind, which is an outflow of ionized gases from the Sun that interacts with its magnetic field and slows its rotation through a magnetic braking process.

As we discussed in Chapter 1, the currently accepted **solar nebula theory** (see Figure 1.11) for the origin of the solar system involves the condensation and collapse of interstellar material in a spiral arm of the Milky Way Galaxy. The collapse of this cloud of gases and small grains into a counterclockwise rotating disk caused about 90% of the material to be concentrated in the central part of the disk and resulted in the formation of an embryonic sun, around which swirled a rotating cloud of material called a *solar nebula*. Within this solar nebula were localized eddies where gases and solid particles condensed. Collisions between the various gases and particles resulted in accretion into **planetesimals** (Figure 20.6). As the planetesimals collided and grew in size and mass, they eventually became planets. According to estimates, a planet the size of Earth formed from planetesimals in as little as 1 million years.

The composition and evolutionary history of the planets are a consequence, in part, of their distance from the Sun. The terrestrial planets are composed of rock and metallic elements that condensed at the high temperatures of the inner nebula. The Jovian planets, all of which have small central rocky cores compared to their overall size, are composed mostly of hydrogen, helium, ammonia, and methane, which condense at low temperatures. Thus, the farther away from the Sun that condensation occurred, the lower the temperature, and hence the higher the percentage of *volatile elements,* which condense at low temperatures, relative to *refractory elements,* which condense at high temperatures.

While the planets were accreting, material that had been pulled into the center of the nebula also condensed, collapsed, and was heated to several million degrees by gravitational compression. The result was the birth of a star, our Sun.

During the early accretionary phase of the solar system's history, collisions between various bodies were common, as indicated by the craters on many planets and moons. An unusually large collision involving Venus could explain why it rotates clockwise rather than counterclockwise, and a collision could also explain why Uranus and Pluto do not rotate nearly perpendicular to the plane of the ecliptic.

Figure 20.6
At the stage of development shown here, planetesimals have formed in the inner solar system, and large eddies of gas and dust remain at great distances from the protosun.

Asteroids probably formed as planetesimals in a localized eddy between what eventually became Mars and Jupiter in much the same way as other planetesimals formed the terrestrial planets. The tremendous gravitational field of Jupiter, however, prevented this material from ever accreting into a planet. *Comets,* interplanetary bodies composed of loosely bound rocky and icy material, are thought to have condensed near the orbits of Uranus and Neptune. Each time comets pass by Jupiter and Saturn, the gravitational effect of those planets increases their speed, forcing them farther out into the solar system.

The solar nebula theory of the formation of the solar system thus accounts for most of the characteristics of the planets and their moons, the differences in composition between the terrestrial and Jovian planets, the slow rotation of the Sun, and the presence of the asteroid belt. Based on the available data, the solar nebula theory best explains the features of the solar system and provides a logical explanation for its evolutionary history.

METEORITES—VISITORS FROM SPACE

Periodic meteor showers afford many people an opportunity to observe a cosmic display during which extraterrestrial pieces of debris flash through the atmosphere. Most of these **meteors,** which are all solid bodies from space entering Earth's atmosphere, are vaporized by frictional heating as they pass through the atmosphere, and thus contribute a tiny quantity of cos-

Figure 20.7
(a) Relative proportions of the three groups of meteorites. (b) Stony meteorite.
(c) Iron meteorite. (d) Stony-iron meteorite.

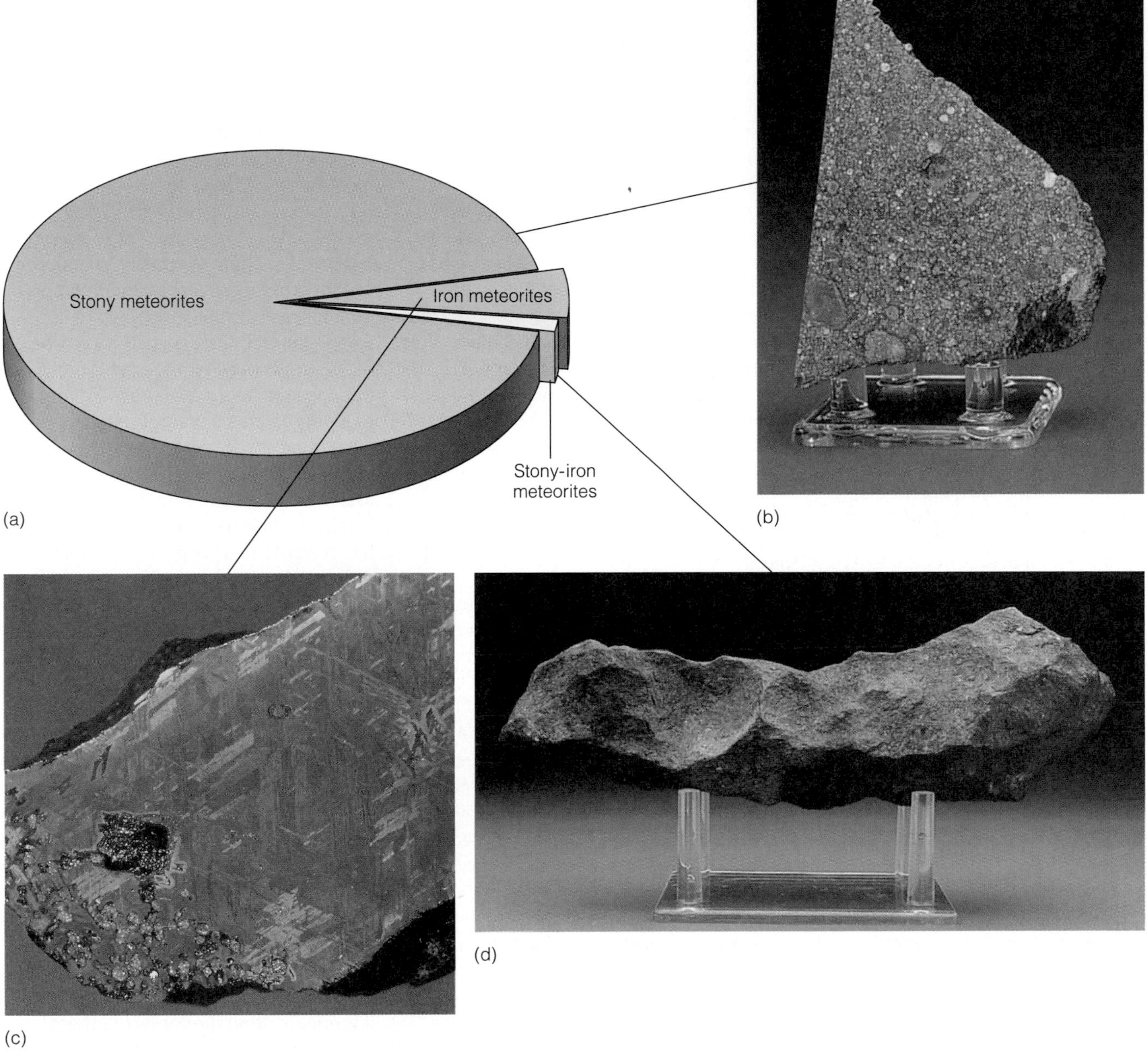

(a)

Stony meteorites

Iron meteorites

Stony-iron
meteorites

(b)

(c)

(d)

mic dust to Earth. But some meteors make it to the surface intact as **meteorites,** all of which are important in considering the solar system's origin and evolution.

Meteorites are probably pieces of material that originated during the formation of the solar system 4.6 billion years ago. Early in the solar system's history, a period of heavy meteorite bombardment occurred as the solar system cleared itself of the many pieces of material that had not yet accreted into planetary bodies or moons. Since then, meteorite activity has greatly diminished. Most meteorites that currently reach Earth are probably fragments resulting from collisions between asteroids.

Meteorites are classified into three broad groups based on their proportions of metals and silicate minerals (Figure 20.7a). About 93% of all meteorites are composed of iron and magnesium silicate minerals and thus are known as **stony meteorites** (Figure 20.7b). The many varieties of stony meteorites provide geologists with much information about the origin and history of the solar system.

Iron meteorites, the second group, accounting for about 6% of all meteorites, are composed primarily of a combination of iron and nickel alloys (Figure 20.7c). Their large crystal size and chemical composition indicate that they cooled very slowly in large objects such as

Guest Essay

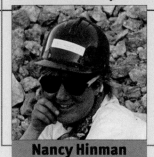

Nancy Hinman

Geochemistry and Life on Mars

Nancy W. Hinman is an assistant professor of geology at the University of Montana. She received her B.S. in chemistry from Reed College and her Ph.D. in Oceanography from Scripp's Institution of Oceanography. She has served as the chair of the Younger Chemist's Committee of the American Chemical Society.

A
s a senior at Reed College, I wanted to understand the elemental cycling through the major aquatic reservoirs, oceans, rivers, and groundwater, so I arranged a senior thesis topic with Jim Pankow at the (then) Oregon Graduate Center to work on the transfer of petroleum-like compounds into seawater. It was a great experience and, as I conducted hourly sampling for 20-hour experiments, totally dispelled the idea that geoscience is done only during daylight hours. From there, my path led me to the study of how the type of silica in diatom shells changes through geologic time. The process occurs in sediments, and many factors are known to affect the rate and sequence of the changes, including temperature/time and water chemistry. Because silica-rich rocks can be both petroleum source and reservoir rocks, the rates of these changes are important in predicting when petroleum is generated and how it moves. So many pieces were involved, and they fit together in such intricate ways, that the more I worked on the problem, the more of a puzzle it became. It is the thrill of finding a key piece that spurs my interest in science. The countless hours spent collecting and analyzing data are erased when the meaning of the data, and the consequence, becomes clear.

I continue to be interested in silica and how it changes, but I no longer work in marine sediments. In what seemed like a sensible move, I shifted my focus to continental silica-rich systems and began working on thermal springs. I thought that the pieces would be more easily identified in what seemed to be a less complex system, that is, one containing fewer elements and mineral phases besides silica; yet another lesson that appearances can be deceiving.

Silica is an excellent medium in which to preserve fossils. What are the chances that organisms become preserved in the geologic record? Very slim but I hope that the results of my recent research will help identify the conditions under which good preservation is achieved. My current research focuses on the processes by which microorganisms become preserved in silica and other minerals, hot springs, and cold seeps. These microorganisms include cyanobacteria, which represent primitive life-forms on Earth and possibly on Mars. It is important to understand the processes by which they are preserved in the geologic record on Earth so that we can understand more about the conditions under which life first appeared here and possibly elsewhere in our solar system. I strive to understand the chemical processes leading to entombment and fossilization so that we can be more selective in choosing the materials to bring back from Mars. I find myself delving into questions involving not only geochemical processes but also microbial processes and hydrogeologic processes. After all, microbes exert a major influence on their chemical surroundings. Further, the supply and properties of water provided to the microorganisms or to their remains determine whether the microbes become preserved in the rock record or lost through weathering.

I find this work fascinating, and it has rewarded me by providing opportunities to work with people in many different fields and to present my results in professional, governmental, and public forums. Recently, I was invited to present testimony at a Senate Subcommittee on Science, Technology, and Space hearing on evidence for life on Mars. This led to several magazine, newspaper, and radio interviews. It seems that the quest to know whether we are alone in our universe has captured the public's imagination. I am excited to be part of the search and am looking forward to finding more pieces of this puzzle and fitting them into their proper place.

Do I look up at the stars and wonder if there are other life-forms out there? Sure, don't you? ∎

---

asteroids where the hot iron–nickel interior was insulated from the cold of space. Collisions between such slowly cooling asteroids produced the iron meteorites that we find today.

Stony-iron meteorites, the third group, are composed of nearly equal amounts of iron and nickel and silicate minerals; they make up less than 1% of all meteorites (Figure 20.7d). Stony-irons are generally thought to represent fragments from the zone between the silicate and metallic portions of large differentiated asteroids.

Astronomers have identified at least 40 asteroids larger than a kilometer in diameter whose orbits cross Earth's and estimate that there may be as many as 1000 such asteroids. Meteor Crater in Arizona formed when

(a)

(b)

Figure 20.8

(a) An Earth-asteroid impact about 25,000 years ago produced Meteor Crater, Arizona. The crater is 1.2 km wide and 180 m deep. (b) If a 48-km-diameter comet hit northern New Jersey, as depicted in this painting, everything seen here including the buildings of lower Manhattan in the foreground would be vaporized, and a plume of material would rise into the atmosphere and circulate around Earth.

an asteroid hit Earth (Figure 20.8a). While asteroid–Earth collisions are rare, they do happen and could have devastating results if they occurred in a populated area (Figure 20.8b). Many scientists think that a collision with a meteorite about 10 km in diameter led to the extinction of dinosaurs and several other groups of animals 66 million years ago (see Perspective 20.1). We know that the ash and gases released into the atmosphere from volcanic eruptions has affected climates, and studies indicate that a collision with a large meteorite could produce enough dust to similarly affect global climate.

Meteorites are important because their age—about 4.6 billion years—and composition provide scientists with information about the origin and history of the solar system. Furthermore, their collisions with Earth and the other planets have affected the topography of the planets and, although there is no unanimity on this point, perhaps the evolution of life on Earth.

PLANETS AND MOONS

A tremendous amount of information about each planet in the solar system has been derived from Earth-based observations and measurements as well as

from the numerous space probes launched during the past 30 years. Information about a planet's size, mass, density, composition, presence of a magnetic field, and atmospheric composition has allowed scientists to formulate hypotheses concerning the origin and history of the planets and their moons.

The Terrestrial Planets

All evidence indicates that the terrestrial planets had similar early histories during which volcanism and cratering from meteorite impacts were common. During the accretionary phase of formation and soon thereafter, each planet seems to have undergone differentiation as a result of heating by radioactive decay. The mass, density, and composition of the planets indicate that each formed a metallic core and a silicate mantle-crust during this phase. Images sent back by these various space probes also clearly show that volcanism and cratering by meteorites continued during the differentiation phase. Volcanic eruptions produced lava flows, and an atmosphere developed on each planet by **outgassing,** a process whereby light gases from the interior rise to the surface during volcanic eruptions (see Perspective 20.2).

Perspective 20.1

Meteorite Impacts and Craters

During the early morning hours of October 20, 1994, many people in southern and central Michigan reported a bright fireball in the sky, followed by a sound like thunder or a sonic boom and then a rumbling or booming noise for up to 1 minute. The flash and sound resulted from a small meteorite detonating in the atmosphere high above Earth's surface, a phenomenon that occurs much more commonly than most people think. Indeed, declassified military satellite data indicate explosions equivalent to small nuclear bombs take place on average about once a month. A small piece of the meteorite disintegrating over Michigan was recovered after it passed through the roof of a house in the town of Coleman (Figure 1a).

Just how common are meteorite impacts, and what is the danger of one hitting a person? Most meteors vaporize in the atmosphere and pose no threat at all, but obviously some make it to the surface. Disregarding unsubstantiated accounts, such as a monk being killed in 1650 in Milan, Italy, there is at least one case of a meteorite striking a person. In 1954, a meteorite hit a woman in Alabama after it passed through her roof and bounced into her, badly bruising her hip. In 1991, a small meteorite narrowly missed two boys in Indiana, and a 1992 meteorite weighing 27 kg struck a car in Peekskill, New York (Figure 1b).

No one can totally disregard the possibility of people being struck by a small meteorite, but a far greater danger is a large meteorite such as the one that formed Meteor Crater in Arizona about 25,000 years ago (see Figure 20.8a). This meteorite was probably only 45 m across, but it released about as much energy as 1000 atomic bombs the size of the one dropped on Hiroshima, Japan, in 1945. The crater it formed is 1.2 km wide and 180 m deep, clearly demonstrating that the size of a meteorite crater is many times larger than the meteorite itself.

On June 30, 1908, a tremendous explosion occurred over the Tunguska River basin in Siberia. A column of incandescent material rose about 20 km into the atmosphere, the shock wave knocked one observer to the ground 80 km away, and more than 1000 km² of forest was leveled (Figure 2). Tens of thousands of animals perished, and according to some accounts several nomadic settlements vanished. The shock wave from the explosion traveled around the world twice, and seismographs everywhere recorded an earthquake.

Because of political turmoil in Russia and the remoteness of the site, investigators did not reach the explosion site

(a)

(b)

Figure 1
(a) Tom Hagon recovered this small meteorite that damaged the roof of his house in Coleman, Michigan, in 1994. (b) In 1992, a 27-kg meteorite damaged Michelle Knapp's car in Peekskill, New York.

for 19 years, and when they did, no crater was found. Subsequently, scientists proposed that a small comet exploded in the atmosphere, thus accounting for the lack of a crater as well as the pattern of destruction. However, scientists now conclude from computer simulations that the Tunguska object was more likely a stony meteorite about 30 m in diameter traveling at 15 km per second that exploded about 8 km above the surface. The energy released by this explosion is estimated at 12.5 megatons (equivalent to 12.5 million tons of TNT). Fortunately, this event took place in a remote, sparsely settled area, or it would have had much more tragic consequences.

Certainly, everyone is familiar with the fact that dinosaurs no longer exist. Numerous hypotheses have been proposed to account for their extinction at the end of the Cretaceous Period along with many other creatures, but most have been dismissed as improbable, untestable, or inconsistent with the available data. A proposal that has become popular since 1980 is based on the discovery of a thin layer of clay with a high concentration of the platinum-group element *iridium* at the boundary between the Cretaceous and following Tertiary Period (Figure 3a). The significance of this iridium-rich clay layer, which is now known from many other areas, is that iridium is rare in crustal rocks but much more abundant in some meteorites. Accordingly, some investigators propose that a 10-km-diameter meteorite hit Earth 66 million years ago, thus accounting for the iridium-rich layer. In addition, they further postulate that the impact set in motion a chain of events leading to extinctions.

According to the impact hypothesis, about 60 times the mass of the meteorite was blasted into the atmosphere, the heat generated started raging fires, and sunlight was blocked for several months, causing a cessation of photosynthesis, the collapse of food chains, and finally, extinctions of many land-dwelling and marine organisms. If sunlight were blocked as proposed, Earth's surface temperatures would have dropped considerably, adding to the catastrophe. In addition, sulfuric acid (H_2SO_4) and nitric acid (HNO_3) from vaporized rock and atmospheric gases

Figure 2
Evidence of the Tunguska event is still apparent in this photograph taken 20 years later. The destruction was caused by some type of explosion in central Siberia in 1908.

would have contributed strongly to acid rain that would have devastated land vegetation as well as a number of marine organisms.

A probable impact site has now been identified, centered on the town of Chicxulub on the Yucatán Peninsula of Mexico (Figure 3b). The structure is deeply buried beneath younger sedimentary rocks, but appears to be about 180 km in diameter and about the right age. However, not all scientists are convinced; some geologists, for instance, think it might be some kind of volcanic feature. Nevertheless, the evidence for an impact origin is mounting, and many scientists now accept that an impact did occur and that if it did not actually cause extinctions, it certainly contributed to them.

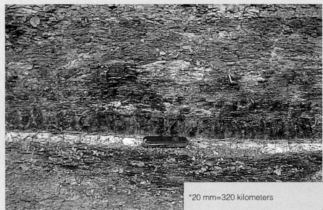

(a)

Figure 3
(a) The thin white band in this image is the iridium-rich clay layer at the boundary between the Cretaceous and Tertiary Periods in Colorado. Many scientists think it represents dust that accumulated following a meteorite impact 66 million years ago. (b) A large circular structure beneath the surface at Chicxulub on the Yucatán Peninsula of Mexico is thought to be a meteorite impact site. It dates from about the end of the Cretaceous Period, when many organisms became extinct.

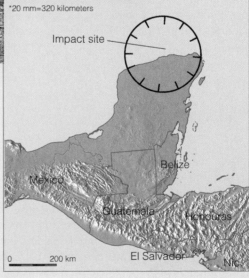

*20 mm=320 kilometers

Impact site

Belize
Mexico
Guatemala
Honduras
El Salvador
Nic.
0   200 km

(b)

Perspective 20.2

The Evolution of Climate on the Terrestrial Planets

The origin and early evolution of the terrestrial planets appear to have been similar, yet each planet has acquired a dramatically different climate. All four planets were initially alike, with atmospheres high in carbon dioxide and water vapor derived by outgassing. Mercury, because of its small size and proximity to the Sun, lost its atmosphere early in its history. Venus, Earth, and Mars, however, all were temperate enough during their early histories to have had liquid water on their surfaces, yet only Earth still has surface water and a climate capable of supporting life.

The reason that these three planets evolved such different climates is related to the recycling of carbon dioxide between the atmosphere and the crust (carbon-silicate geochemical cycle) as well as their distance from the Sun. Carbon dioxide recycling is an important regulator of climates because it, other gases, and water vapor allow sunlight to pass "through" them but trap the heat reflected back from the planet's surface. Heat is thus retained, and the temperature of the atmosphere and surface increases in what is known as the *greenhouse effect*.

On Earth, carbon dioxide combines with water in the atmosphere to form carbonic acid. When this slightly acidic rain falls, it decomposes rocks, releasing calcium and bicarbonate ions into streams and rivers and, ultimately, the oceans (see Chapter 5). In the oceans, marine organisms use some of these ions to construct shells of calcium carbonate. When the organisms die, their shells become part of the total carbonate sediments. During subduction, these carbonate sediments are heated under pressure and release carbon dioxide gas that reenters the atmosphere primarily through volcanic eruptions (Figure 1).

The recycling of carbon dioxide has allowed Earth to maintain a moderate climate throughout its history. For example, when Earth's surface cools, less water vapor is present in the atmosphere, and there is less rain. The amount of carbon dioxide leaving the atmosphere thus decreases, and less decomposition of rocks occurs. In the long term, however, the amount of carbon dioxide returned to the atmosphere does not change because it is continually replenished by plate subduction and volcanism. This leads to a temporary increase in carbon dioxide in the atmosphere, greater greenhouse warming, and higher surface temperatures.

Just the opposite would happen if the surface temperature should increase. Oceanic evaporation would then increase, leading to greater rainfall and more rapid decomposition of rock; and as a result, carbon dioxide would be removed from the atmosphere. Greenhouse warming would then decrease, and surface temperatures would fall.

Venus today is almost completely waterless. Many scientists think that during its early history, when the Sun was dimmer, Venus might have had

---

Mercury Mercury, the closest planet to the Sun, apparently has changed very little since it was heavily cratered during its early history (Figure 20.9a). Most of what we know about this small (4880-km diameter) planet comes from measurements and observations made during the flybys of *Mariner 10* in 1974 and 1975 (Table 20.2). Its high overall density of 5.4 g/cm³ indicates that it has a large metallic core measuring 3600 km in diameter; the core accounts for 80% of Mercury's mass (Figure 20.9b). Furthermore, Mercury has a weak magnetic field (about 1% as strong as Earth's), indicating that the core is probably partially molten.

Images sent back by *Mariner 10* show a heavily cratered surface with the largest impact basins filled with what appear to be lava flows similar to the lava plains on Earth's moon. The lava plains are not deformed, however, indicating that there has been little or no tectonic activity. Another feature of Mercury's

vast oceans. During this time, which probably lasted no longer than the first billion or so years of Venus's existence, water vapor as well as carbon dioxide was being released into the atmosphere by volcanism. The water vapor condensed and formed oceans, while carbon dioxide cycled (by plate tectonics) just as it does on Earth. As the Sun's energy output increased, these oceans eventually evaporated. Once the oceans disappeared, there was no water to return carbon to the crust, and carbon dioxide began accumulating in the atmosphere, creating a greenhouse effect and raising temperatures.

Mars, like Venus and Earth, probably once had a moderate climate and surface water, as indicated by the criss-crossing network of valleys on its oldest terrain. Because Mars is smaller than Earth, it had less internal heat when it formed and hence cooled much more rapidly. Eventually, the interior of Mars became so cold that it no longer released carbon dioxide. As a result, the amount of atmospheric carbon dioxide decreased to its current level. The greenhouse effect was thus weakened, and the Martian atmosphere became thin and cooled to its present low temperature.

If Mars had been the size of Earth or Venus, it very likely would have had enough internal heat to continue recycling carbon dioxide, offsetting the effects of low sunlight levels caused by its distance from the Sun. In other words, Mars would still have enough carbon dioxide in its atmosphere so that it could maintain a "temperate climate."

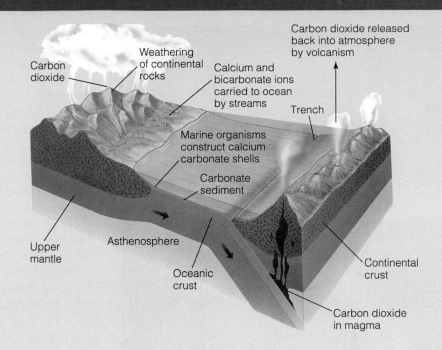

Figure 1
The carbon–silicate geochemical cycle illustrates how carbon dioxide is recycled. Carbon dioxide is removed from the atmosphere by combining with water and forming slightly acidic rain that falls on Earth's surface and decomposes rocks. This decomposition releases calcium and bicarbonate ions that ultimately reach the oceans. Marine organisms use these ions to construct shells of calcium carbonate. When they die, the shells become part of the carbonate sediments that are eventually subducted. As the sediments are subjected to heat and pressure, they release carbon dioxide gas back into the atmosphere primarily through volcanic eruptions.

surface is a large number of scarps (cliffs usually produced by faulting or erosion) (Figure 20.9c). Some scientists think that these scarps formed when Mercury cooled and contracted.

Because Mercury is so small, its gravitational attraction is insufficient to retain atmospheric gases; any atmosphere that it may have held when it formed probably escaped into space quickly. Nevertheless, *Mariner 10* detected small quantities of hydrogen and helium,

thought to have originated from the solar winds that stream by Mercury.

Venus Of all the planets, Venus is the most similar in size and mass to Earth, but it differs in most other respects (Table 20.2). Venus is searingly hot with a surface temperature of 475°C and an oppressively thick atmosphere composed of 96% carbon dioxide and 3.5% nitrogen with traces of sulfur dioxide and sulfuric

(a)

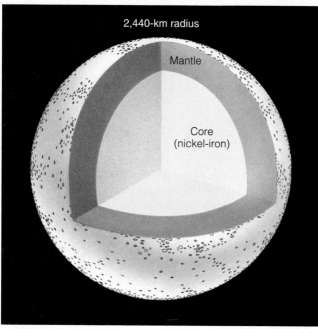

2,440-km radius

Mantle

Core
(nickel-iron)

(b)

North

(c)

Figure 20.9

(a) Mercury has a heavily cratered surface that has changed very little since its early history. (b) Internal structure of Mercury, showing its large solid core relative to its overall size. (c) Seven scarps (indicated by arrows) can clearly be seen in this image. These scarps might have formed when Mercury cooled and contracted early in its history.

and hydrochloric acid (Figure 20.10a). From information obtained by the various space probes that have passed by, orbited, and descended to Venus's surface, we know that three distinct cloud layers composed of droplets of sulfuric acid envelop the planet. Furthermore, winds up to 360 km/hour occur at the top of the clouds, while the planet's surface remains calm.

Radar images from both orbiting spacecraft and the Venusian surface reveal a wide variety of terrains (Figure 20.10b), some of which are unlike anything seen elsewhere in the solar system. Even though no active volcanism has been observed on Venus, the presence of volcanic domes, extensive lava flows, folded mountain ranges, a complex network of faults, and what appear to be deep trenches comparable to the oceanic trenches on Earth indicate that internal and surface activity has occurred during the past (Figure 20.10c). There is, however, no evidence for active plate tectonics as on Earth. As is the case with all terrestrial planets, Venus has a core, mantle, and crust (Figure 20.10d).

Mars Mars, the Red Planet, is differentiated, as are all the terrestrial planets, into a metallic core and a silicate mantle and crust (Figure 20.11a). The thin Martian atmosphere consists of 95% carbon dioxide, 2.7% nitrogen, 1.7% argon, and traces of other gases. Mars has distinct seasons during which its polar ice caps of frozen carbon dioxide expand and recede. The two small moons orbiting Mars are probably captured asteroids (see Perspective 20.3).

Perhaps the most striking aspect of Mars is its surface features, many of which have not yet been satisfactorily explained. Like the surfaces of Mercury and the Moon, the southern hemisphere is heavily cratered, attesting to a period of meteorite bombardment. Hellas, a crater with a diameter of 2000 km, is the largest known impact structure in the solar system and is found in the Martian southern hemisphere.

The northern hemisphere is much different, having large smooth plains, fewer craters, and evidence of extensive volcanism. The largest known volcano in the solar system, Olympus Mons, has a basal diameter of 600 km, rises 27 km above the surrounding plains, and is topped by a huge circular crater 80 km in diameter. Its large size is probably accounted for by the fact that unlike Earth, Mars has no moving plates. Apparently, volcanism simply took place over a hot spot in this one area for a long period, building a gigantic volcano. In contrast, a chain of volcanoes forms on Earth as a plate moves over a hot spot, as in the case of the Hawaiian Islands (see Figure 4.22).

Because NASA has been sending spacecraft to Mars, much has been learned about the Red Planet. However, two recent failures have generated some criticism. In September 1999, the *Mars Climate Orbiter* was lost when it probably crashed into the planet, or burned

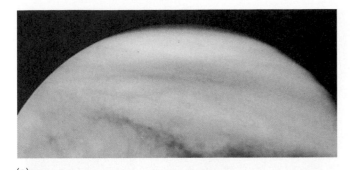

(a)

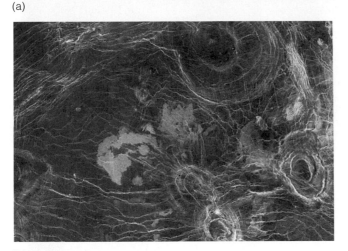

(b)

(c)

Figure 20.10

(a) Venus has a searingly hot surface and is surrounded by an oppressively thick atmosphere composed largely of carbon dioxide. (b) This radar image of Venus made by the *Magellan* spacecraft reveals circular and oval volcanic features. A complex network of cracks and fractures extends outward from the volcanic features. Geologists think these features were created by blobs of magma rising from the interior of Venus, with dikes filling some of the cracks. (c) A nearly complete map of the northern hemisphere of Venus based on radar images beamed back to Earth from the *Magellan* space probe. (d) The internal structure of Venus.

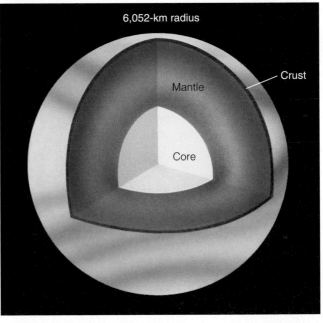

(d)

up in the atmosphere, and in December 1999, the *Mars Polar Lander* fell silent. Despite the loss of these two spacecraft (totaling nearly $300 million) and criticism from some quarters, the Mars missions still enjoy considerable support from the public.

The northern hemisphere is also marked by huge canyons that are essentially parallel to the Martian equator. One of these canyons, Valles Marineris, is at least 4000 km long, 250 km wide, and 7 km deep and is the largest yet discovered in the solar system (Figure 20.11b). If it were present on Earth, it would stretch from San Francisco to New York! It is not yet known how these vast canyons formed, although geologists postulate that they may have started as large rift zones that were subsequently modified by running water and wind erosion. Such hypotheses are based on com-

parison to rift structures found on Earth and topographic features formed by geologic agents of erosion such as water and wind.

Tremendous wind storms have strongly influenced the surface of Mars and led to dramatic dune formations (Figure 20.11c). Even more stunning than the dunes, however, are the outflow channels that appear to be the result of running water (see Figure 15.4). Mars is currently too cold for surface water to exist, yet the channels strongly indicate that there was running water on the planet during the past.

Figure 20.11

(a) The internal structure of Mars. (b) A striking view of Mars is revealed in this mosaic of 102 *Viking* images. The largest canyon known in the solar system, Valles Marineris, can be clearly seen in the center of this image, and three of the planet's volcanoes are visible on the left side of the image. (c) Large dune fields surrounding the north polar ice cap are testimony to the incessant wind action occurring on Mars.

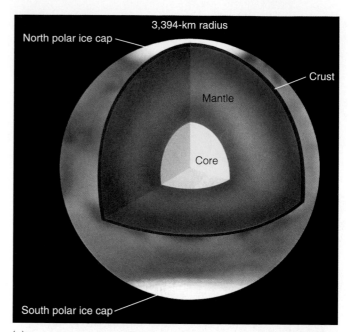

(a)

Further evidence strengthening the case for a warmer and wetter ancient Mars comes from the images and data radioed back by the *Mars Pathfinder,* which landed at Ares Vallis on Mars on July 4, 1997. This evidence includes the numerous rounded pebbles and possible conglomerates, which typically result from the action of running water, as well as abundant sand-sized particles, seen forming numerous small dunes.

The fresh-looking surfaces of its many volcanoes strongly suggest that Mars was tectonically active during the past and may still be. There is, however, no evidence that plate movement, such as occurs on Earth, has ever taken place.

WHAT WOULD YOU DO?

Considering our current knowledge of Mars, do you think it is possible for humans to colonize the planet effectively and become self-sufficient? What problems would our hypothetical colonists face, and how might these problems be solved with currently existing knowledge and technology?

(b)

The Jovian Planets

The Jovian planets are unlike any of the terrestrial planets in size and chemical composition (Table 20.2) and have had completely different evolutionary histories. Although they all apparently contain a small rocky core in relation to their overall size, the bulk of a Jovian planet is composed of volatile elements and compounds, such as hydrogen, helium, methane, and ammonia, that condense at low temperatures.

Jupiter   Jupiter is the largest of the Jovian planets (Table 20.2; Figure 20.12). With its moons, rings, and radiation belts, it is the most complex and varied planet in the solar system. Jupiter's density is only one-fourth that of Earth, but it has 318 times the mass because it is so large (Table 20.2). It is an unusual planet in that it emits almost 2.5 times more energy than it receives from the Sun. One explanation is that most of the excess energy is left over from its time of formation.

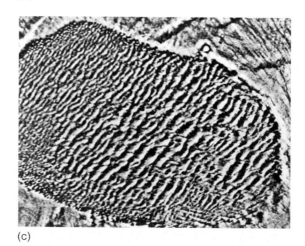

(c)

Perspective 20.3

The Moons of Mars

A persistent myth exists in which the British author Jonathan Swift (1667–1745) somehow knew that Mars had two small moons long before astronomers discovered them in 1877. Swift did in fact mention two moons close to Mars's surface in his work *Gulliver's Travels,* leading some present-day proponents of UFOs to claim that he learned this fact when he was visited by extraterrestrials. Furthermore, Swift wrote that the orbital periods of these two moons were 10 hours and 21.5 hours. But just how did these predictions come about, or were they predictions at all?

Actually, Swift was simply extrapolating on an idea, well known in his time, of the orderly progression in the number of moons possessed by the planets. Johannes Kepler (1571–1630), a famous German astronomer, noted that the closest planets to the Sun, Mercury and Venus, had no moons, Earth had 1, and Jupiter had 4 (actually it has 16, but only 4 were known then). Accordingly, Earth with 1 moon should be followed by Mars with 2, then Jupiter with 4, Saturn with 8, and so on, although the three outermost planets had not been discovered in Kepler's time.

Of course, we now know that this proposed orderly progression in the numbers of moons is incorrect. Jupiter and Saturn have too many moons, Uranus probably has nearly the right number, but Neptune and Pluto have far too few (see Table 20.2). Jonathan Swift was no doubt simply using this presumed knowledge in his reference to the moons of Mars, and just by coincidence happened to be right. But what about Swift's knowledge of the two moons' proximity to Mars's surface and their orbital periods? The fact that they had not been discovered made it apparent that if Mars had any moons at all they must be small and close to the planet. Furthermore, Swift was aware of Kepler's laws of planetary motion, holding that if small moons did exist, they must have short orbital periods. The actual orbital periods are 7.7 hours and 30 hours, not very close to Swift's estimates if he really had some kind of special knowledge.

Kepler's idea and Swift's writings about Mars's moons are nothing more than interesting footnotes in the history of astronomy and literature, but the moons themselves are fascinating objects. Both are irregularly shaped, quite small, and probably represent captured asteroids rather than bodies that formed around Mars when the solar system originated (Figure 1). Their names, Phobos and Deimos, meaning Fear and Terror, respectively, were the names of the horses that pulled the chariot of Mars, the Greek god of war. Phobos measures only 28 km in its greatest dimension, and Deimos is no more than 12 km long. Both moons possess weak gravitational fields, so weak in fact that one could throw a baseball and put it into orbit around Deimos, unless it were thrown at more than 21 km/hr, in which case it would exceed escape velocity and never be seen again.

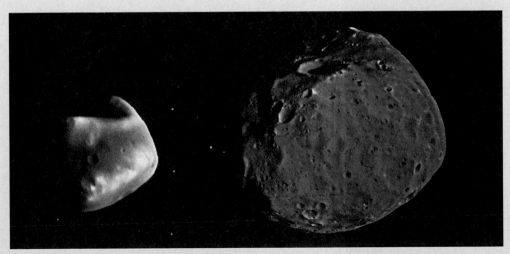

(a) (b)

Figure 1
Both of Mars's moons are small and irregularly shaped. (a) Deimos is smoother than Phobos because it has a layer of dust covering smaller features. (b) Phobos at 28 km long is the larger of the two, whereas Deimos measures only 12 km in its greatest dimension.

Figure 20.12

A composite image showing the Jovian planets and Earth to the same scale. The images of the Jovian planets were all obtained by the *Voyager* spacecraft.

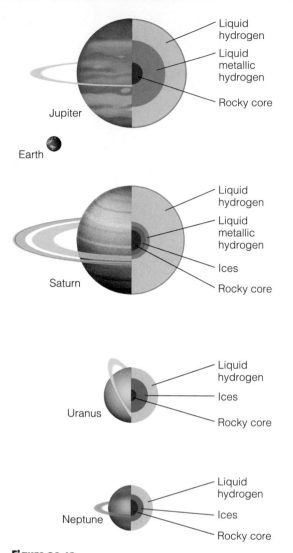

Figure 20.13

Probable internal structure of the Jovian planets. Only Jupiter and Saturn are large enough so that internal pressure is sufficient for liquid metallic hydrogen to exist. All Jovian planets have rings, but they are most obvious around Saturn.

When Jupiter formed, it heated up because of gravitational contraction (as did all the planets) and is still cooling. Jupiter's massive size insulates its interior, and hence it has cooled very slowly.

Jupiter has a relatively small central core of solid rocky material formed by differentiation. Above this core is a thick zone of liquid metallic hydrogen followed by a thicker layer of liquid hydrogen; above that is a thin layer of clouds (Figure 20.13). Surrounding Jupiter are a strong magnetic field and an intense radiation belt.

Jupiter has a dense, hot, and surprisingly dry atmosphere (according to data transmitted by *Galileo*'s atmospheric probe, which plunged about 140 km into Jupiter's atmosphere before melting in December 1995) of hydrogen, helium, methane, and ammonia, which some think are the same gases that composed Earth's first atmosphere. Jupiter's cloudy and violent atmosphere (with winds up to 650 km/hour) is divided into a series of different colored bands as well as a variety of spots (the Great Red Spot) and other features, all interacting in incredibly complex motions.

Revolving around Jupiter are 16 moons varying greatly in geologic activity; one of the most interesting moons is Europa (see the Prologue and chapter opening photo). Also surrounding Jupiter is a thin, faint ring, a feature shared by all the Jovian planets.

Saturn Saturn is slightly smaller than Jupiter, about one-third as massive and about one-half as dense, but has a similar internal structure and atmosphere (Table 20.2). Like Jupiter, Saturn gives off more energy (2.2 times as much) than it gets from the Sun. Saturn's most conspicuous feature is its ring system, consisting of thousands of rippling, spiraling bands of countless particles (Figure 20.12).

The composition of Saturn is similar to Jupiter's but consists of slightly more hydrogen and less helium.

Saturn's core is not as dense as Jupiter's, and as in the case of Jupiter, a layer of liquid metallic hydrogen overlies the core, followed by a zone of liquid hydrogen and helium, and, finally, a layer of clouds (Figure 20.13). Liquid metallic hydrogen can exist only at very high pressures, and because Saturn is smaller than Jupiter, such high pressures are found at greater depths in Saturn. Therefore, there is less of this conducting material than on Jupiter, and as a consequence, Saturn has a weaker magnetic field.

Even though the atmospheres of Saturn and Jupiter are similar, Saturn's atmosphere contains little ammonia because it is farther from the Sun and therefore colder. The cloud layer on Saturn is thicker than on Jupiter, but it lacks the contrast between the different bands. Unlike Jupiter, Saturn has seasons because its axis tilts 27 degrees.

Saturn has 18 known moons, most of which are small with low densities. Titan, with an atmosphere rich in hydrocarbons and nitrogen, and possibly with oceans consisting primarily of lightweight hydrocarbons, is also the most distinctive and perhaps the most interesting moon in the solar system.

Uranus   Uranus is much smaller than Jupiter, but their densities are about the same (Table 20.2). It is the only planet that lies on its side (Figure 20.12); that is, its axis of rotation nearly parallels the plane of the ecliptic. Some scientists think that a collision with an Earth-sized body early in its history may have knocked Uranus on its side.

Data gathered by the flyby of *Voyager 2* indicate that Uranus has a water zone beneath its cloud cover. Because the planet's density is greater than if it were composed entirely of hydrogen and helium, it is thought that Uranus must have a dense, rocky core, and this core may be surrounded by a deep global ocean of liquid water, or ices (Figure 20.13).

The atmosphere of Uranus is similar to that of Jupiter and Saturn, with hydrogen being the dominant gas, followed by helium and some methane. Uranus also has a banded atmosphere and a circulation pattern much like those of Jupiter and Saturn. Surrounding Uranus is a huge corkscrew-shaped magnetic field that stretches for millions of kilometers into space. Uranus has at least nine thin, faint rings and 17 small moons circling it.

Neptune   The flyby of *Voyager 2* in August 1989 provided the first detailed look at Neptune and revealed a dynamic, stormy planet (Figure 20.12). Its atmosphere is similar to those of the other Jovian planets, and it exhibits a pattern of zonal winds and giant storm systems comparable to those of Jupiter. Neptune's internal structure is similar to that of Uranus (Table 20.2) in that it has a rocky core surrounded by a semifrozen slush of water and liquid methane (Figure 20.13). Its atmosphere is composed of hydrogen and helium with some methane. Winds up to 2000 km/hour blow over the planet, creating tremendous storms—the largest of which, the Great Dark Spot, is in Neptune's southern hemisphere. It is nearly as big as Earth and is similar to the Great Red Spot on Jupiter. One of the mysteries raised by *Voyager 2*'s discovery is where Neptune gets the energy to drive such a storm system.

Encircling Neptune are three faint rings and eight moons, the most interesting of which is Triton, Neptune's largest moon. Triton's mottled surface of delicate pinks, reds, and blues consists primarily of water ice, with minor amounts of nitrogen and a methane frost. Geysers erupting carbon-rich material and frozen nitrogen particles some 8 km above its surface have been discovered, making it only the second place other than Earth with active volcanism.

Some areas of Triton are smooth, while others have a very irregular appearance indicating numerous episodes of deformation. Heavily cratered areas bear witness to bombardment by meteorites or the collapse of its surface. Perhaps the most intriguing aspect of Triton is that it may have once been a planet—much like Pluto, which it resembles in size and possibly composition—that was captured by Neptune's gravitational field soon after the formation of the solar system.

Pluto   With a diameter of only 2300 km, Pluto is the smallest planet and, strictly speaking, is not one of the Jovian planets (Table 20.2). Little is known about Pluto, but studies indicate it has a rocky core overlain by a mixture of frozen methane, nitrogen, and carbon monoxide (Figure 20.14). It has a thin, two-layer atmosphere with a clear upper layer overlying a more opaque lower layer. Using the Hubble space telescope to take images of Pluto, researchers have found between seven and nine light and dark patches on its icy surface. These patches, each of which is hundreds of kilometers across, include a polar ice cap bisected by a dark strip.

Pluto differs from all other planets in that its highly elliptic orbit is tilted with respect to the plane of the ecliptic. It has one known moon, Charon, that is nearly half its size, with a surface that appears to differ markedly from Pluto's.

EARTH—ITS ORIGIN AND DIFFERENTIATION

As matter was accreting in the various turbulent eddies that swirled around the early Sun, enough material eventually gathered together in one eddy to

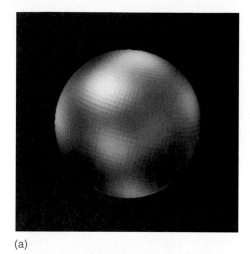

(a)

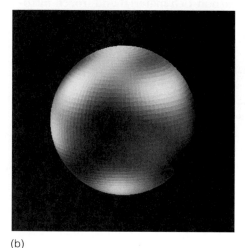

(b)

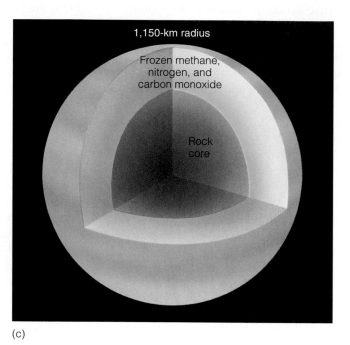

1,150-km radius

Frozen methane, nitrogen, and carbon monoxide

Rock core

(c)

Figure 20.14
(a–b) Even with the Hubbel telescope, surface features on Pluto are barely discernible in these two images. Some of these areas might be darker, older terrain and lighter impact basins. The bright areas at the poles might be frosts. (c) The probable internal structure of Pluto.

form the planet Earth. Recall from Chapter 1 that Earth consists of concentric layers of different composition and densities (Figure 20.15). This differentiation into concentric layers is a fundamental characteristic of all the terrestrial planets and presumably occurred very early in their history.

Geologists know that Earth is 4.6 billion years old. The oldest known rocks, however, are 3.96-billion-year-old metamorphic rocks from Canada. Like younger crustal rocks, these rocks are composed of relatively light silicate minerals. It appears that a crust, a heavier silicate mantle, and an iron–nickel core were already present 3.96 billion years ago, or 640 million years after Earth formed.

Early Earth was probably cool, of generally uniform composition and density throughout, and composed mostly of silicate compounds, iron and magnesium oxides, and smaller amounts of all the other chemical elements (see Figure 1.12a). For the iron and nickel to concentrate in the core, Earth must have heated up

enough for them to melt and sink through the surrounding lighter silicate minerals.

This initial heating could have occurred in three ways. Some heat was no doubt generated by the impact of meteorites; most of this heat was radiated back into space, but some was probably retained by the accreting planet. Heat was also generated within early Earth as gravitational compression reduced it to a smaller volume. Rock is a poor conductor of heat, so this heat accumulated within the evolving planet. The third cause of internal heating was the decay of radioactive elements such as uranium, thorium, and others. Even though these elements form only a very small portion of Earth, the heat generated during radioactive decay was absorbed by the surrounding rock.

The combination of meteorite impacts, gravitational compression, and heat from radioactive decay increased early Earth's temperature enough to melt iron and nickel, which, being denser than silicate minerals, settled to Earth's center and formed the core (see

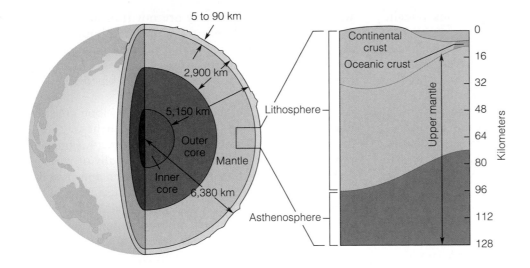

Figure 20.15
Earth's internal structure consisting of a core, mantle, and crust. The expanded view shows details of the upper mantle and crust.

Figure 1.12b). Simultaneously, the lighter silicates slowly flowed upward, beginning the differentiation of the mantle from the core.

Calculations indicate that with a uniform distribution of elements in an early solid Earth, enough heat could be generated to begin melting iron and nickel at depths between 400 and 850 km. Melting would have had to begin at shallow depths because the temperature at which melting begins increases with pressure; therefore, the melting point of any material increases toward Earth's center.

The most significant event in Earth's history was its differentiation into a layered planet, which led to the formation of a crust and eventually to continents. It also led to the outgassing of light volatile elements from the interior, which eventually resulted in the formation of the oceans and atmosphere.

THE EARTH–MOON SYSTEM: HOW DID IT ORIGINATE?

We probably know more about the Moon than any other celestial object except Earth (Figure 20.16a). Nevertheless, even though the Moon has been studied for centuries through telescopes and

(a)

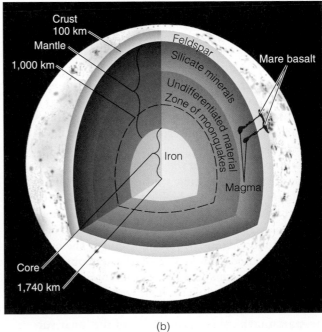

(b)

Figure 20.16
(a) High-quality image of the Moon taken through a telescope on Earth. The light-colored lunar highlands are heavily cratered. The dark-colored areas are maria, which formed when lava flowed onto the surface. (b) The Moon's internal structure. Because seismic waves are not transmitted below 1000 km, it is likely that the innermost mantle is liquid.

has been sampled directly, many questions remain unanswered.

The Moon is one-fourth the diameter of Earth, has a low density (3.3 g/cm³) relative to the terrestrial planets, and exhibits an unusual chemistry in that it is for the most part bone-dry, having been largely depleted of most volatile elements (Table 20.2). However, in 1995 a spacecraft circling the Moon in a vertical orbit discovered ice deposits in the south pole's Aitkin Basin. Although it is uncertain where the water came from, scientists think that a collision with an icy meteor is the most likely source.

The Moon orbits Earth and rotates on its own axis at the same rate, so we always see the same side. Furthermore, the Earth–Moon system is unique among the terrestrial planets. Neither Mercury nor Venus has a moon, and the two small moons of Mars—Phobos and Deimos—are probably captured asteroids (see Perspective 20.3).

The surface of the Moon can be divided into two major parts: the low-lying dark-colored plains, called *maria*, and the light-colored *highlands* (Figure 20.16). The highlands are the oldest parts of the Moon and are heavily cratered, providing striking evidence of the massive meteorite bombardment that occurred in the solar system more than 4 billion years ago.

Study of the several hundred kilograms of rocks returned by the *Apollo* missions indicates that three kinds of materials dominate the lunar surface: igneous rocks, breccias, and dust. Basalt, a common dark-colored igneous rock on Earth, is one of the several different types of igneous rocks on the Moon and makes up the greater part of the maria. The presence of igneous rocks that are essentially the same as those on Earth shows that magmas similar to those on Earth were generated on the Moon long ago.

The lunar surface is covered with a regolith estimated at 3 to 4 m thick. This gray covering, which is composed of breccia, glass spherules, and small particles of dust, is thought to be the result of debris formed by meteorite impacts.

The interior structure of the Moon is quite different from that of Earth, indicating a different evolutionary history (Figure 20.16b). The highland crust is thick (65 to 100 km) and comprises about 12% of the Moon's volume. It was formed about 4.4 billion years ago, immediately following the Moon's accretion. The highlands are composed principally of the igneous rock anorthosite, which is made up of light-colored feldspar minerals that are responsible for their white appearance.

A thin covering (1 to 2 km thick) of basaltic lava fills the maria; lava covers about 17% of the lunar surface, mostly on the side facing Earth. These maria lavas came from partial melting of a thick underlying mantle of silicate composition. Moonquakes occur at a depth of about 1000 km, but below that depth seismic shear waves (S-waves) apparently are not transmitted. Because shear waves do not travel through liquid, their lack of transmission implies that the innermost mantle may be partially molten. There is increasing evidence that the Moon has a small (500–800 km diameter) metallic core.

The origin and earliest history of the Moon are still unclear, but the basic stages in its subsequent development are well understood. It formed some 4.6 billion years ago and shortly thereafter was partially or wholly melted, yielding silicate magma that cooled and crystallized to form the rock anorthosite. Because of the low density of the anorthite crystals and the lack of water in the silicate magma, the thick anorthosite highland crust formed. The remaining silicate magma cooled and crystallized to produce the zoned mantle, while the heavier metallic elements formed the small metallic core.

The formation of the lunar mantle was completed by about 4.4 to 4.3 billion years ago. The maria basalts, derived from partial melting of the upper mantle, were extruded during great lava floods between 3.8 and 3.2 billion years ago.

Numerous models have been proposed for the origin of the Moon, including capture from an independent orbit, formation with Earth as part of an integrated two-planet system, breaking off from Earth during accretion, and formation resulting from a collision between Earth and a large planetesimal. These various models are not mutually exclusive, and elements of some occur in others. At this time, scientists cannot agree on a single model, as each has some inherent problems. However, the model that seems to account best for the Moon's particular composition and structure involves an impact by a large planetesimal with a young Earth (Figure 20.17).

In this model, a giant planetesimal, the size of Mars or larger, crashed into Earth about 4.6 to 4.4 billion years ago, causing the ejection of a large quantity of hot material that formed the Moon. The material that was ejected was mostly in the liquid and vapor phase and came primarily from the mantle of the colliding planetesimal. As it cooled, the various lunar layers crystallized out in the order we have discussed.

THE DYNAMIC EARTH

In Chapter 1, we introduced the concept of *systems* as a combination of related parts that operate in a organized fashion (see Figure 1.3), and gave examples of these interactions while discussing topics such as volcanism, plate tectonics, and running water. Our emphasis in this chapter has obviously been on the origin and evolution of the universe, solar system, and especially Earth. Once Earth formed, its various systems became opera-

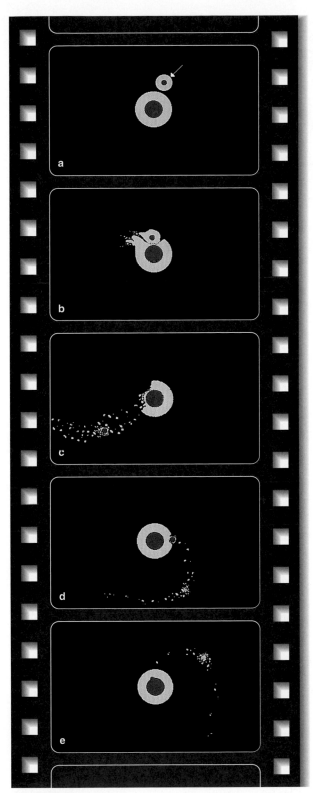

Figure 20.17
According to one hypothesis, the Moon formed from the debris blasted into space during a collision between Earth and a Mars-sized planetesimal 4.4 to 4.6 billion years ago (a–c). The planetesimal's iron-rich core fell back to Earth (d), and the Moon began forming from the debris orbiting Earth (e).

tive, but not all at the same time. For instance, Earth did not become differentiated into a core, mantle, and crust until millions of years after it initially formed. Once it did, though, internal heat was responsible for interactions among lithospheric plates as they began to diverge, converge, and slide past one another at transform plate boundaries. Indeed, the continents grew and continue to grow by accretion along convergent plate boundaries (see Chapter 13).

During its earliest history, Earth's surface was hot and dry, volcanism was ubiquitous and the atmosphere was likely composed largely of carbon dioxide. Meteorites and comets flashed through this primitive atmosphere, and because no ozone layer existed, cosmic radiation was intense. Gases derived from within Earth were released at the surface, a phenomenon known as *outgassing*, accounting for the atmosphere and accumulation of surface waters. However, not until organisms capable of photosynthesis were present in abundance did the atmosphere become more like it is today, that is, composed mostly of nitrogen and oxygen. Nevertheless, the most ancient sedimentary rocks clearly indicate that the atmosphere and hydrosphere had a profound impact on Earth's surface. Only later, but still at least 3.5 billion years ago, did organisms appear, and once they did the biosphere has had a profound effect on several of the other systems.

In marked contrast to Earth, the Moon, Mercury, and Mars show considerably less internal and surface activity. Why, then, is Earth so active? And will it eventually suffer the same fate as these other celestial bodies? Earth is of course much larger than any of the others mentioned. Its greater mass accounts for greater gravitational attraction, thus allowing it to retain an atmosphere and liquid surface water (its distance from the Sun is also important). And among the terrestrial planets and moons of the Jovian planets, it is the only body in the solar system known to support life. Furthermore, Earth remains internally active because it possesses sufficient heat for volcanism, seismic activity, and plate movements. Scientists are convinced that Earth's internally generated heat will diminish in the far distant future and these kinds of activity will eventually cease, but in the meantime Earth remains a dynamic planet (Figure 20.18).

WHAT WOULD YOU DO?

Earth is a planet with considerable internal and surface activity, whereas its Moon shows less activity, and bodies such as Mars's moon Phobos show hardly any. Why are these bodies so different in these aspects, considering that all are rocky bodies? Do you think these types of geologic activities on Earth will diminish and eventually cease? If so, why?

(a)

(b)

(c)

(d)

Figure 20.18
Images showing some aspects of Earth's interacting systems. (a) Plate divergence in the Red Sea (lower right) results in northward movement of the Arabian plate (upper right) against the Eurasian plate. (b) Although several processes account for this spectacular scenery in Yosemite National Park, California, erosion by glaciers and running water, both parts of the hydrosphere, were responsible for most of it. (c) This huge windblown sand dune near Fallon, Nevada, is 180 m high and about 4 km long. (d) Layering in these ancient rocks in Michigan resulted from the activities of cyanobacteria or blue-green algae. Not only have these organisms left their mark in the rocks, they also added oxygen to early Earth's atmosphere during photosynthesis.

Chapter Summary

1. The universe began with a Big Bang approximately 15 billion years ago. Astronomers have deduced this age from the fact that celestial objects are moving away from each other in what appears to be an ever-expanding universe.

2. The universe has a background radiation of 2.7°K above absolute zero, which is interpreted as the faint afterglow of the Big Bang.

3. About 4.6 billion years ago, the solar system formed from a rotating cloud of interstellar matter. As this cloud condensed, it eventually collapsed under the influence of gravity and flattened into a counterclockwise rotating disk. Within this rotating disk, the Sun, planets, and moons formed from the turbulent eddies of nebular gases and solids.

4. Meteorites provide vital information about the age and composition of the solar system. Stony, iron, and stony-iron meteorites are the three main groups of meteorites. Each had a different origin.

5. Temperature as a function of distance from the Sun played a key role in the types of planets that evolved. The terrestrial planets are composed of rock and metallic elements that condense at high temperatures. The Jovian planets plus Pluto are composed mostly of hydrogen, helium, ammonia, and methane, all of which condense at lower temperatures.

6. All terrestrial planets differentiated into a core, mantle, and crust, and all seem to have had a similar early history during which volcanism and cratering from meteorite impacts were common.

7. The Jovian planets differ from the terrestrial planets in size and chemical composition and followed completely different evolutionary histories. All Jovian planets have a small core compared to their overall size; they are mainly composed of volatile elements and compounds.

8. Earth formed from a swirling eddy of nebular material 4.6 billion years ago, and by at least 3.96 billion years ago, it had differentiated into its present-day structure. It accreted as a solid body and then underwent differentiation during a period of internal heating.

9. The Moon probably formed as a result of a Mars-sized planetesimal crashing into Earth 4.6 to 4.4 billion years ago and causing it to eject a large quantity of hot material. As the material cooled, the various lunar layers crystallized, forming a zoned body.

Important Terms

Big Bang
Doppler effect
iron meteorites
Jovian planet

meteor
meteorite
outgassing
planetesimal

solar nebula theory
stony meteorites
stony-iron meteorites
terrestrial planet

Review Questions

1. Which one of the following is one of the four basic forces?

   a. _____ gravity;
   b. _____ meteorite;
   c. _____ solar radiation;
   d. _____ density;
   e. _____ rotation.

2. The most common meteorites are:

   a. _____ Jovian;
   b. _____ stony;
   c. _____ planetesimal;
   d. _____ terrestrial;
   e. _____ solar nebula.

3. The most widely accepted theory for the origin of Earth's moon involves:

   a. _____ capture from an independent orbit;
   b. _____ cooling and condensation of a comet;
   c. _____ a collision between Earth and a large planetesimal;

   d. _____ accretion of asteroids between Earth and Venus;
   e. _____ eruption of a mass of gases from Jupiter.

4. Which one of the following is a terrestrial planet?

   a. _____ Jupiter;
   b. _____ Pluto;
   c. _____ Saturn;
   d. _____ Charon;
   e. _____ Mercury.

5. The composition of the universe has been changing since the Big Bang, yet 98% of it by weight still consists of the elements:

   a. _____ carbon and nitrogen;
   b. _____ silicon and potassium;
   c. _____ oxygen and carbon dioxide;
   d. _____ hydrogen and helium;
   e. _____ asteroids and meteors.

6. The major problem that plagued most early theories of the origin of the solar system involved the:

   a. _____ source of comets and asteroids;
   b. _____ fact that the most common elements in the solar system are silicon and potassium;
   c. _____ slow rotation of the Sun;
   d. _____ number of moons possessed by each planet;
   e. _____ masses of the Jovian planets.

7. Accretion, meteorite impacts, and volcanism are three features common to the early histories of:

   a. _____ the terrestrial planets;
   b. _____ stars;
   c. _____ Saturn and Neptune;
   d. _____ the Asteroid Belt;
   e. _____ comets.

8. The largest volcano in the solar system is on:
   a. _____ Earth;
   b. _____ Triton;
   c. _____ Pluto;
   d. _____ Venus;
   e. _____ Mars.

9. Venus and Earth are similar in that both:
   a. _____ have atmospheres rich in oxygen and nitrogen;
   b. _____ are about the same size and density;
   c. _____ have a large moon;
   d. _____ show evidence of active plate tectonics;
   e. _____ are composed mostly of volatile elements.

10. Which one of the following statements is correct:
   a. _____ The Big Bang occurred about 4.6 million years ago;
   b. _____ All Jovian planets have rings;
   c. _____ Jupiter's surface is the most heavily cratered in the solar system;
   d. _____ Venus's atmosphere is thick and composed of carbon dioxide;
   e. _____ The Moon's density exceeds that of Earth.

11. Both Jupiter and Saturn have a relatively small rocky core overlain by a zone of:
   a. _____ anorthosite and breccia;
   b. _____ semisolid helium and carbon;
   c. _____ gases consisting mostly of carbon dioxide;
   d. _____ rock composed of iron- and magnesium-rich minerals;
   e. _____ liquid metallic hydrogen.

12. What was the main source of heat during Earth's early history?
   a. _____ spontaneous combustion;
   b. _____ comet impacts;
   c. _____ radioactive decay;
   d. _____ volcanism;
   e. _____ the Big Bang.

13. Describe the two phenomena that indicate the Big Bang occurred.

14. Discuss the origin of the Earth–Moon system.

15. Describe the main groups of meteorites. Which is most common, and what do they tell about the origin of the solar system?

16. What are the similarities and differences in the histories of the four terrestrial planets?

17. How and why do the Jovian planets differ from the terrestrial planets?

18. Explain how elements heavier than hydrogen and helium form.

19. Why do you think that the largest volcano in the solar system is present on Mars rather than Earth?

20. How does the solar nebula theory account for the general characteristics of the solar system? Are there any aspects of the solar system that it does not explain?

21. What three sources of heat were important in the differentiation of the terrestrial planets?

22. How was the age of the universe determined?

23. What surface features on Mars indicate that it once had running water and that its surface is currently being modified by wind?

24. How and why do the atmospheres of Earth and Venus differ?

Points to Ponder

1. The Sun is thought to have been much dimmer during the early history of the solar system. Because of this, Venus probably had oceans early in its history, as did Mars. If the Sun was as bright during its early history as it is now, how might this have affected the evolution of the solar system, particularly the evolution of the terrestrial planets?

2. One of the theories proposed to account for the origin of the Moon is that it broke off from Earth during accretion. What information about Earth and the Moon would you need to support this theory? How would you gather such information?

3. It is likely that the atmospheres of Jupiter and Saturn are very similar to that of early Earth. Why have these two planets' atmospheres remained essentially the same during the past 4.6 billion years, while those of the terrestrial planets have evolved?

4. Based on what you know about how volcanic eruptions affect climatic conditions, discuss how the present global ecosystem might be affected by a collision between Earth and a meteorite about 10 km in diameter.

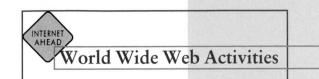
➤ WELCOME TO THE PLANETS

This site, maintained by the California Institute of Technology with U.S. government sponsorship under NASA Contract NAS7-1270, contains a collection of many of the best images of the planets from NASA's planetary exploration program. Each planet is listed, as well as the various spacecraft. In addition, it provides links to other sites around the world and a What's New section. Click on any of the *planets* for a planet profile, as well as spectacular images of the planet and specific regions of it and information about those regions.

➤ NASA HOME PAGE

This site, maintained by NASA, contains a tremendous amount of information about the organization of NASA, what it does, what some of the projects are, as well as a Questions and Answers section.

1. Click on the *Today@NASA* site. This will take you to the latest breaking news about space, as well as sites for information and images from the Hubble space telescope, the current shuttle mission, and much more.
2. Click on the *Mission to Planet Earth* site. This page is dedicated to understanding the various ways Earth is changing and how humans are influencing those changes. What are some of the different ways NASA is acquiring information about Earth systems? What are the different ways this information is used?

➤ VIEWS OF THE SOLAR SYSTEM

This site is maintained by Calvin J. Hamilton and was created as an educational tour of the solar system. It contains information (over 220 pages) and images (over 950 images and animations) about the planets, asteroids, comets, meteoroids and meteorites, and history of space exploration, as well as links to other astronomy Web sites.

1. Click on the *Meteoroids and Meteorites* site. What is the difference between a meteoroid and meteorite? What are the different types of meteorites?
2. Click on the *Mars* site. How do the information and images at this site compare to those from the "Welcome to the Planets" Web site?

➤ SKY ON-LINE

Operated by Sky Publishing, publishers of *Sky and Telescope* magazine, this site offers information on current phenomena in the sky, including comet and asteroid positions, eclipses, and interesting planetary events. There are also many links to other astronomical sites.

➤ NASA PHOTO ARCHIVE

This site is the entry point to NASA's vast photo archives. Planetary photographs and atlases, Hubble space telescope images, plus collections of many other pictures of deep-space objects are just a few of the things available.

➤ EXTRASOLAR PLANETS CATALOG

A French site with references (mostly highly technical), but it also has a table of current data on all known or suspected extrasolar planets.

Appendix A
English-Metric Conversion Chart

	ENGLISH UNIT	CONVERSION FACTOR	METRIC UNIT	CONVERSION FACTOR	ENGLISH UNIT
LENGTH	Inches (in.)	2.54	Centimeters (cm)	0.39	Inches (in.)
	Feet (ft)	0.305	Meters (m)	3.28	Feet (ft)
	Miles (mi)	1.61	Kilometers (km)	0.62	Miles (mi)
AREA	Square inches (in.^{2})	6.45	Square centimeters (cm^2)	0.16	Square inches (in.^{2})
	Square feet (ft^2)	0.093	Square meters (m^2)	10.8	Square feet (ft^2)
	Square miles (mi^2)	2.59	Square kilometers (km^2)	0.39	Square miles (mi^2)
VOLUME	Cubic inches (in.^{3})	16.4	Cubic centimeters (cm^3)	0.061	Cubic inches (in.^{3})
	Cubic feet (ft^3)	0.028	Cubic meters (m^3)	35.3	Cubic feet (ft^3)
	Cubic miles (mi^3)	4.17	Cubic kilometers (km^3)	0.24	Cubic miles (mi^3)
WEIGHT	Ounces (oz)	28.3	Grams (g)	0.035	Ounces (oz)
	Pounds (lb)	0.45	Kilograms (kg)	2.20	Pounds (lb)
	Short tons (st)	0.91	Metric tons (t)	1.10	Short tons (st)
TEMPERATURE	Degrees Fahrenheit (°F)	$-32° \times 0.56$	Degrees centigrade (Celsius)(°C)	$\times 1.80 + 32°$	Degrees Fahrenheit (°F)

Examples:

10 inches = 25.4 centimeters; 10 centimeters = 3.9 inches

100 square feet = 9.3 square meters; 100 square meters = 1080 square feet

50°F = 10.1°C; 50°C = 122°F

Appendix B
Periodic Table of the Elements

47 —— Atomic Number
Ag —— Symbol of Element
silver —— Name of Element
107.9 —— Atomic Mass Number (rounded to three significant figures)

Representative Elements

Transition Elements

Inner-Transition Elements

Noble Gases

Period	(1)* I A	(2) II A	(3) III B	(4) IV B	(5) V B	(6) VI B	(7) VII B	(8)	(9) VIII B
1	1 **H** hydrogen 1.008								
2	3 **Li** lithium 6.941	4 **Be** beryllium 9.012							
3	11 **Na** sodium 22.99	12 **Mg** magnesium 24.31							
4	19 **K** potassium 39.10	20 **Ca** calcium 40.08	21 **Sc** scandium 44.96	22 **Ti** titanium 47.90	23 **V** vanadium 50.94	24 **Cr** chromium 52.00	25 **Mn** manganese 54.94	26 **Fe** iron 55.85	27 **Co** cobalt 58.93
5	37 **Rb** rubidium 85.47	38 **Sr** strontium 87.62	39 **Y** yttrium 88.91	40 **Zr** zirconium 91.22	41 **Nb** niobium 92.91	42 **Mo** molybdenum 95.94	43 **Tc** technetium 98.91	44 **Ru** ruthenium 101.1	45 **Rh** rhodium 102.9
6	55 **Cs** cesium 132.9	56 **Ba** barium 137.3	57 **La** lanthanum 138.9	72 **Hf** hafnium 178.5	73 **Ta** tantalum 180.9	74 **W** tungsten 183.9	75 **Re** rhenium 186.2	76 **Os** osmium 190.2	77 **Ir** iridium 192.2
7	87 **Fr** francium (223)	88 **Ra** radium 226.0	89 **Ac** actinium (227)	104 **Rf** rutherfordium (261)	105 **Db** dubnium (262)	106 **Sg** seaborgium (263)	107 **Bh** bohrium (262)	108 **Hs** hassium (265)	109 **Mt** meitnerium (266)

Lanthanides

58 **Ce** cerium 140.1	59 **Pr** praseodymium 140.9	60 **Nd** neodymium 144.2	61 **Pm** promethium (147)	62 **Sm** samarium 150.4

Actinides

90 **Th** thorium 232.0	91 **Pa** protactinium 231.0	92 **U** uranium 238.0	93 **Np** neptunium 237.0	94 **Pu** plutonium (244)

( ) Indicates mass number of isotope with longest known half-life.

* Number in ( ) heading each column represents the group designation recommended by the American Chemical Society Committee on Nomenclature.

			(13) III A	(14) IV A	(15) V A	(16) VI A	(17) VII A	(18) Noble Gases
								2 **He** helium 4.003
			5 **B** boron 10.81	6 **C** carbon 12.01	7 **N** nitrogen 14.01	8 **O** oxygen 16.00	9 **F** fluorine 19.00	10 **Ne** neon 20.18
(10)	(11) I B	(12) II B	13 **Al** aluminum 26.98	14 **Si** silicon 28.09	15 **P** phosphorus 30.97	16 **S** sulfur 32.06	17 **Cl** chlorine 35.45	18 **Ar** argon 39.95
28 **Ni** nickel 58.71	29 **Cu** copper 63.55	30 **Zn** zinc 65.37	31 **Ga** gallium 69.72	32 **Ge** germanium 72.59	33 **As** arsenic 74.92	34 **Se** selenium 78.96	35 **Br** bromine 79.90	36 **Kr** krypton 83.80
46 **Pd** palladium 106.4	47 **Ag** silver 107.9	48 **Cd** cadmium 112.4	49 **In** indium 114.8	50 **Sn** tin 118.7	51 **Sb** antimony 121.8	52 **Te** tellurium 127.6	53 **I** iodine 126.9	54 **Xe** xenon 131.3
78 **Pt** platinum 195.1	79 **Au** gold 197.0	80 **Hg** mercury 200.6	81 **Tl** thallium 204.4	82 **Pb** lead 207.2	83 **Bi** bismuth 209.0	84 **Po** polonium (210)	85 **At** astatine (210)	86 **Rn** radon (222)
110 **Uun** ununnilium (269)	111 **Uuu** unununium (272)	112 **Uub** ununbium (277)	113	114 **Uuq** ununquadium (289)	115	116 **Uuh** ununhexium (289)	117	118 **Uuo** ununoctium (293)

63 **Eu** europium 152.0	64 **Gd** gadolinium 157.3	65 **Tb** terbium 158.9	66 **Dy** dysprosium 162.5	67 **Ho** holmium 164.9	68 **Er** erbium 167.3	69 **Tm** thulium 168.9	70 **Yb** ytterbium 173.0	71 **Lu** lutetium 175.0

95 **Am** americium (243)	96 **Cm** curium (247)	97 **Bk** berkelium (247)	98 **Cf** californium (251)	99 **Es** einsteinium (254)	100 **Fm** fermium (257)	101 **Md** mendelevium (258)	102 **No** nobelium (255)	103 **Lr** lawrencium (256)

Appendix C
Mineral Identification Tables

Metallic Luster

Mineral	Chemical Composition	Color	Hardness / Specific Gravity	Other Features	Comments
Chalcopyrite	$CuFeS_2$	Brassy yellow	3.5–4 / 4.1–4.3	Usually massive; greenish black streak; iridescent tarnish	Most common copper mineral. Important source of copper. Mostly in hydrothermal rocks.
Galena	PbS	Lead gray	2.5 / 7.6	Cubic crystals; 3 cleavages at right angles	The ore of lead. Mostly in hydrothermal rocks.
Graphite	C	Black	1–2 / 2.09–2.33	Greasy feel; writes on paper; 1 direction of cleavage	Used for pencil "leads" and as a dry lubricant. Mostly in metamorphic rocks.
Hematite	Fe_2O_3	Red brown	6 / 4.8–5.3	Usually granular or massive; reddish brown streak	Most important ore of iron. An accessory mineral in many rocks.
Magnetite	Fe_3O_4	Black	5.5–6.5 / 5.2	Strong magnetism	An important ore of iron. An accessory mineral in many rocks.
Pyrite	FeS_2	Brassy yellow	6.5 / 5.0	Cubic and octahedral crystals	The most common sulfide mineral. Found in some igneous and hydrothermal rocks and in sedimentary rocks associated with coal.

Nonmetallic Luster

Mineral	Chemical Composition	Color	Hardness / Specific Gravity	Other Features	Comments
Anhydrite	$CaSO_4$	White, gray	3.5 / 2.9–3.0	Crystals with 2 cleavages; usually in granular masses	Found in limestones, evaporite deposits, and the cap rocks of salt domes. Used as a soil conditioner.
Apatite	$Ca_5(PO_4)_3F$	Blue, green, brown, yellow, white	5 / 3.1–3.2	6-sided crystals; in massive or granular masses	An accessory mineral in many rocks. The main constituent of bone and dentine. A source of phosphorous for fertilizer.
Augite	$Ca(Mg,Fe,Al)(Al,Si)_2O_6$	Black, dark green	6 / 3.25–3.55	Short 8-sided crystals; 2 cleavages; cleavages nearly at right angles	The most common pyroxene mineral. Found mostly in mafic igneous rocks.
Barite	$BaSO_4$	Colorless, white, gray	3 / 4.5	Tabular crystals; high specific gravity for a nonmetallic mineral	Commonly found with ores of a variety of metals and in limestones and hot spring deposits. A source of barium.
Biotite (mica)	$K(Mg,Fe)_3AlSi_3O_{10}(OH)_2$	Black, brown	2.5 / 2.9–3.4	1 cleavage direction; cleaves into thin sheets	In felsic and mafic igneous rocks, in metamorphic rocks, and in some sedimentary rocks.

Nonmetallic Luster

Mineral	Chemical Composition	Color	Hardness / Specific Gravity	Other Features	Comments
Calcite	$CaCO_3$	Colorless, white	3 / 2.71	3 cleavages at oblique angles; cleaves into rhombs; reacts with dilute hydrochloric acid	The most common carbonate mineral. Main component of limestone and marble. Also common in hydrothermal rocks.
Cassiterite	SnO_2	Brown to black	6.5 / 7.0	High specific gravity for a nonmetallic mineral	The main ore of tin. Most is concentrated in alluvial deposits because of its high specific gravity.
Chlorite	$(Mg,Fe)_3(Si,Al)_4O_{10}$ $(Mg,Fe)_3(OH)_6$	Green	2 / 2.6–3.4	1 cleavage; occurs in scaly masses	Common in low-grade metamorphic rocks such as slate.
Corundum	Al_2O_3	Gray, blue, pink, brown	9 / 4.0	6-sided crystals and great hardness are distinctive	An accessory mineral in some igneous and metamorphic rocks. Used as a gemstone and for abrasives.
Dolomite	$CaMg(CO_3)_2$	White, yellow, gray, pink	3.5–4 / 2.85	Cleavage as in calcite; reacts with dilute hydrochloric acid when powdered	The main constituent of dolostone. Also found associated with calcite in some limestones and marble.
Fluorite	CaF_2	Colorless, purple, green, brown	4 / 3.18	4 cleavage directions; cubic and octahedral crystals	Occurs mostly in hydrothermal rocks and in some limestones and dolostones. Used in the manufacture of steel and the preparation of hydrofluoric acid.
Garnet	$Fe_3Al_2(SiO_4)_3$	Dark red	7–7.5 / 4.32	12-sided crystals common; uneven fracture	Found mostly in gneiss and schist. Used as a semi-precious gemstone and for abrasives.
Gypsum	$CaSO_4 \cdot 2H_2O$	Colorless, white	2 / 2.32	Elongate crystals; fibrous and earthy masses	The most common sulfate mineral. Found mostly in evaporite deposits. Used to manufacture plaster of Paris and cements.
Halite	$NaCl$	Colorless, white	3–4 / 2.2	3 cleavages at right angles; cleaves into cubes; cubic crystals; salty taste	Occurs in evaporite deposits. Used as a source of chlorine and in the manufacture of hydrochloric acid, many sodium compounds, and food seasoning.
Hornblende	$NaCa_2(Mg,Fe,Al)_5$ $(Si,Al)_8O_{22}(OH)_2$	Green, black	6 / 3.0–3.4	Elongate, 6-sided crystals; 2 cleavages intersecting at 56° and 124°	A common rock-forming amphibole mineral in igneous and metamorphic rocks.
Illite	$(Ca,Na,K)(Al,Fe^{+3},Fe^{+2},$ $Mg)_2(Si,Al)_4O_{10}(OH)_2$	White, light gray, buff	1–2 / 2.6–2.9	Earthy masses; particles too small to observe properties	A clay mineral common in soils and clay-rich sedimentary rocks.
Kaolinite	$Al_2Si_4O_{10}(OH)_8$	White	2 / 2.6	Massive; earthy odor; particles too small to observe properties.	A common clay mineral formed by chemical weathering of aluminum-rich silicates. The main ingredient of kaolin clay used for the manufacture of ceramics.

Nonmetallic Luster

Mineral	Chemical Composition	Color	Hardness / Specific Gravity	Other Features	Comments
Muscovite (Mica)	$KAl_2Si_3O_{10}(OH)_2$	Colorless	2–2.5 / 2.7–2.9	1 direction of cleavage; cleaves into thin sheets	Common in felsic igneous rocks, metamorphic rocks, and some sedimentary rocks. Used as an insulator in electrical appliances.
Olivine	$(Fe,Mg)_2SiO_4$	Olive green	6.5 / 3.3–3.6	Small mineral grains in granular masses; conchoidal fracture	Common in mafic igneous rocks.
Plagioclase feldspars	Varies from $CaAl_2Si_2O_8$ to $NaAlSi_3O_8$	White, gray, brown	6 / 2.56	2 cleavages at right angles	Common in igneous rocks and a variety of metamorphic rocks. Also in some arkoses.
Potassium feldspar — Microcline	$KAlSi_3O_8$	White, pink, green	6 / 2.56	2 cleavages at right angles	Common in felsic igneous rocks, some metamorphic rocks, and arkoses. Used in the manufacture of porcelain.
Potassium feldspar — Orthoclase	$KAlSi_3O_8$	White, pink			
Quartz	SiO_2	Colorless, white, gray, pink, green	7 / 2.67	6-sided crystals; no cleavage; conchoidal fracture	A common rock-forming mineral in all rock groups and hydrothermal rocks. Also occurs in varieties known as chert, flint, agate, and chalcedony.
Siderite	$FeCO_3$	Yellow, brown	4 / 3.8–4.0	3 cleavages at oblique angles; cleaves into rhombs	Found mostly in concretions and sedimentary rocks associated with coal.
Smectite	$(Al,Mg)_8(Si_4O_{10})_3(OH)_{10} \cdot 12H_2O$	Gray, buff, white	1–1.5 / 2.5	Earthy masses; particles too small to observe properties	A clay mineral with the unique property of swelling and contracting as it absorbs and releases water.
Sphalerite	ZnS	Yellow, brown, black	3.5–4 / 4.0–4.1	6 cleavages; cleaves into dodecahedra	The most important ore of zinc. Commonly found with galena in hydrothermal rocks.
Talc	$Mg_3Si_4O_{10}(OH)_2$	White, green	1 / 2.82	1 cleavage direction; usually in compact masses	Formed by the alteration of magnesium silicates. Mostly in metamorphic rocks. Used in ceramics, cosmetics, and as a filler in paints.
Topaz	$Al_2SiO_4(OH,F)$	Colorless, white, yellow, blue	8 / 3.5–3.6	High specific gravity; 1 cleavage direction	Found in pegmatites, granites, and hydrothermal rocks. An important gemstone.
Zircon	Zr_2SiO_4	Brown, gray	7.5 / 3.9–4.7	4-sided, elongate crystals	Most common as an accessory in granitic rocks. An ore of zirconium and used as a gemstone.

Appendix D
Topographic Maps

Nearly everyone has used a map of one kind or another and is probably aware that a map is a scaled-down version of the area depicted. For a map to be of any use, however, one must understand what is shown on a map and how to read it. A particularly useful type of map for geologists, and people in many other professions, is a *topographic map,* which shows the three-dimensional configuration of Earth's surface on a two-dimensional sheet of paper.

Maps showing relief—differences in elevation in adjacent areas—are actually models of Earth's surface. Such maps are available for some areas, but they are expensive, difficult to carry, and impossible to record data on. Thus, paper sheets showing relief by using lines of equal elevation known as *contours* are most commonly used. Topographic maps depict (1) relief, which includes hills, mountains, valleys, canyons, and plains; (2) bodies of water such as rivers, lakes, and swamps; (3) natural features such as forests, grasslands, and glaciers; and (4) various cultural features including communities, highways, railroads, land boundaries, canals, and power transmission lines.

Topographic maps known as *quadrangles* are published by the U.S. Geological Survey (USGS). The area depicted on a topographic map can be identified by referring to the map's name in the upper right and lower right corners, which is usually derived from some prominent geographic feature (Lincoln Creek Quadrangle, Idaho) or community (Mt. Pleasant Quadrangle, Michigan). In addition, most maps have a state outline map along the bottom margin, and shown within the outline is a small black rectangle indicating the part of the state represented by the map.

CONTOURS

Contour lines, or simply contours, are lines of equal elevation used to show topography. Think of contours as the lines formed where imaginary horizontal planes intersect Earth's surface at specific elevations. On maps, contours are brown, and every fifth contour, called an *index contour,* is darker than adjacent ones and labeled with its elevation (Figure D1). Elevations on most USGS topographic maps are in feet, although a few use meters; in either case, the specified elevation is above or below mean sea level. Because contours are defined as lines of equal elevation, they cannot divide or cross one another, although they will converge and appear to join in areas with vertical or overhanging cliffs. Notice in Figure D1 that where contours cross a stream they form a V that points upstream toward higher elevations.

The vertical distance between contours is the *contour interval.* If an area has considerable relief, a large contour interval is used, perhaps 80 or 100 feet, whereas a small interval such as 5, 10, or 20 feet is used in areas with little relief. The values recorded on index contours are always multiples of the map's contour interval, shown at the bottom of the map. For instance, if a map has a contour interval of 10 feet, index contour values such as 3600, 3650, and 3700 feet might be shown (Figure D1). In addition to contours, specific elevations are shown at some places on maps and may be indicated by a small ×, next to which is a number. A specific elevation might also be shown adjacent to the designation *BM* (benchmark), a place where the elevation and location are precisely known.

Contour spacing depends on slope, so in areas with steep slopes, contours are closely spaced because there is a considerable increase in elevation in a short distance. In contrast, if slopes are gentle, contours are widely spaced (Figure D1). Furthermore, if contour spacing is uniform the slope angle remains constant, but if spacing changes slope angle changes. However, one must be careful in comparing slopes on maps with different contour intervals or different scales.

Topographic features such as hills, valleys, plains, and so on can easily be shown by contours. For instance, a hill is shown by a concentric pattern of contours with the highest elevation in the central part of the pattern. All contours must close on themselves, but they may do so beyond the confines of a particular map. A concentric contour pattern also might show a closed depression, but in this case special contours with short bars perpendicular to the contour pointing toward the central part of the depression are used (Figure D1).

MAP SCALES

All maps are scaled-down versions of the areas shown, so to be of any use they must have a scale.

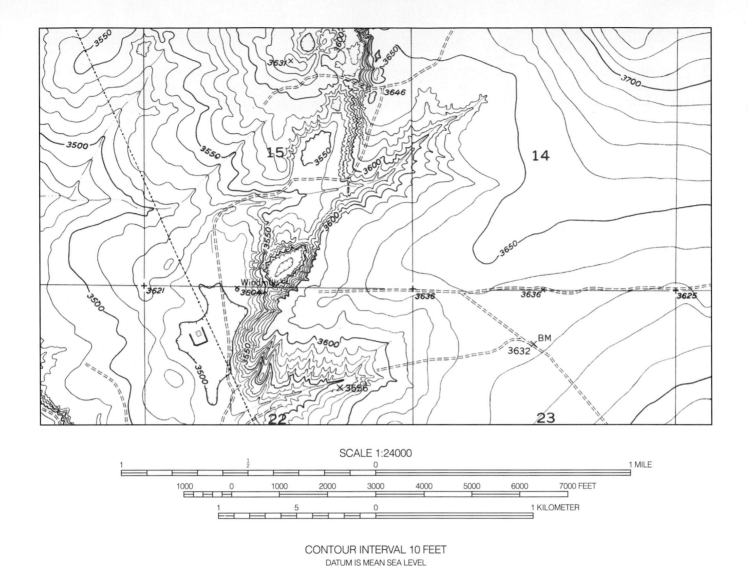

SCALE 1:24000

1   ½   0   1 MILE

1000   0   1000   2000   3000   4000   5000   6000   7000 FEET

1   5   0   1 KILOMETER

CONTOUR INTERVAL 10 FEET
DATUM IS MEAN SEA LEVEL

Figure D1

Part of the Bottomless Lakes Quadrangle, New Mexico, which has a contour interval of 10 feet; every fifth contour is darker and labeled with its elevation. Notice that contours are widely spaced where slopes are gentle and more closely spaced where they are steeper, as in the central part of the map. Hills are shown by contours that close on themselves, whereas depressions are indicated by contours with hachure marks pointing toward the center of the depression. The dashed blue lines on the map represent intermittent streams; notice that where contours cross a stream's channel they form a V that points upstream.

Highway maps, for example, commonly have a scale such as "1 inch equals 10 miles," by which one can readily determine distances. Two types of scales are used on topographic maps. The first and most easily understood is a graphic scale, which is simply a bar subdivided into appropriate units of length (Figure D1). This scale appears at the bottom center of the map and may show miles, feet, kilometers, or meters. Indeed, graphic scales on USGS topographic maps generally show both English and metric distance units.

A ratio or fractional scale, which represents the degree of reduction of the area depicted, appears above the graphic scale. On a map with a ratio scale of 1:24,000, for instance, the area shown is 1/24,000th the size of the actual land area (Figure D1). Another way to express this

relationship is to say that any unit of length on the map equals 24,000 of the same units on the ground. Thus, 1 inch on the map equals 24,000 inches on the ground, which is more meaningful if one converts inches to feet, making 1 inch equal to 2000 feet. A few maps have scales of 1:63,360, which converts to 1 inch equals 5280 feet, or 1 inch equals 1 mile.

USGS topographic maps are published in a variety of scales such as 1:50,000, 1:62,500, 1:125,000, and 1:250,000. One should also realize that large-scale maps cover less area than small-scale maps, and the former show much more detail than the latter. For example, a large-scale map (1:24,000) shows more surface features in greater detail than does a small-scale map (1:125,000) for the same area.

MAP LOCATIONS

Location on topographic maps can be determined in two ways. First, the borders of maps correspond to lines of latitude and longitude. Latitude is measured north and south of the equator in degrees, minutes, and seconds, whereas the same units are used to designate longitude east and west of the prime meridian, which passes through Greenwich, England. Maps depicting all areas within the United States are noted in north latitude and west longitude. Latitude and longitude are noted in degrees and minutes at the corners of maps, but usually only minutes and seconds are shown along the margins. Many USGS topographic maps cover 7½ or 15 minutes of latitude and longitude and are thus referred to as 7½- and 15-minute quadrangles.

Beginning in 1812, the General Land Office (now known as the Bureau of Land Management) developed a standardized method for accurately defining the location of property in the United States. This method, known as the General Land Office Grid System, has been used for all states except those along the eastern seaboard (except Florida), parts of Ohio, Tennessee, Kentucky, West Virginia, and Texas.

As new land acquired by the United States was surveyed, the surveyors laid out north–south lines they called *principal meridians* and east–west lines known as *base lines*. These intersecting lines form a set of coordinates for locating specific pieces of property. The basic unit in the General Land Office Grid System is the *township,* an area measuring 6 miles on a side, and thus covering 36 square miles (Figure D2). Townships are numbered north and south of base lines and are designated as T.1N., T.1S., and so on. Rows of townships known as *ranges* are numbered east and west of principal meridians; R.2W and R.4E, for example. Note in Figure D2 that each township has a unique designation of township and range numbers.

Townships are subdivided into 36 1-square-mile (640-acre) *sections* numbered from 1 to 36. Because of surveying errors and the adjustments necessary to make a grid system conform to Earth's curved surface, not all sections are exactly 1 mile square. Nevertheless, each section can be further subdivided into half sections and

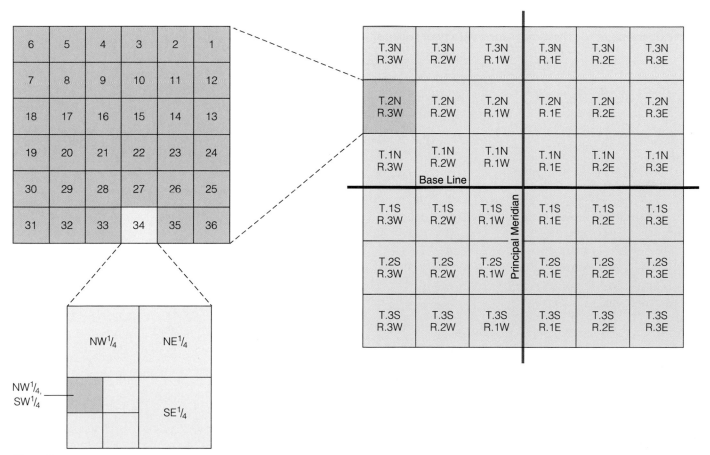

Figure D2
The General Land Office Grid System. Each 36-square-mile township is designated by township and range numbers. Townships are subdivided into sections, which can be further subdivided into quarter sections and quarter-quarter sections.

quarter sections designated NE¼, NW¼, SE¼, and SW¼, and each quarter section can be further divided into quarter-quarter sections. To show the complete designation for an area, the smallest unit is noted first (quarter-quarter section) followed by quarter section, section number, township, and range. For example, the area shown by diagonal lines in Figure D2 is the NW¼, SW¼, Sec. 34, T.2N., R.3W.

Because only a few principal meridians and base lines were established, they do not appear on most topographic maps. Nevertheless, township and range numbers are printed along the margins of 7½- and 15-minute quadrangles, and a grid consisting of red land boundaries depicts sections. In addition, each section number is shown in red within the map. However, small-scale maps show only township and range.

WHERE TO OBTAIN TOPOGRAPHIC MAPS

Many people find topographic maps useful. Land use planners, personnel in various local, state, and federal agencies, as well as engineers and real estate developers might use these maps for a variety of reasons.

In addition, hikers, backpackers, and others interested in exploring undeveloped areas commonly use topographic maps because trails are shown by black dashed lines. Furthermore, map users can readily determine their location by interpreting the topographic features depicted by contours, and they can anticipate the type of terrain they will encounter during off-road excursions.

Topographic maps for local areas are available at some sporting goods stores, at National Park Visitor Centers, and from some state geologic surveys. Free index maps showing the names and locations of all quadrangles for each state are available from the USGS to anyone uncertain of which specific map is needed. Any published topographic map can be purchased from two main sources. For maps of areas east of the Mississippi River, write to

Branch of Distribution
U.S. Geological Survey
1200 S. Eads Street
Arlington, Virginia 22202

Maps for areas west of the Mississippi River can be obtained from

Branch of Distribution
U.S. Geological Survey
Box 25286 Federal Center
Denver, Colorado 80225

Answers
Multiple-Choice Review Questions

Chapter 1
1. d; 2. b; 3. c; 4. a; 5. e; 6. b; 7. a; 8. c; 9. b; 10. d; 11. b

Chapter 2
1. d; 2. a; 3. b; 4. c; 5. e; 6. c; 7. a; 8. c; 9. d; 10. a; 11. c; 12. e

Chapter 3
1. b; 2. b; 3. e; 4. c; 5. d; 6. a; 7. a; 8. e; 9. b; 10. b; 11. a; 12. c

Chapter 4
1. e; 2. c; 3. b; 4. e; 5. a; 6. c; 7. b; 8. c; 9. e; 10. d; 11. b; 12. a

Chapter 5
1. a; 2. d; 3. c; 4. b; 5. a; 6. c; 7. e; 8. c; 9. a; 10. a; 11. e; 12. a

Chapter 6
1. b; 2. c; 3. e; 4. a; 5. d; 6. a; 7. c; 8. c; 9. c; 10. b; 11. d; 12. e

Chapter 7
1. c; 2. b; 3. d; 4. c; 5. a; 6. c; 7. d; 8. b; 9. b; 10. c; 11. b; 12. c; 13. c; 14. b; 15. b

Chapter 8
1. b; 2. b; 3. b; 4. d; 5. b; 6. b; 7. e; 8. a; 9. c; 10. b

Chapter 9
1. b; 2. d; 3. e; 4. d; 5. c; 6. c; 7. a; 8. d; 9. e; 10. e; 11. d; 12. c

Chapter 10
1. c; 2. b; 3. e; 4. a; 5. a; 6. b; 7. c; 8. a; 9. c; 10. d; 11. e; 12. c

Chapter 11
1. b; 2. e; 3. a; 4. b; 5. c; 6. d; 7. a; 8. b; 9. c; 10. b; 11. a; 12. d

Chapter 12
1. a; 2. c; 3. e; 4. d; 5. b; 6. b; 7. c; 8. d; 9. a; 10. d; 11. a; 12. a

Chapter 13
1. a; 2. e; 3. d; 4. c; 5. a; 6. b; 7. e; 8. c; 9. a; 10. d; 11. b; 12. e

Chapter 14
1. e; 2. e; 3. d; 4. a; 5. b; 6. c; 7. b; 8. a; 9. e; 10. b; 11. e; 12. d

Chapter 15
1. b; 2. e; 3. c; 4. a; 5. b; 6. d; 7. e; 8. c; 9. d; 10. a; 11. e; 12. d

Chapter 16
1. b; 2. d; 3. a; 4. e; 5. a; 6. c; 7. d; 8. a; 9. a; 10. c; 11. b; 12. c

Chapter 17
1. a; 2. e; 3. b; 4. c; 5. d; 6. b; 7. e; 8. b; 9. c; 10. c; 11. d; 12. d

Chapter 18
1. d; 2. d; 3. d; 4. e; 5. c; 6. a; 7. c; 8. a; 9. e; 10. c; 11. b; 12. c

Chapter 19
1. b; 2. d; 3. c; 4. a; 5. b; 6. e; 7. a; 8. c; 9. a; 10. d; 11. b; 12. a

Chapter 20
1. a; 2. b; 3. c; 4. e; 5. d; 6. c; 7. a; 8. e; 9. b; 10. d; 11. e; 12. c

Glossary

A

aa A lava flow with a surface of rough, angular blocks and fragments.

abrasion The process whereby exposed rock is worn and scraped by the impact of solid particles.

absolute dating The process of assigning ages in years before the present to geologic events. Various radioactive decay-dating techniques yield absolute ages. See also *relative dating*.

abyssal plain Vast flat area on the seafloor adjacent to the continental rises of passive continental margins.

active continental margin A continental margin characterized by volcanism and seismicity at the leading edge of a continental plate where oceanic lithosphere is subducted. See also *passive continental margin*.

aftershock An earthquake caused by adjustments along a fault following a larger earthquake. Large earthquakes are usually followed by numerous aftershocks, but most are smaller than the main shock.

alluvial fan A cone-shaped alluvial deposit formed where a stream flows from mountains onto an adjacent lowland.

alluvium A general term for all detrital material transported and deposited by streams.

alpha decay A type of radioactive decay involving the emission of a particle consisting of two protons and two neutrons from the nucleus of an atom; decreases the atomic number by 2 and the atomic mass number by 4.

angular unconformity An unconformity below which older strata dip at a different angle (usually steeper) than the overlying strata. See also *disconformity* and *nonconformity*.

anticline An up-arched fold in which the oldest exposed rocks coincide with the fold axis and all strata dip away from the axis.

aphanitic texture An igneous texture in which individual mineral grains are too small to be seen without magnification; results from rapid cooling and generally indicates an extrusive origin.

aquiclude Any material that prevents the movement of groundwater.

aquifer A permeable layer that allows the movement of groundwater.

arête A narrow, serrated ridge separating two glacial valleys or adjacent cirques.

artesian system A confined groundwater system in which high hydrostatic (fluid) pressure builds up causing water to rise above the level of the aquifer.

aseismic ridge A ridge or broad, plateaulike feature rising as much as 2 to 3 km above the surrounding seafloor and lacking seismic activity.

ash Pyroclastic material measuring less than 2 mm.

assimilation A process in which magma changes composition as it reacts with country rock with which it comes in contact.

asthenosphere The part of the mantle that lies below the lithosphere; behaves plastically and flows.

atom The smallest unit of matter that retains the characteristics of an element.

atomic mass number The total number of protons and neutrons in the nucleus of an atom.

atomic number The number of protons in the nucleus of an atom.

aureole A zone surrounding a pluton in which contact metamorphism has taken place.

B

barchan dune A crescent-shaped dune with the tips of the crescent pointing downwind.

barrier island A long, narrow island composed of sand oriented parallel to a shoreline but separated from the mainland by a lagoon.

basal slip A type of glacial movement in which a glacier slides over its underlying surface.

basalt plateau A large area built up by numerous flat-lying lava flows erupted from fissures.

base level The lowest limit to which a stream can erode.

basin The circular equivalent of a syncline. All strata in a basin dip toward a central point, and the youngest exposed rocks are in the center.

batholith A discordant, irregularly shaped pluton composed chiefly of granitic rocks. Has a surface area of at least 100 km².

baymouth bar A spit that has grown until it cuts off a bay from the open sea.

beach A deposit of unconsolidated sediment extending landward from low tide to a change in topography or where permanent vegetation begins.

bed (bedding) A bed is an individual layer of rock, especially sedimentary rock, whereas bedding is the layered arrangement of rocks. See also *stratification*.

bed load The part of a stream's sediment load transported along its bed; consists of sand and gravel.

berm The backshore area of a beach, consisting of a platform composed of sediment deposited by waves. Berms are nearly horizontal or slope gently landward.

beta decay   A type of radioactive decay during which a fast-moving electron emitted from a neutron is converted to a proton; results in an increase of one atomic number but does not change atomic mass number.

Big Bang   A model for the evolution of the universe in which a dense, hot state was followed by expansion, cooling, and a less dense state.

biochemical sedimentary rock   A sedimentary rock resulting from the chemical processes of organisms.

body wave   An earthquake wave that travels through Earth. Both P- and S-waves are body waves.

bonding   The process whereby atoms are joined to other atoms.

Bowen's reaction series   A mechanism accounting for the derivation of intermediate and felsic magmas from a mafic magma. It has a discontinuous branch of ferromagnesian minerals that change from one to another over specific temperature ranges and a continuous branch of plagioclase feldspars whose composition changes as the temperature decreases.

braided stream   A stream possessing an intricate network of dividing and rejoining channels. Braiding occurs when sand and gravel bars are deposited within channels.

breaker   A wave that steepens as it enters shallow water until its crest plunges forward.

butte   An isolated, steep-sided, pinnacle-like erosional feature formed by the breaching of a resistant cap rock, which allows rapid erosion of the less resistant underlying rocks.

C

caldera   A large, steep-sided circular to oval volcanic depression usually formed by summit collapse resulting from partly draining the underlying magma chamber.

carbon 14 dating technique   An absolute dating method that relies on determining the ratio of C^{14} to C^{12} in a sample; useful back to about 70,000 years ago; can be applied only to organic substances.

carbonate mineral   A mineral containing the negatively charged carbonate ion $(CO_3)^{-2}$, e.g., calcite $(CaCO_3)$ and dolomite $[CaMg(CO_3)_2]$.

carbonate rock   A rock containing mostly carbonate minerals (such as limestone and dolostone).

cave   A naturally formed subsurface opening that is generally connected to the surface and is large enough for a person to enter.

cementation   The precipitation of minerals as binding material in and around sediment grains, thus converting sediment to sedimentary rock.

chemical sedimentary rock   Rock formed of minerals derived from materials dissolved during chemical weathering.

chemical weathering   The decomposition of rocks by chemical alteration of parent material.

cinder cone   A small steep-sided volcano composed of pyroclastic materials that accumulate around a vent.

circum-Pacific belt   A zone of seismic and volcanic activity that nearly encircles the margins of the Pacific Ocean basin.

cirque   A steep-walled, bowl-shaped depression formed at the upper end of a glacial valley.

cleavage   The breaking or splitting of mineral crystals along planes of weakness; cleavage is determined by the strength of the bonds within minerals.

columnar joint   Joints in some igneous rocks consisting of six-sided columns that form as a result of shrinkage during cooling.

compaction   A method of lithification whereby the pressure exerted by the weight of overlying sediment reduces the amount of pore space and thus the volume of a deposit.

complex movement   A combination of different types of mass movements in which one type is not dominant; most complex movements involve sliding and flowing.

composite volcano   A volcano composed of pyroclastic layers, lava flows typically of intermediate composition, and mudflows; also called *stratovolcanoes*.

compound   A substance resulting from the bonding of two or more different elements, such as water (H_2O) and quartz (SiO_2).

compression   Stress resulting when rocks are squeezed by external forces directed toward one another.

concordant pluton   Pluton whose boundaries are parallel to the layering in the country rock. See also *discordant pluton*.

cone of depression   A cone-shaped depression in the water table around a well, resulting from pumping water from an aquifer faster than it can be replenished.

contact metamorphism   Metamorphism in which a magma body alters the surrounding country rock.

continental–continental plate boundary   A convergent plate boundary along which two continental lithospheric plates collide, such as the collision of India with Asia.

continental crust   The rocks of continents overlying the upper mantle and consisting of a wide variety of igneous, sedimentary, and metamorphic rocks. It has an overall granitic composition and an average density of about 2.70 g/cm^3.

continental drift   The theory that the continents were once joined into a single landmass that broke apart with the various fragments (continents) moving with respect to one another.

continental glacier   A glacier covering a vast area (at least $50,000 \text{ km}^2$) and unconfined by topography. Also called an *ice sheet*.

continental margin   The area separating the part of a continent above sea level from the deep seafloor.

continental rise   The gently sloping area of the seafloor beyond the base of the continental slope.

continental shelf   The area between the shoreline and the continental slope where the seafloor slopes gently seaward.

continental slope   The relatively steep area between the shelf–slope break (at an average depth of 135 m) and the more gently sloping continental rise or an oceanic trench.

convergent plate boundary   The boundary between two plates that are moving toward one another; three types of convergent plate boundaries are recognized. See also *continental–continental plate boundary, oceanic–continental plate boundary,* and *oceanic–oceanic plate boundary.*

core The interior part of Earth, beginning at a depth of about 2900 km; probably composed mostly of iron and nickel; divided into an outer liquid core and an inner solid core.

Coriolis effect The deflection of winds to the right of their direction of motion (clockwise) in the Northern Hemisphere and to the left of their direction of motion (counterclockwise) in the Southern Hemisphere, due to Earth's rotation.

correlation The demonstration of time equivalency of rock units in different areas.

country rock Any rock that is invaded by and surrounds a pluton.

covalent bond A bond formed by the sharing of electrons between atoms.

crater A circular or oval depression at the summit of a volcano resulting from the extrusion of gases, pyroclastic materials, and lava; connected by a conduit to a magma chamber below Earth's surface.

craton The relatively stable part of a continent; consists of a shield and a platform, a buried extension of a shield; the ancient nucleus of a continent.

creep A type of mass wasting in which soil or rock moves slowly downslope.

crest The highest part of a wave.

cross-bedding Layers in sedimentary rocks deposited at an angle to the surface on which they accumulated.

crust Earth's outermost layer; the upper part of the lithosphere, which is separated from the mantle by the Moho; divided into continental and oceanic crust.

crystal settling The physical separation and concentration of minerals in the lower part of a magma chamber or pluton by crystallization and gravitational settling.

crystalline solid A solid in which the constituent atoms are arranged in a regular, three-dimensional framework.

Curie point The temperature at which iron-bearing minerals in a cooling magma attain their magnetism.

D

debris flow A mass-wasting process involving flowage of a viscous mixture of water, soil, and rocks; much like a mudflow, but at least half the particles are larger than sand.

deflation The removal of loose surface sediment by the wind.

deformation Any change in shape or volume, or both, of rocks in response to stress; deformation involves folding and fracturing.

delta An alluvial deposit formed where a stream flows into a lake or the sea.

depositional environment Any area where sediment is deposited, such as a stream's floodplain or on a beach.

desert Any area that receives less than 25 cm of rain per year and has a high evaporation rate.

desertification The expansion of deserts into formerly productive lands.

desert pavement A surface mosaic of close-fitting pebbles, cobbles, and boulders found in many dry regions; formed by the removal of sand-sized and smaller particles by wind.

detrital sedimentary rock Rock consisting of the solid particles (detritus) of preexisting rocks (such as sandstone and conglomerate).

differential pressure Pressure that is not applied equally to all sides of a rock body.

differential weathering Weathering of rock at different rates, producing an uneven surface.

dike A tabular or sheetlike discordant pluton.

dilatancy model A model used to predict earthquakes based on changes occurring in rocks subjected to very high pressure.

dip A measure of the maximum angular deviation of an inclined plane from horizontal.

dip-slip fault A fault on which all movement is parallel with the dip of the fault plane. See also *normal fault* and *reverse fault.*

discharge The volume of water in a stream moving past a particular point in a given period of time.

disconformity An unconformity above and below which the strata are parallel. See also *angular unconformity* and *nonconformity.*

discontinuity A boundary across which seismic wave velocity or direction changes abruptly, such as the mantle–core boundary.

discordant pluton Pluton whose boundaries cut across the layering in the country rock. See also *concordant pluton.*

dissolved load The part of a stream's load consisting of ions in solution.

divergent plate boundary The boundary between two plates that are moving apart.

divide A topographically high area that separates adjacent drainage basins.

dome A circular equivalent of an anticline; all strata in a dome dip away from a central point, and the oldest exposed rocks are at the dome's center.

Doppler effect The apparent change in frequency of a wave resulting from motion of the source of the wave, the receiver, or both.

drainage basin The surface area drained by a stream and its tributaries.

drainage pattern The regional arrangement of channels in a drainage system.

dripstone Various cave deposits resulting from the deposition of calcite.

drumlin An elongate hill of till formed by the movement of a continental glacier or floods.

dune A mound or ridge of wind-deposited sand.

dynamic metamorphism Metamorphism occurring in fault zones where rocks are subjected to high differential pressure.

E

earthflow A mass-wasting process involving downslope flow of water-saturated soil.

earthquake Vibrations caused by the sudden release of energy, usually as a result of the displacement of rocks along faults.

elastic rebound theory A theory that explains how energy is suddenly released during earthquakes: When rocks are deformed, they store energy and bend; when the inherent strength of the rocks is exceeded, they rupture and release energy, causing earthquakes.

elastic strain A type of deformation in which the material returns to its original shape when stress is relaxed.

electron A negatively charged particle of very little mass that encircles the nucleus of an atom.

electron capture A type of radioactive decay in which an electron is captured by a proton and converted to a neutron; results in a loss of one atomic number but no change in atomic mass number.

electron shell Electrons orbit rapidly around the nuclei of atoms at specific distances known as *electron shells*.

element A substance composed of all the same atoms; it cannot be changed into another element by ordinary chemical means.

emergent coast A coast where the land has risen with respect to sea level.

end moraine A pile of rubble deposited at the terminus of a glacier. See also *recessional moraine* and *terminal moraine*.

epicenter The point on Earth's surface vertically above the focus of an earthquake.

erosion The removal of weathered materials from their source area.

esker A long, sinuous ridge of stratified drift formed by deposition by running water in tunnels beneath stagnant ice.

evaporite A sedimentary rock that formed by inorganic chemical precipitation of minerals from solution (such as rock salt and rock gypsum).

Exclusive Economic Zone An area extending 371 km seaward from the coast of the United States and its territories in which the United States claims all sovereign rights.

exfoliation The process whereby slabs of rock bounded by sheet joints slip or slide off the host rock.

exfoliation dome A large rounded dome of rock resulting from the process of exfoliation.

expansive soil A soil in which the volume increases when water is present.

F

fault A fracture along which movement has occurred parallel to the fracture surface.

fault plane A fracture surface along which blocks of rock on opposite sides have moved relative to one another.

felsic magma A type of magma containing more than 65% silica and considerable sodium, potassium, and aluminum but little calcium, iron, and magnesium. See also *intermediate magma* and *mafic magma*.

ferromagnesian silicate A silicate mineral containing iron or magnesium or both.

fetch The distance the wind blows over a continuous water surface.

fiord An arm of the sea extending into a U-shaped glacial trough eroded below sea level.

firn Granular snow formed by partial melting and refreezing of snow.

fission track dating The process of dating samples by counting the number of small linear tracks (fission tracks) that result when a mineral crystal is damaged by rapidly moving alpha particles generated by radioactive decay of uranium.

fissure eruption An eruption in which lava or pyroclastic material is emitted from a long, narrow fissure or group of fissures.

floodplain A low-lying, relatively flat area adjacent to a stream, which is partly or completely covered with water when the stream overflows its banks.

fluid activity An agent of metamorphism in which water and carbon dioxide promote metamorphism by increasing the rate of chemical reactions.

focus The place within Earth where an earthquake originates and energy is released.

foliated texture A texture of metamorphic rocks in which platy and elongate minerals are arranged in a parallel fashion.

footwall block The block of rock that lies beneath a fault plane.

fossil Remains or traces of prehistoric organisms preserved in rocks of the crust.

fracture A break in a rock resulting from intense applied pressure.

frost action The disaggregation of rocks by repeated freezing and thawing of water in cracks and crevices.

frost heaving The process whereby a mass of sediment or soil undergoes freezing, expansion, and actual lifting, followed by thawing, contraction, and lowering of the mass.

frost wedging The opening and widening of cracks by the repeated freezing and thawing of water.

G

geologic time scale A chart with the designation for the earliest interval of geologic time at the bottom, followed upward by designations for progressively more recent time intervals.

geology The science concerned with the study of Earth; includes studies of Earth materials (minerals and rocks), surface and internal processes, and Earth history.

geothermal energy Energy that comes from the steam and hot water trapped within the crust.

geothermal gradient The temperature increase with depth; it averages 25°C/km near the surface but varies from area to area.

geyser A hot spring that intermittently ejects hot water and steam.

glacial budget The balance between expansion and contraction of a glacier in response to accumulation and wastage.

glacial drift A collective term for all sediment deposited by glaciers or associated processes. Includes till deposited by ice and outwash deposited by streams derived from melting ice.

glacial erratic A rock fragment carried some distance from its source by a glacier and usually deposited on bedrock of a different composition.

glacial ice Water in the solid state within a glacier. Forms as snow partially melts and refreezes and is compacted so that it is transformed first into firn and then into glacial ice.

glacial polish A smooth, glistening rock surface formed by the movement of a sediment-laden glacier over it.

glacial striation A straight scratch rarely more than a few millimeters deep on a rock caused by the movement of sediment-laden glacial ice.

glacier A mass of ice on land that moves by plastic flow and basal slip.

***Glossopteris* flora** A Late Paleozoic association of plants found only on the Southern Hemisphere continents and India.

Gondwana One of six major Paleozoic continents; composed of the present-day continents of South America, Africa, Antarctica, Australia, and India and parts of other continents such as southern Europe, Arabia, and Florida.

graded bedding A type of sedimentary bedding in which an individual bed is characterized by a decrease in grain size from bottom to top.

graded stream A stream possessing an equilibrium profile in which a delicate balance exists between gradient, discharge, flow velocity, channel characteristics, and sediment load such that neither significant erosion nor deposition occurs within its channel.

gradient The slope over which a stream flows; expressed in m/km or ft/mi.

gravity anomaly (positive and negative) A departure from the expected force of gravity; a gravity anomaly might be positive, indicating a mass excess, or negative, indicating a mass deficiency.

ground moraine The layer of sediment liberated from melting ice as a glacier's terminus retreats.

groundwater Underground water stored in the pore spaces of rock, sediment, or soil.

guide fossil Any fossil that can be used to determine the relative geologic ages of rocks and to correlate rocks of the same relative age in different areas.

gullying Erosion of channels too deep (about 30 cm) to be plowed over. Gullies form most actively during and immediately following rainstorms and snowmelt.

guyot A flat-topped seamount of volcanic origin rising more than 1 km above the seafloor.

H

half-life The time required for one-half of the original number of atoms of a radioactive element to decay to a stable daughter product (e.g., the half-life of potassium 40 is 1.3 billion years).

hanging valley A tributary glacial valley whose floor is at a higher level than that of the main glacial valley.

hanging wall block The block of rock that overlies a fault plane.

harmonic tremor Ground motion lasting from minutes to hours resulting from magma moving below the surface, as opposed to the sudden jolts produced by most earthquakes.

heat An agent of metamorphism; heat comes from increasing depth, magma, and applied pressure.

heat flow The flow of heat from Earth's interior to its surface.

historical geology One of the two main subdivisions of geology. The branch of geology that investigates Earth and life history.

horn A steep-walled, pyramidal peak formed by the headward erosion of at least three cirques.

hot spot A localized zone of melting below the lithosphere.

hot spring A spring in which the water temperature is warmer than the temperature of the human body (37°C).

humus The material in soils derived by bacterial decay of organic matter.

hydraulic action The power of moving water.

hydrologic cycle The continuous recycling of water from the oceans, through the atmosphere, to the continents, and back to the oceans.

hydrolysis The chemical reaction between the hydrogen (H^+) ions and hydroxyl (OH^-) ions of water and a mineral's ions.

hydrothermal A term referring to hot water as in hot springs or geysers.

hypothesis A provisional explanation for observations; subject to continual testing and modification. If well supported by evidence, hypotheses are then generally called *theories*.

I

ice sheet See *continental glacier.*

igneous rock Any rock formed by cooling and crystallization of magma or by the accumulation and consolidation of pyroclastic materials such as ash.

incised meander A deep, meandering canyon cut into bedrock by a stream.

index mineral A mineral that forms within specific temperature and pressure ranges during metamorphism.

infiltration capacity The maximum rate at which a soil or sediment can absorb water.

inselberg An isolated steep-sided erosional remnant rising above a surrounding desert plain.

intensity The subjective measure of the kind of damage done by an earthquake, as well as people's reaction to it.

intermediate magma A magma having a silica content between 53% and 65% and an overall composition intermediate between felsic and mafic magmas. See also *felsic magma* and *mafic magma.*

intrusive igneous rock See *plutonic rock.*

ion An electrically charged atom produced by adding or removing electrons from the outermost electron shell.

ionic bond A bond that results from the attraction of positively and negatively charged ions.

iron meteorites A group of meteorites composed primarily of iron and nickel and accounting for about 6% of all meteorites.

isostasy See *principle of isostasy.*

isostatic rebound   The phenomenon in which unloading of Earth's crust causes it to rise upward until equilibrium is again attained. See also *principle of isostasy*.

J

joint   A fracture along which no movement has occurred or where movement has been perpendicular to the fracture surface.

Jovian planet   Any of the four planets (Jupiter, Saturn, Uranus, and Neptune) that resemble Jupiter. All are large and have low mean densities, indicating they are composed mostly of lightweight gases, such as hydrogen and helium, and frozen compounds, such as ammonia and methane.

K

kame   Conical hill of stratified drift originally deposited in a depression on a glacier's surface.

karst topography   Topography consisting of numerous caves, sinkholes, and solution valleys developed by groundwater solution of rocks such as limestone and dolostone. Also characterized by springs, disappearing streams, and underground drainage.

L

laccolith   A concordant pluton with a mushroomlike geometry.

lahar   A mudflow composed of volcanic materials such as ash.

lateral moraine   The sediment deposited as a long ridge of till along the margin of a valley glacier.

laterite   A red soil, rich in iron or aluminum or both, that forms in the tropics by intense chemical weathering.

Laurasia   A Late Paleozoic, Northern Hemisphere continent composed of the present-day continents of North America, Greenland, Europe, and Asia.

lava   Magma at Earth's surface.

lava dome   A bulbous, steep-sided structure formed by viscous magma moving upward through a volcanic conduit.

lava flow   A stream of magma flowing over Earth's surface.

lava tube   A tunnel beneath the solidified surface of a lava flow through which a molten flow continues to move. Also, the hollow space left when the lava within the tube drains away.

lithification   The process of converting sediment into sedimentary rock.

lithosphere   Earth's outer, rigid part consisting of the upper mantle, oceanic crust, and continental crust.

lithostatic pressure   Pressure exerted on rock by the weight of overlying rocks; it is applied equally in all directions.

loess   Windblown silt and clay deposits; derived from three main sources—deserts, Pleistocene glacial outwash deposits, and floodplains of streams in semiarid regions.

longitudinal dune   A long ridge of sand generally parallel to the direction of the prevailing wind.

longshore current   A current between the breaker zone and the beach that flows parallel to the shoreline and is produced by wave refraction.

longshore drift   The movement of sediment along a shoreline by longshore currents.

Love wave (L-wave)   A surface wave in which the individual particles of material only move back and forth in a horizontal plane perpendicular to the direction of wave travel.

low-velocity zone   The zone within the mantle between 100 and 250 km deep where the velocity of both P- and S-waves decreases markedly; it corresponds closely to the asthenosphere.

M

mafic magma   A silica-poor magma containing between 45% and 52% silica and proportionately more calcium, iron, and magnesium than intermediate and felsic magmas. See also *felsic magma* and *intermediate magma*.

magma   Molten rock material generated within Earth.

magma chamber   A reservoir of magma within the upper mantle or lower crust.

magma mixing   The process of mixing magmas of different composition, thereby producing a modified version of the parent magmas.

magnetic anomaly   Any change, such as a change in average strength, of Earth's magnetic field.

magnetic declination   The angle between lines drawn from a compass position to the north magnetic and geographic poles.

magnetic field   The area in which magnetic substances are affected by lines of magnetic force emanating from Earth.

magnetic inclination   The deviation from horizontal of the magnetic lines of force around Earth.

magnetic reversal   The phenomenon in which the north and south magnetic poles are completely reversed.

magnitude   The total amount of energy released by an earthquake at its source. See also *Richter Magnitude Scale*.

mantle   The thick layer between Earth's crust and core.

mantle plume   A stationary column of magma that originates deep within the mantle and slowly rises to Earth's surface to form volcanoes or basalt plateaus.

marine regression   The withdrawal of the sea from a continent or coastal area, resulting in the emergence of the land as sea level falls or the land rises with respect to sea level.

marine terrace   A wave-cut platform now elevated above sea level.

marine transgression   The invasion of coastal areas or much of a continent by the sea, resulting from a rise in sea level or subsidence of the land.

mass wasting   The downslope movement of material under the influence of gravity.

meandering stream   A stream possessing a single, sinuous channel with broadly looping curves.

mechanical weathering Disaggregation of rocks by physical processes that yield smaller pieces retaining the same composition as the parent material.

medial moraine   A moraine formed where two lateral moraines merge.

Mediterranean belt   A zone of seismic and volcanic activity extending westerly from Indonesia through the Himalayas, across Iran and Turkey, and through the Mediterranean region of Europe; about 20% of all active volcanoes and 15% of all earthquakes occur in this belt.

mesa   A broad, flat-topped erosional remnant bounded on all sides by steep slopes; forms when resistant cap rock is breached, allowing rapid erosion of the less resistant underlying sedimentary rock.

metamorphic facies   A group of metamorphic rocks characterized by particular mineral assemblages formed under the same broad temperature–pressure conditions.

metamorphic rock   Any rock type altered by high temperature and pressure and the chemical activities of fluids is said to have been metamorphosed (such as slate, gneiss, or marble).

metamorphic zone   The region between lines of equal metamorphic intensity known as isograds.

meteor   A piece of extraterrestrial debris that enters Earth's atmosphere.

meteorite   A mass of matter of extraterrestrial origin that has fallen to Earth.

Milankovitch theory   A theory that explains cyclic variations in climate and the onset of ice ages as a result of irregularities in Earth's rotation and orbit.

mineral   A naturally occurring, inorganic, crystalline solid having characteristic physical properties and a narrowly defined chemical composition.

Modified Mercalli Intensity Scale   A scale having values ranging from I to XII that is used to characterize earthquake intensity based on damage.

Moho   See *Mohorovičić discontinuity*.

Mohorovičić discontinuity   The boundary between the crust and mantle; also called the Moho.

Moment Magnitude Scale   A scale for expressing the size of an earthquake by considering the area of a fault along which movement took place and the amount of movement of rocks adjacent to a fault.

monocline   A simple bend or flexure in otherwise horizontal or uniformly dipping rock layers.

mud crack   A sedimentary structure found in clay-rich sediment that has dried out. When such sediment dries, it shrinks and forms intersecting fractures.

mudflow   A flow consisting of mostly clay- and silt-sized particles and more than 30% water; most common in semiarid and arid environments.

N

native element   A mineral composed of a single element (such as gold or silver).

natural levee   A ridge of sandy alluvium deposited along the margins of a stream channel during floods.

nearshore sediment budget   The balance between additions and losses of sediment in the nearshore zone.

neutron   An electrically neutral particle found in the nucleus of an atom.

nonconformity   An unconformity in which stratified sedimentary rocks above an erosion surface overlie igneous or metamorphic rocks. See also *angular unconformity* and *disconformity*.

nonferromagnesian silicate   A silicate mineral that has no iron or magnesium.

nonfoliated texture   A metamorphic texture in which there is no discernible preferred orientation of mineral grains.

normal fault   A dip-slip fault on which the hanging wall block has moved downward relative to the footwall block. See also *reverse fault*.

normal polarity   A record of magnetism in which the direction of the magnetic field is the same as it is at the present. See also *reversed polarity*.

nucleus   The central part of an atom consisting of one or more protons and neutrons.

nuée ardente   A mobile dense cloud of hot pyroclastic materials and gases ejected from a volcano.

O

oblique-slip fault   A fault having both dip-slip and strike-slip movement.

oceanic–continental plate boundary   A type of convergent plate boundary along which oceanic lithosphere and continental lithosphere collide; characterized by subduction of the oceanic plate beneath the continental plate and by volcanism and seismicity.

oceanic crust   The crust underlying the ocean basins, it ranges from 5 to 10 km thick, is composed of gabbro and basalt, and has an average density of 3.0 g/cm^3.

oceanic–oceanic plate boundary   A type of convergent plate boundary along which two oceanic lithospheric plates collide and one is subducted beneath the other.

oceanic ridge   A submarine mountain system found in all the oceans, it is composed of volcanic rock (mostly basalt) and displays features produced by tension.

oceanic trench   A long, narrow feature restricted to active continental margins and along which subduction occurs.

ooze   Deep-sea sediment composed mostly of shells of marine animals and plants.

ophiolite   A sequence of igneous rocks thought to represent a fragment of oceanic lithosphere; composed of peridotite overlain successively by gabbro, sheeted basalt dikes, and pillow lavas.

orogeny   The process of forming mountains, especially by folding and thrust faulting; an episode of mountain building.

outgassing   The process whereby gases derived from Earth's interior are released into the atmosphere by volcanic activity.

outwash plain   The sediment deposited by the meltwater discharging from a continental glacier's terminus.

oxbow lake   A cutoff meander filled with water.

oxidation   The reaction of oxygen with other atoms to form oxides or, if water is present, hydroxides.

P

pahoehoe   A type of lava flow with a smooth, ropy surface.

paleocurrent   The direction of an ancient current as indicated by sedimentary structures such as cross-bedding.

paleomagnetism   Remnant magnetism in rocks, studied to determine the intensity and direction of Earth's past magnetic field.

Pangaea   The name Alfred Wegener proposed for a supercontinent consisting of all Earth's landmasses that existed at the end of the Paleozoic Era.

parabolic dune   A crescent-shaped dune in which the tips point upwind.

parent material   The material that is chemically and mechanically weathered to yield sediment and soil.

passive continental margin   The trailing edge of a continental plate consisting of a broad continental shelf and a continental slope and rise. A vast, flat abyssal plain is commonly present adjacent to the rise. See also *active continental margin*.

pedalfer   A soil formed in humid regions with an organic-rich A horizon and aluminum-rich clays and iron oxides in horizon B.

pediment   An erosion surface of low relief gently sloping away from a mountain base.

pedocal   A soil characteristic of arid and semiarid regions with a thin A horizon and a calcium carbonate–rich B horizon.

pelagic clay   Generally brown or reddish deep-sea sediment composed of clay-sized particles derived from the continents and oceanic islands.

permafrost   Ground that remains permanently frozen.

permeability   A material's capacity for transmitting fluids.

phaneritic texture   A coarse-grained texture in igneous rocks in which the mineral grains are easily visible without magnification; results from slow cooling and generally indicates an intrusive origin.

physical geology   One of the two main subdivisions of geology. That branch of geology concerned with Earth materials (minerals and rocks) and Earth's various internal and surface processes.

pillow lava   Bulbous masses of basalt, resembling pillows, formed when lava is rapidly chilled under water.

planetesimal   An asteroid-sized body that along with other planetesimals aggregated to form protoplanets.

plastic flow   The flow that occurs in response to pressure and causes permanent deformation.

plastic strain   The result of stress in which a material cannot recover its original shape and retains the configuration produced by the stress such as folding of rocks.

plate   An individual piece of lithosphere that moves over the asthenosphere.

plate tectonic theory   The theory that large segments of the outer part of Earth (lithospheric plates) move relative to one another.

playa   A dry lakebed found in deserts.

plunging fold   A fold with an inclined axis.

pluton   An intrusive igneous body that forms when magma cools and crystallizes within the crust (such as a batholith and sill).

plutonic (intrusive igneous) rock   Igneous rock that crystallizes from magma intruded into or formed in place within Earth's crust.

point bar   The sediment body deposited on the gently sloping side of a meander loop.

porosity   The percentage of a material's total volume that is pore space.

porphyritic texture   An igneous texture with mineral grains of markedly different sizes.

precursor   Any change within Earth that precedes an earthquake.

pressure release   A mechanical weathering process in which rocks that formed under pressure expand on being exposed at the surface.

pressure ridge   A buckled area on the surface of a lava flow that forms because of pressure on the partly solid crust of a moving flow.

primary wave   See *P-wave*.

principle of cross-cutting relationships   A principle used to determine the relative ages of events; holds that an igneous intrusion or fault must be younger than the rocks it intrudes into or cuts.

principle of fossil succession   A principle holding that fossils, and especially assemblages of fossils, succeed one another through time in a regular and determinable order.

principle of inclusions   A principle holding that inclusions, or fragments, in a rock unit are older than the rock unit itself (for example, granite fragments in a sandstone are older than the sandstone).

principle of isostasy   The theoretical concept of Earth's crust "floating" on a denser underlying layer.

principle of lateral continuity   A principle holding that sediment layers extend outward in all directions until they terminate.

principle of original horizontality   A principle holding that sediment layers are deposited horizontally or very nearly so.

principle of superposition   A principle holding that younger rocks are deposited on top of older layers.

principle of uniformitarianism   A principle holding that we can interpret past events by understanding present-day processes; based on the assumption that natural laws have not changed through time.

proton   A positively charged particle found in the nucleus of an atom.

P-wave   A compressional, or push–pull, wave; the fastest seismic wave and one that can travel through solids, liquids, and gases; also known as a primary wave.

P-wave shadow zone   The area between 103 and 143 degrees from an earthquake focus where little P-wave energy is recorded by seismographs.

pyroclastic materials   Fragmental substances, such as ash explosively ejected from a volcano.

pyroclastic (fragmental) texture   A fragmental texture characteristic of igneous rocks composed of pyroclastic materials.

pyroclastic sheet deposit Vast, sheetlike deposits of felsic pyroclastic materials erupted from fissures.

Q

quick clay A clay that spontaneously liquefies and flows like water when disturbed.

R

radioactive decay The spontaneous change of an atom to an atom of a different element by emission of a particle from its nucleus (alpha and beta decay) or by electron capture.

rainshadow desert A desert found on the lee side of a mountain range; forms because moist marine air moving inland forms clouds and produces precipitation on the windward side of the mountain range so that the air descending on the leeward side is much warmer and drier.

rapid mass movement Any kind of mass movement involving a visible downslope displacement of material.

Rayleigh wave (R-wave) A surface wave in which the individual particles of material move in an elliptic path within a vertical plane oriented in the direction of wave movement.

recessional moraine A type of end moraine formed when a glacier's terminus retreats, then stabilizes and till is deposited. See also *end moraine* and *terminal moraine*.

reef A moundlike, wave-resistant structure composed of the skeletons of organisms.

reflection The return to the surface of some of a seismic wave's energy when it encounters a boundary separating materials of different density or elasticity.

refraction The change in direction and velocity of a seismic wave when it travels from one material into another of different density and elasticity.

regional metamorphism Metamorphism that occurs over a large area, resulting from tremendous temperatures, pressures, and the chemical activity of fluids within the crust.

regolith The layer of unconsolidated rock and mineral fragments and soil that covers most of the surface.

relative dating The process of determining the age of an event relative to other events; involves placing geologic events in their correct chronologic order but involves no consideration of when the events occurred in terms of number of years ago. See also *absolute dating*.

reserve The part of the resource base that can be extracted economically.

resource A concentration of naturally occurring solid, liquid, or gaseous material in or on Earth's crust in such form and amount that economic extraction of a commodity from the concentration is currently or potentially feasible.

reversed polarity A record of magnetism in which the direction of the magnetic field is the opposite of its present orientation. See also *normal polarity*.

reverse fault A dip-slip fault in which the hanging wall block has moved upward relative to the footwall block. See also *normal fault*.

Richter Magnitude Scale An open-ended scale that measures the amount of energy released during an earthquake.

rill erosion Erosion by running water that scours small channels in the ground.

rip current A narrow surface current that flows out to sea through the breaker zone.

ripple mark Wavelike (undulating) structure produced in granular sediment such as sand by unidirectional wind and water currents or by oscillating wave currents.

rock An aggregate of one or more minerals, as in limestone or granite, or a consolidated aggregate of rock fragments, as in conglomerate; although exceptions to this definition, coal and natural glass are considered rocks.

rock cycle A group of processes through which Earth materials may pass as they are transformed from one rock type to another.

rockfall A common type of extremely rapid mass wasting in which rocks fall through the air.

rock-forming mineral A mineral common in rocks, which is important in their identification and classification.

rock slide A type of rapid mass movement in which rocks move downslope along a more or less planar surface.

rounding The process by which the sharp corners and edges of sedimentary particles are abraded during transport.

runoff The surface flow of streams.

R-wave See *Rayleigh wave*.

S

salt crystal growth A mechanical weathering process in which rocks are disaggregated by the growth of salt crystals in crevices and pores.

saltwater incursion The displacement of freshwater by saltwater as a result of excessive pumping in coastal areas.

scientific method A logical, orderly approach that involves gathering data, formulating and testing hypotheses, and proposing theories.

seafloor spreading The theory that the seafloor moves away from spreading ridges and is eventually consumed at subduction zones.

seamount A submarine volcanic mountain rising at least 1 km above the seafloor.

secondary wave See *S-wave*.

sediment Loose aggregate of solids derived from preexisting rocks, or solids precipitated from solution by inorganic chemical processes or extracted from solution by organisms.

sedimentary facies Any aspect of a sedimentary rock unit that makes it recognizably different from adjacent sedimentary rocks of the same, or approximately the same, age (e.g., a sandstone facies).

sedimentary rock Any rock composed of sediment (such as sandstone and limestone).

sedimentary structure Any structure in sedimentary rock formed at or shortly after the time of deposition (such as cross-bedding, mud cracks, and animal burrows).

seismic profiling A method in which strong waves generated at an energy source penetrate the layers beneath the seafloor. Some of the energy is reflected from the various layers to the surface, making it possible to determine the nature of the layers.

seismic risk map   A map based on the distribution and intensity of past earthquakes; such maps indicate the potential severity of future earthquakes and are useful in planning.

seismograph   An instrument that detects, records, and measures the various waves produced by an earthquake.

seismology   The study of earthquakes.

shear strength   The resisting forces helping to maintain slope stability.

shear stress   The result of forces acting parallel to one another but in opposite directions; results in deformation by displacement of adjacent layers along closely spaced planes.

sheet erosion   Erosion that is more or less evenly distributed over the surface and removes thin layers of soil.

sheet joint   A large fracture more or less parallel to a rock surface resulting from pressure released by expansion of the rock.

shield   A vast area of exposed ancient rocks on a continent; the nucleus of a continent around which accretion occurred.

shield volcano   A dome-shaped volcano with a low, rounded profile built up mostly of overlapping basalt lava flows, such as Mauna Loa and Kilauea in Hawaii.

shoreline   The area between mean low tide and the highest level of land affected by storm waves.

silica   A compound of silicon and oxygen atoms.

silicate   A mineral containing silica (such as quartz [SiO_2]).

silica tetrahedron   The basic building block of all silicate minerals, it consists of one silicon atom and four oxygen atoms.

sill   A tabular or sheetlike concordant pluton.

sinkhole   A depression in the ground that forms in karst regions by the solution of the underlying carbonate rocks or by the collapse of a cave roof.

slide   A type of mass wasting involving movement of material along one or more surfaces of failure.

slow mass movement   Mass movement that advances at an imperceptible rate and is usually only detectable by the effects of its movement.

slump   A type of mass wasting that takes place along a curved surface of failure and results in the backward rotation of the slump mass.

soil   Regolith consisting of weathered material, water, air, and humus that can support plants.

soil degradation   Any process leading to a loss of soil productivity; may involve erosion, chemical pollution, or compaction.

soil horizon   A distinct soil layer that differs from other soil layers in texture, structure, composition, and color.

solar nebula theory   A theory for the evolution of the solar system from a rotating cloud of gas.

solifluction   A type of mass wasting involving the slow downslope movement of water-saturated surface materials; especially the flow at high elevations or high latitudes where the flow is underlain by frozen soil.

solution   A reaction in which the ions of a substance become dissociated in a liquid and the solid substance dissolves.

sorting   A term referring to the degree to which all particles of sediment or sedimentary rock are about the same size.

spatter cone   A small, steep-sided cone that forms when gases escaping from a lava flow hurl globs of molten lava into the air that fall back to the surface and adhere to one another.

spheroidal weathering   A type of chemical weathering in which corners and sharp edges of rocks weather more rapidly than flat surfaces, thus yielding spherical shapes.

spit   A continuation of a beach forming a point of land that projects into a body of water, commonly a bay.

spring   A place where groundwater flows or seeps out of the ground.

stock   An irregularly shaped discordant pluton with a surface area less than 100 km².

stony-iron meteorites   A group of meteorites composed of nearly equal amounts of iron and nickel and silicate minerals; they comprise about 1% of all meteorites.

stony meteorites   A group of meteorites composed of iron and magnesium silicate minerals; they comprise about 93% of all meteorites.

stoping   A process in which rising magma detaches and engulfs pieces of the surrounding country rock.

strain   Deformation caused by stress. See also *elastic strain* and *plastic strain*.

strata (stratification)   Strata (singular, *stratum*) are the layers in sedimentary rocks, whereas stratification is the layered aspect of sedimentary rocks. See also *bed (bedding)*.

stratified drift   Glacial drift displaying both sorting and stratification.

stratovolcano   See *composite volcano*.

stream   Runoff confined to channels regardless of size.

stream terrace   An erosional remnant of a floodplain that formed when a stream was flowing at a higher level.

stress   The force per unit area applied to a material such as rock.

strike   The direction of a line formed by the intersection of a horizontal plane with an inclined plane, such as a rock layer.

strike-slip fault   A fault involving horizontal movement so that blocks on opposite sides of a fault plane slide sideways past one another. See *dip-slip fault*.

subduction   The process whereby the leading edge of one plate descends beneath the margin of another plate.

subduction zone   A long, narrow zone at a convergent plate boundary where an oceanic plate descends relative to another plate (for example, the subduction of the Nazca plate beneath the South American plate).

submarine canyon   A steep-walled submarine canyon; best developed on the continental slope but some extend well up onto the continental shelves.

submarine fan   A cone-shaped sedimentary deposit that accumulates on the continental slope and rise.

submarine hydrothermal vent
A crack of fissure in the seafloor through which superheated water issues. Many have a plume of water discolored by dissolved minerals and are called black smokers.

submergent coast   A coast along which sea level rises with respect to the land or the land subsides.

superposed stream   A stream that once flowed on a higher surface and eroded downward into resistant rocks, while still maintaining its course.

surface wave   Earthquake waves that travel along Earth's surface. Rayleigh (R-) and Love (L-) waves are both surface waves.

suspended load   The smallest particles carried by a stream, such as silt and clay, which are kept suspended by fluid turbulence.

sustainable development   The concept of satisfying basic human needs while safeguarding the environment to ensure continued economic development.

S-wave   A shear wave that moves material perpendicular to the direction of travel, thereby producing shear stresses in the material it moves through; also known as a secondary wave; an S-wave travels only through solids.

S-wave shadow zone   Those areas more than 103 degrees from an earthquake focus where no S-waves are recorded.

syncline   A down-arched fold in which the youngest exposed rocks coincide with the fold axis and all strata dip toward the axis.

system   A combination of related parts that interact in an organized fashion. Earth systems include the atmosphere, hydrosphere, biosphere and solid Earth.

T

talus   Weathered material that accumulates at the base of slopes.

tension   A type of stress in which forces act in opposite directions but along the same line, thus tending to stretch an object.

terminal moraine   A type of end moraine; the outermost moraine marking the greatest extent of a glacier. See

also *end moraine* and *recessional moraine*.

terrane   A block of rock with characteristics quite different from those of surrounding rocks. Terranes probably represent seamounts, oceanic rises, and other seafloor features that accreted to the margins of continents during orogenies.

terrestrial planet   Any of the four innermost planets (Mercury, Venus, Earth, and Mars). They are all small and have high mean densities, indicating that they are composed of rock and metallic elements. See also *Jovian planet*.

theory   An explanation for some natural phenomenon that has a large body of supporting evidence; to be considered scientific, a theory must be testable; for example, plate tectonic theory.

thermal convection cell   A type of circulation of material in the asthenosphere during which hot material rises, moves laterally, cools and sinks, and is reheated and continues the cycle.

thermal expansion and contraction   A type of mechanical weathering in which the volume of rock changes in response to heating and cooling.

thrust fault   A type of reverse fault with a fault plane dipping less than 45 degrees.

tide   The regular fluctuation in the sea's surface in response to the gravitational attraction of the Moon and Sun.

till   All sediment deposited directly by glacial ice.

time–distance graph   A graph showing the average travel times for P- and S-waves for any specific distance from an earthquake's focus.

tombolo   A type of spit that extends out into the sea and connects an island to the mainland.

transform fault   A type of fault along which one type of motion is transformed into another. Commonly displaces oceanic ridges, but movement on opposite sides of the fault between displaced ridge segments is the opposite of the apparent displacement. On land recognized as strike-slip faults, such as the San Andreas fault.

transform plate boundary   Plate boundary along which plates slide past

one another and crust is neither produced nor destroyed; on land recognized as strike-slip fault.

transport   The mechanism by which weathered material is moved from one place to another, commonly by running water, wind, or glaciers.

transverse dune   A long ridge of sand perpendicular to the prevailing wind direction.

tree-ring dating   The process of determining the age of a tree or wood in structures by counting the number of annual growth rings.

trough   The lowest point between wave crests.

tsunami   A destructive sea wave that is usually produced by an earthquake but can also be caused by submarine landslides or volcanic eruptions.

turbidity current   A sediment–water mixture, denser than normal seawater, that flows downslope to the deep seafloor.

U

unconformity   An erosion surface separating younger strata from older rocks. See also *angular unconformity*, *disconformity*, and *nonconformity*.

U-shaped glacial trough   A valley with steep or vertical walls and a broad, rather flat floor; formed by the movement of a glacier through a stream valley.

V

valley   A linear depression bounded by higher areas such as ridges, hills, or mountains. Most are eroded by streams, although mass wasting is also important in their origin and evolution.

valley glacier   A glacier confined to a mountain valley or to an interconnected system of mountain valleys.

valley train   A long, narrow deposit of stratified drift confined within a glacial valley.

velocity   A measure of the downstream distance water travels per unit of time. Velocity varies considerably among streams and even within the same stream.

ventifact   A stone whose surface has been polished, pitted, grooved, or faceted by wind abrasion.

vesicle A small hole or cavity formed by gas trapped in cooling lava.

viscosity A fluid's resistance to flow.

volcanic explosivity index (VEI) A semiquantitative scale for the size of a volcanic eruption based on evaluation of such criteria as volume of material explosively ejected and height of eruption cloud.

volcanic island arc A curved chain of volcanic islands parallel to a deep-sea trench where oceanic lithosphere is subducted, causing volcanism and the origin of volcanic islands.

volcanic neck An erosional remnant of the material that solidified in a volcanic pipe.

volcanic pipe The conduit connecting the crater of a volcano with an underlying magma chamber.

volcanic (extrusive igneous) rock An igneous rock formed when magma is extruded onto Earth's surface where it cools and crystallizes, or when pyroclastic materials become consolidated.

volcanism The process whereby magma and its associated gases rise through the crust and are extruded onto the surface or into the atmosphere.

volcano A mountain formed around a vent as a result of the eruption of lava and pyroclastic materials.

W

water table The surface separating the zone of aeration from the underlying zone of saturation.

water well A well made by digging or drilling into the zone of saturation.

wave An undulation on the surface of a body of water, resulting in the water surface rising and falling.

wave base A depth of about one-half wavelength, where the diameter of the orbits of water in waves is essentially zero; the depth below which water is unaffected by surface waves.

wave-cut platform A beveled surface that slopes gently seaward; formed by the retreat of a sea cliff.

wave height The vertical distance from wave trough to wave crest.

wavelength The distance between successive wave crests or troughs.

wave period The time required for two successive wave crests (or troughs) to pass a given point.

wave refraction The bending of waves so that they more nearly parallel the shoreline.

weathering The physical breakdown and chemical alteration of rocks and minerals at or near Earth's surface.

Z

zone of accumulation Part of a glacier where additions exceed losses and the glacier's surface is perennially covered with snow.

zone of aeration The zone above the water table that contains both water and air within the pore spaces of the rock or soil.

zone of leaching Another name for horizon A of a soil. The area in the upper part of a soil where soluble minerals are removed by downward moving water.

zone of saturation The area below the zone of aeration in which all pore spaces are filled with groundwater.

zone of wastage The part of a glacier where losses from melting, sublimation, and calving of icebergs exceed the rate of accumulation.

Credits

This page constitutes an extension of the copyright page. We have made every effort to trace the ownership of all copyrighted material and to secure permission from copyright holders. In the event of any question arising as to the use of any material, we will be pleased to make the necessary corrections in future printings. Thanks are due to the following authors, publishers, and agents for permission to use the material indicated.

Chapter 1
Figure 1.8: THE FAR SIDE. Copyright © 1991. Distributed by Universal Press Syndicate. Reprinted with permission. All rights reserved.

Chapter 4
Figure 4.10 (a–d): From Howell Williams, *Crater Lake: The Story of Its Origin* (Berkeley, Calif. University of California Press): Illustrations from p. 84. © 1941 Regents of the University of California, © renewed 1969, Howell Williams.

Chapter 5
Perspective 5.2, Figure 1(a): From *Dust Bowl: The Southern Plains in the 1930s,* by Donald Worster. Copyright © 1979 by Oxford University Press, Inc. Reprinted by permission.

Chapter 7
Figure 7.3(a): From C. Gillen, *Metamorphic Geology,* Figure 4.4, p. 73. Copyright © 1982. Reprinted with the kind permission of Kluwer Academic Publishers and C. Gillen. **Figure 7.18:** From H. L. James, *G. S. A. Bulletin,* vol. 66, plate 1, page 1454, with permission of the publisher, the Geological Society of America, Boulder, Colorado. USA. Copyright © 1955 Geological Society of America. **Figure 7.19:** From AGI Data Sheet 35.4, *AGI Data Sheets,* 3rd edition (1989) with the kind permission of the American Geological Institute. **Figure 7.21:** From "Effects of Late Jurassic-Early Tertiary Subduction in California," San Joaquin Geological Society Short Course, 1977, 66, Figure 5-9.

Chapter 8
Figure 8.1: Modified from A. R. Palmer, "The Decade of North American Geology, 1983 Geologic Time Scale." *Geology* (Geological Society of America, 1983), p. 504. **Figure 8.6:** From *Giants of Geology,* by C. L. Fenton and M. A. Fenton, copyright 1952 by C. L. Fenton and M.A. Fenton. Used by permission of Doubleday, a division of Random House, Inc. **Figure 8.19:** Based on data from S. M. Richardson and H. Y. McSween, Jr., *Geochemistry—Pathways and Processes,* Prentice-Hall. **Figure 8.25:** From E. K. Ralph, H. N. Michael, and M. C. Han, "Radiocarbon Dates and Reality," *MASCA Newsletter 9* (1973), page 5, figure 8. Used by permission. **Figure 8.26:** From *An Introduction to Tree-Ring Dating,* by Stokes and Smiley, 1968, p. 6. Reprinted by permission of University of Chicago Press.

Chapter 9
Figure 9.7: Data from National Oceanic and Atmospheric Administration. **Figure 9.12 (a, b, c):** From *Nuclear Explosions and Earthquakes: The Parted Veil,* by Bruce A. Bolt. Copyright © 1976 by H. Freeman and Company. Used with permission. **Figure 9.13 (a,**

b): From *Nuclear Explosions and Earthquakes: The Parted Veil,* by Bruce A. Bolt. Copyright © 1976 by W. H. Freeman and Company. Used with permission. **Figure 9.15:** Based on data from C. F. Richter, *Elementary Seismology,* 1958. W. H. Freeman and Company. **Figure 9.19:** From *Earthquakes,* by Bruce A. Bolt. Copyright © 1988 by W. H. Freeman and Company. Used with permission. **Perspective 9.2, Figure 2:** From *Earthquakes,* by Bruce A. Bolt. Copyright © 1988 by W. H. Freeman and Company. Used with permission. **Figure 9.26:** Data from National Oceanic and Atmospheric Administration. **Figure 9.29:** Data from *The Loma Prieta Earthquake of October 17, 1989.* U. S. Geological Survey. **Figure 9.30:** From *Predicting Earthquakes: A Scientific and Technical Evaluation—With Implications for Society,* p. 41. Copyright © 1976 National Academy Press. Reprinted by permission. **Figure 9.31:** From Figure 6, page 17, *Geotimes Vol. 10, No. 9* (1966) with the kind permission of the American Geological Institute.

Chapter 10
Figure 10.6: From G. C. Brown and A. E. Musset, *The Inaccessible Earth* (London: Chapman & Hall, 1981), **Figure 12.7a. Figure 10.11:** From G. C. Brown and A. E. Musset, *The Inaccessible Earth* (London: Chapman & Hall, 1981), Figure 7.11. **Figure 10.12:** From "Journey to the Center of the Earth," by T. A. Heppenheimer, *Discover,* v. 8, no. 11, Nov. 1987. Illustration by Andrew Christie, copyright © 1987. Reprinted with permission of Discover Magazine. **Perspective 10.1, Figure 2:** From Keith G. Cox, "Kimberlite Pipes." Original illustration by Adolph E. Brotman. Copyright © April 1978 by Scientific American. All rights reserved. **Figure 10.20:** From Beno Gutenburg, *Physics of the Earth's Interior* (Orlando, Florida: Academic Press, 1959), 194, Figure 9.1.

Chapter 11
Figure 11.1: From Phyllis Young Forsyth, *Atlantis: The Making of Myth* (Montreal: McGill-Queen's University Press): 13, Figure 2. **Figure 11.11:** From Bruce E. Heezen and Charles D. Hollister, *The Face of the Deep* (New York: Oxford University Press, 1971): Figure 8.15, page 297. **Figure 11.14:** From Alyn and Alison Duxbury, *An Introduction to the World's Oceans.* Copyright © 1984 Addison-Wesley Publishing Company, Inc. **Figure 11.15:** From Bruce E. Heezen and Charles D. Hollister, *The Face of the Deep* (New York: Oxford University Press, 1971): Figure 8.38, page 329. **Figure 11.26:** From U. S. Geological Survey.

Chapter 12
Figure 12.4: From *General Geology,* 5/e by R. J. Foster. Copyright © 1988. Reprinted by permission of Prentice-Hall, Inc., Upper Saddle River, NJ. **Figure 12.7:** Modified from E. H.Colbert, *Wandering Lands and Animals* (1973): 72, Figure 31. **Figure 12.9:** From A. Cox and R. R. Doell, "Review of Paleomagnetism," *G. S. A. Bulletin,* vol. 71, figure 33, page 758, with permission of the publisher, the Geological Society of America, Boulder, Colorado. USA. Copyright © 1955 Geological Society of America. **Figure 12.11:** Reprinted with permission from A. Cox, "Geomagnetic Reversals," *Science 163,* January 17, 1969. Copyright © 1969 American Association for the Advancement of Science. **Figure 12.12:** From Larson, R. L. et al. (1985). *The Bedrock Geology of the World,* W. H. Freeman and Co., New York, NY.

Chapter 13

Figure 13.27: From Peter Molnar, "The Geological History and the Structure of the Himalayas." *American Scientist 74:* 148-149, Figure 4, Journal of Sigma Xi, The Scientific Research Group. **Figure 13.28:** From Zvi Ben-Avraham, "The Movement of Continents." *American Scientist 69:* 291-299, Figure 9, p. 298, Journal of Sigma Xi, The Scientific Research Group. **Figure 13.30:** Reprinted with permission from Kent C. Condie, *Plate Tectonics and Crustal Evolution,* 4d, p. 65 (Fig. 2.26), © 1997, Butterworth-Heinemann. **Perspective 13.3, Figures 1 and 2:** Reprinted with permission from *Geologic Trips: San Francisco Bay Area,* by Ted Konigsmark, © GeoPress, Gualala, Calif. **Figure 13.34:** Reprinted with permission from W. R. Dickinson and Pacific Coast Section SEPM, "Cenozoic Plate Tectonic Setting in the Cordilleran Region in the U. S., in *Cenozoic Paleogeography of the Western U.S.* Pacific Coast Symposium 3, 1979, p. 2.

Chapter 14

Perspective 14.2, Figure 1: From G. A. Kiersch, "Vaiont Reservoir Disaster," *Civil Engineering* 34 (1964). **Perspective 14.2, Figure 3:** From G. A. Kiersch, "Vaiont Reservoir Disaster," *Civil Engineering 34* (1964).

Chapter 16

Figure 16.1 (a, b): From *Trapped* by Robert K. Murray and Roger W. Brucker. Copyright © 1979 by Murray and Brucker, copyright renewed. **Figure 16.2:** Modified from *U.S. News and World Report* (18 March 1991): 72-73. **Figure 16.19:** From J. B. Weeks et al., *U. S. Geological Survey Professional Paper* 1400-A, 1988.

Chapter 17

Figure 17.32: From V. K. Prest, *Economic Geology Report 1,* 5d, 1970, p. 90–91 (fig. 7-6), Geological Survey of Canada. Department of Energy, Mines, and Resources, reproduced with permission of Minister of Supply and Services.

Chapter 18

Perspective 18.1, Figure 1: From C. B. Hunt and D. R. Mabey, *U. S. Geological Survey Professional Paper* 494A (1966): A5, Figure 2.

Chapter 20

Figure 20.2: Reprinted with permission from *The Restless Universe* by N. Henbest and H. Couper, 1982. Published by George Philip and Son, Ltd., © Nigel Henbest and Heather Couper.

PHOTO CREDITS

Chapter opening background photo: PhotoDisc, Inc.

About the Authors
Photo of Reed Wicander by Melanie G. Wicander.

Chapter 1
Chapter Opener: Photo courtesy of Dick Van Effen. **Perspective 1.1 Figure 1:** NASA. **Perspective 1.1 Figure 2:** Black Star Publishing © 1990 David Turnley. **Figure 1.1:** NASA. **Figure 1.2:** © Walt Anderson/Visuals Unlimited. **Figure 1.5:** © Jean Miele/The Stock Market. **Figure 1.6:** Collection of the New York Public Library Astor, Lenox, and Tilden Foundations. **Figure 1.7:** Photo by Reed Wicander. **Figure 1.9:** Photo composite by Elliott Hill. **Figure 1.10:** Superstock. **Figure 1.18 (a-f):** Photos courtesy of Sue Monroe.

Chapter 2
Chapter Opener: Los Angeles County Museum specimen, © Harold and Erica Van Pelt. **Figure 2.1:** National Museum of Natural History (NMNH) Specimen #G7101. Photo by D. Penland, courtesy of Smithsonian Institution. **Figure 2.2:** Jerry Jacka Photography. **Figure 2.3:** © Lanye Kennedy/ CORBIS. **Figure 2.4:** Photo by Stew Monroe. **Figure 2.16 (a-d):** Photos courtesy of Sue Monroe. **Figure 2.17 (a-d):**

Photos courtesy of Sue Monroe. **Figure 2.18 (a-d):** Photos courtesy of Sue Monroe. **Figure 2.19 (a-b):** Photos courtesy of Sue Monroe. **Figure 2.20 (a-c):** Photos courtesy of Sue Monroe. **Figure 2.23:** Photos courtesy of Sue Monroe. **Figure 2.24:** © Gregory G. Dimijian 1992, Photo Researchers, Inc. **Figure 2.25:** Photos by Stew Monroe. **Figure 2.26 (a):** Photo courtesy of Cleveland Cliffs Iron Company. **Figure 2.26 (b):** Photo courtesy of Stew Monroe. **Perspective 2.1, Figure 1(a):** National Museum of Natural History (NMNH) Specimen #R121297. Photo by D. Penland, courtesy of Smithsonian Institution. **Perspective 2.1, Figure 1(b):** © Ken Lucas/Visuals Unlimited. **Perspective 2.1, Figure 1(c):** © E.R. Degginger, Photo Researchers, Inc. **Perspective 2.1, Figure 2(a):** Photo by Stew Monroe. **Perspective 2.1, Figure 2(b):** Photo courtesy of Sue Monroe. **Perspective 2.2, Figure 1(a):** National Museum of Natural History, (NMNH) Specimen #R12804. Photo by Chip Clark, courtesy of Smithsonian Institution. **Perspective 2.2, Figure 1(b-g):** Photos courtesy of Sue Monroe.

Chapter 3
Chapter Opener: J. D. Griggs/USGS. **Figure 3.1 (a):** © Richard Thom/Visuals Unlimited. **Figure 3.1 (b):** Photo courtesy of Steve Stahl. **Figure 3.1 (c):** Photo by Stew Monroe. **Figure 3.1 (d):** © Alissa Crandall/ CORBIS. **Figure 3.2:** P. Mouginis-Mark. **Figure 3.6 (b):** Photo courtesy of Sue Monroe. **Figure 3.8 (b, d-f):** Photos courtesy of Sue Monroe. **Figure 3.9 (a-c):** Photos courtesy of Sue Monroe. **Figure 3.11** Photo courtesy of Sue Monroe. **Figure 3.12 (a-b):** Photos courtesy of Sue Monroe. **Figure 3.13 (a-b):** Photos courtesy of Sue Monroe. **Figure 3.14 (a-b):** Photos courtesy of Sue Monroe. **Figure 3.15:** Photo courtesy of Sue Monroe. **Figure 3.17:** Photo courtesy of David J. Matty. **Perspective 3.1, Figure 1:** Photo courtesy of Wendell E. Wilson. **Perspective 3.1, Figure 2:** W. T. Schaller/USGS. **Figure 3.18 (a-b):** Photos courtesy of Sue Monroe. **Figure 3.20:** © Martin G. Miller/Visuals Unlimited. **Figure 3.21 (a):** Photo courtesy of Richard L. Chambers. **Figure 3.21 (b):** Photo courtesy of Sue Monroe. **Figure 3.22 (c):** Photo by Stew Monroe. **Perspective 3.2, Figure1:** © Michael Nicholson/ CORBIS. **Perspective 3.2, Figure 2 (a):** Photo courtesy of Sue Monroe. **Perspective 3.2, Figure 2 (b):** Photo courtesy of Frank Hanna. **Perspective 3.2, Figure 3:** Photo by Stew Monroe.

Chapter 4
Chapter Opener: 1997 Photo Disc, Inc. **Figure 4.1:** © 1995 Stephen Cottrell. **Figure 4.2 (a):** J.D. Griggs/USGS. **Figure 4.2 (b):** © G. Brad Lewis/Omjalla Images. **Figure 4.2 (c):** University of Colorado. **Figure 4.3 (a-c):** Photos courtesy of Sue Monroe. **Figure 4.4 (a):** J. D. Griggs, U.S. Geological Survey. **Figure 4.4 (b):** J. B. Stokes, U.S. Geological Survey. **Figure 4.5 (a):** Photo courtesy of Hawaii Volcanoes National Park, USGS. **Figure 4.5 (b):** K. V. Cashman, U.S. Geological Survey. **Figure 4.6 (a):** T. J. Takahashi, U.S.Geological Survey. **Figure 4.6 (b):** © G. Brad Lewis/Omjalla Images. **Figure 4.7:** Photo courtesy of Sue Monroe. **Figure 4.8 (a):** Reproduced by permission of Marie Tharp, 1 Washington Avenue, South Nyack, NY 10960. **Figure 4.8 (b):** Photo by Stew Monroe. **Figure 4.9:** Photo courtesy of Sue Monroe. **Perspective 4.1, Figure 1:** © Nik Wheeler/ CORBIS. **Perspective 4.1, Figure 1 (a):** © Richard Cummins/ CORBIS. **Perspective 4.1, Figure 2 (b):** Photo by Stew Monroe. **Figure 4.10 (e):** Photo courtesy of Sue Monroe. **Figure 4.11 (b-c):** Photos by Stew Monroe. **Figure 4.12 (a-b):** Photos by Stew Monroe. **Figure 4.12 (c):** Solarfilma/ GeoScience Features. **Figure 4.13 (b):** R. Solkoski, Consulting Geologists, Vancouver, WA. **Figure 4.13 (c):** D.R. Crandel, U.S.Geological Survey. **Figure 4.14 (a):** Reuters/ CORBIS. **Figure 4.14 (b):** U.S. Department of Interior, U.S.Geological Survey, David A. Johnston Cascades Volcano Observatory, Vancouver, WA. **Perspective 4.2, Figure 1 (b):** Photo courtesy of Keith Ronnholm. **Perspective 4.2, Figure 2:** © Stone /Richard During. **Figure 4.15:** Photo courtesy of Sue Monroe. **Figure 4.16 (a):** Bettman/ CORBIS. **Figure 4.16 (b):** Neg./Transparency no. 256108 (Photo by E.O. Hovey) Courtesy Dept. of Library Services, American Museum of Natural History. **Perspective 4.3, Figure 1:** © Buddy Mays/ CORBIS. **Perspective 4.3, Figure 2:** Photo by Stew Monroe. **Perspective 4.3, Figure 3 (a):** Photo

by C. Nye, Alaska Division of Geological and Geophysical Surveys, USGS. **Perspective 4.3 Figure 3 (b):** U.S.Geological Survey. **Figure 4.17 (a):** Photo courtesy of Ward's Natural Science Inc. **Figure 4.18:** W.E. Scott, U.S. Geological Survey. **Guest Essay:** Stanley N. Williams—Havis Photo.

Chapter 5
Chapter Opener: © Bill Ross/ CORBIS. **Figure 5.1:** Photo courtesy of Sue Monroe. **Figure 5.2 (a):** Photo by Stew Monroe. **Figure 5.2 (b):** Photo courtesy of Sue Monroe. **Figure 5.3 (a-b):** Photos courtesy of Sue Monroe. **Figure 5.4:** Photo by Stew Monroe. **Figure 5.6:** Photo by Stew Monroe. **Figure 5.7:** Photo by Stew Monroe. **Figure 5.8 (a):** Photo by Stew Monroe. **Figure 5.8 (b):** © Mark Gibson/ Visuals Unlimited. **Figure 5.9:** Photo courtesy of W. D. Lowry. **Figure 5.10:** Photo by Stew Monroe. **Figure 5.11:** Photo by Stew Monroe. **Figure 5.13:** Photo by Stew Monroe. **Figure 5.14:** © Bill Beatty/ Visuals Unlimited. **Figure 5.15:** Photo by Stew Monroe. **Figure 5.18 (d):** Photo by Stew Monroe. **Figure 5.19 (a-b):** Photos by Stew Monroe. **Figure 5.21 (a-b):** Photos by Stew Monroe. **Figure 5.22:** © Walt Anderson/Visuals Unlimited. **Figure 5.24 (b):** Photo courtesy of Ed Nuhfer, Director, Teaching Effectiveness & Faculty Development, CU–Denver, Campus Box 137, P.O. Box 173364, Denver, CO 80217-3364; (303) 556-4915, fax (303) 556-5855. **Figure 5.25 (a):** Photo by Stew Monroe. **Figure 5.25 (b):** H.H. Waldron, USGS. **Figure 5.26:** © Frank Lembrecht/Visuals Unlimited. **Perspective 5.2, Figure1 (b):** The Granger Collection, New York. **Perspective 5.2, Figure 2:** Bettman/CORBIS. **Figure 5.27:** ©Science VU/Visuals Unlimited.

Chapter 6
Chapter Opener: Photo by Stew Monroe. **Figure 6.1 (b-c):** Photos courtesy of Sue Monroe. **Figure 6.2:** © Roy Andersen. **Figure 6.4 (a):** Photo courtesy of R. V. Dietrich. **Figure 6.4 (b):** Photo by Stew Monroe. **Figure 6.7:** Photo courtesy of Sue Monroe. **Figure 6.8 (a-b):** Photos by Stew Monroe. **Figure 6.9 (a-b):** Photos courtesy of Sue Monroe. **Figure 6.10 (a-b):** Photos courtesy of Sue Monroe. **Figure 6.11 (a-b):** Photos courtesy of Sue Monroe. **Figure 6.12 (a):** Photo courtesy of Rex Elliot. **Figure 6.12 (b, d):** Photos courtesy of Sue Monroe. **Figure 6.12 (c):** Photo courtesy of Richard V. Dietrich. **Figure 6.13 (a-b):** Photos courtesy of Sue Monroe. **Figure 6.14 (a-b):** Photos courtesy of Sue Monroe. **Figure 6.15 (a-c):** Photos courtesy of Sue Monroe. **Figure 6.19:** Photo courtesy of Sue Monroe. **Figure 6.20 (c):** Photo by Stew Monroe. **Figure 6.21:** Photo by Stew Monroe. **Figure 6.22 (c, e):** Photos by Stew Monroe. **Figure 6.23 (a):** Photo courtesy of R. V. Dietrich. **Figure 6.23 (b):** Photo courtesy of Jeff Mayo, GeoPhoto. **Figure 6.24 (a-c):** Photos courtesy of Sue Monroe. **Figure 6.25 (a):** Photo courtesy of Wayne E. Moore. **Figure 6.25 (b):** Photo courtesy of Sue Monroe. **Figure 6.25 (c-d):** Photos by Stew Monroe. **Figure 6.26 (d):** Photo courtesy of Sue Monroe. **Figure 6.27:** Photo by Stew Monroe. **Figure 6.28:** Alan L. Mayo, GeoPhoto Publishing Company. **Perspective 6.1, Figure 1 (a-b):** Photos courtesy of Sue Monroe. **Perspective 6.1, Figure 2 (a-b):** Photos courtesy of Sue Monroe. **Perspective 6.1, Figure 3:** Reprinted with permission from R.T. Tucker, Human Footprints in Stone, Oklahoma Today, v.25, 1975. **Perspective 6.1, Figure 4:** Martin Land, Science Photo Library/Photo Researchers. **Perspective 6.1, Figure 4 (b):** USDA Photo by Ken Hammond. **Perspective 6.2, Figure 1:** Reproduced from *Bones for Barnum Brown: Adventures of a Dinosaur Hunter* by Roland T. Bird, et al. (Texas Christian University Press, 1985). **Perspective 6.2, Figure 2 (a-b):** Photos courtesy of Sue Monroe. **Perspective 6.2, Figure 3:** University of Nebraska State Museum. **Figure 6.31:** Photo by Stew Monroe.

Chapter 7
Chapter Opener: Garry Hobart/ GEOImagery. **Figure 7.1:** Collection of the J. Paul Getty Museum, Malibu, CA. **Figure 7.3 (b):** Photo courtesy of David J. Matty and Jane M. Matty. **Figure 7.4:** Photo courtesy of Eric Johnson. **Perspective 7.1, Figure1:** Collection of the J. Paul

Getty Museum, Malibu, CA. **Figure 7.6:** Photo courtesy of David J. Matty. **Figure 7.7:** Photo courtesy of Eric Johnson. **Figure 7.9 (b):** Photo by Reed Wicander. **Figure 7.10 (a):** Photo courtesy of Sue Monroe. **Figure 7.10 (b):** Photo by Reed Wicander. **Figure 7.11:** Photo by Brian A. Roberts, courtesy of Reed Wicander. **Figure 7.12 (a):** Photo by Brian A. Roberts, courtesy of Reed Wicander. **Figure 7.12 (b):** Photo courtesy of Sue Monroe. **Figure 7.13:** Photo by Reed Wicander. **Figure 7.14:** Photo by Ed Bartram, courtesy of R. V. Dietrich. **Figure 7.15:** Photo by Reed Wicander. **Figure 7.16:** Photo courtesy of Sue Monroe. **Figure 7.17:** Photo courtesy of Sue Monroe. **Figure 7.10 (b):** Photo courtesy of R. V. Dietrich.

Chapter 8
Chapter Opener: © Bob Krist/ CORBIS. **Figure 8.2:** Photo by Reed Wicander. **Figure 8.3 (a):** Photo by Stew Monroe. **Figure 8.3 (b-c):** Photos by Reed Wicander. **Figure 8.5 (c):** Photo by Stew Monroe. **Figure 8.8 (b):** Photo by Stew Monroe. **Figure 8.9 (b):** Photo courtesy of Dorothy L. Stout. **Figure 8.10 (b):** Photo by Stew Monroe. **Figure 8.14:** Photos by Reed Wicander. **Figure 8.23:** Photo courtesy of Charles W. Naeser, USGS.

Chapter 9
Chapter Opener: © Dave Bartruff/Stock, Boston/PictureQuest. **Figure 9.1 (b):** Mehmet Celebi, U.S.Geological Survey. **Figure 9.1 (c):** Charles Meuller, U.S.Geological Survey. **Figure 9.2:** © Bettmann/ CORBIS. **Figure 9.3 (b):** Photo by Stew Monroe. **Figure 9.4:** Reproduced by the Trustees of the Science Museum, London. **Figure 9.9 (b):** ©Bunyo Ishikawa/ Sygma/CORBIS. **Figure 9.9 (a):** Photo courtesy of Dennis Fox. **Figure 9.10:** ©Cindy Andrews, Liaison Agency, Inc. **Perspective 9.1, Figure 1 (a-b):** Photos by Stew Monroe. **Perspective 9.1, Figure 2:** Steinbrugge Collection, Earthquake Engineering Research Center, University of California, Berkeley. **Perspective 9.1, Figure 3 (a):** Don Bloomer/Time. **Perspective 9.1, Figure 3 (b):** ©Ted Soqui/ Sygma/CORBIS. **Perspective 9.1, Figure 3 (c):** ©Lee Stone/ Sygma/CORBIS. **Figure 9.11:** Courtesy of South Caroliniana Library, University of South Carolina, Columbia. **Guest essay:** Bob Paz/Caltech. **Figure 9.20 (b):** Figures courtesy U.S.Geological Survey; Photo © Lloyd S. Cluff. **Figure 9.21:** National Geophysical Data Center. **Perspective 9.2, Figure 3:** M. Celebi, U.S. Geological Survey. **Figure 9.22:** Photo courtesy of ChinastockPhoto. **Figure 9.23:** Martin E. Klimek, Marin Independent Journal. **Figure 9.24:** Photo courtesy of Bishop Museum. **Figure 9.25:** © Pierre Mion. **Figure 9.27 (a):** Photo by Stew Monroe. **Figure 9.27 (b):** Photo by Reed Wicander.

Chapter 10
Chapter Opener: Photo courtesy of Cornelius Gillen. **Figure 10.8 (a):** Kort-og Matrikelstyrelsen (National Survey and Cadastre-Denmark). **Figure 10.21:** ©Fundamental Photographs.

Chapter 11
Chapter Opener: © Woods Hole Oceanographic Institute. **Figure 11.2 (b):** Painting by Lloyd K. Townsend copyright © National Geographic Society. **Figure 11.5:** World Ocean Floor map by Bruce C. Heezen and Marie Tharp, 1977. Copyright © Marie Tharp 1977. Reproduced by permission of Marie Tharp, 1 Washington Ave., South Nyack, NY 10960. **Figure 11.7 (a):** Ocean Drilling Program, Texas A & M University. **Figure 11.7 (b):** VU/WHOI-R. Catarach. **Figure 11.13:** Copyright 1995, David T. Sandwell, Institute of Geophysics and Planetary Physics, Scripps Institution of Oceanography. Map courtesy of National Oceanic and Atmospheric Administration. **Figure 11.17 (a):** Peter Ryan/Scripps/Science Photo Library/Photo Researchers, Inc. **Figure 11.17 (b):** Visuals Unlimited /WHOI-D. Foster. **Figure 11.21 (a):** Institute of Oceanographic Sciences/Nerc/ Science Library/Photo Researchers, Inc. **Figure 11.21(b):** Dr. James Andrews, University of Hawaii 334: **Figure 11.22 (b):** © A. M. Siegelman/Visuals Unlimited. **Figure 11.22 (c):** © Woods Hole

Oceanographic Institute. **Figure 11.22 (d):** © Roger Klocek/Visuals Unlimited. **Figure 11.22 (e):** © Science VU/Visuals Unlimited. **Figure 11.23 (a):** Photo courtesy of Carl Roessler. **Figure 11.23 (b):** © Douglas Faulkner, Science Source/Photo Researchers, Inc. **Figure 11.27:** © Bruce Hall. **Perspective 11.2, Figure 2:** Photo courtesy of T. Scott, Florida Geological Survey.

Chapter 12
Chapter Opener: NASA. **Figure 12.1 (a):** Photo courtesy of Patricia G. Gensel, University of North Carolina. **Figure 12.1 (b):** Photo courtesy of Patricia G. Gensel, University of North Carolina. **Figure 12.2:** Bildarchiv Preussischer Kulterbesitz. **Figure 12.6 (a):** Photo courtesy Scott Katz. **Figure 12.8:** Photo by Brian A. Roberts, courtesy of Reed Wicander. **Figure 12.10:** Photo courtesy of ALCOA. **Perspective 12.2, Figure 1:** AP/Wide World Photos. **Figure 12.16 (b):** © Robert Caputo/Auora. **Figure 12.17 (b):** Photo courtesy of John M. Faivre. **Figure 12.18 (b):** © Stone/Japan satellite. **Figure 12.19 (b):** NASA. **Figure 12.20 (b):** NASA. **Figure 12.23:** © Stone /James Balogs. **Figure 12.29 (b):** Photo courtesy of R. V. Dietrich.

Chapter 13
Chapter Opener: © Craig Lovell/CORBIS. **Figure 13.1 (a):** Photo courtesy of Steve Stahl. **Figure 13.1 (b):** Photo by Stew Monroe. **Figure 13.5:** Photo courtesy of John S. Shelton. **Figure 13.6 (a-b):** Photos courtesy of Sue Monroe. **Figure 13.7 (b):** Photo courtesy of Sue Monroe. **Figure 13.8 (b):** Photo courtesy of Kevin O'Brien. **Perspective 13.1, Figure 1 (a):** Photo courtesy of Sue Monroe. **Perspective 13.1, Figure 1 (b):** © Jim Winkley; Ecoscene/ CORBIS. **Perspective 13.1, Figure 3 (b):** Photo courtesy Sue Monroe. **Perspective 13.1, Figure 3 (c):** © Yann Arthus-Bertrand/ CORBIS. **Figure 13.9:** Photo by Reed Wicander. **Figure 13.11 (b):** Photo by Martin F. Schmidt. Jr. **Figure 13.14 (c):** Photo courtesy of John S. Shelton. **Figure 13.16 (a):** © Galen Rowell/Peter Arnold, Inc. **Figure 13.16 (b):** Photo courtesy of C.G. Tillman. **Figure 13.16 (c):** Photo by Stew Monroe. **Figure 13.18 (b):** Photo courtesy of David J. Matty. **Figure 13.18 (c):** Photo by Stew Monroe. **Figure 13.20 (a):** Photo by Stew Monroe. **Figure 13.20 (b):** © Martin Miller/Visuals Unlimited **Figure 13.21:** Photo courtesy of John S. Shelton. **Figure 13.23 (b-c):** Photos by Stew Monroe. **Perspective 13.3, Figure 1 (c,d,e):** Photos by Stew Monroe. **Figure 13.32 (b):** Photo courtesy of Dinosaur State Park.

Chapter 14
Chapter Opener: AP/Wide World Photos. **Figure 14.3 (c):** Photo by Reed Wicander. **Figure 14.4** Photo by Stew Monroe. **Figure 14.5 (d):** Photo courtesy of R. V. Dietrich. **Figure 14.6:** Boris Yaro, Los Angeles Times. **Figure 14.8:** ©Rod Rolle/ LiaisonAgency, Inc. **Perspective 14.1, Figure 1:** T. Spencer/Colorific. **Figure 14.10:** W. R. Hansen/ U.S. Geological Survey. **Figure 14.11:** Photo courtesy of Sue Monroe. **Figure 14.13:** Photo courtesy of John S. Shelton. **Figure 14.15 (a):** Photo courtesy of Eleanora I. Robbins, U.S. Geological Survey. **Figure 14.15 (b):** Stephen R. Lower, GeoPhoto Publishing Company. **Figure 14.16 (b):** B. Bradley and the University of Colorado's Geology Department, National Geophysical Data Center, NOAA, Boulder, CO. **Figure 14.17:** Photo by Stew Monroe. **Figure 14.18:** B. Pipkin, University of Southern California. **Figure 14.19 (b):** Photo by Reed Wicander. **Figure 14.20:** Canadian Air Force. **Figure 14.21 (b):** Photograph from "Alaska Earthquake collection," no. 43ct, USGS. **Figure 14.22 (b):** B. Bradley and the University of Colorado's Geology Department. National Geophysical Data Center, NOAA, Boulder, CO. **Figure 14.23:** O. J. Ferrains, Jr., U.S. Geological Survey. **Figure 14.24 (b):** B. Bradley and the University of Colorado's Geology Department. National Geophysical Data Center, NOAA, Boulder, CO. **Figure 14.24 (c):** B. Bradley and the University of Colorado's Geology Department. National Geophysical Data Center, NOAA, Boulder, CO. **Figure 14.24 (d):** Photo courtesy of David J. Matty. **Figure 14.26:** George Plafker, U.S. Geological Survey. **Perspec-**

tive 14.2, Figure 2: Bettmann/CORBIS. **Figure 14.28 (b):** Photo by Reed Wicander. **Figure 14.30 (b):** © John D. Cunningham/Visuals Unlimited. **Figure 14.31 (b):** © Dell R. Foutz/Visuals Unlimited. **Figure 14.32 (b):** Photo by Reed Wicander.

Chapter 15
Chapter Opener: © David Muench/ CORBIS. **Figure 15.1 (a):** © Schenectaday Museum; Hall of Electrical History/ CORBIS. **Figure 15.1 (b):** BettmanCORBIS. **Figure 15.2 (a):** Security Pacific Collection/Los Angeles Public Library. **Figure 15.2 (b):** Security Pacific Collection/Los Angeles Public Library. **Figure 15.3 (a):** ©Lowell Georgia/ Science Source/Photo Researchers. **Figure 15.3 (b):** Photo by Stew Monroe. **Figure 15.4:** NASA. **Perspective 15.1, Figure 2 (b-d):** Photos courtesy of Sue Monroe. **Figure 15.11 (a):** Photo by Stew Monroe. **Figure 15.11 (b):** Photo courtesy of R. V. Dietrich. **Figure 15.12 (a-b):** Photos by Stew Monroe. **Figure 15.15 (a):** Photo courtesy of Sue Monroe. **Figure 15.15 (b):** Photo by Stew Monroe. **Figure 15.17 (b-c):** Photos by Stew Monroe. **Figure 15.21 (b):** Photo by Stew Monroe. **Figure 15.23 (b):** Photo courtesy of John S. Shelton. **Figure 15.24:** © 1997 Jamie Sabau. **Figure 15.16:** Photo courtesy of John S. Shelton. **Perspective 15.2, Figure 1 (a):** Photo courtesy of Michael Lawton. **Perspective 15.2, Figure 1 (b):** Photo courtesy of Sue Monroe. **Figure 15.27 (a):** Photo by Stew Monroe. **Figure 15.29 (b):** Photo courtesy of the Kentucky Department of Parks. **Figure 15.33 (a-b):** Photos by Stew Monroe. **Figure 15.35 (d):** J. R. Stacey, U.S. Geological Survey. **Figure 15.36:** Photo courtesy of John S. Shelton. **Perspective 15.3, Figure 1:** Photo courtesy of Sue Monroe.

Chapter 16
Chapter Opener: © John Elk III/Stock, Boston/PictureQuest. **Figure 16.1 (b):** Brown Brothers. **Figure 16.5:** Photo courtesy of Nassau Department of Public Works. **Figure 16.7 (a-b):** Photos courtesy of Sue Monroe. **Perspective 16.1, Figure 2 (a):** Photo courtesy of Ed Cooper. **Perspective 16.1, Figure 2 (b):** © Chip Clark 1989. **Figure 16.8 (a-b):** Photos courtesy of Sue Monroe. **Figure 16.11:** J. R. Stacey, U.S. Geological Survey. **Figure 16.14 (a):** Frank Kujawa, University of Central Florida, GeoPhoto Publishing Company. **Figure 16.14 (b):** Photo by Stew Monroe. **Figure 16.16 (a):** Photo by Reed Wicander. **Figure 16.16 (b):** Photo courtesy of John S. Shelton. **Figure 16.18:** © Daniel W. Gotshall/ Visuals Unlimited. **Figure 16.22:** U.S. Geological Survey. **Figure 16.23:** ©Stone/Sarah Stone. **Figure 16.24:** Photo courtesy of R. V. Dietrich. **Figure 16.25:** City of Long Beach Department of Oil Properties. **Perspective 16.2, Figure 1:** Courtesy of United States Department of Energy. **Figure 16.27 (a):** Photo by Reed Wicander. **Figure 16.27 (b-d):** Photos by Stew Monroe. **Figure 16.28:** British Tourist Authority. **Figure 16.29(a-b):** Photos by Stew Monroe. **Figure 16.31 (a-b):** Photos by Reed Wicander. **Perspective 16.3, Figure 1:** Julie Donnelly-Nolan, USGS.

Chaper 17
Chapter Opener: © Ric Ergenbright/CORBIS. **Figure 17.1 (a):** Offentliche Kunstammlung, Kupferstichkabinet, Basel. **Figure 17.1 (b):** Reproduced courtesy of the Board of Directory of the Budapest Museum of Fine Arts. **Figure 17.4:** Engineering Mechanics, Virginia Polytechnic Institute and State University. **Figure 17.5 (b):** © Frank Awbrey/Visuals Unlimited. **Figure 17.6 (b):** Photo by Stew Monroe. **Figure 17.10:** National Park Service photograph by Ruth and Louis Kirk. **Perspective 17.1, Figure 1** AP Wide World 544: **Figure 17.11 (b):** Photo by Stew Monroe. **Figure 17.12 (a-c):** Photos by Stew Monroe. **Figure 17.14 (a-b):** Photos by Stew Monroe. **Figure 17.15:** © Brian Vikander/CORBIS. **Figure 17.16:** ©Phil Schermeister /CORBIS. **Figure 17.17 (a):** Photo by Stew Monroe. **Figure 17.17 (b):** Photo courtesy of John S. Shelton. **Figure 17.18:** Photo by Stew Monroe. **Figure 17.19:** Swiss National Tourist Office. **Perspective 17.2, Figure 1 (a):** Photo by Stew Monroe. **Perspective 17.2, Figure 1 (b):** © Ron Watts/CORBIS. **Perspective 17.2, Figure 2 (b-c):** Photos by Stew Monroe. **Figure 17.20:** Alan Kesselhein/Mary Pat Ziter, © JLM

Visuals. **Perspective 17.2, Figure 3:** Carl H. Key, Northern Rocky Mountain Science Center, Glacier Field Station, USGS. **Figure 17.21 (a):** Photo by Richard V. Dietrich. **Figure 17.21 (b):** Photo by Stew Monroe. **Figure 17.23 (a-b):** Photos by Stew Monroe. **Figure 17.24 (a):** Engineering Mechanics, Virginia Polytechnic Institute and State University. **Figure 17.24 (b):** © Peter Kresan. **Figure 17.25:** Photo courtesy Carl Guell Slide Collection, Department of Geography, University of Wisconsin, Oshkosh. **Figure 17.26 (a):** Photo courtesy of David J. Matty. **Figure 17.26 (b):** Photo courtesy of John S. Shelton. **Figure 17.28 (a):** Photo courtesy of B. M. C. Pape. **Figure 17.28 (b):** © Tom Bean/ CORBIS. **Figure 17.29:** Photo courtesy of Canadian Geological Survey. **Perspective 17.3, Figure 2 (a):** Photo courtesy of P. Weiss, U.S. Geological Survey. **Perspective 17.3, Figure 2 (b):** Photo courtesy of P. Weiss, U.S. Geological Survey.

Chaper 18

Chapter Opener: ©George Gerster/The National Audubon Society/ Photo Researchers. **Figure 18.2:** © Martin G. Miller/Visuals Unlimited. **Figure 18.3:** © O. Alamany & E. Vicens/ CORBIS. **Figure 18.4 (b):** © Martin G. Miller/Visuals Unlimited. **Figure 18.5:** Photo courtesy of Marion A. Whitney. **Figure 18.6:** © Martin G. Miller/ VisualsUnlimited. **Figure 18.7 (c)** Photo courtesy of David J. Matty. **Figure 18.8:** Digital imagery ® copyright 1999 PhotoDisc, Inc.. **Figure 18.10:** Photo by Reed Wicander. **Figure 18.11 (b):** Photo courtesy of John S. Shelton. **Figure 18.12 (b):** © 1994 CNES. Provided by SPOT Image Corporation. **Figure 18.13 (b):** © W.J. Weber/Visuals Unlimited. **Figure 18.14:** © Stone /Willard Clay. **Figure 18.15 (b):** Photo by Reed Wicander. **Figure 18.16 (b):** © Nigel J. Dennis /Photo Researchers, Inc. **Figure 18.17:** © Steve McCutcheon/Visuals Unlimited. **Perspective 18.1, Figure 3:** © John D. Cunningham/Visuals Unlimited. **Perspective 18.1, Figure 4:** United States Borax & Chemical Corporation. **Perspective 18.2, Figure 1:** Photo by Reed Wicander. **Perspective 18.2, Figure 2:** Photo by Reed Wicander. **Figure 18.22:** © Charlie Ott,The National Audubon Society Collection/ Photo Researchers. **Figure 18.24 (a-b):** Photos courtesy of John S.Shelton. **Figure 18.25:** © Martin G. Miller/Visuals Unlimited. **Figure 18.26:** Alan L. and Linda D. Mayo, GeoPhoto Publishing Company. **Figure 18.27 (b):** Photo by Reed Wicander. **Perspective 18.3, Figure 2 (a-b):** Photos by Reed Wicander. **Figure 18.28(a):** © Adriel Heisey. **Figure 18.28 (b):** © Tom Bean/ CORBIS.

Chapter 19

Chapter Opener: Photo by Stew Monroe. **Figure 19.1 (a):** Rosenberg Library, Galveston, Texas. **Figure 19.1 (b):** Rosenberg Library, Galveston, Texas. **Figure 19.2 (a):** NOAA/National Hurricane Center. **Figure 19.2 (b):** NOAA/National Hurricane Center. **Figure 19.3:** Photo by Stew Monroe. **Figure 19.4 (a):** Photo courtesy of Karl Kuhn. **Figure 19.4 (b):** Photo courtesy of Karl Kuhn. **Figure 19.6 (b):** Photo by Stew Monroe. **Figure 19.6 (c):** Photo by Stew Monroe.

Perspective 19.1, Figure 2: © La médiathèque EDF/Gerard Halary. **Figure 19.7:** Photo courtesy of Sue Monroe. **Figure 19.8 (a):** Tom Servais, Surfer Magazine. **Figure 19.8 (b):** Tom Servais, Surfer Magazine. **Figure 19.9:** Photo courtesy of John S. Shelton. **Figure 19.10 (a):** Photo courtesy of John S. Shelton. **Figure 19.11 (a):** Photo by Stew Monroe. **Figure 19.11 (b):** Photo courtesy of Michael Slear. **Figure 19.12 (b):** Photo by Stew Monroe. **Figure 19.13 (b):** Photo courtesy of John S. Shelton. **Figure 19.14 (a):** Photo courtesy of Sue Monroe. **Figure 19.15 (b-c):** Photos courtesy of Sue Monroe. **Figure 19.16 (b-c):** Photos by Stew Monroe. **Perspective 19.2, Figure 1** Department of the Interior/USGS, EROS Data Center. **Figure 19.17 (b-c):** Photos by Stew Monroe. **Perspective 19.2, Figure 3:** Photo by Stew Monroe. **Figure 19.18:** Photo courtesy of John S. Shelton. **Figure 19.19 (a-b):** Photos courtesy of Sue Monroe. **Figure 19.21 (a):** Data available from U.S. Geological Survey, EROS Data Center, Sioux Falls, SD. **Figure 19.21 (b):** © Richard Stockton 1998. Duke Photo Department. **Figure 19.24 (b):** Photo courtesy of John S. Shelton. **Figure 19.24 (c):** Photo by Stew Monroe. **Figure 19.25 (b):** Photo courtesy of Nick Harvey. **Figure 19.25 (c):** Photo courtesy of Sue Monroe. **Figure 19.28 (a):** U.S. Army Corps of Engineers. **Figure 19.28 (b):** U.S. Army Corps of Engineers. **Perspective 19.3, Figure 1:** NASA. **Perspective 19.3, Figure 2 (a):** © Tom Till. **Perspective 19.3, Figure 2 (b):** © Myles & Patterson. **Figure 19.29:** Photo courtesy Dr. Stanley R. Riggs, Department of Geology, Graham Bldg. East Carolina University, Greenville, NC 27858 Tel. 252 328-6015 Fax 252 328-4391. **Figure 19.30:** GEOPIC, Earth Satellite Corporation. **Figure 19.31:** Photo courtesy of Sue Monroe. **Figure 19.32:** Photo courtesy of John S. Shelton. **Guest Essay:** Orrin Pilkey—Duke University Photo Department

Chapter 20

Chapter Opener: NASA. **Figure 20.1:** JPL/ NASA. **Figure 20.6:** Photo courtesy of Dana Berry. **Figure 20.7 (b-d):** Fotosmith. **Figure 20.8 (a):** D. J. Roddy, U.S. Geological Survey. **Figure 20.8 (b):** Photo courtesy of Paul Dimare. **Perspective 20.1, Figure 1 (a):** Photographed by Michael Hollenbeck The Saginaw News. **Perspective 20.1, Figure 1 (b):** © John Bortle. **Perspective 20.1, Figure 2:** TASS from Sovfoto. **Perspective 20.1, Figure 3 (a):** Photo courtesy of D. J. Nichols, USGS. **Figure 20.9 (a):** NASA. **Figure 20.9 (c):** Photo courtesy of Victor Royer. **Figure 20.10 (a):** Finley Holiday Films. **Figure 20.10 (b):** NASA. **Figure 20.10 (c):** AP/Wide World Photos. **Figure 20.11 (b):** JPL/ NASA. **Figure 20.11 (c):** NASA. **Perspective 20.3, Figure 1 (a):** Damon Simonelli and Joseph Vererka, Cornell University/ NASA. **Perspective 20.3, Figure 1 (b):** NASA. **Figure 20.12:** Photo courtesy of Dana Berry. **Figure 20.14 (a-b):** NASA. **Figure 20.16 (a):** UCO/Lick Observatory image. **Figure 20.18(a):** NASA. **Figure 20.18 (b-c):** Photos courtesy of Sue Monroe. **Figure 20.18 (d):** Photo by Stew Monroe.

Index

Oxidized ores, 151
Ozone layer, 4, 6, 135

P

Pacific-Farralon ridge, 414, 416, 417
Pacific Ocean, 318, 326, 327, 408
Pacific Palisades (California), 434
Pacific plate, 258, 371
Pahoehoe flow, 96–97
Pakistan, Mount Godwin-Austen (K2), 382
Paleocene Epoch, 158
Paleocurrents, 172
Paleogeographic maps, 356–357
Paleomagnetism, 308, 354–355
Paleontologists, 177, 239
Paleozoic Era, 325, 348, 349, 352, 356, 361, 414
Palisades (New York), 365
Pan American Highway (Mexico), 427
Pangaea, 20, 324, 326, 360–361, 371, 405, 414
Parabolic dunes, 580, 582
Parent element, 234–237
Parent material, 129
  chemical weathering and, 140
  soil formation and, 143–145
Parkfield earthquake prediction program, 279
Partial melting, 67, 68
Particle size, chemical weathering and, 139
Passive continental margins, 324–325
Patagonian region, 587
Patination, 591
Paunsaugunt Plateau (Utah), 127
Peat, 168, 473
Pedalfers, 143
Pediments, 595–596
Pedocals, 143, 144
Pegmatite, 75–77
Pelagic clay, 332, 334
Peninsular Ranges Batholith (California), 198
Pennsylvania Period, 183, 349, 503
Penny Ice Cap, 539
Penzias, Arno, 644
Perched water table, 401, 499
Peridotite, 15, 72, 73, 196, 295, 297, 299
Perihelion, 567
Periodic table, 35, 672–673
Permafrost, 440, 441
Permeability, 182, 497, 498
Persian Gulf region, 182, 183, 374
Peru, earthquake of 1970, 276, 443
Peru-Chile Trench, 324, 326, 405, 412
Petermann Ranges Orogeny, 596
Petrified Forest National Park (Arizona), 590
Petrified wood, 178, 179
Petroglyph National Monument (New Mexico), 591
Petroglyphs, 590–591
Petroleum. *See* Oil
Phaneritic texture, 70
Phenocysts, 70
Philippines
  Mayon volcano, 95, 105, 107, 119, 534
  Mount Pinatubo, 65, 71, 93, 94, 106–107, 112, 115, 116, 119
Phobos, 659, 664, 665
Phosphate, 55
Phosphatization, 340–341
Phosphoria Formation, 340
Phosphorite, 339–341
Photons, 645
pH scale, 135
Phyllite, 202
Physical geology, 7
Picture Gorge Basalt, 157

Piedmont glaciers, 537
Piezoelectricity, 49
Pilkey, Orrin H., 627
Pillow lava, 98, 119, 329
Pipe, volcanic, 80
Piston corers, 320
Pitted outwash plains, 558
Placer deposits, 181
Plagioclase feldspars, 46, 61, 66, 74, 133, 135, 139, 165
Plains
  abyssal, 326–328
  floodplains, 470, 472, 480
  ice-scoured, 548
  outwash, 556
Plane of the ecliptic, 645
Planetesimals, 15, 648
Planets, 645. *See also specific planets*
  characteristics of, 646
  Jovian. *See* Jovian planets
  terrestrial. *See* Terrestrial planets
Plants. *See* Vegetation
Plasma, 34n
Plastic flow, 536, 537
Plastic strain, 385
Plate boundaries
  continental-continental, 366, 367, 405, 407
  convergent. *See* Convergent plate boundaries
  divergent. *See* Divergent plate boundaries
  oceanic-continental, 366, 367, 404–405
  oceanic-oceanic, 365, 366, 404, 406
  orogenesis and, 404–408
  transform, 17, 18, 251, 255, 256, 324, 369
  types of, 361
Plates, 4, 16
  boundaries of, 361
  locations of, 360
  movement/motion, 369–371
Plate tectonics, 118–120
  distribution of natural resources and, 373–375
  driving mechanism of, 372
  metamorphism and, 207–208
  rock cycle and, 22
  theory, 348–349 acceptance of, 16, 17–19, 360 formulation of, 19
Platinum, 42, 43
Plato, 315–317
Playa lakes, 593, 594
Playas, 593–594
Pleistocene Epoch, 113, 240, 323, 484, 535, 538, 542, 544, 551, 593
  climate during, 561
  glaciation, 559, 561–563
  Milankovitch theory and, 566–567
  pluvial lakes in, 562
  sea level in, 563
Plinian eruptions, 91
Pliocene Epoch, 114, 158, 588
Plucking (quarrying), 542, 543
Plunging breakers, 613
Plunging folds, 392, 394
Pluto, 645, 648, 659, 661, 662
  characteristics of, 646
Plutonic rocks, 64, 93. *See also* Intrusive igneous rock
Plutons, 63, 119
  batholiths, 80, 81, 84
  concordant, 78, 79
  dikes, 78–80
  discordant, 78, 79
  laccoliths, 80
  in mountain building, 401, 404, 406
  sills, 78–80

stocks, 80
types of, 78
volcanic pipes and necks, 80
Pluvial lakes, 562
Pocket beaches, 614, 615
Pohnpei (Micronesia), 100, 101
Point bar, 469
Point Fermin (California), 434, 436
Polarity, 308
Polar wandering, continental drift and, 354–355
Pollution, 4, 9
  acid rain, 135–136
  environmental remediation for, 476, 507
  groundwater, 517–518, 521
  soil degradation and, 150
Polonium, 234
Polycyclic aromatic hydrocarbons, 642
Pompeii (Italy), 91–92, 94, 118
Ponds, 620, 621
Popocatépetl volcano (Mexico), 120
Popping, 131, 290
Population growth, 12–13
Pore spaces, 162, 183
Porosity, 497–498
Porphyritic texture, 70
Portage des Sioux (Missouri), 478, 479
Port Royal earthquake (Jamaica), 276
Positive gravity anomaly, 302
Positive magnetic anomaly, 307
Potassium feldspars, 46, 52, 61, 74, 133, 135, 165
Potential energy, 465
Potholes, 466, 467
Potomac River, 489
Pratt, J.H., 303
Precambrian Midcontinent Rift, 185
Precession of the equinoxes, 566–567
Precipitation, 459, 462
Precursors of earthquakes, 278–280
Prediction of earthquakes, 269, 277–280
Pressure
  differential, 194, 195, 201
  lithostatic, 194, 195
  in metamorphic process, 194–195
  release, 130–132
  ridges, 97
Primary sedimentary structures, 174
Primary waves. *See* P-waves
Principal meridians, 680
Principle of
  cross-cutting relationships, 220–221
  fossil succession, 220, 222, 223
  inclusions, 220, 222
  isostasy, 303–305
  lateral continuity, 220
  original horizontality, 220, 385
  superposition, 220
  uniformitarianism, 23, 180, 218–219
Principles of Geology (Lyell), 218
Proglacial lakes, 562
Prograding delta, 472, 473
Proterozoic Belt Supergroup, 550
Proterozoic Eon, 185
Protons, 35, 232
P-S time interval, 263–264
Public policy, geoscience and, 373
Pumice, 75, 78
Purcell Supergroup, 550
P-wave shadow zone, 294
P-waves (primary waves), 260–263, 268, 292–296
Pyrite, 39, 48, 135, 208
Pyroclastic (fragmental) texture, 70
Pyroclastic materials, 63, 91, 93, 94, 96–99, 102, 119